HANDLING HAZARDOUS MATERIALS

This issue includes revisions issued on or before November 14, 2011.

J. J. Keller
& Associates, Inc.®
Since 1953

December 2011

J. J. Keller & Associates, Inc.®
Neenah, Wisconsin

Printed in the U.S.A.

Due to the constantly changing nature of government regulations, it is impossible to guarantee absolute accuracy of the material contained herein. The Publisher and Editors, therefore, cannot assume any responsibility for omissions, errors, misprinting, or ambiguity contained within this publication and shall not be held liable in any degree for any loss or injury caused by such omission, error, misprinting or ambiguity presented in this publication.

This publication is designed to provide reasonably accurate and authoritative information in regard to the subject matter covered. It is sold with the understanding that the Publisher is not engaged in rendering legal, accounting, or other professional service. If legal advice or other expert assistance is required, the services of a competent professional person should be sought.

The Editorial Staff is available to provide information generally associated with this publication to a normal and reasonable extent, and at the option of, and as a courtesy of, the Publisher.

Library of Congress Number: 9772935

Perfect Bound ISBN: 978-1-60287-567-8

Spiral Bound ISBN: 978-1-60287-568-5

Canadian Goods and Services Tax (GST) Number: R123-317687

FOREWORD

The transportation of hazardous materials continues to increase in complexity. Each year new chemicals are introduced into commerce and many of them are classified as hazardous materials. Public interest and pressure for more stringent controls for hazardous materials transportation have also had an impact on government and industry safety programming. A few serious hazardous materials accidents coupled with worldwide publicized accounts of life threatening, environmentally hazardous and costly accidents have produced an atmosphere of public concern about the potential for catastrophic accidents resulting from hazardous materials transportation.

The federal Department of Transportation (DOT) has the responsibility to identify materials moving in commerce that may pose unreasonable risks to health and safety (hazardous materials) and to promulgate regulations for the safe transportation of these hazardous materials. The Environmental Protection Agency (EPA) has developed controls for the proper management and transportation of hazardous waste from "cradle to grave." EPA has established regulations for generators, transporters and treatment, storage, or disposal (TSD) facilities, including significant changes in recordkeeping, reporting, and manifesting requirements. The Clean Water Act was adopted to minimize the adverse affects on the environment by spills of materials identified as hazardous substances. Both DOT and EPA have promulgated requirements under the Act for notification and clean-up of hazardous substance spills.

This book is a guide to the Hazardous Materials Regulations. It should not be used as a substitute. The full text of the Hazardous Materials Regulations should be on hand wherever hazardous materials are handled.

This book contains select portions of the most commonly used parts of the Hazardous Materials Regulations. It does not contain all of the Hazardous Materials Regulations. The regulations are in the front of the book in the same numerical order that they appear in the Code of Federal Regulations. Following the regulations are plain language explanations of the regulations. This gives you the actual regulations and plain language explanations to help you understand the regulations. Regulation numbers are provided in the plain language explanations so you can go to and read the actual regulations.

The Editors & Publisher
J. J. Keller & Associates, Inc.®

Regulation Changes

Changes to the Hazardous Materials Regulations from the following final rules have been incorporated into this edition of *Handling Hazardous Materials*.

- HM-238 - Storage of Explosives During Transportation, published in the *Federal Register* on June 7, 2011.
- HM-145O - Revision to the List of Hazardous Substances and Reportable Quantities, published in the *Federal Register* on June 27, 2011.
- HM-218F - Miscellaneous Amendments, published in the *Federal Register* on July 20, 2011.
- HM-233B - Revisions of Special Permits Procedures, published in the *Federal Register* on July 26, 2011.
- HM-244D - Editorial Corrections and Clarifications, published in the *Federal Register* on September 13, 2011.

Revision bars, like the one next to this paragraph, identify text that has changed in this edition. These revision bars will help you identify changes resulting from the above rulemakings.

REGULATIONS

ENFORCEMENT

§107.307 General. 1
§107.309 Warning Letters. 1
§107.310 Ticketing. 1
§107.311 Notice of Probable Violation. 1
§107.313 Reply. 2
§107.329 Maximum Penalties. 2
§107.333 Criminal Penalties Generally. 2

REGISTRATION

§107.601 Applicability. 3
§107.606 Exceptions. 3
§107.608 General Registration Requirements. 3
§107.612 Amount of Fee. 3
§107.616 Payment Procedures. 4
§107.620 Recordkeeping Requirements. 4

GENERAL

§171.1 Applicability of Hazardous Materials Regulations (HMR) to Persons and Functions. 7
§171.2 General Requirements. 9
§171.3 Hazardous Waste. 10
§171.4 Marine Pollutants. 11

DEFINITIONS

§171.8 Definitions and Abbreviations. 13
§171.9 Rules of Construction. 23

NORTH AMERICAN

§171.12 North American Shipments. 25

INCIDENT REPORTS

§171.15 Immediate Notice of Certain Hazardous Materials Incidents. 27
§171.16 Detailed Hazardous Materials Incident Reports. 27

INTERNATIONAL

§171.22 Authorization and Conditions for the Use of International Standards and Regulations. 29
§171.23 Requirements for Specific Materials and Packagings Transported Under the ICAO Technical Instructions, IMDG Code, Transport Canada TDG Regulations, or the IAEA Regulations. 30
§171.24 Additional Requirements for the Use of the ICAO Technical Instructions. 32
§171.25 Additional Requirements for the Use of the IMDG Code. 32
§171.26 Additional Requirements for the Use of the IAEA Regulations. 33

HAZARDOUS MATERIALS TABLE

§172.1 Purpose and Scope. 35
§172.3 Applicability. 35
§172.101 Purpose and Use of Hazardous Materials Table. 35
§172.101 Hazardous Materials Table. 41

HAZARDOUS SUBSTANCES RQ

Appendix A to §172.101 — List of Hazardous Substances and Reportable Quanties 141

Table 1—Hazardous Substances Other Than Radionuclides 141

RADIONUCLIDES RQ

Table 2—Radionuclides 153

MARINE POLLUTANTS

Appendix B to §172.101 — List of Marine Pollutants 161
List of Marine Pollutants....161

SPECIAL PROVISIONS

§172.102 Special Provisions....167

SHIPPING PAPERS

§172.200 Applicability. 189
§172.201 Preparation and Rentention of Shipping Papers....189
§172.202 Description of Hazardous Material on Shipping Papers....189
§172.203 Additional Description Requirements....190
§172.204 Shipper's Certification. 193
§172.205 Hazardous Waste Manifest. 194

MARKING

§172.300 Applicability. 195
§172.301 General Marking Requirements for Non-Bulk Packagings. 195
§172.302 General Marking Requirements for Bulk Packagings. 195
§172.303 Prohibited Markings....196
§172.304 Marking Requirements. 196
§172.308 Authorized Abbreviations....196
§172.310 Class 7 (Radioactive) Materials....196
§172.312 Liquid Hazardous Materials in Non-Bulk Packagings....196
§172.313 Poisonous Hazardous Materials. 197
§172.315 Limited Quantities....197
§172.316 Packagings Containing Materials Classed as ORM–D. 198
§172.317 KEEP AWAY FROM HEAT Handling Mark....198
§172.320 Explosive Hazardous Materials....198
§172.322 Marine Pollutants....199
§172.323 Infectious Substances. 200
§172.324 Hazardous Substances in Non-Bulk Packagings....200
§172.325 Elevated Temperature Materials. 200
§172.326 Portable Tanks....201
§172.327 Petroleum Sour Crude Oil in Bulk Packaging....201
§172.328 Cargo Tanks....201
§172.330 Tank Cars and Multi-Unit Tank Car Tanks....202
§172.331 Bulk Packagings Other Than Portable Tanks, Cargo Tanks, Tank Cars and Multi-Unit Tank Car Tanks....202
§172.332 Identification Number Markings. 203
§172.334 Identification Numbers; Prohibited Display....203
§172.336 Identification Numbers; Special Provisions. 204
§172.338 Replacement of Identification Numbers....204

TABLE OF CONTENTS

LABELING

§172.400 General Labeling Requirements. 205
§172.400a Exceptions From Labeling. 205
§172.401 Prohibited Labeling. 206
§172.402 Additional Labeling Requirements. 206
§172.403 Class 7 (Radioactive) Material. 207
§172.404 Labels for Mixed and Consolidated Packaging. 208
§172.405 Authorized Label Modifications. 208
§172.406 Placement of Labels. 208
§172.411 EXPLOSIVE 1.1, 1.2, 1.3, 1.4, 1.5 and 1.6 Labels, and EXPLOSIVE Subsidiary Label. 209
§172.415 NON-FLAMMABLE GAS Label. 209
§172.416 POISON GAS Label. 210
§172.417 FLAMMABLE GAS Label. 210
§172.419 FLAMMABLE LIQUID Label. 210
§172.420 FLAMMABLE SOLID Label. 210
§172.422 SPONTANEOUSLY COMBUSTIBLE Label. 210
§172.423 DANGEROUS WHEN WET Label. 211
§172.426 OXIDIZER Label. 211
§172.427 ORGANIC PEROXIDE Label. 211
§172.429 POISON INHALATION HAZARD Label. 211
§172.430 POISON Label. 211
§172.431 [Reserved] 212
§172.432 INFECTIOUS SUBSTANCE Label. 212
§172.436 RADIOACTIVE WHITE–I Label. 212
§172.438 RADIOACTIVE YELLOW–II Label. 212
§172.440 RADIOACTIVE YELLOW–III Label. 212
§172.441 FISSILE Label. 213
§172.442 CORROSIVE Label. 213
§172.446 CLASS 9 Label. 213
§172.448 CARGO AIRCRAFT ONLY Label. 213
§172.450 EMPTY Label. 214

PLACARDING

§172.500 Applicability of Placarding Requirements. 215
§172.502 Prohibited and Permissive Placarding. 215
§172.503 Identification Number Display on Placards. 215
§172.504 General Placarding Requirements. 215
§172.505 Placarding for Subsidiary Hazards. 217
§172.506 Providing and Affixing Placards: Highway. 217
§172.507 Special Placarding Provisions: Highway. 217
§172.508 Providing and Affixing Placards: Rail. 217
§172.510 Special Placarding Provisions: Rail. 217
§172.512 Freight Containers and Aircraft Unit Load Devices. 218
§172.514 Bulk Packagings. 218
§172.516 Visibility and Display of Placards. 218
§172.519 General Specifications for Placards. 219
§172.521 DANGEROUS Placard. 219

§172.522 EXPLOSIVES 1.1, EXPLOSIVES 1.2 and EXPLOSIVES 1.3 Placards..........220
§172.523 EXPLOSIVES 1.4 Placard. 220
§172.524 EXPLOSIVES 1.5 Placard. 220
§172.525 EXPLOSIVES 1.6 Placard. 220
§172.527 Background Requirements for Certain Placards..........221
§172.528 NON-FLAMMABLE GAS Placard..........221
§172.530 OXYGEN Placard..........221
§172.532 FLAMMABLE GAS Placard..........221
§172.540 POISON GAS Placard. 222
§172.542 FLAMMABLE Placard..........222
§172.544 COMBUSTIBLE Placard..........222
§172.546 FLAMMABLE SOLID Placard..........222
§172.547 SPONTANEOUSLY COMBUSTIBLE Placard..........222
§172.548 DANGEROUS WHEN WET Placard. 223
§172.550 OXIDIZER Placard. 223
§172.552 ORGANIC PEROXIDE Placard..........223
§172.553 [Reserved] 223
§172.554 POISON Placard..........223
§172.555 POISON INHALATION HAZARD Placard..........224
§172.556 RADIOACTIVE Placard..........224
§172.558 CORROSIVE Placard. 224
§172.560 CLASS 9 Placard..........224

EMERGENCY RESPONSE

§172.600 Applicability and General Requirements. 227
§172.602 Emergency Response Information..........227
§172.604 Emergency Response Telephone Number..........227
§172.606 Carrier Information Contact..........228

TRAINING

§172.700 Purpose and Scope..........229
§172.701 Federal-State Relationship..........229
§172.702 Applicability and Responsibility for Training and Testing..........229
§172.704 Training Requirements. 229

SECURITY PLANS

§172.800 Purpose and Applicability. 231
§172.802 Components of a Security Plan. 231
§172.804 Relationship to Other Federal Requirements..........232
§172.820 Additional Planning Requirements for Transportation by Rail..........232
§172.822 Limitation on Actions by States, Local Governments, and Indian Tribes. 233

HAZARD PRECEDENCE

§173.2 Hazardous Materials Classes and Index to Hazard Class Definitions. 235
§173.2a Classification of a Material Having More Than One Hazard..........235

SPECIAL EXCEPTIONS

§173.4 Small Quantities for Highway and Rail..........237
§173.4a Excepted Quantities..........237

TABLE OF CONTENTS

§173.4b De Minimis Exceptions. 239
§173.5 Agricultural Operations. 240
§173.6 Materials of Trade Exceptions. 241
§173.8 Exceptions for Non-Specification Packagings Used in Intrastate Transportation. 242
§173.12 Exceptions for Shipment of Waste Materials. 242
§173.13 Exceptions for Class 3, Divisions 4.1, 4.2, 4.3, 5.1, 6.1, and Classes 8 and 9 Materials. 244

PACKAGING PREPARATION

§173.22 Shipper's Responsibility. 245
§173.22a Use of Packagings Authorized Under Special Permits. 245
§173.24 General Requirements for Packagings and Packages. 245
§173.24a Additional General Requirements for Non-Bulk Packagings and Packages. 247
§173.24b Additional General Requirements for Bulk Packagings. 248
§173.25 Authorized Packagings and Overpacks. 249
§173.28 Reuse, Reconditioning and Remanufacture of Packagings. 249
§173.29 Empty Packagings. 251
§173.30 Loading and Unloading of Transport Vehicles. 252

CLASSIFICATION & EXCEPTIONS

§173.50 Class 1—Definitions. 253
§173.115 Class 2, Divisions 2.1, 2.2, and 2.3 — Definitions. 253
§173.116 Class 2—Assignment of Hazard Zone. 254
§173.120 Class 3—Definitions. 254
§173.121 Class 3—Assignment of Packing Group. 255
§173.124 Class 4, Divisions 4.1, 4.2 and 4.3— Definitions. 256
§173.125 Class 4 — Assignment of Packing Group. 257
§173.127 Class 5, Division 5.1 — Definition and Assignment of Packing Groups. 258
§173.128 Class 5, Division 5.2 — Definitions and Types. 258
§173.129 Class 5, Division 5.2 — Assignment of Packing Group. 259
§173.132 Class 6, Division 6.1 — Definitions. 259
§173.133 Assignment of Packing Group and Hazard Zones for Division 6.1 Materials. 260
§173.134 Class 6, Division 6.2—Definitions and Exceptions. 263
§173.136 Class 8—Definitions. 265
§173.137 Class 8—Assignment of Packing Group. 265
§173.140 Class 9—Definitions. 266
§173.141 Class 9—Assignment of Packing Group. 266
§173.144 Other Regulated Materials (ORM)—Definitions. 266
§173.145 Other Regulated Materials—Assignment of Packing Group. 266
§173.150 Exceptions for Class 3 (Flammable and Combustible Liquids). 266
§173.151 Exceptions for Class 4. 267
§173.152 Exceptions for Division 5.1 (Oxidizers) and Division 5.2 (Organic Peroxides). 268
§173.153 Exceptions for Division 6.1 (Poisonous Materials). 268
§173.154 Exceptions for Class 8 (Corrosive Materials). 269
§173.155 Exceptions for Class 9 (Miscellaneous Hazardous Materials). 269
§173.156 Exceptions for Limited Quantity and ORM. 270

AUTHORIZED PACKAGINGS

§173.159 Batteries, Wet....271
§173.159a Exceptions for Non-Spillable Batteries....272
§173.166 Air Bag Inflators, Air Bag Modules and Seat-Belt Pretensioners....273
§173.167 Consumer Commodities....274
§173.171 Smokeless Powder for Small Arms....274
§173.173 Paint, Paint-Related Material, Adhesives, Ink and Resins....274
§173.174 Refrigerating Machines....274
§173.185 Lithium Cells and Batteries....274
§173.186 Matches....275
§173.201 Non-Bulk Packagings for Liquid Hazardous Materials in Packing Group I....276
§173.202 Non-Bulk Packagings for Liquid Hazardous Materials in Packing Group II....276
§173.203 Non-Bulk Packagings for Liquid Hazardous Materials in Packing Group III....276
§173.204 Non-Bulk, Non-Specification Packagings for Certain Hazardous Materials....277
§173.205 Specification Cylinders for Liquid Hazardous Materials....277
§173.206 Packaging Requirements for Chlorosilanes....277
§173.211 Non-Bulk Packagings for Solid Hazardous Materials in Packing Group I....277
§173.212 Non-Bulk Packagings for Solid Hazardous Materials in Packing Group II....278
§173.213 Non-Bulk Packagings for Solid Hazardous Materials in Packing Group III....278
§173.217 Carbon Dioxide, Solid (Dry Ice)....279
§173.220 Internal Combustion Engines, Self-Propelled Vehicles, Mechanical Equipment Containing Internal Combustion Engines, Battery-Powered Equipment or Machinery, Fuel Cell-Powered Equipment or Machinery....279
§173.240 Bulk Packaging for Certain Low Hazard Solid Materials....281
§173.241 Bulk Packagings for Certain Low Hazard Liquid and Solid Materials....282
§173.242 Bulk Packagings for Certain Medium Hazard Liquids and Solids, Including Solids With Dual Hazards....282
§173.243 Bulk Packaging for Certain High Hazard Liquids and Dual Hazard Materials Which Pose a Moderate Hazard....283
§173.244 Bulk Packaging for Certain Pyrophoric Liquids (Division 4.2), Dangerous When Wet (Division 4.3) Materials, and Poisonous Liquids With Inhalation Hazards (Division 6.1)....284
§173.245 Bulk Packaging for Extremely Hazardous Materials Such as Poisonous Gases (Division 2.3)....285
§173.247 Bulk Packaging for Certain Elevated Temperature Materials....285

GASES LTD QTY & EXCEPTIONS

§173.306 Limited Quantities of Compressed Gases....289
§173.307 Exceptions for Compressed Gases....292
§173.309 Fire Extinguishers....293

TANK CAR UNLOADING

§174.67 Tank Car Unloading....295

HIGHWAY TRANSPORT

§177.800 Purpose and Scope of This Part and Responsibility for Compliance and Training....297
§177.801 Unacceptable Hazardous Materials Shipments....297

TABLE OF CONTENTS

§177.802 Inspection. 297
§177.804 Compliance With Federal Motor Carrier Safety Regulations. 297
§177.810 Vehicular Tunnels. 297
§177.816 Driver Training. 297
§177.817 Shipping Papers. 298
§177.823 Movement of Motor Vehicles in Emergency Situations. 298

LOADING & UNLOADING

§177.834 General Requirements. 299
§177.835 Class 1 Materials. 300
§177.837 Class 3 Materials. 301
§177.838 Class 4 (Flammable Solid) Materials, Class 5 (Oxidizing) Materials, and Division 4.2 (Pyroforic Liquid) Materials. 302
§177.839 Class 8 (Corrosive) Materials. 303
§177.840 Class 2 (Gases) Materials. 303
§177.841 Division 6.1 and Division 2.3 Materials. 305
§177.842 Class 7 (Radioactive) Material. 306
§177.843 Contamination of Vehicles. 306

SEGREGATION

§177.848 Segregation of Hazardous Materials. 309
§177.854 Disabled Vehicles and Broken or Leaking Packages; Repairs. 310

PACKAGING CODES

§178.1 Purpose and Scope. 313
§178.2 Applicability and Responsibility. 313
§178.3 Marking of Packagings. 313
§178.500 Purpose, Scope and Definitions. 314
§178.502 Identification Codes for Packagings. 314
§178.503 Marking of Packagings. 314
§178.602 Preparation of Packagings and Packages for Testing. 317

IBC TEST/INSPECTIONS

§180.352 Requirements for Retest and Inspection of IBCs. 319

CARGO TANK TEST/INSPECTIONS

§180.407 Requirements for Test and Inspection of Specification Cargo Tanks. 321
§180.415 Test and Inspection Markings. 326
§180.416 Discharge System Inspection and Maintenance Program for Cargo Tanks Transporting Liquefied Compressed Gases. 326
§180.417 Reporting and Record Retention Requirements. 327

PORTABLE TANK TEST/INSPECTIONS

§180.605 Requirements for Periodic Testing, Inspection and Repair of Portable Tanks. 329

PART 397

§397.1 Application of the Rules in This Part. 333
§397.2 Compliance With Federal Motor Carrier Safety Regulations. 333
§397.3 State and Local Laws, Ordinances and Regulations. 333
§397.5 Attendance and Surveillance of Motor Vehicles. 333

§397.7 Parking. 333
§397.9 [Removed and Reserved] 334
§397.11 Fires. 334
§397.13 Smoking. 334
§397.15 Fueling. 334
§397.17 Tires. 334
§397.19 Instructions and Documents. 334

PLAIN LANGUAGE EXPLANATIONS

TRAINING

Training 335
Training Content 335
Recurrent Training 336
Training Record 336

SECURITY

Security Plans 337
Security Training 339

CLASSIFICATION

Hazard Classifications 341
Other Criteria 342
Hazard of the Material 342

HAZARDOUS MATERIALS TABLE

Column 1: Symbols 343
Column 2: Hazardous Materials Descriptions and Proper Shipping Names 343
Column 3: Hazard Class or Division 345
Column 4: Identification Numbers 345
Column 5: PG (Packing Group) 345
Column 6: Label Codes 346
Column 7: Special Provisions 346
Column 8: Packaging 346
Column 9: Quantity Limitations 346
Column 10: Vessel Stowage 346

PACKAGING

Terms 347
Manufacturer's UN Packaging Marking 348

SHIPPING PAPERS

Exceptions 351
Shipping Paper Retention 351
Display on Papers 351
Description 351
Technical Names 352
Reportable Quantity (RQ) 353
Limited Quantity 353

Elevated Temperature Material....353
Poisons....353
Marine Pollutant....353
Special Permits....353
Empty Packagings....353
Additional Information....353
Prohibited Entries....353
Shipper's Certification....353
Emergency Response Telephone Number....354
Emergency Response Information....355

MARKING

Applicability....357
Responsibility....357
Prohibited Markings....357
Marking Specifications....357
Manufacturer/Specification Packaging Marking....357
Non-Bulk Markings....358
Large Quantities of Non-Bulk Packages....359
Large Quantities of Non-Bulk Poison Inhalation Hazards....360
Bulk Markings....360
Additional Marking Requirements....362

LABELING

Responsibility....369
Applicability....369
Exceptions....369
Limited Quantities....369
Prohibited Labeling....370
General Requirements....370
Placement....370
Duplicate Labeling....371
Consolidated Packaging....371
Modifications....371
Specifications....372
Special Labels....373

PLACARDING

Applicability....375
Responsibility....375
Prohibited Placarding....375
Permissive Placarding....376
Placarding Tables....376
Placarding Exceptions....377
Subsidiary Placards....378
Placard Specifications....379
Placard Placement....379
White Square Background....380
ID Number Display....381
White Bottom Combustible....381

TABLE OF CONTENTS

REGISTRATION

Who Must Register 383
Exceptions 383
Registration Form 383
Fee 383
Recordkeeping 383

HIGHWAY TRANSPORT

Accepting a Shipment 385
CDL Hazardous Materials Endorsement 385
Papers 385
Loading/Unloading 386
On the Road 389

INCIDENT REPORTING

Immediate Notice of Certain Hazardous Materials Incidents 393
Detailed Hazardous Materials Incident Reports 393
Hazardous Substance Discharge Notification 395

PUBLIC RELATIONS

Introduction 397
Responsibility 397
A Written Plan 397
Emergency Information 397
Importance of Attitude 398
One Company's Procedure 398
Positive Actions in Between 398

MATERIALS OF TRADE

Definition 399
Hazard and Size Limitations 399
Gross Weight 399
Markings 400
Packaging 400
Cylinders 400
Drivers 400
Exceptions 400

LIMITED QUANTITIES

Definition 401
Packaging 401
Markings 401
Labels 402
Shipping Papers 402
Training 402
Exceptions 402
Compliance Dates 402

TABLE OF CONTENTS

CONSUMER COMMODITIES

Definition 403
Packaging 403
Markings 403
Training 403
Exceptions 404
Compliance Dates 404

HAZARDOUS WASTES

Identification of Hazardous Waste 405
EPA Transporter Number 405
Waste Manifest 405
Package Marking 408
Transfer Facility 408
Discharges 408
EPA Identified Hazardous Wastes 408
Use of the Hazardous Waste Tables 409

ID CROSS REFERENCE

Identification Number Cross Reference Index to Proper Shipping Names in §172.101 ...411

HAZARDOUS MATERIALS MARKING CHART

HAZARDOUS MATERIALS LABELING CHART

HAZARDOUS MATERIALS PLACARDING CHART

HAZARDOUS MATERIALS SEGREGATION CHART

REGULATIONS

This section contains select portions of the Hazardous Materials Regulations (49 CFR Parts 107-180 and 397). It does not contain all of the Hazardous Materials Regulations. The regulations included are those most commonly used to comply with the Hazardous Materials Regulations.

PART 107

Subpart D—Enforcement

§107.307 General.

(a) When the Associate Administrator and the Office of Chief Counsel have reason to believe that a person is knowingly engaging or has knowingly engaged in conduct which is a violation of the Federal hazardous material transportation law or any provision of this subchapter or subchapter C of this chapter, or any exemption, special permit, or order issued thereunder, for which the Associate Administrator or the Office of Chief Counsel exercise enforcement authority, they may—

(1) Issue a warning letter, as provided in §107.309;

(2) Initiate proceedings to assess a civil penalty, as provided in either §§107.310 or 107.311;

(3) Issue an order directing compliance, regardless of whether a warning letter has been issued or a civil penalty assessed; and

(4) Seek any other remedy available under the Federal hazardous material transportation law.

(b) In the case of a proceeding initiated for failure to comply with an exemption or special permit, the allegation of a violation of a term or condition thereof is considered by the Associate Administrator and the Office of Chief Counsel to constitute an allegation that the special permit holder or party to the special permit is failing, or has failed to comply with the underlying regulations from which relief was granted by the special permit.

§107.309 Warning letters.

(a) The Associate Administrator may issue a warning letter to any person whom the Associate Administrator believes to have committed a probable violation of the Federal hazardous material transportation law or any provision of this subchapter, subchapter C of this chapter, or any special permit issued thereunder.

(b) A warning letter issued under this section includes:

(1) A statement of the facts upon which the Office of Chief Counsel bases its determination that the person has committed a probable violation;

(2) A statement that the recurrence of the probable violations cited may subject the person to enforcement action; and

(3) An opportunity to respond to the warning letter by submitting pertinent information or explanations concerning the probable violations cited therein.

§107.310 Ticketing.

(a) For an alleged violation that does not have a direct or substantial impact on safety, the Associate Administrator may issue a ticket.

(b) The Associate Administrator issues a ticket by mailing it by certified or registered mail to the person alleged to have committed the violation. The ticket includes:

(1) A statement of the facts on which the Associate Administrator bases the conclusion that the person has committed the alleged violation;

(2) The maximum penalty provided for by statute, the proposed full penalty determined according to PHMSA's civil penalty guidelines and the statutory criteria for penalty assessment, and the ticket penalty amount; and

(3) A statement that within 45 days of receipt of the ticket, the person must pay the penalty in accordance with paragraph (d) of this section, make an informal response under §107.317, or request a formal administrative hearing under §107.319.

(c) If the person makes an informal response or requests a formal administrative hearing, the Associate Administrator forwards the inspection report, ticket and response to the Office of the Chief Counsel for processing under §§107.307–107.339, except that the Office of the Chief Counsel will not issue a Notice of Probable Violation under §107.311. The Office of the Chief Counsel may impose a civil penalty that does not exceed the proposed full penalty set forth in the ticket.

(d) Payment of the ticket penalty amount must be made in accordance with the instructions on the ticket.

(e) If within 45 days of receiving the ticket the person does not pay the ticket amount, make an informal response, or request a formal administrative hearing, the person has waived the right to make an informal response or request a hearing, has admitted the violation and owes the ticket penalty amount to PHMSA.

§107.311 Notice of probable violation.

(a) The Office of Chief Counsel may serve a notice of probable violation on a person alleging the violation of one or more provisions of the Federal hazardous material transportation law or any provision of this subchapter or subchapter C of this chapter, or any special permit, or order issued thereunder.

(b) A notice of probable violation issued under this section includes the following information:

(1) A citation of the provisions of the Federal hazardous material transportation law, an order issued thereunder, this subchapter, subchapter C of this chapter, or the terms of any special permit issued thereunder which the Office of Chief Counsel believes the respondent is violating or has violated.

(2) A statement of the factual allegations upon which the demand for remedial action, a civil penalty, or both, is based.

(3) A statement of the respondent's right to present written or oral explanations, information, and arguments in answer to the allegations and in mitigation of the sanction sought in the notice of probable violation.

(4) A statement of the respondent's right to request a hearing and the procedures for requesting a hearing.

(5) In addition, in the case of a notice of probable violation proposing a compliance order, a statement of the proposed actions to be taken by the respondent to achieve compliance.

(6) In addition, in the case of a notice of probable violation proposing a civil penalty:

(i) A statement of the maximum civil penalty for which the respondent may be liable;

(ii) The amount of the preliminary civil penalty being sought by the Office of Chief Counsel constitutes the maximum amount the Chief Counsel may seek throughout the proceedings; and

(iii) A description of the manner in which the respondent makes payment of any money due the United States as a result of the proceeding.

(c) The Office of Chief Counsel may amend a notice of probable violation at any time before issuance of a compliance order or an order assessing a civil penalty. If the Office of Chief Counsel alleges any new material facts or seeks new or additional remedial action or an increase in the amount of the proposed civil penalty, it issues a new notice of probable violation under this section.

§107.313 Reply.

(a) Within 30 days of receipt of a notice of probable violation, the respondent must either:

(1) Admit the violation under §107.315;

(2) Make an informal response under §107.317; or

(3) Request a hearing under §107.319.

(b) Failure of the respondent to file a reply as provided in this section constitutes a waiver of the respondent's right to appear and contest the allegations and authorizes the Chief Counsel, without further notice to the respondent, to find the facts to be as alleged in the notice of probable violation and issue an order directing compliance or assess a civil penalty, or, if proposed in the notice, both. Failure to request a hearing under paragraph (a)(3) of this section constitutes a waiver of the respondent's right to a hearing.

(c) Upon the request of the respondent, the Office of Chief Council may, for good cause shown and filed within the 30 days prescribed in the notice of probable violation, extend the 30-day response period.

§107.329 Maximum penalties.

(a) A person who knowingly violates a requirement of the Federal hazardous material transportation law, an order issued thereunder, this subchapter, subchapter C of this chapter, or a special permit or approval issued under this subchapter applicable to the transportation of hazardous materials or the causing of them to be transported or shipped is liable for a civil penalty of not more than $55,000 and not less than $250 for each violation, except the maximum civil penalty is $110,000 if the violation results in death, serious illness or severe injury to any person or substantial destruction of property, and a minimum $495 civil penalty applies to a violation relating to training. When the violation is a continuing one, each day of the violation constitutes a separate offense.

(b) A person who knowingly violates a requirement of the Federal hazardous material transportation law, an order issued thereunder, this subchapter, subchapter C of this chapter, or a special permit or approval issued under this subchapter applicable to the design, manufacture, fabrication, inspection, marking, maintenance, reconditioning, repair or testing of a package, container, or packaging component which is represented, marked, certified, or sold by that person as qualified for use in the transportation of hazardous materials in commerce is liable for a civil penalty of not more than $55,000 and not less than $250 for each violation, except the maximum civil penalty is $110,000 if the violation results in death, serious illness or severe injury to any person or substantial destruction of property, and a minimum $495 civil penalty applies to a violation relating to training.

§107.333 Criminal penalties generally.

A person who knowingly violates §171.2(l) of this title or willfully or recklessly violates a requirement of the Federal hazardous material transportation law or a regulation, order, special permits, or approval issued thereunder shall be fined under title 18, United States Code, or imprisoned for not more than 5 years, or both, except the maximum amount of imprisonment shall be 10 years in any case in which the violation involves the release of a hazardous material which results in death or bodily injury to any person.

Subpart G—Registration of Persons Who Offer or Transport Hazardous Materials

§107.601 Applicability.

(a) The registration and fee requirements of this subpart apply to any person who offers for transportation, or transports in foreign, interstate or intrastate commerce—

(1) A highway route-controlled quantity of a Class 7 (radioactive) material, as defined in §173.403 of this chapter;

(2) More than 25 kg (55 pounds) of a Division 1.1, 1.2, or 1.3 (explosive) material (see §173.50 of this chapter) in a motor vehicle, rail car or freight container;

(3) More than one L (1.06 quarts) per package of a material extremely toxic by inhalation (i.e., "material poisonous by inhalation," as defined in §171.8 of this chapter, that meets the criteria for "hazard zone A," as specified in §§173.116(a) or 173.133(a) of this chapter);

(4) A shipment of a quantity of hazardous materials in a bulk packaging (see §171.8 of this chapter) having a capacity equal to or greater than 13,248 L (3,500 gallons) for liquids or gases or more than 13.24 cubic meters (468 cubic feet) for solids;

(5) A shipment in other than a bulk packaging of 2,268 kg (5,000 pounds) gross weight or more of one class of hazardous materials for which placarding of a vehicle, rail car, or freight container is required for that class, under the provisions of subpart F of part 172 of this chapter; or

(6) Except as provided in paragraph (b) of this section, a quantity of hazardous material that requires placarding, under provisions of subpart F of part 172 of this chapter.

(b) Paragraph (a)(6) of this section does not apply to those activities of a farmer, as defined in §171.8 of this chapter, that are in direct support of the farmer's farming operations.

(c) In this subpart, the term "shipment" means the offering or loading of hazardous material at one loading facility using one transport vehicle, or the transport of that transport vehicle.

§107.606 Exceptions.

(a) The following are excepted from the requirements of this subpart:

(1) An agency of the Federal government.

(2) A State agency.

(3) An agency of a political subdivision of a State.

(4) An Indian tribe.

(5) An employee of any of those entities in paragraphs (a)(1) through (a)(4) of this section with respect to the employee's official duties.

(6) A hazmat employee (including, for purposes of this subpart, the owner-operator of a motor vehicle that transports in commerce hazardous materials, if that vehicle at the time of those activities, is leased to a registered motor carrier under a 30-day or longer lease as prescribed in 49 CFR part 376 or an equivalent contractual agreement).

(7) A person domiciled outside the United States, who offers solely from a location outside the United States, hazardous materials for transportation in commerce, provided that the country of which such a person is a domiciliary does not require persons domiciled in the United States, who solely offer hazardous materials for transportation to the foreign country from places in the United States, to file a registration statement or to pay a registration fee.

(b) Upon making a determination that persons domiciled in the United States, who offer hazardous materials for transportation to a foreign country solely from places in the United States, must file registration statements or pay fees to that foreign country, the U.S. Competent Authority will provided notice of such determination directly to the Competent Authority of that foreign country and by publication in the *Federal Register*. Persons who offer hazardous materials for transportation to the United States from that foreign country must file a registration statement and pay the required fee no later than 60 days following publication of the determination in the *Federal Register*.

§107.608 General registration requirements.

(a) Each person subject to this subpart must submit a complete and accurate registration statement on DOT Form F 5800.2 not later than June 30 for each registration year, or in time to comply with paragraph (b) of this section, whichever is later. Each registration year begins on July 1 and ends on June 30 of the following year.

(b) No person required to file a registration statement may transport a hazardous material or cause a hazardous material to be transported or shipped, unless such person has on file, in accordance with §107.620, a current Certificate of Registration in accordance with the requirements of this subpart.

(c) A registrant whose name or principal place of business has changed during the year of registration must notify PHMSA of that change by submitting an amended registration statement not later than 30 days after the change.

(d) Copies of DOT Form F 5800.2 and instructions for its completion may be obtained from the Outreach, Training and Grants Division, PHH-50, U.S. Department of Transportation, Washington, DC 20590-0001, by calling 202-366-4109, or via the Internet at *http://phmsa.dot.gov/hazmat/registration*.

(e) If the registrant is not a resident of the United States, the registrant must attach to the registration statement the name and address of a permanent resident of the United States, designated in accordance with §105.40, to serve as agent for service of process.

§107.612 Amount of fee.

(a) For the registration year 2010-2011 and subsequent years, each person offering for transportation or transporting in commerce a material listed in §107.601(a) must pay an annual registration fee, as follows:

(1) *Small business.* Each person that qualifies as a small business, under criteria specified in 13 CFR part 121 applicable to the North American Industry Classification System (NAICS) code that describes that

person's primary commercial activity, must pay an annual registration fee of $250 and the processing fee required by paragraph (a)(4) of this section.

(2) *Not-for-profit organization.* Each not-for-profit organization must pay an annual registration fee of $250 and the processing fee required by paragraph (a)(4) of this section. A not-for-profit organization is an organization exempt from taxation under 26 U.S.C. 501(a).

(3) *Other than a small business or not-for-profit organization.* Each person that does not meet the criteria specified in paragraph (a)(1) or (a)(2) of this section must pay an annual registration fee of $2,575 and the processing fee required by paragraph (a)(4) of this section.

(4) *Processing fee.* The processing fee is $25 for each registration statement filed. A single statement may be filed for one, two, or three registration years as provided in §107.616(c).

(b) For registration years 2009–2010 and prior years, each person that offered for transportation or transported in commerce a material listed in §107.601(a) during that year must pay the annual registration fee, including the processing fee, specified under the requirements of this subchapter in effect for the specific registration year.

(c) *Registration years 2003–2004, 2004–2005 and 2005–2006.* For registration years 2003–2004, 2004–2005, and 2005–2006, each person subject to the requirements of this subpart must pay an annual registration fee as follows:

(1) *Small business.* Each person that qualifies as a small business, under criteria specified in 13 CFR part 121 applicable to the North American Industry Classification System (NAICS) code that describes that person's primary commercial activity, must pay an annual registration fee of $125 and the processing fee required by paragraph (c)(4) of this section.

(2) *Not-for-profit organization.* Each not-for-profit organization must pay an annual registration fee of $125 and the processing fee required by paragraph (c)(4) of this section. A not-for-profit organization is an organization exempt from taxation under 26 U.S.C. 501(a).

(3) *Other than a small business or not-for-profit organization.* Each person that does not meet the criteria specified in paragraph (c)(1) or (c)(2) of this section must pay an annual registration fee of $275 and the processing fee required by paragraph (c)(4) of this section.

(4) *Processing fee.* The processing fee is $25 for each registration statement filed. A single statement may be filed for one, two, or three registration years as provided in §107.616(c).

(d) *Registration years 2006–2007 and following.* For each registration year beginning with 2006–2007, each person subject to the requirements of this subpart must pay an annual fee as follows:

(1) *Small business.* Each person that qualifies as a small business, under criteria specified in 13 CFR part 121 applicable to the North American Industry Classification System (NAICS) code that describes that person's primary commercial activity, must pay an annual registration fee of $250 and the processing fee required by paragraph (d)(4) of this section.

(2) *Not-for-profit organization.* Each not-for-profit organization must pay an annual registration fee of $250 and the processing fee required by paragraph (d)(4) of this section. A not-for-profit organization is an organization exempt from taxation under 26 U.S.C. 501(a).

(3) *Other than a small business or not-for-profit organization.* Each person that does not meet the criteria specified in paragraph (d)(1) or (d)(2) of this section must pay an annual registration fee of $975 and the processing fee required by paragraph (d)(4) of this section.

(4) *Processing fee.* The processing fee is $25 for each registration statement filed. A single statement may be filed for one, two, or three registration years as provided in §107.616(c).

§107.616 Payment procedures.

(a) Each person subject to the requirements of this subpart must mail the registration statement and payment in full to the U.S. Department of Transportation, Hazardous Materials Registration, P.O. Box 530273, Atlanta, GA 30353-0273, or submit the statement and payment electronically through the Department's e-Commerce Internet site. Access to this service is provided at *http://phmsa.dot.gov/hazmat/registration*. A registrant required to file an amended registration statement under §107.608(c) must mail it to the same address or submit it through the same Internet site.

(b) Payment must be made by certified check, cashier's check, personal check, or money order in U.S. funds and drawn on a U.S. bank, payable to the U.S. Department of Transportation and identified as payment for the "Hazmat Registration Fee," or by completing an authorization for payment by credit card or other electronic means of payment acceptable to the Department on the registration statement or as part of an Internet registration as provided in paragraph (a) of this section.

(c) Payment must correspond to the total fees properly calculated in the "Amount Due" block of the DOT form F 5800.2. A person may elect to register and pay the required fees for up to three registration years by filing one complete and accurate registration statement.

§107.620 Recordkeeping requirements.

(a) Each person subject to the requirements of this subpart, or its agent designated under §107.608(e), must maintain at its principal place of business for a period of three years from the date of issuance of each Certificate of Registration:

(1) A copy of the registration statement filed with PHMSA; and

(2) The Certificate of Registration issued to the registrant by PHMSA.

(b) After January 1, 1993, each motor carrier subject to the requirements of this subpart must carry a copy of its current Certificate of Registration issued by PHMSA or another document bearing the registration number identified as the "U.S. DOT Hazmat Reg. No." on board each truck and truck tractor (not including trailers and semi-trailers) used to transport hazardous materials subject to the requirements of this subpart. The Certificate of Registration or document bearing the registration number must be made available, upon request, to enforcement personnel.

(c) In addition to the requirements of paragraph (a) of this section, after January 1, 1995, each person who transports by vessel a hazardous material subject to the requirements of this subpart must carry on board the vessel a copy of its current Certificate of Registration or another document bearing the current registration number identified as the "U.S. DOT Hazmat Reg. No."

(d) Each person subject to this subpart must furnish its Certificate of Registration (or a copy thereof) and all other records and information pertaining to the information contained in the registration statement to an authorized representative or special agent of DOT upon request.

Reserved

PART 171

§171.1 Applicability of Hazardous Materials Regulations (HMR) to persons and functions.

Federal hazardous materials transportation law (49 U.S.C. 5101 *et seq.*) directs the Secretary of Transportation to establish regulations for the safe and secure transportation of hazardous materials in commerce, as the Secretary considers appropriate. The Secretary is authorized to apply these regulations to persons who transport hazardous materials in commerce. In addition, the law authorizes the Secretary to apply these regulations to persons who cause hazardous materials to be transported in commerce. The law also authorizes the Secretary to apply these regulations to persons who manufacture or maintain a packaging or a component of a packaging that is represented, marked, certified, or sold as qualified for use in the transportation of a hazardous material in commerce. Federal hazardous material transportation law also applies to anyone who indicates by marking or other means that a hazardous material being transported in commerce is present in a package or transport conveyance when it is not, and to anyone who tampers with a package or transport conveyance used to transport hazardous materials in commerce or a required marking, label, placard, or shipping description. Regulations prescribed in accordance with Federal hazardous materials transportation law shall govern safety aspects, including security, of the transportation of hazardous materials that the Secretary considers appropriate. In 49 CFR 1.53, the Secretary delegated authority to issue regulations for the safe and secure transportation of hazardous materials in commerce to the Pipeline and Hazardous Materials Safety Administrator. The Administrator issues the Hazardous Materials Regulations (HMR; 49 CFR Parts 171 through 180) under that delegated authority. This section addresses the applicability of the HMR to packagings represented as qualified for use in the transportation of hazardous materials in commerce and to pre-transportation and transportation functions.

(a) *Packagings.* Requirements in the HMR apply to each person who manufactures, fabricates, marks, maintains, reconditions, repairs, or tests a packaging or a component of a packaging that is represented, marked, certified, or sold as qualified for use in the transportation of a hazardous material in commerce, including each person under contract with any department, agency, or instrumentality of the executive, legislative, or judicial branch of the Federal government who manufactures, fabricates, marks, maintains, reconditions, repairs, or tests a packaging or a component of a packaging that is represented, marked, certified, or sold as qualified for use in the transportation of a hazardous material in commerce.

(b) *Pre-transportation functions.* Requirements in the HMR apply to each person who offers a hazardous material for transportation in commerce, causes a hazardous material to be transported in commerce, or transports a hazardous material in commerce and who performs or is responsible for performing a pre-transportation function, including each person performing pre-transportation functions under contract with any department, agency, or instrumentality of the executive, legislative, or judicial branch of the Federal government. Pre-transportation functions include, but are not limited to, the following:

(1) Determining the hazard class of a hazardous material.

(2) Selecting a hazardous materials packaging.

(3) Filling a hazardous materials packaging, including a bulk packaging.

(4) Securing a closure on a filled or partially filled hazardous materials package or container or on a package or container containing a residue of a hazardous material.

(5) Marking a package to indicate that it contains a hazardous material.

(6) Labeling a package to indicate that it contains a hazardous material.

(7) Preparing a shipping paper.

(8) Providing and maintaining emergency response information.

(9) Reviewing a shipping paper to verify compliance with the HMR or international equivalents.

(10) For each person importing a hazardous material into the United States, providing the shipper with timely and complete information as to the HMR requirements that will apply to the transportation of the material within the United States.

(11) Certifying that a hazardous material is in proper condition for transportation in conformance with the requirements of the HMR.

(12) Loading, blocking, and bracing a hazardous materials package in a freight container or transport vehicle.

(13) Segregating a hazardous materials package in a freight container or transport vehicle from incompatible cargo.

(14) Selecting, providing, or affixing placards for a freight container or transport vehicle to indicate that it contains a hazardous material.

(c) *Transportation functions.* Requirements in the HMR apply to transportation of a hazardous material in commerce and to each person who transports a hazardous material in commerce, including each person under contract with any department, agency, or instrumentality of the executive, legislative, or judicial branch of the Federal government who transports a hazardous material in commerce. Transportation of a hazardous material in commerce begins when a carrier takes physical possession of the hazardous material for the purpose of transporting it and continues until the package containing the hazardous material is delivered to the destination indicated on a shipping document, package marking, or other medium, or, in the case of a rail car, until the car is delivered to a private track or siding. For a private motor carrier, transportation of a hazardous material in commerce begins when a motor vehicle driver takes possession of a hazardous material for the purpose of transporting it and continues until the driver relinquishes possession of the package containing the hazardous material at its destination and is no longer responsible for performing functions subject to the HMR with respect to that particular package. Transportation of a hazardous material in commerce includes the following:

(1) *Movement.* Movement of a hazardous material by rail car, aircraft, motor vehicle, or vessel (except as delegated by Department of Homeland Security Delegation No. 0170 at 2(103)).

(2) *Loading incidental to movement of a hazardous material.* Loading of packaged or containerized hazardous material onto a transport vehicle, aircraft, or vessel for the purpose of transporting it, including blocking and bracing a hazardous materials package in a freight container or transport vehicle, and segregating a hazardous materials package in a freight container or transport vehicle from incompatible cargo, when performed by carrier personnel or in the presence of carrier personnel. For a bulk packaging, loading incidental to movement is filling the packaging with a hazardous material for the purpose of transporting it when performed by carrier personnel or in the presence of carrier personnel (except as delegated by Department of Homeland Security Delegation No. 0170 at 2(103)), including transloading.

(3) *Unloading incidental to movement of a hazardous material.* Removing a package or containerized hazardous material from a transport vehicle, aircraft, or vessel; or for a bulk packaging, emptying a hazardous material from the bulk packaging after the hazardous material has been delivered to the consignee when performed by carrier personnel or in the presence of carrier personnel or, in the case of a private motor carrier, while the driver of the motor vehicle from which the hazardous material is being unloaded immediately after movement is completed is present during the unloading operation. (Emptying a hazardous material from a bulk packaging while the packaging is on board a vessel is subject to separate regulations as delegated by Department of Homeland Security Delegation No. 0170 at 2(103).) Unloading incidental to movement includes transloading.

(4) *Storage incidental to movement of a hazardous material.* Storage of a transport vehicle, freight container, or package containing a hazardous material by any person between the time that a carrier takes physical possession of the hazardous material for the purpose of transporting it until the package containing the hazardous material has been delivered to the destination indicated on a shipping document, package marking, or other medium, or, in the case of a private motor carrier, between the time that a motor vehicle driver takes physical possession of the hazardous material for the purpose of transporting it until the driver relinquishes possession of the package at its destination and is no longer responsible for performing functions subject to the HMR with respect to that particular package.

(i) Storage incidental to movement includes—

(A) Storage at the destination shown on a shipping document, including storage at a transloading facility, provided the original shipping documentation identifies the shipment as a through-shipment and identifies the final destination or destinations of the hazardous material; and

(B) A rail car containing a hazardous material that is stored on track that does not meet the definition of "private track or siding" in §171.8, even if the car has been delivered to the destination shown on the shipping document.

(ii) Storage incidental to movement does not include storage of a hazardous material at its final destination as shown on a shipping document.

(d) *Functions not subject to the requirements of the HMR.* The following are examples of activities to which the HMR do not apply:

(1) Storage of a freight container, transport vehicle, or package containing a hazardous material at an offeror facility prior to a carrier taking possession of the hazardous material for movement in transportation in commerce or, for a private motor carrier, prior to a motor vehicle driver taking physical possession of the hazardous material for movement in transportation in commerce.

(2) Unloading of a hazardous material from a transport vehicle or a bulk packaging performed by a person employed by or working under contract to the consignee following delivery of the hazardous material by the carrier to its destination and departure from the consignee's premises of the carrier's personnel or, in the case of a private carrier, departure of the driver from the unloading area.

(3) Storage of a freight container, transport vehicle, or package containing a hazardous material after its delivery by a carrier to the destination indicated on a shipping document, package marking, or other medium, or, in the case of a rail car, storage of a rail car on private track.

(4) Rail and motor vehicle movements of a hazardous material exclusively within a contiguous facility boundary where public access is restricted, except to the extent that the movement is on or crosses a public road or is on track that is part of the general railroad system of transportation, unless access to the public road is restricted by signals, lights, gates, or similar controls.

(5) Transportation of a hazardous material in a motor vehicle, aircraft, or vessel operated by a Federal, state, or local government employee solely for noncommercial Federal, state, or local government purposes.

(6) Transportation of a hazardous material by an individual for non-commercial purposes in a private motor vehicle, including a leased or rented motor vehicle.

(7) Any matter subject to the postal laws and regulations of the United States.

(e) *Requirements of other Federal agencies.* Each facility at which pre-transportation or transportation functions are performed in accordance with the HMR may be subject to applicable standards and regulations of other Federal agencies.

(f) *Requirements of state and local government agencies.* (1) Under 49 U.S.C. 5125, a requirement of a state, political subdivision of a state, or an Indian tribe is preempted, unless otherwise authorized by another Federal statute or DOT issues a waiver of preemption, if—

(i) Complying with both the non-Federal requirement and Federal hazardous materials transportation law, the regulations issued under Federal hazardous material transportation law or a hazardous material transportation security regulation or directive issued by the Secretary of Homeland Security is not possible;

(ii) The non-Federal requirement, as applied or enforced, is an obstacle to accomplishing and carrying out Federal hazardous materials transportation law, the regulations issued under Federal hazardous material transportation law, or a hazardous material transportation security regulation or directive issued by the Secretary of Homeland Security;

(iii) The non-Federal requirement is not substantively the same as a provision of Federal hazardous materials transportation law, the regulations issued under Federal hazardous material transportation law, or a hazardous material transportation security regulation or directive issued by the Secretary of Homeland Security with respect to—

(A) The designation, description, and classification of hazardous material;

(B) The packing, repacking, handling, labeling, marking, and placarding of hazardous material;

(C) The preparation, execution, and use of shipping documents related to hazardous material and requirements related to the number, contents, and placement of those documents;

(D) The written notification, recording, and reporting of the unintentional release of hazardous material; or

(E) The design, manufacturing, fabricating, marking, maintenance, reconditioning, repairing, or testing of a package or container represented, marked, certified, or sold as qualified for use in transporting hazardous material.

(iv) A non-Federal designation, limitation or requirement on highway routes over which hazardous material may or may not be transported does not comply with the regulations in subparts C and D of part 397 of this title; or

(v) A fee related to the transportation of a hazardous material is not fair or is used for a purpose that is not related to transporting hazardous material, including enforcement and planning, developing, and maintaining a capability for emergency response.

(2) Subject to the limitations in paragraph (f)(1) of this section, each facility at which functions regulated under the HMR are performed may be subject to applicable laws and regulations of state and local governments and Indian tribes.

(3) The procedures for DOT to make administrative determinations of preemption are set forth in subpart E of part 397 of this title with respect to non-Federal requirements on highway routing (paragraph (f)(1)(iv) of this section) and in subpart C of part 107 of this chapter with respect to all other non-Federal requirements.

(g) *Penalties for noncompliance.* Each person who knowingly violates a requirement of the Federal hazardous material transportation law, an order issued under Federal hazardous material transportation law, subchapter A of this chapter, or a special permit or approval issued under subchapter A or C of this chapter is liable for a civil penalty of not more than $55,000 and not less than $250 for each violation, except the maximum civil penalty is $110,000 if the violation results in death, serious illness or severe injury to any person or substantial destruction of property, and a minimum $495 civil penalty applies to a violation relating to training. When a violation is a continuing one and involves transporting of hazardous material or causing them to be transported, each day of the violation is a separate offense. Each person who knowingly violates §171.2(l) or willfully or recklessly violates a provision of the Federal hazardous material transportation law, an order issued under Federal hazardous material transportation law, subchapter A of this chapter, or a special permit or approval issued under subchapter A or C of this chapter, shall be fined under title 18, United States Code, or imprisoned for not more than 5 years, or both, except the maximum amount of imprisonment shall be 10 years in any case in which a violation involves the release of a hazardous material which results in death or bodily injury to any person.

§171.2 General requirements.

(a) Each person who performs a function covered by this subchapter must perform that function in accordance with this subchapter.

(b) Each person who offers a hazardous material for transportation in commerce must comply with all applicable requirements of this subchapter, or an exemption or special permit, approval, or registration issued under this subchapter or under subchapter A of this chapter. There may be more than one offeror of a shipment of hazardous materials. Each offeror is responsible for complying with the requirements of this subchapter, or an exemption or special permit, approval, or registration issued under this subchapter or subchapter A of this chapter, with respect to any pre-transportation function that it performs or is required to perform; however, each offeror is responsible only for the specific pre-transportation functions that it performs or is required to perform, and each offeror may rely on information provided by another offeror, unless that offeror knows or, a reasonable person, acting in the circumstances and exercising reasonable care, would have knowledge that the information provided by the other offeror is incorrect.

(c) Each person who performs a function covered by or having an effect on a specification or activity prescribed in part 178, 179, or 180 of this subchapter, an approval issued under this subchapter, or an exemption or special permit issued under subchapter A of this chapter, must perform the function in accordance with that specification, approval, or exemption or special permit, as appropriate.

(d) No person may offer or accept a hazardous material for transportation in commerce or transport a hazardous material in commerce unless that person is registered in conformance with subpart G of part 107 of this chapter, if applicable.

(e) No person may offer or accept a hazardous material for transportation in commerce unless the hazardous material is properly classed, described, packaged, marked, labeled, and in condition for shipment as required or authorized by applicable requirements of this subchapter or an exemption or special permit, approval, or registration issued under this subchapter or subchapter A of this chapter.

(f) No person may transport a hazardous material in commerce unless the hazardous material is transported in accordance with applicable requirements of this subchapter, or an exemption or special permit, approval, or registration issued under this subchapter or subchapter A of this chapter. Each carrier who transports a hazardous material in commerce may rely on information provided by the offeror of the hazardous material or a prior carrier, unless the carrier knows or, a reasonable person, acting in the circumstances and exercising reasonable care, would have knowledge that the information provided by the offeror or prior carrier is incorrect.

(g) No person may represent, mark, certify, sell, or offer a packaging or container as meeting the requirements of this subchapter governing its use in the transportation of a hazardous material in commerce unless the packaging or container is manufactured, fabricated, marked, maintained, reconditioned, repaired, and retested in accordance with the applicable requirements of this subchapter. No person may represent, mark, certify, sell, or offer a packaging or container as meeting the requirements of an exemption, a special permit, approval, or registration issued under this subchapter or subchapter A of this chapter unless the packaging or container is manufactured, fabricated, marked, maintained, reconditioned, repaired, and retested in accordance with the applicable requirements of the exemption, special permit, approval, or registration issued under this subchapter or subchapter A of this chapter. The requirements of this paragraph apply whether or not the packaging or container is used or to be used for the transportation of a hazardous material.

(h) The representations, markings, and certifications subject to the prohibitions of paragraph (g) of this section include:

(1) Specification identifications that include the letters "ICC", "DOT", "CTC", "MC", or "UN";

(2) Exemption, special permit, approval, and registration numbers that include the letters "DOT", "EX", "M", or "R"; and

(3) Test dates associated with specification, registration, approval, retest, exemption, or special permit markings indicating compliance with a test or retest requirement of the HMR, or an exemption, special permit, approval, or registration issued under the HMR or under subchapter A of this chapter.

(i) No person may certify that a hazardous material is offered for transportation in commerce in accordance with the requirements of this subchapter unless the hazardous material is properly classed, described, packaged, marked, labeled, and in condition for shipment as required or authorized by applicable requirements of this subchapter or an exemption or special permit, approval, or registration issued under this subchapter or subchapter A of this chapter. Each person who offers a package containing a hazardous material for transportation in commerce in accordance with the requirements of this subchapter or an exemption or special permit, approval, or registration issued under this subchapter or subchapter A of this chapter, must assure that the package remains in condition for shipment until it is in the possession of the carrier.

(j) No person may, by marking or otherwise, represent that a container or package for transportation of a hazardous material is safe, certified, or in compliance with the requirements of this chapter unless it meets the requirements of all applicable regulations issued under Federal hazardous material transportation law.

(k) No person may, by marking or otherwise, represent that a hazardous material is present in a package, container, motor vehicle, rail car, aircraft, or vessel if the hazardous material is not present.

(l) No person may alter, remove, deface, destroy, or otherwise unlawfully tamper with any marking, label, placard, or description on a document required by Federal hazardous material transportation law or the regulations issued under Federal hazardous material transportation law. No person may alter, deface, destroy, or otherwise unlawfully tamper with a package, container, motor vehicle, rail car, aircraft, or vessel used for the transportation of hazardous materials.

(m) No person may falsify or alter an exemption or special permit, approval, registration, or other grant of authority issued under this subchapter or subchapter A of this chapter. No person may offer a hazardous material for transportation or transport a hazardous material in commerce under an exemption or special permit, approval, registration or other grant of authority issued under this subchapter or subchapter A of this chapter if such grant of authority has been altered without the consent of the issuing authority. No person may represent, mark, certify, or sell a packaging or container under an exemption or special permit, approval, registration or other grant of authority issued under this subchapter or subchapter A of this chapter if such grant of authority has been altered without the consent of the issuing authority.

§171.3 Hazardous waste.

(a) No person may offer for transportation or transport a hazardous waste (as defined in §171.8 of this subchapter) in interstate or intrastate commerce except in accordance with the requirements of this subchapter.

(b) No person may accept for transportation, transport, or deliver a hazardous waste for which a manifest is required unless that person:

(1) Has marked each motor vehicle used to transport hazardous waste in accordance with §390.21 of this title even though placards may not be required;

(2) Complies with the requirements for manifests set forth in §172.205 of this subchapter; and

(3) Delivers, as designated on the manifest by the generator, the entire quantity of the waste received from the generator or a transporter to:

(i) The designated facility or, if not possible, to the designated alternate facility;

(ii) The designated subsequent carrier; or

(iii) A designated place outside the United States.

NOTE: Federal law specifies penalties up to $250,000 fine for an individual and $500,000 for a company and 5 years imprisonment for the willful discharge of hazardous waste at other than designated facilities. 49 U.S.C. 5124.

(c) If a discharge of hazardous waste or other hazardous material occurs during transportation, and an official of a State or local government or a Federal agency, acting within the scope of his official responsibilities, determines that immediate removal of the waste is necessary to prevent further consequence, that official may authorize the removal of the waste without the preparation of a manifest. [NOTE: In such cases, EPA does not require carriers to have EPA identification numbers.]

NOTE 1: EPA requires shippers (generators) and carriers (transporters) of hazardous wastes to have identification numbers which must be displayed on hazardous waste manifests. See 40 CFR parts 262 and 263. (Identification number application forms may be obtained from EPA regional offices.)

NOTE 2: In 40 CFR part 263, the EPA sets forth requirements for the cleanup of releases of hazardous wastes.

§171.4 Marine pollutants.

(a) Except as provided in paragraph (c) of this section, no person may offer for transportation or transport a marine pollutant, as defined in §171.8, in intrastate or interstate commerce except in accordance with the requirements of this subchapter.

(b) The requirements of this subchapter for the transportation of marine pollutants are based on the provisions of Annex III of the 1973 International Convention for Prevention of Pollution from Ships, as modified by the Protocol of 1978 (MARPOL 73/78).

(c) *Exceptions.* Except when all or part of the transportation is by vessel, the requirements of this subchapter specific to marine pollutants do not apply to non-bulk packagings transported by motor vehicle, rail car or aircraft.

Reserved

§171.8 Definitions and abbreviations.

In this subchapter,

Administrator means the Administrator, Pipeline and Hazardous Materials Safety Administration.

Aerosol means any non-refillable receptacle containing a gas compressed, liquefied or dissolved under pressure, the sole purpose of which is to expel a nonpoisonous (other than a Division 6.1 Packing Group III material) liquid, paste, or powder and fitted with a self-closing release device allowing the contents to be ejected by the gas.

Aggregate lithium content means the sum of the grams of lithium content or equivalent lithium content contained by the cells comprising a battery.

Agricultural product means a hazardous material, other than a hazardous waste, whose end use directly supports the production of an agricultural commodity including, but not limited to a fertilizer, pesticide, soil amendment or fuel. An *agricultural product* is limited to a material in Class 3, 8 or 9, Division 2.1, 2.2, 5.1, or 6.1, or an ORM-D material.

Approval means a written authorization, including a competent authority approval, from the Associate Administrator or other designated Department official, to perform a function for which prior authorization by the Associate Administrator is required under subchapter C of this chapter (49 CFR parts 171 through 180.)

Approved means approval issued or recognized by the Department unless otherwise specifically indicated in this subchapter.

Asphyxiant gas means a gas which dilutes or replaces oxygen normally in the atmosphere.

Associate Administrator means the Associate Administrator for Hazardous Materials Safety, Pipeline and Hazardous Materials Safety Administration.

Atmospheric gases means air, nitrogen, oxygen, argon, krypton, neon and xenon.

Authorized Inspection Agency means: (1) A jurisdiction which has adopted and administers one or more sections of the ASME Boiler and Pressure Vessel Code as a legal requirement and has a representative serving as a member of the ASME Conference Committee; or (2) an insurance company which has been licensed or registered by the appropriate authority of a State of the United States or a Province of Canada to underwrite boiler and pressure vessel insurance in such State or Province.

Authorized Inspector means an Inspector who is currently commissioned by the National Board of Boiler and Pressure Vessel Inspectors and employed as an Inspector by an Authorized Inspection Agency.

Bag means a flexible packaging made of paper, plastic film, textiles, woven material or other similar materials.

Bar means 1 BAR = 100 kPa (14.5 psi).

Barge means a non-selfpropelled vessel.

Biological product. See §173.134 of this subchapter.

Biological substances, Category B. See §173.134 of this subchapter.

Bottle means an inner packaging having a neck of relatively smaller cross section than the body and an opening capable of holding a closure for retention of the contents.

Bottom shell means that portion of a tank car tank surface, excluding the head ends of the tank car tank, that lies within two feet, measured circumferentially, of the bottom longitudinal center line of the tank car tank.

Box means a packaging with complete rectangular or polygonal faces, made of metal, wood, plywood, reconstituted wood, fiberboard, plastic, or other suitable material. Holes appropriate to the size and use of the packaging, for purposes such as ease of handling or opening, or to meet classification requirements, are permitted as long as they do not compromise the integrity of the packaging during transportation and are not otherwise prohibited in this subchapter.

Break-bulk means packages of hazardous materials that are handled individually, palletized, or unitized for purposes of transportation as opposed to bulk and containerized freight.

Btu means British thermal unit.

Bulk packaging means a packaging, other than a vessel or a barge, including a transport vehicle or freight container, in which hazardous materials are loaded with no intermediate form of containment. A Large Packaging in which hazardous materials are loaded with an intermediate form of containment, such as one or more articles or inner packagings, is also a bulk packaging. Additionally, a bulk packaging has:

(1) A maximum capacity greater than 450 L (119 gallons) as a receptacle for a liquid;

(2) A maximum net mass greater than 400 kg (882 pounds) and a maximum capacity greater than 450 L (119 gallons) as a receptacle for a solid; or

(3) A water capacity greater than 454 kg (1000 pounds) as a receptacle for a gas as defined in §173.115 of this subchapter.

Bundle of cylinders means assemblies of UN cylinders fastened together and interconnected by a manifold and transported as a unit. The total water capacity for the bundle may not exceed 3,000 L, except that a bundle intended for the transport of gases in Division 2.3 is limited to a water capacity of 1,000 L.

Bureau of Explosives means the Bureau of Explosives (B of E) of the Association of American Railroads.

C means Celsius or Centigrade.

Captain of the Port (COTP) means the officer of the Coast Guard, under the command of a District Commander, so designated by the Commandant for the purpose of giving immediate direction to Coast Guard law enforcement activities within an assigned area. As used in this subchapter, the term *Captain of the Port* includes an authorized representative of the Captain of the Port.

Carfloat means a vessel that operates on a short run on an irregular basis and serves one or more points in a port area as an extension of a rail line or highway over water, and does not operate in ocean, coastwise, or ferry service.

Cargo aircraft only means an aircraft that is used to transport cargo and is not engaged in carrying passengers. For purposes of this subchapter, the terms *cargo aircraft only, cargo-only aircraft and cargo aircraft* have the same meaning.

Cargo tank means a bulk packaging that:

(1) Is a tank intended primarily for the carriage of liquids or gases and includes appurtenances, reinforcements, fittings, and closures (for the definition of a tank,

see 49 CFR 178.320, 178.337-1, or 178.338-1, as applicable);

(2) Is permanently attached to or forms a part of a motor vehicle, or is not permanently attached to a motor vehicle but which, by reason of its size, construction or attachment to a motor vehicle is loaded or unloaded without being removed from the motor vehicle; and

(3) Is not fabricated under a specification for cylinders, intermediate bulk containers, multi-unit tank car tanks, portable tanks, or tank cars.

Cargo tank motor vehicle means a motor vehicle with one or more cargo tanks permanently attached to or forming an integral part of the motor vehicle.

Cargo vessel means:

(1) Any vessel other than a passenger vessel; and

(2) Any ferry being operated under authority of a change of character certificate issued by a Coast Guard Officer-in-Charge, Marine Inspection.

Carrier means a person who transports passengers or property in commerce by rail car, aircraft, motor vehicle, or vessel.

CC means closed-cup.

Character of vessel means the type of service in which the vessel is engaged at the time of carriage of a hazardous material.

Class means hazard class. See *hazard class*.

Class 1. See §173.50 of this subchapter.

Class 2. See §173.115 of this subchapter.

Class 3. See §173.120 of this subchapter.

Class 4. See §173.124 of this subchapter.

Class 5. See §173.128 of this subchapter.

Class 6. See §173.132 of this subchapter.

Class 7. See §173.403 of this subchapter.

Class 8. See §173.136 of this subchapter.

Class 9. See §173.140 of this subchapter.

Closure means a device which closes an opening in a receptacle.

COFC means container-on-flat-car.

Combination packaging means a combination of packaging, for transport purposes, consisting of one or more inner packagings secured in a non-bulk outer packaging. It does not include a composite packaging.

Combustible liquid. See §173.120 of this subchapter.

Commerce means trade or transportation in the jurisdiction of the United States within a single state; between a place in a state and a place outside of the state; that affects trade or transportation between a place in a state and place outside of the state; or on a United States-registered aircraft.

Compatibility group letter means a designated alphabetical letter used to categorize different types of explosive substances and articles for purposes of stowage and segregation. See §173.52 of this subchapter.

Competent Authority means a national agency responsible under its national law for the control or regulation of a particular aspect of the transportation of hazardous materials (dangerous goods). The term *Appropriate Authority*, as used in the ICAO Technical Instructions (IBR, see §171.7), has the same meaning as *Competent Authority*. For purposes of this subchapter, the Associate Administrator is the Competent Authority for the United States.

Composite packaging means a packaging consisting of an outer packaging and an inner receptacle, so constructed that the inner receptacle and the outer packaging form an integral packaging. Once assembled it remains thereafter an integrated single unit; it is filled, stored, shipped and emptied as such.

Compressed gas. See §173.115 of this subchapter.

Consignee means the person or place shown on a shipping document, package marking, or other media as the location to which a carrier is directed to transport a hazardous material.

Consumer commodity means a material that is packaged and distributed in a form intended or suitable for sale through retail sales agencies or instrumentalities for consumption by individuals for purposes of personal care or household use. This term also includes drugs and medicines.

Containership means a cargo vessel designed and constructed to transport, within specifically designed cells, portable tanks and freight containers which are lifted on and off with their contents intact.

Corrosive material. See §173.136 of this subchapter.

Crate means an outer packaging with incomplete surfaces.

Crewmember means a person assigned to perform duty in an aircraft during flight time.

Cryogenic liquid. See §173.115(g) of this subchapter.

Cultures and stocks. See §173.134 of this subchapter.

Cylinder means a pressure vessel designed for pressure higher than 40 psia and having a circular cross section. It does not include a portable tank, multi-unit tank car tank, cargo tank, or tank car.

Dangerous when wet material. See §173.124 of this subchapter.

Design Certifying Engineer means a person registered with the Department in accordance with subpart F of part 107 of this chapter who has the knowledge and ability to perform stress analysis of pressure vessels and otherwise determine whether a cargo tank design and construction meets the applicable DOT specification. A *Design Certifying Engineer* meets the knowledge and ability requirements of this section by meeting any one of the following requirements:

(1) Has an engineering degree and one year of work experience in cargo tank structural or mechanical design;

(2) Is currently registered as a professional engineer by appropriate authority of a state of the United States or a province of Canada; or

(3) Has at least three years' experience in performing the duties of a Design Certifying Engineer prior to September 1, 1991.

Designated facility means a hazardous waste treatment, storage, or disposal facility that has been designated on the manifest by the generator.

District Commander means the District Commander of the Coast guard, or his authorized representative, who has jurisdiction in the particular geographical area.

Division means a subdivision of a hazard class.

DOD means the U.S. Department of Defense.

Domestic transportation means transportation between places within the United States other than through a foreign country.

DOT or Department means U.S. Department of Transportation.

Drum means a flat-ended or convex-ended cylindrical packaging made of metal, fiberboard, plastic, plywood, or other suitable materials. This definition also includes packagings of other shapes made of metal or plastic (e.g., round taper-necked packagings or pail-shaped packagings) but does not include cylinders, jerricans, wooden barrels or bulk packagings.

Elevated temperature material means a material which, when offered for transportation or transported in a bulk packaging:

(1) Is in a liquid phase and at a temperature at or above 100 °C (212 °F);

(2) Is in a liquid phase with a flash point at or above 38 °C (100 °F) that is intentionally heated and offered for transportation or transported at or above its flash point; or

(3) Is in a solid phase and at a temperature at or above 240 °C (464 °F).

Engine means a locomotive propelled by any form of energy and used by a railroad.

EPA means U.S. Environmental Protection Agency.

Equivalent lithium content means, for a lithium-ion cell, the product of the rated capacity, in ampere-hours, of a lithium-ion cell times 0.3, with the result expressed in grams. The equivalent lithium content of a battery equals the sum of the grams of equivalent lithium content contained in the component cells of the battery.

Etilogic agent. See §173.134 of this subchapter.

EX number means a number preceded by the prefix "EX", assigned by the Associate Administrator, to an item that has been evaluated under the provisions of §173.56 of this subchapter.

Explosive. See §173.50 of this subchapter.

F means degree Fahrenheit.

Farmer means a person engaged in the production or raising of crops, poultry, or livestock.

Federal hazardous material transportation law means 49 U.S.C. 5101 *et seq.*

Ferry vessel means a vessel which is limited in its use to the carriage of deck passengers or vehicles or both, operates on a short run on a frequent schedule between two points over the most direct water route, other than in ocean or coastwise service, and is offered as a public service of a type normally attributed to a bridge or tunnel.

Filling density has the following meanings:

(1) For compressed gases in cylinders, see §173.304a(a)(2) table note 1.

(2) For compressed gases in tank cars, see §173.314(c) table note 1.

(3) For compressed gases in cargo tanks and portable tanks, see §173.315(a) table note 1.

(4) For cryogenic liquids in cylinders, except hydrogen, see §173.316(c)(1).

(5) For hydrogen, cryogenic liquid in cylinders, see §173.316(c)(3) table note 1.

(6) For cryogenic liquids in cargo tanks, see §173.318(f)(1).

(7) For cryogenic liquids in tank cars, see §173.319(d)(1).

Flammable gas. See §173.115 of this subchapter.

Flammable liquid. See §173.120 of this subchapter.

Flammable solid. See §173.124 of this subchapter.

Flash point. See §173.120 of this subchapter.

Freight container means a reusable container having a volume of 64 cubic feet or more designed and constructed to permit being lifted with its contents intact and intended primarily for containment of packages (in unit form) during transportation.

Fuel cell means an electrochemical device that converts the energy of the chemical reaction between a fuel, such as hydrogen or hydrogen rich gases, alcohols, or hydrocarbons, and an oxidant, such as air or oxygen, to direct current (d.c.) power, heat, and other reaction products.

Fuel cell cartridge or fuel cartridge means an article that stores fuel for discharge into the fuel cell through a valve(s) that controls the discharge of fuel into the fuel cell.

Fuel cell system means a fuel cell with an installed fuel cell cartridge together with wiring, valves, and other attachments that connect the fuel cell or cartridge to the device it powers. The fuel cell or cartridge may be so constructed that it forms an integral part of the device or may be removed and connected manually to the device.

Fuel tank means a tank other than a cargo tank, used to transport flammable or combustible liquid, or compressed gas for the purpose of supplying fuel for propulsion of the transport vehicle to which it is attached, or for the operation of other equipment on the transport vehicle.

Fumigated lading. See §§172.302(g) and 173.9.

Gas means a material which has a vapor pressure greater than 300 kPa (43.5 psia) at 50°C (122°F) or is completely gaseous at 20°C (68°F) at a standard pressure of 101.3 kPa (14.7 psia).

Gross weight or Gross mass means the weight of a packaging plus the weight of its contents.

Hazard class means the category of hazard assigned to a hazardous material under the definitional criteria of part 173 of this subchapter and the provisions of the §172.101 Table. A material may meet the defining criteria for more than one hazard class but is assigned to only one hazard class.

Hazard zone means one of four levels of hazard (Hazard Zones A through D) assigned to gases, as specified in §173.116(a) of this subchapter, and one of two levels of hazards (Hazard Zones A and B) assigned to liquids that are poisonous by inhalation, as specified in §173.133(a) of this subchapter. A hazard zone is based on the LC50 value for acute inhalation toxicity of gases and vapors, as specified in §173.133(a).

Hazardous material means a substance or material that the Secretary of Transportation has determined is capable of posing an unreasonable risk to health, safety, and property when transported in commerce, and has designated as hazardous under section 5103 of Federal hazardous materials transportation law (49 U.S.C. 5103). The term includes hazardous substances, hazardous wastes, marine pollutants, elevated temperature materials, materials designated as hazardous in the Hazardous Materials Table (see 49 CFR 172.101), and materials that meet the defining criteria for hazard classes and divisions in part 173 of subchapter C of this chapter.

Hazardous substance for the purposes of this subchapter, means a material, including its mixtures and solutions, that—

(1) Is listed in Appendix A to §172.101 of this subchapter;

(2) Is in a quantity, in one package, which equals or exceeds the reportable quantity (RQ) listed in Appendix A to §172.101 of this subchapter; and

(3) When in a mixture or solution—

(i) For radionuclides, conforms to paragraph 7 of Appendix A to §172.101.

(ii) For other than radionuclides, is in a concentration by weight which equals or exceeds the concentration corresponding to the RQ of the material, as shown in the following table:

RQ pounds (kilograms)	Concentration by weight	
	Percent	PPM
5000 (2270)	10	100,000
1000 (454)	2	20,000
100(45.4)	0.2	2,000
10 (4.54)	0.02	200
1 (0.45)	0.002	20

The term does not include petroleum, including crude oil or any fraction thereof which is not otherwise specifically listed or designated as a hazardous substance in Appendix A to §172.101 of this subchapter, and the term does not include natural gas, natural gas liquids, liquefied natural gas, or synthetic gas usable for fuel (or mixtures of natural gas and such synthetic gas).

Hazardous waste, for the purposes of this chapter, means any material that is subject to the Hazardous Waste Manifest Requirements of the U.S. Environmental Protection Agency specified in 40 CFR part 262.

Hazmat means a hazardous material.

Hazmat employee means: (1) A person who is:

(i) Employed on a full-time, part time, or temporary basis by a hazmat employer and who in the course of such full time, part time or temporary employment directly affects hazardous materials transportation safety;

(ii) Self-employed (including an owner-operator of a motor vehicle, vessel, or aircraft) transporting hazardous materials in commerce who in the course of such self-employment directly affects hazardous materials transportation safety;

(iii) A railroad signalman; or

(iv) A railroad maintenance-of-way employee.

(2) This term includes an individual, employed on a full time, part time, or temporary basis by a hazmat employer, or who is self-employed, who during the course of employment:

(i) Loads, unloads, or handles hazardous materials;

(ii) Designs, manufactures, fabricates, inspects, marks, maintains, reconditions, repairs, or tests a package, container or packaging component that is represented, marked, certified, or sold as qualified for use in transporting hazardous material in commerce;

(iii) Prepares hazardous materials for transportation;

(iv) Is responsible for safety of transporting hazardous materials;

(v) Operates a vehicle used to transport hazardous materials.

Hazmat employer means:

(1) A person who employs or uses at least one hazmat employee on a full-time, part time, or temporary basis; and who:

(i) Transports hazardous materials in commerce;

(ii) Causes hazardous materials to be transported in commerce; or

(iii) Designs, manufactures, fabricates, inspects, marks, maintains, reconditions, repairs or tests a package, container, or packaging component that is represented, marked, certified, or sold by that person as qualified for use in transporting hazardous materials in commerce.

(2) A person who is self-employed (including an owner-operator of a motor vehicle, vessel, or aircraft) transporting materials in commerce; and who:

(i) Transports hazardous materials in commerce;

(ii) Causes hazardous materials to be transported in commerce; or

(iii) Designs, manufactures, fabricates, inspects, marks, maintains, reconditions, repairs or tests a package, container, or packaging component that is represented, marked, certified, or sold by that person as qualified for use in transporting hazardous materials in commerce; or

(3) A department, agency, or instrumentality of the United States Government, or an authority of a State, political subdivision of a State, or an Indian tribe; and who:

(i) Transports hazardous materials in commerce;

(ii) Causes hazardous materials to be transported in commerce; or

(iii) Designs, manufactures, fabricates, inspects, marks, maintains, reconditions, repairs or tests a package, container, or packaging component that is represented, marked, certified, or sold by that person as qualified for use in transporting hazardous materials in commerce.

Hermetically sealed means closed by fusion, gasketing, crimping, or equivalent means so that no gas or vapor can enter or escape.

HMR means the Hazardous Materials Regulations, Parts 171 through 180 of this chapter.

Household waste means any solid waste (including garbage, trash, and sanitary waste from septic tanks) derived from households (including single and multiple residences, hotels and motels, bunkhouses, ranger stations, crew quarters, campgrounds, picnic grounds, and day-use recreation areas). This term is not applicable to consolidated shipments of household hazardous materials transported from collection centers. A collection center is a central location where household waste is collected.

IAEA means International Atomic Energy Agency.

IATA means International Air Transport Association.

ICAO means International Civil Aviation Organization.

IMO means International Maritime Organization.

Incorporated by reference or IBR means a publication or a portion of a publication that is made a part of the regulations of this subchapter. See §171.7.

Infectious substance (etiologic agent). See §173.134 of this subchapter.

Inner packaging means a packaging for which an outer packaging is required for transport. It does not include the inner receptacle of a composite packaging.

Inner receptacle means a receptacle which requires an outer packaging in order to perform its containment function. The inner receptacle may be an inner packaging of a combination packaging or the inner receptacle of a composite packaging.

Intermediate bulk container or IBC means a rigid or flexible portable packaging, other than a cylinder or portable tank, which is designed for mechanical handling. Standards for IBCs manufactured in the United States are set forth in subparts N and O of part 178 of this subchapter.

Intermediate packaging means a packaging which encloses an inner packaging or article and is itself enclosed in an outer packaging.

Intermodal container means a freight container designed and constructed to permit it to be used interchangeably in two or more modes of transport.

Intermodal portable tank or IM portable tank means a specific class of portable tanks designed primarily for international intermodal use.

International transportation means transportation—

(1) Between any place in the United States and any place in a foreign country;

(2) Between places in the United States through a foreign country; or

(3) Between places in one or more foreign countries through the United States.

Irritating material. See §173.132(a)(2) of this subchapter.

Jerrican means a metal or plastic packaging of rectangular or polygonal cross-section.

Large packaging means a packaging that—

(1) Consists of an outer packaging that contains articles or inner packagings;

(2) Is designated for mechanical handling;

(3) Exceeds 400 kg net mass or 450 liters (118.9 gallons) capacity;

(4) Has a volume of not more than 3 cubic meters (m^3) (see §178.801(i) of this subchapter); and

(5) Conforms to the requirements for the construction, testing and marking of Large Packagings as specified in subparts P and Q of part 178 of this subchapter.

Lighter means a mechanically operated flame-producing device employing an ignition device and containing a Class 3 or a Division 2.1 material. For design, capacity, and filling density requirements for lighters containing a Division 2.1 material, see §173.308.

Lighter refill means a pressurized container that does not contain an ignition device but does contain a release device and is intended for use as a replacement cartridge in a lighter or to refill a lighter with a Division 2.1 flammable gas fuel. For capacity limits, see §173.306(h) of this subchapter.

Limited quantity, when specified as such in a section applicable to a particular material, means the maximum amount of a hazardous material for which there is a specific labeling or packaging exception.

Liquid means a material, other than an elevated temperature material, with a melting point or initial melting point of 20 °C (68 °F) or lower at a standard pressure of 101.3 kPa (14.7 psia). A viscous material for which a specific melting point cannot be determined must be subjected to the procedures specified in ASTM D 4359 "Standard Test Method for Determining Whether a Material is Liquid or Solid" (IBR, see §171.7).

Liquid phase means a material that meets the definition of liquid when evaluated at the higher of the temperature at which it is offered for transportation or at which it is transported, not at the 38 °C (100 °F) temperature specified in ASTM D 4359 (IBR, see §171.7).

Lithium content means the mass of lithium in the anode of a lithium metal or lithium alloy cell. The lithium content of a battery equals the sum of the grams of lithium content contained in the component cells of the battery. For a lithium-ion cell see the definition for "equivalent lithium content".

Loading incidental to movement means loading by carrier personnel or in the presence of carrier personnel of packaged or containerized hazardous material onto a transport vehicle, aircraft, or vessel for the purpose of transporting it, including the loading, blocking and bracing a hazardous materials package in a freight container or transport vehicle, and segregating a hazardous materials package in a freight container or transport vehicle from incompatible cargo. For a bulk packaging, *loading incidental to movement* means filling the packaging with a hazardous material for the purpose of transporting it. *Loading incidental to movement* includes transloading.

Magazine vessel means a vessel used for the receiving, storing, or dispensing of explosives.

Magnetic material. See §173.21(d) of this subchapter.

Marine pollutant means a material which is listed in appendix B to §172.101 of this subchapter (also see §171.4) and, when in a solution or mixture of one or more marine pollutants, is packaged in a concentration which equals or exceeds:

(1) Ten percent by weight of the solution or mixture for materials listed in the appendix; or

(2) One percent by weight of the solution or mixture for materials that are identified as severe marine pollutants in the appendix.

Marking means a descriptive name, identification number, instructions, cautions, weight, specification, or UN marks, or combinations thereof, required by this subchapter on outer packagings of hazardous materials.

Material of trade means a hazardous material, other than a hazardous waste, that is carried on a motor vehicle—

(1) For the purpose of protecting the health and safety of the motor vehicle operator or passengers;

(2) For the purpose of supporting the operation or maintenance of a motor vehicle (including its auxiliary equipment); or

(3) By a private motor carrier (including vehicles operated by a rail carrier) in direct support of a principal business that is other than transportation by motor vehicle.

Material poisonous by inhalation or Material toxic by inhalation means:

(1) A gas meeting the defining criteria in §173.115(c) of this subchapter and assigned to Hazard Zone A, B, C, or D in accordance with §173.116(a) of this subchapter;

(2) A liquid (other than as a mist) meeting the defining criteria in §173.132(a)(1)(iii) of this subchapter and assigned to Hazard Zone A or B in accordance with §173.133(a) of this subchapter; or

(3)Any material identified as an inhalation hazard by a special provision in column 7 of the §172.101 Table.

Maximum allowable working pressure or MAWP: For DOT specification cargo tanks used to transport liquid hazardous materials, *see* §178.320(a) of this subchapter.

Maximum capacity means the maximum inner volume of receptacles or packagings.

Maximum net mass means the allowable maximum net mass of contents in a single packaging, or as used in subpart M of part 178 of this subchapter, the maximum combined mass of inner packaging, and the contents thereof.

Mechanical displacement meter prover means a mechanical device used in the oilfield service industry consisting of a pipe assembly that is used to calibrate the accuracy and performance of meters that measure the quantities of a product being pumped or transferred at facilities such as drilling locations, refineries, tank farms, and loading racks.

Metal hydride storage system means a single complete hydrogen storage system that includes a receptacle, metal hydride, pressure relief device, shut-off valve, service equipment and internal components used for the transportation of hydrogen only.

Metered delivery service means a cargo tank unloading operation conducted at a metered flow rate of 378.5 L (100 gallons) per minute or less through an attached delivery hose with a nominal inside diameter of 3.175 cm (1¼ inches) or less.

Miscellaneous hazardous material. See §173.140 of this subchapter.

Mixture means a material composed of more than one chemical compound or element.

Mode means any of the following transportation methods; rail, highway, air, or water.

Motor vehicle includes a vehicle, machine, tractor, trailer or semitrailer, or any combination thereof, propelled or drawn by mechanical power and used upon the highways in the transportation of passengers or property. It does not include a vehicle, locomotive, or car operated exclusively on a rail or rails, or a trolley bus operated by electric power derived from a fixed overhead wire, furnishing local passenger transportation similar to street-railway service.

Movement means the physical transfer of a hazardous material from one geographic location to another by rail car, aircraft, motor vehicle, or vessel.

Multiple-element gas container or MEGC means assemblies of UN cylinders, tubes, or bundles of cylinders interconnected by a manifold and assembled within a framework. The term includes all service equipment and structural equipment necessary for the transport of gases.

Name of contents means the proper shipping name as specified in §172.101 of this subchapter.

Navigable waters means, for the purpose of this subchapter, waters of the United States, including the territorial seas.

Non-bulk packaging means a packaging which has:

(1) A maximum capacity of 450 L (119 gallons) or less as a receptacle for a liquid;

(2) A maximum net mass of 400 kg (882 pounds) or less and a maximum capacity of 450 L (119 gallons) or less as a receptacle for a solid; or

(3) A water capacity of 454 kg (1000 pounds) or less as a receptacle for a gas as defined in §173.115 of this subchapter.

Nonflammable gas. See §173.115 of this subchapter.

N.O.S. means not otherwise specified.

N.O.S. description means a shipping description from the §172.101 table which includes the abbreviation *n.o.s.*

NPT means an American Standard taper pipe thread conforming to the requirements of NBS Handbook H-28 (IBR, see §171.7).

NRC (non-reusable container) means a packaging (container) whose reuse is restricted in accordance with the provisions of §173.28 of this subchapter.

Occupied caboose means a rail car being used to transport non-passenger personnel.

Officer in Charge, Marine Inspection means a person from the civilian or military branch of the Coast Guard designated as such by the Commandant and who under the supervision and direction of the Coast Guard District Commander is in charge of a designated inspection zone for the performance of duties with respect to the enforcement and administration of title 52, Revised Statutes, acts amendatory thereof or supplemental thereto, rules and regulations thereunder, and the inspection required thereby.

Offshore supply vessel means a cargo vessel of less than 500 gross tons that regularly transports goods, supplies or equipment in support of exploration or production of offshore mineral or energy resources.

Open cryogenic receptacle means a transportable thermally insulated receptacle for refrigerated liquefied gases maintained at atmospheric pressure by continuous venting of the refrigerated gas.

Operator means a person who controls the use of an aircraft, vessel, or vehicle.

Organic peroxide. See §173.128 of this subchapter.

ORM means other regulated material. See §173.144 of this subchapter.

Outage or ullage means the amount by which a packaging falls short of being liquid full, usually expressed in percent by volume.

Outer packaging means the outermost enclosure of a composite or combination packaging together with any absorbent materials, cushioning and any other components necessary to contain and protect inner receptacles or inner packagings.

Overpack, except as provided in subpart K of part 178 of this subchapter, means an enclosure that is used by a single consignor to provide protection or convenience in handling of a package or to consolidate two or more packages. *Overpack* does not include a transport vehicle, freight container, or aircraft unit load device. Examples of overpacks are one or more packages:

(1) Placed or stacked onto a load board such as a pallet and secured by strapping, shrink wrapping, stretch wrapping, or other suitable means; or

(2) Placed in a protective outer packaging such as a box or crate.

Oxidizer. See §173.127 of this subchapter.

Oxidizing gas means a gas that may, generally by providing oxygen, cause or contribute to the combustion of other material more than air does. Specifically, this means a pure gas or gas mixture with an oxidizing power greater than 23.5% as determined by a method specified in ISO 10156: or 10156–2: (IBR, *see* §171.7 of this subchapter) (*see also* §173.115(k)).

Oxygen generator (chemical) means a device containing chemicals that upon activation release oxygen as a product of chemical reaction.

Package or Outside Package means a packaging plus its contents. For radioactive materials, see §173.403 of this subchapter.

Packaging means a receptacle and any other components or materials necessary for the receptacle to perform its containment function in conformance with the minimum packing requirements of this subchapter. For radioactive materials packaging, see §173.403 of this subchapter.

Packing group means a grouping according to the degree of danger presented by hazardous materials. Packing Group I indicates great danger; Packing Group II, medium danger; Packing Group III, minor danger. See §172.101(f) of this subchapter.

Passenger (With respect to vessels and for the purpose of part 176 only) means a person being carried on a vessel other than:

(1) The owner or his representative;

(2) The operator;

(3) A bona fide member of the crew engaged in the business of the vessel who has contributed no consideration for his carriage and who is paid for his services; or

(4) A guest who has not contributed any consideration directly or indirectly for his carriage.

Passenger-carrying aircraft means an aircraft that carries any person other than a crewmember, company employee, an authorized representative of the United States, or a person accompanying the shipment.

Passenger vessel means—

(1) A vessel subject to any of the requirements of the International Convention for the Safety of Life at Sea, 1974, which carries more than 12 passengers;

(2) A cargo vessel documented under the laws of the United States and not subject to that Convention, which carries more than 16 passengers;

(3) A cargo vessel of any foreign nation that extends reciprocal privileges and is not subject to that Convention and which carries more than 16 passengers; and

(4) A vessel engaged in a ferry operation and which carries passengers.

Person means an individual, corporation, company, association, firm, partnership, society, joint stock company; or a government, Indian Tribe, or authority of a government or Tribe, that offers a hazardous material for transportation in commerce, transports a hazardous material to support a commercial enterprise, or designs, manufactures, fabricates, inspects, marks, maintains, reconditions, repairs, or tests a package, container, or packaging component that is represented, marked, certified, or sold as qualified for use in transporting hazardous material in commerce. This term does not include the United States Postal Service or, for purposes of 49 U.S.C. 5123 and 5124, a Department, agency, or instrumentality of the government.

Person who offers or offeror means:

(1) Any person who does either or both of the following:

(i) Performs, or is responsible for performing, any pre-transportation function required under this subchapter for transportation of the hazardous material in commerce.

(ii) Tenders or makes the hazardous material available to a carrier for transportation in commerce.

(2) A carrier is not an offeror when it performs a function required by this subchapter as a condition of acceptance of a hazardous material for transportation in commerce (*e.g.*, reviewing shipping papers, examining packages to ensure that they are in conformance with this subchapter, or preparing shipping documentation for its own use) or when it transfers a hazardous material to another carrier for continued transportation in commerce without performing a pre-transportation function.

PHMSA means the Pipeline and Hazardous Materials Safety Administration, U.S. Department of Transportation, Washington, DC 20590.

Placarded car means a rail car which is placarded in accordance with the requirements of part 172 of this subchapter.

Poisonous gas. See §173.115 of this subchapter.

Poisonous materials. See §173.132 of this subchapter.

Portable tank means a bulk packaging (except a cylinder having a water capacity of 1000 pounds or less) designed primarily to be loaded onto, or on, or temporarily attached to a transport vehicle or ship and equipped with skids, mountings, or accessories to facilitate handling of the tank by mechanical means. It does not include a cargo tank, tank car, multi-unit tank car tank, or trailer carrying 3AX, 3AAX, or 3T cylinders.

Preferred route or Preferred highway is a highway for shipment of *highway route controlled quantities* of radioactive materials so designated by a State routing agency, and any Interstate System highway for which an alternative highway has not been designated by such State agency as provided by §397.103 of this title.

Pre-transportation function means a function specified in the HMR that is required to assure the safe transportation of a hazardous material in commerce, including—

(1) Determining the hazard class of a hazardous material.

(2) Selecting a hazardous materials packaging.

(3) Filling a hazardous materials packaging, including a bulk packaging.

(4) Securing a closure on a filled or partially filled hazardous materials package or container or on a package or container containing a residue of a hazardous material.

(5) Marking a package to indicate that it contains a hazardous material.

(6) Labeling a package to indicate that it contains a hazardous material.

(7) Preparing a shipping paper.

(8) Providing and maintaining emergency response information.

(9) Reviewing a shipping paper to verify compliance with the HMR or international equivalents.

(10) For each person importing a hazardous material into the United States, providing the shipper with timely and complete information as to the HMR requirements that will apply to the transportation of the material within the United States.

(11) Certifying that a hazardous material is in proper condition for transportation in conformance with the requirements of the HMR.

(12) Loading, blocking, and bracing a hazardous materials package in a freight container or transport vehicle.

(13) Segregating a hazardous materials package in a freight container or transport vehicle from incompatible cargo.

(14) Selecting, providing, or affixing placards for a freight container or transport vehicle to indicate that it contains a hazardous material.

Primary hazard means the hazard class of a material as assigned in the §172.101 Table.

Private track or Private siding means: (i) Track located outside of a carrier's right-of-way, yard, or terminals where the carrier does not own the rails, ties, roadbed, or right-of-way, or

(ii) Track leased by a railroad to a lessee, where the lease provides for, and actual practice entails, exclusive use of that trackage by the lessee and/or a general system railroad for purpose of moving only cars shipped to or by the lessee, and where the lessor otherwise exercises no control over or responsibility for the trackage or the cars on the trackage.

Proper shipping name means the name of the hazardous material shown in Roman print (not italics) in §172.101 of this subchapter.

Psi means pounds per square inch.

Psia means pounds per square inch absolute.

Psig means pounds per square inch gauge.

Public vessel means a vessel owned by and being used in the public service of the United States. It does not include a vessel owned by the United States and engaged in trade or commercial service or a vessel under contract or charter to the United States.

Pyrophoric liquid. See §173.124(b) of this subchapter.

Radioactive materials. See §173.403 of this subchapter for definitions relating to radioactive materials.

Rail car means a car designed to carry freight or non-passenger personnel by rail, and includes a box car, flat car, gondola car, hopper car, tank car, and occupied caboose.

Railroad means a person engaged in transportation by rail.

Receptacle means a containment vessel for receiving and holding materials, including any means of closing.

Reconditioned packaging. See §173.28 of this subchapter.

Registered Inspector means a person registered with the Department in accordance with subpart F of part 107 of this chapter who has the knowledge and ability to determine whether a cargo tank conforms to the applicable DOT specification. A *Registered Inspector* meets the knowledge and ability requirements of this section by meeting any one of the following requirements:

(1) Has an engineering degree and one year of work experience relating to the testing and inspection of cargo tanks;

(2) Has an associate degree in engineering and two years of work experience relating to the testing and inspection of cargo tanks;

(3) Has a high school diploma (or General Equivalency Diploma) and three years of work experience relating to the testing and inspection of cargo tanks; or

(4) Has at least three years' experience performing the duties of a Registered Inspector prior to September 1, 1991.

Regulated medical waste. See §173.134 of this subchapter.

Remanufactured packagings. See §173.28 of this subchapter.

Reportable quantity (RQ) for the purposes of this subchapter means the quantity specified in Column 2 of the appendix to §172.101, for any material identified in Column 1 of the appendix.

Research means investigation or experimentation aimed at the discovery of new theories or laws and the discovery and interpretation of facts or revision of accepted theories or laws in the light of new facts. Research does not include the application of existing technology to industrial endeavors.

Residue means the hazardous material remaining in a packaging, including a tank car, after its contents have been unloaded to the maximum extent practicable and before the packaging is either refilled or cleaned of hazardous materials and purged to remove any hazardous vapors.

Reused packaging. See §173.28 of this subchapter.

SADT means self-accelerated decomposition temperature. See §173.21(f) of this subchapter.

Salvage packaging means a special packaging conforming to §173.3 of this subchapter into which damaged, defective, leaking, or non-conforming hazardous materials packages, or hazardous materials that have spilled or leaked, are placed for purposes of transport for recovery or disposal.

SCF (standard cubic foot) means one cubic foot of gas measured at 60 °F. and 14.7 psia.

Secretary means the Secretary of Transportation.

Self-defense spray means an aerosol or non-pressurized device that:

(1) Is intended to have an irritating or incapacitating effect on a person or animal; and

(2) Meets no hazard criteria other than for Class 9 (for example, a pepper spray; see §173.140(a) of this subchapter) and, for an aerosol, Division 2.1 or 2.2 (see

§173.115 of this subchapter), except that it may contain not more than two percent by mass of a tear gas substance (e.g., chloroacetophenone (CN) or 0-chlorobenzylmalonitrile (CS); see §173.132(a)(2) of this subchapter.)

Settled pressure means the pressure exerted by the contents of a UN pressure receptacle in thermal and diffusive equilibrium.

Sharps. See §173.134 of this subchapter.

Shipping paper means a shipping order, bill of lading, manifest or other shipping document serving a similar purpose and prepared in accordance with subpart C of part 172 of this chapter.

Siftproof packaging means a packaging impermeable to dry contents, including fine solid material produced during transportation.

Single packaging means a non-bulk packaging other than a combination packaging.

Solid means a material which is not a gas or a liquid.

Solution means any homogeneous liquid mixture of two or more chemical compounds or elements that will not undergo any segregation under conditions normal to transportation.

Special permit means a document issued by the Associate Administrator, or other designated Department official, under the authority of 49 U.S.C. 5117 permitting a person to perform a function that is not otherwise permitted under subchapter A or C of this chapter, or other regulations issued under 49 U.S.C. 5101 et seq. (*e.g.*, Federal Motor Carrier Safety routing requirements).

Specification packaging means a packaging conforming to one of the specifications or standards for packagings in part 178 or part 179 of this subchapter.

Spontaneously combustible material. See §173.124(b) of this subchapter.

Stabilized means that the hazardous material is in a condition that precludes uncontrolled reaction. This may be achieved by methods such as adding an inhibiting chemical, degassing the hazardous material to remove dissolved oxygen and inerting the air space in the package, or maintaining the hazardous material under temperature control.

State means a State of the United States, the District of Columbia, the Commonwealth of Puerto Rico, the Commonwealth of the Northern Mariana Islands, the Virgin Islands, American Samoa, Guam, or any other territory or possession of the United States designated by the Secretary.

State-designated route means a preferred route selected in accordance with U.S. DOT “Guidelines for Selecting Preferred Highway Routes for Highway Route Controlled Quantities of Radioactive Materials” or an equivalent routing analysis which adequately considers overall risk to the public.

Storage incidental to movement means storage of a transport vehicle, freight container, or package containing a hazardous material by any person between the time that a carrier takes physical possession of the hazardous material for the purpose of transporting it in commerce until the package containing the hazardous material is physically delivered to the destination indicated on a shipping document, package marking, or other medium, or, in the case of a private motor carrier, between the time that a motor vehicle driver takes physical possession of the hazardous material for the purpose of transporting it in commerce until the driver relinquishes possession of the package at its destination and is no longer responsible for performing functions subject to the HMR with respect to that particular package.

(1) *Storage incidental to movement* includes—

(i) Storage at the destination shown on a shipping document, including storage at a transloading facility, provided the shipping documentation identifies the shipment as a through-shipment and identifies the final destination or destinations of the hazardous material; and

(ii) Rail cars containing hazardous materials that are stored on track that does not meet the definition of “private track or siding” in §171.8, even if those cars have been delivered to the destination shown on the shipping document.

(2) Storage incidental to movement does not include storage of a hazardous material at its final destination as shown on a shipping document.

Stowage means the act of placing hazardous materials on board a vessel.

Strong outer packaging means the outermost enclosure that provides protection against the unintentional release of its contents. It is a packaging that is sturdy, durable, and constructed so that it will retain its contents under normal conditions of transportation. In addition, a strong outer packaging must meet the general packaging requirements of subpart B of part 173 of this subchapter but need not comply with the specification packaging requirements in part 178 of the subchapter. For transport by aircraft, a strong outer packaging is subject to §173.27 of this subchapter. The terms “strong outside container” and “strong outside packaging” are synonymous with “strong outer packaging.”

Subsidiary hazard means a hazard of a material other than the primary hazard. (See *primary hazard*).

Table in §172.101 or §172.101 Table means the Hazardous Materials Table in §172.101 of this subchapter.

Technical name means a recognized chemical name or microbiological name currently used in scientific and technical handbooks, journals, and texts. Generic descriptions are authorized for use as technical names provided they readily identify the general chemical group, or microbiological group. Examples of acceptable generic chemical descriptions are organic phosphate compounds, petroleum aliphatic hydrocarbons and tertiary amines. For proficiency testing only, generic microbiological descriptions such as bacteria, mycobacteria, fungus, and viral samples may be used. Except for names which appear in subpart B of part 172 of this subchapter, trade names may not be used as technical names.

TOFC means trailer-on-flat-car.

Top shell means the tank car tank surface, excluding the head ends and bottom shell of the tank car tank.

Toxin . See §173.134 of this subchapter.

Trailership means a vessel, other than a carfloat, specifically equipped to carry motor transport vehicles and fitted with installed securing devices to tie down each vehicle. The term *trailership* includes *Roll-on/ Roll-off (RO/RO)* vessels.

Train means one or more engines coupled with one or more rail cars, except during switching operations or where the operation is that of classifying and assembling rail cars within a railroad yard for the purpose of making or breaking up trains.

Trainship means a vessel other than a rail car ferry or carfloat, specifically equipped to transport railroad vehicles, and fitted with installed securing devices to tie down each vehicle.

Transloading means the transfer of a hazardous material by any person from one bulk packaging to another bulk packaging, from a bulk packaging to a non-bulk packaging, or from a non-bulk packaging to a bulk packaging for the purpose of continuing the movement of the hazardous material in commerce.

Transport vehicle means a cargo-carrying vehicle such as an automobile, van, tractor, truck, semitrailer, tank car or rail car used for the transportation of cargo by any mode. Each cargo-carrying body (trailer, rail car, etc.) is a separate transport vehicle.

Transportation or transport means the movement of property and loading, unloading, or storage incidental to that movement.

UFC means Uniform Freight Classification.

UN means United Nations.

UN cylinder means a transportable pressure receptacle with a water capacity not exceeding 150 L that has been marked and certified as conforming to the applicable requirements in part 178 of this subchapter.

UN portable tank means a intermodal tank having a capacity of more than 450 liters (118.9 gallons). It includes a shell fitted with service equipment and structural equipment, including stabilizing members external to the shell and skids, mountings or accessories to facilitate mechanical handling. A UN portable tank must be capable of being filled and discharged without the removal of its structural equipment and must be capable of being lifted when full. Cargo tanks, rail tank car tanks, non-metallic tanks, non-specification tanks, bulk bins, and IBCs and packagings made to cylinder specifications are not UN portable tanks.

UN pressure receptacle means a UN cylinder or tube.

U.N. Recommendations means the U.N. Recommendations on the Transport of Dangerous Goods, Model Regulations (IBR, *see* §171.7 of this subchapter).

UN standard packaging means a packaging conforming to standards in the UN Recommendations (IBR, see §171.7).

UN tube means a seamless transportable pressure receptacle with a water capacity exceeding 150 L but not more than 3,000 L that has been marked and certified as conforming to the requirements in part 178 of this subchapter.

Undeclared hazardous material means a hazardous material that is: (1) Subject to any of the hazard communication requirements in subparts C (Shipping Papers), D (Marking), E (Labeling), and F (Placarding) of Part 172 of this subchapter, or an alternative marking requirement in Part 173 of this subchapter (such as §§173.4(a)(10) and 173.6(c)); and (2) offered for transportation in commerce without any visible indication to the person accepting the hazardous material for transportation that a hazardous material is present, on either an accompanying shipping document, or the outside of a transport vehicle, freight container, or package.

Unintentional release means the escape of a hazardous material from a package on an occasion not anticipated or planned. This includes releases resulting from collision, package failures, human error, criminal activity, negligence, improper packing, or unusual conditions such as the operation of pressure relief devices as a result of over-pressurization, overfill or fire exposure. It does not include releases, such as venting of packages, where allowed, and the operational discharge of contents from packages.

Unit load device means any type of freight container, aircraft container, aircraft pallet with a net, or aircraft pallet with a net over an igloo.

United States means a State of the United States, the District of Columbia, the Commonwealth of Puerto Rico, the Commonwealth of the Northern Mariana Islands, the Virgin Islands, American Samoa, Guam, or any other territory or possession of the United States designated by the Secretary.

Unloading incidental to movement means removing a packaged or containerized hazardous material from a transport vehicle, aircraft, or vessel or, for a bulk packaging, emptying a hazardous material from the bulk packaging after the hazardous material has been delivered to the consignee when performed by carrier personnel or in the presence of carrier personnel or, in the case of a private motor carrier, while the driver of the motor vehicle from which the hazardous material is being unloaded immediately after movement is completed is present during the unloading operation. (Emptying a hazardous material from a bulk packaging while the packaging is on board a vessel is subject to separate regulations as delegated by Department of Homeland Security Delegation No. 0170.1 at 2(103).) *Unloading incidental to movement* includes transloading.

Vessel includes every description of watercraft, used or capable of being used as a means of transportation on the water.

Viscous liquid means a liquid material which has a measured viscosity in excess of 2500 centistokes at 25 °C. (77 °F.) when determined in accordance with the procedures specified in ASTM Method D 445-72 "Kinematic Viscosity of Transparent and Opaque Liquids (and the Calculation and Dynamic Viscosity)" or ASTM Method D 1200-70 "Viscosity of Paints, Varnishes, and Lacquers by Ford Viscosity Cup."

Volatility refers to the relative rate of evaporation of materials to assume the vapor state.

Water reactive material . See §173.124(c) of this subchapter.

Water resistant means having a degree of resistance to permeability by and damage caused by water in liquid form.

Wooden barrel means a packaging made of natural wood, of round cross-section, having convex walls, consisting of staves and heads and fitted with hoops.

Working pressure for purposes of UN pressure receptacles, means the settled pressure of a compressed gas at a reference temperature of 15 °C (59 °F).

W.T. means watertight.

§171.9 Rules of construction.

(a) In this subchapter, unless the context requires otherwise:

(1) Words imparting the singular include the plural;

(2) Words imparting the plural include the singular; and

(3) Words imparting the masculine gender include the feminine;

(b) In this subchapter, the word: (1) **Shall** is used in an imperative sense;

(2) **Must** is used in an imperative sense;

(3) **Should** is used in a recommendatory sense;

(4) **May** is used in a permissive sense to state authority or permission to do the act described, and the words "no person may* * *" or "a person may not* * *" means that no person is required, authorized, or permitted to do the act described; and

(5) **Includes** is used as a word of inclusion not limitation.

Reserved

§171.12 North American shipments.

(a) *Requirements for the use of the Transport Canada TDG Regulations.* (1) A hazardous material transported from Canada to the United States, from the United States to Canada, or transiting the United States to Canada or a foreign destination may be offered for transportation or transported by motor carrier and rail in accordance with the Transport Canada TDG Regulations (IBR, see §171.7) as authorized in §171.22, provided the requirements in §§171.22 and 171.23, as applicable, and this section are met. In addition, a cargo tank motor vehicle, portable tank or rail tank car authorized by the Transport Canada TDG Regulations may be used for transportation to, from, or within the United States provided the cargo tank motor vehicle, portable tank or rail tank car conforms to the applicable requirements of this section. Except as otherwise provided in this subpart and subpart C of this part, the requirements in parts 172, 173, and 178 of this subchapter do not apply for a material transported in accordance with the Transport Canada TDG Regulations.

(2) *General packaging requirements.* When the provisions of this subchapter require a DOT specification or UN standard packaging to be used for transporting a hazardous material, a packaging authorized by the Transport Canada TDG Regulations may be used, subject to the limitations of this part, and only if it is equivalent to the corresponding DOT specification or UN packaging (see §173.24(d)(2) of this subchapter) authorized by this subchapter.

(3) *Bulk packagings.* A portable tank, cargo tank motor vehicle or rail tank car equivalent to a corresponding DOT specification and conforming to and authorized by the Transport Canada TDG Regulations may be used provided—

(i) An equivalent type of packaging is authorized for the hazardous material according to the §172.101 table of this subchapter;

(ii) The portable tank, cargo tank motor vehicle or rail tank car conforms to the requirements of the applicable part 173 bulk packaging section specified in the §172.101 table for the material to be transported;

(iii) The portable tank, cargo tank motor vehicle or rail tank car conforms to the requirements of all assigned bulk packaging special provisions (B codes, and T and TP codes) in §172.102 of this subchapter; and

(iv) The bulk packaging conforms to all applicable requirements of §§173.31, 173.32, 173.33 and 173.35 of this subchapter, and parts 177 and 180 of this subchapter. The periodic retests and inspections required by §§173.31, 173.32 and 173.33 of this subchapter may be performed in accordance with part 180 of this subchapter or in accordance with the requirements of the TDG Regulations provided that the intervals prescribed in part 180 of this subchapter are met.

(v) Rail tank cars must conform to the requirements of Canadian General Standards Board standard 43.147 (IBR, see §171.7).

(4) *Cylinders.* When the provisions of this subchapter require that a DOT specification or a UN pressure receptacle must be used for a hazardous material, a packaging authorized by the Transport Canada TDG Regulations may be used only if it corresponds to the DOT specification or UN standard authorized by this subchapter. Unless otherwise excepted in this subchapter, a cylinder (including a UN pressure receptacle) may not be transported unless—

(i) The packaging is a UN pressure receptacle marked with the letters "CAN" for Canada as a country of manufacture or a country of approval or is a cylinder that was manufactured, inspected and tested in accordance with a DOT specification or a UN standard prescribed in part 178 of this subchapter, except that cylinders not conforming to these requirements must meet the requirements in §171.23. Each cylinder must conform to the applicable requirements in part 173 of this subchapter for the hazardous material involved.

(ii) The packaging is a Canadian Transport Commission (CTC) specification cylinder manufactured, originally marked and approved in accordance with the CTC regulations and in full conformance with the Transport Canada TDG Regulations.

(A) The CTC specification corresponds with a DOT specification and the cylinder markings are the same as those specified in this subchapter except that they were originally marked with the letters "CTC" in place of "DOT";

(B) The cylinder has been requalified under a program authorized by the Transport Canada TDG Regulations or requalified in accordance with the requirements in §180.205 within the prescribed requalification period provided for the corresponding DOT specification;

(C) When the regulations authorize a cylinder for a specific hazardous material with a specification marking prefix of "DOT", a cylinder marked "CTC" which otherwise bears the same markings that would be required of the specified "DOT" cylinder may be used; and

(D) Transport of the cylinder and the material it contains is in all other respects in conformance with the requirements of this subchapter (e.g. valve protection, filling requirements, operational requirements, etc.).

(5) *Class 1 (explosive) materials.* When transporting Class 1 (explosive) material, rail and motor carriers must comply with 49 CFR 1572.9 and 1572.11 to the extent the requirements apply.

(6) *Primary lithium batteries and cells.* Packages containing primary lithium batteries and cells that meet the exception in §172.102, Special Provision 188 or 189 of this subchapter must be marked "PRIMARY LITHIUM BATTERIES—FORBIDDEN FOR TRANSPORT ABOARD PASSENGER AIRCRAFT" or "LITHIUM METAL BATTERIES—FORBIDDEN FOR TRANSPORT ABOARD PASSENGER AIRCRAFT." The provisions of this paragraph do not apply to packages that contain 5 kg (11 pounds) net weight or less of primary lithium batteries cells that are contained in or packed with equipment.

(b) *Shipments to or from Mexico* Unless otherwise excepted, hazardous materials shipments from Mexico to the United States or from the United States to Mexico must conform to all applicable requirements of this subchapter. When a hazardous material that is a material poisonous by inhalation (see §171.8) is transported by highway or rail from Mexico to the United States, or from the United States to Mexico, the following requirements apply:

(1) The shipping description must include the words "Toxic Inhalation Hazard" or "Poison-Inhalation Hazard" or "Inhalation Hazard", as required in §172.203(m) of this subchapter.

(2) The material must be packaged in accordance with requirements of this subchapter.

(3) The package must be marked in accordance with §172.313 of this subchapter.

(4) Except as provided in paragraph (e)(5) of this section, the package must be labeled or placarded POISON GAS or POISON INHALATION HAZARD, as appropriate, in accordance with subparts E and F of this subchapter.

(5) A label or placard that conforms to the UN Recommendations (IBR, see §171.7) specifications for a "Division 2.3" or "Division 6.1" label or placard may be substituted for the POISON GAS or POISON INHALATION HAZARD label or placard required by §§172.400(a) and 172.504(e) of this subchapter on a package transported in a closed transport vehicle or freight container. The transport vehicle or freight container must be marked with identification numbers for the material, regardless of the total quantity contained in the transport vehicle or freight container, in the manner specified in §172.313(c) of this subchapter and placarded as required by subpart F of this subchapter.

§171.15 Immediate notice of certain hazardous materials incidents.

(a) *General*. As soon as practical but no later than 12 hours after the occurrence of any incident described in paragraph (b) of this section, each person in physical possession of the hazardous material must provide notice by telephone to the National Response Center (NRC) on 800-424-8802 (toll free) or 202-267-2675 (toll call) or online at *http://www.nrc.uscg.mil*. Each notice must include the following information:

(1) Name of reporter;

(2) Name and address of person represented by reporter;

(3) Phone number where reporter can be contacted;

(4) Date, time, and location of incident;

(5) The extent of injury, if any;

(6) Class or division, proper shipping name, and quantity of hazardous materials involved, if such information is available; and

(7) Type of incident and nature of hazardous material involvement and whether a continuing danger to life exists at the scene.

(b) *Reportable incident*. A telephone report is required whenever any of the following occurs during the course of transportation in commerce (including loading, unloading, and temporary storage):

(1) As a direct result of a hazardous material—

(i) A person is killed;

(ii) A person receives an injury requiring admittance to a hospital;

(iii) The general public is evacuated for one hour or more;

(iv) A major transportation artery or facility is closed or shut down for one hour or more; or

(v) The operational flight pattern or routine of an aircraft is altered;

(2) Fire, breakage, spillage, or suspected radioactive contamination occurs involving a radioactive material (see also §176.48 of this subchapter);

(3) Fire, breakage, spillage, or suspected contamination occurs involving an infectious substance other than a regulated medical waste;

(4) A release of a marine pollutant occurs in a quantity exceeding 450 L (119 gallons) for a liquid or 400 kg (882 pounds) for a solid;

(5) A situation exists of such a nature (*e.g.*, a continuing danger to life exists at the scene of the incident) that, in the judgment of the person in possession of the hazardous material, it should be reported to the NRC even though it does not meet the criteria of paragraphs (b)(1), (2), (3) or (4) of this section; or

(6) During transportation by aircraft, a fire, violent rupture, explosion or dangerous evolution of heat (*i.e.*, an amount of heat sufficient to be dangerous to packaging or personal safety to include charring of packaging, melting of packaging, scorching of packaging, or other evidence) occurs as a direct result of a battery or battery-powered device.

(c) *Written report*. Each person making a report under this section must also make the report required by §171.16 of this subpart.

NOTE to §171.15: Under 40 CFR 302.6, EPA requires persons in charge of facilities (including transport vehicles, vessels, and aircraft) to report any release of a hazardous substance in a quantity equal to or greater than its reportable quantity, as soon as that person has knowledge of the release, to DOT's National Response Center at (toll free) 800-424-8802 or (toll) 202-267-2675.

§171.16 Detailed hazardous materials incident reports.

(a) *General*. Each person in physical possession of a hazardous material at the time that any of the following incidents occurs during transportation (including loading, unloading, and temporary storage) must submit a Hazardous Materials Incident Report on DOT Form F 5800.1 (01/2004) within 30 days of discovery of the incident:

(1) Any of the circumstances set forth in §171.15(b);

(2) An unintentional release of a hazardous material or the discharge of any quantity of hazardous waste;

(3) A specification cargo tank with a capacity of 1,000 gallons or greater containing any hazardous material suffers structural damage to the lading retention system or damage that requires repair to a system intended to protect the lading retention system, even if there is no release of hazardous material;

(4) An undeclared hazardous material is discovered; or

(5) A fire, violent rupture, explosion or dangerous evolution of heat (*i.e.*, an amount of heat sufficient to be dangerous to packaging or personal safety to include charring of packaging, melting of packaging, scorching of packaging, or other evidence) occurs as a direct result of a battery or battery-powered device.

(b) *Providing and retaining copies of the report.* Each person reporting under this section must—

(1) Submit a written Hazardous Materials Incident Report to the Information Systems Manager, PHH-60, Pipeline and Hazardous Materials Safety Administration, Department of Transportation, East Building, 1200 New Jersey Ave., SE., Washington, DC 20590-0001, or an electronic Hazardous Material Incident Report to the Information System Manager, PHH-60, Pipeline and Hazardous Materials Safety Administration, Department of Transportation, Washington, DC 20590-0001 at *http://hazmat.dot.gov*;

(2) For an incident involving transportation by aircraft, submit a written or electronic copy of the Hazardous Materials Incident Report to the FAA Security Field Office nearest the location of the incident; and

(3) Retain a written or electronic copy of the Hazardous Materials Incident Report for a period of two years at the reporting person's principal place of business. If the written or electronic Hazardous Materials Incident Report is maintained at other than the reporting person's principal place of business, the report must be made available at the reporting person's principal place of business within 24 hours of a request for the report by an authorized representative or special agent of the Department of Transportation.

(c) *Updating the incident report*. A Hazardous Materials Incident Report must be updated within one year of the date of occurrence of the incident whenever:

(1) A death results from injury caused by a hazardous material;

(2) There was a misidentification of the hazardous material or package information on a prior incident report;

(3) Damage, loss or related cost that was not known when the initial incident report was filed becomes known; or

(4) Damage, loss, or related cost changes by $25,000 or more, or 10% of the prior total estimate, whichever is greater.

(d) *Exceptions*. Unless a telephone report is required under the provisions of §171.15 of this part, the requirements of paragraphs (a), (b), and (c) of this section do not apply to the following incidents:

(1) A release of a minimal amount of material from—

(i) A vent, for materials for which venting is authorized;

(ii) The routine operation of a seal, pump, compressor, or valve; or

(iii) Connection or disconnection of loading or unloading lines, provided that the release does not result in property damage.

(2) An unintentional release of hazardous material when:

(i) The material is properly classed as—

(A) ORM-D; or

(B) a Packing Group III material in Class or Division 3, 4, 5, 6.1, 8, or 9;

(ii) Each package has a capacity of less than 20 liters (5.2 gallons) for liquids or less than 30 kg (66 pounds) for solids;

(iii) The total aggregate release is less than 20 liters (5.2 gallons) for liquids or less than 30 kg (66 pounds) for solids; and

(iv) The material is not—

(A) Offered for transportation or transported by aircraft,

(B) A hazardous waste, or

(C) An undeclared hazardous material.

(3) An undeclared hazardous material discovered in an air passenger's checked or carry-on baggage during the airport screening process. (For discrepancy reporting by carriers, see §175.31 of this subchapter.)

§171.22 Authorization and conditions for the use of international standards and regulations.

(a) *Authorized international standards and regulations.* This subpart authorizes, with certain conditions and limitations, the offering for transportation and the transportation in commerce of hazardous materials in accordance with the International Civil Aviation Organization's Technical Instructions for the Safe Transport of Dangerous Goods by Air (ICAO Technical Instructions), the International Maritime Dangerous Goods Code (IMDG Code), Transport Canada's Transportation of Dangerous Goods Regulations (Transport Canada TDG Regulations), and the International Atomic Energy Agency Regulations for the Safe Transport of Radioactive Material (IAEA Regulations) (IBR, see §171.7).

(b) *Limitations on the use of international standards and regulations.* A hazardous material that is offered for transportation or transported in accordance with the international standards and regulations authorized in paragraph (a) of this section—

(1) Is subject to the requirements of the applicable international standard or regulation and must be offered for transportation or transported in conformance with the applicable standard or regulation; and

(2) Must conform to all applicable requirements of this subpart.

(c) *Materials excepted from regulation under international standards and regulations.* A material designated as a hazardous material under this subchapter, but excepted from or not subject to the international transport standards and regulations authorized in paragraph (a) of this section (e.g., paragraph 1.16 of the Transport Canada TDG Regulations excepts from regulation quantities of hazardous materials less than or equal to 500 kg gross transported by rail) must be transported in accordance with all applicable requirements of this subchapter.

(d) *Materials not regulated under this subchapter.* Materials not designated as hazardous materials under this subchapter but regulated by an international transport standard or regulation authorized in paragraph (a) of this section may be offered for transportation and transported in the United States in full compliance (i.e., packaged, marked, labeled, classed, described, stowed, segregated, secured) with the applicable international transport standard or regulation.

(e) *Forbidden materials.* No person may offer for transportation or transport a hazardous material that is a forbidden material or package as designated in—

(1) Section 173.21 of this subchapter;

(2) Column (3) of the §172.101 Table of this subchapter;

(3) Column (9A) of the §172.101 Table of this subchapter when offered for transportation or transported on passenger aircraft or passenger railcar; or

(4) Column (9B) of the §172.101 Table of this subchapter when offered for transportation or transported by cargo aircraft.

(f) *Complete information and certification.* (1) Except for shipments into the United States from Canada conforming to §171.12, each person importing a hazardous material into the United States must provide the forwarding agent at the place of entry into the United States timely and complete written information as to the requirements of this subchapter applicable to the particular shipment.

(2) The shipper, directly or through the forwarding agent at the place of entry, must provide the initial U.S. carrier with the shipper's certification required by §172.204 of this subchapter, unless the shipment is otherwise excepted from the certification requirement. Except for shipments for which the certification requirement does not apply, a carrier may not accept a hazardous material for transportation unless provided a shipper's certification.

(3) All shipping paper information and package markings required in accordance with this subchapter must be in English. The use of shipping papers and a package marked with both English and a language other than English, in order to dually comply with this subchapter and the regulations of a foreign entity, is permitted under this subchapter.

(4) Each person who provides for transportation or receives for transportation (see §§174.24, 175.30, 176.24 and 177.817 of this subchapter) a shipping paper must retain a copy of the shipping paper or an electronic image thereof that is accessible at or through its principal place of business in accordance with §172.201(e) of this part.

(g) *Additional requirements for the use of international standards and regulations.* All shipments offered for transportation or transported in the United States in accordance with this subpart must conform to the following requirements of this subchapter, as applicable:

(1) The emergency response information requirements in subpart G of part 172 of this subchapter;

(2) The training requirements in subpart H of part 172 of this subchapter, including function-specific training in the use of the international transport standards and regulations authorized in paragraph (a) of this section, as applicable;

(3) The security requirements in subpart I of part 172 of this subchapter;

(4) The incident reporting requirements in §§171.15 and 171.16 of this part for incidents occurring within the jurisdiction of the United States including on board vessels in the navigable waters of the United States and aboard aircraft of United States registry anywhere in air commerce;

(5) For export shipments, the general packaging requirements in §§173.24 and 173.24a of this subchapter;

(6) For export shipments, the requirements for the reuse, reconditioning, and remanufacture of packagings in §173.28 of this subchapter; and

(7) The registration requirements in subpart G of part 107 of this chapter.

§171.23 Requirements for specific materials and packagings transported under the ICAO technical instructions, IMDG code, Transport Canada TDG regulations, or the IAEA regulations.

All shipments offered for transportation or transported in the United States under the ICAO Technical Instructions, IMDG Code, Transport Canada TDG Regulations, or the IAEA Regulations (IBR, see §171.7) must conform to the requirements of this section, as applicable.

(a) *Conditions and requirements for cylinders*—(1) Except as provided in this paragraph, a filled cylinder (pressure receptacle) manufactured to other than a DOT specification or a UN standard in accordance with part 178 of this subchapter, or a DOT exemption or special permit cylinder or a cylinder used as a fire extinguisher in conformance with §173.309(a) of this subchapter, may not be transported to, from, or within the United States.

(2) Cylinders (including UN pressure receptacles) transported to, from, or within the United States must conform to the applicable requirements of this subchapter. Unless otherwise excepted in this subchapter, a cylinder must not be transported unless—

(i) The cylinder is manufactured, inspected and tested in accordance with a DOT specification or a UN standard prescribed in part 178 of this subchapter, except that cylinders not conforming to these requirements must meet the requirements in paragraphs (a)(3), (a)(4) or (a)(5) of this section;

(ii) The cylinder is equipped with a pressure relief device in accordance with §173.301(f) of this subchapter and conforms to the applicable requirements in part 173 of this subchapter for the hazardous material involved;

(iii) The openings on an aluminum cylinder in oxygen service conform to the requirements of this paragraph, except when the cylinder is used for aircraft parts or used aboard an aircraft in accordance with the applicable airworthiness requirements and operating regulations. An aluminum DOT specification cylinder must have an opening configured with straight (parallel) threads. A UN pressure receptacle may have straight (parallel) or tapered threads provided the UN pressure receptacle is marked with the thread type, e.g. "17E, 25E, 18P, or 25P" and fitted with the properly marked valve; and

(iv) A UN pressure receptacle is marked with "USA" as a country of approval in conformance with §§178.69 and 178.70 of this subchapter.

(3) Importation of cylinders for discharge within a single port area: A cylinder manufactured to other than a DOT specification or UN standard in accordance with part 178 of this subchapter and certified as being in conformance with the transportation regulations of another country may be authorized, upon written request to and approval by the Associate Administrator, for transportation within a single port area, provided—

(i) The cylinder is transported in a closed freight container;

(ii) The cylinder is certified by the importer to provide a level of safety at least equivalent to that required by the regulations in this subchapter for a comparable DOT specification or UN cylinder; and

(iii) The cylinder is not refilled for export unless in compliance with paragraph (a)(4) of this section.

(4) Filling of cylinders for export or for use on board a vessel: A cylinder not manufactured, inspected, tested and marked in accordance with part 178 of this subchapter, or a cylinder manufactured to other than a UN standard, DOT specification, exemption or special permit, may be filled with a gas in the United States and offered for transportation and transported for export or alternatively, for use on board a vessel, if the following conditions are met:

(i) The cylinder has been requalified and marked with the month and year of requalification in accordance with subpart C of part 180 of this subchapter, or has been requalified as authorized by the Associate Administrator;

(ii) In addition to other requirements of this subchapter, the maximum filling density, service pressure, and pressure relief device for each cylinder conform to the requirements of this part for the gas involved; and

(iii) The bill of lading or other shipping paper identifies the cylinder and includes the following certification: "This cylinder has (These cylinders have) been qualified, as required, and filled in accordance with the DOT requirements for export."

(5) Cylinders not equipped with pressure relief devices: A DOT specification or a UN cylinder manufactured, inspected, tested and marked in accordance with part 178 of this subchapter and otherwise conforms to the requirements of part 173 for the gas involved, except that the cylinder is not equipped with a pressure relief device may be filled with a gas and offered for transportation and transported for export if the following conditions are met:

(i) Each DOT specification cylinder or UN pressure receptacle must be plainly and durably marked "For Export Only";

(ii) The shipping paper must carry the following certification: "This cylinder has (These cylinders have) been retested and refilled in accordance with the DOT requirements for export."; and

(iii) The emergency response information provided with the shipment and available from the emergency response telephone contact person must indicate that the pressure receptacles are not fitted with pressure relief devices and provide appropriate guidance for exposure to fire.

(b) *Conditions and requirements specific to certain materials*—(1) Aerosols. Except for a limited quantity of a compressed gas in a container of not more than 4 fluid ounces capacity meeting the requirements in §173.306(a)(1) of this subchapter, the proper shipping name "Aerosol," UN1950, may be used only for a non-refillable receptacle containing a gas compressed, liquefied, or dissolved under pressure the sole purpose of which is to expel a nonpoisonous (other than Division 6.1, Packing Group III material) liquid, paste, or powder and fitted with a self-closing release device (see §171.8). In addition, an aerosol must be in a metal packaging when the packaging exceeds 7.22 cubic inches.

(2) *Air bag inflator, air bag module and seat-belt pretensioner.* For each approved air bag inflator, air bag module and seat-belt pretensioner, the shipping paper description must conform to the requirements in §173.166(c) of this subchapter.

(i) The EX number or product code must be included in association with the basic shipping description. When a product code is used, it must be traceable to the specific EX number assigned to the inflator, module or seatbelt pretensioner by the Associate Administrator. The EX number or product code is not required to be marked on the outside package.

(ii) The proper shipping name "Articles, pyrotechnic for technical purposes, UN0431" must be used for all air bag inflators, air bag modules, and seat-belt pretensioners meeting the criteria for a Division 1.4G material.

(3) *Chemical oxygen generators.* Chemical oxygen generators must be approved, classed, described, packaged, and transported in accordance with the requirements of this subchapter.

(4) *Class 1 (explosive) materials.* Prior to being transported, Class 1 (explosive) materials must be approved by the Associate Administrator in accordance with §173.56 of this subchapter. Each package containing a Class 1 (explosive) material must conform to the marking requirements in §172.320 of this subchapter.

(5) *Hazardous substances.* A material meeting the definition of a hazardous substance as defined in §171.8, must conform to the shipping paper requirements in §172.203(c) of this subchapter and the marking requirements in §172.324 of this subchapter:

(i) The proper shipping name must identify the hazardous substance by name, or the name of the substance must be entered in parentheses in association with the basic description and marked on the package in association with the proper shipping name. If the hazardous substance meets the definition for a hazardous waste, the waste code (for example, D001), may be used to identify the hazardous substance;

(ii) The shipping paper and the package markings must identify at least two hazardous substances with the lowest reportable quantities (RQs) when the material contains two or more hazardous substances; and

(iii) The letters "RQ" must be entered on the shipping paper either before or after the basic description, and marked on the package in association with the proper shipping name for each hazardous substance listed.

(6) *Hazardous wastes.* A material meeting the definition of a hazardous waste (see §171.8) must conform to the following:

(i) The shipping paper and the package markings must include the word "Waste" immediately preceding the proper shipping name;

(ii) The shipping paper must be retained by the shipper and by each carrier for three years after the material is accepted by the initial carrier (see §172.205(e)(5)); and

(iii) A hazardous waste manifest must be completed in accordance with §172.205 of this subchapter.

(7) *Marine pollutants.* Except for marine pollutants (see §171.8) transported in accordance with the IMDG Code, marine pollutants transported in bulk packages must meet the shipping paper requirements in §172.203(l) of this subchapter and the package marking requirements in §172.322 of this subchapter.

(8) *Organic peroxides.* Organic peroxides not identified by technical name in the Organic Peroxide Table in §173.225(b) of this subchapter must be approved by the Associate Administrator in accordance with §173.128(d) of this subchapter.

(9) [Reseved]

(10) *Poisonous by inhalation materials.* A material poisonous by inhalation (see §171.8) must conform to the following requirements:

(i) The words "Poison-Inhalation Hazard" or "Toxic-Inhalation Hazard" and the words "Zone A," "Zone B," "Zone C," or "Zone D" for gases, or "Zone A" or "Zone B" for liquids, as appropriate, must be entered on the shipping paper immediately following the basic shipping description. The word "Poison" or "Toxic" or the phrase "Poison-Inhalation Hazard" or "Toxic- Inhalation Hazard" need not be repeated if it otherwise appears in the shipping description;

(ii) The material must be packaged in accordance with the requirements of this subchapter;

(iii) The package must be marked in accordance with §172.313 of this subchapter; and

(iv) Except as provided in subparagraph (B) of this paragraph (b)(10)(iv) and for a package containing anhydrous ammonia prepared in accordance with the Transport Canada TDG Regulations, the package must be labeled or placarded with POISON INHALATION HAZARD or POISON GAS, as appropriate, in accordance with Subparts E and F of part 172 of this subchapter.

(A) For a package transported in accordance with the IMDG Code in a closed transport vehicle or freight container, a label or placard conforming to the IMDG Code specifications for a "Class 2.3" or "Class 6.1" label or placard may be substituted for the POISON GAS or POISON INHALATION HAZARD label or placard, as appropriate. The transport vehicle or freight container must be marked with the identification numbers for the hazardous material, regardless of the total quantity contained in the transport vehicle or freight container, in the manner specified in §172.313(c) of this subchapter and placarded as required by subpart F of part 172 of this subchapter.

(B) For a package transported in accordance with the Transport Canada TDG Regulations in a closed transport vehicle or freight container, a label or placard conforming to the TDG Regulations specifications for a "Class 2.3" or "Class 6.1" label or placard may be substituted for the POISON GAS or POISON INHALATION HAZARD label or placard, as appropriate. The transport vehicle or freight container must be marked with the identification numbers for the hazardous material, regardless of the total quantity contained in the transport vehicle or freight container, in the manner specified in §172.313(c) of this subchapter and placarded as required by subpart F of part 172 of this subchapter. While in transportation in the United States, the transport vehicle or freight container may also be placarded in accordance with the appropriate Transport Canada TDG Regulations in addition to being placarded with the POISON GAS or POISON INHALATION HAZARD placards.

(11) *Class 7 (radioactive) materials.* (i) Highway route

controlled quantities (see §173.403 of this subchapter) must be shipped in accordance with §§172.203(d)(4) and (d)(10); 172.507, and 173.22(c) of this subchapter;

(ii) For fissile materials and Type B, Type B(U), and Type B(M) packagings, the competent authority certification and any necessary revalidation must be obtained from the appropriate competent authorities as specified in §§173.471, 173.472, and 173.473 of this subchapter, and all requirements of the certificates and revalidations must be met;

(iii) Type A package contents are limited in accordance with §173.431 of this subchapter;

(iv) The country of origin for the shipment must have adopted the edition of TS–R–1 of the IAEA Regulations referenced in §171.7;

(v) The shipment must conform to the requirements of §173.448, when applicable;

(vi) The definition for "radioactive material" in §173.403 of this subchapter must be applied to radioactive materials transported under the provisions of this subpart;

(vii) Except for limited quantities, the shipment must conform to the requirements of §172.204(c)(4) of this subchapter; and

(viii) Excepted packages of radioactive material, instruments or articles, or articles containing natural uranium or thorium must conform to the requirements of §§173.421, 173.424, or 173.426 of this subchapter, as appropriate.

(12) *Self-reactive materials.* Selfreactive materials not identified by technical name in the Self-reactive Materials Table in §173.224(b) of this subchapter must be approved by the Associate Administrator in accordance with §173.124(a)(2)(iii) of this subchapter.

§171.24 Additional requirements for the use of the ICAO technical instructions.

(a) A hazardous material that is offered for transportation or transported within the United States by aircraft, and by motor vehicle or rail either before or after being transported by aircraft in accordance with the ICAO Technical Instructions (IBR, see §171.7), as authorized in paragraph (a) of §171.22, must conform to the requirements in §171.22, as applicable, and this section.

(b) Any person who offers for transportation or transports a hazardous material in accordance with the ICAO Technical Instructions must comply with the following additional conditions and requirements:

(1) All applicable requirements in parts 171 and 175 of this subchapter (also see 14 CFR 121.135, 121.401, 121.433a, 135.323, 135.327 and 135.333);

(2) The quantity limits prescribed in the ICAO Technical Instructions for transportation by passenger-carrying or cargo aircraft, as applicable;

(3) The conditions or requirements of a United States variation, when specified in the ICAO Technical Instructions.

(c) *Highway transportation.* For transportation by highway prior to or after transportation by aircraft, a shipment must conform to the applicable requirements of part 177 of this subchapter, and the motor vehicle must be placarded in accordance with subpart F of part 172.

(d) *Conditions and requirements specific to certain materials.* Hazardous materials offered for transportation or transported in accordance with the ICAO Technical Instructions must conform to the following specific conditions and requirements, as applicable:

(1) *Batteries*—(i) *Nonspillable wet electric storage batteries.* Nonspillable wet electric storage batteries are not subject to the requirements of this subchapter provided—

(A) The battery meets the conditions specified in Special Provision 67 of the ICAO Technical Instructions;

(B) The battery, its outer packaging, and any overpack are plainly and durably marked "NONSPILLABLE" or "NONSPILLABLE BATTERY"; and

(C) The batteries or battery assemblies are offered for transportation or transported in a manner that prevents short circuiting or forced discharge, including, but not limited to, protection of exposed terminals.

(ii) *Primary lithium batteries and cells.* Primary lithium batteries and cells are forbidden for transportation aboard passenger-carrying aircraft. Equipment containing or packed with primary lithium batteries or cells are forbidden for transport aboard passenger-carrying aircraft except as provided in §172.102, Special Provision A101 of this subchapter. When transported aboard cargo-only aircraft, packages containing primary lithium batteries and cells transported in accordance with Special Provision A45 of the ICAO Technical Instructions must be marked "PRIMARY LITHIUM BATTERIES—FORBIDDEN FOR TRANSPORT ABOARD PASSENGER AIRCRAFT" or "LITHIUM METAL BATTERIES—FORBIDDEN FOR TRANSPORT ABOARD PASSENGER AIRCRAFT." This marking is not required on packages that contain 5 kg (11 pounds) net weight or less of primary lithium batteries or cells that are contained in or packed with equipment.

(iii) *Prototype lithium batteries and cells.* Prototype lithium batteries and cells are forbidden for transport aboard passenger aircraft and must be approved by the Associate Administrator prior to transportation aboard cargo aircraft, in accordance with the requirements of Special Provision A55 in §172.102 of this subchapter.

(2) A package containing Oxygen, compressed, or any of the following oxidizing gases must be packaged as required by Parts 173 and 178 of this subchapter: carbon dioxide and oxygen mixtures, compressed; compressed gas, oxidizing, n.o.s.; liquefied gas, oxidizing, n.o.s.; nitrogen trifluoride; and nitrous oxide.

§171.25 Additional requirements for the use of the IMDG code.

(a) A hazardous material may be offered for transportation or transported to, from or within the United States by vessel, and by motor carrier and rail in accordance with the IMDG Code (IBR, see §171.7), as authorized in §171.22, provided all or part of the movement is by vessel. Such shipments must conform to the requirements in §171.22, as applicable, and this section.

(b) Any person who offers for transportation or transports a hazardous material in accordance with the IMDG Code must conform to the following additional conditions and requirements:

(1) Unless specified otherwise in this subchapter, a shipment must conform to the requirements in part 176 of this subchapter. For transportation by rail or highway prior to or subsequent to transportation by vessel, a shipment must conform to the applicable requirements of parts 174 and 177 respectively, of this subchapter, and the motor vehicle or rail car must be placarded in accordance with subpart F of part 172 of this subchapter. When a hazardous material regulated by this subchapter for transportation by highway is transported by motor vehicle on a public highway or by rail under the provisions of subpart C of part 171, the segregation requirements of Part 7, Chapter 7.2 of the IMDG Code are authorized.

(2) For transportation by vessel, the stowage and segregation requirements in Part 7 of the IMDG Code may be substituted for the stowage and segregation requirements in part 176 of this subchapter.

(3) Packages containing primary lithium batteries and cells that are transported in accordance with Special Provision 188 of the IMDG Code must be marked "PRIMARY LITHIUM BATTERIES—FORBIDDEN FOR TRANSPORT ABOARD PASSENGER AIRCRAFT" or "LITHIUM METAL BATTERIES—FORBIDDEN FOR TRANSPORT ABOARD PASSENGER AIRCRAFT." This marking is not required on packages that contain 5 kg (11 pounds) net weight or less of primary lithium batteries and cells that are contained in or packed with equipment.

(4) Material consigned under UN3166 and UN3171 (*e.g.*, Engines, internal combustion, *etc.*, Vehicles, *etc.* and Battery-powered equipment) may be prepared in accordance with the IMDG Code or this subchapter.

(c) *Conditions and requirements for bulk packagings.* Except for IBCs and UN portable tanks used for the transportation of liquids or solids, bulk packagings must conform to the requirements of this subchapter. Additionally, the following requirements apply:

(1) UN portable tanks must conform to the requirements in Special Provisions TP37, TP38, TP44 and TP45 when applicable, and any applicable bulk special provisions assigned to the hazardous material in the Hazardous Materials Table in §172.101 of this subchapter;

(2) IMO Type 5 portable tanks must conform to DOT Specification 51 or UN portable tank requirements, unless specifically authorized in this subchapter or approved by the Associate Administrator;

(3) Except as specified in this subpart, for a material poisonous (toxic) by inhalation, the T Codes specified in Column 13 of the Dangerous Goods List in the IMDG Code may be applied to the transportation of those materials in IM, IMO and DOT Specification 51 portable tanks, when these portable tanks are authorized in accordance with the requirements of this subchapter; and

(4) No person may offer an IM or UN portable tank containing liquid hazardous materials of Class 3, PG I or II, or PG III with a flash point less than 100 °F (38 °C); Division 5.1, PG I or II; or Division 6.1, PG I or II, for unloading while it remains on a transport vehicle with the motive power unit attached, unless it conforms to the requirements in §177.834(o) of this subchapter.

(d) *Use of IMDG Code in port areas.* (1) Except for Division 1.1, 1.2, and Class 7 materials, a hazardous material being imported into or exported from the United States or passing through the United States in the course of being shipped between locations outside the United States may be offered and accepted for transportation and transported by motor vehicle within a single port area, including contiguous harbors, when packaged, marked, classed, labeled, stowed and segregated in accordance with the IMDG Code, offered and accepted in accordance with the requirements of subparts C and F of part 172 of this subchapter pertaining to shipping papers and placarding, and otherwise conforms to the applicable requirements of part 176 of this subchapter.

(2) The requirement in §172.201(d) of this subchapter for an emergency telephone number does not apply to shipments made in accordance with the IMDG Code if the hazardous material is not offloaded from the vessel, or is offloaded between ocean vessels at a U.S. port facility without being transported by public highway.

§171.26 Additional requirements for the use of the IAEA regulations.

A Class 7 (radioactive) material being imported into or exported from the United States or passing through the United States in the course of being shipped between places outside the United States may be offered for transportation or transported in accordance with the IAEA Regulations (IBR, see §171.7) as authorized in paragraph (a) of §171.22, provided the requirements in §171.22, as applicable, are met.

Reserved

PART 172

Subpart A—General

§172.1 Purpose and scope.

This Part lists and classifies those materials which the Department has designated as hazardous materials for purposes of transportation and prescribes the requirements for shipping papers, package marking, labeling, and transport vehicle placarding applicable to the shipment and transportation of those hazardous materials.

§172.3 Applicability.

(a) This Part applies to —

(1) Each person who offers a hazardous material for transportation, and

(2) Each carrier by air, highway, rail, or water who transports a hazardous material.

(b) When a person, other than one of those provided for in paragraph (a) of this section, performs a packaging labeling or marking function required by this part, that person shall perform the function in accordance with this part.

Subpart B — Table of Hazardous Materials and Special Provisions

§172.101 Purpose and use of hazardous materials table.

(a) The Hazardous Materials Table (Table) in this section designates the materials listed therein as hazardous materials for the purpose of transportation of those materials. For each listed material, the Table identifies the hazard class or specifies that the material is forbidden in transportation, and gives the proper shipping name or directs the user to the preferred proper shipping name. In addition, the Table specifies or references requirements in this subchapter pertaining to labeling, packaging, quantity limits aboard aircraft and stowage of hazardous materials aboard vessels.

(b) **Column 1: Symbols.** Column 1 of the Table contains six symbols ("+", "A", "D", "G", "I", and "W"), as follows:

(1) The plus (+) sign fixes the proper shipping name, hazard class and packing group for that entry without regard to whether the material meets the definition of that class, packing group or any other hazard class definition. When the plus sign is assigned to a proper shipping name in Column (1) of the §172.101 Table, it means that the material is known to pose a risk to humans. When a plus sign is assigned to mixtures or solutions containing a material where the hazard to humans is significantly different from that of the pure material or where no hazard to humans is posed, the material may be described using an alternative shipping name that represents the hazards posed by the material. An appropriate alternate proper shipping name and hazard class may be authorized by the Associate Administrator.

(2) The letter "A" denotes a material that is subject to the requirements of this subchapter only when offered or intended for transportation by aircraft, unless the material is a hazardous substance or a hazardous waste. A shipping description entry preceded by an "A" may be used to describe a material for other modes of transportation provided all applicable requirements for the entry are met.

(3) The letter "D" identifies proper shipping names which are appropriate for describing materials for domestic transportation but may be inappropriate for international transportation under the provisions of international regulations (e.g., IMO, ICAO). An alternate proper shipping name may be selected when either domestic or international transportation is involved.

(4) The letter "G" identifies proper shipping names for which one or more technical names of the hazardous material must be entered in parentheses, in association with the basic description. (See §172.203(k).)

(5) The letter "I" identifies proper shipping names which are appropriate for describing materials in international transportation. An alternate proper shipping name may be selected when only domestic transportation is involved.

(6) The letter "W" restricts the application of requirements of this subchapter to materials offered or intended for transportation by vessel, unless the material is a hazardous substance or a hazardous waste.

(c) **Column 2: Hazardous materials descriptions and proper shipping names.** Column 2 lists the hazardous materials descriptions and proper shipping names of materials designated as hazardous materials. Modification of a proper shipping name may otherwise be required or authorized by this section. Proper shipping names are limited to those shown in Roman type (not italics).

(1) Proper shipping names may be used in the singular or plural and in either capital or lower case letters. Words may be alternatively spelled in the same manner as they appear in the ICAO Technical Instructions or the IMDG Code. For example "aluminum" may be spelled "aluminium" and "sulfur" may be spelled "sulphur". However, the word "inflammable" may not be used in place of the word "flammable".

(2) Punctuation marks and words in italics are not part of the proper shipping name, but may be used in addition to the proper shipping name. The word "or" in italics indicates that there is a choice of terms in the sequence that may alternately be used as the proper shipping name or as part of the proper shipping name, as appropriate. For example, for the hazardous materials description "Carbon dioxide, solid or Dry ice" either "Carbon dioxide, solid" or "Dry ice" may be used as the proper shipping name; and for the hazardous materials description "Articles, pressurized pneumatic or hydraulic," either "Articles, pressurized pneumatic" or "Articles, pressurized hydraulic" may be used as the proper shipping name.

(3) The word "poison" or "poisonous" may be used interchangeably with the word "toxic" when only domestic transportation is involved. The abbreviation "n.o.i." or "n.o.i.b.n." may be used interchangeably with "n.o.s.".

(4) Except for hazardous wastes, when qualifying words are used as part of the proper shipping name, their sequence in the package markings and shipping paper description is optional. However, the entry in the Table reflects the preferred sequence.

(5) When one entry references another entry by use of the word "see", if both names are in Roman type, either name may be used as the proper shipping name (e.g., Ethyl alcohol, see Ethanol).

(6) When a proper shipping name includes a concentration range as part of the shipping description, the actual concentration, if it is within the range stated, may be used in place of the concentration range. For example, an aqueous solution of hydrogen peroxide containing 30 percent peroxide may be described as "Hydrogen peroxide, aqueous solution *with not less than 20 percent but not more than 40 percent hydrogen peroxide*" or "Hydrogen peroxide, aqueous solution *with 30 percent hydrogen peroxide*".

(7) Use of the prefix "mono" is optional in any shipping name, when appropriate. Thus, Iodine monochloride may be used interchangeably with Iodine chloride. In "Glycerol alphamonochlorohydrin" the term "mono" is considered a prefix to the term "chlorohydrin" and may be deleted.

(8) Use of the word "liquid" or "solid". The word "liquid" or "solid" may be added to a proper shipping name when a hazardous material specifically listed by name may, due to differing physical states, be a liquid or solid. When the packaging specified in Column 8 is inappropriate for the physical state of the material, the table provided in paragraph (i)(4) of this section should be used to determine the appropriate packaging section.

(9) *Hazardous wastes.* If the word "waste" is not included in the hazardous material description in Column 2 of the Table, the proper shipping name for a hazardous waste (as defined in §171.8 of this subchapter), shall include the word "Waste" preceding the proper shipping name of the material. For example: Waste acetone.

(10) *Mixtures and solutions.* (i) A mixture or solution not identified specifically by name, comprised of a single predominant hazardous material identified in the Table by technical name and one or more hazardous and/or non-hazardous material, must be described using the proper shipping name of the hazardous material and the qualifying word "mixture" or "solution", as appropriate, unless—

(A) Except as provided in §172.101(i)(4) the packaging specified in Column 8 is inappropriate to the physical state of the material;

(B) The shipping description indicates that the proper shipping name applies only to the pure or technically pure hazardous material;

(C) The hazard class, packing group, or subsidiary hazard of the mixture or solution is different from that specified for the entry;

(D) There is a significant change in the measures to be taken in emergencies;

(E) The material is identified by special provision in Column 7 of the §172.101 Table as a material poisonous by inhalation; however, it no longer meets the definition of poisonous by inhalation or it falls within a different hazard zone than that specified in the special provision; or

(F) The material can be appropriately described by a shipping name that describes its intended application, such as "Coating solution", "Extracts, flavoring" or "Compound, cleaning liquid."

(ii) If one or more of the conditions specified in paragraph (c)(10)(i) of this section is satisfied then a proper shipping name shall be selected as prescribed in paragraph (c)(12)(ii) of this section.

(iii) A mixture or solution not identified in the Table specifically by name, comprised of two or more hazardous materials in the same hazard class, shall be described using an appropriate shipping description (e.g., "Flammable liquid, n.o.s."). The name that most appropriately describes the material shall be used; e.g., an alcohol not listed by its technical name in the Table shall be described as "Alcohol, n.o.s." rather than "Flammable liquid, n.o.s.". Some mixtures may be more appropriately described according to their application, such as "Coating solution" or "Extracts, flavoring liquid" rather than by an n.o.s. entry. Under the provisions of subparts C and D of this part, the technical names of at least two components most predominately contributing to the hazards of the mixture or solution may be required in association with the proper shipping name.

(11) Except for a material subject to or prohibited by §§173.21, 173.54, 173.56(d), 173.56(e), 173.224(c) or 173.225(b) of this subchapter, a material that is considered to be a hazardous waste or a sample of a material for which the hazard class is uncertain and must be determined by testing may be assigned a tentative proper shipping name, hazard class, identification number and packing group, if applicable, based on the shipper's tentative determination according to:

(i) Defining criteria in this subchapter;

(ii) The hazard precedence prescribed in §173.2a of this subchapter;

(iii) The shippers knowledge of the material;

(iv) In addition to paragraphs (c)(11)(i) through (iii) of this section, for a sample of a material other than a waste, the following must be met:

(A) Except when the word "Sample" already appears in the proper shipping name, the word "Sample" must appear as part of the proper shipping name or in association with the basic description on the shipping paper;

(B) When the proper shipping description for a sample is assigned a "G" in Column (1) of the §172.101 Table, and the primary constituent(s) for which the tentative classification is based are not known, the provisions requiring a technical name for the constituent(s) do not apply; and

(C) A sample must be transported in a combination packaging which conforms to the requirements of this subchapter that are applicable to the tentative packing group assigned, and may not exceed a net mass of 2.5 kg. (5.5 pounds) per package.

Note to Paragraph (c)(11): For the transportation of samples of self-reactive materials, organic peroxides, explosives or lighters see §§173.224(c)(3), 173.225(c)(2), 173.56(d) or 173.308(b)(2) of this subchapter, respectively.

(12) Except when the proper shipping name in the Table is preceded by a plus (+)—

(i) If it is specifically determined that a material meets the definition of a hazard class or packing group, other than the class or packing group shown in association with the proper shipping name, or does not meet the defining criteria for a subsidiary hazard shown in Column 6 of the Table, the material shall be described by an appropriate proper shipping name listed in association with the correct hazard class, packing group, or subsidiary hazard for the material.

(ii) *Generic or n.o.s. descriptions.* If an appropriate technical name is not shown in the Table, selection of a proper shipping name shall be made from the generic or n.o.s. descriptions corresponding to the specific hazard class, packing group, or subsidiary hazard, if any, for the material. The name that most appropriately describes the material shall be used; e.g, an alcohol not listed by its technical name in the Table shall be described as "Alcohol, n.o.s." rather than "Flammable liquid, n.o.s.". Some mixtures may be more appropriately described according to their application, such as "Coating solution" or "Extracts, flavoring, liquid", rather than by an n.o.s. entry, such as "Flammable liquid, n.o.s." It should be noted, however, that an n.o.s. description as a proper shipping name may not provide sufficient information for shipping papers and package marking. Under the provisions of subparts C and D of this part, the technical name of one or more constituents which makes the product a hazardous material may be required in association with the proper shipping name.

(iii) *Multiple hazard materials.* If a material meets the definition of more than one hazard class, and is not identified in the Table specifically by name (e.g., acetyl chloride), the hazard class of the material shall be determined by using the precedence specified in §173.2a of this subchapter, and an appropriate shipping description (e.g., "Flammable liquid, corrosive n.o.s.") shall be selected as described in paragraph (c)(12)(ii) of this section.

(iv) If it is specifically determined that a material is not a forbidden material and does not meet the definition of any hazard class, the material is not a hazardous material.

(13) *Self-reactive materials and organic peroxides.* A generic proper shipping name for a self-reactive material or an organic peroxide, as listed in Column 2 of the Table, must be selected based on the material's technical name and concentration, in accordance with the provisions of §§173.224 or 173.225 of this subchapter, respectively.

(14) A proper shipping name that describes all isomers of a material may be used to identify any isomer of that material if the isomer meets criteria for the same hazard class or division, subsidiary risk(s) and packing group, unless the isomer is specifically identified in the Table.

(15) Unless a hydrate is specifically listed in the Table, a proper shipping name for the equivalent anhydrous substance may be used, if the hydrate meets the same hazard class or division, subsidiary risk(s) and packing group.

(16) Unless it is already included in the proper shipping name in the §172.101 Table, the qualifying words "liquid" or "solid" may be added in association with the proper shipping name when a hazardous material specifically listed by name in the §172.101 Table may, due to the differing physical states of the various isomers of the material, be either a liquid or a solid (for example "Dinitrotoluenes, liquid" and "Dinitrotoluenes, solid"). Use of the words "liquid" or "solid" is subject to the limitations specified for the use of the words "mixture" or "solution" in paragraph (c)(10) of this section. The qualifying word "molten" may be added in association with the proper shipping name when a hazardous material, which is a solid in accordance with the definition in §171.8 of this subchapter, is offered for transportation in the molten state (for example, "Alkylphenols, solid, n.o.s., molten").

(d) **Column 3: Hazard class or Division.** Column 3 contains a designation of the hazard class or division corresponding to each proper shipping name, or the word "Forbidden".

(1) A material for which the entry in this column is "Forbidden" may not be offered for transportation or transported. This prohibition does not apply if the material is diluted, stabilized or incorporated in a device and it is classed in accordance with the definitions of hazardous materials contained in part 173 of this subchapter.

(2) When a reevaluation of test data or new data indicates a need to modify the "Forbidden" designation or the hazard class or packing group specified for a material specifically identified in the Table, this data should be submitted to the Associate Administrator.

(3) A basic description of each hazard class and the section reference for class definitions appear in §173.2 of this subchapter.

(4) Each reference to a Class 3 material is modified to read "Combustible liquid" when that material is reclassified in accordance with §173.150(e) or (f) of this subchapter or has a flash point above 60°C (140°F) but below 93°C (200°F).

(e) **Column 4: Identification number.** Column 4 lists the identification number assigned to each proper shipping name. Those preceded by the letter "UN" are associated with proper shipping names considered appropriate for international transportation as well as domestic transportation. Those preceded by the letters "NA" are associated with proper shipping names not recognized for international transportation, except to and from Canada. Identification numbers in the "NA9000" series are associated with proper shipping names not appropriately covered by international hazardous materials (dangerous goods) transportation standards, or not appropriately addressed by the international transportation standards for emergency response information purposes, except for transportation between the United States and Canada. Those preceded by the letters "ID" are associated with proper shipping names recognized by the ICAO Technical Instructions (IBR, *see* §171.7 of this subchapter).

(f) **Column 5: Packing group.** Column 5 specifies one or more packing groups assigned to a material corresponding to the proper shipping name and hazard class for that material. Class 2, Class 7, Division 6.2 (other than regulated medical wastes), and ORM-D materials, do not have packing groups, Packing Groups I, II, and III indicate the degree of danger presented by the material is either great, medium or minor, respectively. If more than one packing group is indicated for an entry, the packing group for the hazardous material is determined using the criteria for assignment of packing groups specified in subpart D of part 173. When a reevaluation of test data or new data indicates a need to modify the specified packing group(s), the data should be submitted to the Associate Administrator. Each reference in this column to a material which is a hazardous waste or a hazardous substance, and whose proper shipping name is preceded in Column 1 of the Table by the letter "A" or "W", is modified to read "III" on those occasions when the material is offered for transportation or transported by a mode in which its transportation is not otherwise subject to requirements of this subchapter.

(g) **Column 6: Labels.** Column 6 specifies codes which represent the hazard warning labels required for a package filled with a material conforming to the associated hazard class and proper shipping name, unless the package is otherwise excepted from labeling by a provision in subpart E of this part, or part 173 of this subchapter. The first code is indicative of the primary hazard of the material. Additional label codes are indicative of subsidiary hazards. Provisions in §172.402 may require that a label other than that specified in Column 6 be affixed to the package in addition to that specified in Column 6. No label is required for a material classed as a combustible liquid or for a Class 3 material that is reclassed as a combustible liquid. The codes contained in Column 6 are defined according to the following table:

LABEL SUBSTITUTION TABLE

Label code	Label name
1	Explosive.
1.1[1]	Explosive 1.1.[1]
1.2[1]	Explosive 1.2.[1]
1.3[1]	Explosive 1.3.[1]
1.4[1]	Explosive 1.4.[1]
1.5[1]	Explosive 1.5.[1]
1.6[1]	Explosive 1.6.[1]
2.1	Flammable Gas.
2.2	Non-Flammable Gas.
2.3	Poison Gas.
3	Flammable Liquid.
4.1	Flammable Solid.
4.2	Spontaneously Combustible.
4.3	Dangerous When Wet.
5.1	Oxidizer.
5.2	Organic Peroxide.
6.1 (inhalation hazard, Zone A or B)	Poison Inhalation Hazard.
6.1 (other than inhalation hazard, Zone A or B)[2]	Poison.
6.2	Infectious Substance.
7	Radioactive.
8	Corrosive.
9	Class 9.

[1]Refers to the appropriate compatibility group letter.

[2]The packing group for a material is indicated in column 5 of the table.

(h) **Column 7: Special provisions.** Column 7 specifies codes for special provisions applicable to hazardous materials. When Column 7 refers to a special provision for a hazardous material, the meaning and requirements of that special provision are as set forth in §172.102 of this subpart.

(i) **Column 8: Packaging authorizations.** Columns 8A, 8B and 8C specify the applicable sections for exceptions, non-bulk packaging requirements and bulk packaging requirements, respectively, in part 173 of this subchapter. Columns 8A, 8B and 8C are completed in a manner which indicates that "§173." precedes the designated numerical entry. For example, the entry "202" in Column 8B associated with the proper shipping name "Gasoline" indicates that for this material conformance to non-bulk packaging requirements prescribed in §173.202 of this subchapter is required. When packaging requirements are specified, they are in addition to the standard requirements for all packagings prescribed in §173.24 of this subchapter and any other applicable requirements in subparts A and B of part 173 of this subchapter.

(1) Exceptions. Column 8A contains exceptions from some of the requirements of this subchapter. The referenced exceptions are in addition to those specified in subpart A of part 173 and elsewhere in this subchapter. A "None" in this column means no packaging exceptions are authorized, except as may be provided by special provisions in Column 7.

(2) Non-bulk packaging. Column 8B references the section in part 173 of this subchapter which prescribes packaging requirements for non-bulk packagings. A "None" in this column means non-bulk packagings are not authorized, except as may be provided by special provisions in Column 7. Each reference in this column to a material which is a hazardous waste or a hazardous substance, and whose proper shipping name is preceded in Column 1 of the Table by the letter "A" or "W", is modified to include "§173.203" or "§173.213", as appropriate for liquids and solids, respectively, on those occasions when the material is offered for transportation or transported by a mode in which its transportation is not otherwise subject to the requirements of this subchapter.

(3) *Bulk packaging.* Column (8C) specifies the section in part 173 of this subchapter that prescribes packaging requirements for bulk packagings, subject to the limitations, requirements, and additional authorizations of Columns (7) and (8B). A "None" in Column (8C) means bulk packagings are not authorized, except as may be provided by special provisions in Column (7) and in packaging authorizations Column (8B). Additional authorizations and limitations for use of UN portable tanks are set forth in Column 7. For each reference in this column to a material that is a hazardous waste or a hazardous substance, and whose proper shipping name is preceded in Column 1 of the Table by the letter "A" or "W" and that is offered for transportation or transported by a mode in which its transportation is not otherwise subject to the requirements of this subchapter:

(i) The column reference is §173.240 or §173.241, as appropriate.

(ii) For a solid material, the exception provided in Special provision B54 is applicable.

(iii) For a Class 9 material which meets the definition of an elevated temperature material, the column reference is §173.247.

(4) For a hazardous material which is specifically named in the Table and whose packaging sections specify packagings not applicable to the form of the material (e.g., packaging specified is for solid material and the material is being offered for transportation in a liquid form) the following Table should be used to determine the appropriate packaging section:

Packaging section reference for solid materials	Corresponding packaging section for liquid materials
§173.187	§173.181
§173.211	§173.201
§173.212	§173.202
§173.213	§173.203
§173.240	§173.241
§173.242	§173.243

(5) *Cylinders.* For cylinders, both non-bulk and bulk packaging authorizations are set forth in Column (8B). Notwithstanding a designation of "None" in Column (8C), a bulk cylinder may be used when specified through the section reference in Column (8B).

(j) **Column 9: Quantity limitations.** Columns 9A and 9B specify the maximum quantities that may be offered for transportation in one package by passenger-carrying aircraft or passenger-carrying rail car (Column 9A) or by cargo aircraft only (Column 9B), subject to the following:

(1) "Forbidden" means the material may not be offered for transportation or transported in the applicable mode of transport.

(2) The quantity limitation is "net" except where otherwise specified, such as for "Consumer commodity" which specifies "30 kg gross."

(3) When articles or devices are specifically listed by name, the net quantity limitation applies to the entire article or device (less packaging and packaging materials) rather than only to its hazardous components.

(4) A package offered or intended for transportation by aircraft and which is filled with a material forbidden on passenger-carrying aircraft but permitted on cargo aircraft only, or which exceeds the maximum net quantity authorized on passenger-carrying aircraft, shall be labelled with the CARGO AIRCRAFT ONLY label specified in §172.448 of this part.

(5) The total net quantity of hazardous material for an outer non- bulk packaging that contains more than one hazardous material may not exceed the lowest permitted maximum net quantity per package as shown in Column 9A or 9B, as appropriate. If one material is a liquid and one is a solid, the maximum net quantity must be calculated in kilograms. *See* § 173.24a(c)(1)(iv).

(k) **Column 10: Vessel Stowage.** Column 10A [Vessel stowage] specifies the authorized stowage locations on board cargo and passenger vessels. Column 10B [Other provisions] specifies codes for stowage requirements for specific hazardous materials. The meaning of each code in Column 10B is set forth in §176.84 of this subchapter. Section §176.63 of this subchapter sets forth the physical requirements for each of the authorized locations listed in Column 10A. (For bulk transportation by vessel, see 46 CFR parts 30 to 40, 70, 98,148, 151, 153 and 154.) The authorized stowage locations specified in Column 10A are defined as follows:

(1) Stowage category "A" means the material may be stowed "on deck" or "under deck" on a cargo vessel and on a passenger vessel.

(2) Stowage category "B" means-

(i) The material may be stowed "on deck" or "under deck" on a cargo vessel and on a passenger vessel carrying a number of passengers limited to not more than the larger of 25 passengers, or one passenger per each 3 m of overall vessel length; and

(ii) "On deck only" on passenger vessels in which the number of passengers specified in paragraph (k)(2)(i) of this section is exceeded.

(3) Stowage category "C" means the material must be stowed "on deck only" on a cargo vessel and on a passenger vessel.

(4) Stowage category "D" means the material must be stowed "on deck only" on a cargo vessel and on a passenger vessel carrying a number of passengers limited to not more than the larger of 25 passengers or one passenger per each 3 m of overall vessel length, but the material is prohibited on passenger vessels in which the limiting number of passengers is exceeded.

(5) Stowage category "E" means the material may be stowed "on deck" or "under deck" on a cargo vessel and on a passenger vessel carrying a number of passengers limited to not more than the larger of 25 passengers, or one passenger per each 3 m of overall vessel length, but is prohibited from carriage on passenger vessels in which the limiting number of passengers is exceeded.

(6) Stowage category "01" means the material may be stowed "on deck" or "under deck" on a cargo vessel (up to 12 passengers) and on a passenger vessel.

(7) Stowage category "02" means the material may be stowed "on deck" or "under deck" on a cargo vessel (up to 12 passengers) and "on deck" in closed cargo transport units or "under deck" in closed cargo transport units on a passenger vessel.

(8) Stowage category "03" means the material may be stowed "on deck" or "under deck" on a cargo vessel (up to 12 passengers) and "on deck" in closed cargo transport units on a passenger vessel.

(9) Stowage category "04" means the material may be stowed "on deck" or "under deck" on a cargo vessel (up to 12 passengers) but the material is prohibited on a passenger vessel.

(10) Stowage category "05" means the material may be stowed "on deck" in closed cargo transport units or "under deck" on a cargo vessel (up to 12 passengers) and on a passenger vessel.

(11) Stowage category "06" means the material may be stowed "on deck" in closed cargo transport units or "under deck" on a cargo vessel (up to 12 passengers) and "on deck" in closed cargo transport units or "under deck" in closed cargo transport units on a passenger vessel.

(12) Stowage category "07" means the material may be stowed "on deck" in closed cargo transport units or "under deck" on a cargo vessel (up to 12 passengers) and "on deck" only in closed cargo transport units on a passenger vessel.

(13) Stowage category "08" means the material may be stowed "on deck" in closed cargo transport units or "under deck" on a cargo vessel (up to 12 passengers) but the material is prohibited on a passenger vessel.

(14) Stowage category "09" means the material may be stowed "on deck only" in closed cargo transport units or "under deck" in closed cargo transport units on a cargo vessel (up to 12 passengers) and on a passenger vessel.

(15) Stowage category "10" means the material may be stowed "on deck" in closed cargo transport units or "under deck" in closed cargo transport units on a cargo vessel (up to 12 passengers) and "on deck" only in closed cargo transport units on a passenger vessel.

(16) Stowage category "11" means the material may be stowed "on deck" in closed cargo transport units or "under deck" in magazine stowage type "c" on a cargo vessel (up to 12 passengers) and "on deck" only in closed cargo transport units on a passenger vessel.

(17) Stowage category "12" means the material may be stowed "on deck" in closed cargo transport units or "under deck" in magazine stowage type "c" on a cargo vessel (up to 12 passengers) but the material is prohibited on a passenger vessel.

(18) Stowage category "13" means the material may be stowed "on deck" in closed cargo transport units or "under deck" in magazine stowage type "A" on a cargo vessel (up to 12 passengers) and "on deck" only in closed cargo transport units on a passenger vessel.

(19) Stowage category "14" means the material may be stowed "on deck" in closed cargo transport units on a cargo vessel (up to 12 passengers) but the material is prohibited on a passenger vessel.

(20) Stowage category "15... means the material may be stowed "on deck" in closed cargo transport units or "under deck" in closed cargo transport units on a cargo vessel (up to 12 passengers) but the material is prohibited on a passenger vessel.

(l) *Changes to the Table.*(1) Unless specifically stated otherwise in a rule document published in the *Federal Register* amending the Table-

(i) Such a change does not apply to the shipment of any package filled prior to the effective date of the amendment; and

(ii) Stocks of preprinted shipping papers and package marking may be continued in use, in the manner previously authorized, until depleted or for a one-year period, subsequent to the effective date of the amendment, whichever is less.

(2) Except as otherwise provided in this section any alteration of a shipping description or associated entry which is listed in the §172.101 Table must receive prior written approval from the Associate Administrator.

(3) Cylinders used for chlorine (UN1017) with preprinted markings conforming to §172.400a(a)(1)(ii) without the Division 5.1 subsidiary hazard number may continue to be used until January 1, 2011.

ERG Guide No. The ERG Guide No. column is not part of the §172.101 Table in the Hazardous Materials Regulations. This column has been added for your convenience. This column provides the three-digit emergency response guide page number assigned to the name of the material in the 2004 Emergency Response Guidebook.

The letter "P" following some guide numbers indicates that the material may undergo violent polymerization if subjected to heat or contamination.

ERG Guide numbers that are shaded indicate that the Table of Initial Isolation and Protective Action Distances (the green pages of the ERG) should be consulted. These materials are either TIH (Toxic Inhalation Hazard) material, or a chemical warfare agent, or a Dangerous Water Reactive Material (produces toxic gas upon contact with water).

Placard Advisory. The placard advisory column is not part of the §172.101 Table in the Hazardous Materials Regulations. This column has been added for your convenience. It provides a quick indication of the placard required for the shipping description, based only on the information in the other columns of the §172.101 Table. To determine the actual placards required for a specific shipment, consult the Hazardous Materials Regulations (§172.504 & §172.505).

* An asterisk after the placard name indicates that placarding is required for any quantity of material being transported. If multiple placards are listed, all placards are required for any quantity.

Symbols (1)	Hazardous materials descriptions and proper shipping names (2)	Hazard class or Division (3)	Identification Numbers (4)	PG (5)	Label Codes (6)	Special Provisions (§172.102) (7)	(8) Packaging (§173.***) Exceptions (8A)	Non-bulk (8B)	Bulk (8C)	(9) Quantity limitations (see §§173.27 and 175.75) Passenger aircraft/rail (9A)	Cargo aircraft only (9B)	(10) Vessel stowage Location (10A)	Other (10B)	ERG Guide No.	Placard Advisory Consult §172.504 & §172.505 to determine placard *Denotes placard for any quantity
	Accellerene, see **p-Nitrosodimethylaniline**														
	Accumulators, electric, see **Batteries, wet** *etc.*														
	Accumulators, pressurized, pneumatic or hydraulic (containing non-flammable gas) see **Articles, pressurized pneumatic** *or* **hydraulic** *(containing non-flammable gas)*														
	Acetal	3	UN1088	II	3	IB2, T4, TP1	150	202	242	5 L	60 L	E		127	FLAMMABLE
	Acetaldehyde	3	UN1089	I	3	A3, B16, T11, TP2, TP7	None	201	243	Forbidden	30 L	E		129	FLAMMABLE
A	**Acetaldehyde ammonia**	9	UN1841	III	9	IB8, IP3, IP7, T1, TP33	155	204	240	200 kg	200 kg	A	34	171	CLASS 9
	Acetaldehyde oxime	3	UN2332	III	3	B1, IB3, T4, TP1	150	203	242	60 L	220 L	A		129	FLAMMABLE
	Acetic acid, glacial *or* **Acetic acid solution,** *with more than 80 percent acid, by mass*	8	UN2789	II	8, 3	A3, A6, A7, A10, B2, IB2, T7, TP2	154	202	243	1 L	30 L	A		132	CORROSIVE
	Acetic acid solution, *not less than 50 percent but not more than 80 percent acid, by mass*	8	UN2790	II	8	A3, A6, A7, A10, B2, IB2, T7, TP2	154	202	242	1 L	30 L	A		153	CORROSIVE
	Acetic acid solution, *with more than 10 percent and less than 50 percent acid, by mass.*	8	UN2790	III	8	IB3, T4, TP1	154	203	242	5 L	60 L	A		153	CORROSIVE
	Acetic anhydride	8	UN1715	II	8, 3	A3, A6, A7, A10, B2, IB2, T7, TP2	154	202	243	1 L	30 L	A	40	137	CORROSIVE
	Acetone	3	UN1090	II	3	IB2, T4, TP1	150	202	242	5 L	60 L	B		127	FLAMMABLE
	Acetone cyanohydrin, stabilized	6.1	UN1541	I	6.1	2, B9, B14, B32, B76, B77, N34, T20, TP2, TP13, TP38, TP45	None	227	244	Forbidden	Forbidden	D	25, 40, 52, 53	155	POISON INHALATION HAZARD*
	Acetone oils	3	UN1091	II	3	IB2, T4, TP1, TP8	150	202	242	5 L	60 L	B		127	FLAMMABLE
	Acetonitrile	3	UN1648	II	3	IB2, T7, TP2	150	202	242	5 L	60 L	B	40	127	FLAMMABLE
	Acetyl acetone peroxide with more than 9 percent by mass active oxygen	Forbidden													
	Acetyl benzoyl peroxide, solid, or with more than 40 percent in solution	Forbidden													
	Acetyl bromide	8	UN1716	II	8	B2, IB2, T8, TP2	154	202	242	1 L	30 L	C	40	156	CORROSIVE
	Acetyl chloride	3	UN1717	II	3, 8	A3, A6, A7, IB1, N34, T8, TP2	150	202	243	1 L	5 L	B	40	155	FLAMMABLE
	Acetyl cyclohexanesulfonyl peroxide, with more than 82 percent wetted with less than 12 percent water	Forbidden													
	Acetyl iodide	8	UN1898	II	8	B2, IB2, T7, TP2, TP13	154	202	242	1 L	30 L	C	40	156	CORROSIVE
	Acetyl methyl carbinol	3	UN2621	III	3	B1, IB3, T2, TP1	150	203	242	60 L	220 L	A		127	FLAMMABLE
	Acetyl peroxide, solid, or with more than 25 percent in solution	Forbidden													
	Acetylene, dissolved	2.1	UN1001		2.1	N86, N88	None	303	None	Forbidden	15 kg	D	25, 40, 57	116	FLAMMABLE GAS
	Acetylene (liquefied)	Forbidden													
	Acetylene silver nitrate	Forbidden													
	Acetylene, solvent free	Forbidden													
	Acetylene tetrabromide, see **Tetrabromoethane**														
	Acid butyl phosphate, see **Butyl acid phosphate**														
	Acid, sludge, *see* **Sludge acid**														
	Acridine	6.1	UN2713	III	6.1	IB8, IP3, T1, TP33	153	213	240	100 kg	200 kg	A		153	POISON
	Acrolein dimer, stabilized	3	UN2607	III	3	B1, IB3, T2, TP1	150	203	242	60 L	220 L	A	40	129P	FLAMMABLE
	Acrolein, stabilized	6.1	UN1092	I	6.1, 3	1, B9, B14, B30, B42, B77, T22, TP2, TP7, TP13, TP38, TP44	None	226	244	Forbidden	Forbidden	D	40	131P	POISON INHALATION HAZARD*
	Acrylamide, solid	6.1	UN2074	III	6.1	IB8, IP3, T1, TP33	153	213	240	100 kg	200 kg	A	12	153P	POISON
	Acrylamide solution	6.1	UN3426	III	6.1	IB3, T4, TP1	153	203	241	60 L	220 L	A	12	153P	POISON
	Acrylic acid, stabilized	8	UN2218	II	8, 3	B2, IB2, T7, TP2	154	202	243	1 L	30 L	C	25, 40	132P	CORROSIVE
	Acrylonitrile, stabilized	3	UN1093	I	3, 6.1	B9, T14, TP2, TP13	None	201	243	Forbidden	30 L	E	40	131P	FLAMMABLE
	Actuating cartridge, explosive, see **Cartridges, power device**														

Symbols (1)	Hazardous materials descriptions and proper shipping names (2)	Hazard class or Division (3)	Identification Numbers (4)	PG (5)	Label Codes (6)	Special Provisions (§172.102) (7)	(8) Packaging (§173.***) Exceptions (8A)	Non-bulk (8B)	Bulk (8C)	(9) Quantity limitations (see §§173.27 and 175.75) Passenger aircraft/rail (9A)	Cargo aircraft only (9B)	(10) Vessel stowage Location (10A)	Other (10B)	ERG Guide No.	Placard Advisory Consult §172.504 & §172.505 to determine placard *Denotes placard for any quantity
	Adhesives, *containing a flammable liquid*	3	UN1133	I	3	T11, TP1, TP8, TP27	150	201	243	1 L	30 L	B		128	FLAMMABLE
				II	3	149, B52, IB2, T4, TP1, TP8	150	173	242	5 L	60 L	B		128	FLAMMABLE
				III	3	B1, B52, IB3, T2, TP1	150	173	242	60 L	220 L	A		128	FLAMMABLE
	Adiponitrile	6.1	UN2205	III	6.1	IB3, T3, TP1	153	203	241	60 L	220 L	A		153	POISON
	Aerosols, *corrosive, Packing Group II or III, (each not exceeding 1 L capacity)*	2.2	UN1950		2.2, 8	A34	306	None	None	75 kg	150 kg	A	48, 87, 126	126	NONFLAMMABLE GAS
	Aerosols, *flammable, (each not exceeding 1 L capacity)*	2.1	UN1950		2.1	N82	306	None	None	75 kg	150 kg	A	48, 87, 126	126	FLAMMABLE GAS
	Aerosols, flammable, n.o.s. *(engine starting fluid) (each not exceeding 1 L capacity)*	2.1	UN1950		2.1	N82	306	304	None	Forbidden	150 kg	A	48, 87, 126	126	FLAMMABLE GAS
	Aerosols, *non-flammable, (each not exceeding 1 L capacity)*	2.2	UN1950		2.2		306	None	None	75 kg	150 kg	A	48, 87, 126	126	NONFLAMMABLE GAS
	Aerosols, *poison, Packing Group III (each not exceeding 1 L capacity)*	2.2	UN1950		2.2, 6.1		306	None	None	Forbidden	Forbidden	A	48, 87, 126	126	NONFLAMMABLE GAS
I	**Air bag inflators,** *or* **Air bag modules,** *or* **Seat-belt pretensioners**	1.4G	UN0503	II	1.4G	161	None	62	None	Forbidden	75 kg	02		114	EXPLOSIVES 1.4
	Air bag inflators, *or* **Air bag modules,** *or* **Seat-belt pretensioners**	9	UN3268	III	9	160	166	166	166	25 kg	100 kg	A		171	CLASS 9
	Air, compressed	2.2	UN1002		2.2	78	306, 307	302	302	75 kg	150 kg	A		122	NONFLAMMABLE GAS
	Air, refrigerated liquid, *(cryogenic liquid)*	2.2	UN1003		2.2, 5.1	T75, TP5, TP22	320	316	318, 319	Forbidden	Forbidden	D	51	122	NONFLAMMABLE GAS
	Air, refrigerated liquid, *(cryogenic liquid) non-pressurized*	2.2	UN1003		2.2, 5.1	T75, TP5, TP22	320	316	318, 319	Forbidden	Forbidden	D	51	122	NONFLAMMABLE GAS
	Aircraft engines (including turbines), see **Engines, internal combustion**														
	Aircraft evacuation slides, see **Lifesaving appliances** *etc.*														
	Aircraft hydraulic power unit fuel tank *(containing a mixture of anhydrous hydrazine and monomethyl hydrazine) (M86 fuel)*	3	UN3165	I	3, 6.1, 8		None	172	None	Forbidden	42 L	E		131	FLAMMABLE
	Aircraft survival kits, see **Life saving appliances** *etc.*														
G	**Alcoholates solution, n.o.s.,** *in alcohol*	3	UN3274	II	3, 8	IB2	150	202	243	1 L	5 L	B		132	FLAMMABLE
	Alcoholic beverages	3	UN3065	II	3	24, 149, B1, IB2, T4, TP1	150	202	242	5 L	60 L	A		127	FLAMMABLE
				III	3	24, B1, IB3, N11, T2, TP1	150	203	242	60 L	220 L	A		127	FLAMMABLE
	Alcohols, n.o.s.	3	UN1987	I	3	172, T11, TP1, TP8, TP27	4b	201	243	1 L	30 L	E		127	FLAMMABLE
				II	3	172, IB2, T7, TP1, TP8, TP28	4b, 150	202	242	5 L	60 L	B		127	FLAMMABLE
				III	3	172, B1, IB3, T4, TP1, TP29	4b, 150	203	242	60 L	220 L	A		127	FLAMMABLE
G	**Alcohols, flammable, toxic, n.o.s.**	3	UN1986	I	3, 6.1	T14, TP2, TP13, TP27	None	201	243	Forbidden	30 L	E	40	131	FLAMMABLE
				II	3, 6.1	IB2, T11, TP2, TP27	150	202	243	1 L	60 L	B	40	131	FLAMMABLE
				III	3, 6.1	B1, IB3, T7, TP1, TP28	150	203	242	60 L	220 L	A		131	FLAMMABLE
	Aldehydes, n.o.s.	3	UN1989	I	3	T11, TP1, TP27	None	201	243	1 L	30 L	E		129	FLAMMABLE
				II	3	IB2, T7, TP1, TP8, TP28	150	202	242	5 L	60 L	B		129	FLAMMABLE
				III	3	B1, IB3, T4, TP1, TP29	150	203	242	60 L	220 L	A		129	FLAMMABLE
G	**Aldehydes, flammable, toxic, n.o.s.**	3	UN1988	I	3, 6.1	T14, TP2, TP13, TP27	None	201	243	Forbidden	30 L	E	40	131	FLAMMABLE
				II	3, 6.1	IB2, T11, TP2, TP27	150	202	243	1 L	60 L	B	40	131	FLAMMABLE
				III	3, 6.1	B1, IB3, T7, TP1, TP28	150	203	242	60 L	220 L	A		131	FLAMMABLE
	Aldol	6.1	UN2839	II	6.1	IB2, T7, TP2	153	202	243	5 L	60 L	A	12	153	POISON

Symbols (1)	Hazardous materials descriptions and proper shipping names (2)	Hazard class or Division (3)	Identification Numbers (4)	PG (5)	Label Codes (6)	Special Provisions (§172.102) (7)	(8) Packaging (§173.***) Exceptions (8A)	Non-bulk (8B)	Bulk (8C)	(9) Quantity limitations (see §§173.27 and 175.75) Passenger aircraft/rail (9A)	Cargo aircraft only (9B)	(10) Vessel stowage Location (10A)	Other (10B)	ERG Guide No.	Placard Advisory Consult §172.504 & §172.505 to determine placard *Denotes placard for any quantity
G	**Alkali metal alcoholates, self-heating, corrosive, n.o.s.**	4.2	UN3206	II	4.2, 8	64, A7, IB5, IP2, T3, TP33	None	212	242	15 kg	50 kg	B		136	SPONTANEOUSLY COMBUSTIBLE
				III	4.2, 8	64, A7, IB8, IP3, T1, TP33	None	213	242	25 kg	100 kg	B		136	SPONTANEOUSLY COMBUSTIBLE
	Alkali metal alloys, liquid, n.o.s.	4.3	UN1421	I	4.3	A2, A3, A7, B48, N34	None	201	244	Forbidden	1 L	D	52	138	DANGEROUS WHEN WET*
	Alkali metal amalgam, liquid	4.3	UN1389	I	4.3	A2, A3, A7, N34	None	201	244	Forbidden	1 L	D	40, 52	138	DANGEROUS WHEN WET*
	Alkali metal amalgam, solid	4.3	UN3401	I	4.3	IB4, IP1, N40, T9, TP7, TP33	None	211	242	Forbidden	15 kg	D	52	138	DANGEROUS WHEN WET*
	Alkali metal amides	4.3	UN1390	II	4.3	A6, A7, A8, A19, A20, IB7, IP2	151	212	241	15 kg	50 kg	E	40, 52	139	DANGEROUS WHEN WET*
	Alkali metal dispersions, flammable *or* **Alkaline earth metal dispersions, flammable**	4.3	UN3482	I	4.3, 3	A2, A3, A7	None	201	244	Forbidden	1 L	D	52	—	DANGEROUS WHEN WET*
	Alkali metal dispersions, *or* **Alkaline earth metal dispersions**	4.3	UN1391	I	4.3	A2, A3, A7	None	201	244	Forbidden	1 L	D	52	138	DANGEROUS WHEN WET*
	Alkaline corrosive liquids, n.o.s., see **Caustic alkali liquids, n.o.s.**														
G	**Alkaline earth metal alcoholates, n.o.s.**	4.2	UN3205	II	4.2	65, A7, IB6, IP2, T3, TP33	None	212	241	15 kg	50 kg	B		135	SPONTANEOUSLY COMBUSTIBLE
				III	4.2	65, A7, IB8, IP3, T1, TP33	None	213	241	25 kg	100 kg	B		135	SPONTANEOUSLY COMBUSTIBLE
	Alkaline earth metal alloys, n.o.s.	4.3	UN1393	II	4.3	A19, IB7, IP2, T3, TP33	151	212	241	15 kg	50 kg	E	52	138	DANGEROUS WHEN WET*
	Alkaline earth metal amalgams, liquid	4.3	UN1392	I	4.3	A19, N34, N40	None	201	244	Forbidden	1 L	E	40, 52	138	DANGEROUS WHEN WET*
	Alkaline earth metal amalgams, solid	4.3	UN3402	I	4.3	A19, N34, N40, T9, TP7, TP33	None	211	242	Forbidden	15 kg	D	52	138	DANGEROUS WHEN WET*
G	**Alkaloids, liquid, n.o.s.,** *or* **Alkaloid salts, liquid, n.o.s.**	6.1	UN3140	I	6.1	A4, T14, TP2, TP27	None	201	243	1 L	30 L	A		151	POISON
				II	6.1	IB2, T11, TP2, TP27	153	202	243	5 L	60 L	A		151	POISON
				III	6.1	IB3, T7, TP1, TP28	153	203	241	60 L	220 L	A		151	POISON
G	**Alkaloids, solid, n.o.s.** *or* **Alkaloid salts, solid, n.o.s.** *poisonous*	6.1	UN1544	I	6.1	IB7, IP1, T6, TP33	None	211	242	5 kg	50 kg	A		151	POISON
				II	6.1	IB8, IP2, IP4, T3, TP33	153	212	242	25 kg	100 kg	A		151	POISON
				III	6.1	IB8, IP3, T1, TP33	153	213	240	100 kg	200 kg	A		151	POISON
	Alkyl sulfonic acids, liquid *or* **Aryl sulfonic acids, liquid** *with more than 5 percent free sulfuric acid*	8	UN2584	II	8	B2, IB2, T8, TP2, TP13	154	202	242	1 L	30 L	B		153	CORROSIVE
	Alkyl sulfonic acids, liquid *or* **Aryl sulfonic acids, liquid** *with not more than 5 percent free sulfuric acid*	8	UN2586	III	8	IB3, T4, TP1	154	203	241	5 L	60 L	B		153	CORROSIVE
	Alkyl sulfonic acids, solid *or* **Aryl sulfonic acids, solid,** *with more than 5 percent free sulfuric acid*	8	UN2583	II	8	IB8, IP2, IP4, T3, TP33	154	212	240	15 kg	50 kg	A		153	CORROSIVE
	Alkyl sulfonic acids, solid *or* **Aryl sulfonic acids, solid** *with not more than 5 percent free sulfuric acid*	8	UN2585	III	8	IB8, IP3, T1, TP33	154	213	240	25 kg	100 kg	A		153	CORROSIVE
	Alkylphenols, liquid, n.o.s. *(including C2-C12 homologues)*	8	UN3145	I	8	A6, T14, TP2	None	201	243	0.5 L	2.5 L	B		153	CORROSIVE
				II	8	IB2, T11, TP2, TP27	154	202	242	1 L	30 L	B		153	CORROSIVE
				III	8	IB3, T7, TP1, TP28	154	203	241	5 L	60 L	A		153	CORROSIVE
	Alkylphenols, solid, n.o.s. *(including C2-C12 homologues)*	8	UN2430	I	8	IB7, IP1, T6, TP33	None	211	242	1 kg	25 kg	B		153	CORROSIVE
				II	8	IB8, IP2, IP4, T3, TP33	154	212	240	15 kg	50 kg	B		153	CORROSIVE
				III	8	IB8, IP3, T1, TP33	154	213	240	25 kg	100 kg	A		153	CORROSIVE
	Alkylsulfuric acids	8	UN2571	II	8	B2, IB2, T8, TP2, TP13, TP28	154	202	242	1 L	30 L	C	14	156	CORROSIVE
	Allethrin, see **Pesticides, liquid, toxic, n.o.s.**														
	Allyl acetate	3	UN2333	II	3, 6.1	IB2, T7, TP1, TP13	150	202	243	1 L	60 L	E	40	131	FLAMMABLE
	Allyl alcohol	6.1	UN1098	I	6.1, 3	2, B9, B14, B32, B77, T20, TP13, TP38, TP45	None	227	244	Forbidden	Forbidden	D	40	131	POISON INHALATION HAZARD*

Sym-bols (1)	Hazardous materials descriptions and proper shipping names (2)	Hazard class or Division (3)	Identifi-cation Numbers (4)	PG (5)	Label Codes (6)	Special Provisions (§172.102) (7)	(8) Packaging (§173.***) Excep-tions (8A)	Non-bulk (8B)	Bulk (8C)	(9) Quantity limitations (see §§173.27 and 175.75) Passenger aircraft/rail (9A)	Cargo aircraft only (9B)	(10) Vessel stowage Location (10A)	Other (10B)	ERG Guide No.	Placard Advisory Consult §172.504 & §172.505 to determine placard *Denotes placard for any quantity
	Allyl bromide	3	UN1099	I	3, 6.1	T14, TP2, TP13	None	201	243	Forbidden	30 L	B	40	131	FLAMMABLE
	Allyl chloride	3	UN1100	I	3, 6.1	T14, TP2, TP13	None	201	243	Forbidden	30 L	E	40	131	FLAMMABLE
	Allyl chlorocarbonate, see **Allyl chloroformate**														
	Allyl chloroformate	6.1	UN1722	I	6.1, 3, 8	2, B9, B14, B32, N41, T20, TP2, TP13, TP38, TP45	None	227	244	Forbidden	Forbidden	D	40	155	POISON INHALATION HAZARD*
	Allyl ethyl ether	3	UN2335	II	3, 6.1	IB2, T7, TP1, TP13	150	202	243	1 L	60 L	E	40	131	FLAMMABLE
	Allyl formate	3	UN2336	I	3, 6.1	T14, TP2, TP13	None	201	243	Forbidden	30 L	E	40	131	FLAMMABLE
	Allyl glycidyl ether	3	UN2219	III	3	B1, IB3, T2, TP1	150	203	242	60 L	220 L	A		129	FLAMMABLE
	Allyl iodide	3	UN1723	II	3, 8	A3, A6, IB1, N34, T7, TP2, TP13	150	202	243	1 L	5 L	B	40	132	FLAMMABLE
	Allyl isothiocyanate, stabilized	6.1	UN1545	II	6.1, 3	A3, A7, IB2, T7, TP2	None	202	243	Forbidden	60 L	D	40	155	POISON
	Allylamine	6.1	UN2334	I	6.1, 3	2, B9, B14, B32, T20, TP2, TP13, TP38, TP45	None	227	244	Forbidden	Forbidden	D	40	131	POISON INHALATION HAZARD*
	Allyltrichlorosilane, stabilized	8	UN1724	II	8, 3	A7, B2, B6, N34, T10, TP2, TP7, TP13	None	206	243	Forbidden	30 L	C	40	155	CORROSIVE
	Aluminum borohydride *or* **Aluminum borohydride in devices**	4.2	UN2870	I	4.2, 4.3	B11, T21, TP7, TP33	None	181	244	Forbidden	Forbidden	D		135	SPONTANEOUSLY COMBUSTIBLE, DANGEROUS WHEN WET*
	Aluminum bromide, anhydrous	8	UN1725	II	8	IB8, IP2, IP4, T3, TP33	154	212	240	15 kg	50 kg	A	40	137	CORROSIVE
	Aluminum bromide, solution	8	UN2580	III	8	IB3, T4, TP1	154	203	241	5 L	60 L	A		154	CORROSIVE
	Aluminum carbide	4.3	UN1394	II	4.3	A20, IB7, IP2, N41, T3, TP33	151	212	242	15 kg	50 kg	A	52	138	DANGEROUS WHEN WET*
	Aluminum chloride, anhydrous	8	UN1726	II	8	IB8, IP2, IP4, T3, TP33	154	212	240	15 kg	50 kg	A	40	137	CORROSIVE
	Aluminum chloride, solution	8	UN2581	III	8	IB3, T4, TP1	154	203	241	5 L	60 L	A		154	CORROSIVE
	Aluminum dross, wet or hot	Forbidden													
	Aluminum ferrosilicon powder	4.3	UN1395	II	4.3, 6.1	A19, IB5, IP2, T3, TP33	151	212	242	15 kg	50 kg	A	39, 40, 52, 53, 85, 103	139	DANGEROUS WHEN WET*
				III	4.3, 6.1	A19, A20, IB4	151	213	241	25 kg	100 kg	A	39, 40, 52, 53, 85, 103	139	DANGEROUS WHEN WET*
	Aluminum hydride	4.3	UN2463	I	4.3	A19, N40	None	211	242	Forbidden	15 kg	E		138	DANGEROUS WHEN WET*
D	**Aluminum, molten**	9	NA9260	III	9	IB3, T1, TP3	None	None	247	Forbidden	Forbidden	D		169	CLASS 9
	Aluminum nitrate	5.1	UN1438	III	5.1	A1, A29, IB8, IP3, T1, TP33	152	213	240	25 kg	100 kg	A		140	OXIDIZER
	Aluminum phosphate solution, see **Corrosive liquids,** *etc.*														
	Aluminum phosphide	4.3	UN1397	I	4.3, 6.1	A8, A19, N40	None	211	242	Forbidden	15 kg	E	40, 52, 85	139	DANGEROUS WHEN WET*
	Aluminum phosphide pesticides	6.1	UN3048	I	6.1	A8, IB7, IP1, T6, TP33	None	211	242	Forbidden	15 kg	E	40, 85	157	POISON
	Aluminum powder, coated	4.1	UN1309	II	4.1	IB8, IP2, IP4, T3, TP33	151	212	240	15 kg	50 kg	A	13, 39, 52, 53, 74, 101	170	FLAMMABLE SOLID
				III	4.1	IB8, IP3, T1, TP33	151	213	240	25 kg	100 kg	A	13, 39, 52, 53, 74, 101	170	FLAMMABLE SOLID
	Aluminum powder, uncoated	4.3	UN1396	II	4.3	A19, A20, IB7, IP2, T3, TP33	151	212	242	15 kg	50 kg	A	39, 52, 53	138	DANGEROUS WHEN WET*
				III	4.3	A19, A20, IB8, IP4, T1, TP33	151	213	241	25 kg	100 kg	A	39, 52, 53	138	DANGEROUS WHEN WET*
	Aluminum resinate	4.1	UN2715	III	4.1	IB6, T1, TP33	151	213	240	25 kg	100 kg	A		133	FLAMMABLE SOLID
	Aluminum silicon powder, uncoated	4.3	UN1398	III	4.3	A1, A19, IB8, IP4, T1, TP33	151	213	241	25 kg	100 kg	A	39, 40, 52, 53, 85, 103	138	DANGEROUS WHEN WET*
	Aluminum smelting by-products *or* **Aluminum remelting by-products**	4.3	UN3170	II	4.3	128, B115, IB7, IP2, T3, TP33	None	212	242	15 kg	50 kg	B	85, 103	138	DANGEROUS WHEN WET*
				III	4.3	128, B115, IB8, IP4, T1, TP33	None	213	241	25 kg	100 kg	B	85, 103	138	DANGEROUS WHEN WET*

Symbols (1)	Hazardous materials descriptions and proper shipping names (2)	Hazard class or Division (3)	Identification Numbers (4)	PG (5)	Label Codes (6)	Special Provisions (§172.102) (7)	(8) Packaging (§173.***) Exceptions (8A)	Non-bulk (8B)	Bulk (8C)	(9) Quantity limitations (see §§173.27 and 175.75) Passenger aircraft/rail (9A)	Cargo aircraft only (9B)	(10) Vessel stowage Location (10A)	Other (10B)	ERG Guide No.	Placard Advisory Consult §172.504 & §172.505 to determine placard *Denotes placard for any quantity
	Amatols, see **Explosives, blasting, type B**														
G	**Amine, flammable, corrosive, n.o.s.** *or* **Polyamines, flammable, corrosive, n.o.s.**	3	UN2733	I	3, 8	T14, TP1, TP27	None	201	243	0.5 L	2.5 L	D	40, 52	132	FLAMMABLE
				II	3, 8	IB2, T11, TP1, TP27	150	202	243	1 L	5 L	B	40, 52	132	FLAMMABLE
				III	3, 8	B1, IB3, T7, TP1, TP28	150	203	242	5 L	60 L	A	40, 52	132	FLAMMABLE
G	**Amine, liquid, corrosive, flammable, n.o.s.** *or* **Polyamines, liquid, corrosive, flammable, n.o.s.**	8	UN2734	I	8, 3	A3, A6, N34, T14, TP2, TP27	None	201	243	0.5 L	2.5 L	A	52	132	CORROSIVE
				II	8, 3	IB2, T11, TP2, TP27	None	202	243	1 L	30 L	A	52	132	CORROSIVE
G	**Amines, liquid, corrosive, n.o.s.,** *or* **Polyamines, liquid, corrosive, n.o.s.**	8	UN2735	I	8	A3, A6, B10, N34, T14, TP2, TP27	None	201	243	0.5 L	2.5 L	A	52	153	CORROSIVE
				II	8	B2, IB2, T11, TP1, TP27	154	202	242	1 L	30 L	A	52	153	CORROSIVE
				III	8	IB3, T7, TP1, TP28	154	203	241	5 L	60 L	A	52	153	CORROSIVE
G	**Amines, solid, corrosive, n.o.s.,** *or* **Polyamines, solid, corrosive n.o.s.**	8	UN3259	I	8	IB7, IP1, T6, TP33	None	211	242	1 kg	25 kg	A	52	154	CORROSIVE
				II	8	IB8, IP2, IP4, T3, TP33	154	212	240	15 kg	50 kg	A	52	154	CORROSIVE
				III	8	IB8, IP3, T1, TP33	154	213	240	25 kg	100 kg	A	52	154	CORROSIVE
	2-Amino-4-chlorophenol	6.1	UN2673	II	6.1	IB8, IP2, IP4, T3, TP33	153	212	242	25 kg	100 kg	A		151	POISON
	2-Amino-5-diethylaminopentane	6.1	UN2946	III	6.1	IB3, T4, TP1	153	203	241	60 L	220 L	A		153	POISON
	2-Amino-4, 6-Dinitrophenol, wetted *with not less than 20 percent water by mass*	4.1	UN3317	I	4.1	23, A8, A19, A20, N41	None	211	None	1 kg	15 kg	E	28, 36	113	FLAMMABLE SOLID
	2-(2-Aminoethoxy) ethanol	8	UN3055	III	8	IB3, T4, TP1	154	203	241	5 L	60 L	A		154	CORROSIVE
	N-Aminoethylpiperazine	8	UN2815	III	8	IB3, T4, TP1	154	203	241	5 L	60 L	A	12	153	CORROSIVE
+	**Aminophenols (o-; m-; p-)**	6.1	UN2512	III	6.1	IB8, IP3, T1, TP33	153	213	240	100 kg	200 kg	A		152	POISON
	Aminopropyldiethanolamine, see **Amines,** *etc.*														
	n-Aminopropylmorpholine, see **Amines,** *etc.*														
	Aminopyridines *(o-; m-; p-)*	6.1	UN2671	II	6.1	IB8, IP2, IP4, T3, TP33	153	212	242	25 kg	100 kg	B	12, 40, 52	153	POISON
I	**Ammonia, anhydrous**	2.3	UN1005		2.3, 8	4, N87, T50	None	304	314, 315	Forbidden	Forbidden	D	40, 52, 57	125	POISON GAS*
D	**Ammonia, anhydrous**	2.2	UN1005		2.2	13, T50	None	304	314, 315	Forbidden	Forbidden	D	40, 52, 57	125	NONFLAMMABLE GAS
D	**Ammonia solution,** *relative density less than 0.880 at 15 degrees C in water, with more than 50 percent ammonia.*	2.2	UN3318		2.2	13, T50	None	304	314, 315	Forbidden	Forbidden	D	40, 52, 57	125	NONFLAMMABLE GAS
I	**Ammonia solution,** *relative density less than 0.880 at 15 degrees C in water, with more than 50 percent ammonia.*	2.3	UN3318		2.3, 8	4, N87, T50	None	304	314, 315	Forbidden	Forbidden	D	40, 52, 57	125	POISON GAS*
	Ammonia solution, *relative density between 0.880 and 0.957 at 15 degrees C in water, with more than 10 percent but not more than 35 percent ammonia*	8	UN2672	III	8	IB3, IP8, T7, TP1	154	203	241	5 L	60 L	A	40, 52, 85	154	CORROSIVE
	Ammonia solutions, *relative density less than 0.880 at 15 degrees C in water, with more than 35 percent but not more than 50 percent ammonia*	2.2	UN2073		2.2	N87	306	304	314, 315	Forbidden	150 kg	E	40, 52, 57	125	NONFLAMMABLE GAS
	Ammonium arsenate	6.1	UN1546	II	6.1	IB8, IP2, IP4, T3, TP33	153	212	242	25 kg	100 kg	A	53	151	POISON
	Ammonium azide	Forbidden													
	Ammonium bifluoride, solid, see **Ammonium hydrogendifluoride, solid**														
	Ammonium bifluoride solution, see **Ammonium hydrogendifluoride, solution**														
	Ammonium bromate	Forbidden													
	Ammonium chlorate	Forbidden													
	Ammonium dichromate	5.1	UN1439	II	5.1	IB8, IP2, IP4, T3, TP33	152	212	242	5 kg	25 kg	A	52	141	OXIDIZER
	Ammonium dinitro-o-cresolate, solid	6.1	UN1843	II	6.1	IB8, IP2, IP4, T3, TP33	153	212	242	25 kg	100 kg	B	36, 65, 66, 77	141	POISON

Symbols (1)	Hazardous materials descriptions and proper shipping names (2)	Hazard class or Division (3)	Identification Numbers (4)	PG (5)	Label Codes (6)	Special Provisions (§172.102) (7)	(8) Packaging (§173.***) Exceptions (8A)	Non-bulk (8B)	Bulk (8C)	(9) Quantity limitations (see §§173.27 and 175.75) Passenger aircraft/rail (9A)	Cargo aircraft only (9B)	(10) Vessel stowage Location (10A)	Other (10B)	ERG Guide No.	Placard Advisory Consult §172.504 & §172.505 to determine placard *Denotes placard for any quantity
	Ammonium dinitro-o-cresolate solution	6.1	UN3424	II	6.1	IB2, T7, TP2	153	202	243	5 L	60 L	B	36, 66, 78, 91	141	POISON
				III	6.1	IB2, T7, TP2	153	203	241	60 L	220 L	A	36, 66, 78, 91	141	POISON
	Ammonium fluoride	6.1	UN2505	III	6.1	IB8, IP3, T1, TP33	153	213	240	100 kg	200 kg	A	52	154	POISON
	Ammonium fluorosilicate	6.1	UN2854	III	6.1	IB8, IP3, T1, TP33	153	213	240	100 kg	200 kg	A	52	151	POISON
	Ammonium fulminate	Forbidden													
	Ammonium hydrogen sulfate	8	UN2506	II	8	IB8, IP2, IP4, T3, TP33	154	212	240	15 kg	50 kg	A	40	154	CORROSIVE
	Ammonium hydrogendifluoride, solid	8	UN1727	II	8	IB8, IP2, IP4, N34, T3, TP33	154	212	240	15 kg	50 kg	A	25, 40, 52	154	CORROSIVE
	Ammonium hydrogendifluoride, solution	8	UN2817	II	8, 6.1	IB2, N34, T8, TP2, TP13	154	202	243	1 L	30 L	B	40	154	CORROSIVE
				III	8, 6.1	IB3, N3, T4, TP1, TP13	154	203	241	5 L	60 L	B	40, 95	154	CORROSIVE
	Ammonium hydrosulfide, solution, see **Ammonium sulfide solution**														
D	**Ammonium hydroxide,** *see* **Ammonia solutions,** *etc.*														
	Ammonium metavanadate	6.1	UN2859	II	6.1	IB8, IP2, IP4, T3, TP33	153	212	242	25 kg	100 kg	A	44, 89, 100, 141	154	POISON
	Ammonium nitrate based fertilizer	5.1	UN2067	III	5.1	52, 150, IB8, IP3, T1, TP33	152	213	240	25 kg	100 kg	B	48, 59, 60, 66, 117	140	OXIDIZER
A W	**Ammonium nitrate based fertilizer**	9	UN2071	III	9	132, IB8, IP3	155	213	240	200 kg	200 kg	A		140	CLASS 9
	Ammonium nitrate emulsion *or* **Ammonium nitrate suspension** *or* **Ammonium nitrate gel,** *intermediate for blasting explosives*	5.1	UN3375	II	5.1	147, 163	None	214	214	Forbidden	Forbidden	D	48, 59, 60, 66, 124	140	OXIDIZER
D	**Ammonium nitrate-fuel oil mixture** *containing only prilled ammonium nitrate and fuel oil*	1.5D	NA0331	II	1.5D		None	62	None	Forbidden	Forbidden	10	19E	112	EXPLOSIVES 1.5
	Ammonium nitrate, liquid *(hot concentrated solution)*	5.1	UN2426		5.1	B5, T7	None	None	243	Forbidden	Forbidden	D	59, 60	140	OXIDIZER
	Ammonium nitrate, *with more than 0.2 percent combustible substances, including any organic substance calculated as carbon, to the exclusion of any other added substance*	1.1D	UN0222	II	1.1D		None	62	None	Forbidden	Forbidden	10	19E	112	EXPLOSIVES 1.1*
	Ammonium nitrate, *with not more than 0.2% total combustible material, including any organic substance, calculated as carbon to the exclusion of any other added substance*	5.1	UN1942	III	5.1	A1, A29, IB8, IP3, T1, TP33	152	213	240	25 kg	100 kg	A	48, 59, 60, 116	140	OXIDIZER
	Ammonium nitrite	Forbidden													
	Ammonium perchlorate	1.1D	UN0402	II	1.1D	107	None	62	None	Forbidden	Forbidden	10	19E	112	EXPLOSIVES 1.1*
	Ammonium perchlorate	5.1	UN1442	II	5.1	107, A9, IB6, IP2, T3, TP33	152	212	242	5 kg	25 kg	E	58, 69	143	OXIDIZER
	Ammonium permanganate	Forbidden													
	Ammonium persulfate	5.1	UN1444	III	5.1	A1, A29, IB8, IP3, T1, TP33	152	213	240	25 kg	100 kg	A		140	OXIDIZER
	Ammonium picrate, *dry or wetted with less than 10 percent water, by mass*	1.1D	UN0004	II	1.1D		None	62	None	Forbidden	Forbidden	10	5E, 19E	112	EXPLOSIVES 1.1*
	Ammonium picrate, wetted *with not less than 10 percent water, by mass*	4.1	UN1310	I	4.1	23, A2, N41	None	211	None	0.5 kg	0.5 kg	D	28, 36	113	FLAMMABLE SOLID
	Ammonium polysulfide, solution	8	UN2818	II	8, 6.1	IB2, T7, TP2, TP13	154	202	243	1 L	30 L	B	12, 40, 52	154	CORROSIVE
				III	8, 6.1	IB3, T4, TP1, TP13	154	203	241	5 L	60 L	B	12, 40, 52	154	CORROSIVE
	Ammonium polyvanadate	6.1	UN2861	II	6.1	IB8, IP2, IP4, T3, TP33	153	212	242	25 kg	100 kg	A	44, 89, 100, 141	151	POISON
	Ammonium silicofluoride, see **Ammonium fluorosilicate**														
	Ammonium sulfide solution	8	UN2683	II	8, 6.1, 3	IB1, T7, TP2, TP13	154	202	243	1 L	30 L	B	12, 22, 52, 100	132	CORROSIVE
	Ammunition, blank, see **Cartridges for weapons, blank**														
	Ammunition, illuminating *with or without burster, expelling charge or propelling charge*	1.2G	UN0171	II	1.2G			62	62	Forbidden	Forbidden	03		112	EXPLOSIVES 1.2*
	Ammunition, illuminating *with or without burster, expelling charge or propelling charge*	1.3G	UN0254	II	1.3G			62	62	Forbidden	Forbidden	03		112	EXPLOSIVES 1.3*

Symbols (1)	Hazardous materials descriptions and proper shipping names (2)	Hazard class or Division (3)	Identification Numbers (4)	PG (5)	Label Codes (6)	Special Provisions (§172.102) (7)	(8) Packaging (§173.***) Exceptions (8A)	Non-bulk (8B)	Bulk (8C)	(9) Quantity limitations (see §§173.27 and 175.75) Passenger aircraft/rail (9A)	Cargo aircraft only (9B)	(10) Vessel stowage Location (10A)	Other (10B)	ERG Guide No.	Placard Advisory Consult §172.504 & §172.505 to determine placard *Denotes placard for any quantity
	Ammunition, illuminating *with or without burster, expelling charge or propelling charge*	1.4G	UN0297	II	1.4G			62	62	Forbidden	75 kg	02		114	EXPLOSIVES 1.4
	Ammunition, incendiary *liquid or gel, with burster, expelling charge or propelling charge*	1.3J	UN0247	II	1.3J			62	None	Forbidden	Forbidden	04	23E	112	EXPLOSIVES 1.3*
	Ammunition, incendiary *(water-activated contrivances) with burster, expelling charge or propelling charge, see* **Contrivances, water-activated, etc.**														
	Ammunition, incendiary, white phosphorus, *with burster, expelling charge or propelling charge*	1.2H	UN0243	II	1.2H			62	62	Forbidden	Forbidden	08	8E, 14E, 15E, 17E	112	EXPLOSIVES 1.2*
	Ammunition, incendiary, white phosphorus, *with burster, expelling charge or propelling charge*	1.3H	UN0244	II	1.3H			62	62	Forbidden	Forbidden	08	8E, 14E, 15E, 17E	112	EXPLOSIVES 1.3*
	Ammunition, incendiary *with or without burster, expelling charge, or propelling charge*	1.2G	UN0009	II	1.2G			62	62	Forbidden	Forbidden	03		112	EXPLOSIVES 1.2*
	Ammunition, incendiary *with or without burster, expelling charge, or propelling charge*	1.3G	UN0010	II	1.3G			62	62	Forbidden	Forbidden	03		112	EXPLOSIVES 1.3*
	Ammunition, incendiary *with or without burster, expelling charge or propelling charge*	1.4G	UN0300	II	1.4G			62	62	Forbidden	75 kg	02		114	EXPLOSIVES 1.4
	Ammunition, practice	1.4G	UN0362	II	1.4G			62	62	Forbidden	75 kg	02		114	EXPLOSIVES 1.4
	Ammunition, practice	1.3G	UN0488	II	1.3G			62	62	Forbidden	Forbidden	03		112	EXPLOSIVES 1.3*
	Ammunition, proof	1.4G	UN0363	II	1.4G			62	62	Forbidden	75 kg	02		114	EXPLOSIVES 1.4
	Ammunition, rocket, see **Warheads, rocket** *etc.*														
	Ammunition, SA (small arms), see **Cartridges for weapons,** *etc.*														
	Ammunition, smoke (water-activated contrivances), white phosphorus, with burster, expelling charge or propelling charge, see **Contrivances, water-activated,** *etc. (UN0248)*														
	Ammunition, smoke (water-activated contrivances), without white phosphorus or phosphides, with burster, expelling charge or propelling charge, see **Contrivances, water-activated,** *etc. (UN0249)*														
	Ammunition smoke, white phosphorus *with burster, expelling charge, or propelling charge*	1.2H	UN0245	II	1.2H			62	62	Forbidden	Forbidden	08	8E, 14E, 15E, 17E	112	EXPLOSIVES 1.2*
	Ammunition, smoke, white phosphorus *with burster, expelling charge, or propelling charge*	1.3H	UN0246	II	1.3H			62	62	Forbidden	Forbidden	08	8E, 14E, 15E, 17E	112	EXPLOSIVES 1.3*
	Ammunition, smoke *with or without burster, expelling charge or propelling charge*	1.2G	UN0015	II	1.2G			62	62	Forbidden	Forbidden		8E, 17E, 20E	112	EXPLOSIVES 1.2*
	Ammunition, smoke *with or without burster, expelling charge or propelling charge*	1.3G	UN0016	II	1.3G			62	62	Forbidden	Forbidden		8E, 17E, 20E	112	EXPLOSIVES 1.3*
	Ammunition, smoke *with or without burster, expelling charge or propelling charge*	1.4G	UN0303	II	1.4G			62	62	Forbidden	75 kg		7E, 8E, 14E, 15E, 17E	114	EXPLOSIVES 1.4
	Ammunition, sporting, see **Cartridges for weapons,** *etc. (UN0012; UN0328; UN0339)*														
	Ammunition, tear-producing, non-explosive, *without burster or expelling charge, non-fuzed*	6.1	UN2017	II	6.1, 8		None	212	None	Forbidden	50 kg	E	13, 40	159	POISON
	Ammunition, tear-producing *with burster, expelling charge or propelling charge*	1.2G	UN0018	II	1.2G, 8, 6.1			62	62	Forbidden	Forbidden		8E, 17E, 20E	112	EXPLOSIVES 1.2*
	Ammunition, tear-producing *with burster, expelling charge or propelling charge*	1.3G	UN0019	II	1.3G, 8, 6.1			62	62	Forbidden	Forbidden		8E, 17E, 20E	112	EXPLOSIVES 1.3*
	Ammunition, tear-producing *with burster, expelling charge or propelling charge*	1.4G	UN0301	II	1.4G, 8, 6.1			62	62	Forbidden	75 kg		7E, 8E, 14E, 15E, 17E	114	EXPLOSIVES 1.4
	Ammunition, toxic, non-explosive, *without burster or expelling charge, non-fuzed*	6.1	UN2016	II	6.1		None	212	None	Forbidden	100 kg	E	13, 40	151	POISON
	Ammunition, toxic (water-activated contrivances), with burster, expelling charge or propelling charge, see **Contrivances, water-activated,***etc.*														
G	**Ammunition, toxic** *with burster, expelling charge, or propelling charge*	1.2K	UN0020	II	1.2K, 6.1			62	None	Forbidden	Forbidden	08	8E, 14E, 15E, 17E	112	EXPLOSIVES 1.2*
G	**Ammunition, toxic** *with burster, expelling charge, or propelling charge*	1.3K	UN0021	II	1.3K, 6.1			62	None	Forbidden	Forbidden	08	8E, 14E, 15E, 17E	112	EXPLOSIVES 1.3*

Symbols (1)	Hazardous materials descriptions and proper shipping names (2)	Hazard class or Division (3)	Identification Numbers (4)	PG (5)	Label Codes (6)	Special Provisions (§172.102) (7)	(8) Packaging (§173.***) Exceptions (8A)	Non-bulk (8B)	Bulk (8C)	(9) Quantity limitations (see §§173.27 and 175.75) Passenger aircraft/rail (9A)	Cargo aircraft only (9B)	(10) Vessel stowage Location (10A)	Other (10B)	ERG Guide No.	Placard Advisory Consult §172.504 & §172.505 to determine placard *Denotes placard for any quantity
	Amyl acetates	3	UN1104	III	3	B1, IB3, T2, TP1	150	203	242	60 L	220 L	A		129	FLAMMABLE
	Amyl acid phosphate	8	UN2819	III	8	IB3, T4, TP1	154	203	241	5 L	60 L	A		153	CORROSIVE
	Amyl butyrates	3	UN2620	III	3	B1, IB3, T2, TP1	150	203	242	60 L	220 L	A		130	FLAMMABLE
	Amyl chlorides	3	UN1107	II	3	IB2, T4, TP1	150	202	242	5 L	60 L	B		129	FLAMMABLE
	Amyl formates	3	UN1109	III	3	B1, IB3, T2, TP1	150	203	242	60 L	220 L	A		129	FLAMMABLE
	Amyl mercaptans	3	UN1111	II	3	A3, A6, IB2, T4, TP1	None	202	242	5 L	60 L	B	95, 102	130	FLAMMABLE
	n-Amyl methyl ketone	3	UN1110	III	3	B1, IB3, T2, TP1	150	203	242	60 L	220 L	A		127	FLAMMABLE
	Amyl nitrate	3	UN1112	III	3	B1, IB3, T2, TP1	150	203	242	60 L	220 L	A	40	140	FLAMMABLE
	Amyl nitrites	3	UN1113	II	3	IB2, T4, TP1	150	202	242	5 L	60 L	E	40	129	FLAMMABLE
	Amylamines	3	UN1106	II	3, 8	IB2, T7, TP1	150	202	243	1 L	5 L	B		132	FLAMMABLE
				III	3, 8	B1, IB3, T4, TP1	150	203	242	5 L	60 L	A		132	FLAMMABLE
	Amyltrichlorosilane	8	UN1728	II	8	A7, B2, B6, N34, T10, TP2, TP7, TP13	None	206	242	Forbidden	30 L	C	40	155	CORROSIVE
	Anhydrous ammonia, *see* **Ammonia, anhydrous**														
	Anhydrous hydrofluoric acid, see **Hydrogen fluoride, anhydrous**														
+	**Aniline**	6.1	UN1547	II	6.1	IB2, T7, TP2	153	202	243	5 L	60 L	A	40, 52	153	POISON
	Aniline hydrochloride	6.1	UN1548	III	6.1	IB8, IP3, T1, TP33	153	213	240	100 kg	200 kg	A		153	POISON
	Aniline oil, see **Aniline**														
	Anisidines	6.1	UN2431	III	6.1	IB3, T4, TP1	153	203	241	60 L	220 L	A		153	POISON
	Anisole	3	UN2222	III	3	B1, IB3, T2, TP1	150	203	242	60 L	220 L	A		128	FLAMMABLE
	Anisoyl chloride	8	UN1729	II	8	B2, B4, IB8, IP2, IP4, T3, TP33	154	212	240	15 kg	50 kg	A	40	156	CORROSIVE
	Anti-freeze, liquid, see **Flammable liquids, n.o.s.**														
	Antimonous chloride, see **Antimony trichloride**														
G	**Antimony compounds, inorganic, liquid, n.o.s.**	6.1	UN3141	III	6.1	35, IB3, T7, TP1, TP28	153	203	241	60 L	220 L	A		157	POISON
G	**Antimony compounds, inorganic, solid, n.o.s.**	6.1	UN1549	III	6.1	35, IB8, IP3, T1, TP33	153	213	240	100 kg	200 kg	A		157	POISON
	Antimony lactate	6.1	UN1550	III	6.1	IB8, IP3, T1, TP33	153	213	240	100 kg	200 kg	A		151	POISON
	Antimony pentachloride, liquid	8	UN1730	II	8	B2, IB2, T7, TP2	None	202	242	1 L	30 L	C	40	157	CORROSIVE
	Antimony pentachloride, solutions	8	UN1731	II	8	B2, IB2, T7, TP2	154	202	242	1 L	30 L	C	40	157	CORROSIVE
				III	8	IB3, T4, TP1	154	203	241	5 L	60 L	C	40	157	CORROSIVE
	Antimony pentafluoride	8	UN1732	II	8, 6.1	A3, A6, A7, A10, IB2, N3, N36, T7, TP2	None	202	243	Forbidden	30 L	D	40, 89, 100, 141	157	CORROSIVE
	Antimony potassium tartrate	6.1	UN1551	III	6.1	IB8, IP3, T1, TP33	153	213	240	100 kg	200 kg	A		151	POISON
	Antimony powder	6.1	UN2871	III	6.1	IB8, IP3, T1, TP33	153	213	240	100 kg	200 kg	A		170	POISON
	Antimony sulfide and a chlorate, mixtures of	Forbidden													
	Antimony sulfide, solid, see **Antimony compounds, inorganic, n.o.s.**														
	Antimony trichloride, liquid	8	UN1733	II	8	B2, IB2	154	202	242	1 L	30 L	C	40	157	CORROSIVE
	Antimony trichloride, solid	8	UN1733	II	8	IB8, IP2, IP4, T3, TP33	154	212	240	15 kg	50 kg	A	40	157	CORROSIVE
	Aqua ammonia, see **Ammonia solution,** *etc.*														
	Argon, *compressed*	2.2	UN1006		2.2		306, 307	302	314, 315	75 kg	150 kg	A		121	NONFLAMMABLE GAS
	Argon, refrigerated liquid *(cryogenic liquid)*	2.2	UN1951		2.2	T75, TP5	320	316	318	50 kg	500 kg	D		120	NONFLAMMABLE GAS
	Arsenic	6.1	UN1558	II	6.1	IB8, IP2, IP4, T3, TP33	153	212	242	25 kg	100 kg	A		152	POISON
	Arsenic acid, liquid	6.1	UN1553	I	6.1	T20, TP2, TP7, TP13	None	201	243	1 L	30 L	B	46	154	POISON
	Arsenic acid, solid	6.1	UN1554	II	6.1	IB8, IP2, IP4, T3, TP33	153	212	242	25 kg	100 kg	A		154	POISON
	Arsenic bromide	6.1	UN1555	II	6.1	IB8, IP2, IP4, T3, TP33	153	212	242	25 kg	100 kg	A	12, 40	151	POISON
	Arsenic chloride, see **Arsenic trichloride**														

Symbols (1)	Hazardous materials descriptions and proper shipping names (2)	Hazard class or Division (3)	Identification Numbers (4)	PG (5)	Label Codes (6)	Special Provisions (§172.102) (7)	(8) Packaging (§173.***) Exceptions (8A)	Non-bulk (8B)	Bulk (8C)	(9) Quantity limitations (see §§173.27 and 175.75) Passenger aircraft/rail (9A)	Cargo aircraft only (9B)	(10) Vessel stowage Location (10A)	Other (10B)	ERG Guide No.	Placard Advisory Consult §172.504 & §172.505 to determine placard *Denotes placard for any quantity
G	**Arsenic compounds, liquid, n.o.s.** *inorganic, including arsenates, n.o.s.; arsenites, n.o.s.; arsenic sulfides, n.o.s.; and organic compounds of arsenic, n.o.s.*	6.1	UN1556	I	6.1	T14, TP2, TP13, TP27	None	201	243	1 L	30 L	B	40, 137	152	POISON
				II	6.1	IB2, T11, TP2, TP13, TP27	153	202	243	5 L	60 L	B	40, 137	152	POISON
				III	6.1	IB3, T7, TP2, TP28	153	203	241	60 L	220 L	B	40, 137	152	POISON
G	**Arsenic compounds, solid, n.o.s.** *inorganic, including arsenates, n.o.s.; arsenites, n.o.s.; arsenic sulfides, n.o.s.; and organic compounds of arsenic, n.o.s.*	6.1	UN1557	I	6.1	IB7, IP1, T6, TP33	None	211	242	5 kg	50 kg	A	137	152	POISON
				II	6.1	IB8, IP2, IP4, T3, TP33	153	212	242	25 kg	100 kg	A	137	152	POISON
				III	6.1	IB8, IP3, T1, TP33	153	213	240	100 kg	200 kg	A	137	152	POISON
	Arsenic pentoxide	6.1	UN1559	II	6.1	IB8, IP2, IP4, T3, TP33	153	212	242	25 kg	100 kg	A		151	POISON
	Arsenic sulfide and a chlorate, mixtures of	Forbidden													
	Arsenic trichloride	6.1	UN1560	I	6.1	2, B9, B14, B32, T20, TP2, TP13, TP38, TP45	None	227	244	Forbidden	Forbidden	B	40	157	POISON INHALATION HAZARD*
	Arsenic trioxide	6.1	UN1561	II	6.1	IB8, IP2, IP4, T3, TP33	153	212	242	25 kg	100 kg	A		151	POISON
	Arsenic, white, solid, see **Arsenic trioxide**														
	Arsenical dust	6.1	UN1562	II	6.1	IB8, IP2, IP4, T3, TP33	153	212	242	25 kg	100 kg	A		152	POISON
	Arsenical pesticides, liquid, flammable, toxic, *flash point less than 23 degrees C*	3	UN2760	I	3, 6.1	T14, TP2, TP13, TP27	None	201	243	Forbidden	30 L	B	40	131	FLAMMABLE
				II	3, 6.1	IB2, T11, TP2, TP13, TP27	150	202	243	1 L	60 L	B	40	131	FLAMMABLE
	Arsenical pesticides, liquid, toxic	6.1	UN2994	I	6.1	T14, TP2, TP13, TP27	None	201	243	1 L	30 L	B	40	151	POISON
				II	6.1	IB2, T11, TP2, TP13, TP27	153	202	243	5 L	60 L	B	40	151	POISON
				III	6.1	IB3, T7, TP2, TP28	153	203	241	60 L	220 L	A	40	151	POISON
	Arsenical pesticides, liquid, toxic, flammable, *flash point not less than 23 degrees C*	6.1	UN2993	I	6.1, 3	T14, TP2, TP13, TP27	None	201	243	1 L	30 L	B	40	131	POISON
				II	6.1, 3	IB2, T11, TP2, TP13, TP27	153	202	243	5 L	60 L	B	40	131	POISON
				III	6.1, 3	B1, IB3, T7, TP2, TP28	153	203	242	60 L	220 L	A	40	131	POISON
	Arsenical pesticides, solid, toxic	6.1	UN2759	I	6.1	IB7, IP1, T6, TP33	None	211	242	5 kg	50 kg	A	40	151	POISON
				II	6.1	IB8, IP2, IP4, T3, TP33	153	212	242	25 kg	100 kg	A	40	151	POISON
				III	6.1	IB8, IP3, T1, TP33	153	213	240	100 kg	200 kg	A	40	151	POISON
	Arsenious acid, solid, see **Arsenic trioxide**														
	Arsenious and mercuric iodide solution, see **Arsenic compounds, liquid, n.o.s.**														
	Arsine	2.3	UN2188		2.3, 2.1	1	None	192	245	Forbidden	Forbidden	D	40	119	POISON GAS*
	Articles, explosive, extremely insensitive *or* **Articles, EEI**	1.6N	UN0486	II	1.6N		None	62	None	Forbidden	Forbidden	07		112	EXPLOSIVES 1.6
G	**Articles, explosive, n.o.s.**	1.4S	UN0349	II	1.4S		None	62	None	25 kg	100 kg	05		114	EXPLOSIVES 1.4
G	**Articles, explosive, n.o.s.**	1.4B	UN0350	II	1.4B		None	62	None	Forbidden	Forbidden	06		114	EXPLOSIVES 1.4
G	**Articles, explosive, n.o.s.**	1.4C	UN0351	II	1.4C		None	62	None	Forbidden	75 kg	06		114	EXPLOSIVES 1.4
G	**Articles, explosive, n.o.s.**	1.4D	UN0352	II	1.4D		None	62	None	Forbidden	75 kg	06		114	EXPLOSIVES 1.4
G	**Articles, explosive, n.o.s.**	1.4G	UN0353	II	1.4G		None	62	None	Forbidden	75 kg	06		114	EXPLOSIVES 1.4
G	**Articles, explosive, n.o.s.**	1.1L	UN0354	II	1.1L		None	62	None	Forbidden	Forbidden	08	8E, 14E, 15E, 17E	112	EXPLOSIVES 1.1*
G	**Articles, explosive, n.o.s.**	1.2L	UN0355	II	1.2L		None	62	None	Forbidden	Forbidden	08	8E, 14E, 15E, 17E	112	EXPLOSIVES 1.2*

Symbols (1)	Hazardous materials descriptions and proper shipping names (2)	Hazard class or Division (3)	Identification Numbers (4)	PG (5)	Label Codes (6)	Special Provisions (§172.102) (7)	(8) Packaging (§173.***) Exceptions (8A)	Non-bulk (8B)	Bulk (8C)	(9) Quantity limitations (see §§173.27 and 175.75) Passenger aircraft/rail (9A)	Cargo aircraft only (9B)	(10) Vessel stowage Location (10A)	Other (10B)	ERG Guide No.	Placard Advisory Consult §172.504 & §172.505 to determine placard *Denotes placard for any quantity
G	**Articles, explosive, n.o.s.**	1.3L	UN0356	II	1.3L		None	62	None	Forbidden	Forbidden	08	8E, 14E, 15E, 17E	112	EXPLOSIVES 1.3*
G	**Articles, explosive, n.o.s.**	1.1C	UN0462	II	1.1C		None	62	None	Forbidden	Forbidden	07		112	EXPLOSIVES 1.1*
G	**Articles, explosive, n.o.s.**	1.1D	UN0463	II	1.1D		None	62	None	Forbidden	Forbidden	07		112	EXPLOSIVES 1.1*
G	**Articles, explosive, n.o.s.**	1.1E	UN0464	II	1.1E		None	62	None	Forbidden	Forbidden	07		112	EXPLOSIVES 1.1*
G	**Articles, explosive, n.o.s.**	1.1F	UN0465	II	1.1F		None	62	None	Forbidden	Forbidden	08		112	EXPLOSIVES 1.1*
G	**Articles, explosive, n.o.s.**	1.2C	UN0466	II	1.2C		None	62	None	Forbidden	Forbidden	07		112	EXPLOSIVES 1.2*
G	**Articles, explosive, n.o.s.**	1.2D	UN0467	II	1.2D		None	62	None	Forbidden	Forbidden	07		112	EXPLOSIVES 1.2*
G	**Articles, explosive, n.o.s.**	1.2E	UN0468	II	1.2E		None	62	None	Forbidden	Forbidden	07		112	EXPLOSIVES 1.2*
G	**Articles, explosive, n.o.s.**	1.2F	UN0469	II	1.2F		None	62	None	Forbidden	Forbidden	08		112	EXPLOSIVES 1.2*
G	**Articles, explosive, n.o.s.**	1.3C	UN0470	II	1.3C		None	62	None	Forbidden	Forbidden	07		112	EXPLOSIVES 1.3*
G	**Articles, explosive, n.o.s.**	1.4E	UN0471	II	1.4E		None	62	None	Forbidden	75 kg	06		114	EXPLOSIVES 1.4
G	**Articles, explosive, n.o.s.**	1.4F	UN0472	II	1.4F		None	62	None	Forbidden	Forbidden	08		114	EXPLOSIVES 1.4
	Articles, pressurized pneumatic *or* **hydraulic***containing non-flammable gas*	2.2	UN3164		2.2		306	302, 304	None	No limit	No limit	A		126	NONFLAMMABLE GAS
	Articles, pyrophoric	1.2L	UN0380	II	1.2L		None	62	None	Forbidden	Forbidden	08	8E, 14E, 15E, 17E	112	EXPLOSIVES 1.2*
	Articles, pyrotechnic *for technical purposes*	1.1G	UN0428	II	1.1G		None	62	None	Forbidden	Forbidden	07		112	EXPLOSIVES 1.1*
	Articles, pyrotechnic *for technical purposes*	1.2G	UN0429	II	1.2G		None	62	None	Forbidden	Forbidden	07		112	EXPLOSIVES 1.2*
	Articles, pyrotechnic *for technical purposes*	1.3G	UN0430	II	1.3G		None	62	None	Forbidden	Forbidden	07		112	EXPLOSIVES 1.3*
	Articles, pyrotechnic *for technical purposes*	1.4G	UN0431	II	1.4G		None	62	None	Forbidden	75 kg	06		114	EXPLOSIVES 1.4
	Articles, pyrotechnic *for technical purposes*	1.4S	UN0432	II	1.4S		None	62	None	25 kg	100 kg	05		114	EXPLOSIVES 1.4
D	**Asbestos**	9	NA2212	III	9	156, IB8, IP2, IP4	155	216	240	200 kg	200 kg	A	34, 40	171	CLASS 9
	Ascaridole (organic peroxide)	Forbidden													
D	**Asphalt,***at or above its flashpoint*	3	NA1999	III	3	IB3, T1, TP3	150	203	247	Forbidden	Forbidden	D		130	FLAMMABLE
D	**Asphalt, cut back,** *see* **Tars, liquid,** *etc.*														
	Automobile, motorcycle, tractor, other self-propelled vehicle, engine, or other mechanical apparatus, see **Vehicles** *or* **Battery** *etc.*														
A G	**Aviation regulated liquid, n.o.s.**	9	UN3334		9	A35	155	204		No limit	No limit	A		171	CLASS 9
A G	**Aviation regulated solid, n.o.s.**	9	UN3335		9	A35	155	204		No limit	No limit	A		171	CLASS 9
	Azaurolic acid (salt of) (dry)	Forbidden													
	Azido guanidine picrate (dry)	Forbidden													
	5-Azido-1-hydroxy tetrazole	Forbidden													
	Azido hydroxy tetrazole (mercury and silver salts)	Forbidden													
	3-Azido-1,2-Propylene glycol dinitrate	Forbidden													
	Azidodithiocarbonic acid	Forbidden													
	Azidoethyl nitrate	Forbidden													
	1-Aziridinylphosphine oxide-(tris), see **Tris-(1-aziridinyl)phosphine oxide, solution**														
	Azodicarbonamide	4.1	UN3242	II	4.1	38, IB8, T3, TP33	151	223	240	Forbidden	Forbidden	D	2, 52, 53, 74	149	FLAMMABLE SOLID
	Azotetrazole (dry)	Forbidden													
	Barium	4.3	UN1400	II	4.3	A19, IB7, IP2, T3, TP33	151	212	241	15 kg	50 kg	E	52	138	DANGEROUS WHEN WET*
	Barium alloys, pyrophoric	4.2	UN1854	I	4.2	T21, TP7, TP33	None	181	None	Forbidden	Forbidden	D		135	SPONTANEOUSLY COMBUSTIBLE
	Barium azide, *dry or wetted with less than 50 percent water, by mass*	1.1A	UN0224	II	1.1A, 6.1	111, 117	None	62	None	Forbidden	Forbidden	12		112	EXPLOSIVES 1.1*
	Barium azide, wetted *with not less than 50 percent water, by mass*	4.1	UN1571	I	4.1, 6.1	162, A2	None	182	None	Forbidden	0.5 kg	D	28	113	FLAMMABLE SOLID
	Barium bromate	5.1	UN2719	II	5.1, 6.1	IB8, IP2, IP4, T3, TP33	152	212	242	5 kg	25 kg	A	56, 58	141	OXIDIZER
	Barium chlorate, solid	5.1	UN1445	II	5.1, 6.1	A9, IB6, IP2, N34, T3, TP33	152	212	242	5 kg	25 kg	A	56, 58	141	OXIDIZER

Symbols (1)	Hazardous materials descriptions and proper shipping names (2)	Hazard class or Division (3)	Identification Numbers (4)	PG (5)	Label Codes (6)	Special Provisions (§172.102) (7)	(8) Packaging (§173.***) Exceptions (8A)	Non-bulk (8B)	Bulk (8C)	(9) Quantity limitations (see §§173.27 and 175.75) Passenger aircraft/rail (9A)	Cargo aircraft only (9B)	(10) Vessel stowage Location (10A)	Other (10B)	ERG Guide No.	Placard Advisory Consult §172.504 & §172.505 to determine placard *Denotes placard for any quantity
	Barium chlorate, solution	5.1	UN3405	II	5.1, 6.1	A9, IB2, N34, T4, TP1	152	202	243	1 L	5 L	A	56, 58, 133	141	OXIDIZER
				III	5.1, 6.1	A9, IB2, N34, T4, TP1	152	203	242	2.5 L	30 L	A	56, 58, 133	141	OXIDIZER
G	**Barium compounds, n.o.s.**	6.1	UN1564	II	6.1	IB8, IP2, IP4, T3, TP33	153	212	242	25 kg	100 kg	A		154	POISON
				III	6.1	IB8, IP3, T1, TP33	153	213	240	100 kg	200 kg	A		154	POISON
	Barium cyanide	6.1	UN1565	I	6.1	IB7, IP1, N74, N75, T6, TP33	None	211	242	5 kg	50 kg	A	40, 52	157	POISON
	Barium hypochlorite *with more than 22 percent available chlorine*	5.1	UN2741	II	5.1, 6.1	A7, A9, IB8, IP2, IP4, N34, T3, TP33	152	212	None	5 kg	25 kg	B	4, 52, 56, 58, 106	141	OXIDIZER
	Barium nitrate	5.1	UN1446	II	5.1, 6.1	IB8, IP2, IP4, T3, TP33	152	212	242	5 kg	25 kg	A		141	OXIDIZER
	Barium oxide	6.1	UN1884	III	6.1	IB8, IP3, T1, TP33	153	213	240	100 kg	200 kg	A		157	POISON
	Barium perchlorate, solid	5.1	UN1447	II	5.1, 6.1	IB6, IP2, T3, TP33	152	212	242	5 kg	25 kg	A	56, 58	141	OXIDIZER
	Barium perchlorate, solution	5.1	UN3406	II	5.1, 6.1	IB2, T4, TP1	152	202	243	1 L	5 L	A	56, 58, 133	141	OXIDIZER
				III	5.1, 6.1	IB2, T4, TP1	152	203	242	2.5 L	30 L	A	56, 58, 133	141	OXIDIZER
	Barium permanganate	5.1	UN1448	II	5.1, 6.1	IB6, IP2, T3, TP33	152	212	242	5 kg	25 kg	D	56, 58, 138	141	OXIDIZER
	Barium peroxide	5.1	UN1449	II	5.1, 6.1	A9, IB6, IP2, T3, TP33	152	212	242	5 kg	25 kg	A	13, 52, 56, 75	141	OXIDIZER
	Barium selenate, see **Selenates** *or* **Selenites**														
	Barium selenite, see **Selenates** *or* **Selenites**														
	Batteries, containing sodium	4.3	UN3292	II	4.3		189	189	189	Forbidden	No limit	A		138	DANGEROUS WHEN WET*
	Batteries, dry, containing potassium hydroxide solid, *electric, storage*	8	UN3028	III	8	237	None	213	None	25 kg gross	230 kg gross	A	52	154	CORROSIVE
	Batteries, dry, sealed, n.o.s.					130								—	NONE
W	**Batteries, nickel-metal hydride** *see* **Batteries, dry, sealed, n.o.s.** *for nickel-metal hydride batteries transported by modes other than vessel*	9	UN3496		9	340						A	48	—	CLASS 9
	Batteries, wet, filled with acid, *electric storage*	8	UN2794	III	8		159	159	159	30 kg gross	No limit	A	146	154	CORROSIVE
	Batteries, wet, filled with alkali, *electric storage*	8	UN2795	III	8		159	159	159	30 kg gross	No limit	A	52, 146	154	CORROSIVE
	Batteries, wet, non-spillable, *electric storage*	8	UN2800	III	8		159a	159	159	No Limit	No Limit	A		154	CORROSIVE
	Battery fluid, acid	8	UN2796	II	8	A3, A7, B2, B15, IB2, N6, N34, T8, TP2	154	202	242	1 L	30 L	B		157	CORROSIVE
	Battery fluid, alkali	8	UN2797	II	8	B2, IB2, N6, T7, TP2, TP28	154	202	242	1 L	30 L	A	29, 52	154	CORROSIVE
	Battery lithium type, see **Lithium batteries** *etc.*														
	Battery-powered vehicle *or* **Battery-powered equipment**	9	UN3171		9	134	220	220	None	No limit	No limit	A		154	CLASS 9
	Battery, wet, filled with acid or alkali with vehicle or mechanical equipment containing an internal combustion engine, see **Vehicle,** *etc. or* **Engines, Internal combustion,** *etc.*														
+	**Benzaldehyde**	9	UN1990	III	9	IB3, T2, TP1	155	203	241	100 L	220 L	A		129	CLASS 9
	Benzene	3	UN1114	II	3	IB2, T4, TP1	150	202	242	5 L	60 L	B	40	130	FLAMMABLE
	Benzene diazonium chloride (dry)	Forbidden													
	Benzene diazonium nitrate (dry)	Forbidden													
	Benzene phosphorus dichloride, see **Phenyl phosphorus dichloride**														
	Benzene phosphorus thiodichloride, see **Phenyl phosphorus thiodichloride**														
	Benzene sulfonyl chloride	8	UN2225	III	8	IB3, T4, TP1	154	203	241	5 L	60 L	A	40	156	CORROSIVE
	Benzene triozonide	Forbidden													
	Benzenethiol, see **Phenyl mercaptan**														

HAZARDOUS MATERIALS TABLE

Symbols (1)	Hazardous materials descriptions and proper shipping names (2)	Hazard class or Division (3)	Identification Numbers (4)	PG (5)	Label Codes (6)	Special Provisions (§172.102) (7)	(8) Packaging (§173.***) Exceptions (8A)	Non-bulk (8B)	Bulk (8C)	(9) Quantity limitations (see §§173.27 and 175.75) Passenger aircraft/rail (9A)	Cargo aircraft only (9B)	(10) Vessel stowage Location (10A)	Other (10B)	ERG Guide No.	Placard Advisory Consult §172.504 & §172.505 to determine placard *Denotes placard for any quantity
	Benzidine	6.1	UN1885	II	6.1	IB8, IP2, IP4, T3, TP33	153	212	242	25 kg	100 kg	A		153	POISON
	Benzol, see **Benzene**														
	Benzonitrile	6.1	UN2224	II	6.1	IB2, T7, TP2	153	202	243	5 L	60 L	A	40, 52	152	POISON
	Benzoquinone	6.1	UN2587	II	6.1	IB8, IP2, IP4, T3, TP33	153	212	242	25 kg	100 kg	A		153	POISON
	Benzotrichloride	8	UN2226	II	8	B2, IB2, T7, TP2	154	202	242	1 L	30 L	A	40	156	CORROSIVE
	Benzotrifluoride	3	UN2338	II	3	IB2, T4, TP1	150	202	242	5 L	60 L	B	40	127	FLAMMABLE
	Benzoxidiazoles (dry)	Forbidden													
	Benzoyl azide	Forbidden													
	Benzoyl chloride	8	UN1736	II	8	B2, IB2, T8, TP2, TP13	154	202	242	1 L	30 L	C	40	137	CORROSIVE
	Benzyl bromide	6.1	UN1737	II	6.1, 8	A3, A7, IB2, N33, N34, T8, TP2, TP13	None	202	243	1 L	30 L	D	13, 40	156	POISON
	Benzyl chloride	6.1	UN1738	II	6.1, 8	A3, A7, B70, IB2, N33, N42, T8, TP2, TP13	None	202	243	1 L	30 L	D	13, 40	156	POISON
	Benzyl chloride *unstabilized*	6.1	UN1738	II	6.1, 8	A3, A7, B8, B11, IB2, N33, N34, N43, T8, TP2, TP13	153	202	243	1 L	30 L	D	13, 40	156	POISON
	Benzyl chloroformate	8	UN1739	I	8	A3, A6, B4, N41, T10, TP2, TP13	None	201	243	Forbidden	2.5 L	D	40	137	CORROSIVE
	Benzyl iodide	6.1	UN2653	II	6.1	IB2, T7, TP2	153	202	243	5 L	60 L	B	12, 40	156	POISON
	Benzyldimethylamine	8	UN2619	II	8, 3	B2, IB2, T7, TP2	154	202	243	1 L	30 L	A	40, 48	132	CORROSIVE
	Benzylidene chloride	6.1	UN1886	II	6.1	IB2, T7, TP2	153	202	243	5 L	60 L	D	40	156	POISON
G	**Beryllium compounds, n.o.s.**	6.1	UN1566	II	6.1	IB8, IP2, IP4, T3, TP33	153	212	242	25 kg	100 kg	A		154	POISON
				III	6.1	IB8, IP3, T1, TP33	153	213	240	100 kg	200 kg	A		154	POISON
	Beryllium nitrate	5.1	UN2464	II	5.1, 6.1	IB8, IP2, IP4, T3, TP33	152	212	242	5 kg	25 kg	A		141	OXIDIZER
	Beryllium, powder	6.1	UN1567	II	6.1, 4.1	IB8, IP2, IP4, T3, TP33	153	212	242	15 kg	50 kg	A		134	POISON
	Bicyclo [2,2,1] hepta-2,5-diene, stabilized *or* **2,5-Norbornadiene, stabilized**	3	UN2251	II	3	IB2, T7, TP2	150	202	242	5 L	60 L	D		128P	FLAMMABLE
	Biological substance, Category B	6.2	UN3373			A82	134	199	None	4 L or 4 kg	4 L or 4 kg	A	40	158	NONE
	Biphenyl triozonide	Forbidden													
	Bipyridilium pesticides, liquid, flammable, toxic, *flash point less than 23 degrees C*	3	UN2782	I	3, 6.1	T14, TP2, TP13, TP27	None	201	243	Forbidden	30 L	E		131	FLAMMABLE
				II	3, 6.1	IB2, T11, TP2, TP13, TP27	150	202	243	1 L	60 L	B	40	131	FLAMMABLE
	Bipyridilium pesticides, liquid, toxic	6.1	UN3016	I	6.1	T14, TP2, TP13, TP27	None	201	243	1 L	30 L	B	40	151	POISON
				II	6.1	IB2, T11, TP2, TP13, TP27	153	202	243	5 L	60 L	B	40	151	POISON
				III	6.1	IB3, T7, TP2, TP28	153	203	241	60 L	220 L	A	40	151	POISON
	Bipyridilium pesticides, liquid, toxic, flammable, *flash point not less than 23 degrees C*	6.1	UN3015	I	6.1, 3	T14, TP2, TP13, TP27	None	201	243	1 L	30 L	B	21, 40	131	POISON
				II	6.1, 3	IB2, T11, TP2, TP13, TP27	153	202	243	5 L	60 L	B	21, 40	131	POISON
				III	6.1, 3	B1, IB3, T7, TP2, TP28	153	203	242	60 L	220 L	A	21, 40	131	POISON
	Bipyridilium pesticides, solid, toxic	6.1	UN2781	I	6.1	IB7, IP1, T6, TP33	None	211	242	5 kg	50 kg	A	40	151	POISON
				II	6.1	IB8, IP2, IP4, T3, TP33	153	212	242	25 kg	100 kg	A	40	151	POISON
				III	6.1	IB8, IP3, T1, TP33	153	213	240	100 kg	200 kg	A	40	151	POISON
	Bis (Aminopropyl) piperazine, see **Corrosive liquid, n.o.s.**														
	Bisulfate, aqueous solution	8	UN2837	II	8	A7, B2, IB2, N34, T7, TP2	154	202	242	1 L	30 L	A		154	CORROSIVE

Symbols (1)	Hazardous materials descriptions and proper shipping names (2)	Hazard class or Division (3)	Identification Numbers (4)	PG (5)	Label Codes (6)	Special Provisions (§172.102) (7)	(8) Packaging (§173.***) Exceptions (8A)	Non-bulk (8B)	Bulk (8C)	(9) Quantity limitations (see §§173.27 and 175.75) Passenger aircraft/rail (9A)	Cargo aircraft only (9B)	(10) Vessel stowage Location (10A)	Other (10B)	ERG Guide No.	Placard Advisory Consult §172.504 & §172.505 to determine placard *Denotes placard for any quantity
				III	8	A7, IB3, N34, T4, TP1	154	203	241	5 L	60 L	A		154	CORROSIVE
	Bisulfites, aqueous solutions, n.o.s.	8	UN2693	III	8	IB3, T7, TP1, TP28	154	203	241	5 L	60 L	A	40, 52	154	CORROSIVE
	Black powder, compressed *or* **Gunpowder, compressed** *or* **Black powder, in pellets** *or* **Gunpowder, in pellets**	1.1D	UN0028	II	1.1D		None	62	None	Forbidden	Forbidden	10		112	EXPLOSIVES 1.1*
	Black powder *or* **Gunpowder,** *granular or as a meal*	1.1D	UN0027	II	1.1D		None	62	None	Forbidden	Forbidden	10		112	EXPLOSIVES 1.1*
D	**Black powder for small arms**	4.1	NA0027	I	4.1	70	None	170	None	Forbidden	Forbidden	E		—	FLAMMABLE SOLID
	Blasting agent, n.o.s., see **Explosives, blasting** *etc.*														
	Blasting cap assemblies, see **Detonator assemblies, non-electric,** *for blasting*														
	Blasting caps, electric, see **Detonators, electric** *for blasting*														
	Blasting caps, non-electric, see **Detonators, non-electric,** *for blasting*														
	Bleaching powder, see **Calcium hypochlorite mixtures,** *etc.*														
I	**Blue asbestos** *(Crocidolite) or* **Brown asbestos** *(amosite, mysorite)*	9	UN2212	II	9	156, IB8, IP2, IP4, T3, TP33	155	216	240	Forbidden	Forbidden	A	34, 40	171	CLASS 9
	Bombs, photo-flash	1.1F	UN0037	II	1.1F			62	None	Forbidden	Forbidden	08		112	EXPLOSIVES 1.1*
	Bombs, photo-flash	1.1D	UN0038	II	1.1D			62	62	Forbidden	Forbidden	03		112	EXPLOSIVES 1.1*
	Bombs, photo-flash	1.2G	UN0039	II	1.2G			62	62	Forbidden	Forbidden	03		112	EXPLOSIVES 1.2*
	Bombs, photo-flash	1.3G	UN0299	II	1.3G			62	62	Forbidden	Forbidden	03		112	EXPLOSIVES 1.3*
	Bombs, smoke, non-explosive, *with corrosive liquid, without initiating device*	8	UN2028	II	8		None	160	None	Forbidden	50 kg	E	40	153	CORROSIVE
	Bombs, *with bursting charge*	1.1F	UN0033	II	1.1F			62	None	Forbidden	Forbidden	08		112	EXPLOSIVES 1.1*
	Bombs, *with bursting charge*	1.1D	UN0034	II	1.1D			62	62	Forbidden	Forbidden	03		112	EXPLOSIVES 1.1*
	Bombs, *with bursting charge*	1.2D	UN0035	II	1.2D			62	62	Forbidden	Forbidden	03		112	EXPLOSIVES 1.2*
	Bombs, *with bursting charge*	1.2F	UN0291	II	1.2F			62	None	Forbidden	Forbidden	08		112	EXPLOSIVES 1.2*
	Bombs with flammable liquid, *with bursting charge*	1.1J	UN0399	II	1.1J			62	None	Forbidden	Forbidden	04	23E	112	EXPLOSIVES 1.1*
	Bombs with flammable liquid, *with bursting charge*	1.2J	UN0400	II	1.2J			62	None	Forbidden	Forbidden	04	23E	112	EXPLOSIVES 1.2*
	Boosters with detonator	1.1B	UN0225	II	1.1B		None	62	None	Forbidden	Forbidden	11		112	EXPLOSIVES 1.1*
	Boosters with detonator	1.2B	UN0268	II	1.2B		None	62	None	Forbidden	Forbidden	07		112	EXPLOSIVES 1.2*
	Boosters, *without detonator*	1.1D	UN0042	II	1.1D		None	62	None	Forbidden	Forbidden	07		112	EXPLOSIVES 1.1*
	Boosters, *without detonator*	1.2D	UN0283	II	1.2D		None	62	None	Forbidden	Forbidden	07		112	EXPLOSIVES 1.2*
	Borate and chlorate mixtures, see **Chlorate and borate mixtures**														
	Borneol	4.1	UN1312	III	4.1	A1, IB8, IP3, T1, TP33	None	213	240	25 kg	100 kg	A		133	FLAMMABLE SOLID
+	**Boron tribromide**	8	UN2692	I	8, 6.1	2, B9, B14, B32, N34, T20, TP2, TP13, TP38, TP45	None	227	244	Forbidden	Forbidden	C	12	157	CORROSIVE, POISON INHALATION HAZARD*
	Boron trichloride	2.3	UN1741		2.3, 8	3, B9, B14	None	304	314	Forbidden	Forbidden	D	25, 40	125	POISON GAS*
	Boron trifluoride	2.3	UN1008		2.3, 8	2, B9, B14	None	302	314, 315	Forbidden	Forbidden	D	40	125	POISON GAS*
	Boron trifluoride acetic acid complex, liquid	8	UN1742	II	8	B2, B6, IB2, T8, TP2	154	202	242	1 L	30 L	A		157	CORROSIVE
	Boron trifluoride acetic acid complex, solid	8	UN3419	II	8	B2, B6, IB8, IP2, IP4, T3, TP33	154	212	240	15 kg	50 kg	A		157	CORROSIVE
	Boron trifluoride diethyl etherate	8	UN2604	I	8, 3	A3, A19, T10, TP2	None	201	243	0.5 L	2.5 L	D	40	132	CORROSIVE
	Boron trifluoride dihydrate	8	UN2851	II	8	IB2, T7, TP2	154	212	240	15 kg	50 kg	B	12, 40	157	CORROSIVE
	Boron trifluoride dimethyl etherate	4.3	UN2965	I	4.3, 8, 3	A19, T10, TP2, TP7	None	201	243	Forbidden	1 L	D	21, 28, 40, 49, 100	139	DANGEROUS WHEN WET*
	Boron trifluoride propionic acid complex, liquid	8	UN1743	II	8	B2, IB2, T8, TP2, TP12	154	202	242	1 L	30 L	A		157	CORROSIVE
	Boron trifluoride proponic acid complex, solid	8	UN3420	II	8	B2, IB8, IP2, IP4, T3, TP33	154	212	240	15 kg	50 kg	A		157	CORROSIVE
	Box toe gum, see **Nitrocellulose** *etc.*														
G	**Bromates, inorganic, aqueous solution, n.o.s.**	5.1	UN3213	II	5.1	350, IB2, T4, TP1	152	202	242	1 L	5 L	B	56, 58, 133	140	OXIDIZER
				III	5.1	350, IB2, T4, TP1	152	203	241	2.5 L	30 L	B	56, 58, 133	140	OXIDIZER

Symbols (1)	Hazardous materials descriptions and proper shipping names (2)	Hazard class or Division (3)	Identification Numbers (4)	PG (5)	Label Codes (6)	Special Provisions (§172.102) (7)	(8) Packaging (§173.***) Exceptions (8A)	Non-bulk (8B)	Bulk (8C)	(9) Quantity limitations (see §§173.27 and 175.75) Passenger aircraft/rail (9A)	Cargo aircraft only (9B)	(10) Vessel stowage Location (10A)	Other (10B)	ERG Guide No.	Placard Advisory Consult §172.504 & §172.505 to determine placard *Denotes placard for any quantity
G	**Bromates, inorganic, n.o.s.**	5.1	UN1450	II	5.1	350, IB8, IP2, IP4, T3, TP33	152	212	242	5 kg	25 kg	A	56, 58	141	OXIDIZER
+	**Bromine**	8	UN1744	I	8, 6.1	1, B9, B85, N34, N43, T22, TP2, TP10, TP13	None	226	249	Forbidden	Forbidden	D	12, 40, 66, 74, 89, 90	154	CORROSIVE, POISON INHALATION HAZARD*
	Bromine azide	Forbidden													
	Bromine chloride	2.3	UN2901		2.3, 8, 5.1	2, B9, B14, N86	None	304	314, 315	Forbidden	Forbidden	D	40, 89, 90	124	POISON GAS*
+	**Bromine pentafluoride**	5.1	UN1745	I	5.1, 6.1, 8	1, B9, B14, B30, T22, TP2, TP13, TP38, TP44	None	228	244	Forbidden	Forbidden	D	25, 40, 66, 90	144	OXIDIZER, POISON INHALATION HAZARD*
+	**Bromine solutions**	8	UN1744	I	8, 6.1	1, B9, B85, N34, N43, T22, TP2, TP10, TP13	None	226	249	Forbidden	Forbidden	D	12, 40, 66, 74, 89, 90	154	CORROSIVE, POISON INHALATION HAZARD*
+	**Bromine solutions**	8	UN1744	I	8, 6.1	2, B9, B85, N34, N43, T22, TP2, TP10, TP13	None	227	249	Forbidden	Forbidden	D	12, 40, 66, 74, 89, 90	154	CORROSIVE, POISON INHALATION HAZARD*
+	**Bromine trifluoride**	5.1	UN1746	I	5.1, 6.1, 8	2, B9, B14, B32, T22, TP2, TP13, TP38, TP45	None	228	244	Forbidden	Forbidden	D	25, 40, 66, 90	144	OXIDIZER, POISON INHALATION HAZARD*
	4-Bromo-1,2-dinitrobenzene	Forbidden													
	4-Bromo-1,2-dinitrobenzene (unstable at 59 degrees C.)	Forbidden													
	1-Bromo-3-chloropropane	6.1	UN2688	III	6.1	IB3, T4, TP1	153	203	241	60 L	220 L	A		159	POISON
	1-Bromo-3-methylbutane	3	UN2341	III	3	B1, IB3, T2, TP1	150	203	242	60 L	220 L	A		130	FLAMMABLE
	1-Bromo-3-nitrobenzene (unstable at 56 degrees C)	Forbidden													
	2-Bromo-2-nitropropane-1,3-diol	4.1	UN3241	III	4.1	46, IB8, IP3	151	213	None	25 kg	50 kg	C	12, 25, 40	133	FLAMMABLE SOLID
	Bromoacetic acid solid	8	UN3425	II	8	A7, IB8, IP2, IP4, N34, T3, TP33	154	212	240	15 kg	50 kg	A		156	CORROSIVE
	Bromoacetic acid solution	8	UN1938	II	8	A7, B2, IB2, T7, TP2	154	202	242	1 L	30 L	A	40	156	CORROSIVE
				III	8	B2, IB3, T7, TP2	154	203	241	5 L	60 L	A	40	156	CORROSIVE
+	**Bromoacetone**	6.1	UN1569	II	6.1, 3	2, T20, TP2, TP13	None	193	245	Forbidden	Forbidden	D	40	131	POISON INHALATION HAZARD*
	Bromoacetyl bromide	8	UN2513	II	8	B2, IB2, T8, TP2	154	202	242	1 L	30 L	C	40, 53	156	CORROSIVE
	Bromobenzene	3	UN2514	III	3	B1, IB3, T2, TP1	150	203	242	60 L	220 L	A		130	FLAMMABLE
	Bromobenzyl cyanides, liquid	6.1	UN1694	I	6.1	T14, TP2, TP13	None	201	243	Forbidden	30 L	D	12, 40, 52	159	POISON
	Bromobenzyl cyanides, solid	6.1	UN3449	I	6.1	T6, TP33	None	211	242	5 kg	50 kg	D	12, 40, 52	159	POISON
	1-Bromobutane	3	UN1126	II	3	IB2, T4, TP1	150	202	242	5 L	60 L	B	40	130	FLAMMABLE
	2-Bromobutane	3	UN2339	II	3	B1, IB2, T4, TP1	150	202	242	5 L	60 L	B	40	130	FLAMMABLE
	Bromochloromethane	6.1	UN1887	III	6.1	IB3, T4, TP1	153	203	241	60 L	220 L	A		160	POISON
	2-Bromoethyl ethyl ether	3	UN2340	II	3	IB2, T4, TP1	150	202	242	5 L	60 L	B	40	130	FLAMMABLE
	Bromoform	6.1	UN2515	III	6.1	IB3, T4, TP1	153	203	241	60 L	220 L	A	12, 40	159	POISON
	Bromomethylpropanes	3	UN2342	II	3	IB2, T4, TP1	150	202	242	5 L	60 L	B		130	FLAMMABLE
	2-Bromopentane	3	UN2343	II	3	IB2, T4, TP1	150	202	242	5 L	60 L	B		130	FLAMMABLE
	Bromopropanes	3	UN2344	II	3	IB2, T4, TP1	150	202	242	5 L	60 L	B	40	129	FLAMMABLE
				III	3	IB3, T2, TP1	150	203	242	60 L	220 L	A		129	FLAMMABLE
	3-Bromopropyne	3	UN2345	II	3	IB2, T4, TP1	150	202	242	5 L	60 L	D	40	130	FLAMMABLE
	Bromosilane	Forbidden													
	Bromotoluene-alpha, see **Benzyl bromide**														
	Bromotrifluoroethylene	2.1	UN2419		2.1		None	304	314, 315	Forbidden	150 kg	B	40	116	FLAMMABLE GAS
	Bromotrifluoromethane *or* **Refrigerant gas, R13B1**	2.2	UN1009		2.2	T50	306	304	314, 315	75 kg	150 kg	A		126	NONFLAMMABLE GAS
	Brucine	6.1	UN1570	I	6.1	IB7, IP1, T6, TP33	None	211	242	5 kg	50 kg	A		152	POISON
	Bursters, *explosive*	1.1D	UN0043	II	1.1D		None	62	None	Forbidden	Forbidden	07		112	EXPLOSIVES 1.1*
	Butadienes, stabilized *or* **Butadienes and Hydrocarbon mixture, stabilized** *containing more than 40% Butadienes*	2.1	UN1010		2.1	T50	306	304	314, 315	Forbidden	150 kg	B	40	116P	FLAMMABLE GAS

Symbols	Hazardous materials descriptions and proper shipping names	Hazard class or Division	Identification Numbers	PG	Label Codes	Special Provisions (§172.102)	(8) Packaging (§173.***) Exceptions	Non-bulk	Bulk	(9) Quantity limitations (see §§173.27 and 175.75) Passenger aircraft/rail	Cargo aircraft only	(10) Vessel stowage Location	Other	ERG Guide No.	Placard Advisory Consult §172.504 & §172.505 to determine placard *Denotes placard for any quantity
(1)	(2)	(3)	(4)	(5)	(6)	(7)	(8A)	(8B)	(8C)	(9A)	(9B)	(10A)	(10B)		
	Butane *see also* **Petroleum gases, liquefied**	2.1	UN1011		2.1	19, T50	306	304	314, 315	Forbidden	150 kg	E	40	115	FLAMMABLE GAS
	Butane, butane mixtures and mixtures having similar properties in cartridges each not exceeding 500 grams, see **Receptacles,** *etc.*														
	Butanedione	3	UN2346	II	3	IB2, T4, TP1	150	202	242	5 L	60 L	B		127	FLAMMABLE
	1,2,4-Butanetriol trinitrate	Forbidden													
	Butanols	3	UN1120	II	3	IB2, T4, TP1, TP29	150	202	242	5 L	60 L	B		129	FLAMMABLE
				III	3	B1, IB3, T2, TP1	150	203	242	60 L	220 L	A		129	FLAMMABLE
	tert-Butoxycarbonyl azide	Forbidden													
	Butyl acetates	3	UN1123	II	3	IB2, T4, TP1	150	202	242	5 L	60 L	B		129	FLAMMABLE
				III	3	B1, IB3, T2, TP1	150	203	242	60 L	220 L	A		129	FLAMMABLE
	Butyl acid phosphate	8	UN1718	III	8	IB3, T4, TP1	154	203	241	5 L	60 L	A		153	CORROSIVE
	Butyl acrylates, stabilized	3	UN2348	III	3	B1, IB3, T2, TP1	150	203	242	60 L	220 L	A		129P	FLAMMABLE
	Butyl alcohols, see **Butanols**														
	Butyl benzenes	3	UN2709	III	3	B1, IB3, T2, TP1	150	203	242	60 L	220 L	A		128	FLAMMABLE
	n-Butyl bromide, see **1-Bromobutane**														
	n-Butyl chloride, see **Chlorobutanes**														
	n-Butyl chloroformate	6.1	UN2743	I	6.1, 8, 3	2, B9, B14, B32, T20, TP2, TP13, TP38, TP45	None	227	244	Forbidden	Forbidden	A	12, 13, 21, 25, 40, 100	155	POISON INHALATION HAZARD*
	Butyl ethers, see **Dibutyl ethers**														
	Butyl ethyl ether, see **Ethyl butyl ether**														
	n-Butyl formate	3	UN1128	II	3	IB2, T4, TP1	150	202	242	5 L	60 L	B		129	FLAMMABLE
	tert-Butyl hydroperoxide, with more than 90 percent with water	Forbidden													
	tert-Butyl hypochlorite	4.2	UN3255	I	4.2, 8		None	211	243	Forbidden	Forbidden	D		135	SPONTANEOUSLY COMBUSTIBLE
	N-n-Butyl imidazole	6.1	UN2690	II	6.1	IB2, T7, TP2	153	202	243	5 L	60 L	A		152	POISON
	tert-Butyl isocyanate	6.1	UN2484	I	6.1, 3	1, B9, B14, B30, B72, T20, TP2, TP13, TP38, TP44	None	226	244	Forbidden	Forbidden	D	40	155	POISON INHALATION HAZARD*
	n-Butyl isocyanate	6.1	UN2485	I	6.1, 3	2, B9, B14, B32, B77, T20, TP2, TP13, TP38, TP45	None	227	244	Forbidden	Forbidden	D	40	155	POISON INHALATION HAZARD*
	Butyl mercaptans	3	UN2347	II	3	A3, A6, IB2, T4, TP1	150	202	242	5 L	60 L	D	26, 95	130	FLAMMABLE
	n-Butyl methacrylate, stabilized	3	UN2227	III	3	B1, IB3, T2, TP1	150	203	242	60 L	220 L	A		130P	FLAMMABLE
	Butyl methyl ether	3	UN2350	II	3	IB2, T4, TP1	150	202	242	5 L	60 L	B		127	FLAMMABLE
	Butyl nitrites	3	UN2351	I	3	T11, TP1, TP8, TP27	150	201	243	1 L	30 L	E	40	129	FLAMMABLE
				II	3	IB2, T4, TP1	150	202	242	5 L	60 L	B	40	129	FLAMMABLE
				III	3	B1, IB3, T2, TP1	150	203	242	60 L	220 L	A	40	129	FLAMMABLE
	tert-Butyl peroxyacetate, with more than 76 percent in solution	Forbidden													
	n-Butyl peroxydicarbonate, with more than 52 percent in solution	Forbidden													
	tert-Butyl peroxyisobutyrate, with more than 77 percent in solution	Forbidden													
	Butyl phosphoric acid, see **Butyl acid phosphate**														
	Butyl propionates	3	UN1914	III	3	B1, IB3, T2, TP1	150	203	242	60 L	220 L	A		130	FLAMMABLE
	5-tert-Butyl-2,4,6-trinitro-m-xylene *or* **Musk xylene**	4.1	UN2956	III	4.1	159	None	223	None	Forbidden	Forbidden	D	12, 25, 48, 127	149	FLAMMABLE SOLID
	Butyl vinyl ether, stabilized	3	UN2352	II	3	IB2, T4, TP1	150	202	242	5 L	60 L	B	40	127P	FLAMMABLE
	n-Butylamine	3	UN1125	II	3, 8	IB2, T7, TP1	150	202	242	1 L	5 L	B	40	132	FLAMMABLE
	N-Butylaniline	6.1	UN2738	II	6.1	IB2, T7, TP2	153	202	243	5 L	60 L	A	74	153	POISON
	tert-Butylcyclohexylchloroformate	6.1	UN2747	III	6.1	IB3, T4, TP1	153	203	241	60 L	220 L	A	12, 13, 25	156	POISON
	Butylene *see also* **Petroleum gases, liquefied**	2.1	UN1012		2.1	19, T50	306	304	314, 315	Forbidden	150 kg	E	40	115	FLAMMABLE GAS

Symbols (1)	Hazardous materials descriptions and proper shipping names (2)	Hazard class or Division (3)	Identification Numbers (4)	PG (5)	Label Codes (6)	Special Provisions (§172.102) (7)	(8) Packaging (§173.***) Exceptions (8A)	Non-bulk (8B)	Bulk (8C)	(9) Quantity limitations (see §§173.27 and 175.75) Passenger aircraft/rail (9A)	Cargo aircraft only (9B)	(10) Vessel stowage Location (10A)	Other (10B)	ERG Guide No.	Placard Advisory Consult §172.504 & §172.505 to determine placard *Denotes placard for any quantity
	1,2-Butylene oxide, stabilized	3	UN3022	II	3	IB2, T4, TP1	150	202	242	5 L	60 L	B	27, 49	127P	FLAMMABLE
	Butyltoluenes	6.1	UN2667	III	6.1	IB3, T4, TP1	153	203	241	60 L	220 L	A		152	POISON
	Butyltrichlorosilane	8	UN1747	II	8, 3	A7, B2, B6, N34, T10, TP2, TP7, TP13	None	206	243	Forbidden	30 L	C	40	155	CORROSIVE
	1,4-Butynediol	6.1	UN2716	III	6.1	A1, IB8, IP3, T1, TP33	None	213	240	100 kg	200 kg	C	52, 53, 70	153	POISON
	Butyraldehyde	3	UN1129	II	3	IB2, T4, TP1	150	202	242	5 L	60 L	B		129	FLAMMABLE
	Butyraldoxime	3	UN2840	III	3	B1, IB3, T2, TP1	150	203	242	60 L	220 L	A		129	FLAMMABLE
	Butyric acid	8	UN2820	III	8	IB3, T4, TP1	154	203	241	5 L	60 L	A	12	153	CORROSIVE
	Butyric anhydride	8	UN2739	III	8	IB3, T4, TP1	154	203	241	5 L	60 L	A		156	CORROSIVE
	Butyronitrile	3	UN2411	II	3, 6.1	IB2, T7, TP1, TP13	150	202	243	1 L	60 L	E	40	131	FLAMMABLE
	Butyryl chloride	3	UN2353	II	3, 8	IB2, T8, TP2, TP13	150	202	243	1 L	5 L	C	40	132	FLAMMABLE
	Cacodylic acid	6.1	UN1572	II	6.1	IB8, IP2, IP4, T3, TP33	153	212	242	25 kg	100 kg	E	52	151	POISON
G	**Cadmium compounds**	6.1	UN2570	I	6.1	IB7, IP1, T6, TP33	None	211	242	5 kg	50 kg	A		154	POISON
				II	6.1	IB8, IP2, IP4, T3, TP33	153	212	242	25 kg	100 kg	A		154	POISON
				III	6.1	IB8, IP3, T1, TP33	153	213	240	100 kg	200 kg	A		154	POISON
	Caesium hydroxide	8	UN2682	II	8	IB8, IP2, IP4, T3, TP33	154	212	240	15 kg	50 kg	A	29, 52	157	CORROSIVE
	Caesium hydroxide solution	8	UN2681	II	8	B2, IB2, T7, TP2	154	202	242	1 L	30 L	A	29, 52	154	CORROSIVE
				III	8	IB3, T4, TP1	154	203	241	5 L	60 L	A	29, 52	154	CORROSIVE
	Calcium	4.3	UN1401	II	4.3	IB7, IP2, T3, TP33	151	212	241	15 kg	50 kg	E	52	138	DANGEROUS WHEN WET*
	Calcium arsenate	6.1	UN1573	II	6.1	IB8, IP2, IP4, T3, TP33	153	212	242	25 kg	100 kg	A		151	POISON
	Calcium arsenate and calcium arsenite, mixtures, solid	6.1	UN1574	II	6.1	IB8, IP2, IP4, T3, TP33	153	212	242	25 kg	100 kg	A		151	POISON
	Calcium bisulfite solution, see **Bisulfites, aqueous solutions, n.o.s.**														
	Calcium carbide	4.3	UN1402	I	4.3	A1, A8, B55, B59, IB4, IP1, N34, T9, TP7, TP33	None	211	242	Forbidden	15 kg	B	52	138	DANGEROUS WHEN WET*
				II	4.3	A1, A8, B55, B59, IB7, IP2, N34, T3, TP33	151	212	241	15 kg	50 kg	B	52	138	DANGEROUS WHEN WET*
	Calcium chlorate	5.1	UN1452	II	5.1	A9, IB8, IP2, IP4, N34, T3, TP33	152	212	242	5 kg	25 kg	A	56, 58	140	OXIDIZER
	Calcium chlorate aqueous solution	5.1	UN2429	II	5.1	A2, IB2, N41, T4, TP1	152	202	242	1 L	5 L	B	56, 58, 133	140	OXIDIZER
				III	5.1	A2, IB2, N41, T4, TP1	152	203	241	2.5 L	30 L	B	56, 68, 133	140	OXIDIZER
	Calcium chlorite	5.1	UN1453	II	5.1	A9, IB8, IP2, IP4, N34, T3, TP33	152	212	242	5 kg	25 kg	A	56, 58	140	OXIDIZER
	Calcium cyanamide *with more than 0.1 percent of calcium carbide*	4.3	UN1403	III	4.3	A1, A19, IB8, IP4, T1, TP33	151	213	241	25 kg	100 kg	A	52	138	DANGEROUS WHEN WET*
	Calcium cyanide	6.1	UN1575	I	6.1	IB7, IP1, N79, N80, T6, TP33	None	211	242	5 kg	50 kg	A	40, 52	157	POISON
	Calcium dithionite *or* **Calcium hydrosulfite**	4.2	UN1923	II	4.2	A19, A20, IB6, IP2, T3, TP33	None	212	241	15 kg	50 kg	E	13	135	SPONTANEOUSLY COMBUSTIBLE
	Calcium hydride	4.3	UN1404	I	4.3	A19, N40	None	211	242	Forbidden	15 kg	E	52	138	DANGEROUS WHEN WET*
	Calcium hydrosulfite, *see* **Calcium dithionite**														
	Calcium hypochlorite, dry, corrosive *or* **Calcium hypochlorite mixtures, dry, corrosive** *with more than 39% available chlorine (8.8% available oxygen)*	5.1	UN3485	II	5.1, 8	165, 166, A7, A9, IB8, IP2, IP4, IP13, N34, W9	152	212	None	5 kg	25 kg	D	4, 48, 52, 56, 58, 69, 142	—	OXIDIZER
	Calcium hypochlorite, dry *or* **Calcium hypochlorite mixtures dry** *with more than 39 percent available chlorine (8.8 percent available oxygen)*	5.1	UN1748	II	5.1	165, 166, A7, A9, IB8, IP2, IP4, IP13, N34, W9	152	212	None	5 kg	25 kg	D	4, 25, 48, 52, 56, 58, 69, 142	140	OXIDIZER

Symbols (1)	Hazardous materials descriptions and proper shipping names (2)	Hazard class or Division (3)	Identification Numbers (4)	PG (5)	Label Codes (6)	Special Provisions (§172.102) (7)	(8) Packaging (§173.***) Exceptions (8A)	Non-bulk (8B)	Bulk (8C)	(9) Quantity limitations (see §§173.27 and 175.75) Passenger aircraft/rail (9A)	Cargo aircraft only (9B)	(10) Vessel stowage Location (10A)	Other (10B)	ERG Guide No.	Placard Advisory Consult §172.504 & §172.505 to determine placard *Denotes placard for any quantity
				III	5.1	165, 171, A7, A9, IB8, IP4, IP13, N34, W9	152	213	240	25 kg	100 kg	D	4, 25, 48, 52, 56, 58, 69, 142	140	OXIDIZER
	Calcium hypochlorite, hydrated, corrosive *or* **Calcium hypochlorite, hydrated mixture, corrosive** *with not less than 5.5% but not more than 16% water*	5.1	UN3487	II	5.1, 8	165, IB8, IP2, IP4, IP13, W9	152	212	240	5 kg	25 kg	D	4, 48, 52, 56, 58, 69, 142	—	OXIDIZER
				III	5.1, 8	165, IB8, IP4, W9	152	213	240	25 kg	100 kg	D	4, 48, 52, 56, 58, 69, 142	—	OXIDIZER
	Calcium hypochlorite, hydrated *or* **Calcium hypochlorite, hydrated mixtures,** *with not less than 5.5 percent but not more than 16 percent water*	5.1	UN2880	II	5.1	165, IB8, IP2, IP4, IP13, W9	152	212	240	5 kg	25 kg	D	4, 25, 48, 52, 56, 58, 69, 142	140	OXIDIZER
				III	5.1	165, 171, IB8, IP4, IP13, W9	152	213	240	25 kg	100 kg	D	4, 25, 48, 52, 56, 58, 69, 142	140	OXIDIZER
	Calcium hypochlorite mixture, dry, corrosive *with more than 10% but not more than 39% available chlorine*	5.1	UN3486	III	5.1, 8	165, A1, A29, IB8, IP3, IP13, N34, W9	152	213	240	5 kg	25 kg	D	4, 48, 52, 56, 58, 69, 142	—	OXIDIZER
	Calcium hypochlorite mixtures, dry, *with more than 10 percent but not more than 39 percent available chlorine*	5.1	UN2208	III	5.1	165, A1, A29, IB8, IP3, IP13, N34, W9	152	213	240	25 kg	100 kg	D	4, 25, 48, 52, 56, 58, 69, 142	140	OXIDIZER
	Calcium manganese silicon	4.3	UN2844	III	4.3	A1, A19, IB8, IP4, T1, TP33	151	213	241	25 kg	100 kg	A	52, 85, 103	138	DANGEROUS WHEN WET*
	Calcium nitrate	5.1	UN1454	III	5.1	34, IB8, IP3, T1, TP33	152	213	240	25 kg	100 kg	A		140	OXIDIZER
A	**Calcium oxide**	8	UN1910	III	8	IB8, IP3, T1, TP33	154	213	240	25 kg	100 kg	A		157	CORROSIVE
	Calcium perchlorate	5.1	UN1455	II	5.1	IB6, IP2, T3, TP33	152	212	242	5 kg	25 kg	A	56, 58	140	OXIDIZER
	Calcium permanganate	5.1	UN1456	II	5.1	IB6, IP2, T3, TP33	152	212	242	5 kg	25 kg	D	56, 58, 138	140	OXIDIZER
	Calcium peroxide	5.1	UN1457	II	5.1	IB6, IP2, T3, TP33	152	212	242	5 kg	25 kg	A	13, 52, 56, 75	140	OXIDIZER
	Calcium phosphide	4.3	UN1360	I	4.3, 6.1	A8, A19, N40	None	211	242	Forbidden	15 kg	E	40, 52, 85	139	DANGEROUS WHEN WET*
	Calcium, pyrophoric *or* **Calcium alloys, pyrophoric**	4.2	UN1855	I	4.2		None	187	None	Forbidden	Forbidden	D		135	SPONTANEOUSLY COMBUSTIBLE
	Calcium resinate	4.1	UN1313	III	4.1	A1, A19, IB6, T1, TP33	None	213	240	25 kg	100 kg	A		133	FLAMMABLE SOLID
	Calcium resinate, fused	4.1	UN1314	III	4.1	A1, A19, IB4, T1, TP33	None	213	240	25 kg	100 kg	A		133	FLAMMABLE SOLID
	Calcium selenate, see **Selenates** *or* **Selenites**														
	Calcium silicide	4.3	UN1405	II	4.3	A19, IB7, IP2, T3, TP33	151	212	241	15 kg	50 kg	B	52, 85, 103	138	DANGEROUS WHEN WET*
				III	4.3	A1, A19, IB8, IP4, T1, TP33	151	213	241	25 kg	100 kg	B	52, 85, 103	138	DANGEROUS WHEN WET*
	Camphor oil	3	UN1130	III	3	B1, IB3, T2, TP1	150	203	242	60 L	220 L	A		128	FLAMMABLE
	Camphor, *synthetic*	4.1	UN2717	III	4.1	A1, IB8, IP3, T1, TP33	None	213	240	25 kg	100 kg	A		133	FLAMMABLE SOLID
	Cannon primers, see **Primers, tubular**														
	Caproic acid	8	UN2829	III	8	IB3, T4, TP1	154	203	241	5 L	60 L	A		153	CORROSIVE
	Caps, blasting, see **Detonators,** *etc.*														
	Carbamate pesticides, liquid, flammable, toxic, *flash point less than 23 degrees C*	3	UN2758	I	3, 6.1	T14, TP2, TP13, TP27	None	201	243	Forbidden	30 L	B	40	131	FLAMMABLE
				II	3, 6.1	IB2, T11, TP2, TP13, TP27	150	202	243	1 L	60 L	B	40	131	FLAMMABLE
	Carbamate pesticides, liquid, toxic	6.1	UN2992	I	6.1	T14, TP2, TP13, TP27	None	201	243	1 L	30 L	B	40	151	POISON
				II	6.1	IB2, T11, TP2, TP13, TP27	153	202	243	5 L	60 L	B	40	151	POISON
				III	6.1	IB3, T7, TP2, TP28	153	203	241	60 L	220 L	A	40	151	POISON
	Carbamate pesticides, liquid, toxic, flammable, *flash point not less than 23 degrees C*	6.1	UN2991	I	6.1, 3	T14, TP2, TP13, TP27	None	201	243	1 L	30 L	B	40	131	POISON

HAZARDOUS MATERIALS TABLE

Symbols (1)	Hazardous materials descriptions and proper shipping names (2)	Hazard class or Division (3)	Identification Numbers (4)	PG (5)	Label Codes (6)	Special Provisions (§172.102) (7)	(8) Packaging (§173.***) Exceptions (8A)	Non-bulk (8B)	Bulk (8C)	(9) Quantity limitations (see §§173.27 and 175.75) Passenger aircraft/rail (9A)	Cargo aircraft only (9B)	(10) Vessel stowage Location (10A)	Other (10B)	ERG Guide No.	Placard Advisory Consult §172.504 & §172.505 to determine placard *Denotes placard for any quantity
				II	6.1, 3	IB2, T11, TP2, TP13, TP27	153	202	243	5 L	60 L	B	40	131	POISON
				III	6.1, 3	B1, IB3, T7, TP2, TP28	153	203	242	60 L	220 L	A	40	131	POISON
	Carbamate pesticides, solid, toxic	6.1	UN2757	I	6.1	IB7, IP1, T6, TP33	None	211	242	5 kg	50 kg	A	40	151	POISON
				II	6.1	IB8, IP2, IP4, T3, TP33	153	212	242	25 kg	100 kg	A	40	151	POISON
				III	6.1	IB8, IP3, T1, TP33	153	213	240	100 kg	200 kg	A	40	151	POISON
	Carbolic acid, see **Phenol, solid** *or* **Phenol, molten**														
	Carbolic acid solutions, see **Phenol solutions**														
I	**Carbon, activated**	4.2	UN1362	III	4.2	IB8, IP3, T1, TP33	None	213	241	0.5 kg	0.5 kg	A	12	133	SPONTANEOUSLY COMBUSTIBLE
I	**Carbon,** *animal or vegetable origin*	4.2	UN1361	II	4.2	IB6, T3, TP33	None	212	242	Forbidden	Forbidden	A	12	133	SPONTANEOUSLY COMBUSTIBLE
				III	4.2	IB8, IP3, T1, TP33	None	213	241	Forbidden	Forbidden	A	12	133	SPONTANEOUSLY COMBUSTIBLE
	Carbon bisulfide, see **Carbon disulfide**														
	Carbon dioxide	2.2	UN1013		2.2		306	302, 304	302, 314, 315	75 kg	150 kg	A		120	NONFLAMMABLE GAS
	Carbon dioxide, refrigerated liquid	2.2	UN2187		2.2	T75, TP5	306	304	314, 315	50 kg	500 kg	D		120	NONFLAMMABLE GAS
A W	**Carbon dioxide, solid** *or* **Dry ice**	9	UN1845		None		217	217	240	200 kg	200 kg	C	40	120	CLASS 9
	Carbon disulfide	3	UN1131	I	3, 6.1	B16, T14, TP2, TP7, TP13	None	201	243	Forbidden	Forbidden	D	40, 78, 115	131	FLAMMABLE
	Carbon monoxide, compressed	2.3	UN1016		2.3, 2.1	4	None	302	314, 315	Forbidden	25 kg	D	40	119	POISON GAS*
D	**Carbon monoxide, refrigerated liquid** *(cryogenic liquid)*	2.3	NA9202		2.3, 2.1	4, T75, TP5	None	316	318	Forbidden	Forbidden	D		168	POISON GAS*
	Carbon tetrabromide	6.1	UN2516	III	6.1	IB8, IP3, T1, TP33	153	213	240	100 kg	200 kg	A	25	151	POISON
	Carbon tetrachloride	6.1	UN1846	II	6.1	IB2, N36, T7, TP2	153	202	243	5 L	60 L	A	40	151	POISON
	Carbonyl chloride, see **Phosgene**														
	Carbonyl fluoride	2.3	UN2417		2.3, 8	2	None	302	None	Forbidden	Forbidden	D	40	125	POISON GAS*
	Carbonyl sulfide	2.3	UN2204		2.3, 2.1	3, B14	None	304	314, 315	Forbidden	Forbidden	D	40	119	POISON GAS*
	Cartridge cases, empty primed, see **Cases, cartridge, empty, with primer**														
	Cartridges, actuating, for aircraft ejector seat catapult, fire extinguisher, canopy removal or apparatus, see **Cartridges, power device**														
	Cartridges, explosive, see **Charges, demolition**														
	Cartridges, flash	1.1G	UN0049	II	1.1G		None	62	None	Forbidden	Forbidden	07		112	EXPLOSIVES 1.1*
	Cartridges, flash	1.3G	UN0050	II	1.3G		None	62	None	Forbidden	75 kg	07		112	EXPLOSIVES 1.3*
	Cartridges for weapons, blank	1.1C	UN0326	II	1.1C		None	62	None	Forbidden	Forbidden	07		112	EXPLOSIVES 1.1*
	Cartridges for weapons, blank	1.2C	UN0413	II	1.2C		None	62	None	Forbidden	Forbidden	07		112	EXPLOSIVES 1.2*
	Cartridges for weapons, blank *or* **Cartridges, small arms, blank**	1.4S	UN0014	II	None		63	62	None	25 kg	100 kg	05		114	EXPLOSIVES 1.4
	Cartridges for weapons, blank *or* **Cartridges, small arms, blank**	1.3C	UN0327	II	1.3C		None	62	None	Forbidden	Forbidden	07		112	EXPLOSIVES 1.3*
	Cartridges for weapons, blank *or* **Cartridges, small arms, blank**	1.4C	UN0338	II	1.4C		None	62	None	Forbidden	75 kg	06		114	EXPLOSIVES 1.4
	Cartridges for weapons, inert projectile	1.2C	UN0328	II	1.2C		None	62	62	Forbidden	Forbidden	03		112	EXPLOSIVES 1.2*
	Cartridges for weapons, inert projectile *or* **Cartridges, small arms**	1.4S	UN0012	II	None		63	62	None	25 kg	100 kg	05		114	EXPLOSIVES 1.4
	Cartridges for weapons, inert projectile *or* **Cartridges, small arms**	1.4C	UN0339	II	1.4C		None	62	None	Forbidden	75 kg	06		114	EXPLOSIVES 1.4
	Cartridges for weapons, inert projectile *or* **Cartridges, small arms**	1.3C	UN0417	II	1.3C		None	62	None	Forbidden	Forbidden	06		112	EXPLOSIVES 1.3*

Sym-bols (1)	Hazardous materials descriptions and proper shipping names (2)	Hazard class or Division (3)	Identification Numbers (4)	PG (5)	Label Codes (6)	Special Provisions (§172.102) (7)	(8) Packaging (§173.***) Exceptions (8A)	Non-bulk (8B)	Bulk (8C)	(9) Quantity limitations (see §§173.27 and 175.75) Passenger aircraft/rail (9A)	Cargo aircraft only (9B)	(10) Vessel stowage Location (10A)	Other (10B)	ERG Guide No.	Placard Advisory Consult §172.504 & §172.505 to determine placard *Denotes placard for any quantity
	Cartridges for weapons, *with bursting charge*	1.1F	UN0005	II	1.1F		None	62	None	Forbidden	Forbidden	08		112	EXPLOSIVES 1.1*
	Cartridges for weapons, *with bursting charge*	1.1E	UN0006	II	1.1E		None	62	62	Forbidden	Forbidden	03		112	EXPLOSIVES 1.1*
	Cartridges for weapons, *with bursting charge*	1.2F	UN0007	II	1.2F		None	62	None	Forbidden	Forbidden	08		112	EXPLOSIVES 1.2*
	Cartridges for weapons, *with bursting charge*	1.2E	UN0321	II	1.2E		None	62	62	Forbidden	Forbidden	03		112	EXPLOSIVES 1.2*
	Cartridges for weapons, *with bursting charge*	1.4F	UN0348	II	1.4F		None	62	None	Forbidden	Forbidden	08		114	EXPLOSIVES 1.4
	Cartridges for weapons, *with bursting charge*	1.4E	UN0412	II	1.4E		None	62	62	Forbidden	75 kg	02		114	EXPLOSIVES 1.4
	Cartridges, oil well	1.3C	UN0277	II	1.3C		None	62	62	Forbidden	Forbidden	07		112	EXPLOSIVES 1.3*
	Cartridges, oil well	1.4C	UN0278	II	1.4C		None	62	62	Forbidden	75 kg	06		114	EXPLOSIVES 1.4
	Cartridges, power device	1.3C	UN0275	II	1.3C		None	62	62	Forbidden	75 kg	07		112	EXPLOSIVES 1.3*
	Cartridges, power device	1.4C	UN0276	II	1.4C	110	None	62	62	Forbidden	75 kg	06		114	EXPLOSIVES 1.4
	Cartridges, power device	1.4S	UN0323	II	1.4S	110, 347	63	62	62	25 kg	100 kg	05		114	EXPLOSIVES 1.4
	Cartridges, power device	1.2C	UN0381	II	1.2C		None	62	62	Forbidden	Forbidden	07		112	EXPLOSIVES 1.2*
D	**Cartridges power device** *(used to project fastening devices)*	ORM-D	None		None	347	63	None	None	30 kg gross	30 kg gross	A		—	NONE
	Cartridges, safety, blank, see **Cartridges for weapons, blank** *(UN 0014)*														
	Cartridges, safety, see **Cartridges for weapons, inert projectile,** *or* **Cartriges, small arms** *or* **Cartridges, power device** *(UN 0323)*														
	Cartridges, signal	1.3G	UN0054	II	1.3G		None	62	None	Forbidden	75 kg	07		112	EXPLOSIVES 1.3*
	Cartridges, signal	1.4G	UN0312	II	1.4G		None	62	None	Forbidden	75 kg	06		114	EXPLOSIVES 1.4
	Cartridges, signal	1.4S	UN0405	II	1.4S		None	62	None	25 kg	100 kg	05		114	EXPLOSIVES 1.4
D	**Cartridges, small arms**	ORM-D			None		63	None	None	30 kg gross	30 kg gross	A		—	NONE
	Cartridges, sporting, see **Cartridges for weapons, inert projectile,** *or* **Cartridges, small arms**														
	Cartridges, starter, jet engine, see **Cartridges, power device**														
	Cases, cartridge, empty with primer	1.4S	UN0055	II	1.4S	50	None	62	None	25 kg	100 kg	05		114	EXPLOSIVES 1.4
	Cases, cartridges, empty with primer	1.4C	UN0379	II	1.4C	50	None	62	None	Forbidden	75 kg	06		114	EXPLOSIVES 1.4
	Cases, combustible, empty, without primer	1.4C	UN0446	II	1.4C		None	62	None	Forbidden	75 kg	06		114	EXPLOSIVES 1.4
	Cases, combustible, empty, without primer	1.3C	UN0447	II	1.3C		None	62	None	Forbidden	Forbidden	07		112	EXPLOSIVES 1.3*
	Casinghead gasoline see **Gasoline**														
A W	**Castor beans** *or* **Castor meal** *or* **Castor pomace** *or* **Castor flake**	9	UN2969	II	None	IB8, IP2, IP4, T3, TP33	155	204	240	No limit	No limit	E	34, 40	171	CLASS 9
G	**Caustic alkali liquids, n.o.s.**	8	UN1719	II	8	B2, IB2, T11, TP2, TP27	154	202	242	1 L	30 L	A	29, 52	154	CORROSIVE
				III	8	IB3, T7, TP1, TP28	154	203	241	5 L	60 L	A	29, 52	154	CORROSIVE
	Caustic potash, see **Potassium hydroxide** *etc.*														
	Caustic soda, (etc.), see **Sodium hydroxide** *etc.*														
	Cells, containing sodium	4.3	UN3292	II	4.3		189	189	189	25 kg gross	No limit	A		138	DANGEROUS WHEN WET*
	Celluloid, *in block, rods, rolls, sheets, tubes, etc., except scrap*	4.1	UN2000	III	4.1		None	213	240	25 kg	100 kg	A		133	FLAMMABLE SOLID
	Celluloid, scrap	4.2	UN2002	III	4.2	IB8, IP3	None	213	241	Forbidden	Forbidden	D		135	SPONTANEOUSLY COMBUSTIBLE
	Cement, see **Adhesives** *containing flammable liquid*														
	Cerium, *slabs, ingots, or rods*	4.1	UN1333	II	4.1	IB8, IP2, IP4, N34	None	212	240	15 kg	50 kg	A	74, 91	170	FLAMMABLE SOLID
	Cerium, *turnings or gritty powder*	4.3	UN3078	II	4.3	A1, IB7, IP2, T3, TP33	151	212	242	15 kg	50 kg	E	52	138	DANGEROUS WHEN WET*
	Cesium *or* **Caesium**	4.3	UN1407	I	4.3	A7, A19, IB4, IP1, N34, N40	None	211	242	Forbidden	15 kg	D	52	138	DANGEROUS WHEN WET*
	Cesium nitrate *or* **Caesium nitrate**	5.1	UN1451	III	5.1	A1, A29, IB8, IP3, T1, TP33	152	213	240	25 kg	100 kg	A		140	OXIDIZER
D	**Charcoal** *briquettes, shell, screenings, wood, etc.*	4.2	NA1361	III	4.2	IB8, T1, TP33	151	213	240	25 kg	100 kg	A	12	133	SPONTANEOUSLY COMBUSTIBLE
	Charges, bursting, plastics bonded	1.1D	UN0457	II	1.1D		None	62	None	Forbidden	Forbidden	07		112	EXPLOSIVES 1.1*
	Charges, bursting, plastics bonded	1.2D	UN0458	II	1.2D		None	62	None	Forbidden	Forbidden	07		112	EXPLOSIVES 1.2*
	Charges, bursting, plastics bonded	1.4D	UN0459	II	1.4D		None	62	None	Forbidden	75 kg	06		114	EXPLOSIVES 1.4

Symbols (1)	Hazardous materials descriptions and proper shipping names (2)	Hazard class or Division (3)	Identification Numbers (4)	PG (5)	Label Codes (6)	Special Provisions (§172.102) (7)	(8) Packaging (§173.***) Exceptions (8A)	Non-bulk (8B)	Bulk (8C)	(9) Quantity limitations (see §§173.27 and 175.75) Passenger aircraft/rail (9A)	Cargo aircraft only (9B)	(10) Vessel stowage Location (10A)	Other (10B)	ERG Guide No.	Placard Advisory Consult §172.504 & §172.505 to determine placard *Denotes placard for any quantity
	Charges, bursting, plastics bonded	1.4S	UN0460	II	1.4S	347	None	62	None	25 kg	100 kg	05		114	EXPLOSIVES 1.4
	Charges, demolition	1.1D	UN0048	II	1.1D		None	62	62	Forbidden	Forbidden	03		112	EXPLOSIVES 1.1*
	Charges, depth	1.1D	UN0056	II	1.1D		None	62	62	Forbidden	Forbidden	03		112	EXPLOSIVES 1.1*
	Charges, expelling, explosive, for fire extinguishers, see **Cartridges, power device**														
	Charges, explosive, commercial *without detonator*	1.1D	UN0442	II	1.1D		None	62	None	Forbidden	Forbidden	07		112	EXPLOSIVES 1.1*
	Charges, explosive, commercial *without detonator*	1.2D	UN0443	II	1.2D		None	62	None	Forbidden	Forbidden	07		112	EXPLOSIVES 1.2*
	Charges, explosive, commercial *without detonator*	1.4D	UN0444	II	1.4D		None	62	None	Forbidden	75 kg	06		114	EXPLOSIVES 1.4
	Charges, explosive, commercial *without detonator*	1.4S	UN0445	II	1.4S	347	None	62	None	25 kg	100 kg	05		114	EXPLOSIVES 1.4
	Charges, propelling	1.1C	UN0271	II	1.1C		None	62	None	Forbidden	Forbidden	07		112	EXPLOSIVES 1.1*
	Charges, propelling	1.3C	UN0272	II	1.3C		None	62	None	Forbidden	Forbidden	07		112	EXPLOSIVES 1.3*
	Charges, propelling	1.2C	UN0415	II	1.2C		None	62	None	Forbidden	Forbidden	07		112	EXPLOSIVES 1.2*
	Charges, propelling	1.4C	UN0491	II	1.4C		None	62	None	Forbidden	75 kg	06		114	EXPLOSIVES 1.4
	Charges, propelling, for cannon	1.3C	UN0242	II	1.3C		None	62	None	Forbidden	Forbidden	10		112	EXPLOSIVES 1.3*
	Charges, propelling, for cannon	1.1C	UN0279	II	1.1C		None	62	None	Forbidden	Forbidden	10		112	EXPLOSIVES 1.1*
	Charges, propelling, for cannon	1.2C	UN0414	II	1.2C		None	62	None	Forbidden	Forbidden	10		112	EXPLOSIVES 1.2*
	Charges, shaped, flexible, linear	1.4D	UN0237	II	1.4D		None	62	None	Forbidden	75 kg	06		114	EXPLOSIVES 1.4
	Charges, shaped, flexible, linear	1.1D	UN0288	II	1.1D		None	62	None	Forbidden	Forbidden	07		112	EXPLOSIVES 1.1*
	Charges, shaped, *without detonator*	1.1D	UN0059	II	1.1D		None	62	None	Forbidden	Forbidden	07		112	EXPLOSIVES 1.1*
	Charges, shaped, *without detonator*	1.2D	UN0439	II	1.2D		None	62	None	Forbidden	Forbidden	07		112	EXPLOSIVES 1.2*
	Charges, shaped, *without detonator*	1.4D	UN0440	II	1.4D		None	62	None	Forbidden	75 kg	06		114	EXPLOSIVES 1.4
	Charges, shaped, *without detonator*	1.4S	UN0441	II	1.4S	347	None	62	None	25 kg	100 kg	05		114	EXPLOSIVES 1.4
	Charges, supplementary explosive	1.1D	UN0060	II	1.1D		None	62	None	Forbidden	Forbidden	10		112	EXPLOSIVES 1.1*
D	**Chemical kit**	8	NA1760	II	8		154	161	None	1 L	30 L	B	40	154	CORROSIVE
	Chemical kits	9	UN3316		9	15	161	161	None	10 kg	10 kg	A		171	CLASS 9
	Chloral, anhydrous, stabilized	6.1	UN2075	II	6.1	IB2, T7, TP2	153	202	243	5 L	60 L	D	40	153	POISON
	Chlorate and borate mixtures	5.1	UN1458	II	5.1	A9, IB8, IP2, IP4, N34, T3, TP33	152	212	240	5 kg	25 kg	A	56, 58	140	OXIDIZER
				III	5.1	A9, IB8, IP3, N34, T1, TP33	152	213	240	25 kg	100 kg	A	56, 58	140	OXIDIZER
	Chlorate and magnesium chloride mixture solid	5.1	UN1459	II	5.1	A9, IB8, IP2, IP4, N34, T3, TP33	152	212	240	5 kg	25 kg	A	56, 58	140	OXIDIZER
				III	5.1	A9, IB8, IP3, N34, T1, TP33	152	213	240	25 kg	100 kg	A	56, 58	140	OXIDIZER
	Chlorate and magnesium chloride mixture solution	5.1	UN3407	II	5.1	A9, IB2, N34, T4, TP1	152	202	242	1 L	5 L	A	56, 58, 133	140	OXIDIZER
				III	5.1	A9, IB2, N34, T4, TP1	152	203	241	2.5 L	30 L	A	56, 58, 133	140	OXIDIZER
	Chlorate of potash, see **Potassium chlorate**														
	Chlorate of soda, see **Sodium chlorate**														
G	**Chlorates, inorganic, aqueous solution, n.o.s.**	5.1	UN3210	II	5.1	351, IB2, T4, TP1	152	202	242	1 L	5 L	B	56, 58, 133	140	OXIDIZER
				III	5.1	351, IB2, T4, TP1	152	203	241	2.5 L	30 L	B	56, 58, 133	140	OXIDIZER
G	**Chlorates, inorganic, n.o.s.**	5.1	UN1461	II	5.1	351, A9, IB6, IP2, N34, T3, TP33	152	212	242	5 kg	25 kg	A	56, 58	140	OXIDIZER
	Chloric acid aqueous solution, *with not more than 10 percent chloric acid*	5.1	UN2626	II	5.1	IB2, T4, TP1	None	229	None	Forbidden	Forbidden	D	56, 58	140	OXIDIZER
	Chloride of phosphorus, see **Phosphorus trichloride**														
	Chloride of sulfur, see **Sulfur chloride**														
	Chlorinated lime, see **Calcium hypochlorite mixtures,** *etc.*														
	Chlorine	2.3	UN1017		2.3, 5.1, 8	2, B9, B14, N86, T50, TP19	None	304	314, 315	Forbidden	Forbidden	D	40, 51, 55, 62, 68, 89, 90	124	POISON GAS*
	Chlorine azide	Forbidden													
D	**Chlorine dioxide, hydrate, frozen**	5.1	NA9191	II	5.1, 6.1		None	229	None	Forbidden	Forbidden	E		143	OXIDIZER

Symbols (1)	Hazardous materials descriptions and proper shipping names (2)	Hazard class or Division (3)	Identification Numbers (4)	PG (5)	Label Codes (6)	Special Provisions (§172.102) (7)	(8) Packaging (§173.***) Exceptions (8A)	Non-bulk (8B)	Bulk (8C)	(9) Quantity limitations (see §§173.27 and 175.75) Passenger aircraft/rail (9A)	Cargo aircraft only (9B)	(10) Vessel stowage Location (10A)	Other (10B)	ERG Guide No.	Placard Advisory Consult §172.504 & §172.505 to determine placard *Denotes placard for any quantity
	Chlorine dioxide (not hydrate)	Forbidden													
	Chlorine pentafluoride	2.3	UN2548		2.3, 5.1, 8	1, B7, B9, B14, N86	None	304	314	Forbidden	Forbidden	D	40, 89, 90	124	POISON GAS*
	Chlorine trifluoride	2.3	UN1749		2.3, 5.1, 8	2, B7, B9, B14, N86	None	304	314	Forbidden	Forbidden	D	40, 89, 90	124	POISON GAS*
	Chlorite solution	8	UN1908	II	8	A3, A6, A7, B2, IB2, N34, T7, TP2, TP24	154	202	242	1 L	30 L	B	26, 44, 89, 100, 141	154	CORROSIVE
				III	8	A3, A6, A7, B2, IB3, N34, T4, TP2, TP24	154	203	241	5 L	60 L	B	26, 44, 89, 100, 141	154	CORROSIVE
G	**Chlorites, inorganic, n.o.s.**	5.1	UN1462	II	5.1	352, A7, IB6, IP2, N34, T3, TP33	152	212	242	5 kg	25 kg	A	56, 58	143	OXIDIZER
	1-Chloro-1,1-difluoroethane *or* **Refrigerant gas R 142b**	2.1	UN2517		2.1	T50	306	304	314, 315	Forbidden	150 kg	B	40	115	FLAMMABLE GAS
	3-Chloro-4-methylphenyl isocyanate, liquid	6.1	UN2236	II	6.1	IB2	153	202	243	5 L	60 L	B	40	156	POISON
	3-Chloro-4-methylphenyl isocyanate, solid	6.1	UN3428	II	6.1	IB8, IP2, IP4, T3, TP33	153	212	242	25 kg	100 kg	B	40	156	POISON
	1-Chloro-1,2,2,2-tetrafluoroethane *or* **Refrigerant gas R 124**	2.2	UN1021		2.2	T50	306	304	314, 315	75 kg	150 kg	A		126	NONFLAMMABLE GAS
	4-Chloro-o-toluidine hydrochloride, solid	6.1	UN1579	III	6.1	IB8, IP3, T1, TP33	153	213	240	100 kg	200 kg	A		153	POISON
	4-Chloro-o-toluidine hydrochloride, solution	6.1	UN3410	III	6.1	IB3, T4, TP1	153	203	241	60 L	220 L	A		153	POISON
	1-Chloro-2,2,2-trifluoroethane *or* **Refrigerant gas R 133a**	2.2	UN1983		2.2	T50	306	304	314, 315	75 kg	150 kg	A		126	NONFLAMMABLE GAS
	Chloroacetic acid, molten	6.1	UN3250	II	6.1, 8	IB1, T7, TP3, TP28	None	202	243	Forbidden	Forbidden	C	40	153	POISON
	Chloroacetic acid, solid	6.1	UN1751	II	6.1, 8	A3, A7, IB8, IP2, IP4, N34, T3, TP33	153	212	242	15 kg	50 kg	C	40	153	POISON
	Chloroacetic acid, solution	6.1	UN1750	II	6.1, 8	A7, IB2, N34, T7, TP2	153	202	243	1 L	30 L	C	40	153	POISON
	Chloroacetone, stabilized	6.1	UN1695	I	6.1, 3, 8	2, B9, B14, B32, N12, N32, N34, T20, TP2, TP13, TP38, TP45	None	227	244	Forbidden	Forbidden	D	21, 40, 100	131	POISON INHALATION HAZARD*
	Chloroacetone (unstabilized)	Forbidden													
+	**Chloroacetonitrile**	6.1	UN2668	I	6.1, 3	2, B9, B14, B32, IB9, T20, TP2, TP13, TP38, TP45	None	227	244	Forbidden	Forbidden	A	12, 40, 52	131	POISON INHALATION HAZARD*
	Chloroacetophenone, liquid, (*CN*)	6.1	UN3416	II	6.1	A3, IB2, N12, N32, N33, T7, TP2, TP13	None	202	243	Forbidden	60 L	D	12, 40	153	POISON
	Chloroacetophenone, solid, (*CN*)	6.1	UN1697	II	6.1	A3, IB8, IP2, IP4, N12, N32, N33, N34, T3, TP2, TP13, TP33	None	212	None	Forbidden	100 kg	D	12, 40	153	POISON
	Chloroacetyl chloride	6.1	UN1752	I	6.1, 8	2, B3, B8, B9, B14, B32, B77, N34, N43, T20, TP2, TP13, TP38, TP45	None	227	244	Forbidden	Forbidden	D	40	156	POISON INHALATION HAZARD*
	Chloroanilines, liquid	6.1	UN2019	II	6.1	IB2, T7, TP2	153	202	243	5 L	60 L	A	52	152	POISON
	Chloroanilines, solid	6.1	UN2018	II	6.1	IB8, IP2, IP4, T3, TP33	153	212	242	25 kg	100 kg	A		152	POISON
	Chloroanisidines	6.1	UN2233	III	6.1	IB8, IP3, T1, TP33	153	213	240	100 kg	200 kg	A		152	POISON
	Chlorobenzene	3	UN1134	III	3	B1, IB3, T2, TP1	150	203	242	60 L	220 L	A		130	FLAMMABLE
	Chlorobenzol, see **Chlorobenzene**														
	Chlorobenzotrifluorides	3	UN2234	III	3	B1, IB3, T2, TP1	150	203	242	60 L	220 L	A	40	130	FLAMMABLE
	Chlorobenzyl chlorides, liquid	6.1	UN2235	III	6.1	IB3, T4, TP1	153	203	241	60 L	220 L	A		153	POISON
	Chlorobenzyl chlorides, solid	6.1	UN3427	III	6.1	IB8, IP3, T1, TP33	153	213	240	100 kg	200 kg	A		153	POISON
	Chlorobutanes	3	UN1127	II	3	IB2, T4, TP1	150	202	242	5 L	60 L	B		130	FLAMMABLE
	Chlorocresols solution	6.1	UN2669	II	6.1	IB2, T7, TP2	153	202	243	5 L	60 L	A	12	152	POISON
				III	6.1	IB3, T7, TP2	153	203	241	60 L	220 L	A	12	152	POISON
	Chlorocresols, solid	6.1	UN3437	II	6.1	IB8, IP2, IP4, T3, TP33	153	212	242	25 kg	100 kg	A	12	152	POISON

Symbols (1)	Hazardous materials descriptions and proper shipping names (2)	Hazard class or Division (3)	Identification Numbers (4)	PG (5)	Label Codes (6)	Special Provisions (§172.102) (7)	(8) Packaging (§173.***) Exceptions (8A)	Non-bulk (8B)	Bulk (8C)	(9) Quantity limitations (see §§173.27 and 175.75) Passenger aircraft/rail (9A)	Cargo aircraft only (9B)	(10) Vessel stowage Location (10A)	Other (10B)	ERG Guide No.	Placard Advisory Consult §172.504 & §172.505 to determine placard *Denotes placard for any quantity
	Chlorodifluorobromomethane *or* **Refrigerant gas R 12B1**	2.2	UN1974		2.2	T50	306	304	314, 315	75 kg	150 kg	A		126	NONFLAMMABLE GAS
	Chlorodifluoromethane and chloropentafluoroethane mixture *or* **Refrigerant gas R 502** *with fixed boiling point, with approximately 49 percent chlorodifluoromethane*	2.2	UN1973		2.2	T50	306	304	314, 315	75 kg	150 kg	A		126	NONFLAMMABLE GAS
	Chlorodifluoromethane *or* **Refrigerant gas R 22**	2.2	UN1018		2.2	T50	306	304	314, 315	75 kg	150 kg	A		126	NONFLAMMABLE GAS
+	**Chlorodinitrobenzenes, liquid**	6.1	UN1577	II	6.1	IB2, T7, TP2	153	202	243	5 L	60 L	B	91	153	POISON
+	**Chlorodinitrobenzenes, solid**	6.1	UN3441	II	6.1	IB8, IP2, IP4, T3, TP33	153	212	242	25 kg	100 kg	A	91	153	POISON
	2-Chloroethanal	6.1	UN2232	I	6.1	2, B9, B14, B32, T20, TP2, TP13, TP38, TP45	None	227	244	Forbidden	Forbidden	D	40	153	POISON INHALATION HAZARD*
	Chloroform	6.1	UN1888	III	6.1	IB3, N36, T7, TP2	153	203	241	60 L	220 L	A	40	151	POISON
G	**Chloroformates, toxic, corrosive, flammable, n.o.s.**	6.1	UN2742	II	6.1, 8, 3	5, IB1, T7, TP2	153	202	243	1 L	30 L	A	12, 13, 21, 25, 40, 100	155	POISON
G	**Chloroformates, toxic, corrosive, n.o.s.**	6.1	UN3277	II	6.1, 8	IB8, T8, TP2, TP13, TP28	153	202	243	1 L	30 L	A	12, 13, 25, 40	154	POISON
	Chloromethyl chloroformate	6.1	UN2745	II	6.1, 8	IB2, T7, TP2, TP13	153	202	243	1 L	30 L	A	12, 13, 21, 25, 40, 100	157	POISON
	Chloromethyl ethyl ether	3	UN2354	II	3, 6.1	IB2, T7, TP1, TP13	150	202	243	1 L	60 L	E	40	131	FLAMMABLE
	Chloronitroanilines	6.1	UN2237	III	6.1	IB8, IP3, T1, TP33	153	213	240	100 kg	200 kg	A		153	POISON
+	**Chloronitrobenzenes, liquid**	6.1	UN3409	II	6.1	IB2, T7, TP2	153	202	243	5 L	60 L	A		152	POISON
+	**Chloronitrobenzenes, solid**	6.1	UN1578	II	6.1	IB8, IP2, IP4, T3, TP33	153	212	242	25 kg	100 kg	A		152	POISON
	Chloronitrotoluenes, liquid	6.1	UN2433	III	6.1	IB3, T4, TP1	153	203	241	60 L	220 L	A	44, 89, 100, 141	152	POISON
	Chloronitrotoluenes, solid	6.1	UN3457	III	6.1	IB8, IP3, T1, TP33	153	213	240	25 kg	200 kg	A		152	POISON
	Chloropentafluoroethane *or* **Refrigerant gas R 115**	2.2	UN1020		2.2	T50	306	304	314, 315	75 kg	150 kg	A		126	NONFLAMMABLE GAS
	Chlorophenolates, liquid *or* **Phenolates, liquid**	8	UN2904	III	8	IB3	154	203	241	5 L	60 L	A		154	CORROSIVE
	Chlorophenolates, solid *or* **Phenolates, solid**	8	UN2905	III	8	IB8, IP3, T1, TP33	154	213	240	25 kg	100 kg	A		154	CORROSIVE
	Chlorophenols, liquid	6.1	UN2021	III	6.1	IB3, T4, TP1	153	203	241	60 L	220 L	A		153	POISON
	Chlorophenols, solid	6.1	UN2020	III	6.1	IB8, IP3, T1, TP1, TP33	153	213	240	100 kg	200 kg	A		153	POISON
	Chlorophenyltrichlorosilane	8	UN1753	II	8	A7, B2, B6, N34, T10, TP2, TP7	None	206	242	Forbidden	30 L	C	40	156	CORROSIVE
+	**Chloropicrin**	6.1	UN1580	I	6.1	2, B7, B9, B14, B32, B46, T22, TP2, TP13, TP38, TP45	None	227	244	Forbidden	Forbidden	D	40	154	POISON INHALATION HAZARD*
	Chloropicrin and methyl bromide mixtures	2.3	UN1581		2.3	2, B9, B14, N86, T50	None	193	314, 315	Forbidden	Forbidden	D	25, 40	123	POISON GAS*
	Chloropicrin and methyl chloride mixtures	2.3	UN1582		2.3	2, N86, T50	None	193	245	Forbidden	Forbidden	D	25, 40	119	POISON GAS*
	Chloropicrin mixture, flammable (pressure not exceeding 14.7 psia at 115 degrees F flash point below 100 degrees F) see **Toxic liquids, flammable,** *etc.*														
G	**Chloropicrin mixtures, n.o.s.**	6.1	UN1583	I	6.1	5	None	201	243	Forbidden	Forbidden	C	40	154	POISON INHALATION HAZARD*
				II	6.1	IB2	153	202	243	Forbidden	Forbidden	C	40	154	POISON
				III	6.1	IB3	153	203	241	Forbidden	Forbidden	C	40	154	POISON
D	**Chloropivaloyl chloride**	6.1	NA9263	I	6.1, 8	2, B9, B14, B32, T20, TP4, TP13, TP38, TP45	None	227	244	Forbidden	Forbidden	B	40	156	POISON INHALATION HAZARD*
	Chloroplatinic acid, solid	8	UN2507	III	8	IB8, IP3, T1, TP33	154	213	240	25 kg	100 kg	A		154	CORROSIVE
	Chloroprene, stabilized	3	UN1991	I	3, 6.1	B57, T14, TP2, TP13	None	201	243	Forbidden	30 L	D	40	131P	FLAMMABLE
	Chloroprene, uninhibited	Forbidden													
	1-Chloropropane	3	UN1278	II	3	IB2, IP8, N34, T7, TP2	None	202	242	Forbidden	60 L	E		129	FLAMMABLE

Symbols (1)	Hazardous materials descriptions and proper shipping names (2)	Hazard class or Division (3)	Identification Numbers (4)	PG (5)	Label Codes (6)	Special Provisions (§172.102) (7)	(8) Packaging (§173.***) Exceptions (8A)	Non-bulk (8B)	Bulk (8C)	(9) Quantity limitations (see §§173.27 and 175.75) Passenger aircraft/rail (9A)	Cargo aircraft only (9B)	(10) Vessel stowage Location (10A)	Other (10B)	ERG Guide No.	Placard Advisory Consult §172.504 & §172.505 to determine placard *Denotes placard for any quantity
	2-Chloropropane	3	UN2356	I	3	N36, T11, TP2, TP13	150	201	243	1 L	30 L	E		129	FLAMMABLE
	3-Chloropropanol-1	6.1	UN2849	III	6.1	IB3, T4, TP1	153	203	241	60 L	220 L	A		153	POISON
	2-Chloropropene	3	UN2456	I	3	A3, N36, T11, TP2	150	201	243	1 L	30 L	E		130P	FLAMMABLE
	2-Chloropropionic acid	8	UN2511	III	8	IB3, T4, TP2	154	203	241	5 L	60 L	A	8	153	CORROSIVE
	2-Chloropyridine	6.1	UN2822	II	6.1	IB2, T7, TP2	153	202	243	5 L	60 L	A	40	153	POISON
	Chlorosilanes, corrosive, flammable, n.o.s.	8	UN2986	II	8, 3	T14, TP2, TP7, TP13, TP27	None	206	243	Forbidden	30 L	C	40	155	CORROSIVE
	Chlorosilanes, corrosive, n.o.s.	8	UN2987	II	8	B2, T14, TP2, TP7, TP13, TP27	None	206	242	Forbidden	30 L	C	40	156	CORROSIVE
	Chlorosilanes, flammable, corrosive, n.o.s.	3	UN2985	II	3, 8	T14, TP2, TP7, TP13, TP27	None	206	243	1 L	5 L	B	40	155	FLAMMABLE
G	**Chlorosilanes, toxic, corrosive, flammable, n.o.s.**	6.1	UN3362	II	6.1, 3, 8	T14, TP2, TP7, TP13, TP27	None	206	243	1 L	30 L	C	40, 125	155	POISON
G	**Chlorosilanes, toxic, corrosive, n.o.s.**	6.1	UN3361	II	6.1, 8	T14, TP2, TP7, TP13, TP27	None	206	243	1 L	30 L	C	40	156	POISON
	Chlorosilanes, water-reactive, flammable, corrosive, n.o.s.	4.3	UN2988	I	4.3, 3, 8	A2, T14, TP2, TP7, TP13	None	201	244	Forbidden	1 L	D	21, 28, 40, 49, 100	139	DANGEROUS WHEN WET*
+	**Chlorosulfonic acid** *(with or without sulfur trioxide)*	8	UN1754	I	8, 6.1	2, B9, B10, B14, B32, T20, TP2, TP38, TP45	None	227	244	Forbidden	Forbidden	C	40	137	CORROSIVE, POISON INHALATION HAZARD*
	Chlorotoluenes	3	UN2238	III	3	B1, IB3, T2, TP1	150	203	242	60 L	220 L	A		129	FLAMMABLE
	Chlorotoluidines, liquid	6.1	UN3429	III	6.1	IB3, T4, TP1	153	203	241	60 L	220 L	A		153	POISON
	Chlorotoluidines, solid	6.1	UN2239	III	6.1	IB8, IP3, T1, TP33	153	213	240	100 kg	200 kg	A		153	POISON
	Chlorotrifluoromethane and trifluoromethane azeotropic mixture *or* **Refrigerant gas R 503** *with approximately 60 percent chlorotrifluoromethane*	2.2	UN2599		2.2		306	304	314, 315	75 kg	150 kg	A		126	NONFLAMMABLE GAS
	Chlorotrifluoromethane *or* **Refrigerant gas R 13**	2.2	UN1022		2.2		306	304	314, 315	75 kg	150 kg	A		126	NONFLAMMABLE GAS
	Chromic acid solution	8	UN1755	II	8	B2, IB2, T8, TP2	154	202	242	1 L	30 L	C	40, 44, 89, 100, 141	154	CORROSIVE
				III	8	IB3, T4, TP1	154	203	241	5 L	60 L	C	40, 44, 89, 100, 141	154	CORROSIVE
	Chromic anhydride, see **Chromium trioxide, anhydrous**														
	Chromic fluoride, solid	8	UN1756	II	8	IB8, IP2, IP4, T3, TP33	154	212	240	15 kg	50 kg	A	26	154	CORROSIVE
	Chromic fluoride, solution	8	UN1757	II	8	B2, IB2, T7, TP2	154	202	242	1 L	30 L	A		154	CORROSIVE
				III	8	IB3, T4, TP1	154	203	241	5 L	60 L	A		154	CORROSIVE
	Chromium nitrate	5.1	UN2720	III	5.1	A1, A29, IB8, IP3, T1, TP33	152	213	240	25 kg	100 kg	A		141	OXIDIZER
	Chromium oxychloride	8	UN1758	I	8	A3, A6, A7, B10, N34, T10, TP2	None	201	243	0.5 L	2.5 L	C	40, 66, 74, 89, 90	137	CORROSIVE
	Chromium trioxide, anhydrous	5.1	UN1463	II	5.1, 6.1, 8	IB8, IP2, IP4, T3, TP33	None	212	242	5 kg	25 kg	A	66, 90	141	OXIDIZER
	Chromosulfuric acid	8	UN2240	I	8	A3, A6, A7, B4, B6, N34, T10, TP2, TP13	None	201	243	0.5 L	2.5 L	B	40, 66, 74, 89, 90	154	CORROSIVE
	Chromyl chloride, see **Chromium oxychloride**														
	Cigar and cigarette lighters, charged with fuel, see **Lighters** *or* **Lighter refills** *containing flammable gas*														
	Coal briquettes, hot	Forbidden													
	Coal gas, compressed	2.3	UN1023		2.3, 2.1	3	None	302	314, 315	Forbidden	Forbidden	D	40	119	POISON GAS*
	Coal tar distillates, flammable	3	UN1136	II	3	IB2, T4, TP1	150	202	242	5 L	60 L	B		128	FLAMMABLE
				III	3	B1, IB3, T4, TP1, TP29	150	203	242	60 L	220 L	A		128	FLAMMABLE
	Coal tar dye, corrosive, liquid, n.o.s, see **Dyes, liquid** *or* **solid, n.o.s.** *or* **Dye intermediates, liquid** *or* **solid, n.o.s.,** *corrosive*														

Symbols (1)	Hazardous materials descriptions and proper shipping names (2)	Hazard class or Division (3)	Identification Numbers (4)	PG (5)	Label Codes (6)	Special Provisions (§172.102) (7)	(8) Packaging (§173.***) Exceptions (8A)	Non-bulk (8B)	Bulk (8C)	(9) Quantity limitations (see §§173.27 and 175.75) Passenger aircraft/rail (9A)	Cargo aircraft only (9B)	(10) Vessel stowage Location (10A)	Other (10B)	ERG Guide No.	Placard Advisory Consult §172.504 & §172.505 to determine placard *Denotes placard for any quantity
	Coating solution *(includes surface treatments or coatings used for industrial or other purposes such as vehicle under-coating, drum or barrel lining)*	3	UN1139	I	3	T11, TP1, TP8, TP27	150	201	243	1 L	30 L	E		127	FLAMMABLE
				II	3	149, IB2, T4, TP1, TP8	150	202	242	5 L	60 L	B		127	FLAMMABLE
				III	3	B1, IB3, T2, TP1	150	203	242	60 L	220 L	A		127	FLAMMABLE
	Cobalt naphthenates, powder	4.1	UN2001	III	4.1	A19, IB8, IP3, T1, TP33	151	213	240	25 kg	100 kg	A		133	FLAMMABLE SOLID
	Cobalt resinate, precipitated	4.1	UN1318	III	4.1	A1, A19, IB6, T1, TP33	151	213	240	25 kg	100 kg	A		133	FLAMMABLE SOLID
	Coke, hot	Forbidden													
	Collodion, see **Nitrocellulose** *etc.*														
D G	**Combustible liquid, n.o.s.**	Comb liq	NA1993	III	None	IB3, T1, T4, TP1	150	203	241	60 L	220 L	A		128	COMBUSTIBLE (BULK ONLY)
G	**Components, explosive train, n.o.s.**	1.2B	UN0382	II	1.2B		None	62	None	Forbidden	Forbidden	11		112	EXPLOSIVES 1.2*
G	**Components, explosive train, n.o.s.**	1.4B	UN0383	II	1.4B		None	62	None	Forbidden	75 kg	06		114	EXPLOSIVES 1.4
G	**Components, explosive train, n.o.s.**	1.4S	UN0384	II	1.4S		None	62	None	25 kg	100 kg	05		114	EXPLOSIVES 1.4
G	**Components, explosive train, n.o.s.**	1.1B	UN0461	II	1.1B		None	62	None	Forbidden	Forbidden	11		112	EXPLOSIVES 1.1*
	Composition B, see **Hexolite**, *etc.*														
D G	**Compounds, cleaning liquid**	8	NA1760	I	8	A7, B10, T14, TP2, TP27	None	201	243	0.5 L	2.5 L	B	40	154	CORROSIVE
				II	8	B2, IB2, N37, T11, TP2, TP27	154	202	242	1 L	30 L	B	40	154	CORROSIVE
				III	8	IB3, N37, T7, TP1, TP28	154	203	241	5 L	60 L	A	40	154	CORROSIVE
D G	**Compounds, cleaning liquid**	3	NA1993	I	3	T11, TP1	150	201	243	1 L	30 L	E		128	FLAMMABLE
				II	3	IB2, T7, TP1, TP8, TP28	150	202	242	5 L	60 L	B		128	FLAMMABLE
				III	3	B1, B52, IB3, T4, TP1, TP29	150	203	242	60 L	220 L	A		128	FLAMMABLE
D G	**Compounds, tree killing, liquid** *or* **Compounds, weed killing, liquid**	8	NA1760	I	8	A7, B10, T14, TP2, TP27	None	201	243	0.5 L	2.5 L	B	40	154	CORROSIVE
				II	8	B2, IB2, N37, T11, TP2, TP27	154	202	242	1 L	30 L	B	40	154	CORROSIVE
				III	8	IB3, N37, T7, TP1, TP28	154	203	241	5 L	60 L	A	40	154	CORROSIVE
D G	**Compounds, tree killing, liquid** *or* **Compounds, weed killing, liquid**	3	NA1993	I	3	T11, TP1	150	201	243	1 L	30 L	E		128	FLAMMABLE
				II	3	IB2, T7, TP1, TP8, TP28	150	202	242	5 L	60 L	B		128	FLAMMABLE
				III	3	B1, B52, IB3, T4, TP1, TP29	150	203	242	60 L	220 L	A		128	FLAMMABLE
D G	**Compounds, tree killing, liquid** *or* **Compounds, weed killing, liquid**	6.1	NA2810	I	6.1	T14, TP2, TP13, TP27	None	201	243	1 L	30 L	B	40	153	POISON
				II	6.1	IB2, T11, TP2, TP27	153	202	243	5 L	60 L	B	40	153	POISON
				III	6.1	IB3, T7, TP1, TP28	153	203	241	60 L	220 L	A	40	153	POISON
G	**Compressed gas, flammable, n.o.s.**	2.1	UN1954		2.1		306	302, 305	314, 315	Forbidden	150 kg	D	40	115	FLAMMABLE GAS
G	**Compressed gas, n.o.s.**	2.2	UN1956		2.2		306, 307	302, 305	314, 315	75 kg	150 kg	A		126	NONFLAMMABLE GAS
G	**Compressed gas, oxidizing, n.o.s.**	2.2	UN3156		2.2, 5.1	A14	306	302	314, 315	75 kg	150 kg	D		122	NONFLAMMABLE GAS
G I	**Compressed gas, toxic, corrosive, n.o.s.** *Inhalation Hazard Zone A*	2.3	UN3304		2.3, 8	1	None	192	245	Forbidden	Forbidden	D	40	123	POISON GAS*
G I	**Compressed gas, toxic, corrosive, n.o.s.** *Inhalation Hazard Zone B*	2.3	UN3304		2.3, 8	2, B9, B14	None	302, 305	314, 315	Forbidden	Forbidden	D	40	123	POISON GAS*
G I	**Compressed gas, toxic, corrosive, n.o.s.** *Inhalation Hazard Zone C*	2.3	UN3304		2.3, 8	3, B14	None	302, 305	314, 315	Forbidden	Forbidden	D	40	123	POISON GAS*

Symbols (1)	Hazardous materials descriptions and proper shipping names (2)	Hazard class or Division (3)	Identification Numbers (4)	PG (5)	Label Codes (6)	Special Provisions (§172.102) (7)	(8) Packaging (§173.***) Exceptions (8A)	Non-bulk (8B)	Bulk (8C)	(9) Quantity limitations (see §§173.27 and 175.75) Passenger aircraft/rail (9A)	Cargo aircraft only (9B)	(10) Vessel stowage Location (10A)	Other (10B)	ERG Guide No.	Placard Advisory Consult §172.504 & §172.505 to determine placard *Denotes placard for any quantity
G I	**Compressed gas, toxic, corrosive, n.o.s.** *Inhalation Hazard Zone D*	2.3	UN3304		2.3, 8	4	None	302, 305	314, 315	Forbidden	Forbidden	D	40	123	POISON GAS*
G I	**Compressed gas, toxic, flammable, corrosive, n.o.s** *Inhalation Hazard Zone A*	2.3	UN3305		2.3, 2.1, 8	1	None	192	245	Forbidden	Forbidden	D	17, 40	119	POISON GAS*
G I	**Compressed gas, toxic, flammable, corrosive, n.o.s** *Inhalation Hazard Zone B*	2.3	UN3305		2.3, 2.1, 8	2, B9, B14	None	302, 305	314, 315	Forbidden	Forbidden	D	17, 40	119	POISON GAS*
G I	**Compressed gas, toxic, flammable, corrosive, n.o.s** *Inhalation Hazard Zone C*	2.3	UN3305		2.3, 2.1, 8	3, B14	None	302, 305	314, 315	Forbidden	Forbidden	D	17, 40	119	POISON GAS*
G I	**Compressed gas, toxic, flammable, corrosive, n.o.s** *Inhalation Hazard Zone D*	2.3	UN3305		2.3, 2.1, 8	4	None	302, 305	314, 315	Forbidden	Forbidden	D	17, 40	119	POISON GAS*
G	**Compressed gas, toxic, flammable, n.o.s.** *Inhalation hazard Zone A*	2.3	UN1953		2.3, 2.1	1	None	192	245	Forbidden	Forbidden	D	40	119	POISON GAS*
G	**Compressed gas, toxic, flammable, n.o.s.** *Inhalation hazard Zone B*	2.3	UN1953		2.3, 2.1	2, B9, B14	None	302, 305	314, 315	Forbidden	Forbidden	D	40	119	POISON GAS*
G	**Compressed gas, toxic, flammable, n.o.s.** *Inhalation Hazard Zone C*	2.3	UN1953		2.3, 2.1	3, B14	None	302, 305	314, 315	Forbidden	Forbidden	D	40	119	POISON GAS*
G	**Compressed gas, toxic, flammable, n.o.s.** *Inhalation Hazard Zone D*	2.3	UN1953		2.3, 2.1	4	None	302, 305	314, 315	Forbidden	Forbidden	D	40	119	POISON GAS*
G	**Compressed gas, toxic, n.o.s.** *Inhalation Hazard Zone A*	2.3	UN1955		2.3	1	None	192	245	Forbidden	Forbidden	D	40	123	POISON GAS*
G	**Compressed gas, toxic, n.o.s.** *Inhalation Hazard Zone B*	2.3	UN1955		2.3	2, B9, B14	None	302, 305	314, 315	Forbidden	Forbidden	D	40	123	POISON GAS*
G	**Compressed gas, toxic, n.o.s.** *Inhalation Hazard Zone C*	2.3	UN1955		2.3	3, B14	None	302, 305	314, 315	Forbidden	Forbidden	D	40	123	POISON GAS*
G	**Compressed gas, toxic, n.o.s.** *Inhalation Hazard Zone D*	2.3	UN1955		2.3	4	None	302, 305	314, 315	Forbidden	Forbidden	D	40	123	POISON GAS*
G I	**Compressed gas, toxic, oxidizing, corrosive, n.o.s** *Inhalation Hazard Zone A*	2.3	UN3306		2.3, 5.1, 8	1	None	192	244	Forbidden	Forbidden	D	40, 89, 90	124	POISON GAS*
G I	**Compressed gas, toxic, oxidizing, corrosive, n.o.s** *Inhalation Hazard Zone B*	2.3	UN3306		2.3, 5.1, 8	2, B9, B14	None	302, 305	314, 315	Forbidden	Forbidden	D	40, 89, 90	124	POISON GAS*
G I	**Compressed gas, toxic, oxidizing, corrosive, n.o.s** *Inhalation Hazard Zone C*	2.3	UN3306		2.3, 5.1, 8	3, B14	None	302, 305	314, 315	Forbidden	Forbidden	D	40, 89, 90	124	POISON GAS*
G I	**Compressed gas, toxic, oxidizing, corrosive, n.o.s** *Inhalation Hazard Zone D*	2.3	UN3306		2.3, 5.1, 8	4	None	302, 305	314, 315	Forbidden	Forbidden	D	40, 89, 90	124	POISON GAS*
G	**Compressed gas, toxic, oxidizing, n.o.s** *Inhalation Hazard Zone A*	2.3	UN3303		2.3, 5.1	1	None	192	245	Forbidden	Forbidden	D	40	124	POISON GAS*
G	**Compressed gas, toxic, oxidizing, n.o.s** *Inhalation Hazard Zone B*	2.3	UN3303		2.3, 5.1	2, B9, B14	None	302, 305	314, 315	Forbidden	Forbidden	D	40	124	POISON GAS*
G	**Compressed gas, toxic, oxidizing, n.o.s** *Inhalation Hazard Zone C*	2.3	UN3303		2.3, 5.1	3, B14	None	302, 305	314, 315	Forbidden	Forbidden	D	40	124	POISON GAS*
G	**Compressed gas, toxic, oxidizing, n.o.s** *Inhalation Hazard Zone D*	2.3	UN3303		2.3, 5.1	4	None	302, 305	314, 315	Forbidden	Forbidden	D	40	124	POISON GAS*
D	**Consumer commodity**	ORM-D			None		156, 306	156, 306	None	30 kg gross	30 kg gross	A		—	NONE
	Consumer commodity	9	ID8000		9		167	167	None	30 kg gross	30 kg gross			171	CLASS 9
G	**Contrivances, water-activated,** *with burster, expelling charge or propelling charge*	1.2L	UN0248	II	1.2L		None	62	None	Forbidden	Forbidden	08	8E, 14E, 15E, 17E	112	EXPLOSIVES 1.2*
G	**Contrivances, water-activated,** *with burster, expelling charge or propelling charge*	1.3L	UN0249	II	1.3L		None	62	None	Forbidden	Forbidden	08	8E, 14E, 15E, 17E	112	EXPLOSIVES 1.3*
	Copper acetoarsenite	6.1	UN1585	II	6.1	IB8, IP2, IP4, T3, TP33	None	212	242	25 kg	100 kg	A		151	POISON
	Copper acetylide	Forbidden													
	Copper amine azide	Forbidden													
	Copper arsenite	6.1	UN1586	II	6.1	IB8, IP2, IP4, T3, TP33	153	212	242	25 kg	100 kg	A		151	POISON
	Copper based pesticides, liquid, flammable, toxic, *flash point less than 23 degrees C*	3	UN2776	I	3, 6.1	T14, TP2, TP13, TP27	None	201	243	Forbidden	30 L	B	40	131	FLAMMABLE
				II	3, 6.1	IB2, T11, TP2, TP13, TP27	150	202	243	1 L	60 L	B	40	131	FLAMMABLE
	Copper based pesticides, liquid, toxic	6.1	UN3010	I	6.1	T14, TP2, TP13, TP27	None	201	243	1 L	30 L	B	40	151	POISON

Symbols (1)	Hazardous materials descriptions and proper shipping names (2)	Hazard class or Division (3)	Identification Numbers (4)	PG (5)	Label Codes (6)	Special Provisions (§172.102) (7)	(8) Packaging (§173.***) Exceptions (8A)	Non-bulk (8B)	Bulk (8C)	(9) Quantity limitations (see §§173.27 and 175.75) Passenger aircraft/rail (9A)	Cargo aircraft only (9B)	(10) Vessel stowage Location (10A)	Other (10B)	ERG Guide No.	Placard Advisory Consult §172.504 & §172.505 to determine placard *Denotes placard for any quantity
				II	6.1	IB2, T11, TP2, TP13, TP27	153	202	243	5 L	60 L	B	40	151	POISON
				III	6.1	IB3, T7, TP2, TP28	153	203	241	60 L	220 L	A	40	151	POISON
	Copper based pesticides, liquid, toxic, flammable *flashpoint not less than 23 degrees C*	6.1	UN3009	I	6.1, 3	T14, TP2, TP13, TP27	None	201	243	1 L	30 L	B	40	131	POISON
				II	6.1, 3	IB2, T11, TP2, TP13, TP27	153	202	243	5 L	60 L	B	40	131	POISON
				III	6.1, 3	B1, IB3, T7, TP2, TP28	153	203	242	60 L	220 L	A	40	131	POISON
	Copper based pesticides, solid, toxic	6.1	UN2775	I	6.1	IB7, IP1, T6, TP33	None	211	242	5 kg	50 kg	A	40	151	POISON
				II	6.1	IB8, IP2, IP4, T3, TP33	153	212	242	25 kg	100 kg	A	40	151	POISON
				III	6.1	IB8, IP3, T1, TP33	153	213	240	100 kg	200 kg	A	40	151	POISON
	Copper chlorate	5.1	UN2721	II	5.1	A1, IB8, IP2, IP4, T3, TP33	152	212	242	5 kg	25 kg	A	56, 58	141	OXIDIZER
	Copper chloride	8	UN2802	III	8	IB8, IP3, T1, TP33	154	213	240	25 kg	100 kg	A		154	CORROSIVE
	Copper cyanide	6.1	UN1587	II	6.1	IB8, IP2, IP4, T3, TP33	153	204	242	25 kg	100 kg	A	52	151	POISON
	Copper selenate, see **Selenates** *or* **Selenites**														
	Copper selenite, see **Selenates** *or* **Selenites**														
	Copper tetramine nitrate	Forbidden													
A W	**Copra**	4.2	UN1363	III	4.2	IB8, IP3, IP7	None	213	241	Forbidden	Forbidden	A	13, 19, 48, 119	135	SPONTANEOUSLY COMBUSTIBLE
	Cord, detonating, *flexible*	1.1D	UN0065	II	1.1D	102	63(a)	62	None	Forbidden	Forbidden	07		112	EXPLOSIVES 1.1*
	Cord, detonating, *flexible*	1.4D	UN0289	II	1.4D		None	62	None	Forbidden	75 kg	06		114	EXPLOSIVES 1.4
	Cord detonating *or* **Fuse detonating** *metal clad*	1.2D	UN0102	II	1.2D		None	62	None	Forbidden	Forbidden	07		112	EXPLOSIVES 1.2*
	Cord, detonating *or* **Fuse, detonating** *metal clad*	1.1D	UN0290	II	1.1D		None	62	None	Forbidden	Forbidden	07		112	EXPLOSIVES 1.1*
	Cord, detonating, mild effect *or* **Fuse, detonating, mild effect** *metal clad*	1.4D	UN0104	II	1.4D		None	62	None	Forbidden	75 kg	06		114	EXPLOSIVES 1.4
	Cord, igniter	1.4G	UN0066	II	1.4G		None	62	None	Forbidden	75 kg	06		114	EXPLOSIVES 1.4
	Cordeau detonant fuse, see **Cord, detonating,** *etc;* **Cord, detonating,** *flexible*														
	Cordite, see **Powder, smokeless**														
G	**Corrosive liquid, acidic, inorganic, n.o.s.**	8	UN3264	I	8	A6, B10, T14, TP2, TP27	None	201	243	0.5 L	2.5 L	B	40	154	CORROSIVE
				II	8	B2, IB2, T11, TP2, TP27	154	202	242	1 L	30 L	B	40	154	CORROSIVE
				III	8	IB3, T7, TP1, TP28	154	203	241	5 L	60 L	A	40	154	CORROSIVE
G	**Corrosive liquid, acidic organic, n.o.s.**	8	UN3265	I	8	A6, B10, T14, TP2, TP27	None	201	243	0.5 L	2.5 L	B	40	153	CORROSIVE
				II	8	B2, IB2, T11, TP2, TP27	154	202	242	1 L	30 L	B	40	153	CORROSIVE
				III	8	IB3, T7, TP1, TP28	154	203	241	5 L	60 L	A	40	153	CORROSIVE
G	**Corrosive liquid, basic, inorganic, n.o.s.**	8	UN3266	I	8	A6, T14, TP2, TP27	None	201	243	0.5 L	2.5 L	B	40, 52	154	CORROSIVE
				II	8	B2, IB2, T11, TP2, TP27	154	202	242	1 L	30 L	B	40, 52	154	CORROSIVE
				III	8	IB3, T7, TP1, TP28	154	203	241	5 L	60 L	A	40, 52	154	CORROSIVE
G	**Corrosive liquid, basic, organic, n.o.s.**	8	UN3267	I	8	A6, B10, T14, TP2, TP27	None	201	243	0.5 L	2.5 L	B	40, 52	153	CORROSIVE
				II	8	B2, IB2, T11, TP2, TP27	154	202	242	1 L	30 L	B	40, 52	153	CORROSIVE
				III	8	IB3, T7, TP1, TP28	154	203	241	5 L	60 L	A	40, 52	153	CORROSIVE
G	**Corrosive liquid, self-heating, n.o.s.**	8	UN3301	I	8, 4.2	A6, B10	None	201	243	0.5 L	2.5 L	D		136	CORROSIVE
				II	8, 4.2	B2, IB1	154	202	242	1 L	30 L	D		136	CORROSIVE
G	**Corrosive liquids, flammable, n.o.s.**	8	UN2920	I	8, 3	A6, B10, T14, TP2, TP27	None	201	243	0.5 L	2.5 L	C	25, 40	132	CORROSIVE

Symbols (1)	Hazardous materials descriptions and proper shipping names (2)	Hazard class or Division (3)	Identification Numbers (4)	PG (5)	Label Codes (6)	Special Provisions (§172.102) (7)	(8) Packaging (§173.***) Exceptions (8A)	Non-bulk (8B)	Bulk (8C)	(9) Quantity limitations (see §§173.27 and 175.75) Passenger aircraft/rail (9A)	Cargo aircraft only (9B)	(10) Vessel stowage Location (10A)	Other (10B)	ERG Guide No.	Placard Advisory Consult §172.504 & §172.505 to determine placard *Denotes placard for any quantity
				II	8, 3	B2, IB2, T11, TP2, TP27	None	202	243	1 L	30 L	C	25, 40	132	CORROSIVE
G	**Corrosive liquids, n.o.s.**	8	UN1760	I	8	A6, A7, B10, T14, TP2, TP27	None	201	243	0.5 L	2.5 L	B	40	154	CORROSIVE
				II	8	B2, IB2, T11, TP2, TP27	154	202	242	1 L	30 L	B	40	154	CORROSIVE
				III	8	IB3, T7, TP1, TP28	154	203	241	5 L	60 L	A	40	154	CORROSIVE
G	**Corrosive liquids, oxidizing, n.o.s.**	8	UN3093	I	8, 5.1	A6, A7	None	201	243	Forbidden	2.5 L	C	89	140	CORROSIVE
				II	8, 5.1	A6, A7, IB2	None	202	243	1 L	30 L	C	89	140	CORROSIVE
G	**Corrosive liquids, toxic, n.o.s.**	8	UN2922	I	8, 6.1	A6, A7, B10, T14, TP2, TP13, TP27	None	201	243	0.5 L	2.5 L	B	40	154	CORROSIVE
				II	8, 6.1	B3, IB2, T7, TP2	154	202	243	1 L	30 L	B	40	154	CORROSIVE
				III	8, 6.1	IB3, T7, TP1, TP28	154	203	241	5 L	60 L	B	40	154	CORROSIVE
G	**Corrosive liquids, water-reactive, n.o.s.**	8	UN3094	I	8, 4.3	A6, A7	None	201	243	Forbidden	1 L	E		138	CORROSIVE, DANGEROUS WHEN WET*
				II	8, 4.3	A6, A7	None	202	243	1 L	5 L	E		138	CORROSIVE, DANGEROUS WHEN WET*
G	**Corrosive solid, acidic, inorganic, n.o.s.**	8	UN3260	I	8	IB7, IP1, T6, TP33	None	211	242	1 kg	25 kg	B		154	CORROSIVE
				II	8	IB8, IP2, IP4, T3, TP33	154	212	240	15 kg	50 kg	B		154	CORROSIVE
				III	8	IB8, IP3, T1, TP33	154	213	240	25 kg	100 kg	A		154	CORROSIVE
G	**Corrosive solid, acidic, organic, n.o.s.**	8	UN3261	I	8	IB7, IP1, T6, TP33	None	211	242	1 kg	25 kg	B		154	CORROSIVE
				II	8	IB8, IP2, IP4, T3, TP33	154	212	240	15 kg	50 kg	B		154	CORROSIVE
				III	8	IB8, IP3, T1, TP33	154	213	240	25 kg	100 kg	A		154	CORROSIVE
G	**Corrosive solid, basic, inorganic, n.o.s.**	8	UN3262	I	8	IB7, IP1, T6, TP33	None	211	242	1 kg	25 kg	B	52	154	CORROSIVE
				II	8	IB8, IP2, IP4, T3, TP33	154	212	240	15 kg	50 kg	B	52	154	CORROSIVE
				III	8	IB8, IP3, T1, TP33	154	213	240	25 kg	100 kg	A	52	154	CORROSIVE
G	**Corrosive solid, basic, organic, n.o.s.**	8	UN3263	I	8	IB7, IP1, T6, TP33	None	211	242	1 kg	25 kg	B	52	154	CORROSIVE
				II	8	IB8, IP2, IP4, T3, TP33	154	212	240	15 kg	50 kg	B	52	154	CORROSIVE
				III	8	IB8, IP3, T1, TP33	154	213	240	25 kg	100 kg	A	52	154	CORROSIVE
G	**Corrosive solids, flammable, n.o.s.**	8	UN2921	I	8, 4.1	IB6, T6, TP33	None	211	242	1 kg	25 kg	B	12, 25	134	CORROSIVE
				II	8, 4.1	IB8, IP2, IP4, T3, TP33	None	212	242	15 kg	50 kg	B	12, 25	134	CORROSIVE
G	**Corrosive solids, n.o.s.**	8	UN1759	I	8	IB7, IP1, T6, TP33	None	211	242	1 kg	25 kg	B		154	CORROSIVE
				II	8	128, IB8, IP2, IP4, T3, TP33	154	212	240	15 kg	50 kg	A		154	CORROSIVE
				III	8	128, IB8, IP3, T1, TP33	154	213	240	25 kg	100 kg	A		154	CORROSIVE
G	**Corrosive solids, oxidizing, n.o.s.**	8	UN3084	I	8, 5.1	T6, TP33	None	211	242	1 kg	25 kg	C		140	CORROSIVE
				II	8, 5.1	IB6, IP2, T3, TP33	None	212	242	15 kg	50 kg	C		140	CORROSIVE
G	**Corrosive solids, self-heating, n.o.s.**	8	UN3095	I	8, 4.2	T6, TP33	None	211	243	1 kg	25 kg	C		136	CORROSIVE
				II	8, 4.2	IB6, IP2, T3, TP33	None	212	242	15 kg	50 kg	C		136	CORROSIVE
G	**Corrosive solids, toxic, n.o.s.**	8	UN2923	I	8, 6.1	IB7, T6, TP33	None	211	242	1 kg	25 kg	B	40	154	CORROSIVE
				II	8, 6.1	IB8, IP2, IP4, T3, TP33	154	212	240	15 kg	50 kg	B	40	154	CORROSIVE
				III	8, 6.1	IB8, IP3, T1, TP33	154	213	240	25 kg	100 kg	B	40, 95	154	CORROSIVE
G	**Corrosive solids, water-reactive, n.o.s.**	8	UN3096	I	8, 4.3	IB4, IP1, T6, TP33	None	211	243	1 kg	25 kg	D		138	CORROSIVE, DANGEROUS WHEN WET*
				II	8, 4.3	IB6, IP2, T3, TP33	None	212	242	15 kg	50 kg	D		138	CORROSIVE, DANGEROUS WHEN WET*

Symbols (1)	Hazardous materials descriptions and proper shipping names (2)	Hazard class or Division (3)	Identification Numbers (4)	PG (5)	Label Codes (6)	Special Provisions (§172.102) (7)	(8) Packaging (§173.***) Exceptions (8A)	Non-bulk (8B)	Bulk (8C)	(9) Quantity limitations (see §§173.27 and 175.75) Passenger aircraft/rail (9A)	Cargo aircraft only (9B)	(10) Vessel stowage Location (10A)	Other (10B)	ERG Guide No.	Placard Advisory Consult §172.504 & §172.505 to determine placard *Denotes placard for any quantity
D W	**Cotton**	9	NA1365		9	137, IB8, IP2, IP4, W41	None	None	None	No limit	No limit	A		133	CLASS 9
A W	**Cotton waste, oily**	4.2	UN1364	III	4.2	IB8, IP3, IP7	None	213	None	Forbidden	Forbidden	A	54	133	SPONTANEOUSLY COMBUSTIBLE
A I W	**Cotton, wet**	4.2	UN1365	III	4.2	IB8, IP3, IP7	None	204	241	Forbidden	Forbidden	A		133	SPONTANEOUSLY COMBUSTIBLE
	Coumarin derivative pesticides, liquid, flammable, toxic, *flashpoint less than 23 degrees C*	3	UN3024	I	3, 6.1	T14, TP2, TP13, TP27	None	201	243	Forbidden	30 L	B	40	131	FLAMMABLE
				II	3, 6.1	IB2, T11, TP2, TP13, TP27	150	202	243	1 L	60 L	B	40	131	FLAMMABLE
	Coumarin derivative pesticides, liquid, toxic	6.1	UN3026	I	6.1	T14, TP2, TP13, TP27	None	201	243	1 L	30 L	B	40	151	POISON
				II	6.1	IB2, T11, TP2, TP27	153	202	243	5 L	60 L	B	40	151	POISON
				III	6.1	IB3, T7, TP1, TP28	153	203	241	60 L	220 L	A	40	151	POISON
	Coumarin derivative pesticides, liquid, toxic, flammable, *flash point not less than 23 degrees C*	6.1	UN3025	I	6.1, 3	T14, TP2, TP13, TP27	None	201	243	1 L	30 L	B	40	131	POISON
				II	6.1, 3	IB2, T11, TP2, TP13, TP27	153	202	243	5 L	60 L	B	40	131	POISON
				III	6.1, 3	B1, IB3, T7, TP1, TP28	153	203	242	60 L	220 L	A	40	131	POISON
	Coumarin derivative pesticides, solid, toxic	6.1	UN3027	I	6.1	IB7, IP1, T6, TP33	None	211	242	5 kg	50 kg	A	40	151	POISON
				II	6.1	IB8, IP2, IP4, T3, TP33	153	212	242	25 kg	100 kg	A	40	151	POISON
				III	6.1	IB8, IP3, T1, TP33	153	213	240	100 kg	200 kg	A	40	151	POISON
	Cresols, liquid	6.1	UN2076	II	6.1, 8	IB8, IP2, IP4, T7, TP2	153	202	243	1 L	30 L	B		153	POISON
	Cresols, solid	6.1	UN3455	II	6.1, 8	IB8, IP2, IP4, T3, TP33	153	212	242	15 kg	50 kg	B		153	POISON
	Cresylic acid	6.1	UN2022	II	6.1, 8	IB2, T7, TP2, TP13	153	202	243	1 L	30 L	B		153	POISON
	Crotonaldehyde *or* **Crotonaldehyde, stabilized**	6.1	UN1143	I	6.1, 3	2, 175, B9, B14, B32, B77, T20, TP2, TP13, TP38, TP45	None	227	244	Forbidden	Forbidden	D	40	131P	POISON INHALATION HAZARD*
	Crotonic acid, liquid	8	UN3472	III	8	IB8, T1	154	203	241	5 L	60 L	A	12	153	CORROSIVE
	Crotonic acid, solid	8	UN2823	III	8	IB8, IP3, T1, TP33	154	213	240	25 kg	100 kg	A	12	153	CORROSIVE
	Crotonylene	3	UN1144	I	3	T11, TP2	150	201	243	1 L	30 L	E		128	FLAMMABLE
	Cupriethylenediamine solution	8	UN1761	II	8, 6.1	IB2, T7, TP2	154	202	243	1 L	30 L	A		154	CORROSIVE
				III	8, 6.1	IB3, T7, TP1, TP28	154	203	242	5 L	60 L	A	95	154	CORROSIVE
	Cutters, cable, explosive	1.4S	UN0070	II	1.4S		None	62	62	25 kg	100 kg	05		112	EXPLOSIVES 1.4
	Cyanide or cyanide mixtures, dry, see **Cyanides, inorganic, solid, n.o.s.**														
G	**Cyanide solutions, n.o.s.**	6.1	UN1935	I	6.1	B37, T14, TP2, TP13, TP27	None	201	243	1 L	30 L	B	40, 52	157	POISON
				II	6.1	IB2, T11, TP2, TP13, TP27	153	202	243	5 L	60 L	A	40, 52	157	POISON
				III	6.1	IB3, T7, TP2, TP13, TP28	153	203	241	60 L	220 L	A	40, 52	157	POISON
	Cyanides, inorganic, solid, n.o.s.	6.1	UN1588	I	6.1	IB7, IP1, N74, N75, T6, TP33	None	211	242	5 kg	50 kg	A	52	157	POISON
				II	6.1	IB8, IP2, IP4, N74, N75, T3, TP33	153	212	242	25 kg	100 kg	A	52	157	POISON
				III	6.1	IB8, IP3, N74, N75, T1, TP33	153	213	240	100 kg	200 kg	A	52	157	POISON
	Cyanogen	2.3	UN1026		2.3, 2.1	2	None	304	245	Forbidden	Forbidden	D	40	119	POISON GAS*
	Cyanogen bromide	6.1	UN1889	I	6.1, 8	A6, A8, T6, TP33	None	211	242	1kg	15kg	D	40	157	POISON
	Cyanogen chloride, stabilized	2.3	UN1589		2.3, 8	1	None	192	245	Forbidden	Forbidden	D	40	125	POISON GAS*
	Cyanuric chloride	8	UN2670	II	8	IB8, IP2, IP4, T3, TP33	None	212	240	15 kg	50 kg	A	12, 40	157	CORROSIVE

Symbols (1)	Hazardous materials descriptions and proper shipping names (2)	Hazard class or Division (3)	Identification Numbers (4)	PG (5)	Label Codes (6)	Special Provisions (§172.102) (7)	(8) Packaging (§173.***) Exceptions (8A)	Non-bulk (8B)	Bulk (8C)	(9) Quantity limitations (see §§173.27 and 175.75) Passenger aircraft/rail (9A)	Cargo aircraft only (9B)	(10) Vessel stowage Location (10A)	Other (10B)	ERG Guide No.	Placard Advisory Consult §172.504 & §172.505 to determine placard *Denotes placard for any quantity
	Cyanuric triazide	Forbidden													
	Cyclobutane	2.1	UN2601		2.1		306	304	314, 315	Forbidden	150 kg	B	40	115	FLAMMABLE GAS
	Cyclobutyl chloroformate	6.1	UN2744	II	6.1, 8, 3	IB1, T7, TP2, TP13	None	202	243	1 L	30 L	A	12, 13, 21, 25, 40, 100	155	POISON
	1,5,9-Cyclododecatriene	6.1	UN2518	III	6.1	IB3, T4, TP1	153	203	241	60 L	220 L	A	40	153	POISON
	Cycloheptane	3	UN2241	II	3	IB2, T4, TP1	150	202	242	5 L	60 L	B	40	128	FLAMMABLE
	Cycloheptatriene	3	UN2603	II	3, 6.1	IB2, T7, TP1, TP13	150	202	243	1 L	60 L	E	40	131	FLAMMABLE
	Cycloheptene	3	UN2242	II	3	B1, IB2, T4, TP1	150	202	242	5 L	60 L	B		128	FLAMMABLE
	Cyclohexane	3	UN1145	II	3	IB2, T4, TP1	150	202	242	5 L	60 L	E		128	FLAMMABLE
	Cyclohexanone	3	UN1915	III	3	B1, IB3, T2, TP1	150	203	242	60 L	220 L	A		127	FLAMMABLE
	Cyclohexene	3	UN2256	II	3	IB2, T4, TP1	150	202	242	5 L	60 L	E		130	FLAMMABLE
	Cyclohexenyltrichlorosilane	8	UN1762	II	8	A7, B2, N34, T10, TP2, TP7, TP13	None	206	242	Forbidden	30 L	C	40	156	CORROSIVE
	Cyclohexyl acetate	3	UN2243	III	3	B1, IB3, T2, TP1	150	203	242	60 L	220 L	A		130	FLAMMABLE
	Cyclohexyl isocyanate	6.1	UN2488	I	6.1, 3	2, B9, B14, B32, B77, T20, TP2, TP13, TP38, TP45	None	227	244	Forbidden	Forbidden	D	40	155	POISON INHALATION HAZARD*
	Cyclohexyl mercaptan	3	UN3054	III	3	B1, IB3, T2, TP1	150	203	242	60 L	220 L	A	40, 95	129	FLAMMABLE
	Cyclohexylamine	8	UN2357	II	8, 3	IB2, T7, TP2	None	202	243	1 L	30 L	A	40	132	CORROSIVE
	Cyclohexyltrichlorosilane	8	UN1763	II	8	A7, B2, N34, T10, TP2, TP7, TP13	None	206	242	Forbidden	30 L	C	40	156	CORROSIVE
	Cyclonite and cyclotetramethylenetetranitramine mixtures, wetted *or* **desensitized** *see* **RDX and HMX mixtures, wetted** *or* **desensitized** *etc.*														
	Cyclonite and HMX mixtures, wetted *or* **desensitized** *see* **RDX and HMX mixtures, wetted** *or* **desensitized** *etc.*														
	Cyclonite and octogen mixtures, wetted *or* **desensitized** *see* **RDX and HMX mixtures, wetted** *or* **desensitized** *etc.*														
	Cyclonite, *see* **Cyclotrimethylenetrinitramine,** *etc.*														
	Cyclooctadiene phosphines, *see* **9-Phosphabicyclononanes**														
	Cyclooctadienes	3	UN2520	III	3	B1, IB3, T2, TP1	150	203	242	60 L	220 L	A		130P	FLAMMABLE
	Cyclooctatetraene	3	UN2358	II	3	IB2, T4, TP1	150	202	242	5 L	60 L	B		128P	FLAMMABLE
	Cyclopentane	3	UN1146	II	3	IB2, T7, TP1	150	202	242	5 L	60 L	E		128	FLAMMABLE
	Cyclopentane, methyl, see **Methylcyclopentane**														
	Cyclopentanol	3	UN2244	III	3	B1, IB3, T2, TP1	150	203	242	60 L	220 L	A		129	FLAMMABLE
	Cyclopentanone	3	UN2245	III	3	B1, IB3, T2, TP1	150	203	242	60 L	220 L	A		128	FLAMMABLE
	Cyclopentene	3	UN2246	II	3	IB2, IP8, T7, TP2	150	202	242	5 L	60 L	E		128	FLAMMABLE
	Cyclopropane	2.1	UN1027		2.1	T50	306	304	314, 315	Forbidden	150 kg	E	40	115	FLAMMABLE GAS
	Cyclotetramethylene tetranitramine (dry or unphlegmatized) (HMX)	Forbidden													
	Cyclotetramethylenetetranitramine, desensitized *or* **Octogen, desensitized** *or* **HMX, desensitized**	1.1D	UN0484	II	1.1D		None	62	None	Forbidden	Forbidden	10		112	EXPLOSIVES 1.1*
	Cyclotetramethylenetetranitramine, wetted *or* **HMX, wetted** *or* **Octogen, wetted** *with not less than 15 percent water, by mass*	1.1D	UN0226	II	1.1D		None	62	None	Forbidden	Forbidden	10		112	EXPLOSIVES 1.1*
	Cyclotrimethylenenitramine and octogen, mixtures, wetted *or* **desensitized** *see* **RDX and HMX mixtures, wetted** *or* **desensitized** *etc.*														
	Cyclotrimethylenetrinitramine and cyclotetramethylenetetranitramine mixtures, wetted *or* **desensitized** *see* **RDX and HMX mixtures, wetted** *or* **desensitized** *etc.*														
	Cyclotrimethylenenitramine and HMX mixtures, wetted *or* **desensitized** *see* **RDX and HMX mixtures, wetted** *or* **desensitized** *etc.*														

Symbols (1)	Hazardous materials descriptions and proper shipping names (2)	Hazard class or Division (3)	Identification Numbers (4)	PG (5)	Label Codes (6)	Special Provisions (§172.102) (7)	(8) Packaging (§173.***) Exceptions (8A)	Non-bulk (8B)	Bulk (8C)	(9) Quantity limitations (see §§173.27 and 175.75) Passenger aircraft/rail (9A)	Cargo aircraft only (9B)	(10) Vessel stowage Location (10A)	Other (10B)	ERG Guide No.	Placard Advisory Consult §172.504 & §172.505 to determine placard *Denotes placard for any quantity
	Cyclotrimethylenetrinitramine, desensitized *or* **Cyclonite, desensitized** *or* **Hexogen, desensitized** *or* **RDX, desensitized**	1.1D	UN0483	II	1.1D		None	62	None	Forbidden	Forbidden	10		112	EXPLOSIVES 1.1*
	Cyclotrimethylenetrinitramine, wetted *or* **Cyclonite, wetted** *or* **Hexogen, wetted** *or* **RDX, wetted** *with not less than 15 percent water by mass*	1.1D	UN0072	II	1.1D		None	62	None	Forbidden	Forbidden	10		112	EXPLOSIVES 1.1*
	Cymenes	3	UN2046	III	3	B1, IB3, T2, TP1	150	203	242	60 L	220 L	A		130	FLAMMABLE
	Dangerous Goods in Machinery *or* **Dangerous Goods in Apparatus**	9	UN3363			136, A105	None	222	None	See A105	See A105	A		171	CLASS 9
	Decaborane	4.1	UN1868	II	4.1, 6.1	A19, A20, IB6, IP2, T3, TP33	None	212	None	Forbidden	50 kg	A	74	134	FLAMMABLE SOLID
	Decahydronaphthalene	3	UN1147	III	3	B1, IB3, T2, TP1	150	203	242	60 L	220 L	A		130	FLAMMABLE
	n-Decane	3	UN2247	III	3	B1, IB3, T2, TP1	150	203	242	60 L	220 L	A		128	FLAMMABLE
	Deflagrating metal salts of aromatic nitroderivatives, n.o.s	1.3C	UN0132	II	1.3C		None	62	None	Forbidden	Forbidden	10	5E	112	EXPLOSIVES 1.3*
	Delay electric igniter, see **Igniters**														
D	**Denatured alcohol**	3	NA1987	II	3	172, T8	150	202	242	5 L	60 L	B		127	FLAMMABLE
				III	3	172, B1, T7	150	203	242	60 L	220 L	A		127	FLAMMABLE
	Depth charges, see **Charges, depth**														
G	**Desensitized explosive, liquid, n.o.s.**	3	UN3379	I	3	164	None	201	None	Forbidden	Forbidden	D	36	128	FLAMMABLE
G	**Desensitized explosive, solid, n.o.s.**	4.1	UN3380	I	4.1	164	None	211	None	Forbidden	Forbidden	D	28, 36	133	FLAMMABLE SOLID
	Detonating relays, see **Detonators,** *etc.*														
	Detonator assemblies, non-electric *for blasting*	1.1B	UN0360	II	1.1B		None	62	None	Forbidden	Forbidden	11		112	EXPLOSIVES 1.1*
	Detonator assemblies, non-electric, *for blasting*	1.4B	UN0361	II	1.4B	103	63(f), 63(g)	62	None	Forbidden	75 kg	06		114	EXPLOSIVES 1.4
	Detonator assemblies, non-electric, *for blasting*	1.4S	UN0500	II	1.4S	347	63(f), 63(g)	62	None	25 kg	100 kg	05		114	EXPLOSIVES 1.4
	Detonators, electric, *for blasting*	1.1B	UN0030	II	1.1B		63(f), 63(g)	62	None	Forbidden	Forbidden	11		112	EXPLOSIVES 1.1*
	Detonators, electric, *for blasting*	1.4B	UN0255	II	1.4B	103	63(f), 63(g)	62	None	Forbidden	75 kg	06		114	EXPLOSIVES 1.4
	Detonators, electric *for blasting*	1.4S	UN0456	II	1.4S	347	63(f), 63(g)	62	None	25 kg	100 kg	05		114	EXPLOSIVES 1.4
	Detonators for ammunition	1.1B	UN0073	II	1.1B		None	62	None	Forbidden	Forbidden	11		112	EXPLOSIVES 1.1*
	Detonators for ammunition	1.2B	UN0364	II	1.2B		None	62	None	Forbidden	Forbidden	11		112	EXPLOSIVES 1.2*
	Detonators for ammunition	1.4B	UN0365	II	1.4B	103	None	62	None	Forbidden	75 kg	06		114	EXPLOSIVES 1.4
	Detonators for ammunition	1.4S	UN0366	II	1.4S	347	None	62	None	25 kg	100 kg	05		114	EXPLOSIVES 1.4
	Detonators, non-electric, *for blasting*	1.1B	UN0029	II	1.1B		None	62	None	Forbidden	Forbidden	11		112	EXPLOSIVES 1.1*
	Detonators, non-electric, *for blasting*	1.4B	UN0267	II	1.4B	103	63(f), 63(g)	62	None	Forbidden	75 kg	06		114	EXPLOSIVES 1.4
	Detonators, non-electric *for blasting*	1.4S	UN0455	II	1.4S	347	63(f), 63(g)	62	None	25 kg	100 kg	05		114	EXPLOSIVES 1.4
	Deuterium, compressed	2.1	UN1957		2.1	N89	306	302	None	Forbidden	150 kg	E	40	115	FLAMMABLE GAS
	Devices, small, hydrocarbon gas powered *or* **Hydrocarbon gas refills for small devices** *with release device*	2.1	UN3150		2.1		306	304	None	1 kg	15 kg	B	40	115	FLAMMABLE GAS
	Di-n-amylamine	3	UN2841	III	3, 6.1	B1, IB3, T4, TP1	150	203	242	60 L	220 L	A		131	FLAMMABLE
	Di-n-butyl peroxydicarbonate, with more than 52 percent in solution	Forbidden													
	Di-n-butylamine	8	UN2248	II	8, 3	IB2, T7, TP2	None	202	243	1 L	30 L	A		132	CORROSIVE
	2,2-Di-(tert-butylperoxy) butane, with more than 55 percent in solution	Forbidden													
	Di-(tert-butylperoxy) phthalate, with more than 55 percent in solution	Forbidden													
	2,2-Di-(4,4-di-tert-butylperoxycyclohexyl) propane, with more than 42 percent with inert solid	Forbidden													
	Di-2, 4-dichlorobenzoyl peroxide, with more than 75 percent with water	Forbidden													
	1,2-Di-(dimethylamino)ethane	3	UN2372	II	3	IB2, T4, TP1	150	202	242	5 L	60 L	B		129	FLAMMABLE

Symbols (1)	Hazardous materials descriptions and proper shipping names (2)	Hazard class or Division (3)	Identification Numbers (4)	PG (5)	Label Codes (6)	Special Provisions (§172.102) (7)	(8) Packaging (§173.***) Exceptions (8A)	Non-bulk (8B)	Bulk (8C)	(9) Quantity limitations (see §§173.27 and 175.75) Passenger aircraft/rail (9A)	Cargo aircraft only (9B)	(10) Vessel stowage Location (10A)	Other (10B)	ERG Guide No.	Placard Advisory Consult §172.504 & §172.505 to determine placard *Denotes placard for any quantity
	Di-2-ethylhexyl phosphoric acid, see **Diisooctyl acid phosphate**														
	Di-(1-hydroxytetrazole) (dry)	Forbidden													
	Di-(1-naphthoyl) peroxide	Forbidden													
	a,a'-Di-(nitroxy) methylether	Forbidden													
	Di-(beta-nitroxyethyl) ammonium nitrate	Forbidden													
	Diacetone alcohol	3	UN1148	II	3	IB2, T4, TP1	150	202	242	5 L	60 L	B		129	FLAMMABLE
				III	3	B1, IB3, T2, TP1	150	203	242	60 L	220 L	A		129	FLAMMABLE
	Diacetone alcohol peroxides, with more than 57 percent in solution with more than 9 percent hydrogen peroxide, less than 26 percent diacetone alcohol and less than 9 percent water; total active oxygen content more than 9 percent by mass	Forbidden													
	Diacetyl, see **Butanedione**														
	Diacetyl peroxide, solid, or with more than 25 percent in solution	Forbidden													
	Diallylamine	3	UN2359	II	3, 6.1, 8	IB2, T7, TP1	150	202	243	1 L	5 L	B	21, 40, 100	132	FLAMMABLE
	Diallylether	3	UN2360	II	3, 6.1	IB2, N12, T7, TP1, TP13	150	202	243	1 L	60 L	E	40	131P	FLAMMABLE
	4,4'-Diaminodiphenyl methane	6.1	UN2651	III	6.1	IB8, IP3, T1, TP33	153	213	240	100 kg	200 kg	A		153	POISON
	p-Diazidobenzene	Forbidden													
	1,2-Diazidoethane	Forbidden													
	1,1'-Diazoaminonaphthalene	Forbidden													
	Diazoaminotetrazole (dry)	Forbidden													
	Diazodinitrophenol (dry)	Forbidden													
	Diazodinitrophenol, wetted *with not less than 40 percent water or mixture of alcohol and water, by mass*	1.1A	UN0074	II	1.1A	111, 117	None	62	None	Forbidden	Forbidden	12		112	EXPLOSIVES 1.1*
	Diazodiphenylmethane	Forbidden													
	Diazonium nitrates (dry)	Forbidden													
	Diazonium perchlorates (dry)	Forbidden													
	1,3-Diazopropane	Forbidden													
	Dibenzyl peroxydicarbonate, with more than 87 percent with water	Forbidden													
	Dibenzyldichlorosilane	8	UN2434	II	8	B2, T10, TP2, TP7, TP13	154	206	242	Forbidden	30 L	C	40	156	CORROSIVE
	Diborane	2.3	UN1911		2.3, 2.1	1, N89	None	302	None	Forbidden	Forbidden	D	40, 57	119	POISON GAS*
D	**Diborane mixtures**	2.1	NA1911		2.1	5	None	302	245	Forbidden	Forbidden	D	40, 57	119	FLAMMABLE GAS
	Dibromoacetylene	Forbidden													
	1,2-Dibromobutan-3-one	6.1	UN2648	II	6.1	IB2	153	202	243	5 L	60 L	B	40	154	POISON
	Dibromochloropropane	6.1	UN2872	II	6.1	IB2, T7, TP2	153	202	243	5 L	60 L	A		159	POISON
				III	6.1	IB3, T4, TP1	153	203	241	60 L	220 L	A		159	POISON
A	**Dibromodifluoromethane,** *R12B2*	9	UN1941	III	None	T11, TP2	155	203	241	100 L	220 L	A	25	171	CLASS 9
	1,2-Dibromoethane, see **Ethylene dibromide**														
	Dibromomethane	6.1	UN2664	III	6.1	IB3, T4, TP1	153	203	241	60 L	220 L	A		160	POISON
	Dibutyl ethers	3	UN1149	III	3	B1, IB3, T2, TP1	150	203	242	60 L	220 L	A		128	FLAMMABLE
	Dibutylaminoethanol	6.1	UN2873	III	6.1	IB3, T4, TP1	153	203	241	60 L	220 L	A		153	POISON
	N,N'-Dichlorazodicarbonamidine (salts of) (dry)	Forbidden													
	1,1-Dichloro-1-nitroethane	6.1	UN2650	II	6.1	IB2, T7, TP2	153	202	243	5 L	60 L	A	12, 40, 74	153	POISON
D	**3,5-Dichloro-2,4,6-trifluoropyridine**	6.1	NA9264	I	6.1	2, B9, B14, B32, T20, TP4, TP13, TP38, TP45	None	227	244	Forbidden	Forbidden	A	40	151	POISON INHALATION HAZARD*
	Dichloroacetic acid	8	UN1764	II	8	A3, A6, A7, B2, IB2, N34, T8, TP2	154	202	242	1 L	30 L	A		153	CORROSIVE
	1,3-Dichloroacetone	6.1	UN2649	II	6.1	IB8, IP2, IP4, T3, TP33	153	212	242	25 kg	100 kg	B	12, 40	153	POISON

Symbols (1)	Hazardous materials descriptions and proper shipping names (2)	Hazard class or Division (3)	Identification Numbers (4)	PG (5)	Label Codes (6)	Special Provisions (§172.102) (7)	(8) Packaging (§173.***) Exceptions (8A)	Non-bulk (8B)	Bulk (8C)	(9) Quantity limitations (see §§173.27 and 175.75) Passenger aircraft/rail (9A)	Cargo aircraft only (9B)	(10) Vessel stowage Location (10A)	Other (10B)	ERG Guide No.	Placard Advisory Consult §172.504 & §172.505 to determine placard *Denotes placard for any quantity
	Dichloroacetyl chloride	8	UN1765	II	8	A3, A6, A7, B2, B6, IB2, N34, T7, TP2	154	202	242	1 L	30 L	D	40	156	CORROSIVE
	Dichloroacetylene	Forbidden													
+	**Dichloroanilines, liquid**	6.1	UN1590	II	6.1	IB2, T7, TP2	153	202	243	5 L	60 L	A	40	153	POISON
	Dichloroanilines, solid	6.1	UN3442	II	6.1	IB8, IP2, IP4, T3, TP33	153	212	242	25 kg	100 kg	A	40	153	POISON
+	**o-Dichlorobenzene**	6.1	UN1591	III	6.1	IB3, T4, TP1	153	203	241	60 L	220 L	A		152	POISON
	2,2'-Dichlorodiethyl ether	6.1	UN1916	II	6.1, 3	IB2, N33, N34, T7, TP2	153	202	243	5 L	60 L	A		152	POISON
	Dichlorodifluoromethane and difluoroethane azeotropic mixture *or* **Refrigerant gas R 500** *with approximately 74 per-cent dichlorodifluoromethane*	2.2	UN2602		2.2	T50	306	304	314, 315	75 kg	150 kg	A		126	NONFLAMMABLE GAS
	Dichlorodifluoromethane *or* **Refrigerant gas R 12**	2.2	UN1028		2.2	T50	306	304	314, 315	75 kg	150 kg	A		126	NONFLAMMABLE GAS
	Dichlorodimethyl ether, symmetrical	6.1	UN2249	I	6.1, 3		None	201	243	Forbidden	Forbidden		40	131	POISON
	1,1-Dichloroethane	3	UN2362	II	3	IB2, T4, TP1	150	202	242	5 L	60 L	B	40	130	FLAMMABLE
	1,2-Dichloroethane, see **Ethylene dichloride**														
	Dichloroethyl sulfide	Forbidden													
	1,2-Dichloroethylene	3	UN1150	II	3	IB2, T7, TP2	150	202	242	5 L	60 L	B		130P	FLAMMABLE
	Dichlorofluoromethane *or* **Refrigerant gas R 21**	2.2	UN1029		2.2	T50	306	304	314, 315	75 kg	150 kg	A		126	NONFLAMMABLE GAS
	Dichloroisocyanuric acid, dry *or* **Dichloroisocyanuric acid salts**	5.1	UN2465	II	5.1	28, IB8, IP2, IP4, T3, TP33	152	212	240	5 kg	25 kg	A	13	140	OXIDIZER
	Dichloroisopropyl ether	6.1	UN2490	II	6.1	IB2, T7, TP2	153	202	243	5 L	60 L	B		153	POISON
	Dichloromethane	6.1	UN1593	III	6.1	IB3, IP8, N36, T7, TP2	153	203	241	60 L	220 L	A		160	POISON
	Dichloropentanes	3	UN1152	III	3	B1, IB3, T2, TP1	150	203	242	60 L	220 L	A		130	FLAMMABLE
	Dichlorophenyl isocyanates	6.1	UN2250	II	6.1	IB8, IP2, IP4, T3, TP33	153	212	242	25 kg	100 kg	B	25, 40, 48	156	POISON
	Dichlorophenyltrichlorosilane	8	UN1766	II	8	A7, B2, B6, N34, T10, TP2, TP7, TP13	None	206	242	Forbidden	30 L	C	40	156	CORROSIVE
	1,2-Dichloropropane	3	UN1279	II	3	IB2, N36, T4, TP1	150	202	242	5 L	60 L	B		130	FLAMMABLE
	1,3-Dichloropropanol-2	6.1	UN2750	II	6.1	IB2, T7, TP2	153	202	243	5 L	60 L	A	12, 40	153	POISON
	Dichloropropene and propylene dichloride mixture, see **1,2-Dichloropropane**														
	Dichloropropenes	3	UN2047	II	3	IB2, T4, TP1	150	202	242	5 L	60 L	B		129	FLAMMABLE
				III	3	B1, IB3, T2, TP1	150	203	242	60 L	220 L	A		129	FLAMMABLE
	Dichlorosilane	2.3	UN2189		2.3, 2.1, 8	2, B9, B14	None	304	314, 315	Forbidden	Forbidden	D	17, 40	119	POISON GAS*
	1,2 Dichloro-1,1,2,2-tetrafluoroethane *or* **Refrigerant gas R 114**	2.2	UN1958		2.2	T50	306	304	314, 315	75 kg	150 kg	A		126	NONFLAMMABLE GAS
	Dichlorovinylchloroarsine	Forbidden													
	Dicycloheptadiene, see **Bicyclo [2,2,1] hepta-2,5-diene, stabilized**														
	Dicyclohexylamine	8	UN2565	III	8	IB3, T4, TP1	154	203	241	5 L	60 L	A		153	CORROSIVE
	Dicyclohexylammonium nitrite	4.1	UN2687	III	4.1	IB8, IP3, T1, TP33	151	213	240	25 kg	100 kg	A	48	133	FLAMMABLE SOLID
	Dicyclopentadiene	3	UN2048	III	3	B1, IB3, T2, TP1	150	203	242	60 L	220 L	A		130	FLAMMABLE
	Didymium nitrate	5.1	UN1465	III	5.1	A1, IB8, IP3, T1, TP33	152	213	240	25 kg	100 kg	A		140	OXIDIZER
D	**Diesel fuel**	3	NA1993	III	None	144, B1, IB3, T4, TP1, TP29	150	203	242	60 L	220 L	A		128	COMBUSTIBLE (BULK ONLY)
I	**Diesel fuel**	3	UN1202	III	3	144, B1, IB3, T2, TP1	150	203	242	60 L	220 L	A		128	FLAMMABLE
	Diethanol nitrosamine dinitrate (dry)	Forbidden													
	Diethoxymethane	3	UN2373	II	3	IB2, T4, TP1	150	202	242	5 L	60 L	E		127	FLAMMABLE
	3,3-Diethoxypropene	3	UN2374	II	3	IB2, T4, TP1	150	202	242	5 L	60 L	B		127	FLAMMABLE

Symbols (1)	Hazardous materials descriptions and proper shipping names (2)	Hazard class or Division (3)	Identification Numbers (4)	PG (5)	Label Codes (6)	Special Provisions (§172.102) (7)	(8) Packaging (§173.***) Exceptions (8A)	Non-bulk (8B)	Bulk (8C)	(9) Quantity limitations (see §§173.27 and 175.75) Passenger aircraft/rail (9A)	Cargo aircraft only (9B)	(10) Vessel stowage Location (10A)	Other (10B)	ERG Guide No.	Placard Advisory Consult §172.504 & §172.505 to determine placard *Denotes placard for any quantity
	Diethyl carbonate	3	UN2366	III	3	B1, IB3, T2, TP1	150	203	242	60 L	220 L	A		128	FLAMMABLE
	Diethyl cellosolve, see **Ethylene glycol diethyl ether**														
	Diethyl ether *or* **Ethyl ether**	3	UN1155	I	3	T11, TP2	150	201	243	1 L	30 L	E	40	127	FLAMMABLE
	Diethyl ketone	3	UN1156	II	3	IB2, T4, TP1	150	202	242	5 L	60 L	B		127	FLAMMABLE
	Diethyl peroxydicarbonate, with more than 27 percent in solution	Forbidden													
	Diethyl sulfate	6.1	UN1594	II	6.1	IB2, T7, TP2	153	202	243	5 L	60 L	C		152	POISON
	Diethyl sulfide	3	UN2375	II	3	IB2, T7, TP1, TP13	None	202	243	5 L	60 L	E		129	FLAMMABLE
	Diethylamine	3	UN1154	II	3, 8	A3, IB2, N34, T7, TP1	150	202	243	1 L	5 L	E	40	132	FLAMMABLE
	2-Diethylaminoethanol	8	UN2686	II	8, 3	B2, IB2, T7, TP2	None	202	243	1 L	30 L	A		132	CORROSIVE
	3-Diethylamino-propylamine	3	UN2684	III	3, 8	B1, IB3, T4, TP1	150	203	242	5 L	60 L	A		132	FLAMMABLE
+	**N,N-Diethylaniline**	6.1	UN2432	III	6.1	IB3, T4, TP1	153	203	241	60 L	220 L	A		153	POISON
	Diethylbenzene	3	UN2049	III	3	B1, IB3, T2, TP1	150	203	242	60 L	220 L	A		130	FLAMMABLE
	Diethyldichlorosilane	8	UN1767	II	8, 3	A7, B6, N34, T10, TP2, TP7, TP13	None	206	243	Forbidden	30 L	C	40	155	CORROSIVE
	Diethylene glycol dinitrate	Forbidden													
	Diethyleneglycol dinitrate, desensitized *with not less than 25 percent non-volatile water-insoluble phlegmatizer, by mass*	1.1D	UN0075	II	1.1D		None	62	None	Forbidden	Forbidden	13	21E	112	EXPLOSIVES 1.1*
	Diethylenetriamine	8	UN2079	II	8	B2, IB2, T7, TP2	154	202	242	1 L	30 L	A	40, 52	154	CORROSIVE
	N,N-Diethylethylenediamine	8	UN2685	II	8, 3	IB2, T7, TP2	None	202	243	1 L	30 L	A		132	CORROSIVE
	Diethylgold bromide	Forbidden													
	Diethylthiophosphoryl chloride	8	UN2751	II	8	B2, IB2, T7, TP2	None	212	240	15 kg	50 kg	D	12, 40	155	CORROSIVE
	Difluorochloroethanes, see **1-Chloro-1,1-difluoroethanes**														
	1,1-Difluoroethane *or* **Refrigerant gas R 152a**	2.1	UN1030		2.1	T50	306	304	314, 315	Forbidden	150 kg	B	40	115	FLAMMABLE GAS
	1,1-Difluoroethylene *or* **Refrigerant gas R 1132a**	2.1	UN1959		2.1		306	304	None	Forbidden	150 kg	E	40	116P	FLAMMABLE GAS
	Difluoromethane *or* **Refrigerant gas R 32**	2.1	UN3252		2.1	T50	306	302	314, 315	Forbidden	150 kg	D	40	115	FLAMMABLE GAS
	Difluorophosphoric acid, anhydrous	8	UN1768	II	8	A6, A7, B2, IB2, N5, N34, T8, TP2	None	202	242	1 L	30 L	A	40	154	CORROSIVE
	2,3-Dihydropyran	3	UN2376	II	3	IB2, T4, TP1	150	202	242	5 L	60 L	B		127	FLAMMABLE
	1,8-Dihydroxy-2,4,5,7-tetranitroanthraquinone (chrysamminic acid)	Forbidden													
	Diiodoacetylene	Forbidden													
	Diisobutyl ketone	3	UN1157	III	3	B1, IB3, T2, TP1	150	203	242	60 L	220 L	A		128	FLAMMABLE
	Diisobutylamine	3	UN2361	III	3, 8	B1, IB3, T4, TP1	150	203	242	5 L	60 L	A		132	FLAMMABLE
	Diisobutylene, isomeric compounds	3	UN2050	II	3	IB2, T4, TP1	150	202	242	5 L	60 L	B		128	FLAMMABLE
	Diisooctyl acid phosphate	8	UN1902	III	8	IB3, T4, TP1	154	203	241	5 L	60 L	A		153	CORROSIVE
	Diisopropyl ether	3	UN1159	II	3	IB2, T4, TP1	150	202	242	5 L	60 L	E	40	127	FLAMMABLE
	Diisopropylamine	3	UN1158	II	3, 8	IB2, T7, TP1	150	202	243	1 L	5 L	B		132	FLAMMABLE
	Diisopropylbenzene hydroperoxide, with more than 72 percent in solution	Forbidden													
	Diketene, stabilized	6.1	UN2521	I	6.1, 3	2, B9, B14, B32, T20, TP2, TP13, TP38, TP45	None	227	244	Forbidden	Forbidden	D	26, 27, 40	131P	POISON INHALATION HAZARD*
	1,2-Dimethoxyethane	3	UN2252	II	3	IB2, T4, TP1	150	202	242	5 L	60 L	B		127	FLAMMABLE
	1,1-Dimethoxyethane	3	UN2377	II	3	IB2, T7, TP1	150	202	242	5 L	60 L	B		127	FLAMMABLE
	Dimethyl carbonate	3	UN1161	II	3	IB2, T4, TP1	150	202	242	5 L	60 L	B		129	FLAMMABLE
	Dimethyl chlorothiophosphate, see **Dimethyl thiophosphoryl chloride**														
	2,5-Dimethyl-2,5-dihydroperoxy hexane, with more than 82 percent with water	Forbidden													
	Dimethyl disulfide	3	UN2381	II	3	IB2, T4, TP1	150	202	242	5 L	60 L	B	40	130	FLAMMABLE

Symbols (1)	Hazardous materials descriptions and proper shipping names (2)	Hazard class or Division (3)	Identification Numbers (4)	PG (5)	Label Codes (6)	Special Provisions (§172.102) (7)	(8) Packaging (§173.***) Exceptions (8A)	Non-bulk (8B)	Bulk (8C)	(9) Quantity limitations (see §§173.27 and 175.75) Passenger aircraft/rail (9A)	Cargo aircraft only (9B)	(10) Vessel stowage Location (10A)	Other (10B)	ERG Guide No.	Placard Advisory Consult §172.504 & §172.505 to determine placard *Denotes placard for any quantity
	Dimethyl ether	2.1	UN1033		2.1	T50	306	304	314, 315	Forbidden	150 kg	B	40	115	FLAMMABLE GAS
	Dimethyl-N-propylamine	3	UN2266	II	3, 8	IB2, T7, TP2, TP13	150	202	243	1 L	5 L	B	40	132	FLAMMABLE
	Dimethyl sulfate	6.1	UN1595	I	6.1, 8	2, B9, B14, B32, B77, T20, TP2, TP13, TP38, TP45	None	227	244	Forbidden	Forbidden	D	40	156	POISON INHALATION HAZARD*
	Dimethyl sulfide	3	UN1164	II	3	IB2, IP8, T7, TP2	150	202	242	5 L	60 L	E	40	130	FLAMMABLE
	Dimethyl thiophosphoryl chloride	6.1	UN2267	II	6.1, 8	IB2, T7, TP2	153	202	243	1 L	30 L	B	25	156	POISON
	Dimethylamine, anhydrous	2.1	UN1032		2.1	N87, T50	None	304	314, 315	Forbidden	150 kg	D	40	118	FLAMMABLE GAS
	Dimethylamine solution	3	UN1160	II	3, 8	IB2, T7, TP1	150	202	243	1 L	5 L	B	52	132	FLAMMABLE
	2-Dimethylaminoacetonitrile	3	UN2378	II	3, 6.1	IB2, T7, TP1	150	202	243	1 L	60 L	A	40, 52	131	FLAMMABLE
	2-Dimethylaminoethanol	8	UN2051	II	8, 3	B2, IB2, T7, TP2	154	202	243	1 L	30 L	A		132	CORROSIVE
	2-Dimethylaminoethyl acrylate	6.1	UN3302	II	6.1	IB2, T7, TP2	153	202	243	5 L	60 L	D	25	152	POISON
	2-Dimethylaminoethyl methacrylate	6.1	UN2522	II	6.1	IB2, T7, TP2	153	202	243	5 L	60 L	B	40	153P	POISON
	N,N-Dimethylaniline	6.1	UN2253	II	6.1	IB1, T7, TP2	153	202	243	5 L	60 L	A		153	POISON
	2,3-Dimethylbutane	3	UN2457	II	3	IB2, T7, TP1	150	202	242	5 L	60 L	E		128	FLAMMABLE
	1,3-Dimethylbutylamine	3	UN2379	II	3, 8	IB2, T7, TP1	150	202	243	1 L	5 L	B	52	132	FLAMMABLE
	Dimethylcarbamoyl chloride	8	UN2262	II	8	B2, IB2, T7, TP2	154	202	242	1 L	30 L	A	40	156	CORROSIVE
	Dimethylcyclohexanes	3	UN2263	II	3	IB2, T4, TP1	150	202	242	5 L	60 L	B		128	FLAMMABLE
	N,N-Dimethylcyclohexylamine	8	UN2264	II	8, 3	B2, IB2, T7, TP2	154	202	243	1 L	30 L	A	40	132	CORROSIVE
	Dimethyldichlorosilane	3	UN1162	II	3, 8	B77, T10, TP2, TP7, TP13	None	206	243	Forbidden	Forbidden	B	40	155	FLAMMABLE
	Dimethyldiethoxysilane	3	UN2380	II	3	IB2, T4, TP1	150	202	242	5 L	60 L	B		127	FLAMMABLE
	Dimethyldioxanes	3	UN2707	II	3	IB2, T4, TP1	150	202	242	5 L	60 L	B		127	FLAMMABLE
				III	3	B1, IB3, T2, TP1	150	203	242	60 L	220 L	A		127	FLAMMABLE
	N,N-Dimethylformamide	3	UN2265	III	3	B1, IB3, T2, TP2	150	203	242	60 L	220 L	A		129	FLAMMABLE
	Dimethylhexane dihydroperoxide (dry)	Forbidden													
	Dimethylhydrazine, symmetrical	6.1	UN2382	I	6.1, 3	2, B9, B14, B32, B77, T20, TP2, TP13, TP38, TP45	None	227	244	Forbidden	Forbidden	D	40, 52, 74	131	POISON INHALATION HAZARD*
	Dimethylhydrazine, unsymmetrical	6.1	UN1163	I	6.1, 3, 8	2, B7, B9, B14, B32, T20, TP2, TP13, TP38, TP45	None	227	244	Forbidden	Forbidden	D	21, 38, 40, 52, 100	131	POISON INHALATION HAZARD*
	2,2-Dimethylpropane	2.1	UN2044		2.1		306	304	314, 315	Forbidden	150 kg	E	40	115	FLAMMABLE GAS
	Dinitro-o-cresol	6.1	UN1598	II	6.1	IB8, IP2, IP4, T3, TP33	153	212	242	25 kg	100 kg	A		153	POISON
	1,3-Dinitro-5,5-dimethyl hydantoin	Forbidden													
	Dinitro-7,8-dimethylglycoluril (dry)	Forbidden													
	1,3-Dinitro-4,5-dinitrosobenzene	Forbidden													
	1,4-Dinitro-1,1,4,4-tetramethylolbutanetetranitrate (dry)	Forbidden													
	2,4-Dinitro-1,3,5-trimethylbenzene	Forbidden													
	Dinitroanilines	6.1	UN1596	II	6.1	IB8, IP2, IP4, T3, TP33	153	212	242	25 kg	100 kg	A	91	153	POISON
	Dinitrobenzenes, liquid	6.1	UN1597	II	6.1	11, IB2, T7, TP2	153	202	243	5 L	60 L	A	91	152	POISON
				III	6.1	11, IB3, T7, TP2	153	203	241	60 L	220 L	A	91	152	POISON
	Dinitrobenzenes, solid	6.1	UN3443	II	6.1	IB8, IP2, IP4, T3, TP33	153	212	242	25 kg	100 kg	A	91	152	POISON
	Dinitrochlorobenzene, see **Chlorodinitrobenzene**														
	1,2-Dinitroethane	Forbidden													
	1,1-Dinitroethane (dry)	Forbidden													
	Dinitrogen tetroxide	2.3	UN1067		2.3, 5.1, 8	1, B7, B14, B45, B46, B61, B66, B67, B77, T50, TP21	None	336	314	Forbidden	Forbidden	D	40, 89, 90	124	POISON GAS*
	Dinitroglycoluril *or* **Dingu**	1.1D	UN0489	II	1.1D		None	62	None	Forbidden	Forbidden	10		112	EXPLOSIVES 1.1*

Symbols (1)	Hazardous materials descriptions and proper shipping names (2)	Hazard class or Division (3)	Identification Numbers (4)	PG (5)	Label Codes (6)	Special Provisions (§172.102) (7)	(8) Packaging (§173.***) Exceptions (8A)	Non-bulk (8B)	Bulk (8C)	(9) Quantity limitations (see §§173.27 and 175.75) Passenger aircraft/rail (9A)	Cargo aircraft only (9B)	(10) Vessel stowage Location (10A)	Other (10B)	ERG Guide No.	Placard Advisory Consult §172.504 & §172.505 to determine placard *Denotes placard for any quantity
	Dinitromethane	Forbidden													
	Dinitrophenol, *dry or wetted with less than 15 percent water, by mass*	1.1D	UN0076	II	1.1D, 6.1		None	62	None	Forbidden	Forbidden	10	5E	112	EXPLOSIVES 1.1*
	Dinitrophenol solutions	6.1	UN1599	II	6.1	IB2, T7, TP2	153	202	243	5 L	60 L	A	36	153	POISON
				III	6.1	IB2, T4, TP1	153	203	241	60 L	220 L	A	36	153	POISON
	Dinitrophenol, wetted *with not less than 15 percent water, by mass*	4.1	UN1320	I	4.1, 6.1	23, A8, A19, A20, N41	None	211	None	1 kg	15 kg	E	28, 36	113	FLAMMABLE SOLID
	Dinitrophenolates *alkali metals, dry or wetted with less than 15 percent water, by mass*	1.3C	UN0077	II	1.3C, 6.1		None	62	None	Forbidden	Forbidden	10	5E	112	EXPLOSIVES 1.3*
	Dinitrophenolates, wetted *with not less than 15 percent water, by mass*	4.1	UN1321	I	4.1, 6.1	23, A8, A19, A20, N41	None	211	None	1 kg	15 kg	E	28, 36	113	FLAMMABLE SOLID
	Dinitropropylene glycol	Forbidden													
	Dinitroresorcinol, *dry or wetted with less than 15 percent water, by mass*	1.1D	UN0078	II	1.1D		None	62	None	Forbidden	Forbidden	10	5E	112	EXPLOSIVES 1.1*
	2,4-Dinitroresorcinol (heavy metal salts of) (dry)	Forbidden													
	4,6-Dinitroresorcinol (heavy metal salts of) (dry)	Forbidden													
	Dinitroresorcinol, wetted *with not less than 15 percent water, by mass*	4.1	UN1322	I	4.1	23, A8, A19, A20, N41	None	211	None	1 kg	15 kg	E	28, 36	113	FLAMMABLE SOLID
	3,5-Dinitrosalicylic acid (lead salt) (dry)	Forbidden													
	Dinitrosobenzene	1.3C	UN0406	II	1.3C		None	62	None	Forbidden	Forbidden	10		112	EXPLOSIVES 1.3*
	Dinitrosobenzylamidine and salts of (dry)	Forbidden													
	2,2-Dinitrostilbene	Forbidden													
	Dinitrotoluenes, *liquid*	6.1	UN2038	II	6.1	IB2, T7, TP2	153	202	243	5 L	60 L	A		152	POISON
	Dinitrotoluenes, molten	6.1	UN1600	II	6.1	T7, TP3	None	202	243	Forbidden	Forbidden	C		152	POISON
	Dinitrotoluenes, solid	6.1	UN3454	II	6.1	IB8, IP2, IP4, T3, TP33	153	212	242	25 kg	100 kg	A		152	POISON
	1,9-Dinitroxy pentamethylene-2,4, 6,8-tetramine (dry)	Forbidden													
	Dioxane	3	UN1165	II	3	IB2, T4, TP1	150	202	242	5 L	60 L	B		127	FLAMMABLE
	Dioxolane	3	UN1166	II	3	IB2, T4, TP1	150	202	242	5 L	60 L	B	40	127	FLAMMABLE
	Dipentene	3	UN2052	III	3	B1, IB3, T2, TP1	150	203	242	60 L	220 L	A		128	FLAMMABLE
	Diphenylamine chloroarsine	6.1	UN1698	I	6.1	T6, TP33	None	201	None	Forbidden	Forbidden	D	40	154	POISON
	Diphenylchloroarsine, liquid	6.1	UN1699	I	6.1	A8, B14, B32, N33, N34, T14, TP2, TP13, TP27	None	201	243	Forbidden	30 L	D	40	151	POISON
	Diphenylchloroarsine, solid	6.1	UN3450	I	6.1	IB7, IP1, T6, TP33	None	211	242	5 kg	50 kg	D	40	151	POISON
	Diphenyldichlorosilane	8	UN1769	II	8	A7, B2, N34, T10, TP2, TP7, TP13	None	206	242	Forbidden	30 L	C	40	156	CORROSIVE
	Diphenylmethyl bromide	8	UN1770	II	8	IB8, IP2, IP4, T3, TP33	154	212	240	15 kg	50 kg	D	40	153	CORROSIVE
	Dipicryl sulfide, *dry or wetted with less than 10 percent water, by mass*	1.1D	UN0401	II	1.1D		None	62	None	Forbidden	Forbidden	10		112	EXPLOSIVES 1.1*
	Dipicryl sulfide, wetted *with not less than 10 percent water, by mass*	4.1	UN2852	I	4.1	162, A2, N41, N84	None	211	None	Forbidden	0.5 kg	D	28	113	FLAMMABLE SOLID
	Dipicrylamine, *see* **Hexanitrodiphenylamine**														
	Dipropionyl peroxide, with more than 28 percent in solution	Forbidden													
	Di-n-propyl ether	3	UN2384	II	3	IB2, T4, TP1	150	202	242	5 L	60 L	B		127	FLAMMABLE
	Dipropyl ketone	3	UN2710	III	3	B1, IB3, T2, TP1	150	203	242	60 L	220 L	A		128	FLAMMABLE
	Dipropylamine	3	UN2383	II	3, 8	IB2, T7, TP1	150	202	243	1 L	5 L	B		132	FLAMMABLE
G	**Disinfectant, liquid, corrosive, n.o.s**	8	UN1903	I	8	A6, A7, B10, T14, TP2, TP27	None	201	243	0.5 L	2.5 L	B		153	CORROSIVE
G	**Disinfectants, liquid, corrosive, n.o.s**	8	UN1903	II	8	B2, IB2, T7, TP2	154	202	242	1 L	30 L	B		153	CORROSIVE
				III	8	IB3, T4, TP1	154	203	241	5 L	60 L	A		153	CORROSIVE
G	**Disinfectants, liquid, toxic, n.o.s.**	6.1	UN3142	I	6.1	A4, T14, TP2, TP27	None	201	243	1 L	30 L	A	40	151	POISON
				II	6.1	IB2, T11, TP2, TP27	153	202	243	5 L	60 L	A	40	151	POISON
				III	6.1	IB3, T7, TP1, TP28	153	203	241	60 L	220 L	A	40	151	POISON

Symbols (1)	Hazardous materials descriptions and proper shipping names (2)	Hazard class or Division (3)	Identification Numbers (4)	PG (5)	Label Codes (6)	Special Provisions (§172.102) (7)	(8) Packaging (§173.***) Exceptions (8A)	Non-bulk (8B)	Bulk (8C)	(9) Quantity limitations (see §§173.27 and 175.75) Passenger aircraft/rail (9A)	Cargo aircraft only (9B)	(10) Vessel stowage Location (10A)	Other (10B)	ERG Guide No.	Placard Advisory Consult §172.504 & §172.505 to determine placard *Denotes placard for any quantity
G	**Disinfectants, solid, toxic, n.o.s.**	6.1	UN1601	I	6.1	IB7, IP1, T6, TP33	None	211	242	5 kg	50 kg	A	40	151	POISON
				II	6.1	IB8, IP2, IP4, T3, TP33	153	212	242	25 kg	100 kg	A	40	151	POISON
				III	6.1	IB8, IP3, T1, TP33	153	213	240	100 kg	200 kg	A	40	151	POISON
	Disodium trioxosilicate	8	UN3253	III	8	IB8, IP3, T1, TP33	154	213	240	25 kg	100 kg	A	52	154	CORROSIVE
G	**Dispersant gases, n.o.s.** *see* **Refrigerant gases, n.o.s.**														
	Divinyl ether, stabilized	3	UN1167	I	3	A7, T11, TP2	None	201	243	1 L	30 L	E	40	128P	FLAMMABLE
	Dodecyltrichlorosilane	8	UN1771	II	8	A7, B2, B6, N34, T10, TP2, TP7, TP13	None	206	242	Forbidden	30 L	C	40	156	CORROSIVE
	Dry ice, *see* **Carbon dioxide, solid**														
G	**Dyes, liquid, corrosive n.o.s.** *or* **Dye intermediates, liquid, corrosive, n.o.s.**	8	UN2801	I	8	11, A6, B10, T14, TP2, TP27	None	201	243	0.5 L	2.5 L	A		154	CORROSIVE
				II	8	11, B2, IB2, T11, TP2, TP27	154	202	242	1 L	30 L	A		154	CORROSIVE
				III	8	11, IB3, T7, TP1, TP28	154	203	241	5 L	60 L	A		154	CORROSIVE
G	**Dyes, liquid, toxic, n.o.s.** *or* **Dye intermediates, liquid, toxic, n.o.s.**	6.1	UN1602	I	6.1		None	201	243	1 L	30 L	A		151	POISON
				II	6.1	IB2	153	202	243	5 L	60 L	A		151	POISON
				III	6.1	IB3	153	203	241	60 L	220	A		151	POISON
G	**Dyes, solid, corrosive, n.o.s.** *or* **Dye intermediates, solid, corrosive, n.o.s.**	8	UN3147	I	8	IB7, IP1, T6, TP33	None	211	242	1 kg	25 kg	A		154	CORROSIVE
				II	8	IB8, IP2, IP4, T3, TP33	154	212	240	15 kg	50 kg	A		154	CORROSIVE
				III	8	IB8, IP3, T1, TP33	154	213	240	25 kg	100 kg	A		154	CORROSIVE
G	**Dyes, solid, toxic, n.o.s.** *or* **Dye intermediates, solid, toxic, n.o.s.**	6.1	UN3143	I	6.1	A5, IB7, IP1, T6, TP33	None	211	242	5 kg	50 kg	A		151	POISON
				II	6.1	IB8, IP2, IP4, T3, TP33	153	212	242	25 kg	100 kg	A		151	POISON
				III	6.1	IB8, IP3, T1, TP33	153	213	240	100 kg	200 kg	A		151	POISON
	Dynamite, see **Explosive, blasting, type A**														
	Electrolyte (acid or alkali) for batteries, see **Battery fluid, acid** *or* **Battery fluid, alkali**														
G	**Elevated temperature liquid, flammable, n.o.s.,** *with flash point above 37.8 C, at or above its flash point*	3	UN3256	III	3	IB1, T3, TP3, TP29	None	None	247	Forbidden	Forbidden	A		128	FLAMMABLE
G	**Elevated temperature liquid, n.o.s.,** *at or above 100 C and below its flash point (including molten metals, molten salts, etc.)*	9	UN3257	III	9	IB1, T3, TP3, TP29	None	None	247	Forbidden	Forbidden	A	85	128	CLASS 9
G	**Elevated temperature solid, n.o.s.,** *at or above 240 C, see §173.247(h)(4)*	9	UN3258	III	9		247(h)(4)	None	247	Forbidden	Forbidden	A	85	171	CLASS 9
	Engines, internal combustion, *or* **Engines, fuel cell,** *flammable gas powered*	9	UN3166		9	135	220	220	220	Forbidden	No limit	A		128	CLASS 9
	Engines, internal combustion, *or* **Engines, fuel cell,** *flammable liquid powered*	9	UN3166		9	135	220	220	220	No limit	No limit	A		128	CLASS 9
G	**Environmentally hazardous substance, liquid, n.o.s.**	9	UN3082	III	9	8, 146, 173, 335, IB3, T4, TP1, TP29	155	203	241	No limit	No limit	A		171	CLASS 9
G	**Environmentally hazardous substance, solid, n.o.s.**	9	UN3077	III	9	8, 146, 335, A112, B54, IB8, IP3, N20, T1, TP33	155	213	240	No limit	No limit	A		171	CLASS 9
	Epibromoyhydrin	6.1	UN2558	I	6.1, 3	T14, TP2, TP13	None	201	243	Forbidden	Forbidden	D	40	131	POISON
+	**Epichlorohydrin**	6.1	UN2023	II	6.1, 3	IB2, T7, TP2, TP13	153	202	243	5 L	60 L	A	40	131P	POISON
	1,2-Epoxy-3-ethoxypropane	3	UN2752	III	3	B1, IB3, T2, TP1	150	203	242	60 L	220 L	A		127	FLAMMABLE
	Esters, n.o.s.	3	UN3272	II	3	IB2, T7, TP1, TP8, TP28	150	202	242	5 L	60 L	B		127	FLAMMABLE
				III	3	B1, IB3, T4, TP1, TP29	150	203	242	60 L	220 L	A		127	FLAMMABLE
	Etching acid, liquid, n.o.s., see **Hydrofluoric acid,** *etc.*														
	Ethane	2.1	UN1035		2.1		306	304	302	Forbidden	150 kg	E	40	115	FLAMMABLE GAS

Symbols (1)	Hazardous materials descriptions and proper shipping names (2)	Hazard class or Division (3)	Identification Numbers (4)	PG (5)	Label Codes (6)	Special Provisions (§172.102) (7)	(8) Packaging (§173.***) Exceptions (8A)	Non-bulk (8B)	Bulk (8C)	(9) Quantity limitations (see §§173.27 and 175.75) Passenger aircraft/rail (9A)	Cargo aircraft only (9B)	(10) Vessel stowage Location (10A)	Other (10B)	ERG Guide No.	Placard Advisory Consult §172.504 & §172.505 to determine placard *Denotes placard for any quantity
D	**Ethane-Propane mixture, refrigerated liquid**	2.1	NA1961		2.1	T75, TP5	None	316	314, 315	Forbidden	Forbidden	D	40	115	FLAMMABLE GAS
	Ethane, refrigerated liquid	2.1	UN1961		2.1	T75, TP5	None	None	315	Forbidden	Forbidden	D	40	115	FLAMMABLE GAS
	Ethanol amine dinitrate	Forbidden													
	Ethanol and gasoline mixture *or* **Ethanol and motor spirit mixture** *or* **Ethanol and petrol mixture,** *with more than 10% ethanol*	3	UN3475	II	3	144, 177, IB2, T4, TP1	150	202	242	5 L	60 L	E		127	FLAMMABLE
	Ethanol *or* **Ethyl alcohol** *or* **Ethanol solutions** *or* **Ethyl alcohol solutions**	3	UN1170	II	3	24, IB2, T4, TP1	4b, 150	202	242	5 L	60 L	A		127	FLAMMABLE
				III	3	24, B1, IB3, T2, TP1	4b, 150	203	242	60 L	220 L	A		127	FLAMMABLE
	Ethanolamine *or* **Ethanolamine solutions**	8	UN2491	III	8	IB3, T4, TP1	154	203	241	5 L	60 L	A	52	153	CORROSIVE
	Ether, see **Diethyl ether**														
	Ethers, n.o.s.	3	UN3271	II	3	IB2, T7, TP1, TP8, TP28	150	202	242	5 L	60 L	B		127	FLAMMABLE
				III	3	B1, IB3, T4, TP1, TP29	150	203	242	60 L	220 L	A		127	FLAMMABLE
	Ethyl acetate	3	UN1173	II	3	IB2, T4, TP1	150	202	242	5 L	60 L	B		129	FLAMMABLE
	Ethyl acrylate, stabilized	3	UN1917	II	3	IB2, T4, TP1, TP13	150	202	242	5 L	60 L	B	40	129P	FLAMMABLE
	Ethyl alcohol, *see* **Ethanol**														
	Ethyl aldehyde, see **Acetaldehyde**														
	Ethyl amyl ketone	3	UN2271	III	3	B1, IB3, T2, TP1	150	203	242	60 L	220 L	A		128	FLAMMABLE
	N-Ethyl-N-benzylaniline	6.1	UN2274	III	6.1	IB3, T4, TP1	153	203	241	60 L	220 L	A		153	POISON
	Ethyl borate	3	UN1176	II	3	IB2, T4, TP1	150	202	242	5 L	60 L	B		129	FLAMMABLE
	Ethyl bromide	6.1	UN1891	II	6.1	IB2, IP8, T7, TP2, TP13	None	202	243	5 L	60 L	B	40, 85	131	POISON
	Ethyl bromoacetate	6.1	UN1603	II	6.1, 3	IB2, T7, TP3	None	202	243	Forbidden	Forbidden	D	40	155	POISON
	Ethyl butyl ether	3	UN1179	II	3	B1, IB2, T4, TP1	150	202	242	5 L	60 L	B		127	FLAMMABLE
	Ethyl butyrate	3	UN1180	III	3	B1, IB3, T2, TP1	150	203	242	60 L	220 L	A		130	FLAMMABLE
	Ethyl chloride	2.1	UN1037		2.1	B77, N86, T50	None	322	314, 315	Forbidden	150 kg	B	40	115	FLAMMABLE GAS
	Ethyl chloroacetate	6.1	UN1181	II	6.1, 3	IB2, T7, TP2	153	202	243	5 L	60 L	A		155	POISON
	Ethyl chloroformate	6.1	UN1182	I	6.1, 3, 8	2, B9, B14, B32, N34, T20, TP2, TP13, TP38, TP45	None	227	244	Forbidden	Forbidden	D	21, 40, 100	155	POISON INHALATION HAZARD*
	Ethyl 2-chloropropionate	3	UN2935	III	3	B1, IB3, T2, TP1	150	203	242	60 L	220 L	A		129	FLAMMABLE
+	**Ethyl chlorothioformate**	8	UN2826	II	8, 6.1, 3	2, B9, B14, B32, T20, TP2, TP38, TP45	None	227	244	Forbidden	Forbidden	A	40	155	CORROSIVE, POISON INHALATION HAZARD*
	Ethyl crotonate	3	UN1862	II	3	IB2, T4, TP2	150	202	242	5 L	60 L	B		130	FLAMMABLE
	Ethyl ether, *see* **Diethyl ether**														
	Ethyl fluoride *or* **Refrigerant gas R161**	2.1	UN2453		2.1		306	304	314, 315	Forbidden	150 kg	E	40	115	FLAMMABLE GAS
	Ethyl formate	3	UN1190	II	3	IB2, T4, TP1	150	202	242	5 L	60 L	E		129	FLAMMABLE
	Ethyl hydroperoxide	Forbidden													
	Ethyl isobutyrate	3	UN2385	II	3	IB2, T4, TP1	150	202	242	5 L	60 L	B		129	FLAMMABLE
+	**Ethyl isocyanate**	6.1	UN2481	I	6.1, 3	1, B9, B14, B30, T20, TP2, TP13, TP38, TP44	None	226	244	Forbidden	Forbidden	D	40, 52	155	POISON INHALATION HAZARD*
	Ethyl lactate	3	UN1192	III	3	B1, IB3, T2, TP1	150	203	242	60 L	220 L	A		129	FLAMMABLE
	Ethyl mercaptan	3	UN2363	I	3	A6, T11, TP2, TP13	None	201	243	Forbidden	30 L	E	95, 102	129	FLAMMABLE
	Ethyl methacrylate, stabilized	3	UN2277	II	3	IB2, T4, TP1	150	202	242	5 L	60 L	B		130P	FLAMMABLE
	Ethyl methyl ether	2.1	UN1039		2.1		None	201	314, 315	Forbidden	150 kg	B	40	115	FLAMMABLE GAS
	Ethyl methyl ketone *or* **Methyl ethyl ketone**	3	UN1193	II	3	IB2, T4, TP1	150	202	242	5 L	60 L	B		127	FLAMMABLE
	Ethyl nitrite solutions	3	UN1194	I	3, 6.1		None	201	None	Forbidden	Forbidden	E	40, 105	131	FLAMMABLE

Symbols (1)	Hazardous materials descriptions and proper shipping names (2)	Hazard class or Division (3)	Identification Numbers (4)	PG (5)	Label Codes (6)	Special Provisions (§172.102) (7)	(8) Packaging (§173.***) Exceptions (8A)	Non-bulk (8B)	Bulk (8C)	(9) Quantity limitations (see §§173.27 and 175.75) Passenger aircraft/rail (9A)	Cargo aircraft only (9B)	(10) Vessel stowage Location (10A)	Other (10B)	ERG Guide No.	Placard Advisory Consult §172.504 & §172.505 to determine placard *Denotes placard for any quantity
	Ethyl orthoformate	3	UN2524	III	3	B1, IB3, T2, TP1	150	203	242	60 L	220 L	A		129	FLAMMABLE
	Ethyl oxalate	6.1	UN2525	III	6.1	IB3, T4, TP1	153	203	241	60 L	220 L	A		156	POISON
	Ethyl perchlorate	Forbidden													
D	**Ethyl phosphonothioic dichloride, anhydrous**	6.1	NA2927	I	6.1, 8	2, B9, B14, B32, B74, T20, TP4, TP13, TP38, TP45	None	227	244	Forbidden	Forbidden	D	40	154	POISON INHALATION HAZARD*
D	**Ethyl phosphonous dichloride, anhydrous** *pyrophoric liquid*	6.1	NA2845	I	6.1, 4.2	2, B9, B14, B32, B74, T20, TP4, TP13, TP38, TP45	None	227	244	Forbidden	Forbidden	D	18	135	POISON INHALATION HAZARD*
D	**Ethyl phosphorodichloridate**	6.1	NA2927	I	6.1, 8	2, B9, B14, B32, B74, T20, TP4, TP13, TP38, TP45	None	227	244	Forbidden	Forbidden	D	40	154	POISON INHALATION HAZARD*
	Ethyl propionate	3	UN1195	II	3	IB2, T4, TP1	150	202	242	5 L	60 L	B		129	FLAMMABLE
	Ethyl propyl ether	3	UN2615	II	3	IB2, T4, TP1	150	202	242	5 L	60 L	E		127	FLAMMABLE
	Ethyl silicate, see **Tetraethyl silicate**														
	Ethylacetylene, stabilized	2.1	UN2452		2.1	N88	None	304	314, 315	Forbidden	150 kg	B	40	116P	FLAMMABLE GAS
	Ethylamine	2.1	UN1036		2.1	B77, N87, T50	None	321	314, 315	Forbidden	150 kg	D	40	118	FLAMMABLE GAS
	Ethylamine, aqueous solution *with not less than 50 percent but not more than 70 percent ethylamine*	3	UN2270	II	3, 8	IB2, T7, TP1	150	202	243	1 L	5 L	B	40, 52	132	FLAMMABLE
	N-Ethylaniline	6.1	UN2272	III	6.1	IB3, T4, TP1	153	203	241	60 L	220 L	A	52, 74	153	POISON
	2-Ethylaniline	6.1	UN2273	III	6.1	IB3, T4, TP1	153	203	241	60 L	220 L	A	52, 74	153	POISON
	Ethylbenzene	3	UN1175	II	3	IB2, T4, TP1	150	202	242	5 L	60 L	B		130	FLAMMABLE
	N-Ethylbenzyltoluidines liquid	6.1	UN2753	III	6.1	IB3, T7, TP1	153	203	241	60 L	220 L	A		153	POISON
	N-Ethylbenzyltoluidines, solid	6.1	UN3460	III	6.1	IB8, IP3, T1, TP33	153	213	240	100 kg	200 kg	A		153	POISON
	2-Ethylbutanol	3	UN2275	III	3	B1, IB3, T2, TP1	150	203	242	60 L	220 L	A		129	FLAMMABLE
	2-Ethylbutyl acetate	3	UN1177	III	3	B1, IB3, T2, TP1	150	203	242	60 L	220 L	A		130	FLAMMABLE
	2-Ethylbutyraldehyde	3	UN1178	II	3	B1, IB2, T4, TP1	150	202	242	5 L	60 L	B		130	FLAMMABLE
	Ethyldichloroarsine	6.1	UN1892	I	6.1	2, B9, B14, B32, T20, TP2, TP13, TP38, TP45	None	227	244	Forbidden	Forbidden	D	40	151	POISON INHALATION HAZARD*
	Ethyldichlorosilane	4.3	UN1183	I	4.3, 8, 3	A2, A3, A7, N34, T14, TP2, TP7, TP13	None	201	244	Forbidden	1 L	D	21, 28, 40, 49, 100	139	DANGEROUS WHEN WET*
	Ethylene, acetylene and propylene in mixture, refrigerated liquid *with at least 71.5 percent ethylene with not more than 22.5 percent acetylene and not more than 6 percent propylene*	2.1	UN3138		2.1	T75, TP5	None	304	314, 315	Forbidden	Forbidden	D	40, 57	115	FLAMMABLE GAS
	Ethylene chlorohydrin	6.1	UN1135	I	6.1, 3	2, B9, B14, B32, T20, TP2, TP13, TP38, TP45	None	227	244	Forbidden	Forbidden	D	40	131	POISON INHALATION HAZARD*
	Ethylene	2.1	UN1962		2.1		306	304	302	Forbidden	150 kg	E	40	116P	FLAMMABLE GAS
	Ethylene diamine diperchlorate	Forbidden													
	Ethylene dibromide	6.1	UN1605	I	6.1	2, B9, B14, B32, B77, T20, TP2, TP13, TP38, TP45	None	227	244	Forbidden	Forbidden	D	40	154	POISON INHALATION HAZARD*
	Ethylene dibromide and methyl bromide liquid mixtures, see **Methyl bromide and ethylene dibromide liquid mixtures**														
	Ethylene dichloride	3	UN1184	II	3, 6.1	IB2, N36, T7, TP1	150	202	243	1 L	60 L	B	40	131	FLAMMABLE
	Ethylene glycol diethyl ether	3	UN1153	II	3	IB2, T4, TP1	150	202	242	5 L	60 L	A		127	FLAMMABLE
				III	3	B1, IB3, T2, TP1	150	203	242	60 L	220 L	A		127	FLAMMABLE
	Ethylene glycol dinitrate	Forbidden													
	Ethylene glycol monoethyl ether	3	UN1171	III	3	B1, IB3, T2, TP1	150	203	242	60 L	220 L	A		127	FLAMMABLE
	Ethylene glycol monoethyl ether acetate	3	UN1172	III	3	B1, IB3, T2, TP1	150	203	242	60 L	220 L	A		129	FLAMMABLE
	Ethylene glycol monomethyl ether	3	UN1188	III	3	B1, IB3, T2, TP1	150	203	242	60 L	220 L	A		127	FLAMMABLE
	Ethylene glycol monomethyl ether acetate	3	UN1189	III	3	B1, IB3, T2, TP1	150	203	242	60 L	220 L	A		129	FLAMMABLE
	Ethylene oxide and carbon dioxide mixture *with more than 87 percent ethylene oxide*	2.3	UN3300		2.3, 2.1	4	None	304	314, 315	Forbidden	Forbidden	D	40	119P	POISON GAS*

Symbols (1)	Hazardous materials descriptions and proper shipping names (2)	Hazard class or Division (3)	Identification Numbers (4)	PG (5)	Label Codes (6)	Special Provisions (§172.102) (7)	(8) Packaging (§173.***) Exceptions (8A)	Non-bulk (8B)	Bulk (8C)	(9) Quantity limitations (see §§173.27 and 175.75) Passenger aircraft/rail (9A)	Cargo aircraft only (9B)	(10) Vessel stowage Location (10A)	Other (10B)	ERG Guide No.	Placard Advisory Consult §172.504 & §172.505 to determine placard *Denotes placard for any quantity
	Ethylene oxide and carbon dioxide mixtures *with more than 9 percent but not more than 87 percent ethylene oxide*	2.1	UN1041		2.1	T50	306	304	314, 315	Forbidden	25 kg	B	40	115	FLAMMABLE GAS
	Ethylene oxide and carbon dioxide mixtures *with not more than 9 percent ethylene oxide*	2.2	UN1952		2.2		306	304	314, 315	75 kg	150 kg	A		126	NONFLAMMABLE GAS
	Ethylene oxide and chlorotetrafluoroethane mixture *with not more than 8.8 percent ethylene oxide*	2.2	UN3297		2.2	T50	306	304	314, 315	75 kg	150 kg	A		126	NONFLAMMABLE GAS
	Ethylene oxide and dichlorodifluoromethane mixture, *with not more than 12.5 percent ethylene oxide*	2.2	UN3070		2.2	T50	306	304	314, 315	75 kg	150 kg	A		126	NONFLAMMABLE GAS
	Ethylene oxide and pentafluoroethane mixture *with not more than 7.9 percent ethylene oxide*	2.2	UN3298		2.2	T50	306	304	314, 315	75 kg	150 kg	A		126	NONFLAMMABLE GAS
	Ethylene oxide and propylene oxide mixtures, *with not more than 30 percent ethylene oxide*	3	UN2983	I	3, 6.1	5, A11, N4, N34, T14, TP2, TP7, TP13	None	201	243	Forbidden	30 L	E	40	129P	FLAMMABLE
	Ethylene oxide and tetrafluoroethane mixture *with not more than 5.6 percent ethylene oxide*	2.2	UN3299		2.2	T50	306	304	314, 315	75 kg	150 kg	A		126	NONFLAMMABLE GAS
	Ethylene oxide *or* **Ethylene oxide with nitrogen** *up to a total pressure of 1 MPa (10 bar) at 50 degrees C*	2.3	UN1040		2.3, 2.1	4, 342, T50, TP20	None	323	323	Forbidden	Forbidden	D	40	119P	POISON GAS*
	Ethylene, refrigerated liquid *(cryogenic liquid)*	2.1	UN1038		2.1	T75, TP5	None	316	318, 319	Forbidden	Forbidden	D	40	115	FLAMMABLE GAS
	Ethylenediamine	8	UN1604	II	8, 3	IB2, T7, TP2	154	202	243	1 L	30 L	A	40, 52	132	CORROSIVE
	Ethyleneimine, stabilized	6.1	UN1185	I	6.1, 3	1, B9, B14, B30, B77, N25, N32, T22, TP2, TP13, TP38, TP44	None	226	244	Forbidden	Forbidden	D	40	131P	POISON INHALATION HAZARD*
	Ethylhexaldehyde, see **Octyl aldehydes** *etc.*														
	2-Ethylhexyl chloroformate	6.1	UN2748	II	6.1, 8	IB2, T7, TP2, TP13	153	202	243	1 L	30 L	A	12, 13, 21, 25, 40, 100	156	POISON
	2-Ethylhexylamine	3	UN2276	III	3, 8	B1, IB3, T4, TP1	150	203	242	5 L	60 L	A	40	132	FLAMMABLE
	Ethylphenyldichlorosilane	8	UN2435	II	8	A7, B2, N34, T10, TP2, TP7, TP13	None	206	242	Forbidden	30 L	C		156	CORROSIVE
	1-Ethylpiperidine	3	UN2386	II	3, 8	IB2, T7, TP1	150	202	243	1 L	5 L	B	52	132	FLAMMABLE
	N-Ethyltoluidines	6.1	UN2754	II	6.1	IB2, T7, TP2	153	202	243	5 L	60 L	A		153	POISON
	Ethyltrichlorosilane	3	UN1196	II	3, 8	A7, N34, T10, TP2, TP7, TP13	None	206	243	1 L	5 L	B	40	155	FLAMMABLE
	Etiologic agent, see **Infectious substances,** *etc.*														
	Explosive articles, see **Articles, explosive, n.o.s.** *etc.*														
	Explosive, blasting, type A	1.1D	UN0081	II	1.1D		None	62	None	Forbidden	Forbidden	10	19E, 21E	112	EXPLOSIVES 1.1*
	Explosive, blasting, type B	1.1D	UN0082	II	1.1D		None	62	None	Forbidden	Forbidden	10	19E	112	EXPLOSIVES 1.1*
	Explosive, blasting, type B *or* **Agent blasting, Type B**	1.5D	UN0331	II	1.5D	105, 106	None	62	None	Forbidden	Forbidden	10	19E	112	EXPLOSIVES 1.5
	Explosive, blasting, type C	1.1D	UN0083	II	1.1D	123	None	62	None	Forbidden	Forbidden	10	22E	112	EXPLOSIVES 1.1*
	Explosive, blasting, type D	1.1D	UN0084	II	1.1D		None	62	None	Forbidden	Forbidden	10		112	EXPLOSIVES 1.1*
	Explosive, blasting, type E	1.1D	UN0241	II	1.1D		None	62	None	Forbidden	Forbidden	10	19E	112	EXPLOSIVES 1.1*
	Explosive, blasting, type E *or* **Agent blasting, Type E**	1.5D	UN0332	II	1.5D	105, 106	None	62	None	Forbidden	Forbidden	10	19E	112	EXPLOSIVES 1.5
	Explosive, forbidden. See §173.54	Forbidden													
	Explosive substances, see **Substances, explosive, n.o.s.** *etc.*														
	Explosives, slurry, see **Explosive, blasting, type E**														
	Explosives, water gels, see **Explosive, blasting, type E**														
	Extracts, aromatic, liquid	3	UN1169	II	3	149, IB2, T4, TP1, TP8	150	202	242	5 L	60 L	B		127	FLAMMABLE
				III	3	B1, IB3, T2, TP1	150	203	242	60 L	220 L	A		127	FLAMMABLE
	Extracts, flavoring, liquid	3	UN1197	II	3	149, IB2, T4, TP1, TP8	150	202	242	5 L	60 L	B		127	FLAMMABLE
				III	3	B1, IB3, T2, TP1	150	203	242	60 L	220 L	A		127	FLAMMABLE
	Fabric with animal or vegetable oil, see **Fibers** *or* **Fabrics,** *etc.*														
	Ferric arsenate	6.1	UN1606	II	6.1	IB8, IP2, IP4, T3, TP33	153	212	242	25 kg	100 kg	A		151	POISON
	Ferric arsenite	6.1	UN1607	II	6.1	IB8, IP2, IP4, T3, TP33	153	212	242	25 kg	100 kg	A		151	POISON

Symbols (1)	Hazardous materials descriptions and proper shipping names (2)	Hazard class or Division (3)	Identification Numbers (4)	PG (5)	Label Codes (6)	Special Provisions (§172.102) (7)	(8) Packaging (§173.***) Exceptions (8A)	Non-bulk (8B)	Bulk (8C)	(9) Quantity limitations (see §§173.27 and 175.75) Passenger aircraft/rail (9A)	Cargo aircraft only (9B)	(10) Vessel stowage Location (10A)	Other (10B)	ERG Guide No.	Placard Advisory Consult §172.504 & §172.505 to determine placard *Denotes placard for any quantity
	Ferric chloride, anhydrous	8	UN1773	III	8	IB8, IP3, T1, TP33	154	213	240	25 kg	100 kg	A		157	CORROSIVE
	Ferric chloride, solution	8	UN2582	III	8	B15, IB3, T4, TP1	154	203	241	5 L	60 L	A		154	CORROSIVE
	Ferric nitrate	5.1	UN1466	III	5.1	A1, A29, IB8, IP3, T1, TP33	152	213	240	25 kg	100 kg	A		140	OXIDIZER
	Ferrocerium	4.1	UN1323	II	4.1	59, A19, IB8, IP2, IP4, T3, TP33	151	212	240	15 kg	50 kg	A		170	FLAMMABLE SOLID
	Ferrosilicon, *with 30 percent or more but less than 90 percent silicon*	4.3	UN1408	III	4.3, 6.1	A1, A19, B6, IB8, IP4, IP7, T1, TP33	151	213	240	25 kg	100 kg	A	13, 40, 52, 53, 85, 103	139	DANGEROUS WHEN WET*
	Ferrous arsenate	6.1	UN1608	II	6.1	IB8, IP2, IP4, T3, TP33	153	212	242	25 kg	100 kg	A		151	POISON
D	**Ferrous chloride, solid**	8	NA1759	II	8	IB8, IP2, IP4, T3, TP33	154	212	240	15 kg	50 kg	A		154	CORROSIVE
D	**Ferrous chloride, solution**	8	NA1760	II	8	B3, IB2, T11, TP2, TP27	154	202	242	1 L	30 L	B	40	154	CORROSIVE
	Ferrous metal borings *or* **Ferrous metal shavings** *or* **Ferrous metal turnings** *or* **Ferrous metal cuttings** *in a form liable to self-heating*	4.2	UN2793	III	4.2	A1, A19, IB8, IP3, IP7	None	213	241	25 kg	100 kg	A		170	SPONTANEOUSLY COMBUSTIBLE
	Fertilizer ammoniating solution *with free ammonia*	2.2	UN1043		2.2	N87	306	304	314, 315	Forbidden	150 kg	E	40	125	NONFLAMMABLE GAS
A I W	**Fibers, animal** *or* **Fibers, vegetable** *burnt, wet or damp*	4.2	UN1372	III	4.2		151	213	240	Forbidden	Forbidden	A		133	SPONTANEOUSLY COMBUSTIBLE
I W	**Fibers, vegetable, dry**	4.1	UN3360	III	4.1	137	151	213	240	No limit	No limit	A		133	FLAMMABLE SOLID
A W	**Fibers** *or* **Fabrics, animal** *or* **vegetable** *or* **Synthentic, n.o.s.** *with animal or vegetable oil*	4.2	UN1373	III	4.2	137, IB8, IP3, T1, TP33	None	213	241	Forbidden	Forbidden	A		133	SPONTANEOUSLY COMBUSTIBLE
	Fibers *or* **Fabrics impregnated with weakly nitrated nitrocellulose, n.o.s.**	4.1	UN1353	III	4.1	A1, IB8, IP3	None	213	240	25 kg	100 kg	D		133	FLAMMABLE SOLID
	Films, nitrocellulose base, from which gelatine has been removed; film scrap, see **Celluloid, scrap**														
	Films, nitrocellulose base, *gelatine coated (except scrap)*	4.1	UN1324	III	4.1		None	183	None	25 kg	100 kg	D	28	133	FLAMMABLE SOLID
	Fire extinguisher charges, *corrosive liquid*	8	UN1774	II	8	N41	154	202	None	1 L	30 L	A		154	CORROSIVE
	Fire extinguisher charges, expelling, explosive, see **Cartridges, power device**														
	Fire extinguishers *containing compressed or liquefied gas*	2.2	UN1044		2.2	18, 110	309	309	None	75 kg	150 kg	A		126	NONFLAMMABLE GAS
	Firelighters, solid *with flammable liquid*	4.1	UN2623	III	4.1	A1, A19	None	213	None	25 kg	100 kg	A	52	133	FLAMMABLE SOLID
	Fireworks	1.1G	UN0333	II	1.1G	108	None	62	None	Forbidden	Forbidden	07		112	EXPLOSIVES 1.1*
	Fireworks	1.2G	UN0334	II	1.2G	108	None	62	None	Forbidden	Forbidden	07		112	EXPLOSIVES 1.2*
	Fireworks	1.3G	UN0335	II	1.3G	108	None	62	None	Forbidden	Forbidden	07		112	EXPLOSIVES 1.3*
	Fireworks	1.4G	UN0336	II	1.4G	108	None	62	None	Forbidden	75 kg	06		114	EXPLOSIVES 1.4
	Fireworks	1.4S	UN0337	II	1.4S	108	None	62	None	25 kg	100 kg	05		114	EXPLOSIVES 1.4
	First aid kits	9	UN3316		9	15	161	161	None	10 kg	10 kg	A		171	CLASS 9
W	**Fish meal, stabilized** *or* **Fish scrap, stabilized**	9	UN2216	III	None	155, IB8, IP3, T1, TP33	155	218	218	No limit	No limit	B	88, 122, 128	171	CLASS 9
	Fish meal, unstablized *or* **Fish scrap, unstabilized**	4.2	UN1374	II	4.2	155, A1, A19, IB8, IP2, IP4, T3, TP33	None	212	241	15 kg	50 kg	B	18, 128	133	SPONTANEOUSLY COMBUSTIBLE
	Flammable compressed gas, see **Compressed** *or* **Liquefied gas, flammable,** *etc.*														
	Flammable compressed gas (small receptacles not fitted with a dispersion device, not refillable), see **Receptacles,** *etc.*														
	Flammable gas in lighters, see **Lighters** *or* **lighter refills,** *cigarettes, containing flammable gas*														
G	**Flammable liquid, toxic, corrosive, n.o.s.**	3	UN3286	I	3, 6.1, 8	T14, TP2, TP13, TP27	None	201	243	Forbidden	2.5 L	E	21, 40, 100	131	FLAMMABLE
				II	3, 6.1, 8	IB2, T11, TP2, TP13, TP27	150	202	243	1 L	5 L	B	21, 40, 100	131	FLAMMABLE
G	**Flammable liquids, corrosive, n.o.s.**	3	UN2924	I	3, 8	T14, TP2	None	201	243	0.5 L	2.5 L	E	40	132	FLAMMABLE
				II	3, 8	IB2, T11, TP2, TP27	150	202	243	1 L	5 L	B	40	132	FLAMMABLE
				III	3, 8	B1, IB3, T7, TP1, TP28	150	203	242	5 L	60 L	A	40	132	FLAMMABLE

Symbols (1)	Hazardous materials descriptions and proper shipping names (2)	Hazard class or Division (3)	Identification Numbers (4)	PG (5)	Label Codes (6)	Special Provisions (§172.102) (7)	(8) Packaging (§173.***) Exceptions (8A)	Non-bulk (8B)	Bulk (8C)	(9) Quantity limitations (see §§173.27 and 175.75) Passenger aircraft/rail (9A)	Cargo aircraft only (9B)	(10) Vessel stowage Location (10A)	Other (10B)	ERG Guide No.	Placard Advisory Consult §172.504 & §172.505 to determine placard *Denotes placard for any quantity
G	**Flammable liquids, n.o.s.**	3	UN1993	I	3	T11, TP1, TP27	150	201	243	1 L	30 L	E		128	FLAMMABLE
				II	3	IB2, T7, TP1, TP8, TP28	150	202	242	5 L	60 L	B		128	FLAMMABLE
				III	3	B1, B52, IB3, T4, TP1, TP29	150	203	242	60 L	220 L	A		128	FLAMMABLE
G	**Flammable liquids, toxic, n.o.s.**	3	UN1992	I	3, 6.1	T14, TP2, TP13, TP27	None	201	243	Forbidden	30 L	E	40	131	FLAMMABLE
				II	3, 6.1	IB2, T7, TP2, TP13	150	202	243	1 L	60 L	B	40	131	FLAMMABLE
				III	3, 6.1	B1, IB3, T7, TP1, TP28	150	203	242	60 L	220 L	A		131	FLAMMABLE
G	**Flammable solid, corrosive, inorganic, n.o.s.**	4.1	UN3180	II	4.1, 8	A1, IB6, IP2, T3, TP33	151	212	242	15 kg	50 kg	D	40	134	FLAMMABLE SOLID
				III	4.1, 8	A1, IB6, T1, TP33	151	213	242	25 kg	100 kg	D	40	134	FLAMMABLE SOLID
G	**Flammable solid, inorganic, n.o.s.**	4.1	UN3178	II	4.1	A1, IB8, IP2, IP4, T3, TP33	151	212	240	15 kg	50 kg	B		133	FLAMMABLE SOLID
				III	4.1	A1, IB8, IP3, T1, TP33	151	213	240	25 kg	100 kg	B		133	FLAMMABLE SOLID
G	**Flammable solid, organic, molten, n.o.s.**	4.1	UN3176	II	4.1	IB1, T3, TP3, TP26	151	212	240	Forbidden	Forbidden	C		133	FLAMMABLE SOLID
				III	4.1	IB1, T1, TP3, TP26	151	213	240	Forbidden	Forbidden	C		133	FLAMMABLE SOLID
G	**Flammable solid, oxidizing, n.o.s.**	4.1	UN3097	II	4.1, 5.1	131	None	214	214	Forbidden	Forbidden	E	40	140	FLAMMABLE SOLID
				III	4.1, 5.1	131, T1, TP33	None	214	214	Forbidden	Forbidden	D	40	140	FLAMMABLE SOLID
G	**Flammable solid, toxic, inorganic, n.o.s.**	4.1	UN3179	II	4.1, 6.1	A1, IB6, IP2, T3, TP33	151	212	242	15 kg	50 kg	B	40	134	FLAMMABLE SOLID
				III	4.1, 6.1	A1, IB6, T1, TP33	151	213	242	25 kg	100 kg	B	40	134	FLAMMABLE SOLID
G	**Flammable solids, corrosive, organic, n.o.s.**	4.1	UN2925	II	4.1, 8	A1, IB6, IP2, T3, TP33	None	212	242	15 kg	50 kg	D	40	134	FLAMMABLE SOLID
				III	4.1, 8	A1, IB6, T1, TP33	151	213	242	25 kg	100 kg	D	40	134	FLAMMABLE SOLID
G	**Flammable solids, organic, n.o.s.**	4.1	UN1325	II	4.1	A1, IB8, IP2, IP4, T3, TP33	151	212	240	15 kg	50 kg	B		133	FLAMMABLE SOLID
				III	4.1	A1, IB8, IP3, T1, TP33	151	213	240	25 kg	100 kg	B		133	FLAMMABLE SOLID
G	**Flammable solids, toxic, organic, n.o.s.**	4.1	UN2926	II	4.1, 6.1	A1, IB6, IP2, T3, TP33	151	212	242	15 kg	50 kg	B	40	134	FLAMMABLE SOLID
				III	4.1, 6.1	A1, IB6, T1, TP33	151	213	242	25 kg	100 kg	B	40	134	FLAMMABLE SOLID
	Flares, aerial	1.3G	UN0093	II	1.3G		None	62	None	Forbidden	75 kg	07		112	EXPLOSIVES 1.3*
	Flares, aerial	1.4G	UN0403	II	1.4G		None	62	None	Forbidden	75 kg	06		114	EXPLOSIVES 1.4
	Flares, aerial	1.4S	UN0404	II	1.4S		None	62	None	25 kg	100 kg	05		114	EXPLOSIVES 1.4
	Flares, aerial	1.1G	UN0420	II	1.1G		None	62	None	Forbidden	Forbidden	07		112	EXPLOSIVES 1.1*
	Flares, aerial	1.2G	UN0421	II	1.2G		None	62	None	Forbidden	Forbidden	07		112	EXPLOSIVES 1.2*
	Flares, airplane, see **Flares, aerial**														
	Flares, signal, see **Cartridges, signal**														
	Flares, surface	1.3G	UN0092	II	1.3G		None	62	None	Forbidden	75 kg	07		112	EXPLOSIVES 1.3*
	Flares, surface	1.1G	UN0418	II	1.1G		None	62	None	Forbidden	Forbidden	07		112	EXPLOSIVES 1.1*
	Flares, surface	1.2G	UN0419	II	1.2G		None	62	None	Forbidden	Forbidden	07		112	EXPLOSIVES 1.2*
	Flares, water-activated, see **Contrivances, water-activated,** *etc.*														
	Flash powder	1.1G	UN0094	II	1.1G		None	62	None	Forbidden	Forbidden	15		112	EXPLOSIVES 1.1*
	Flash powder	1.3G	UN0305	II	1.3G		None	62	None	Forbidden	Forbidden	15		112	EXPLOSIVES 1.3*
	Flue dusts, poisonous, see **Arsenical dust**														
	Fluoric acid, see **Hydrofluoric acid,** *etc.*														
	Fluorine, compressed	2.3	UN1045		2.3, 5.1, 8	1, N86	None	302	None	Forbidden	Forbidden	D	40, 89, 90	124	POISON GAS*

Symbols (1)	Hazardous materials descriptions and proper shipping names (2)	Hazard class or Division (3)	Identification Numbers (4)	PG (5)	Label Codes (6)	Special Provisions (§172.102) (7)	(8) Packaging (§173.***) Exceptions (8A)	Non-bulk (8B)	Bulk (8C)	(9) Quantity limitations (see §§173.27 and 175.75) Passenger aircraft/rail (9A)	Cargo aircraft only (9B)	(10) Vessel stowage Location (10A)	Other (10B)	ERG Guide No.	Placard Advisory Consult §172.504 & §172.505 to determine placard *Denotes placard for any quantity
	Fluoroacetic acid	6.1	UN2642	I	6.1	IB7, IP1, T6, TP33	None	211	242	1 kg	15 kg	E		154	POISON
	Fluoroanilines	6.1	UN2941	III	6.1	IB3, T4, TP1	153	203	241	60 L	220 L	A		153	POISON
	Fluorobenzene	3	UN2387	II	3	IB2, T4, TP1	150	202	242	5 L	60 L	B		130	FLAMMABLE
	Fluoroboric acid	8	UN1775	II	8	A6, A7, B2, B15, IB2, N3, N34, T7, TP2	154	202	242	1 L	30 L	A		154	CORROSIVE
	Fluorophosphoric acid anhydrous	8	UN1776	II	8	A6, A7, B2, IB2, N3, N34, T8, TP2	None	202	242	1 L	30 L	A		154	CORROSIVE
G	**Fluorosilicates, n.o.s.**	6.1	UN2856	III	6.1	IB8, IP3, T1, TP33	153	213	240	100 kg	200 kg	A	52	151	POISON
	Fluorosilicic acid	8	UN1778	II	8	A6, A7, B2, B15, IB2, N3, N34, T8, TP2	None	202	242	1 L	30 L	A		154	CORROSIVE
	Fluorosulfonic acid	8	UN1777	I	8	A3, A6, A7, A10, B6, B10, N3, N36, T10, TP2	None	201	243	0.5 L	2.5 L	D	40	137	CORROSIVE
	Fluorotoluenes	3	UN2388	II	3	IB2, T4, TP1	150	202	242	5 L	60 L	B	40	130	FLAMMABLE
	Forbidden materials. See §173.21	Forbidden													
	Formaldehyde solutions, flammable	3	UN1198	III	3, 8	176, B1, IB3, T4, TP1	150	203	242	5 L	60 L	A	40	132	FLAMMABLE
	Formaldehyde solutions, *with not less than 25 percent formaldehyde*	8	UN2209	III	8	IB3, T4, TP1	154	203	241	5 L	60 L	A		132	CORROSIVE
	Formaldehyde solutions (with not less than 10% and less than 25% formaldehyde), see **Aviation regulated liquid, n.o.s.** *or* **Other regulated substances, liquid, n.o.s.**														
	Formalin, see **Formaldehyde solutions**														
	Formic acid *with more than 85% acid by mass*	8	UN1779	II	8, 3	B2, B28, IB2, T7, TP2	154	202	242	1 L	30 L	A	40	153	CORROSIVE
	Formic acid *with not less than 10% but not more than 85% acid by mass*	8	UN3412	II	8	IB2, T7, TP2	154	202	242	1 L	30 L	A	40	153	CORROSIVE
	Formic acid *with not less than 5% but less than 10% acid by mass*	8	UN3412	III	8	IB3, T4, TP1	154	203	241	5 L	60 L	A	40	153	CORROSIVE
	Fracturing devices, explosive, *without detonators for oil wells*	1.1D	UN0099	II	1.1D		None	62	62	Forbidden	Forbidden	07		112	EXPLOSIVES 1.1*
	Fuel, aviation, turbine engine	3	UN1863	I	3	144, T11, TP1, TP8, TP28	150	201	243	1 L	30 L	E		128	FLAMMABLE
				II	3	144, IB2, T4, TP1, TP8	150	202	242	5 L	60 L	B		128	FLAMMABLE
				III	3	144, B1, IB3, T2, TP1	150	203	242	60 L	220 L	A		128	FLAMMABLE
	Fuel cell cartridges *or* **Fuel cell cartridges contained in equipment** *or* **Fuel cell cartridges packed with equipment,** *containing corrosive substances*	8	UN3477		8		230	230	230	5 kg	50 kg	A		153	CORROSIVE
	Fuel cell cartridges *or* **Fuel cell cartridges contained in equipment** *or* **Fuel cell cartridges packed with equipment** *containing flammable liquids*	3	UN3473		3		230	230	230	5 kg	50 kg	A		128	FLAMMABLE
	Fuel cell cartridges *or* **Fuel cell cartridges contained in equipment** *or* **Fuel cell cartridges packed with equipment,** *containing hydrogen in metal hydride*	2.1	UN3479		2.1		230	230	230	1 kg	15 kg	B		115	FLAMMABLE GAS
	Fuel cell cartridges *or* **Fuel cell cartridges contained in equipment** *or* **Fuel cell cartridges packed with equipment,** *containing liquefied flammable gas*	2.1	UN3478		2.1		230	230	230	1 kg	15 kg	B		115	FLAMMABLE GAS
	Fuel cell cartridges *or* **Fuel cell cartridges contained in equipment** *or* **Fuel cell cartridges packed with equipment,** *containing water-reactive substances*	4.3	UN3476		4.3		230	230	230	5 kg	50 kg	A		138	DANGEROUS WHEN WET
D	**Fuel oil** *(No. 1, 2, 4, 5, or 6)*	3	NA1993	III	3	144, B1, IB3, T4, TP1, TP29	150	203	242	60 L	220 L	A		128	FLAMMABLE
	Fuel system components (including fuel control units (FCU), carburetors, fuel lines, fuel pumps) see **Dangerous Goods in Apparatus** *or* **Dangerous Goods in Machinery**														
	Fulminate of mercury (dry)	Forbidden													
	Fulminate of mercury, wet, see **Mercury fulminate,** *etc.*														
	Fulminating gold	Forbidden													
	Fulminating mercury	Forbidden													

Symbols (1)	Hazardous materials descriptions and proper shipping names (2)	Hazard class or Division (3)	Identification Numbers (4)	PG (5)	Label Codes (6)	Special Provisions (§172.102) (7)	(8) Packaging (§173.***) Exceptions (8A)	Non-bulk (8B)	Bulk (8C)	(9) Quantity limitations (see §§173.27 and 175.75) Passenger aircraft/rail (9A)	Cargo aircraft only (9B)	(10) Vessel stowage Location (10A)	Other (10B)	ERG Guide No.	Placard Advisory Consult §172.504 & §172.505 to determine placard *Denotes placard for any quantity
	Fulminating platinum	Forbidden													
	Fulminating silver	Forbidden													
	Fulminic acid	Forbidden													
	Fumaryl chloride	8	UN1780	II	8	B2, IB2, T7, TP2	154	202	242	1 L	30 L	C	8, 40	156	CORROSIVE
	Fumigated lading, see §§172.302(g), 173.9 and 176.76(h)														
	Fumigated transport vehicle or freight container see §173.9														
	Furaldehydes	6.1	UN1199	II	6.1, 3	IB2, T7, TP2	153	202	243	5 L	60 L	A		132P	POISON
	Furan	3	UN2389	I	3	T12, TP2, TP13	None	201	243	1 L	30 L	E	40	128	FLAMMABLE
	Furfuryl alcohol	6.1	UN2874	III	6.1	IB3, T4, TP1	153	203	241	60 L	220 L	A	52, 74	153	POISON
	Furfurylamine	3	UN2526	III	3, 8	B1, IB3, T4, TP1	150	203	242	5 L	60 L	A	40	132	FLAMMABLE
	Fuse, detonating, *metal clad, see* **Cord, detonating,** *metal clad*														
	Fuse, detonating, mild effect, *metal clad, see* **Cord, detonating, mild effect,** *metal clad*														
	Fuse, igniter *tubular metal clad*	1.4G	UN0103	II	1.4G		None	62	None	Forbidden	75 kg	06		114	EXPLOSIVES 1.4
	Fuse, non-detonating *instantaneous or quickmatch*	1.3G	UN0101	II	1.3G		None	62	None	Forbidden	Forbidden	07		112	EXPLOSIVES 1.3*
	Fuse, safety	1.4S	UN0105	II	1.4S		None	62	None	25 kg	100 kg	05		114	EXPLOSIVES 1.4
D	**Fusee** *(railway or highway)*	4.1	NA1325	II	4.1		None	184	None	15 kg	50 kg	B		133	FLAMMABLE SOLID
	Fusel oil	3	UN1201	II	3	IB2, T4, TP1	150	202	242	5 L	60 L	B		127	FLAMMABLE
				III	3	B1, IB3, T2, TP1	150	203	242	60 L	220 L	A		127	FLAMMABLE
	Fuses, tracer, see **Tracers for ammunition**														
	Fuzes, combination, percussion and time, see **Fuzes, detonating** *(UN0257, UN0367);* **Fuzes, igniting** *(UN0317, UN0368)*														
	Fuzes, detonating	1.1B	UN0106	II	1.1B		None	62	None	Forbidden	Forbidden	11		112	EXPLOSIVES 1.1*
	Fuzes, detonating	1.2B	UN0107	II	1.2B		None	62	None	Forbidden	Forbidden	11		112	EXPLOSIVES 1.2*
	Fuzes, detonating	1.4B	UN0257	II	1.4B	116	None	62	None	Forbidden	75 kg	06		114	EXPLOSIVES 1.4
	Fuzes, detonating	1.4S	UN0367	II	1.4S	116	None	62	None	25 kg	100 kg	05		114	EXPLOSIVES 1.4
	Fuzes, detonating, *with protective features*	1.1D	UN0408	II	1.1D		None	62	None	Forbidden	Forbidden	07		112	EXPLOSIVES 1.1*
	Fuzes, detonating, *with protective features*	1.2D	UN0409	II	1.2D		None	62	None	Forbidden	Forbidden	07		112	EXPLOSIVES 1.2*
	Fuzes, detonating, *with protective features*	1.4D	UN0410	II	1.4D	116	None	62	None	Forbidden	75 kg	06		114	EXPLOSIVES 1.4
	Fuzes, igniting	1.3G	UN0316	II	1.3G		None	62	None	Forbidden	Forbidden	07		112	EXPLOSIVES 1.3*
	Fuzes, igniting	1.4G	UN0317	II	1.4G		None	62	None	Forbidden	75 kg	06		114	EXPLOSIVES 1.4
	Fuzes, igniting	1.4S	UN0368	II	1.4S		None	62	None	25 kg	100 kg	05		114	EXPLOSIVES 1.4
	Galactsan trinitrate	Forbidden													
	Gallium	8	UN2803	III	8	T1, TP33	None	162	240	20 kg	20 kg	B	48	172	CORROSIVE
	Gas cartridges, *(flammable) without a release device, non-refillable*	2.1	UN2037		2.1		306	304	None	1 kg	15 kg	B	40	115	FLAMMABLE GAS
D	**Gas identification set**	2.3	NA9035		2.3	6	None	194	None	Forbidden	Forbidden	D		123	POISON GAS*
	Gas oil	3	UN1202	III	3	144, B1, IB3, T2, TP1	150	203	242	60 L	220 L	A		128	FLAMMABLE
G	**Gas, refrigerated liquid, flammable, n.o.s.** *(cryogenic liquid)*	2.1	UN3312		2.1	T75, TP5	None	316	318	Forbidden	Forbidden	D	40	115	FLAMMABLE GAS
G	**Gas, refrigerated liquid, n.o.s.** *(cryogenic liquid)*	2.2	UN3158		2.2	T75, TP5	320	316	318	50 kg	500 kg	D		120	NONFLAMMABLE GAS
G	**Gas, refrigerated liquid, oxidizing, n.o.s.** *(cryogenic liquid)*	2.2	UN3311		2.2, 5.1	T75, TP5, TP22	320	316	318	Forbidden	Forbidden	D		122	NONFLAMMABLE GAS
	Gas sample, non-pressurized, flammable, n.o.s., *not refrigerated liquid*	2.1	UN3167		2.1		306	302, 304	None	1 L	5 L	D		115	FLAMMABLE GAS
	Gas sample, non-pressurized, toxic, flammable, n.o.s., *not refrigerated liquid*	2.3	UN3168		2.3, 2.1	6	306	302	None	Forbidden	1 L	D		119	POISON GAS*
	Gas sample, non-pressurized, toxic, n.o.s., *not refrigerated liquid*	2.3	UN3169		2.3	6	306	302, 304	None	Forbidden	1 L	D		123	POISON GAS*
D	**Gasohol** *gasoline mixed with ethyl alcohol, with not more than 10% alcohol*	3	NA1203	II	3	144, 177	150	202	242	5 L	60 L	E		128	FLAMMABLE
	Gasoline, casinghead, see **Gasoline**														

Symbols (1)	Hazardous materials descriptions and proper shipping names (2)	Hazard class or Division (3)	Identification Numbers (4)	PG (5)	Label Codes (6)	Special Provisions (§172.102) (7)	(8) Packaging (§173.***) Exceptions (8A)	Non-bulk (8B)	Bulk (8C)	(9) Quantity limitations (see §§173.27 and 175.75) Passenger aircraft/rail (9A)	Cargo aircraft only (9B)	(10) Vessel stowage Location (10A)	Other (10B)	ERG Guide No.	Placard Advisory Consult §172.504 & §172.505 to determine placard *Denotes placard for any quantity
	Gasoline *includes gasoline mixed with ethyl alcohol, with not more than 10% alcohol*	3	UN1203	II	3	144, 177, B1, B33, IB2, T8	150	202	242	5 L	60 L	E		128	FLAMMABLE
	Gelatine, blasting, see **Explosive, blasting, type A**														
	Gelatine dynamites, see **Explosive, blasting, type A**														
	Germane	2.3	UN2192		2.3, 2.1	2	None	302	245	Forbidden	Forbidden	D	40	119	POISON GAS*
	Glycerol-1,3-dinitrate	Forbidden													
	Glycerol gluconate trinitrate	Forbidden													
	Glycerol lactate trinitrate	Forbidden													
	Glycerol alpha-monochlorohydrin	6.1	UN2689	III	6.1	IB3, T4, TP1	153	203	241	60 L	220 L	A		153	POISON
	Glyceryl trinitrate, see **Nitroglycerin,** *etc.*														
	Glycidaldehyde	3	UN2622	II	3, 6.1	IB2, IP8, T7, TP1	150	202	243	1 L	60 L	A	40	131P	FLAMMABLE
	Grenades, *hand or rifle, with bursting charge*	1.1D	UN0284	II	1.1D			62	None	Forbidden	Forbidden	07		112	EXPLOSIVES 1.1*
	Grenades, *hand or rifle, with bursting charge*	1.2D	UN0285	II	1.2D			62	None	Forbidden	Forbidden	07		112	EXPLOSIVES 1.2*
	Grenades, *hand or rifle, with bursting charge*	1.1F	UN0292	II	1.1F			62	None	Forbidden	Forbidden	08		112	EXPLOSIVES 1.1*
	Grenades, *hand or rifle, with bursting charge*	1.2F	UN0293	II	1.2F			62	None	Forbidden	Forbidden	08		112	EXPLOSIVES 1.2*
	Grenades, illuminating, see **Ammunition, illuminating,** *etc.*														
	Grenades, practice, *hand or rifle*	1.4S	UN0110	II	1.4S			62	None	25 kg	100 kg	05		114	EXPLOSIVES 1.4
	Grenades, practice, *hand or rifle*	1.3G	UN0318	II	1.3G			62	None	Forbidden	Forbidden	07		112	EXPLOSIVES 1.3*
	Grenades, practice, *hand or rifle*	1.2G	UN0372	II	1.2G			62	None	Forbidden	Forbidden	07		112	EXPLOSIVES 1.2*
	Grenades practice *Hand or rifle*	1.4G	UN0452	II	1.4G			62	None	Forbidden	75 kg	06		114	EXPLOSIVES 1.4
	Grenades, smoke, see **Ammunition, smoke,** *etc.*														
	Guanidine nitrate	5.1	UN1467	III	5.1	A1, IB8, IP3, T1, TP33	152	213	240	25 kg	100 kg	A	73	143	OXIDIZER
	Guanyl nitrosaminoguanylidene hydrazine (dry)	Forbidden													
	Guanyl nitrosaminoguanylidene hydrazine, wetted *with not less than 30 percent water, by mass*	1.1A	UN0113	II	1.1A	111, 117	None	62	None	Forbidden	Forbidden	12		112	EXPLOSIVES 1.1*
	Guanyl nitrosaminoguanyltetrazene (dry)	Forbidden													
	Guanyl nitrosaminoguanyltetrazene, wetted *or* **Tetrazene, wetted** *with not less than 30 percent water or mixture of alcohol and water, by mass*	1.1A	UN0114	II	1.1A	111, 117	None	62	None	Forbidden	Forbidden	12		112	EXPLOSIVES 1.1*
	Gunpowder, compressed *or* **Gunpowder in pellets,** *see* **Black powder** *(UN0028)*														
	Gunpowder, *granular or as a meal, see* **Black powder** *(UN0027)*														
	Hafnium powder, dry	4.2	UN2545	I	4.2		None	211	242	Forbidden	Forbidden	D		135	SPONTANEOUSLY COMBUSTIBLE
				II	4.2	A19, A20, IB6, IP2, N34, T3, TP33	None	212	241	15 kg	50 kg	D		135	SPONTANEOUSLY COMBUSTIBLE
				III	4.2	IB8, IP3, T1, TP33	None	213	241	25 kg	100 kg	D		135	SPONTANEOUSLY COMBUSTIBLE
	Hafnium powder, wetted *with not less than 25 percent water (a visible excess of water must be present) (a) mechanically produced, particle size less than 53 microns; (b) chemically produced, particle size less than 840 microns*	4.1	UN1326	II	4.1	A6, A19, A20, IB6, IP2, N34, T3, TP33	None	212	241	15 kg	50 kg	E	74	170	FLAMMABLE SOLID
	Hand signal device, see **Signal devices, hand**														
	Hazardous substances, liquid or solid, n.o.s., see **Environmentally hazardous substances,** *etc.*														
D G	**Hazardous waste, liquid, n.o.s.**	9	NA3082	III	9	IB3, T2, TP1	155	203	241	No limit	No limit	A		171	CLASS 9
D G	**Hazardous waste, solid, n.o.s.**	9	NA3077	III	9	B54, IB8, IP2, T1, TP33	155	213	240	No limit	No limit	A		171	CLASS 9
	Heating oil, light	3	UN1202	III	3	B1, IB3, T2, TP1	150	203	242	60 L	220 L	A		128	FLAMMABLE
	Helium, *compressed*	2.2	UN1046		2.2		306	302	302, 314	75 kg	150 kg	A	85	121	NONFLAMMABLE GAS
	Helium, refrigerated liquid *(cryogenic liquid)*	2.2	UN1963		2.2	T75, TP5	320	316	318	50 kg	500 kg	D		120	NONFLAMMABLE GAS

Symbols (1)	Hazardous materials descriptions and proper shipping names (2)	Hazard class or Division (3)	Identification Numbers (4)	PG (5)	Label Codes (6)	Special Provisions (§172.102) (7)	(8) Packaging (§173.***) Exceptions (8A)	Non-bulk (8B)	Bulk (8C)	(9) Quantity limitations (see §§173.27 and 175.75) Passenger aircraft/rail (9A)	Cargo aircraft only (9B)	(10) Vessel stowage Location (10A)	Other (10B)	ERG Guide No.	Placard Advisory Consult §172.504 & §172.505 to determine placard *Denotes placard for any quantity
	Heptafluoropropane *or* **Refrigerant gas R 227**	2.2	UN3296		2.2	T50	306	304	314, 315	75 kg	150 kg	A		126	NONFLAMMABLE GAS
	n-Heptaldehyde	3	UN3056	III	3	B1, IB3, T2, TP1	150	203	242	60 L	220 L	A		129	FLAMMABLE
	Heptanes	3	UN1206	II	3	IB2, T4, TP1	150	202	242	5 L	60 L	B		128	FLAMMABLE
	n-Heptene	3	UN2278	II	3	IB2, T4, TP1	150	202	242	5 L	60 L	B		128	FLAMMABLE
	Hexachloroacetone	6.1	UN2661	III	6.1	IB3, T4, TP1	153	203	241	60 L	220 L	B	12, 40	153	POISON
	Hexachlorobenzene	6.1	UN2729	III	6.1	B3, IB8, IP3, T1, TP33	153	203	241	60 L	220 L	A		152	POISON
	Hexachlorobutadiene	6.1	UN2279	III	6.1	IB3, T4, TP1	153	203	241	60 L	220 L	A		151	POISON
	Hexachlorocyclopentadiene	6.1	UN2646	I	6.1	2, B9, B14, B32, B77, T20, TP2, TP13, TP38, TP45	None	227	244	Forbidden	Forbidden	D	40	151	POISON INHALATION HAZARD*
	Hexachlorophene	6.1	UN2875	III	6.1	IB8, IP3, T1, TP33	153	213	240	100 kg	200 kg	A		151	POISON
	Hexadecyltrichlorosilane	8	UN1781	II	8	A7, B2, B6, N34, T10, TP2, TP7, TP13	None	206	242	Forbidden	30 L	C	40	156	CORROSIVE
	Hexadienes	3	UN2458	II	3	IB2, T4, TP1	None	202	242	5 L	60 L	B		130	FLAMMABLE
	Hexaethyl tetraphosphate and compressed gas mixtures	2.3	UN1612		2.3	3	None	334	None	Forbidden	Forbidden	D	40	123	POISON GAS*
	Hexaethyl tetraphosphate, *liquid*	6.1	UN1611	II	6.1	IB2, N76, T7, TP2	153	202	243	5 L	60 L	E	40	151	POISON
	Hexaethyl tetraphosphate, *solid*	6.1	UN1611	II	6.1	IB8, IP2, IP4, N76	153	212	242	25 kg	100 kg	E	40	151	POISON
	Hexafluoroacetone	2.3	UN2420		2.3, 8	2, B9, B14	None	304	314, 315	Forbidden	Forbidden	D	40	125	POISON GAS*
	Hexafluoroacetone hydrate, liquid	6.1	UN2552	II	6.1	IB2, T7, TP2	153	202	243	5 L	60 L	B	40	151	POISON
	Hexafluoroacetone hydrate, solid	6.1	UN3436	II	6.1	IB8, IP2, IP4, T3, TP33	153	212	242	25 kg	100 kg	B	40	151	POISON
	Hexafluoroethane, *or* **Refrigerant gas R 116**	2.2	UN2193		2.2		306	304	314, 315	75 kg	150 kg	A		126	NONFLAMMABLE GAS
	Hexafluorophosphoric acid	8	UN1782	II	8	A6, A7, B2, IB2, N3, N34, T8, TP2	None	202	242	1 L	30 L	A		154	CORROSIVE
	Hexafluoropropylene, compressed *or* **Refrigerant gas R 1216**	2.2	UN1858		2.2	T50	306	304	314, 315	75 kg	150 kg	A		126	NONFLAMMABLE GAS
	Hexaldehyde	3	UN1207	III	3	B1, IB3, T2, TP1	150	203	242	60 L	220 L	A		130	FLAMMABLE
	Hexamethylene diisocyanate	6.1	UN2281	II	6.1	IB2, T7, TP2, TP13	153	202	243	5 L	60 L	C	13, 40	156	POISON
	Hexamethylene triperoxide diamine (dry)	Forbidden													
	Hexamethylenediamine, solid	8	UN2280	III	8	IB8, IP3, T1, TP33	154	213	240	25 kg	100 kg	A	12	153	CORROSIVE
	Hexamethylenediamine solution	8	UN1783	II	8	IB2, T7, TP2	None	202	242	1 L	30 L	A		153	CORROSIVE
				III	8	IB3, T4, TP1	154	203	241	5 L	60 L	A		153	CORROSIVE
	Hexamethyleneimine	3	UN2493	II	3, 8	IB2, T7, TP1	150	202	243	1 L	5 L	B	40	132	FLAMMABLE
	Hexamethylenetetramine	4.1	UN1328	III	4.1	A1, IB8, IP3, T1, TP33	151	213	240	25 kg	100 kg	A		133	FLAMMABLE SOLID
	Hexamethylol benzene hexanitrate	Forbidden													
	Hexanes	3	UN1208	II	3	IB2, T4, TP1	150	202	242	5 L	60 L	E		128	FLAMMABLE
	2,2',4,4',6,6'-Hexanitro-3,3'-dihydroxyazobenzene (dry)	Forbidden													
	Hexanitroazoxy benzene	Forbidden													
	N,N'-(hexanitrodiphenyl) ethylene dinitramine (dry)	Forbidden													
	Hexanitrodiphenyl urea	Forbidden													
	2,2',3',4,4',6-Hexanitrodiphenylamine	Forbidden													
	Hexanitrodiphenylamine *or* **Dipicrylamine** *or* **Hexyl**	1.1D	UN0079	II	1.1D		None	62	None	Forbidden	Forbidden	10		112	EXPLOSIVES 1.1*
	2,3',4,4',6,6'-Hexanitrodiphenylether	Forbidden													
	Hexanitroethane	Forbidden													
	Hexanitrooxanilide	Forbidden													
	Hexanitrostilbene	1.1D	UN0392	II	1.1D		None	62	None	Forbidden	Forbidden	10		112	EXPLOSIVES 1.1*
	Hexanoic acid, see **Corrosive liquids, n.o.s.**														
	Hexanols	3	UN2282	III	3	B1, IB3, T2, TP1	150	203	242	60 L	220 L	A	74	129	FLAMMABLE
	1-Hexene	3	UN2370	II	3	IB2, T4, TP1	150	202	242	5 L	60 L	E		128	FLAMMABLE

Symbols	Hazardous materials descriptions and proper shipping names	Hazard class or Division	Identification Numbers	PG	Label Codes	Special Provisions (§172.102)	(8) Packaging (§173.***)			(9) Quantity limitations (see §§173.27 and 175.75)		(10) Vessel stowage		ERG Guide No.	Placard Advisory Consult §172.504 & §172.505 to determine placard *Denotes placard for any quantity
							Exceptions	Non-bulk	Bulk	Passenger aircraft/rail	Cargo aircraft only	Location	Other		
(1)	(2)	(3)	(4)	(5)	(6)	(7)	(8A)	(8B)	(8C)	(9A)	(9B)	(10A)	(10B)		
	Hexogen and cyclotetramethylenetetranitramine mixtures, wetted *or* **desensitized** *see* **RDX and HMX mixtures, wetted** *or* **desensitized** *etc.*														
	Hexogen and HMX mixtures, wetted *or* **desensitized** *see* **RDX and HMX mixtures, wetted** *or* **desensitized** *etc.*														
	Hexogen and octogen mixtures, wetted *or* **desensitized** *see* **RDX and HMX mixtures, wetted** *or* **desensitized** *etc.*														
	Hexogen, *see* **Cyclotrimethylenetrinitramine,** *etc.*														
	Hexolite, *or* **Hexotol** *dry or wetted with less than 15 percent water, by mass*	1.1D	UN0118	II	1.1D		None	62	None	Forbidden	Forbidden	10		112	EXPLOSIVES 1.1*
	Hexotonal	1.1D	UN0393	II	1.1D		None	62	None	Forbidden	Forbidden	10		112	EXPLOSIVES 1.1*
	Hexyl, *see* **Hexanitrodiphenylamine**														
	Hexyltrichlorosilane	8	UN1784	II	8	A7, B2, B6, N34, T10, TP2, TP7, TP13	None	206	242	Forbidden	30 L	C	40	156	CORROSIVE
	High explosives, see individual explosives' entries														
	HMX, *see* **Cyclotetramethylenete tranitramine,** *etc.*														
	Hydrazine, anhydrous	8	UN2029	I	8, 3, 6.1	A3, A6, A7, A10, B7, B16, B53	None	201	243	Forbidden	2.5 L	D	40, 52, 125	132	CORROSIVE
	Hydrazine aqueous solution, flammable *with more than 37% hydrazine, by mass*	8	UN3484	I	8, 3, 6.1	B16, B53, T10, TP2, TP13	None	201	243	Forbidden	2.5 L	D	40, 52, 125	—	CORROSIVE
	Hydrazine aqueous solution, *with more than 37% hydrazine, by mass*	8	UN2030	I	8, 6.1	B16, B53, T10, TP2, TP13	None	201	243	Forbidden	2.5 L	D	40, 52	153	CORROSIVE
				II	8, 6.1	B16, B53, IB2, T7, TP2, TP13	None	202	243	Forbidden	30 L	D	40, 52	153	CORROSIVE
				III	8, 6.1	B16, B53, IB3, T4, TP1	154	203	241	5 L	60 L	D	40, 52	153	CORROSIVE
	Hydrazine, aqueous solution, *with not more than 37 percent hydrazine, by mass*	6.1	UN3293	III	6.1	IB3, T4, TP1	153	203	241	60 L	220 L	A	52	152	POISON
	Hydrazine azide	Forbidden													
	Hydrazine chlorate	Forbidden													
	Hydrazine dicarbonic acid diazide	Forbidden													
	Hydrazine perchlorate	Forbidden													
	Hydrazine selenate	Forbidden													
	Hydriodic acid, anhydrous, see **Hydrogen iodide, anhydrous**														
	Hydriodic acid	8	UN1787	II	8	A3, A6, B2, IB2, N41, T7, TP2	154	202	242	1 L	30 L	C		154	CORROSIVE
				III	8	IB3, T4, TP1	154	203	241	5 L	60 L	C	8	154	CORROSIVE
	Hydrobromic acid, anhydrous, see **Hydrogen bromide, anhydrous**														
	Hydrobromic acid, *with more than 49 percent hydrobromic acid*	8	UN1788	II	8	B2, B15, IB2, N41, T7, TP2	154	202	242	Forbidden	Forbidden	C		154	CORROSIVE
				III	8	IB3, T4, TP1	154	203	241	Forbidden	Forbidden	C	8	154	CORROSIVE
	Hydrobromic acid, *with not more than 49 percent hydrobromic acid*	8	UN1788	II	8	A3, A6, B2, B15, IB2, N41, T7, TP2	154	202	242	1 L	30 L	C		154	CORROSIVE
				III	8	A3, IB3, T4, TP1	154	203	241	5 L	60 L	C	8	154	CORROSIVE
	Hydrocarbon gas mixture, compressed, n.o.s.	2.1	UN1964		2.1		306	302	314, 315	Forbidden	150 kg	E	40	115	FLAMMABLE GAS
	Hydrocarbon gas mixture, liquefied, n.o.s.	2.1	UN1965		2.1	T50	306	304	314, 315	Forbidden	150 kg	E	40	115	FLAMMABLE GAS
	Hydrocarbons, liquid, n.o.s.	3	UN3295	I	3	144, T11, TP1, TP8, TP28	150	201	243	1 L	30 L	E		128	FLAMMABLE
				II	3	144, IB2, T7, TP1, TP8, TP28	150	202	242	5 L	60 L	B		128	FLAMMABLE
				III	3	144, B1, IB3, T4, TP1, TP29	150	203	242	60 L	220 L	A		128	FLAMMABLE
	Hydrochloric acid, anhydrous, see **Hydrogen chloride, anhydrous**														

Symbols	Hazardous materials descriptions and proper shipping names	Hazard class or Division	Identification Numbers	PG	Label Codes	Special Provisions (§172.102)	(8) Packaging (§173.***) Exceptions	Non-bulk	Bulk	(9) Quantity limitations (see §§173.27 and 175.75) Passenger aircraft/rail	Cargo aircraft only	(10) Vessel stowage Location	Other	ERG Guide No.	Placard Advisory Consult §172.504 & §172.505 to determine placard *Denotes placard for any quantity
(1)	(2)	(3)	(4)	(5)	(6)	(7)	(8A)	(8B)	(8C)	(9A)	(9B)	(10A)	(10B)		
	Hydrochloric acid	8	UN1789	II	8	A3, A6, B3, B15, IB2, N41, T8, TP2	154	202	242	1 L	30 L	C		157	CORROSIVE
				III	8	A3, IB3, T4, TP1	154	203	241	5 L	60 L	C	8	157	CORROSIVE
	Hydrocyanic acid, anhydrous, see **Hydrogen cyanide** *etc.*														
	Hydrocyanic acid, aqueous solutions *or* **Hydrogen cyanide, aqueous solutions** *with not more than 20 percent hydrogen cyanide*	6.1	UN1613	I	6.1	2, B61, B65, B77, B82, T20, TP2, TP13	None	195	244	Forbidden	Forbidden	D	40	154	POISON INHALATION HAZARD*
D	**Hydrocyanic acid, aqueous solutions** *with less than 5 percent hydrogen cyanide*	6.1	NA1613	II	6.1	IB1, T14, TP2, TP13, TP27	None	195	243	Forbidden	5 L	D	40	154	POISON
	Hydrocyanic acid, liquefied, see **Hydrogen cyanide,** *etc.*														
	Hydrocyanic acid (prussic), unstabilized	Forbidden													
	Hydrofluoric acid and Sulfuric acid mixtures	8	UN1786	I	8, 6.1	A6, A7, B15, B23, N5, N34, T10, TP2, TP13	None	201	243	Forbidden	2.5 L	D	40	157	CORROSIVE
	Hydrofluoric acid, anhydrous, see **Hydrogen fluoride, anhydrous**														
	Hydrofluoric acid, *with more than 60 percent strength*	8	UN1790	I	8, 6.1	A6, A7, B4, B15, B23, N5, N34, T10, TP2, TP13	None	201	243	0.5 L	2.5 L	D	12, 40	157	CORROSIVE
	Hydrofluoric acid, *with not more than 60 percent strength*	8	UN1790	II	8, 6.1	A6, A7, B15, IB2, N5, N34, T8, TP2	154	202	243	1 L	30 L	D	12, 40	157	CORROSIVE
	Hydrofluoroboric acid, see **Fluoroboric acid**														
	Hydrofluorosilicic acid, see **Fluorosilicic acid**														
	Hydrogen and Methane mixtures, compressed	2.1	UN2034		2.1	N89	306	302	302, 314, 315	Forbidden	150 kg	E	40, 57	115	FLAMMABLE GAS
	Hydrogen bromide, anhydrous	2.3	UN1048		2.3, 8	3, B14, N86, N89	None	304	314, 315	Forbidden	Forbidden	D	40	125	POISON GAS*
	Hydrogen chloride, anhydrous	2.3	UN1050		2.3, 8	3, N86, N89	None	304	None	Forbidden	Forbidden	D	40	125	POISON GAS*
	Hydrogen chloride, refrigerated liquid	2.3	UN2186		2.3, 8	3, B6	None	None	314, 315	Forbidden	Forbidden	B	40	125	POISON GAS*
	Hydrogen, compressed	2.1	UN1049		2.1	N89	306	302	302, 314	Forbidden	150 kg	E	40, 57	115	FLAMMABLE GAS
	Hydrogen cyanide, solution in alcohol *with not more than 45 percent hydrogen cyanide*	6.1	UN3294	I	6.1, 3	2, B9, B14, B32, T20, TP2, TP13, TP38, TP45	None	227	244	Forbidden	Forbidden	D	40	131	POISON INHALATION HAZARD*
	Hydrogen cyanide, stabilized *with less than 3 percent water*	6.1	UN1051	I	6.1, 3	1, B35, B61, B65, B77, B82	None	195	244	Forbidden	Forbidden	D	40	117	POISON INHALATION HAZARD*
	Hydrogen cyanide, stabilized, *with less than 3 percent water and absorbed in a porous inert material*	6.1	UN1614	I	6.1	5	None	195	None	Forbidden	Forbidden	D	25, 40	152	POISON
	Hydrogen fluoride, anhydrous	8	UN1052	I	8, 6.1	3, B7, B46, B77, N86, T10, TP2	None	163	244	Forbidden	Forbidden	D	40	125	CORROSIVE, POISON INHALATION HAZARD*
	Hydrogen in a metal hydride storage system *or* **Hydrogen in a metal hydride storage system contained in equipment** *or* **Hydrogen in a metal hydride storage system packed with equipment**	2.1	UN3468		2.1	167	None	311	None	Forbidden	100 kg gross	D		115	FLAMMABLE GAS
	Hydrogen iodide, anhydrous	2.3	UN2197		2.3, 8	3, B14, N86, N89	None	304	314, 315	Forbidden	Forbidden	D	40	125	POISON GAS*
	Hydrogen iodide solution, see **Hydriodic acid**														
	Hydrogen peroxide and peroxyacetic acid mixtures, stabilized *with acids, water, and not more than 5 percent peroxyacetic acid*	5.1	UN3149	II	5.1, 8	145, A2, A3, A6, B53, IB2, IP5, T7, TP2, TP6, TP24	None	202	243	1 L	5 L	D	25, 66, 75	140	OXIDIZER
	Hydrogen peroxide, aqueous solutions *with more than 40 percent but not more than 60 percent hydrogen peroxide (stabilized as necessary)*	5.1	UN2014	II	5.1, 8	12, A60, B53, B80, B81, B85, IB2, IP5, T7, TP2, TP6, TP24, TP37	None	202	243	Forbidden	Forbidden	D	25, 66, 75	140	OXIDIZER
	Hydrogen peroxide, aqueous solutions *with not less than 20 percent but not more than 40 percent hydrogen peroxide (stabilized as necessary)*	5.1	UN2014	II	5.1, 8	A2, A3, A6, B53, IB2, IP5, T7, TP2, TP6, TP24, TP37	None	202	243	1 L	5 L	D	25, 66, 75	140	OXIDIZER

Symbols (1)	Hazardous materials descriptions and proper shipping names (2)	Hazard class or Division (3)	Identification Numbers (4)	PG (5)	Label Codes (6)	Special Provisions (§172.102) (7)	(8) Packaging (§173.***) Exceptions (8A)	Non-bulk (8B)	Bulk (8C)	(9) Quantity limitations (see §§173.27 and 175.75) Passenger aircraft/rail (9A)	Cargo aircraft only (9B)	(10) Vessel stowage Location (10A)	Other (10B)	ERG Guide No.	Placard Advisory Consult §172.504 & §172.505 to determine placard *Denotes placard for any quantity
	Hydrogen peroxide, aqueous solutions *with not less than 8 percent but less than 20 percent hydrogen peroxide (stabilized as necessary)*	5.1	UN2984	III	5.1	A1, IB2, IP5, T4, TP1, TP6, TP24, TP37	152	203	241	2.5 L	30 L	B	25, 66, 75	140	OXIDIZER
	Hydrogen peroxide, stabilized *or* **Hydrogen peroxide aqueous solutions, stabilized** *with more than 60 percent hydrogen peroxide*	5.1	UN2015	I	5.1, 8	12, B53, B80, B81, B85, T9, TP2, TP6, TP24, TP37	None	201	243	Forbidden	Forbidden	D	25, 66, 75	143	OXIDIZER
	Hydrogen, refrigerated liquid *(cryogenic liquid)*	2.1	UN1966		2.1	T75, TP5	None	316	318, 319	Forbidden	Forbidden	D	40	115	FLAMMABLE GAS
	Hydrogen selenide, anhydrous	2.3	UN2202		2.3, 2.1	1	None	192	245	Forbidden	Forbidden	D	40	117	POISON GAS*
	Hydrogen sulfate, see **Sulfuric acid**														
	Hydrogen sulfide	2.3	UN1053		2.3, 2.1	2, B9, B14, N89	None	304	314, 315	Forbidden	Forbidden	D	40	117	POISON GAS*
	Hydrogendifluoride, solid, n.o.s.	8	UN1740	II	8	IB8, IP2, IP4, N3, N34, T3, TP33	None	212	240	15 kg	50 kg	A	25, 40, 52	154	CORROSIVE
				III	8	IB8, IP3, N3, N34, T1, TP33	154	213	240	25 kg	100 kg	A	25, 40, 52	154	CORROSIVE
	Hydrogendifluorides, solution, n.o.s.	8	UN3471	II	8, 6.1	IB2, T7, TP2	154	202	242	1 L	30 L	A	25, 40, 52	154	CORROSIVE
				III	8, 6.1	IB3, T4, TP1	154	203	241	5 L	60 L	A	25, 40, 52	154	CORROSIVE
	Hydrosilicofluoric acid, see **Fluorosilicic acid**														
	1-Hydroxybenzotriazole, anhydrous, *dry or wetted with less than 20 percent water, by mass*	1.3C	UN0508		1.3C		None	62	None	Forbidden	Forbidden	10		112	EXPLOSIVES 1.3*
	1–Hydroxybenzotriazole, monohydrate	4.1	UN3474	I	4.1	N90	None	211	None	0.5 kg	0.5 kg	D	28, 36	113	FLAMMABLE SOLID
	Hydroxyl amine iodide	Forbidden													
	Hydroxylamine sulfate	8	UN2865	III	8	IB8, IP3, T1, TP33	154	213	240	25 kg	100 kg	A		154	CORROSIVE
	Hypochlorite solutions	8	UN1791	II	8	A7, B2, B15, IB2, IP5, N34, T7, TP2, TP24	154	202	242	1 L	30 L	B	26	154	CORROSIVE
				III	8	IB3, N34, T4, TP2, TP24	154	203	241	5 L	60 L	B	26	154	CORROSIVE
G	**Hypochlorites, inorganic, n.o.s.**	5.1	UN3212	II	5.1	349, A9, IB8, IP2, IP4, T3, TP33	152	212	240	5 kg	25 kg	D	4, 48, 52, 56, 58, 69, 106, 116, 118	140	OXIDIZER
	Hyponitrous acid	Forbidden													
	Igniter fuse, metal clad, see **Fuse,** *igniter, tubular, metal clad*														
	Igniters	1.1G	UN0121	II	1.1G		None	62	None	Forbidden	Forbidden	07		112	EXPLOSIVES 1.1*
	Igniters	1.2G	UN0314	II	1.2G		None	62	None	Forbidden	Forbidden	07		112	EXPLOSIVES 1.2*
	Igniters	1.3G	UN0315	II	1.3G		None	62	None	Forbidden	Forbidden	07		112	EXPLOSIVES 1.3*
	Igniters	1.4G	UN0325	II	1.4G		None	62	None	Forbidden	75 kg	06		114	EXPLOSIVES 1.4
	Igniters	1.4S	UN0454	II	1.4S		None	62	None	25 kg	100 kg	05		114	EXPLOSIVES 1.4
	3,3'-Iminodipropylamine	8	UN2269	III	8	IB3, T4, TP2	154	203	241	5 L	60 L	A		153	CORROSIVE
G	**Infectious substances, affecting animals** *only*	6.2	UN2900		6.2	A82	134	196	None	50 mL or 50 g	4 L or 4 kg	B	40	158	NONE
G	**Infectious substances, affecting humans**	6.2	UN2814		6.2	A82	134	196	None	50 mL or 50 g	4 L or 4 kg	B	40	158	NONE
	Inflammable, see Flammable														
	Initiating explosives (dry)	Forbidden													
	Inositol hexanitrate (dry)	Forbidden													
G	**Insecticide gases, n.o.s.**	2.2	UN1968		2.2		306	304	314, 315	75 kg	150 kg	A		126	NONFLAMMABLE GAS
G	**Insecticide gases, flammable, n.o.s.**	2.1	UN3354		2.1	T50	306	304	314, 315	Forbidden	150 kg	D	40	115	FLAMMABLE GAS
G	**Insecticide gases, toxic, flammable, n.o.s.** *Inhalation hazard Zone A*	2.3	UN3355		2.3, 2.1	1	None	192	245	Forbidden	Forbidden	D	40	119	POISON GAS*
G	**Insecticide gases, toxic, flammable, n.o.s.** *Inhalation hazard Zone B*	2.3	UN3355		2.3, 2.1	2, B9, B14	None	302, 305	314, 315	Forbidden	Forbidden	D	40	119	POISON GAS*

Symbols (1)	Hazardous materials descriptions and proper shipping names (2)	Hazard class or Division (3)	Identification Numbers (4)	PG (5)	Label Codes (6)	Special Provisions (§172.102) (7)	(8) Packaging (§173.***) Exceptions (8A)	Non-bulk (8B)	Bulk (8C)	(9) Quantity limitations (see §§173.27 and 175.75) Passenger aircraft/rail (9A)	Cargo aircraft only (9B)	(10) Vessel stowage Location (10A)	Other (10B)	ERG Guide No.	Placard Advisory Consult §172.504 & §172.505 to determine placard *Denotes placard for any quantity
G	**Insecticide gases, toxic, flammable, n.o.s.** *Inhalation hazard Zone C*	2.3	UN3355		2.3, 2.1	3, B14	None	302, 305	314, 315	Forbidden	Forbidden	D		119	POISON GAS*
G	**Insecticide gases, toxic, flammable, n.o.s.** *Inhalation hazard Zone D*	2.3	UN3355		2.3, 2.1	4	None	302, 305	314, 315	Forbidden	Forbidden	D		119	POISON GAS*
G	**Insecticide gases, toxic, n.o.s.**	2.3	UN1967		2.3	3	None	193, 334	245	Forbidden	Forbidden	D	40	123	POISON GAS*
	Inulin trinitrate (dry)	Forbidden													
+	**Iodine**	8	UN3495	III	8, 6.1	IB8, IP3, T1, TP33	154	213	240	25 kg	100 kg	B	40, 55	—	CORROSIVE
	Iodine azide (dry)	Forbidden													
	Iodine monochloride	8	UN1792	II	8	B6, IB8, IP2, IP4, N41, T7, TP2	None	212	240	Forbidden	50 kg	D	40, 66, 74, 89, 90	157	CORROSIVE
	Iodine pentafluoride	5.1	UN2495	I	5.1, 6.1, 8		None	205	243	Forbidden	Forbidden	D	25, 40, 52, 66, 90	144	OXIDIZER
	2-Iodobutane	3	UN2390	II	3	IB2, T4, TP1	150	202	242	5 L	60 L	B		129	FLAMMABLE
	Iodomethylpropanes	3	UN2391	II	3	IB2, T4, TP1	150	202	242	5 L	60 L	B		129	FLAMMABLE
	Iodopropanes	3	UN2392	III	3	B1, IB3, T2, TP1	150	203	242	60 L	220 L	A		129	FLAMMABLE
	Iodoxy compounds (dry)	Forbidden													
	Iridium nitratopentamine iridium nitrate	Forbidden													
	Iron chloride, see **Ferric chloride**														
	Iron oxide, spent, *or* **Iron sponge, spent** *obtained from coal gas purification*	4.2	UN1376	III	4.2	B18, IB8, IP3, T1, TP33	None	213	240	Forbidden	Forbidden	E		135	SPONTANEOUSLY COMBUSTIBLE
	Iron pentacarbonyl	6.1	UN1994	I	6.1, 3	1, B9, B14, B30, B77, T22, TP2, TP13, TP38, TP44	None	226	244	Forbidden	Forbidden	D	40	131	POISON INHALATION HAZARD*
	Iron sesquichloride, see **Ferric chloride**														
	Irritating material, see **Tear gas substances,** *etc.*														
	Isobutane, *see also* **Petroleum gases, liquefied**	2.1	UN1969		2.1	19, T50	306	304	314, 315	Forbidden	150 kg	E	40	115	FLAMMABLE GAS
	Isobutanol *or* **Isobutyl alcohol**	3	UN1212	III	3	B1, IB3, T2, TP1	150	203	242	60 L	220 L	A		129	FLAMMABLE
	Isobutyl acetate	3	UN1213	II	3	IB2, T4, TP1	150	202	242	5 L	60 L	B		129	FLAMMABLE
	Isobutyl acrylate, inhibited	3	UN2527	III	3	B1, IB3, T2, TP1	150	203	242	60 L	220 L	A		129P	FLAMMABLE
	Isobutyl alcohol, *see* **Isobutanol**														
	Isobutyl aldehyde, *see* **Isobutyraldehyde**														
	Isobutyl formate	3	UN2393	II	3	IB2, T4, TP1	150	202	242	5 L	60 L	B		129	FLAMMABLE
	Isobutyl isobutyrate	3	UN2528	III	3	B1, IB3, T2, TP1	150	203	242	60 L	220 L	A		130	FLAMMABLE
+	**Isobutyl isocyanate**	6.1	UN2486	I	6.1, 3	1, B9, B14, B30, T20, TP2, TP13, TP27	None	226	244	Forbidden	Forbidden	D	40	155	POISON INHALATION HAZARD*
	Isobutyl methacrylate, stabilized	3	UN2283	III	3	B1, IB3, T2, TP1	150	203	242	60 L	220 L	A		130P	FLAMMABLE
	Isobutyl propionate	3	UN2394	III	3	B1, IB3, T2, TP1	150	203	242	60 L	220 L	B		129	FLAMMABLE
	Isobutylamine	3	UN1214	II	3, 8	IB2, T7, TP1	150	202	243	1 L	5 L	B	40	132	FLAMMABLE
	Isobutylene, *see also* **Petroleum gases, liquefied**	2.1	UN1055		2.1	19, T50	306	304	314, 315	Forbidden	150 kg	E	40	115	FLAMMABLE GAS
	Isobutyraldehyde *or* **Isobutyl aldehyde**	3	UN2045	II	3	IB2, T4, TP1	150	202	242	5 L	60 L	E	40	130	FLAMMABLE
	Isobutyric acid	3	UN2529	III	3, 8	B1, IB3, T4, TP1	150	203	242	5 L	60 L	A		132	FLAMMABLE
	Isobutyronitrile	3	UN2284	II	3, 6.1	IB2, T7, TP2, TP13	150	202	243	1 L	60 L	E	40	131	FLAMMABLE
	Isobutyryl chloride	3	UN2395	II	3, 8	IB1, T7, TP2	150	202	243	1 L	5 L	C	40	132	FLAMMABLE
G	**Isocyanates, flammable, toxic, n.o.s.** *or* **Isocyanate solutions, flammable, toxic, n.o.s.** *flash point less than 23 degrees C*	3	UN2478	II	3, 6.1	5, A3, A7, IB2, T11, TP2, TP13, TP27	150	202	243	1 L	60 L	D	40	155	FLAMMABLE, POISON INHALATION HAZARD*
				III	3, 6.1	5, A3, A7, IB3, T7, TP1, TP13, TP28	150	203	242	60 L	220 L	A		155	FLAMMABLE, POISON INHALATION HAZARD*
G	**Isocyanates, toxic, flammable, n.o.s.** *or* **Isocyanate solutions, toxic, flammable, n.o.s.,** *flash point not less than 23 degrees C but not more than 61 degrees C and boiling point less than 300 degrees C*	6.1	UN3080	II	6.1, 3	IB2, T11, TP2, TP13, TP27	153	202	243	5 L	60 L	B	25, 40, 48	155	POISON

Symbols (1)	Hazardous materials descriptions and proper shipping names (2)	Hazard class or Division (3)	Identification Numbers (4)	PG (5)	Label Codes (6)	Special Provisions (§172.102) (7)	(8) Packaging (§173.***) Exceptions (8A)	Non-bulk (8B)	Bulk (8C)	(9) Quantity limitations (see §§173.27 and 175.75) Passenger aircraft/rail (9A)	Cargo aircraft only (9B)	(10) Vessel stowage Location (10A)	Other (10B)	ERG Guide No.	Placard Advisory Consult §172.504 & §172.505 to determine placard *Denotes placard for any quantity
G	**Isocyanates, toxic, n.o.s.** *or* **Isocyanate, solutions, toxic, n.o.s.,** *flash point more than 61 degrees C and boiling point less than 300 degrees C*	6.1	UN2206	II	6.1	IB2, T11, TP2, TP13, TP27	153	202	243	5 L	60 L	E	25, 40, 48	155	POISON
				III	6.1	IB3, T7, TP1, TP13, TP28	153	203	241	60 L	220 L	E	25, 40, 48	155	POISON
	Isocyanatobenzotrifluorides	6.1	UN2285	II	6.1, 3	5, IB2, T7, TP2	153	202	243	5 L	60 L	D	25, 40, 48	156	POISON INHALATION HAZARD*
	Isoheptenes	3	UN2287	II	3	IB2, T4, TP1	150	202	242	5 L	60 L	B		128	FLAMMABLE
	Isohexenes	3	UN2288	II	3	IB2, IP8, T11, TP1	150	202	242	5 L	60 L	E		128	FLAMMABLE
	Isooctane, see **Octanes**														
	Isooctenes	3	UN1216	II	3	IB2, T4, TP1	150	202	242	5 L	60 L	B		128	FLAMMABLE
	Isopentane, see **Pentane**														
	Isopentanoic acid, see **Corrosive liquids, n.o.s.**														
	Isopentenes	3	UN2371	I	3	T11, TP2	150	201	243	1 L	30 L	E		128	FLAMMABLE
	Isophorone diisocyanate	6.1	UN2290	III	6.1	IB3, T4, TP2	153	203	241	60 L	220 L	B	40	156	POISON
	Isophoronediamine	8	UN2289	III	8	IB3, T4, TP1	154	203	241	5 L	60 L	A		153	CORROSIVE
	Isoprene, stabilized	3	UN1218	I	3	T11, TP2	150	201	243	1 L	30 L	E		130P	FLAMMABLE
	Isopropanol *or* **Isopropyl alcohol**	3	UN1219	II	3	IB2, T4, TP1	4b, 150	202	242	5 L	60 L	B		129	FLAMMABLE
	Isopropenyl acetate	3	UN2403	II	3	IB2, T4, TP1	150	202	242	5 L	60 L	B		129P	FLAMMABLE
	Isopropenylbenzene	3	UN2303	III	3	B1, IB3, T2, TP1	150	203	242	60 L	220 L	A		128	FLAMMABLE
	Isopropyl acetate	3	UN1220	II	3	IB2, T4, TP1	150	202	242	5 L	60 L	B		129	FLAMMABLE
	Isopropyl acid phosphate	8	UN1793	III	8	IB2, T4, TP1	154	213	240	25 kg	100 kg	A		153	CORROSIVE
	Isopropyl alcohol, *see* **Isopropanol**														
	Isopropyl butyrate	3	UN2405	III	3	B1, IB3, T2, TP1	150	203	242	60 L	220 L	A		129	FLAMMABLE
	Isopropyl chloroacetate	3	UN2947	III	3	B1, IB3, T2, TP1	150	203	242	60 L	220 L	A		155	FLAMMABLE
	Isopropyl chloroformate	6.1	UN2407	I	6.1, 3, 8	2, B9, B14, B32, B77, T20, TP2, TP13, TP38, TP44	None	227	244	Forbidden	Forbidden	B	40	155	POISON INHALATION HAZARD*
	Isopropyl 2-chloropropionate	3	UN2934	III	3	B1, IB3, T2, TP1	150	203	242	60 L	220 L	A		129	FLAMMABLE
	Isopropyl isobutyrate	3	UN2406	II	3	IB2, T4, TP1	150	202	242	5 L	60 L	B		127	FLAMMABLE
+	**Isopropyl isocyanate**	6.1	UN2483	I	6.1, 3	1, B9, B14, B30, T20, TP2, TP13, TP38, TP44	None	226	244	Forbidden	Forbidden	D	40	155	POISON INHALATION HAZARD*
	Isopropyl mercaptan, see **Propanethiols**														
	Isopropyl nitrate	3	UN1222	II	3	IB9	150	202	None	5 L	60 L	D		130	FLAMMABLE
	Isopropyl phosphoric acid, see **Isopropyl acid phosphate**														
	Isopropyl propionate	3	UN2409	II	3	IB2, T4, TP1	150	202	242	5 L	60 L	B		129	FLAMMABLE
	Isopropylamine	3	UN1221	I	3, 8	T11, TP2	None	201	243	0.5 L	2.5 L	E		132	FLAMMABLE
	Isopropylbenzene	3	UN1918	III	3	B1, IB3, T2, TP1	150	203	242	60 L	220 L	A		130	FLAMMABLE
	Isopropylcumyl hydroperoxide, with more than 72 percent in solution	Forbidden													
	Isosorbide dinitrate mixture *with not less than 60 percent lactose, mannose, starch or calcium hydrogen phosphate*	4.1	UN2907	II	4.1	IB6, IP2, N85	None	212	None	15 kg	50 kg	E	28, 36	133	FLAMMABLE SOLID
	Isosorbide-5-mononitrate	4.1	UN3251	III	4.1	66, 159, IB8	151	223	240	Forbidden	Forbidden	D	12	133	FLAMMABLE SOLID
	Isothiocyanic acid	Forbidden													
	Jet fuel, see **Fuel aviation, turbine engine**														
D	**Jet perforating guns, charged oil well, with detonator**	1.1D	NA0124	II	1.1D	55, 56	None	62	None	Forbidden	Forbidden	07		112	EXPLOSIVES 1.1*
D	**Jet perforating guns, charged oil well, with detonator**	1.4D	NA0494	II	1.4D	55, 56	None	62	None	Forbidden	Forbidden	06		114	EXPLOSIVES 1.4
	Jet perforating guns, charged *oil well, without detonator*	1.1D	UN0124	II	1.1D	55	None	62	None	Forbidden	Forbidden	07		112	EXPLOSIVES 1.1*
	Jet perforating guns, charged, *oil well, without detonator*	1.4D	UN0494	II	1.4D	55, 114	None	62	None	Forbidden	300 kg	06		114	EXPLOSIVES 1.4
	Jet perforators, see **Charges, shaped,** *etc.*														
	Jet tappers, without detonator, see **Charges, shaped,** *etc.*														
	Jet thrust igniters, for rocket motors or Jato, see **Igniters**														

Symbols (1)	Hazardous materials descriptions and proper shipping names (2)	Hazard class or Division (3)	Identification Numbers (4)	PG (5)	Label Codes (6)	Special Provisions (§172.102) (7)	(8) Packaging (§173.***) Exceptions (8A)	Non-bulk (8B)	Bulk (8C)	(9) Quantity limitations (see §§173.27 and 175.75) Passenger aircraft/rail (9A)	Cargo aircraft only (9B)	(10) Vessel stowage Location (10A)	Other (10B)	ERG Guide No.	Placard Advisory Consult §172.504 & §172.505 to determine placard *Denotes placard for any quantity
	Jet thrust unit (Jato), see **Rocket motors**														
	Kerosene	3	UN1223	III	3	144, B1, IB3, T2, TP2	150	203	242	60 L	220 L	A		128	FLAMMABLE
G	**Ketones, liquid, n.o.s.**	3	UN1224	I	3	T11, TP1, TP8, TP27	None	201	243	1 L	30 L	E		127	FLAMMABLE
				II	3	IB2, T7, TP1, TP8, TP28	150	202	242	5 L	60 L	B		127	FLAMMABLE
				III	3	B1, IB3, T4, TP1, TP29	150	203	242	60 L	220 L	A		127	FLAMMABLE
	Krypton, compressed	2.2	UN1056		2.2		306, 307	302	None	75 kg	150 kg	A		121	NONFLAMMABLE GAS
	Krypton, refrigerated liquid *(cryogenic liquid)*	2.2	UN1970		2.2	T75, TP5	320	None	None	50 kg	500 kg	D		120	NONFLAMMABLE GAS
	Lacquer base or lacquer chips, nitrocellulose, dry, see **Nitrocellulose,** *etc. (UN2557)*														
	Lacquer base or lacquer chips, plastic, wet with alcohol or solvent, see **Nitrocellulose** *(UN2059, UN2555, UN2556, UN2557) or* **Paint** *etc. (UN1263)*														
	Lead acetate	6.1	UN1616	III	6.1	IB8, IP3, T1, TP33	153	213	240	100 kg	200 kg	A		151	POISON
	Lead arsenates	6.1	UN1617	II	6.1	IB8, IP2, IP4, T3, TP33	153	212	242	25 kg	100 kg	A		151	POISON
	Lead arsenites	6.1	UN1618	II	6.1	IB8, IP2, IP4, T3, TP33	153	212	242	25 kg	100 kg	A		151	POISON
	Lead azide (dry)	Forbidden													
	Lead azide, wetted *with not less than 20 percent water or mixture of alcohol and water, by mass*	1.1A	UN0129	II	1.1A	111, 117	None	62	None	Forbidden	Forbidden	12		112	EXPLOSIVES 1.1*
G	**Lead compounds, soluble, n.o.s.**	6.1	UN2291	III	6.1	138, IB8, IP3, T1, TP33	153	213	240	100 kg	200 kg	A		151	POISON
	Lead cyanide	6.1	UN1620	II	6.1	IB8, IP2, IP4, T3, TP33	153	212	242	25 kg	100 kg	A	52	151	POISON
	Lead dioxide	5.1	UN1872	III	5.1	A1, IB8, IP3, T1, TP33	152	213	240	25 kg	100 kg	A		141	OXIDIZER
	Lead dross, see **Lead sulfate,** *with more than 3 percent free acid*														
	Lead nitrate	5.1	UN1469	II	5.1, 6.1	IB8, IP2, IP4, T3, TP33	152	212	242	5 kg	25 kg	A		141	OXIDIZER
	Lead nitroresorcinate (dry)	Forbidden													
	Lead perchlorate, solid	5.1	UN1470	II	5.1, 6.1	IB6, IP2, T3, TP33	152	212	242	5 kg	25 kg	A	56, 58	141	OXIDIZER
	Lead perchlorate, solution	5.1	UN3408	II	5.1, 6.1	IB2, T4, TP1	152	202	243	1 L	5 L	A	56, 58	141	OXIDIZER
				III	5.1, 6.1	IB2, T4, TP1	152	203	242	2.5 L	30 L	A	56, 58	141	OXIDIZER
	Lead peroxide, see **Lead dioxide**														
	Lead phosphite, dibasic	4.1	UN2989	II	4.1	IB8, IP2, IP4, T3, TP33	None	212	240	15 kg	50 kg	B	34	133	FLAMMABLE SOLID
				III	4.1	IB8, IP3, T1, TP33	151	213	240	25 kg	100 kg	B	34	133	FLAMMABLE SOLID
	Lead picrate (dry)	Forbidden													
	Lead styphnate (dry)	Forbidden													
	Lead styphnate, wetted *or* **Lead trinitroresorcinate, wetted** *with not less than 20 percent water or mixture of alcohol and water, by mass*	1.1A	UN0130	II	1.1A	111, 117	None	62	None	Forbidden	Forbidden	12		112	EXPLOSIVES 1.1*
	Lead sulfate *with more than 3 percent free acid*	8	UN1794	II	8	IB8, IP2, IP4, T3, T33	154	212	240	15 kg	50 kg	A		154	CORROSIVE
	Lead trinitroresorcinate, *see* **Lead styphnate,** *etc.*														
	Life-saving appliances, not self inflating *containing dangerous goods as equipment*	9	UN3072		None		None	219	None	No limit	No limit	A		171	CLASS 9
	Life-saving appliances, self inflating	9	UN2990		None		None	219	None	No limit	No limit	A		171	CLASS 9
	Lighter refills *containing flammable gas not exceeding 4 fluid ounces (7.22 cubic inches) and 65 grams of flammable gas*	2.1	UN1057		2.1	169	306	306	None	1 kg	15 kg	B	40	115	FLAMMABLE GAS

Symbols (1)	Hazardous materials descriptions and proper shipping names (2)	Hazard class or Division (3)	Identification Numbers (4)	PG (5)	Label Codes (6)	Special Provisions (§172.102) (7)	(8) Packaging (§173.***) Exceptions (8A)	Non-bulk (8B)	Bulk (8C)	(9) Quantity limitations (see §§173.27 and 175.75) Passenger aircraft/rail (9A)	Cargo aircraft only (9B)	(10) Vessel stowage Location (10A)	Other (10B)	ERG Guide No.	Placard Advisory Consult §172.504 & §172.505 to determine placard *Denotes placard for any quantity
	Lighter replacement cartridges containing liquefied petroleum gases see **Lighter** *refills containing flammable gas, Etc.*														
	Lighters *containing flammable gas*	2.1	UN1057		2.1	168	21, 308	21, 308	None	1 kg	15 kg	B	40	115	FLAMMABLE GAS
	Lighters, fuse	1.4S	UN0131	II	1.4S		None	62	None	25 kg	100 kg	05		114	EXPLOSIVES 1.4
	Lighters, new or empty, purged of all residual fuel and vapors					168									
	Lighters, *non-pressurized, containing flammable liquid*	3	NA1057	II	3	168	21	None	None	Forbidden	Forbidden	B	40	128	FLAMMABLE
	Lime, unslaked, see **Calcium oxide**														
G	**Liquefied gas, flammable, n.o.s.**	2.1	UN3161		2.1	T50	306	304	314, 315	Forbidden	150 kg	D	40	115	FLAMMABLE GAS
G	**Liquefied gas, n.o.s.**	2.2	UN3163		2.2	T50	306	304	314, 315	75 kg	150 kg	A		126	NONFLAMMABLE GAS
G	**Liquefied gas, oxidizing, n.o.s.**	2.2	UN3157		2.2, 5.1	A14	306	304	314, 315	75 kg	150 kg	D		122	NONFLAMMABLE GAS
G I	**Liquefied gas, toxic, corrosive, n.o.s.** *Inhalation Hazard Zone A*	2.3	UN3308		2.3, 8	1	None	192	245	Forbidden	Forbidden	D	40	123	POISON GAS*
G I	**Liquefied gas, toxic, corrosive, n.o.s.** *Inhalation Hazard Zone B*	2.3	UN3308		2.3, 8	2, B9, B14	None	304	314, 315	Forbidden	Forbidden	D	40	123	POISON GAS*
G I	**Liquefied gas, toxic, corrosive, n.o.s.** *Inhalation Hazard Zone C*	2.3	UN3308		2.3, 8	3, B14	None	304	314, 315	Forbidden	Forbidden	D	40	123	POISON GAS*
G I	**Liquefied gas, toxic, corrosive, n.o.s.** *Inhalation Hazard Zone D*	2.3	UN3308		2.3, 8	4	None	304	314, 315	Forbidden	Forbidden	D	40	123	POISON GAS*
G I	**Liquefied gas, toxic, flammable, corrosive, n.o.s.** *Inhalation Hazard Zone A*	2.3	UN3309		2.3, 2.1, 8	1	None	192	245	Forbidden	Forbidden	D	17, 40	119	POISON GAS*
G I	**Liquefied gas, toxic, flammable, corrosive, n.o.s.** *Inhalation Hazard Zone B*	2.3	UN3309		2.3, 2.1, 8	2, B9, B14	None	304	314, 315	Forbidden	Forbidden	D	17, 40	119	POISON GAS*
G I	**Liquefied gas, toxic, flammable, corrosive, n.o.s.** *Inhalation Hazard Zone C*	2.3	UN3309		2.3, 2.1, 8	3, B14	None	304	314, 315	Forbidden	Forbidden	D	17, 40	119	POISON GAS*
G I	**Liquefied gas, toxic, flammable, corrosive, n.o.s.** *Inhalation Hazard Zone D*	2.3	UN3309		2.3, 2.1, 8	4	None	304	314, 315	Forbidden	Forbidden	D	17, 40	119	POISON GAS*
G	**Liquefied gas, toxic, flammable, n.o.s.** *Inhalation Hazard Zone A*	2.3	UN3160		2.3, 2.1	1	None	192	245	Forbidden	Forbidden	D	40	119	POISON GAS*
G	**Liquefied gas, toxic, flammable, n.o.s.** *Inhalation Hazard Zone B*	2.3	UN3160		2.3, 2.1	2, B9, B14	None	304	314, 315	Forbidden	Forbidden	D	40	119	POISON GAS*
G	**Liquefied gas, toxic, flammable, n.o.s.** *Inhalation Hazard Zone C*	2.3	UN3160		2.3, 2.1	3, B14	None	304	314, 315	Forbidden	Forbidden	D	40	119	POISON GAS*
G	**Liquefied gas, toxic, flammable, n.o.s.** *Inhalation Hazard Zone D*	2.3	UN3160		2.3, 2.1	4	None	304	314, 315	Forbidden	Forbidden	D	40	119	POISON GAS*
G	**Liquefied gas, toxic, n.o.s.** *Inhalation Hazard Zone A*	2.3	UN3162		2.3	1	None	192	245	Forbidden	Forbidden	D	40	123	POISON GAS*
G	**Liquefied gas, toxic, n.o.s.** *Inhalation Hazard Zone B*	2.3	UN3162		2.3	2, B9, B14	None	304	314, 315	Forbidden	Forbidden	D	40	123	POISON GAS*
G	**Liquefied gas, toxic, n.o.s.** *Inhalation Hazard Zone C*	2.3	UN3162		2.3	3, B14	None	304	314, 315	Forbidden	Forbidden	D	40	123	POISON GAS*
G	**Liquefied gas, toxic, n.o.s.** *Inhalation Hazard Zone D*	2.3	UN3162		2.3	4	None	304	314, 315	Forbidden	Forbidden	D	40	123	POISON GAS*
G I	**Liquefied gas, toxic, oxidizing, corrosive, n.o.s.** *Inhalation Hazard Zone A*	2.3	UN3310		2.3, 5.1, 8	1	None	192	245	Forbidden	Forbidden	D	40, 89, 90	124	POISON GAS*
G I	**Liquefied gas, toxic, oxidizing, corrosive, n.o.s.** *Inhalation Hazard Zone B*	2.3	UN3310		2.3, 5.1, 8	2, B9, B14	None	304	314, 315	Forbidden	Forbidden	D	40, 89, 90	124	POISON GAS*
G I	**Liquefied gas, toxic, oxidizing, corrosive, n.o.s.** *Inhalation Hazard Zone C*	2.3	UN3310		2.3, 5.1, 8	3, B14	None	304	314, 315	Forbidden	Forbidden	D	40, 89, 90	124	POISON GAS*
G I	**Liquefied gas, toxic, oxidizing, corrosive, n.o.s.** *Inhalation Hazard Zone D*	2.3	UN3310		2.3, 5.1, 8	4	None	304	314, 315	Forbidden	Forbidden	D	40, 89, 90	124	POISON GAS*
G	**Liquefied gas, toxic, oxidizing, n.o.s.** *Inhalation Hazard Zone A*	2.3	UN3307		2.3, 5.1	1	None	192	245	Forbidden	Forbidden	D	40	124	POISON GAS*
G	**Liquefied gas, toxic, oxidizing, n.o.s.** *Inhalation Hazard Zone B*	2.3	UN3307		2.3, 5.1	2, B9, B14	None	304	314, 315	Forbidden	Forbidden	D	40	124	POISON GAS*
G	**Liquefied gas, toxic, oxidizing, n.o.s.** *Inhalation Hazard Zone C*	2.3	UN3307		2.3, 5.1	3, B14	None	304	314, 315	Forbidden	Forbidden	D	40	124	POISON GAS*

Symbols (1)	Hazardous materials descriptions and proper shipping names (2)	Hazard class or Division (3)	Identification Numbers (4)	PG (5)	Label Codes (6)	Special Provisions (§172.102) (7)	(8) Packaging (§173.***) Exceptions (8A)	Non-bulk (8B)	Bulk (8C)	(9) Quantity limitations (see §§173.27 and 175.75) Passenger aircraft/rail (9A)	Cargo aircraft only (9B)	(10) Vessel stowage Location (10A)	Other (10B)	ERG Guide No.	Placard Advisory Consult §172.504 & §172.505 to determine placard *Denotes placard for any quantity
G	**Liquefied gas, toxic, oxidizing, n.o.s.** *Inhalation Hazard Zone D*	2.3	UN3307		2.3, 5.1	4	None	304	314, 315	Forbidden	Forbidden	D	40	124	POISON GAS*
	Liquefied gases, *non-flammable charged with nitrogen, carbon dioxide or air*	2.2	UN1058		2.2		306	304	None	75 kg	150 kg	A		120	NONFLAMMABLE GAS
	Liquefied hydrocarbon gas, see **Hydrocarbon gas mixture, liquefied, n.o.s.**														
	Liquefied natural gas, see **Methane,** *etc. (UN1972)*														
	Liquefied petroleum gas, *see* **Petroleum gases, liquefied**														
	Lithium	4.3	UN1415	I	4.3	A7, A19, IB4, IP1, N45	None	211	244	Forbidden	15 kg	E	52	138	DANGEROUS WHEN WET*
	Lithium acetylide ethylenediamine complex, see **Water reactive solid,** *etc.*														
	Lithium aluminum hydride	4.3	UN1410	I	4.3	A19	None	211	242	Forbidden	15 kg	E	52	138	DANGEROUS WHEN WET*
	Lithium aluminum hydride, ethereal	4.3	UN1411	I	4.3, 3	A2, A3, A11, N34	None	201	244	Forbidden	1 L	D	40	138	DANGEROUS WHEN WET*
	Lithium batteries, contained in equipment	9	UN3091	II	9	29, 188, 189, 190, A54, A55, A101, A104	185	185	None	See A101, A104	35 kg	A		138	CLASS 9
	Lithium batteries packed with equipment	9	UN3091	II	9	29, 188, 189, 190, A54, A55, A101, A103	185	185	None	See A101, A103	35 kg gross	A		138	CLASS 9
	Lithium battery	9	UN3090	II	9	29, 188, 189, 190, A54, A55, A100	185	185	None	See A100	35 kg gross	A		138	CLASS 9
	Lithium borohydride	4.3	UN1413	I	4.3	A19, N40	None	211	242	Forbidden	15 kg	E	52	138	DANGEROUS WHEN WET*
	Lithium ferrosilicon	4.3	UN2830	II	4.3	A19, IB7, IP2, T3, TP33	151	212	241	15 kg	50 kg	E	40, 85, 103	139	DANGEROUS WHEN WET*
	Lithium hydride	4.3	UN1414	I	4.3	A19, N40	None	211	242	Forbidden	15 kg	E	52	138	DANGEROUS WHEN WET*
	Lithium hydride, fused solid	4.3	UN2805	II	4.3	A8, A19, A20, IB4, T3, TP33	151	212	241	15 kg	50 kg	E	52	138	DANGEROUS WHEN WET*
	Lithium hydroxide	8	UN2680	II	8	IB8, IP2, IP4, T3, TP33	154	212	240	15 kg	50 kg	A	52	154	CORROSIVE
	Lithium hydroxide, solution	8	UN2679	II	8	B2, IB2, T7, TP2	154	202	242	1 L	30 L	A	29, 52	154	CORROSIVE
				III	8	IB3, T4, TP2	154	203	241	5 L	60 L	A	29, 52, 96	154	CORROSIVE
	Lithium hypochlorite, dry *or* **Lithium hypochlorite mixture**	5.1	UN1471	II	5.1	A9, IB8, IP2, IP4, N34, T3, TP33	152	212	240	5 kg	25 kg	A	4, 48, 52, 56, 58, 69, 106, 116	140	OXIDIZER
				III	5.1	IB8, IP3, N34, T1, TP33	152	213	240	25 kg	100 kg	A	4, 48, 52, 56, 58, 69, 106, 116	140	OXIDIZER
	Lithium in cartridges, see **Lithium**														
	Lithium nitrate	5.1	UN2722	III	5.1	A1, IB8, IP3, T1, TP33	152	213	240	25 kg	100 kg	A		140	OXIDIZER
	Lithium nitride	4.3	UN2806	I	4.3	A19, IB4, IP1, N40	None	211	242	Forbidden	15 kg	E		138	DANGEROUS WHEN WET*
	Lithium peroxide	5.1	UN1472	II	5.1	A9, IB6, IP2, N34, T3, TP33	152	212	None	5 kg	25 kg	A	13, 52, 66, 75	143	OXIDIZER
	Lithium silicon	4.3	UN1417	II	4.3	A19, A20, IB7, IP2, T3, TP33	151	212	241	15 kg	50 kg	A	85, 103	138	DANGEROUS WHEN WET*
	LNG, see **Methane** *etc. (UN1972)*														
	London purple	6.1	UN1621	II	6.1	IB8, IP2, IP4, T3, TP33	153	212	242	25 kg	100 kg	A		151	POISON
	LPG, see **Petroleum gases, liquefied**														
	Lye, see **Sodium hydroxide solutions**														
	Magnesium aluminum phosphide	4.3	UN1419	I	4.3, 6.1	A19, N34, N40	None	211	242	Forbidden	15 kg	E	40, 52, 85	139	DANGEROUS WHEN WET*
+	**Magnesium arsenate**	6.1	UN1622	II	6.1	IB8, IP2, IP4, T3, TP33	153	212	242	25 kg	100 kg	A		151	POISON

Symbols (1)	Hazardous materials descriptions and proper shipping names (2)	Hazard class or Division (3)	Identification Numbers (4)	PG (5)	Label Codes (6)	Special Provisions (§172.102) (7)	(8) Packaging (§173.***) Exceptions (8A)	Non-bulk (8B)	Bulk (8C)	(9) Quantity limitations (see §§173.27 and 175.75) Passenger aircraft/rail (9A)	Cargo aircraft only (9B)	(10) Vessel stowage Location (10A)	Other (10B)	ERG Guide No.	Placard Advisory Consult §172.504 & §172.505 to determine placard *Denotes placard for any quantity
	Magnesium bisulfite solution, see **Bisulfites, aqueous solutions, n.o.s.**														
	Magnesium bromate	5.1	UN1473	II	5.1	A1, IB8, IP2, IP4, T3, TP33	152	212	242	5 kg	25 kg	A	56, 58	140	OXIDIZER
	Magnesium chlorate	5.1	UN2723	II	5.1	IB8, IP2, IP4, T3, TP33	152	212	242	5 kg	25 kg	A	56, 58	140	OXIDIZER
	Magnesium diamide	4.2	UN2004	II	4.2	A8, A19, A20, IB6, T3, TP33	None	212	241	15 kg	50 kg	C		135	SPONTANEOUSLY COMBUSTIBLE
	Magnesium dross, wet or hot	Forbidden													
	Magnesium fluorosilicate	6.1	UN2853	III	6.1	IB8, IP3, T1, TP33	153	213	240	100 kg	200 kg	A	52	151	POISON
	Magnesium granules, coated, *particle size not less than 149 microns*	4.3	UN2950	III	4.3	A1, A19, IB8, IP4, T1, TP33	151	213	240	25 kg	100 kg	A	52	138	DANGEROUS WHEN WET*
	Magnesium hydride	4.3	UN2010	I	4.3	A19, N40	None	211	242	Forbidden	15 kg	E	52	138	DANGEROUS WHEN WET*
	Magnesium *or* **Magnesium alloys** *with more than 50 percent magnesium in pellets, turnings or ribbons*	4.1	UN1869	III	4.1	A1, IB8, IP3, T1, TP33	151	213	240	25 kg	100 kg	A	39, 52, 53, 74, 101	138	FLAMMABLE SOLID
	Magnesium nitrate	5.1	UN1474	III	5.1	332, A1, IB8, IP3, T1, TP33	152	213	240	25 kg	100 kg	A		140	OXIDIZER
	Magnesium perchlorate	5.1	UN1475	II	5.1	IB6, IP2, T3, TP33	152	212	242	5 kg	25 kg	A	56, 58	140	OXIDIZER
	Magnesium peroxide	5.1	UN1476	II	5.1	IB6, IP2, T3, TP33	152	212	242	5 kg	25 kg	A	13, 52, 66, 75	140	OXIDIZER
	Magnesium phosphide	4.3	UN2011	I	4.3, 6.1	A19, N40	None	211	None	Forbidden	15 kg	E	40, 52, 85	139	DANGEROUS WHEN WET*
	Magnesium, powder *or* **Magnesium alloys, powder**	4.3	UN1418	I	4.3, 4.2	A19, B56	None	211	244	Forbidden	15 kg	A	39, 52	138	DANGEROUS WHEN WET*
				II	4.3, 4.2	A19, B56, IB5, IP2, T3, TP33	None	212	241	15 kg	50 kg	A	39, 52	138	DANGEROUS WHEN WET*
				III	4.3, 4.2	A19, B56, IB8, IP4, T1, TP33	None	213	241	25 kg	100 kg	A	39, 52	138	DANGEROUS WHEN WET*
	Magnesium scrap, see **Magnesium,** *etc. (UN 1869)*														
	Magnesium silicide	4.3	UN2624	II	4.3	A19, A20, IB7, IP2, T3, TP33	151	212	241	15 kg	50 kg	B	85, 103	138	DANGEROUS WHEN WET*
	Magnetized material, see §173.21														
	Maleic anhydride	8	UN2215	III	8	IB8, IP3, T1, TP33	154	213	240	25 kg	100 kg	A		156	CORROSIVE
	Maleic anhydride, molten	8	UN2215	III	8	T4, TP3	None	213	240	Forbidden	Forbidden	A		156	CORROSIVE
	Malononitrile	6.1	UN2647	II	6.1	IB8, IP2, IP4, T3, TP33	153	212	242	25 kg	100 kg	A	12	153	POISON
	Mancozeb (manganese ethylenebisdithiocarbamate complex with zinc), see **Maneb**														
	Maneb *or* **Maneb preparations** *with not less than 60 percent maneb*	4.2	UN2210	III	4.2, 4.3	57, A1, A19, IB6, T1, TP33	None	213	242	25 kg	100 kg	A	34	135	SPONTANEOUSLY COMBUSTIBLE, DANGEROUS WHEN WET*
	Maneb stabilized *or* **Maneb preparations, stabilized** *against self-heating*	4.3	UN2968	III	4.3	54, A1, A19, IB8, IP4, T1, TP33	151	213	242	25 kg	100 kg	B	34, 52	135	DANGEROUS WHEN WET*
	Manganese nitrate	5.1	UN2724	III	5.1	A1, IB8, IP3, T1, TP33	152	213	240	25 kg	100 kg	A		140	OXIDIZER
	Manganese resinate	4.1	UN1330	III	4.1	A1, IB6, T1, TP33	151	213	240	25 kg	100 kg	A		133	FLAMMABLE SOLID
	Mannitan tetranitrate	Forbidden													
	Mannitol hexanitrate (dry)	Forbidden													
	Mannitol hexanitrate, wetted *or* **Nitromannite, wetted** *with not less than 40 percent water, or mixture of alcohol and water, by mass*	1.1D	UN0133	II	1.1D	121	None	62	None	Forbidden	Forbidden	10		112	EXPLOSIVES 1.1*
	Marine pollutants, liquid or solid, n.o.s., see **Environmentally hazardous substances, liquid** *or* **solid, n.o.s.**														
	Matches, block, see **Matches, 'strike anywhere'**														
	Matches, fusee	4.1	UN2254	III	4.1		186	186	None	Forbidden	Forbidden	A		133	FLAMMABLE SOLID
	Matches, safety *(book, card or strike on box)*	4.1	UN1944	III	4.1		186	186	None	25 kg	100 kg	A		133	FLAMMABLE SOLID

Symbols (1)	Hazardous materials descriptions and proper shipping names (2)	Hazard class or Division (3)	Identification Numbers (4)	PG (5)	Label Codes (6)	Special Provisions (§172.102) (7)	(8) Packaging (§173.***) Exceptions (8A)	Non-bulk (8B)	Bulk (8C)	(9) Quantity limitations (see §§173.27 and 175.75) Passenger aircraft/rail (9A)	Cargo aircraft only (9B)	(10) Vessel stowage Location (10A)	Other (10B)	ERG Guide No.	Placard Advisory Consult §172.504 & §172.505 to determine placard *Denotes placard for any quantity
	Matches, strike anywhere	4.1	UN1331	III	4.1		186	186	None	Forbidden	Forbidden	B		133	FLAMMABLE SOLID
	Matches, wax, Vesta	4.1	UN1945	III	4.1		186	186	None	25 kg	100 kg	B		133	FLAMMABLE SOLID
	Matting acid, see **Sulfuric acid**														
	Medicine, liquid, flammable, toxic, n.o.s.	3	UN3248	II	3, 6.1	IB2	150	202	243	1 L	60 L	B	40	131	FLAMMABLE
				III	3, 6.1	IB3	150	203	242	60 L	220 L	A		131	FLAMMABLE
	Medicine, liquid, toxic, n.o.s.	6.1	UN1851	II	6.1		153	202	243	5 L	60 L	C	40	151	POISON
				III	6.1		153	203	241	60 L	220 L	C	40	151	POISON
	Medicine, solid, toxic, n.o.s.	6.1	UN3249	II	6.1	T3, TP33	153	212	242	25 kg	100 kg	C	40	151	POISON
				III	6.1	T3, TP33	153	213	240	100 kg	200 kg	C	40	151	POISON
	Memtetrahydrophthalic anhydride, see **Corrosive liquids, n.o.s.**														
	Mercaptans, liquid, flammable, n.o.s. *or* **Mercaptan mixture, liquid, flammable, n.o.s.**	3	UN3336	I	3	T11, TP2	150	201	243	1 L	30 L	E	95	130	FLAMMABLE
				II	3	IB2, T7, TP1, TP8, TP28	150	202	242	5 L	60 L	B	95	130	FLAMMABLE
				III	3	B1, B52, IB3, T4, TP1, TP29	150	203	241	60 L	220 L	B	95	130	FLAMMABLE
	Mercaptans, liquid, flammable, toxic, n.o.s *or* **Mercaptan mixtures, liquid, flammable, toxic, n.o.s.**	3	UN1228	II	3, 6.1	IB2, T11, TP2, TP27	None	202	243	Forbidden	60 L	B	40, 95	131	FLAMMABLE
				III	3, 6.1	A6, B1, IB3, T7, TP1, TP28	150	203	242	5 L	220 L	A	40, 95	131	FLAMMABLE
	Mercaptans, liquid, toxic, flammable, n.o.s. *or* **Mercaptan mixtures, liquid, toxic, flammable, n.o.s.,** *flash point not less than 23 degrees C*	6.1	UN3071	II	6.1, 3	A6, IB2, T11, TP2, TP13, TP27	153	202	243	5 L	60 L	C	40, 121	131	POISON
	5-Mercaptotetrazol-1-acetic acid	1.4C	UN0448	II	1.4C		None	62	None	Forbidden	75 kg	09		114	EXPLOSIVES 1.4
	Mercuric arsenate	6.1	UN1623	II	6.1	IB8, IP2, IP4, T3, TP33	153	212	242	25 kg	100 kg	A		151	POISON
	Mercuric chloride	6.1	UN1624	II	6.1	IB8, IP2, IP4, T3, TP33	153	212	242	25 kg	100 kg	A		154	POISON
	Mercuric compounds, see **Mercury compounds,** *etc.*														
	Mercuric nitrate	6.1	UN1625	II	6.1	IB8, IP2, IP4, N73, T3, TP33	153	212	242	25 kg	100 kg	A		141	POISON
+	**Mercuric potassium cyanide**	6.1	UN1626	I	6.1	IB7, IP1, N74, N75, T6, TP33	None	211	242	5 kg	50 kg	A	52	157	POISON
	Mercuric sulfocyanate, see **Mercury thiocyanate**														
	Mercurol, see **Mercury nucleate**														
	Mercurous azide	Forbidden													
	Mercurous compounds, see **Mercury compounds,** *etc.*														
	Mercurous nitrate	6.1	UN1627	II	6.1	IB8, IP2, IP4, T3, TP33	153	212	242	25 kg	100 kg	A		141	POISON
A W	**Mercury**	8	UN2809	III	8		164	164	240	35 kg	35 kg	B	40, 97	172	CORROSIVE
	Mercury acetate	6.1	UN1629	II	6.1	IB8, IP2, IP4, T3, TP33	153	212	242	25 kg	100 kg	A		151	POISON
	Mercury acetylide	Forbidden													
	Mercury ammonium chloride	6.1	UN1630	II	6.1	IB8, IP2, IP4, T3, TP33	153	212	242	25 kg	100 kg	A		151	POISON
	Mercury based pesticides, liquid, flammable, toxic, *flash point less than 23 degrees C*	3	UN2778	I	3, 6.1	T14, TP2, TP13, TP27	None	201	243	Forbidden	30 L	B	40	131	FLAMMABLE
				II	3, 6.1	IB2, T11, TP2, TP13, TP27	150	202	243	1 L	60 L	B	40	131	FLAMMABLE
	Mercury based pesticides, liquid, toxic	6.1	UN3012	I	6.1	T14, TP2, TP13, TP27	None	201	243	1 L	30 L	B	40	151	POISON
				II	6.1	IB2, T11, TP2, TP13, TP27	153	202	243	5 L	60 L	B	40	151	POISON
				III	6.1	IB3, T7, TP2, TP28	153	203	241	60 L	220 L	A	40	151	POISON
	Mercury based pesticides, liquid, toxic, flammable, *flash point not less than 23 degrees C*	6.1	UN3011	I	6.1, 3	T14, TP2, TP13, TP27	None	201	243	1 L	30 L	B	40	131	POISON

HAZARDOUS MATERIALS TABLE

Symbols (1)	Hazardous materials descriptions and proper shipping names (2)	Hazard class or Division (3)	Identification Numbers (4)	PG (5)	Label Codes (6)	Special Provisions (§172.102) (7)	(8) Packaging (§173.***) Exceptions (8A)	Non-bulk (8B)	Bulk (8C)	(9) Quantity limitations (see §§173.27 and 175.75) Passenger aircraft/rail (9A)	Cargo aircraft only (9B)	(10) Vessel stowage Location (10A)	Other (10B)	ERG Guide No.	Placard Advisory Consult §172.504 & §172.505 to determine placard *Denotes placard for any quantity
				II	6.1, 3	IB2, T11, TP2, TP13, TP27	153	202	243	5 L	60 L	B	40	131	POISON
				III	6.1, 3	IB3, T7, TP2, TP28	153	203	242	60 L	220 L	A	40	131	POISON
	Mercury based pesticides, solid, toxic	6.1	UN2777	I	6.1	IB7, IP1, T6, TP33	None	211	242	5 kg	50 kg	A	40	151	POISON
				II	6.1	IB8, IP2, IP4, T3, TP33	153	212	242	25 kg	100 kg	A	40	151	POISON
				III	6.1	IB8, IP3, T1, TP33	153	213	240	100 kg	200 kg	A	40	151	POISON
	Mercury benzoate	6.1	UN1631	II	6.1	IB8, IP2, IP4, T3, TP33	153	212	242	25 kg	100 kg	A		154	POISON
	Mercury bromides	6.1	UN1634	II	6.1	IB8, IP2, IP4, T3, TP33	153	212	242	25 kg	100 kg	A		154	POISON
G	**Mercury compound, liquid, n.o.s.**	6.1	UN2024	I	6.1		None	201	243	1 L	30 L	B	40	151	POISON
				II	6.1	IB2	153	202	243	5 L	60 L	B	40	151	POISON
				III	6.1	IB3	153	203	241	60 L	220 L	B	40	151	POISON
G	**Mercury compound, solid, n.o.s.**	6.1	UN2025	I	6.1	IB7, IP1, T6, TP33	None	211	242	5 kg	50 kg	A		151	POISON
				II	6.1	IB8, IP2, IP4, T3, TP33	153	212	242	25 kg	100 kg	A		151	POISON
				III	6.1	IB8, IP3, T1, TP33	153	213	240	100 kg	200 kg	A		151	POISON
A	**Mercury** *contained in manufactured articles*	8	UN2809	III	8		None	164	None	No limit	No limit	B	40, 97	172	CORROSIVE
	Mercury cyanide	6.1	UN1636	II	6.1	IB8, IP2, IP4, N74, N75, T3, TP33	153	212	242	25 kg	100 kg	A	52	154	POISON
	Mercury fulminate, wetted *with not less than 20 percent water, or mixture of alcohol and water, by mass*	1.1A	UN0135	II	1.1A	111, 117	None	62	None	Forbidden	Forbidden	12		112	EXPLOSIVES 1.1*
	Mercury gluconate	6.1	UN1637	II	6.1	IB8, IP2, IP4, T3, TP33	153	212	242	25 kg	100 kg	A		151	POISON
	Mercury iodide	6.1	UN1638	II	6.1	IB2, IP2, IP4, T3, TP33	153	212	242	25 kg	100 kg	A		151	POISON
	Mercury iodide aquabasic ammonobasic (Iodide of Millon's base)	Forbidden													
	Mercury nitride	Forbidden													
	Mercury nucleate	6.1	UN1639	II	6.1	IB8, IP2, IP4, T3, TP33	153	212	242	25 kg	100 kg	A		151	POISON
	Mercury oleate	6.1	UN1640	II	6.1	IB8, IP2, IP4, T3, TP33	153	212	242	25 kg	100 kg	A		151	POISON
	Mercury oxide	6.1	UN1641	II	6.1	IB8, IP2, IP4, T3, TP33	153	212	242	25 kg	100 kg	A		151	POISON
	Mercury oxycyanide	Forbidden													
	Mercury oxycyanide, desensitized	6.1	UN1642	II	6.1	IB8, IP2, IP4, T3, TP33	153	212	242	25 kg	100 kg	A	52, 91	151	POISON
	Mercury potassium iodide	6.1	UN1643	II	6.1	IB8, IP2, IP4, T3, TP33	153	212	242	25 kg	100 kg	A		151	POISON
	Mercury salicylate	6.1	UN1644	II	6.1	IB8, IP2, IP4, T3, TP33	153	212	242	25 kg	100 kg	A		151	POISON
+	**Mercury sulfates**	6.1	UN1645	II	6.1	IB8, IP2, IP4, T3, TP33	153	212	242	25 kg	100 kg	A		151	POISON
	Mercury thiocyanate	6.1	UN1646	II	6.1	IB8, IP2, IP4, T3, TP33	153	212	242	25 kg	100 kg	A		151	POISON
	Mesityl oxide	3	UN1229	III	3	B1, IB3, T2, TP1	150	203	242	60 L	220 L	A		129	FLAMMABLE
G	**Metal carbonyls, liquid, n.o.s.**	6.1	UN3281	I	6.1	5, T14, TP2, TP13, TP27	None	201	243	1 L	30 L	B	40	151	POISON
				II	6.1	IB2, T11, TP2, TP27	153	202	243	5 L	60 L	B	40	151	POISON
				III	6.1	IB3, T7, TP1, TP28	153	203	241	60 L	220 L	A	40	151	POISON
G	**Metal carbonyls, solid, n.o.s.**	6.1	UN3466	I	6.1	IB7, IP1, T6, TP33	None	211	242	5 kg	50 kg	D	40	151	POISON
				II	6.1	IB8, IP2, IP4, T3, TP33	153	212	242	25 kg	100 kg	B	40	151	POISON
				III	6.1	IB8, IP3, T1, TP33	153	213	240	100 kg	200 kg	B	40	151	POISON

Symbols (1)	Hazardous materials descriptions and proper shipping names (2)	Hazard class or Division (3)	Identification Numbers (4)	PG (5)	Label Codes (6)	Special Provisions (§172.102) (7)	(8) Packaging (§173.***) Exceptions (8A)	Non-bulk (8B)	Bulk (8C)	(9) Quantity limitations (see §§173.27 and 175.75) Passenger aircraft/rail (9A)	Cargo aircraft only (9B)	(10) Vessel stowage Location (10A)	Other (10B)	ERG Guide No.	Placard Advisory Consult §172.504 & §172.505 to determine placard *Denotes placard for any quantity
G	**Metal catalyst, dry**	4.2	UN2881	I	4.2	N34, T21, TP7, TP33	None	187	None	Forbidden	Forbidden	C		135	SPONTANEOUSLY COMBUSTIBLE
				II	4.2	IB6, IP2, N34, T3, TP33	None	187	242	Forbidden	50 kg	C		135	SPONTANEOUSLY COMBUSTIBLE
				III	4.2	IB8, IP3, N34, T1, TP33	None	187	241	25 kg	100 kg	C		135	SPONTANEOUSLY COMBUSTIBLE
G	**Metal catalyst, wetted** *with a visible excess of liquid*	4.2	UN1378	II	4.2	A2, A8, IB1, N34, T3, TP33	None	212	None	Forbidden	50 kg	C		170	SPONTANEOUSLY COMBUSTIBLE
	Metal hydrides, flammable, n.o.s.	4.1	UN3182	II	4.1	A1, IB4, T3, TP33	151	212	240	15 kg	50 kg	E		170	FLAMMABLE SOLID
				III	4.1	A1, IB4, T1, TP33	151	213	240	25 kg	100 kg	E		170	FLAMMABLE SOLID
	Metal hydrides, water reactive n.o.s.	4.3	UN1409	I	4.3	A19, N34, N40	None	211	242	Forbidden	15 kg	D	52	138	DANGEROUS WHEN WET*
				II	4.3	A19, IB4, N34, N40, T3, TP33	151	212	242	15 kg	50 kg	D	52	138	DANGEROUS WHEN WET*
	Metal powder, self-heating, n.o.s.	4.2	UN3189	II	4.2	IB6, IP2, T3, TP33	None	212	241	15 kg	50 kg	C		135	SPONTANEOUSLY COMBUSTIBLE
				III	4.2	IB8, IP3, T1, TP33	None	213	241	25 kg	100 kg	C		135	SPONTANEOUSLY COMBUSTIBLE
	Metal powders, flammable, n.o.s.	4.1	UN3089	II	4.1	IB8, IP2, IP4, T3, TP33	151	212	240	15 kg	50 kg	B		170	FLAMMABLE SOLID
				III	4.1	IB6, T1, TP33	151	213	240	25 kg	100 kg	B		170	FLAMMABLE SOLID
	Metal salts of methyl nitramine (dry)	Forbidden													
G	**Metal salts of organic compounds, flammable, n.o.s.**	4.1	UN3181	II	4.1	A1, IB8, IP2, IP4, T3, TP33	151	212	240	15 kg	50 kg	B	40	133	FLAMMABLE SOLID
				III	4.1	A1, IB8, IP3, T1, TP33	151	213	240	25 kg	100 kg	B	40	133	FLAMMABLE SOLID
	Metaldehyde	4.1	UN1332	III	4.1	A1, IB8, IP3, T1, TP33	151	213	240	25 kg	100 kg	A		133	FLAMMABLE SOLID
G	**Metallic substance, water-reactive, n.o.s.**	4.3	UN3208	I	4.3	A7, IB4	None	211	242	Forbidden	15 kg	E	40	138	DANGEROUS WHEN WET*
				II	4.3	A7, IB7, IP2, T3, TP33	151	212	242	15 kg	50 kg	E	40	138	DANGEROUS WHEN WET*
				III	4.3	A7, IB8, IP4, T1, TP33	151	213	241	25 kg	100 kg	E	40	138	DANGEROUS WHEN WET*
G	**Metallic substance, water-reactive, self-heating, n.o.s.**	4.3	UN3209	I	4.3, 4.2	A7	None	211	242	Forbidden	15 kg	E	40	138	DANGEROUS WHEN WET*
				II	4.3, 4.2	A7, IB5, IP2, T3, TP33	None	212	242	15 kg	50 kg	E	40	138	DANGEROUS WHEN WET*
				III	4.3, 4.2	A7, IB8, IP4, T1, TP33	None	213	242	25 kg	100 kg	E	40	138	DANGEROUS WHEN WET*
	Methacrylaldehyde, stabilized	3	UN2396	II	3, 6.1	45, IB2, T7, TP1, TP13	150	202	243	1 L	60 L	E	40	131P	FLAMMABLE
	Methacrylic acid, stabilized	8	UN2531	II	8	41, IB2, T7, TP1, TP18, TP30	154	202	242	1 L	30 L	C	40	153P	CORROSIVE
+	**Methacrylonitrile, stabilized**	6.1	UN3079	I	6.1, 3	2, B9, B14, B32, T20, TP2, TP13, TP38, TP45	None	227	244	Forbidden	Forbidden	D	12, 40, 48	131P	POISON INHALATION HAZARD*
	Methallyl alcohol	3	UN2614	III	3	B1, IB3, T2, TP1	150	203	242	60 L	220 L	A		129	FLAMMABLE
	Methane and hydrogen, mixtures, see **Hydrogen and Methane mixtures,** *etc.*														
	Methane, compressed *or* **Natural gas, compressed** *(with high methane content)*	2.1	UN1971		2.1		306	302	302	Forbidden	150 kg	E	40	115	FLAMMABLE GAS
	Methane, refrigerated liquid *(cryogenic liquid) or* **Natural gas, refrigerated liquid** *(cryogenic liquid), with high methane content)*	2.1	UN1972		2.1	T75, TP5	None	None	318	Forbidden	Forbidden	D	40	115	FLAMMABLE GAS
	Methanesulfonyl chloride	6.1	UN3246	I	6.1, 8	2, B9, B14, B32, T20, TP2, TP13, TP38, TP45	None	227	244	Forbidden	Forbidden	D	40	156	POISON INHALATION HAZARD*
+ I	**Methanol**	3	UN1230	II	3, 6.1	IB2, T7, TP2	150	202	242	1 L	60 L	B	40	131	FLAMMABLE
D	**Methanol**	3	UN1230	II	3	IB2, T7, TP2	150	202	242	1 L	60 L	B	40	131	FLAMMABLE

Symbols	Hazardous materials descriptions and proper shipping names	Hazard class or Division	Identification Numbers	PG	Label Codes	Special Provisions (§172.102)	(8) Packaging (§173.***)			(9) Quantity limitations (see §§173.27 and 175.75)		(10) Vessel stowage		ERG Guide No.	Placard Advisory Consult §172.504 & §172.505 to determine placard *Denotes placard for any quantity
							Exceptions	Non-bulk	Bulk	Passenger aircraft/rail	Cargo aircraft only	Location	Other		
(1)	(2)	(3)	(4)	(5)	(6)	(7)	(8A)	(8B)	(8C)	(9A)	(9B)	(10A)	(10B)		
	Methazoic acid	Forbidden													
	4-Methoxy-4-methylpentan-2-one	3	UN2293	III	3	B1, IB3, T2, TP1	150	203	242	60 L	220 L	A		128	FLAMMABLE
	1-Methoxy-2-propanol	3	UN3092	III	3	B1, IB3, T2, TP1	150	203	242	60 L	220 L	A		129	FLAMMABLE
+	**Methoxymethyl isocyanate**	6.1	UN2605	I	6.1, 3	1, B9, B14, B30, T20, TP2, TP13, TP38, TP44	None	226	244	Forbidden	Forbidden	D	40	155	POISON INHALATION HAZARD*
	Methyl acetate	3	UN1231	II	3	IB2, T4, TP1	150	202	242	5 L	60 L	B		129	FLAMMABLE
	Methyl acetylene and propadiene mixtures, stabilized	2.1	UN1060		2.1	N88, T50	306	304	314, 315	Forbidden	150 kg	B	40	116P	FLAMMABLE GAS
	Methyl acrylate, stabilized	3	UN1919	II	3	IB2, T4, TP1, TP13	150	202	242	5 L	60 L	B		129P	FLAMMABLE
	Methyl alcohol, see **Methanol**														
	Methyl allyl chloride	3	UN2554	II	3	IB2, T4, TP1, TP13	150	202	242	5 L	60 L	E		130P	FLAMMABLE
	Methyl amyl ketone, see **Amyl methyl ketone**														
	Methyl bromide	2.3	UN1062		2.3	3, B14, N86, T50	None	193	314, 315	Forbidden	Forbidden	D	40	123	POISON GAS*
	Methyl bromide and chloropicrin mixtures with more than 2 percent chloropicrin, see **Chloropicrin and methyl bromide mixtures**														
	Methyl bromide and chloropicrin mixtures with not more than 2 percent chloropicrin, see **Methyl bromide**														
	Methyl bromide and ethylene dibromide mixtures, liquid	6.1	UN1647	I	6.1	2, B9, B14, B32, N65, T20, TP2, TP13, TP38, TP44	None	227	244	Forbidden	Forbidden	D	40	151	POISON INHALATION HAZARD*
	Methyl bromoacetate	6.1	UN2643	II	6.1	IB2, T7, TP2	153	202	243	5 L	60 L	D	40	155	POISON
	2-Methyl-1-butene	3	UN2459	I	3	T11, TP2	None	201	243	1 L	30 L	E		128	FLAMMABLE
	2-Methyl-2-butene	3	UN2460	II	3	IB2, IP8, T7, TP1	None	202	242	5 L	60 L	E		128	FLAMMABLE
	3-Methyl-1-butene	3	UN2561	I	3	T11, TP2	None	201	243	1 L	30 L	E		128	FLAMMABLE
	Methyl tert-butyl ether	3	UN2398	II	3	IB2, T7, TP1	150	202	242	5 L	60 L	E		127	FLAMMABLE
	Methyl butyrate	3	UN1237	II	3	IB2, T4, TP1	150	202	242	5 L	60 L	B		129	FLAMMABLE
	Methyl chloride *or* **Refrigerant gas R 40**	2.1	UN1063		2.1	N86, T50	306	304	314, 315	5 kg	100 kg	D	40	115	FLAMMABLE GAS
	Methyl chloride and chloropicrin mixtures, see **Chloropicrin and methyl chloride mixtures**														
	Methyl chloride and methylene chloride mixtures	2.1	UN1912		2.1	N86, T50	306	304	314, 315	Forbidden	150 kg	D	40	115	FLAMMABLE GAS
	Methyl chloroacetate	6.1	UN2295	I	6.1, 3	T14, TP2, TP13	None	201	243	1 L	30 L	D		155	POISON
	Methyl chlorocarbonate, see **Methyl chloroformate**														
	Methyl chloroform, see **1,1,1-Trichloroethane**														
	Methyl chloroformate	6.1	UN1238	I	6.1, 3, 8	1, B9, B14, B30, N34, T22, TP2, TP13, TP38, TP44	None	226	244	Forbidden	Forbidden	D	21, 40, 100	155	POISON INHALATION HAZARD*
	Methyl chloromethyl ether	6.1	UN1239	I	6.1, 3	1, B9, B14, B30, T22, TP2, TP13, TP38, TP44	None	226	244	Forbidden	Forbidden	D	40	131	POISON INHALATION HAZARD*
	Methyl 2-chloropropionate	3	UN2933	III	3	B1, IB3, T2, TP1	150	203	242	60 L	220 L	A		129	FLAMMABLE
	Methyl dichloroacetate	6.1	UN2299	III	6.1	IB3, T4, TP1	153	203	241	60 L	220 L	A		155	POISON
	Methyl ethyl ether, see **Ethyl methyl ether**														
	Methyl ethyl ketone, *see* **Ethyl methyl ketone** *or* **Methyl ethyl ketone**														
	Methyl ethyl ketone peroxide, in solution with more than 9 percent by mass active oxygen	Forbidden													
	2-Methyl-5-ethylpyridine	6.1	UN2300	III	6.1	IB3, T4, TP1	153	203	241	60 L	220 L	A		153	POISON
	Methyl fluoride *or* **Refrigerant gas R 41**	2.1	UN2454		2.1		306	304	314, 315	Forbidden	150 kg	E	40	115	FLAMMABLE GAS
	Methyl formate	3	UN1243	I	3	T11, TP2	150	201	243	1 L	30 L	E		129	FLAMMABLE

Symbols (1)	Hazardous materials descriptions and proper shipping names (2)	Hazard class or Division (3)	Identification Numbers (4)	PG (5)	Label Codes (6)	Special Provisions (§172.102) (7)	(8) Packaging (§173.***) Exceptions (8A)	Non-bulk (8B)	Bulk (8C)	(9) Quantity limitations (see §§173.27 and 175.75) Passenger aircraft/rail (9A)	Cargo aircraft only (9B)	(10) Vessel stowage Location (10A)	Other (10B)	ERG Guide No.	Placard Advisory Consult §172.504 & §172.505 to determine placard *Denotes placard for any quantity
	2-Methyl-2-heptanethiol	6.1	UN3023	I	6.1, 3	2, B9, B14, B32, T20, TP2, TP13, TP38, TP45	None	227	244	Forbidden	Forbidden	D	40, 102	131	POISON INHALATION HAZARD*
	Methyl iodide	6.1	UN2644	I	6.1	2, B9, B14, B32, T20, TP2, TP13, TP38, TP45	None	227	244	Forbidden	Forbidden	D	12, 40	151	POISON INHALATION HAZARD*
	Methyl isobutyl carbinol	3	UN2053	III	3	B1, IB3, T2, TP1	150	203	242	60 L	220 L	A		129	FLAMMABLE
	Methyl isobutyl ketone	3	UN1245	II	3	IB2, T4, TP1	150	202	242	5 L	60 L	B		127	FLAMMABLE
	Methyl isobutyl ketone peroxide, in solution with more than 9 percent by mass active oxygen	Forbidden													
	Methyl isocyanate	6.1	UN2480	I	6.1, 3	1, B9, B14, B30, T22, TP2, TP13, TP38, TP44	None	226	244	Forbidden	Forbidden	D	40, 52	155	POISON INHALATION HAZARD*
	Methyl isopropenyl ketone, stabilized	3	UN1246	II	3	IB2, T4, TP1	150	202	242	5 L	60 L	B		127P	FLAMMABLE
	Methyl isothiocyanate	6.1	UN2477	I	6.1, 3	2, B9, B14, B32, T20, TP2, TP13, TP38, TP45	None	227	244	Forbidden	Forbidden	D	40	131	POISON INHALATION HAZARD*
	Methyl isovalerate	3	UN2400	II	3	IB2, T4, TP1	150	202	242	5 L	60 L	B		130	FLAMMABLE
	Methyl magnesium bromide, in ethyl ether	4.3	UN1928	I	4.3, 3		None	201	243	Forbidden	1 L	D		135	DANGEROUS WHEN WET*
	Methyl mercaptan	2.3	UN1064		2.3, 2.1	3, B7, B9, B14, N89, T50	None	304	314, 315	Forbidden	Forbidden	D	40	117	POISON GAS*
	Methyl mercaptopropionaldehyde, see **4-Thiapentanal**														
	Methyl methacrylate monomer, stabilized	3	UN1247	II	3	IB2, T4, TP1	150	202	242	5 L	60 L	B	40	129P	FLAMMABLE
	Methyl nitramine (dry)	Forbidden													
	Methyl nitrate	Forbidden													
	Methyl nitrite	Forbidden													
	Methyl norbornene dicarboxylic anhydride, see **Corrosive liquids, n.o.s.**														
	Methyl orthosilicate	6.1	UN2606	I	6.1, 3	2, B9, B14, B32, T20, TP2, TP13, TP38, TP45	None	227	244	Forbidden	Forbidden	D	40	155	POISON INHALATION HAZARD*
D	**Methyl phosphonic dichloride**	6.1	NA9206	I	6.1, 8	2, B9, B14, B32, N34, N43, T20, TP4, TP13, TP38, TP45	None	227	244	Forbidden	Forbidden	C		137	POISON INHALATION HAZARD*
	Methyl phosphonothioic dichloride, anhydrous, see **Corrosive liquids, n.o.s.**														
D	**Methyl phosphonous dichloride,** *pyrophoric liquid*	6.1	NA2845	I	6.1, 4.2	2, B9, B14, B16, B32, B74, T20, TP4, TP13, TP38, TP45	None	227	244	Forbidden	Forbidden	D	18	135	POISON INHALATION HAZARD*
	Methyl picric acid (heavy metal salts of)	Forbidden													
	Methyl propionate	3	UN1248	II	3	IB2, T4, TP1	150	202	242	5 L	60 L	B		129	FLAMMABLE
	Methyl propyl ether	3	UN2612	II	3	IB2, IP8, T7, TP2	150	202	242	5 L	60 L	E	40	127	FLAMMABLE
	Methyl propyl ketone	3	UN1249	II	3	IB2, T4, TP1	150	202	242	5 L	60 L	B		127	FLAMMABLE
	Methyl sulfate, see **Dimethyl sulfate**														
	Methyl sulfide, see **Dimethyl sulfide**														
	Methyl trichloroacetate	6.1	UN2533	III	6.1	IB3, T4, TP1	153	203	241	60 L	220 L	A		156	POISON
	Methyl trimethylol methane trinitrate	Forbidden													
	Methyl vinyl ketone, stabilized	6.1	UN1251	I	6.1, 3, 8	1, B9, B14, B30, T22, TP2, TP13, TP38, TP44	None	226	244	Forbidden	Forbidden	B	40	131P	POISON INHALATION HAZARD*
	Methylal	3	UN1234	II	3	IB2, IP8, T7, TP2	None	202	242	5 L	60 L	E		127	FLAMMABLE
	Methylamine, anhydrous	2.1	UN1061		2.1	N87, T50	306	304	314, 315	Forbidden	150 kg	B	40	118	FLAMMABLE GAS
	Methylamine, aqueous solution	3	UN1235	II	3, 8	B1, IB2, T7, TP1	150	202	243	1 L	5 L	E	52, 135	132	FLAMMABLE
	Methylamine dinitramine and dry salts thereof	Forbidden													
	Methylamine nitroform	Forbidden													
	Methylamine perchlorate (dry)	Forbidden													

Symbols (1)	Hazardous materials descriptions and proper shipping names (2)	Hazard class or Division (3)	Identification Numbers (4)	PG (5)	Label Codes (6)	Special Provisions (§172.102) (7)	(8) Packaging (§173.***) Exceptions (8A)	Non-bulk (8B)	Bulk (8C)	(9) Quantity limitations (see §§173.27 and 175.75) Passenger aircraft/rail (9A)	Cargo aircraft only (9B)	(10) Vessel stowage Location (10A)	Other (10B)	ERG Guide No.	Placard Advisory Consult §172.504 & §172.505 to determine placard *Denotes placard for any quantity
	Methylamyl acetate	3	UN1233	III	3	B1, IB3, T2, TP1	150	203	242	60 L	220 L	A		130	FLAMMABLE
	N-Methylaniline	6.1	UN2294	III	6.1	IB3, T4, TP1	153	203	241	60 L	220 L	A		153	POISON
	alpha-Methylbenzyl alcohol, liquid	6.1	UN2937	III	6.1	IB3, T4, TP1	153	203	241	60 L	220 L	A		153	POISON
	alpha-Methylbenzyl alcohol, solid	6.1	UN3438	III	6.1	IB8, IP3, T1, TP33	153	213	240	100 kg	200 kg	A		153	POISON
	3-Methylbutan-2-one	3	UN2397	II	3	IB2, T4, TP1	150	202	242	5 L	60 L	B		127	FLAMMABLE
	2-Methylbutanal	3	UN3371	II	3	IB2, T4, TP1	150	202	242	5 L	60 L	B		129	FLAMMABLE
	N-Methylbutylamine	3	UN2945	II	3, 8	IB2, T7, TP1	150	202	243	1 L	5 L	B	40	132	FLAMMABLE
	Methylchlorosilane	2.3	UN2534		2.3, 2.1, 8	2, B9, B14, N34	None	226	314, 315	Forbidden	Forbidden	D	17, 40	119	POISON GAS*
	Methylcyclohexane	3	UN2296	II	3	B1, IB2, T4, TP1	150	202	242	5 L	60 L	B		128	FLAMMABLE
	Methylcyclohexanols, *flammable*	3	UN2617	III	3	B1, IB3, T2, TP1	150	203	242	60 L	220 L	A		129	FLAMMABLE
	Methylcyclohexanone	3	UN2297	III	3	B1, IB3, T2, TP1	150	203	242	60 L	220 L	A		128	FLAMMABLE
	Methylcyclopentane	3	UN2298	II	3	IB2, T4, TP1	150	202	242	5 L	60 L	B		128	FLAMMABLE
D	**Methyldichloroarsine**	6.1	NA1556	I	6.1	2, T20, TP4, TP13, TP38, TP45	None	192	None	Forbidden	Forbidden	D	40	152	POISON INHALATION HAZARD*
	Methyldichlorosilane	4.3	UN1242	I	4.3, 8, 3	A2, A3, A7, B6, B77, N34, T14, TP2, TP7, TP13	None	201	243	Forbidden	1 L	D	21, 28, 40, 49, 100	139	DANGEROUS WHEN WET*
	Methylene chloride, see **Dichloromethane**														
	Methylene glycol dinitrate	Forbidden													
	2-Methylfuran	3	UN2301	II	3	IB2, T4, TP1	150	202	242	5 L	60 L	E		128	FLAMMABLE
	a-Methylglucoside tetranitrate	Forbidden													
	a-Methylglycerol trinitrate	Forbidden													
	5-Methylhexan-2-one	3	UN2302	III	3	B1, IB3, T2, TP1	150	203	242	60 L	220 L	A		127	FLAMMABLE
	Methylhydrazine	6.1	UN1244	I	6.1, 3, 8	1, B7, B9, B14, B30, B77, N34, T22, TP2, TP13, TP38, TP44	None	226	244	Forbidden	Forbidden	D	21, 40, 49, 52, 100	131	POISON INHALATION HAZARD*
	4-Methylmorpholine *or* **n-methylmorpholine**	3	UN2535	II	3, 8	B6, IB2, T7, TP1	150	202	243	1 L	5 L	B	40	132	FLAMMABLE
	Methylpentadienes	3	UN2461	II	3	IB2, T4, TP1	150	202	242	5 L	60 L	E		128	FLAMMABLE
	2-Methylpentan-2-ol	3	UN2560	III	3	B1, IB3, T2, TP1	150	203	242	60 L	220 L	A		129	FLAMMABLE
	Methylpentanes, see **Hexanes**														
	Methylphenyldichlorosilane	8	UN2437	II	8	T10, TP2, TP7, TP13	None	206	242	Forbidden	30 L	C	40	156	CORROSIVE
	1-Methylpiperidine	3	UN2399	II	3, 8	IB2, T7, TP1	150	202	243	1 L	5 L	B	52	132	FLAMMABLE
	Methyltetrahydrofuran	3	UN2536	II	3	IB2, T4, TP1	150	202	242	5 L	60 L	B		127	FLAMMABLE
	Methyltrichlorosilane	3	UN1250	II	3, 8	A7, B6, B77, N34, T10, TP2, TP7, TP13	None	206	243	1 L	5 L	B	40	155	FLAMMABLE
	alpha-Methylvaleraldehyde	3	UN2367	II	3	B1, IB2, T4, TP1	150	202	242	5 L	60 L	B		130	FLAMMABLE
	Mine rescue equipment containing carbon dioxide, see **Carbon dioxide**														
	Mines *with bursting charge*	1.1F	UN0136	II	1.1F			62	None	Forbidden	Forbidden	08		112	EXPLOSIVES 1.1*
	Mines *with bursting charge*	1.1D	UN0137	II	1.1D			62	62	Forbidden	Forbidden	03		112	EXPLOSIVES 1.1*
	Mines *with bursting charge*	1.2D	UN0138	II	1.2D			62	62	Forbidden	Forbidden	03		112	EXPLOSIVES 1.2*
	Mines *with bursting charge*	1.2F	UN0294	II	1.2F			62	None	Forbidden	Forbidden	08		112	EXPLOSIVES 1.2*
	Mixed acid, see **Nitrating acid mixtures** *etc.*														
	Mobility aids, see **Battery powered equipment** *or* **Battery powered vehicle**														
D	**Model rocket motor**	1.4C	NA0276	II	1.4C	51	None	62	None	Forbidden	75 kg	06		114	EXPLOSIVES 1.4*
D	**Model rocket motor**	1.4S	NA0323	II	1.4S	51	None	62	None	25 kg	100 kg	05		114	EXPLOSIVES 1.4*
	Molybdenum pentachloride	8	UN2508	III	8	IB8, IP3, T1, TP33	154	213	240	25 kg	100 kg	C	40	156	CORROSIVE
	Monochloroacetone (unstabilized)	Forbidden													
	Monochloroethylene, see **Vinyl chloride, stabilized**														
	Monoethanolamine, see **Ethanolamine, solutions**														
	Monoethylamine, see **Ethylamine**														
	Morpholine	8	UN2054	I	8, 3	A6, T10, TP2	None	201	243	0.5 L	2.5 L	A		132	CORROSIVE

Sym-bols (1)	Hazardous materials descriptions and proper shipping names (2)	Hazard class or Division (3)	Identifi-cation Numbers (4)	PG (5)	Label Codes (6)	Special Provisions (§172.102) (7)	(8) Packaging (§173.***) Excep-tions (8A)	Non-bulk (8B)	Bulk (8C)	(9) Quantity limitations (see §§173.27 and 175.75) Passenger aircraft/rail (9A)	Cargo aircraft only (9B)	(10) Vessel stowage Location (10A)	Other (10B)	ERG Guide No.	Placard Advisory Consult §172.504 & §172.505 to determine placard *Denotes placard for any quantity
	Morpholine, aqueous, mixture, see **Corrosive liquids, n.o.s.**														
	Motor fuel anti-knock compounds, *see* **Motor fuel anti-knock mixtures**														
+	**Motor fuel anti-knock mixture, flammable**	6.1	UN3483	I	6.1, 3	14, T14, TP2, TP13	None	201	244	Forbidden	Forbidden	D	25, 40	—	POISON
+	**Motor fuel anti-knock mixtures**	6.1	UN1649	I	6.1	14, B9, B90, T14, TP2, TP13	None	201	244	Forbidden	30 L	D	25, 40	131	POISON
	Motor spirit, *see* **Gasoline**														
	Muriatic acid, see **Hydrochloric acid**														
	Musk xylene, *see* **5-tert-Butyl-2,4,6-trinitro-m-Xylene**														
	Naphtha see **Petroleum distillate n.o.s.**														
	Naphthalene, crude *or* **Naphthalene, refined**	4.1	UN1334	III	4.1	A1, IB8, IP3, T1, TP33	151	213	240	25 kg	100 kg	A		133	FLAMMABLE SOLID
	Naphthalene diozonide	Forbidden													
	beta-Naphthylamine, solid	6.1	UN1650	II	6.1	IB8, IP2, IP4, T3, TP33	153	212	242	25 kg	100 kg	A		153	POISON
	beta-Naphthylamine solution	6.1	UN3411	II	6.1	IB2, T7, TP2	153	202	243	5 L	60 L	A		153	POISON
				III	6.1	IB2, T7, TP2	153	203	241	60 L	220 L	A		153	POISON
	alpha-Naphthylamine	6.1	UN2077	III	6.1	IB8, IP3, T1, TP33	153	213	240	100 kg	200 kg	A		153	POISON
	Naphthalene, molten	4.1	UN2304	III	4.1	IB1, T1, TP3	151	213	241	Forbidden	Forbidden	C		133	FLAMMABLE SOLID
	Naphthylamineperchlorate	Forbidden													
	Naphthylthiourea	6.1	UN1651	II	6.1	IB8, IP2, IP4, T3, TP33	153	212	242	25 kg	100 kg	A		153	POISON
	Naphthylurea	6.1	UN1652	II	6.1	IB8, IP2, IP4, T3, TP33	153	212	242	25 kg	100 kg	A		153	POISON
	Natural gases (with high methane content), see **Methane** *,etc. (UN 1971, UN 1972)*														
	Neohexane, see **Hexanes**														
	Neon, compressed	2.2	UN1065		2.2		306, 307	302	None	75 kg	150 kg	A		121	NONFLAMMABLE GAS
	Neon, refrigerated liquid *(cryogenic liquid)*	2.2	UN1913		2.2	T75, TP5	320	316	None	50 kg	500 kg	D		120	NONFLAMMABLE GAS
	New explosive or explosive device, see §§ 173.51 and *173.56*														
	Nickel carbonyl	6.1	UN1259	I	6.1, 3	1	None	198	None	Forbidden	Forbidden	D	40, 78	131	POISON INHALATION HAZARD*
	Nickel cyanide	6.1	UN1653	II	6.1	IB8, IP2, IP4, N74, N75, T3, TP33	153	212	242	25 kg	100 kg	A	52	151	POISON
	Nickel nitrate	5.1	UN2725	III	5.1	A1, IB8, IP3, T1, TP33	152	213	240	25 kg	100 kg	A		140	OXIDIZER
	Nickel nitrite	5.1	UN2726	III	5.1	A1, IB8, IP3, T1, TP33	152	213	240	25 kg	100 kg	A	56, 58	140	OXIDIZER
	Nickel picrate	Forbidden													
	Nicotine	6.1	UN1654	II	6.1	IB2	153	202	243	5 L	60 L	A		151	POISON
G	**Nicotine compounds, liquid, n.o.s.** *or* **Nicotine preparations, liquid, n.o.s.**	6.1	UN3144	I	6.1	A4	None	201	243	1 L	30 L	B	40	151	POISON
				II	6.1	IB2, T11, TP2, TP27	153	202	243	5 L	60 L	B	40	151	POISON
				III	6.1	IB3, T7, TP1, TP28	153	203	241	60 L	220 L	B	40	151	POISON
G	**Nicotine compounds, solid, n.o.s.** *or* **Nicotine preparations, solid, n.o.s.**	6.1	UN1655	I	6.1	IB7, IP1, T6, TP33	None	211	242	5 kg	50 kg	B		151	POISON
				II	6.1	IB8, IP2, IP4, T3, TP33	153	212	242	25 kg	100 kg	A		151	POISON
				III	6.1	IB8, IP3, T1, TP33	153	213	240	100 kg	200 kg	A		151	POISON
	Nicotine hydrochloride liquid *or* **solution**	6.1	UN1656	II	6.1	IB2	153	202	243	5 L	60 L	A		151	POISON
				III	6.1	IB3	153	203	241	60 L	220 L	A		151	POISON
	Nicotine hydrochloride, solid	6.1	UN3444	II	6.1	IB8, IP2, IP4, T3, TP33	153	212	242	25 kg	100 kg	A		151	POISON
	Nicotine salicylate	6.1	UN1657	II	6.1	IB8, IP2, IP4, T3, TP33	153	212	242	25 kg	100 kg	A		151	POISON

Symbols (1)	Hazardous materials descriptions and proper shipping names (2)	Hazard class or Division (3)	Identification Numbers (4)	PG (5)	Label Codes (6)	Special Provisions (§172.102) (7)	(8) Packaging (§173.***) Exceptions (8A)	Non-bulk (8B)	Bulk (8C)	(9) Quantity limitations (see §§173.27 and 175.75) Passenger aircraft/rail (9A)	Cargo aircraft only (9B)	(10) Vessel stowage Location (10A)	Other (10B)	ERG Guide No.	Placard Advisory Consult §172.504 & §172.505 to determine placard *Denotes placard for any quantity
	Nicotine sulfate solution	6.1	UN1658	II	6.1	IB2, T7, TP2	153	202	243	5 L	60 L	A		151	POISON
				III	6.1	IB3, T7, TP2	153	203	241	60 L	220 L	A		151	POISON
	Nicotine sulphate, solid	6.1	UN3445	II	6.1	IB8, IP2, IP4, T3, TP33	153	212	242	25 kg	100 kg	A		151	POISON
	Nicotine tartrate	6.1	UN1659	II	6.1	IB8, IP2, IP4, T3, TP33	153	212	242	25 kg	100 kg	A		151	POISON
	Nitrated paper (unstable)	Forbidden													
	Nitrates, inorganic, aqueous solution, n.o.s.	5.1	UN3218	II	5.1	58, IB2, T4, TP1	152	202	242	1 L	5 L	B	56, 58, 133	140	OXIDIZER
				III	5.1	58, IB2, T4, TP1	152	203	241	2.5 L	30 L	B	56, 58, 133	140	OXIDIZER
	Nitrates, inorganic, n.o.s.	5.1	UN1477	II	5.1	IB8, IP2, IP4, T3, TP33	152	212	240	5 kg	25 kg	A	56, 58	140	OXIDIZER
				III	5.1	IB8, IP3, T1, TP33	152	213	240	25 kg	100 kg	A	56, 58	140	OXIDIZER
	Nitrates of diazonium compounds	Forbidden													
	Nitrating acid mixtures, spent *with more than 50 percent nitric acid*	8	UN1826	I	8, 5.1	A7, T10, TP2, TP13	None	158	243	Forbidden	2.5 L	D	40, 66	157	CORROSIVE
	Nitrating acid mixtures spent *with not more than 50 percent nitric acid*	8	UN1826	II	8	A7, B2, IB2, T8, TP2	None	158	242	Forbidden	30 L	D	40	157	CORROSIVE
	Nitrating acid mixtures *with more than 50 percent nitric acid*	8	UN1796	I	8, 5.1	A7, T10, TP2, TP13	None	158	243	Forbidden	2.5 L	D	40, 66	157	CORROSIVE
	Nitrating acid mixtures *with not more than 50 percent nitric acid*	8	UN1796	II	8	A7, B2, IB2, T8, TP2, TP13	None	158	242	Forbidden	30 L	D	40	157	CORROSIVE
	Nitric acid *other than red fuming, with at least 65 percent, but not more than 70 percent nitric acid*	8	UN2031	II	8, 5.1	A6, B2, B47, B53, IB2, IP15, T8, TP2	None	158	242	Forbidden	30 L	D	66, 74, 89, 90	157	CORROSIVE
	Nitric acid *other than red fuming, with more than 20 percent and less than 65 percent nitric acid*	8	UN2031	II	8	A6, B2, B47, B53, IB2, IP15, T8, TP2	None	158	242	Forbidden	30 L	D	44, 66, 74, 89, 90	157	CORROSIVE
	Nitric acid *other than red fuming, with more than 70 percent nitric acid*	8	UN2031	I	8, 5.1	A3, B47, B53, T10, TP2, TP12, TP13	None	158	243	Forbidden	2.5 L	D	44, 66, 89, 90, 110, 111	157	CORROSIVE
	Nitric acid *other than red fuming with not more than 20 percent nitric acid*	8	UN2031	II	8	A6, B2, B47, B53, IB2, T8, TP2	None	158	242	1 L	30 L	D		157	CORROSIVE
+	**Nitric acid, red fuming**	8	UN2032	I	8, 5.1, 6.1	2, B9, B32, T20, TP2, TP13, TP38, TP45	None	227	244	Forbidden	Forbidden	D	40, 66, 74, 89, 90	157	CORROSIVE, POISON INHALATION HAZARD*
	Nitric oxide, compressed	2.3	UN1660		2.3, 5.1, 8	1, B77	None	337	None	Forbidden	Forbidden	D	40, 89, 90	124	POISON GAS*
	Nitric oxide and dinitrogen tetroxide mixtures *or* **Nitric oxide and nitrogen dioxide mixtures**	2.3	UN1975		2.3, 5.1, 8	1, B77	None	337	None	Forbidden	Forbidden	D	40, 89, 90	124	POISON GAS*
G	**Nitriles, flammable, toxic, n.o.s.**	3	UN3273	I	3, 6.1	T14, TP2, TP13, TP27	None	201	243	Forbidden	30 L	E	40, 52	131	FLAMMABLE
				II	3, 6.1	IB2, T11, TP2, TP13, TP27	150	202	243	1 L	60 L	B	40, 52	131	FLAMMABLE
G	**Nitriles, toxic, flammable, n.o.s.**	6.1	UN3275	I	6.1, 3	5, T14, TP2, TP13, TP27	None	201	243	1 L	30 L	B	40, 52	131	POISON
				II	6.1, 3	IB2, T11, TP2, TP13, TP27	153	202	243	5 L	60 L	B	40, 52	131	POISON
G	**Nitriles, toxic, liquid, n.o.s.**	6.1	UN3276	I	6.1	5, T14, TP2, TP13, TP27	None	201	243	1 L	30 L	B	52	151	POISON
				II	6.1	IB2, T11, TP2, TP27	153	202	243	5 L	60 L	B	52	151	POISON
				III	6.1	IB3, T7, TP1, TP28	153	203	241	60 L	220 L	A	52	151	POISON
G	**Nitriles, toxic, solid, n.o.s.**	6.1	UN3439	I	6.1	IB7, IP1, T6, TP33	None	211	242	5 kg	50 kg	D	52	151	POISON
				II	6.1	IB8, IP2, IP4, T3, TP33	153	212	242	25 kg	100 kg	B	52	151	POISON
				III	6.1	IB8, IP3, T1, TP33	153	213	240	100 kg	200 kg	A	52	151	POISON
G	**Nitrites, inorganic, aqueous solution, n.o.s.**	5.1	UN3219	II	5.1	IB1, T4, TP1	152	202	242	1 L	5 L	B	46, 56, 58, 133	140	OXIDIZER
				III	5.1	IB2, T4, TP1	152	203	241	2.5 L	30 L	B	46, 56, 58, 133	140	OXIDIZER
G	**Nitrites, inorganic, n.o.s.**	5.1	UN2627	II	5.1	33, IB8, IP2, IP4, T3, TP33	152	212	None	5 kg	25 kg	A	46, 56, 58, 133	140	OXIDIZER
	3-Nitro-4-chlorobenzotrifluoride	6.1	UN2307	II	6.1	IB2, T7, TP2	153	202	243	5 L	60 L	A	40	152	POISON

Symbols (1)	Hazardous materials descriptions and proper shipping names (2)	Hazard class or Division (3)	Identification Numbers (4)	PG (5)	Label Codes (6)	Special Provisions (§172.102) (7)	(8) Packaging (§173.***) Exceptions (8A)	Non-bulk (8B)	Bulk (8C)	(9) Quantity limitations (see §§173.27 and 175.75) Passenger aircraft/rail (9A)	Cargo aircraft only (9B)	(10) Vessel stowage Location (10A)	Other (10B)	ERG Guide No.	Placard Advisory Consult §172.504 & §172.505 to determine placard *Denotes placard for any quantity
	6-Nitro-4-diazotoluene-3-sulfonic acid (dry)	Forbidden													
	Nitro isobutane triol trinitrate	Forbidden													
	N-Nitro-N-methylglycolamide nitrate	Forbidden													
	2-Nitro-2-methylpropanol nitrate	Forbidden													
	Nitro urea	1.1D	UN0147	II	1.1D		None	62	None	Forbidden	Forbidden	10		112	EXPLOSIVES 1.1*
	N-Nitroaniline	Forbidden													
+	**Nitroanilines** *(o-; m-; p-;)*	6.1	UN1661	II	6.1	IB8, IP2, IP4, T3, TP33	153	212	242	25 kg	100 kg	A		153	POISON
	Nitroanisole, liquid	6.1	UN2730	III	6.1	IB3, T4, TP1	153	203	241	60 L	220 L	A		152	POISON
	Nitroanisoles, solid	6.1	UN3458	III	6.1	IB8, IP3, T1, TP33	153	213	240	100 kg	200 kg	A		152	POISON
+	**Nitrobenzene**	6.1	UN1662	II	6.1	IB2, T7, TP2	153	202	243	5 L	60 L	A	40	152	POISON
	m-Nitrobenzene diazonium perchlorate	Forbidden													
	Nitrobenzenesulfonic acid	8	UN2305	II	8	B2, B4, IB8, IP2, IP4, T3, TP33	154	202	242	1 L	30 L	A		153	CORROSIVE
	Nitrobenzol, see **Nitrobenzene**														
	5-Nitrobenzotriazol	1.1D	UN0385	II	1.1D		None	62	None	Forbidden	Forbidden	10		112	EXPLOSIVES 1.1*
	Nitrobenzotrifluorides, liquid	6.1	UN2306	II	6.1	IB2, T7, TP2	153	202	243	5 L	60 L	A	40	152	POISON
	Nitrobenzotrifluorides, solid	6.1	UN3431	II	6.1	IB8, IP2, IP4, T3, TP33	153	212	242	25 kg	100 kg	A	40	152	POISON
	Nitrobromobenzenes, liquid	6.1	UN2732	III	6.1	IB3, T4, TP1	153	203	241	60 L	220 L	A		152	POISON
	Nitrobromobenzenes, solid	6.1	UN3459	III	6.1	IB8, IP3, T1, TP33	153	213	240	100 kg	200 kg	A		152	POISON
	Nitrocellulose, *dry or wetted with less than 25 percent water (or alcohol), by mass*	1.1D	UN0340	II	1.1D		None	62	None	Forbidden	Forbidden	13	27E	112	EXPLOSIVES 1.1*
	Nitrocellulose membrane filters, *with not more than 12.6% nitrogen, by dry mass*	4.1	UN3270	II	4.1	43, A1	151	212	240	1 kg	15 kg	D		133	FLAMMABLE SOLID
	Nitrocellulose, plasticized *with not less than 18 percent plasticizing substance, by mass*	1.3C	UN0343	II	1.3C		None	62	None	Forbidden	Forbidden	10		112	EXPLOSIVES 1.3*
	Nitrocellulose, solution, flammable *with not more than 12.6 percent nitrogen, by mass, and not more than 55 percent nitrocellulose*	3	UN2059	I	3	198, T11, TP1, TP8, TP27	None	201	243	1 L	30 L	E		127	FLAMMABLE
				II	3	198, IB2, T4, TP1, TP8	150	202	242	5 L	60 L	B		127	FLAMMABLE
				III	3	198, B1, IB3, T2, TP1	150	203	242	60 L	220 L	A		127	FLAMMABLE
	Nitrocellulose, *unmodified or plasticized with less than 18 percent plasticizing substance, by mass*	1.1D	UN0341	II	1.1D		None	62	None	Forbidden	Forbidden	13	27E	112	EXPLOSIVES 1.1*
	Nitrocellulose, wetted *with not less than 25 percent alcohol, by mass*	1.3C	UN0342	II	1.3C		None	62	None	Forbidden	Forbidden	10		112	EXPLOSIVES 1.3*
	Nitrocellulose with alcohol *with not less than 25 percent alcohol by mass, and with not more than 12.6 percent nitrogen, by dry mass*	4.1	UN2556	II	4.1		151	212	None	1 kg	15 kg	D	28, 36	113	FLAMMABLE SOLID
	Nitrocellulose, *with not more than 12.6 percent, by dry mass* **mixture with** *or* **without plasticizer, with** *or* **without pigment**	4.1	UN2557	II	4.1	44	151	212	None	1 kg	15 kg	D	28, 36	133	FLAMMABLE SOLID
	Nitrocellulose with water *with not less than 25 percent water by mass*	4.1	UN2555	II	4.1		151	212	None	15 kg	50 kg	E	28, 36	113	FLAMMABLE SOLID
	Nitrochlorobenzene, see **Chloronitrobenzenes** *etc.*														
	Nitrocresols, liquid	6.1	UN3434	III	6.1	IB3, T4, TP1	153	203	241	60 L	220 L	A		153	POISON
	Nitrocresols, solid	6.1	UN2446	III	6.1	IB8, TP3, T1, TP33	153	213	240	100 kg	200 kg	A		153	POISON
	Nitroethane	3	UN2842	III	3	B1, IB3, T2, TP1	150	203	242	60 L	220 L	A		129	FLAMMABLE
	Nitroethyl nitrate	Forbidden													
	Nitroethylene polymer	Forbidden													
	Nitrogen, compressed	2.2	UN1066		2.2		306, 307	302	314, 315	75 kg	150 kg	A		121	NONFLAMMABLE GAS
	Nitrogen dioxide, *see* **Dinitrogen tetroxide**														
	Nitrogen fertilizer solution, see **Fertilizer ammoniating solution** *etc.*														

Symbols (1)	Hazardous materials descriptions and proper shipping names (2)	Hazard class or Division (3)	Identification Numbers (4)	PG (5)	Label Codes (6)	Special Provisions (§172.102) (7)	(8) Packaging (§173.***) Exceptions (8A)	Non-bulk (8B)	Bulk (8C)	(9) Quantity limitations (see §§173.27 and 175.75) Passenger aircraft/rail (9A)	Cargo aircraft only (9B)	(10) Vessel stowage Location (10A)	Other (10B)	ERG Guide No.	Placard Advisory Consult §172.504 & §172.505 to determine placard *Denotes placard for any quantity
	Nitrogen peroxide, see **Dinitrogen tetroxide**														
	Nitrogen, refrigerated liquid *cryogenic liquid*	2.2	UN1977		2.2	345, 346, T75, TP5	320	316	318	50 kg	500 kg	D		120	NONFLAMMABLE GAS
	Nitrogen tetroxide and nitric oxide mixtures, see **Nitric oxide and nitrogen tetroxide mixtures**														
	Nitrogen tetroxide, see **Dinitrogen tetroxide**														
	Nitrogen trichloride	Forbidden													
	Nitrogen trifluoride	2.2	UN2451		2.2, 5.1		None	302	None	75 kg	150 kg	D	40	122	NONFLAMMABLE GAS
	Nitrogen triiodide	Forbidden													
	Nitrogen triiodide monoamine	Forbidden													
	Nitrogen trioxide	2.3	UN2421		2.3, 5.1, 8	1	None	336	245	Forbidden	Forbidden	D	40, 89, 90	124	POISON GAS*
	Nitroglycerin, desensitized *with not less than 40 percent non-volatile water insoluble phlegmatizer, by mass*	1.1D	UN0143	II	1.1D, 6.1	125	None	62	None	Forbidden	Forbidden	13	21E	112	EXPLOSIVES 1.1*
	Nitroglycerin, liquid, not desensitized	Forbidden													
	Nitroglycerin mixture, desensitized, liquid, flammable, n.o.s. *with not more than 30 percent nitroglycerin, by mass*	3	UN3343		3	129	None	214	None	Forbidden	Forbidden	D		113	FLAMMABLE
	Nitroglycerin mixture, desensitized, liquid, n.o.s. *with not more than 30% nitroglycerin, by mass*	3	UN3357	II	3	142	None	202	243	5 L	60 L	E		113	FLAMMABLE
	Nitroglycerin mixture, desensitized, solid, n.o.s. *with more than 2 percent but not more than 10 percent nitroglycerin, by mass*	4.1	UN3319	II	4.1	118	None	None	None	Forbidden	0.5 kg	E		113	FLAMMABLE SOLID
	Nitroglycerin, solution in alcohol, *with more than 1 percent but not more than 5 percent nitroglycerin*	3	UN3064	II	3	N8	None	202	None	Forbidden	5 L	E		127	FLAMMABLE
	Nitroglycerin, solution in alcohol, *with more than 1 percent but not more than 10 percent nitrogylcerin*	1.1D	UN0144	II	1.1D		None	62	None	Forbidden	Forbidden	10	21E	112	EXPLOSIVES 1.1*
	Nitroglycerin solution in alcohol *with not more than 1 percent nitroglycerin*	3	UN1204	II	3	IB2, N34	150	202	None	5 L	60 L	B		127	FLAMMABLE
	Nitroguanidine nitrate	Forbidden													
	Nitroguanidine *or* **Picrite,** *dry or wetted with less than 20 percent water, by mass*	1.1D	UN0282	II	1.1D		None	62	None	Forbidden	Forbidden	10		112	EXPLOSIVES 1.1*
	Nitroguanidine, wetted *or* **Picrite, wetted** *with not less than 20 percent water, by mass*	4.1	UN1336	I	4.1	23, A8, A19, A20, N41	None	211	None	1 kg	15 kg	E	28, 36	113	FLAMMABLE SOLID
	1-Nitrohydantoin	Forbidden													
	Nitrohydrochloric acid	8	UN1798	I	8	A3, B10, N41, T10, TP2, TP13	None	201	243	Forbidden	2.5 L	D	40, 66, 74, 89, 90	157	CORROSIVE
	Nitromannite (dry)	Forbidden													
	Nitromannite, wetted, *see* **Mannitol hexanitrate,** *etc.*														
	Nitromethane	3	UN1261	II	3		150	202	None	Forbidden	60 L	A		129	FLAMMABLE
	Nitromuriatic acid, see **Nitrohydrochloric acid**														
	Nitronaphthalene	4.1	UN2538	III	4.1	A1, IB8, IP3, T1, TP33	151	213	240	25 kg	100 kg	A		133	FLAMMABLE SOLID
+	**Nitrophenols** *(o-; m-; p-;)*	6.1	UN1663	III	6.1	IB8, IP3, T1, TP33	153	213	240	100 kg	200 kg	A		153	POISON
	m-Nitrophenyldinitro methane	Forbidden													
	4-Nitrophenylhydrazine, *with not less than 30 percent water, by mass*	4.1	UN3376	I	4.1	162, A8, A19, A20, N41	None	211	None	Forbidden	15 kg	E	28, 36	113	FLAMMABLE SOLID
	Nitropropanes	3	UN2608	III	3	B1, IB3, T2, TP1	150	203	242	60 L	220 L	A		129	FLAMMABLE
	p-Nitrosodimethylaniline	4.2	UN1369	II	4.2	A19, A20, IB6, IP2, N34, T3, TP33	None	212	241	15 kg	50 kg	D	34	135	SPONTANEOUSLY COMBUSTIBLE
	Nitrostarch, *dry or wetted with less than 20 percent water, by mass*	1.1D	UN0146	II	1.1D		None	62	None	Forbidden	Forbidden	10		112	EXPLOSIVES 1.1*
	Nitrostarch, wetted *with not less than 20 percent water, by mass*	4.1	UN1337	I	4.1	23, A8, A19, A20, N41	None	211	None	1 kg	15 kg	D	28, 36	113	FLAMMABLE SOLID
	Nitrosugars (dry)	Forbidden													
	Nitrosyl chloride	2.3	UN1069		2.3, 8	3, B14	None	304	314, 315	Forbidden	Forbidden	D	40	125	POISON GAS*

Symbols (1)	Hazardous materials descriptions and proper shipping names (2)	Hazard class or Division (3)	Identification Numbers (4)	PG (5)	Label Codes (6)	Special Provisions (§172.102) (7)	(8) Packaging (§173.***) Exceptions (8A)	Non-bulk (8B)	Bulk (8C)	(9) Quantity limitations (see §§173.27 and 175.75) Passenger aircraft/rail (9A)	Cargo aircraft only (9B)	(10) Vessel stowage Location (10A)	Other (10B)	ERG Guide No.	Placard Advisory Consult §172.504 & §172.505 to determine placard *Denotes placard for any quantity
	Nitrosylsulfuric acid, liquid	8	UN2308	II	8	A3, A6, A7, B2, IB2, N34, T8, TP2	154	202	242	1 L	30 L	D	40, 66, 74, 89, 90	157	CORROSIVE
	Nitrosylsulphuric acid, solid	8	UN3456	II	8	IB8, IP2, IP4, T3, TP33	154	212	240	15 kg	50 kg	D	40, 66, 74, 89, 90	157	CORROSIVE
	Nitrotoluenes, liquid	6.1	UN1664	II	6.1	IB2, T7, TP2	153	202	243	5 L	60 L	A		152	POISON
	Nitrotoluenes, solid	6.1	UN3446	II	6.1	IB8, IP2, IP4, T3, TP33	153	212	242	25 kg	100 kg	A		152	POISON
	Nitrotoluidines (mono)	6.1	UN2660	III	6.1	IB8, IP3, T1, TP33	153	213	240	100 kg	200 kg	A		153	POISON
	Nitrotriazolone *or* **NTO**	1.1D	UN0490	II	1.1D		None	62	None	Forbidden	Forbidden	10		112	EXPLOSIVES 1.1*
	Nitrous oxide	2.2	UN1070		2.2, 5.1	A14	306	304	314, 315	75 kg	150 kg	A	40	122	NONFLAMMABLE GAS
	Nitrous oxide, refrigerated liquid	2.2	UN2201		2.2, 5.1	B6, T75, TP5, TP22	None	304	314, 315	Forbidden	Forbidden	D	40	122	NONFLAMMABLE GAS
	Nitroxylenes, liquid	6.1	UN1665	II	6.1	IB2, T7, TP2	153	202	243	5 L	60 L	A		152	POISON
	Nitroxylenes, solid	6.1	UN3447	II	6.1	IB8, IP2, IP4, T3, TP33	153	212	242	25 kg	100 kg	A		152	POISON
	Nitroxylol, see **Nitroxylenes**														
	Nonanes	3	UN1920	III	3	B1, IB3, T2, TP1	150	203	242	60 L	220 L	A		128	FLAMMABLE
	Nonflammable gas, n.o.s., see **Compressed gas,** *etc. or* **Liquefied gas** *etc.*														
	Nonliquefied gases, see **Compressed gases,** *etc.*														
	Nonliquefied hydrocarbon gas, see **Hydrocarbon gas mixture, compressed, n.o.s.**														
	Nonyltrichlorosilane	8	UN1799	II	8	A7, B2, B6, N34, T10, TP2, TP7, TP13	None	206	242	Forbidden	30 L	C	40	156	CORROSIVE
	2,5-Norbornadiene, stabilized, see **Bicyclo [2,2,1] hepta-2,5-diene, stabilized**														
	Nordhausen acid, see **Sulfuric acid, fuming** *etc.*														
	Octadecyltrichlorosilane	8	UN1800	II	8	A7, B2, B6, N34, T10, TP2, TP7, TP13	None	206	242	Forbidden	30 L	C	40	156	CORROSIVE
	Octadiene	3	UN2309	II	3	B1, IB2, T4, TP1	150	202	242	5 L	60 L	B		128P	FLAMMABLE
	1,7-Octadine-3,5-diyne-1,8-dimethoxy-9-octadecynoic acid	Forbidden													
	Octafluorobut-2-ene *or* **Refrigerant gas R 1318**	2.2	UN2422		2.2		None	304	314, 315	75 kg	150 kg	A		126	NONFLAMMABLE GAS
	Octafluorocyclobutane *or* **Refrigerant gas RC 318**	2.2	UN1976		2.2	T50	None	304	314, 315	75 kg	150 kg	A		126	NONFLAMMABLE GAS
	Octafluoropropane, *or* **Refrigerant gas, R 218**	2.2	UN2424		2.2	T50	None	304	314, 315	75 kg	150 kg	A		126	NONFLAMMABLE GAS
	Octanes	3	UN1262	II	3	IB2, T4, TP1	150	202	242	5 L	60 L	B		128	FLAMMABLE
	Octogen, *etc. see* **Cyclotetramethylene tetranitramine,** *etc.*														
	Octolite *or* **Octol,** *dry or wetted with less than 15 percent water, by mass*	1.1D	UN0266	II	1.1D		None	62	None	Forbidden	Forbidden	10		112	EXPLOSIVES 1.1*
	Octonal	1.1D	UN0496		1.1D		None	62	None	Forbidden	Forbidden	10		112	EXPLOSIVES 1.1*
	Octyl aldehydes	3	UN1191	III	3	B1, IB3, T2, TP1	150	203	242	60 L	220 L	A		129	FLAMMABLE
	Octyltrichlorosilane	8	UN1801	II	8	A7, B2, B6, N34, T10, TP2, TP7, TP13	None	206	242	Forbidden	30 L	C	40	156	CORROSIVE
	Oil gas, compressed	2.3	UN1071		2.3, 2.1	6	None	304	314, 315	Forbidden	25 kg	D	40	119	POISON GAS*
	Oleum, see **Sulfuric acid, fuming**														
	Organic peroxide type A, liquid or solid	Forbidden													
G	**Organic peroxide type B, liquid**	5.2	UN3101	II	5.2, 1	53	152	225	None	Forbidden	Forbidden	D	12, 40, 52, 53	146	ORGANIC PEROXIDE
G	**Organic peroxide type B, liquid, temperature controlled**	5.2	UN3111	II	5.2, 1	53	None	225	None	Forbidden	Forbidden	D	2, 40, 52, 53	148	ORGANIC PEROXIDE*
G	**Organic peroxide type B, solid**	5.2	UN3102	II	5.2, 1	53	152	225	None	Forbidden	Forbidden	D	12, 40, 52, 53	146	ORGANIC PEROXIDE

Symbols (1)	Hazardous materials descriptions and proper shipping names (2)	Hazard class or Division (3)	Identification Numbers (4)	PG (5)	Label Codes (6)	Special Provisions (§172.102) (7)	(8) Packaging (§173.***) Exceptions (8A)	Non-bulk (8B)	Bulk (8C)	(9) Quantity limitations (see §§173.27 and 175.75) Passenger aircraft/rail (9A)	Cargo aircraft only (9B)	(10) Vessel stowage Location (10A)	Other (10B)	ERG Guide No.	Placard Advisory Consult §172.504 & §172.505 to determine placard *Denotes placard for any quantity
G	**Organic peroxide type B, solid, temperature controlled**	5.2	UN3112	II	5.2, 1	53	None	225	None	Forbidden	Forbidden	D	2, 40, 52, 53	148	ORGANIC PEROXIDE*
G	**Organic peroxide type C, liquid**	5.2	UN3103	II	5.2		152	225	None	5 L	10 L	D	12, 40, 52, 53	146	ORGANIC PEROXIDE
G	**Organic peroxide type C, liquid, temperature controlled**	5.2	UN3113	II	5.2		None	225	None	Forbidden	Forbidden	D	2, 40, 52, 53	148	ORGANIC PEROXIDE
G	**Organic peroxide type C, solid**	5.2	UN3104	II	5.2		152	225	None	5 kg	10 kg	D	12, 40, 52, 53	146	ORGANIC PEROXIDE
G	**Organic peroxide type C, solid, temperature controlled**	5.2	UN3114	II	5.2		None	225	None	Forbidden	Forbidden	D	2, 40, 52, 53	148	ORGANIC PEROXIDE
G	**Organic peroxide type D, liquid**	5.2	UN3105	II	5.2		152	225	None	5 L	10 L	D	12, 40, 52, 53	145	ORGANIC PEROXIDE
G	**Organic peroxide type D, liquid, temperature controlled**	5.2	UN3115	II	5.2		None	225	None	Forbidden	Forbidden	D	2, 40, 52, 53	148	ORGANIC PEROXIDE
G	**Organic peroxide type D, solid**	5.2	UN3106	II	5.2		152	225	None	5 kg	10 kg	D	12, 40, 52, 53	145	ORGANIC PEROXIDE
G	**Organic peroxide type D, solid, temperature controlled**	5.2	UN3116	II	5.2		None	225	None	Forbidden	Forbidden	D	2, 40, 52, 53	148	ORGANIC PEROXIDE
G	**Organic peroxide type E, liquid**	5.2	UN3107	II	5.2		152	225	None	10 L	25 L	D	12, 40, 52, 53	145	ORGANIC PEROXIDE
G	**Organic peroxide type E, liquid, temperature controlled**	5.2	UN3117	II	5.2		None	225	None	Forbidden	Forbidden	D	2, 40, 52, 53	148	ORGANIC PEROXIDE
G	**Organic peroxide type E, solid**	5.2	UN3108	II	5.2		152	225	None	10 kg	25 kg	D	12, 40, 52, 53	145	ORGANIC PEROXIDE
G	**Organic peroxide type E, solid, temperature controlled**	5.2	UN3118	II	5.2		None	225	None	Forbidden	Forbidden	D	2, 40, 52, 53	148	ORGANIC PEROXIDE
G	**Organic peroxide type F, liquid**	5.2	UN3109	II	5.2	IP5	152	225	225	10 L	25 L	D	12, 40, 52, 53	145	ORGANIC PEROXIDE
G	**Organic peroxide type F, liquid, temperature controlled**	5.2	UN3119	II	5.2	IP5	None	225	225	Forbidden	Forbidden	D	2, 40, 52, 53	148	ORGANIC PEROXIDE
G	**Organic peroxide type F, solid**	5.2	UN3110	II	5.2	TP33	152	225	225	10 kg	25 kg	D	12, 40, 52, 53	145	ORGANIC PEROXIDE
G	**Organic peroxide type F, solid, temperature controlled**	5.2	UN3120	II	5.2	TP33	None	225	225	Forbidden	Forbidden	D	2, 52, 53	148	ORGANIC PEROXIDE
D	**Organic phosphate, mixed with compressed gas** *or* **Organic phosphate compound, mixed with compressed gas** *or* **Organic phosphorus compound, mixed with compressed gas**	2.3	NA1955		2.3	3	None	334	None	Forbidden	Forbidden	D	40	123	POISON GAS*
	Organic pigments, self-heating	4.2	UN3313	II	4.2	IB8, IP2, IP4, T3, TP33	None	212	241	15 kg	50 kg	C		135	SPONTANEOUSLY COMBUSTIBLE
				III	4.2	IB8, IP3, T1, TP33	None	213	241	25 kg	100 kg	C		135	SPONTANEOUSLY COMBUSTIBLE
G	**Organoarsenic compound, liquid, n.o.s.**	6.1	UN3280	I	6.1	5, T14, TP2, TP13, TP27	None	201	242	1 L	30 L	B		151	POISON
				II	6.1	IB2, T11, TP2, TP27	153	202	242	5 L	60 L	B		151	POISON
				III	6.1	IB3, T7, TP1, TP28	153	203	241	60 L	220 L	A		151	POISON
G	**Organoarsenic compound, solid, n.o.s.**	6.1	UN3465	I	6.1	IB7, IP1, T6, TP33	None	211	242	5 kg	50 kg	B		151	POISON
				II	6.1	IB8, IP2, IP4, T3, TP33	153	212	242	25 kg	100 kg	B		151	POISON
				III	6.1	IB8, IP3, T1, TP33	153	213	240	100 kg	200 kg	A		151	POISON
	Organochlorine pesticides liquid, flammable, toxic, *flash point less than 23 degrees C*	3	UN2762	I	3, 6.1	T14, TP2, TP13, TP27	None	201	243	Forbidden	30 L	B	40	131	FLAMMABLE
				II	3, 6.1	IB2, T11, TP2, TP13, TP27	150	202	243	1 L	60 L	B	40	131	FLAMMABLE
	Organochlorine pesticides, liquid, toxic	6.1	UN2996	I	6.1	T14, TP2, TP13, TP27	None	201	243	1 L	30 L	B	40	151	POISON
				II	6.1	IB2, T11, TP2, TP13, TP27	153	202	243	5 L	60 L	B	40	151	POISON
				III	6.1	IB3, T7, TP2, TP28	153	203	241	60 L	220 L	A	40	151	POISON
	Organochlorine pesticides, liquid, toxic, flammable, *flash point not less than 23 degrees C*	6.1	UN2995	I	6.1, 3	T14, TP2, TP13, TP27	None	201	243	1 L	30 L	B	40	131	POISON

Symbols (1)	Hazardous materials descriptions and proper shipping names (2)	Hazard class or Division (3)	Identification Numbers (4)	PG (5)	Label Codes (6)	Special Provisions (§172.102) (7)	(8) Packaging (§173.***) Exceptions (8A)	Non-bulk (8B)	Bulk (8C)	(9) Quantity limitations (see §§173.27 and 175.75) Passenger aircraft/rail (9A)	Cargo aircraft only (9B)	(10) Vessel stowage Location (10A)	Other (10B)	ERG Guide No.	Placard Advisory Consult §172.504 & §172.505 to determine placard *Denotes placard for any quantity
				II	6.1, 3	IB2, T11, TP2, TP13, TP27	153	202	243	5 L	60 L	B	40	131	POISON
				III	6.1, 3	B1, IB3, T7, TP2, TP28	153	203	242	60 L	220 L	A	40	131	POISON
	Organochlorine, pesticides, solid, toxic	6.1	UN2761	I	6.1	IB7, IP1, T6, TP33	None	211	242	5 kg	50 kg	A	40	151	POISON
				II	6.1	IB8, IP2, IP4, T3, TP33	None	212	242	25 kg	100 kg	A	40	151	POISON
				III	6.1	IB8, IP3, T1, TP33	153	213	240	100 kg	200 kg	A	40	151	POISON
G	**Organometallic compound, toxic, liquid, n.o.s.**	6.1	UN3282	I	6.1	T14, TP2, TP13, TP27	None	201	242	1 L	30 L	B		151	POISON
				II	6.1	IB2, T11, TP2, TP27	153	202	242	5 L	60 L	B		151	POISON
				III	6.1	IB3, T7, TP1, TP28	153	203	241	60 L	220 L	A		151	POISON
G	**Organometallic compound, toxic, solid, n.o.s.**	6.1	UN3467	I	6.1	IB7, IP1, T6, TP33	None	211	242	5 kg	50 kg	B		151	POISON
				II	6.1	IB8, IP2, IP4, T3, TP33	153	212	242	25 kg	100 kg	B		151	POISON
				III	6.1	IB8, IP3, T1, TP33	153	213	240	100 kg	200 kg	A		151	POISON
G	**Organometallic substance, liquid, pyrophoric**	4.2	UN3392	I	4.2	B11, T21, TP2, TP7, TP36	None	181	244	Forbidden	Forbidden	D	78	135	SPONTANEOUSLY COMBUSTIBLE
G	**Organometallic substance, liquid, pyrophoric, water-reactive**	4.2	UN3394	I	4.2, 4.3	B11, T21, TP2, TP7, TP36	None	181	244	Forbidden	Forbidden	D	78	135	SPONTANEOUSLY COMBUSTIBLE, DANGEROUS WHEN WET*
G	**Organometallic substance, liquid, water-reactive**	4.3	UN3398	I	4.3	T13, TP2, TP7, TP36	None	201	244	Forbidden	1 L	E	40, 52	135	DANGEROUS WHEN WET*
				II	4.3	IB1, T7, TP2, TP7, TP36	None	202	243	1 L	5 L	E	40, 52	135	DANGEROUS WHEN WET*
				III	4.3	IB2, T7, TP2, TP7, TP36	None	203	242	5 L	60 L	E	40, 52	135	DANGEROUS WHEN WET*
G	**Organometallic substance, liquid, water-reactive, flammable**	4.3	UN3399	I	4.3, 3	T13, TP2, TP7, TP36	None	201	244	Forbidden	1 L	D	40, 52	138	DANGEROUS WHEN WET*
				II	4.3, 3	IB1, IP2, T7, TP2, TP7, TP36	None	202	243	1 L	5 L	D	40, 52	138	DANGEROUS WHEN WET*
				III	4.3, 3	IB2, IP4, T7, TP2, TP7, TP36	None	203	242	5 L	60 L	E	40, 52	138	DANGEROUS WHEN WET*
G	**Organometallic substance, solid, pyrophoric**	4.2	UN3391	I	4.2	T21, TP7, TP33, TP36	None	187	244	Forbidden	Forbidden	D		135	SPONTANEOUSLY COMBUSTIBLE
G	**Organometallic substance, solid, pyrophoric, water-reactive**	4.2	UN3393	I	4.2, 4.3	B11, T21, TP7, TP33, TP36	None	187	244	Forbidden	Forbidden	D	52	135	SPONTANEOUSLY COMBUSTIBLE, DANGEROUS WHEN WET*
G	**Organometallic substance, solid, self-heating**	4.2	UN3400	II	4.2	IB6, T3, TP33, TP36	None	212	242	15 kg	50 kg	C		138	SPONTANEOUSLY COMBUSTIBLE
				III	4.2	IB8, T1, TP33, TP36	None	203	242	25 kg	100 kg	C		138	SPONTANEOUSLY COMBUSTIBLE
G	**Organometallic substance, solid, water-reactive**	4.3	UN3395	I	4.3	N40, T9, TP7, TP33, TP36	None	211	242	Forbidden	15 kg	E	40, 52	135	DANGEROUS WHEN WET*
				II	4.3	IB4, T3, TP33, TP36	151	212	242	15 kg	50 kg	E	40, 52	135	DANGEROUS WHEN WET*
				III	4.3	IB6, T1, TP33, TP36	151	213	241	25 kg	100 kg	E	40, 52	135	DANGEROUS WHEN WET*
G	**Organometallic substance, solid, water-reactive, flammable**	4.3	UN3396	I	4.3, 4.1	N40, T9, TP7, TP33, TP36	None	211	242	Forbidden	15 kg	E	40, 52	138	DANGEROUS WHEN WET*
				II	4.3, 4.1	IB4, T3, TP33, TP36	151	212	242	15 kg	50 kg	E	40, 52	138	DANGEROUS WHEN WET*
				III	4.3, 4.1	IB6, T1, TP33, TP36	151	213	241	25 kg	100 kg	E	40, 52	138	DANGEROUS WHEN WET*
G	**Organometallic substance, solid, water-reactive, self-heating**	4.3	UN3397	I	4.3, 4.2	N40, T9, TP7, TP33, TP36	None	211	242	Forbidden	15 kg	E	40, 52	138	DANGEROUS WHEN WET*
				II	4.3, 4.2	IB4, T3, TP33, TP36	None	212	242	15 kg	50 kg	E	40, 52	138	DANGEROUS WHEN WET*

Symbols (1)	Hazardous materials descriptions and proper shipping names (2)	Hazard class or Division (3)	Identification Numbers (4)	PG (5)	Label Codes (6)	Special Provisions (§172.102) (7)	(8) Packaging (§173.***) Exceptions (8A)	Non-bulk (8B)	Bulk (8C)	(9) Quantity limitations (see §§173.27 and 175.75) Passenger aircraft/rail (9A)	Cargo aircraft only (9B)	(10) Vessel stowage Location (10A)	Other (10B)	ERG Guide No.	Placard Advisory Consult §172.504 & §172.505 to determine placard *Denotes placard for any quantity
				III	4.3, 4.2	IB6, T1, TP33, TP36	None	213	241	25 kg	100 kg	E	40, 52	138	DANGEROUS WHEN WET*
	Organophosphorus compound, toxic, flammable, n.o.s.	6.1	UN3279	I	6.1, 3	5, T14, TP2, TP13, TP27	None	201	243	1 L	30 L	B	40	131	POISON
				II	6.1, 3	IB2, T11, TP2, TP13, TP27	153	202	243	5 L	60 L	B	40	131	POISON
G	**Organophosphorus compound, toxic, liquid, n.o.s.**	6.1	UN3278	I	6.1	5, T14, TP2, TP13, TP27	None	201	243	1 L	30 L	B		151	POISON
				II	6.1	IB2, T11, TP2, TP27	153	202	243	5 L	60 L	B		151	POISON
				III	6.1	IB3, T7, TP1, TP28	153	203	241	60 L	220 L	A		151	POISON
G	**Organophosphorus compound, toxic, solid, n.o.s.**	6.1	UN3464	I	6.1	IB7, IP1, T6, TP33	None	211	242	5 kg	50 kg	B		151	POISON
				II	6.1	IB8, IP2, IP4, T3, TP33	153	212	242	25 kg	100 kg	B		151	POISON
				III	6.1	IB8, IP3, T1, TP33	153	213	240	100 kg	200 kg	A		151	POISON
	Organophosphorus pesticides, liquid, flammable, toxic, *flash point less than 23 degrees C*	3	UN2784	I	3, 6.1	T14, TP2, TP13, TP27	None	201	243	Forbidden	30 L	B	40	131	FLAMMABLE
				II	3, 6.1	IB2, T11, TP2, TP13, TP27	150	202	243	1 L	60 L	B	40	131	FLAMMABLE
	Organophosphorus pesticides, liquid, toxic	6.1	UN3018	I	6.1	N76, T14, TP2, TP13, TP27	None	201	243	1 L	30 L	B	40	152	POISON
				II	6.1	IB2, N76, T11, TP2, TP13, TP27	153	202	243	5 L	60 L	B	40	152	POISON
				III	6.1	IB3, N76, T7, TP2, TP28	153	203	241	60 L	220 L	A	40	152	POISON
	Organophosphorus pesticides, liquid, toxic, flammable, *flashpoint not less than 23 degrees C*	6.1	UN3017	I	6.1, 3	N76, T14, TP2, TP13, TP27	None	201	243	1 L	30 L	B	40	131	POISON
				II	6.1, 3	IB2, N76, T11, TP2, TP13, TP27	153	202	243	5 L	60 L	B	40	131	POISON
				III	6.1, 3	B1, IB3, N76, T7, TP2, TP28	153	203	242	60 L	220 L	A	40	131	POISON
	Organophosphorus pesticides, solid, toxic	6.1	UN2783	I	6.1	IB7, IP1, N77, T6, TP33	None	211	242	5 kg	50 kg	A	40	152	POISON
				II	6.1	IB8, IP2, IP4, N77, T3, TP33	153	212	242	25 kg	100 kg	A	40	152	POISON
				III	6.1	IB8, IP3, N77, T1, TP33	153	213	240	100 kg	200 kg	A	40	152	POISON
	Organotin compounds, liquid, n.o.s.	6.1	UN2788	I	6.1	A3, N33, N34, T14, TP2, TP13, TP27	None	201	243	1 L	30 L	B	40	153	POISON
				II	6.1	A3, IB2, N33, N34, T11, TP2, TP13, TP27	153	202	243	5 L	60 L	A	40	153	POISON
				III	6.1	IB3, T7, TP2, TP28	153	203	241	60 L	220 L	A	40	153	POISON
	Organotin compounds, solid, n.o.s.	6.1	UN3146	I	6.1	A5, IB7, IP1, T6, TP33	None	211	242	5 kg	50 kg	B	40	153	POISON
				II	6.1	IB8, IP2, IP4, T3, TP33	153	212	242	25 kg	100 kg	A	40	153	POISON
				III	6.1	IB8, IP3, T1, TP33	153	213	240	100 kg	200 kg	A	40	153	POISON
	Organotin pesticides, liquid, flammable, toxic, *flash point less than 23 degrees C*	3	UN2787	I	3, 6.1	T14, TP2, TP13, TP27	None	201	243	Forbidden	30 L	B	40	131	FLAMMABLE
				II	3, 6.1	IB2, T11, TP2, TP13, TP27	150	202	243	1 L	60 L	B	40	131	FLAMMABLE
	Organotin pesticides, liquid, toxic	6.1	UN3020	I	6.1	T14, TP2, TP13, TP27	None	201	243	1 L	30 L	B	40	153	POISON
				II	6.1	IB2, T11, TP2, TP13, TP27	153	202	243	5 L	60 L	B	40	153	POISON
				III	6.1	IB3, T7, TP2, TP28	153	203	241	60 L	220 L	A	40	153	POISON
	Organotin pesticides, liquid, toxic, flammable, *flashpoint not less than 23 degrees C*	6.1	UN3019	I	6.1, 3	T14, TP2, TP13, TP27	None	201	243	1 L	30 L	B	40	131	POISON

Symbols (1)	Hazardous materials descriptions and proper shipping names (2)	Hazard class or Division (3)	Identification Numbers (4)	PG (5)	Label Codes (6)	Special Provisions (§172.102) (7)	(8) Packaging (§173.***) Exceptions (8A)	Non-bulk (8B)	Bulk (8C)	(9) Quantity limitations (see §§173.27 and 175.75) Passenger aircraft/rail (9A)	Cargo aircraft only (9B)	(10) Vessel stowage Location (10A)	Other (10B)	ERG Guide No.	Placard Advisory Consult §172.504 & §172.505 to determine placard *Denotes placard for any quantity
				II	6.1, 3	IB2, T11, TP2, TP13, TP27	153	202	243	5 L	60 L	B	40	131	POISON
				III	6.1, 3	B1, IB3, T7, TP2, TP28	153	203	242	60 L	220 L	A	40	131	POISON
	Organotin pesticides, solid, toxic	6.1	UN2786	I	6.1	IB7, IP1, T6. TP33	None	211	242	5 kg	50 kg	A	40	153	POISON
				II	6.1	IB8, IP2, IP4, T3, TP33	153	212	242	25 kg	100 kg	A	40	153	POISON
				III	6.1	IB8, IP3, T1, TP33	153	213	240	100 kg	200 kg	A	40	153	POISON
	Orthonitroaniline, see **Nitroanilines** *etc.*														
	Osmium tetroxide	6.1	UN2471	I	6.1	A8, IB7, IP1, N33, N34, T6, TP33	None	211	242	5 kg	50 kg	B	40	154	POISON
D G	**Other regulated substances, liquid, n.o.s.**	9	NA3082	III	9	IB3, T2, TP1	155	203	241	No Limit	No Limit	A		171	CLASS 9
D G	**Other regulated substances, solid, n.o.s.**	9	NA3077	III	9	B54, IB8, IP2	155	213	240	No Limit	No Limit	A		171	CLASS 9
G	**Oxidizing liquid, corrosive, n.o.s.**	5.1	UN3098	I	5.1, 8	62, A6	None	201	244	Forbidden	2.5 L	D	13, 56, 58, 106, 138	140	OXIDIZER
				II	5.1, 8	62, IB1	None	202	243	1 L	5 L	B	13, 34, 56, 58, 106, 138	140	OXIDIZER
				III	5.1, 8	62, IB2	152	203	242	2.5 L	30 L	B	13, 34, 56, 58, 106, 138	140	OXIDIZER
G	**Oxidizing liquid, n.o.s.**	5.1	UN3139	I	5.1	62, 127, A2, A6	None	201	243	Forbidden	2.5 L	D	56, 58, 106, 138	140	OXIDIZER
				II	5.1	62, 127, A2, IB2	152	202	242	1 L	5 L	B	56, 58, 106, 138	140	OXIDIZER
				III	5.1	62, 127, A2, IB2	152	203	241	2.5 L	30 L	B	56, 58, 106, 138	140	OXIDIZER
G	**Oxidizing liquid, toxic, n.o.s.**	5.1	UN3099	I	5.1, 6.1	62, A6	None	201	244	Forbidden	2.5 L	D	56, 58, 106, 138	142	OXIDIZER
				II	5.1, 6.1	62, IB1	152	202	243	1 L	5 L	B	56, 58, 95, 106, 138	142	OXIDIZER
				III	5.1, 6.1	62, IB2	152	203	242	2.5 L	30 L	B	56, 58, 95, 106, 138	142	OXIDIZER
G	**Oxidizing solid, corrosive, n.o.s.**	5.1	UN3085	I	5.1, 8	62	None	211	242	1 kg	15 kg	D	13, 56, 58, 106, 138	140	OXIDIZER
				II	5.1, 8	62, IB6, IP2, T3, TP33	None	212	242	5 kg	25 kg	B	13, 34, 56, 58, 106, 138	140	OXIDIZER
				III	5.1, 8	62, IB8, IP3, T1, TP33	152	213	240	25 kg	100 kg	B	13, 34, 56, 58, 106, 138	140	OXIDIZER
G	**Oxidizing solid, flammable, n.o.s.**	5.1	UN3137	I	5.1, 4.1	62	None	214	214	Forbidden	Forbidden			140	OXIDIZER
G	**Oxidizing solid, n.o.s.**	5.1	UN1479	I	5.1	62, IB5, IP1	None	211	242	1 kg	15 kg	D	56, 58, 106, 138	140	OXIDIZER
				II	5.1	62, IB8, IP2, IP4, T3, TP33	152	212	240	5 kg	25 kg	B	56, 58, 106, 138	140	OXIDIZER
				III	5.1	62, IB8, IP3, T1, TP33	152	213	240	25 kg	100 kg	B	56, 58, 106, 138	140	OXIDIZER
G	**Oxidizing solid, self-heating, n.o.s.**	5.1	UN3100	I	5.1, 4.2	62	None	214	214	Forbidden	Forbidden			135	OXIDIZER
				II	5.1, 4.2	62	None	214	214	Forbidden	Forbidden			135	OXIDIZER
G	**Oxidizing solid, toxic, n.o.s.**	5.1	UN3087	I	5.1, 6.1	62	None	211	242	1 kg	15 kg	D	56, 58, 106, 138	141	OXIDIZER
				II	5.1, 6.1	62, IB6, IP2, T3, TP33	152	212	242	5 kg	25 kg	B	56, 58, 95, 106, 138	141	OXIDIZER
				III	5.1, 6.1	62, IB8, IP3, T1, TP33	152	213	240	25 kg	100 kg	B	56, 58, 95, 106, 138	141	OXIDIZER
G	**Oxidizing solid, water-reactive, n.o.s.**	5.1	UN3121		5.1, 4.3	62	None	214	214	Forbidden	Forbidden			144	OXIDIZER, DANGEROUS WHEN WET*
	Oxygen, compressed	2.2	UN1072		2.2, 5.1	110, A14	306	302	314, 315	75 kg	150 kg	A		122	NONFLAMMABLE GAS

Symbols (1)	Hazardous materials descriptions and proper shipping names (2)	Hazard class or Division (3)	Identification Numbers (4)	PG (5)	Label Codes (6)	Special Provisions (§172.102) (7)	(8) Packaging (§173.***) Exceptions (8A)	Non-bulk (8B)	Bulk (8C)	(9) Quantity limitations (see §§173.27 and 175.75) Passenger aircraft/rail (9A)	Cargo aircraft only (9B)	(10) Vessel stowage Location (10A)	Other (10B)	ERG Guide No.	Placard Advisory Consult §172.504 & §172.505 to determine placard *Denotes placard for any quantity
	Oxygen difluoride, compressed	2.3	UN2190		2.3, 5.1, 8	1, N86	None	304	None	Forbidden	Forbidden	D	13, 40, 89, 90	124	POISON GAS*
	Oxygen generator, chemical *(including when contained in associated equipment, e.g., passenger service units (PSUs), portable breathing equipment (PBE), etc.)*	5.1	UN3356	II	5.1		None	168	None	Forbidden	25 kg	D	56, 58, 69, 106	140	OXIDIZER
+	**Oxygen generator, chemical, spent**	9	NA3356	III	9	61	None	213	None	Forbidden	Forbidden	A		140	CLASS 9
	Oxygen, refrigerated liquid *(cryogenic liquid)*	2.2	UN1073		2.2, 5.1	T75, TP5, TP22	320	316	318	Forbidden	Forbidden	D		122	NONFLAMMABLE GAS
	Paint, corrosive, flammable *(including paint, lacquer, enamel, stain, shellac, varnish, polish, liquid filler and liquid lacquer base)*	8	UN3470	II	8, 3	IB2, T7, TP2, TP8, TP28	154	202	243	1 L	30 L	B	40	132	CORROSIVE
	Paint, flammable, corrosive *(including paint, lacquer, enamel, stain, shellac, varnish, polish, liquid filler and liquid lacquer base)*	3	UN3469	I	3, 8	T11, TP2, TP27	None	201	243	0.5 L	2.5 L	E	40	132	FLAMMABLE
				II	3, 8	IB2, T7, TP2, TP8, TP28	150	202	243	1 L	5 L	B	40	132	FLAMMABLE
				III	3, 8	IB3, T4, TP1, TP29	150	203	242	5 L	60 L	A	40	132	FLAMMABLE
	Paint *including paint, lacquer, enamel, stain, shellac solutions, varnish, polish, liquid filler, and liquid lacquer base*	3	UN1263	I	3	T11, TP1, TP8, TP27	150	201	243	1 L	30 L	E		128	FLAMMABLE
				II	3	149, B52, IB2, T4, TP1, TP8, TP28	150	173	242	5 L	60 L	B		128	FLAMMABLE
				III	3	B1, B52, IB3, T2, TP1, TP29	150	173	242	60 L	220 L	A		128	FLAMMABLE
	Paint *or* **Paint related material**	8	UN3066	II	8	B2, IB2, T7, TP2, TP28	154	173	242	1 L	30 L	A	40	153	CORROSIVE
				III	8	B52, IB3, T4, TP1, TP29	154	173	241	5 L	60 L	A	40	153	CORROSIVE
	Paint related material corrosive, flammable *(including paint thinning or reducing compound)*	8	UN3470	II	8, 3	IB2, T7, TP2, TP8, TP28	154	202	243	1 L	30 L	B	40	132	CORROSIVE
	Paint related material, flammable, corrosive *(including paint thinning or reducing compound)*	3	UN3469	I	3, 8	T11, TP2, TP27	None	201	243	0.5 L	2.5 L	E	40	132	FLAMMABLE
				II	3, 8	IB2, T7, TP2, TP8, TP28	150	202	243	1 L	5 L	B	40	132	FLAMMABLE
				III	3, 8	IB3, T4, TP1, TP29	150	203	242	5 L	60 L	A	40	132	FLAMMABLE
	Paint related material *including paint thinning, drying, removing, or reducing compound*	3	UN1263	I	3	T11, TP1, TP8, TP27	150	201	243	1 L	30 L	E		128	FLAMMABLE
				II	3	149, B52, IB2, T4, TP1, TP8, TP28	150	173	242	5 L	60 L	B		128	FLAMMABLE
				III	3	B1, B52, IB3, T2, TP1, TP29	150	173	242	60 L	220 L	A		128	FLAMMABLE
	Paper, unsaturated oil treated *incompletely dried (including carbon paper)*	4.2	UN1379	III	4.2	IB8, IP3	None	213	241	Forbidden	Forbidden	A		133	SPONTANEOUSLY COMBUSTIBLE
	Paraformaldehyde	4.1	UN2213	III	4.1	A1, IB8, IP3, T1, TP33	151	213	240	25 kg	100 kg	A		133	FLAMMABLE SOLID
	Paraldehyde	3	UN1264	III	3	B1, IB3, T2, TP1	150	203	242	60 L	220 L	A		129	FLAMMABLE
	Paranitroaniline, solid, see **Nitroanilines** *etc.*														
D	**Parathion and compressed gas mixture**	2.3	NA1967		2.3	3	None	334	245	Forbidden	Forbidden	E	40	123	POISON GAS*
	Paris green, solid, see **Copper acetoarsenite**														
	PCB, *see* **Polychlorinated biphenyls**														
+	**Pentaborane**	4.2	UN1380	I	4.2, 6.1	1	None	205	245	Forbidden	Forbidden	D		135	SPONTANEOUSLY COMBUSTIBLE, POISON INHALATION HAZARD*
	Pentachloroethane	6.1	UN1669	II	6.1	IB2, T7, TP2	153	202	243	5 L	60 L	A	40	151	POISON
	Pentachlorophenol	6.1	UN3155	II	6.1	IB8, IP2, IP4, T3, TP33	153	212	242	25 kg	100 kg	A		154	POISON
	Pentaerythrite tetranitrate (dry)	Forbidden													

Symbols (1)	Hazardous materials descriptions and proper shipping names (2)	Hazard class or Division (3)	Identification Numbers (4)	PG (5)	Label Codes (6)	Special Provisions (§172.102) (7)	(8) Packaging (§173.***) Exceptions (8A)	Non-bulk (8B)	Bulk (8C)	(9) Quantity limitations (see §§173.27 and 175.75) Passenger aircraft/rail (9A)	Cargo aircraft only (9B)	(10) Vessel stowage Location (10A)	Other (10B)	ERG Guide No.	Placard Advisory Consult §172.504 & §172.505 to determine placard *Denotes placard for any quantity
	Pentaerythrite tetranitrate mixture, desensitized, solid, n.o.s. *or* **Pentaerythritol tetranitrate mixture, desensitized, solid, n.o.s.** *or* **PETN mixture, desensitized, solid, n.o.s.,** *with more than 10 percent but not more than 20 percent PETN, by mass*	4.1	UN3344	II	4.1	118, N85	None	214	None	Forbidden	Forbidden	E		113	FLAMMABLE SOLID
	Pentaerythrite tetranitrate *or* **Pentaerythritol tetranitrate** *or* **PETN,** *with not less than 7 percent wax by mass*	1.1D	UN0411	II	1.1D	120	None	62	None	Forbidden	Forbidden	10		122	EXPLOSIVES 1.1*
	Pentaerythrite tetranitrate, wetted *or* **Pentaerythritol tetranitrate, wetted,** *or* **PETN, wetted** *with not less than 25 percent water, by mass, or* **Pentaerythrite tetranitrate,** *or* **Pentaerythritol tetranitrate** *or* **PETN, desensitized** *with not less than 15 percent phlegmatizer by mass*	1.1D	UN0150	II	1.1D	121	None	62	None	Forbidden	Forbidden	10		112	EXPLOSIVES 1.1*
	Pentaerythritol tetranitrate, *see* **Pentaerythrite tetranitrate,** *etc.*														
	Pentafluoroethane *or* **Refrigerant gas R 125**	2.2	UN3220		2.2	T50	306	304	314, 315	75 kg	150 kg	A		126	NONFLAMMABLE GAS
	Pentamethylheptane	3	UN2286	III	3	B1, IB3, T2, TP1	150	203	242	60 L	220 L	A		128	FLAMMABLE
	Pentane-2,4-dione	3	UN2310	III	3, 6.1	B1, IB3, T4, TP1	150	203	242	60 L	220 L	A		131	FLAMMABLE
	Pentanes	3	UN1265	I	3	T11, TP2	150	201	243	1 L	30 L	E		128	FLAMMABLE
				II	3	IB2, IP8, T4, TP1	150	202	242	5 L	60 L	E		128	FLAMMABLE
	Pentanitroaniline (dry)	Forbidden													
	Pentanols	3	UN1105	II	3	IB2, T4, TP1, TP29	150	202	242	5 L	60 L	B		129	FLAMMABLE
				III	3	B1, B3, IB3, T2, TP1	150	203	242	60 L	220 L	A		129	FLAMMABLE
	1-Pentene *(n-amylene)*	3	UN1108	I	3	T11, TP2	150	201	243	1 L	30 L	E		128	FLAMMABLE
	1-Pentol	8	UN2705	II	8	B2, IB2, T7, TP2	154	202	242	1 L	30 L	B	26, 27	153P	CORROSIVE
	Pentolite, *dry or wetted with less than 15 percent water, by mass*	1.1D	UN0151	II	1.1D		None	62	None	Forbidden	Forbidden	10		112	EXPLOSIVES 1.1*
	Pepper spray, see **Aerosols,** *etc. or* **Self-defense spray, non-pressurized**														
	Perchlorates, inorganic, aqueous solution, n.o.s.	5.1	UN3211	II	5.1	IB2, T4, TP1	152	202	242	1 L	5 L	B	56, 58, 133	140	OXIDIZER
				III	5.1	IB2, T4, TP1	152	202	241	2.5 L	30 L	B	56, 58, 69, 133	140	OXIDIZER
	Perchlorates, inorganic, n.o.s.	5.1	UN1481	II	5.1	IB6, IP2, T3, TP33	152	212	242	5 kg	25 kg	A	56, 58	140	OXIDIZER
				III	5.1	IB8, IP3, T1, TP33	152	213	240	25 kg	100 kg	A	56, 58	140	OXIDIZER
	Perchloric acid, with more than 72 percent acid by mass	Forbidden													
	Perchloric acid *with more than 50 percent but not more than 72 percent acid, by mass*	5.1	UN1873	I	5.1, 8	A2, A3, N41, T10, TP1	None	201	243	Forbidden	2.5 L	D	66	143	OXIDIZER
	Perchloric acid *with not more than 50 percent acid by mass*	8	UN1802	II	8, 5.1	IB2, N41, T7, TP2	None	202	243	Forbidden	30 L	C	66	140	CORROSIVE
	Perchloroethylene, see **Tetrachloroethylene**														
	Perchloromethyl mercaptan	6.1	UN1670	I	6.1	2, B9, B14, B32, N34, T20, TP2, TP13, TP38, TP45	None	227	244	Forbidden	Forbidden	D	40	157	POISON INHALATION HAZARD*
	Perchloryl fluoride	2.3	UN3083		2.3, 5.1	2, B9, B14	None	302	314, 315	Forbidden	Forbidden	D	40	124	POISON GAS*
	Percussion caps, see **Primers, cap type**														
	Perfluoro-2-butene, see **Octafluorobut-2-ene**														
	Perfluoro(ethyl vinyl ether)	2.1	UN3154		2.1		306	302, 304, 305	314, 315	Forbidden	150 kg	E	40	115	FLAMMABLE GAS
	Perfluorom(ethyl vinyl ether)	2.1	UN3153		2.1	T50	306	302, 304, 305	314, 315	Forbidden	150 kg	E	40	115	FLAMMABLE GAS
	Perfumery products *with flammable solvents*	3	UN1266	II	3	149, IB2, T4, TP1, TP8	150	202	242	15 L	60 L	B		127	FLAMMABLE
				III	3	B1, IB3, T2, TP1	150	203	242	60 L	220 L	A		127	FLAMMABLE
G	**Permanganates, inorganic, aqueous solution, n.o.s.**	5.1	UN3214	II	5.1	26, 353, IB2, T4, TP1	152	202	242	1 L	5 L	D	56, 58, 133, 138	140	OXIDIZER
G	**Permanganates, inorganic, n.o.s.**	5.1	UN1482	II	5.1	26, 353, A30, IB6, IP2, T3, TP33	152	212	242	5 kg	25 kg	D	56, 58, 138	140	OXIDIZER

Symbols (1)	Hazardous materials descriptions and proper shipping names (2)	Hazard class or Division (3)	Identification Numbers (4)	PG (5)	Label Codes (6)	Special Provisions (§172.102) (7)	(8) Packaging (§173.***) Exceptions (8A)	Non-bulk (8B)	Bulk (8C)	(9) Quantity limitations (see §§173.27 and 175.75) Passenger aircraft/rail (9A)	Cargo aircraft only (9B)	(10) Vessel stowage Location (10A)	Other (10B)	ERG Guide No.	Placard Advisory Consult §172.504 & §172.505 to determine placard *Denotes placard for any quantity
				III	5.1	26, 353, A30, IB8, IP3, T1, TP33	152	213	240	25 kg	100 kg	D	56, 58, 138	140	OXIDIZER
	Permeation devices *for calibrating air quality monitoring equipment See §173.175*														
	Peroxides, inorganic, n.o.s.	5.1	UN1483	II	5.1	A7, A20, IB6, IP2, N34, T3, TP33	None	212	242	5 kg	25 kg	A	13, 52, 66, 75	140	OXIDIZER
				III	5.1	A7, A20, IB8, IP3, N34, T1, TP33	152	213	240	25 kg	100 kg	A	13, 52, 66, 75	140	OXIDIZER
	Peroxyacetic acid, with more than 43 percent and with more than 6 percent hydrogen peroxide	Forbidden													
	Persulfates, inorganic, aqueous solution, n.o.s.	5.1	UN3216	III	5.1	IB2, T4, TP1, TP29	152	203	241	2.5 L	30 L	A	56, 133	140	OXIDIZER
	Persulfates, inorganic, n.o.s.	5.1	UN3215	III	5.1	IB8, IP3, T1, TP33	152	213	240	25 kg	100 kg	A	56, 58	140	OXIDIZER
G	**Pesticides, liquid, flammable, toxic,** *flash point less than 23 degrees C*	3	UN3021	I	3, 6.1	B5, T14, TP2, TP13, TP27	None	201	243	Forbidden	30 L	B		131	FLAMMABLE
				II	3, 6.1	IB2, T11, TP2, TP13, TP27	150	202	243	1 L	60 L	B		131	FLAMMABLE
G	**Pesticides, liquid, toxic, flammable, n.o.s.** *flash point not less than 23 degrees C*	6.1	UN2903	I	6.1, 3	T14, TP2, TP13, TP27	None	201	243	1 L	30 L	B	40	131	POISON
				II	6.1, 3	IB2, T11, TP2, TP13, TP27	153	202	243	5 L	60 L	B	40	131	POISON
				III	6.1, 3	B1, IB3, T7, TP2	153	203	242	60 L	220 L	A	40	131	POISON
G	**Pesticides, liquid, toxic, n.o.s.**	6.1	UN2902	I	6.1	T14, TP2, TP13, TP27	None	201	243	1 L	30 L	B	40	151	POISON
				II	6.1	IB2, T11, TP2, TP13, TP27	153	202	243	5 L	60 L	B	40	151	POISON
				III	6.1	IB3, T7, TP2, TP28	153	203	241	60 L	220 L	A	40	151	POISON
G	**Pesticides, solid, toxic, n.o.s.**	6.1	UN2588	I	6.1	IB7, T6, TP33	None	211	242	5 kg	50 kg	A	40	151	POISON
				II	6.1	IB8, IP2, IP4, T3, TP33	153	212	242	25 kg	100 kg	A	40	151	POISON
				III	6.1	IB8, IP3, T1, TP33	153	213	240	100 kg	200 kg	A	40	151	POISON
	PETN, *see* **Pentaerythrite tetranitrate**														
	PETN/TNT, *see* **Pentolite** *etc.*														
	Petrol, *see* **Gasoline**														
	Petroleum crude oil	3	UN1267	I	3	144, 357, T11, TP1, TP8	150	201	243	1 L	30 L	E		128	FLAMMABLE
				II	3	144, 357, IB2, T4, TP1, TP8	150	202	242	5 L	60 L	B		128	FLAMMABLE
				III	3	144, 357, B1, IB3, T2, TP1	150	203	242	60 L	220 L	A		128	FLAMMABLE
	Petroleum distillates, n.o.s. *or* **Petroleum products, n.o.s.**	3	UN1268	I	3	144, T11, TP1, TP8	150	201	243	1 L	30 L	E		128	FLAMMABLE
				II	3	144, IB2, T7, TP1, TP8, TP28	150	202	242	5 L	60 L	B		128	FLAMMABLE
				III	3	144, B1, IB3, T4, TP1, TP29	150	203	242	60 L	220 L	A		128	FLAMMABLE
	Petroleum gases, liquefied *or* **Liquefied petroleum gas**	2.1	UN1075		2.1	T50	306	304	314, 315	Forbidden	150 kg	E	40	115	FLAMMABLE GAS
D	**Petroleum oil**	3	NA1270	I	3	144, T11, TP1	None	201	243	1 L	30 L	E		128	FLAMMABLE
				II	3	144, IB2, T7, TP1, TP8, TP28	150	202	242	5 L	60 L	B		128	FLAMMABLE
				III	3	144, B1, IB3, T4, TP1, TP29	150	203	242	60 L	220 L	A		128	FLAMMABLE
I	**Petroleum sour crude oil, flammable, toxic**	3	UN3494	I	3, 6.1	343, T14, TP2, TP13	None	201	243	Forbidden	30 L	D	40	—	FLAMMABLE
				II	3, 6.1	343, IB2, T7, TP2	150	202	243	1 L	60 L	D	40	—	FLAMMABLE
				III	3, 6.1	343, IB3, T4, TP1	150	203	242	60 L	220 L	C	40	—	FLAMMABLE
	Phenacyl bromide	6.1	UN2645	II	6.1	IB8, IP2, IP4, T3, TP33	153	212	242	25 kg	100 kg	B	40	153	POISON
+	**Phenetidines**	6.1	UN2311	III	6.1	IB3, T4, TP1	153	203	241	60 L	220 L	A		153	POISON

Symbols (1)	Hazardous materials descriptions and proper shipping names (2)	Hazard class or Division (3)	Identification Numbers (4)	PG (5)	Label Codes (6)	Special Provisions (§172.102) (7)	(8) Packaging (§173.***) Exceptions (8A)	Non-bulk (8B)	Bulk (8C)	(9) Quantity limitations (see §§173.27 and 175.75) Passenger aircraft/rail (9A)	Cargo aircraft only (9B)	(10) Vessel stowage Location (10A)	Other (10B)	ERG Guide No.	Placard Advisory Consult §172.504 & §172.505 to determine placard *Denotes placard for any quantity
	Phenol, molten	6.1	UN2312	II	6.1	B14, T7, TP3	None	202	243	Forbidden	Forbidden	B	40	153	POISON
+	**Phenol, solid**	6.1	UN1671	II	6.1	IB8, IP2, IP4, N78, T3, TP33	153	212	242	25 kg	100 kg	A		153	POISON
	Phenol solutions	6.1	UN2821	II	6.1	IB2, T7, TP2	153	202	243	5 L	60 L	A		153	POISON
				III	6.1	IB3, T4, TP1	153	203	241	60 L	220 L	A		153	POISON
	Phenolsulfonic acid, liquid	8	UN1803	II	8	B2, IB2, N41, T7, TP2	154	202	242	1 L	30 L	C	14	153	CORROSIVE
	Phenoxyacetic acid derivative pesticide, liquid, flammable, toxic, *flash point less than 23 degrees C*	3	UN3346	I	3, 6.1	T14, TP2, TP13, TP27	None	201	243	Forbidden	30 L	B	40	131	FLAMMABLE
				II	3, 6.1	IB2, T11, TP2, TP13, TP27	150	202	243	1 L	60 L	B	40	131	FLAMMABLE
	Phenoxyacetic acid derivative pesticide, liquid, toxic	6.1	UN3348	I	6.1	T14, TP2, TP13, TP27	None	201	243	1 L	30 L	B	40	153	POISON
				II	6.1	IB2, T11, TP2, TP27	153	202	243	5 L	60 L	B	40	153	POISON
				III	6.1	IB3, T7, TP2, TP28	153	203	241	60 L	220 L	A	40	153	POISON
	Phenoxyacetic acid derivative pesticide, liquid, toxic, flammable, *flash point not less than 23 degrees C*	6.1	UN3347	I	6.1, 3	T14, TP2, TP13, TP27	None	201	243	1 L	30 L	B	40	131	POISON
				II	6.1, 3	IB2, T11, TP2, TP13, TP27	153	202	243	5 L	60 L	B	40	131	POISON
				III	6.1, 3	IB3, T7, TP2, TP28	153	203	241	60 L	220 L	A	40	131	POISON
	Phenoxyacetic acid derivative pesticide, solid, toxic	6.1	UN3345	I	6.1	IB7, IP1, T6, TP33	None	211	242	5 kg	50 kg	A	40	153	POISON
				II	6.1	IB8, IP2, IP4, T3, TP33	153	212	242	25 kg	100 kg	A	40	153	POISON
				III	6.1	IB8, IP3, T1, TP33	153	213	240	100 kg	200 kg	A	40	153	POISON
	Phenyl chloroformate	6.1	UN2746	II	6.1, 8	IB2, T7, TP2, TP13	153	202	243	1 L	30 L	A	12, 13, 21, 25, 40, 100	156	POISON
	Phenyl isocyanate	6.1	UN2487	I	6.1, 3	2, B9, B14, B32, B77, N33, N34, T20, TP2, TP13, TP38, TP45	None	227	244	Forbidden	Forbidden	D	40	155	POISON INHALATION HAZARD*
	Phenyl mercaptan	6.1	UN2337	I	6.1, 3	2, B9, B14, B32, B77, T20, TP2, TP13, TP38, TP45	None	227	244	Forbidden	Forbidden	D	40, 52	131	POISON INHALATION HAZARD*
	Phenyl phosphorus dichloride	8	UN2798	II	8	B2, B15, IB2, T7, TP2	154	202	242	Forbidden	30 L	B	40	137	CORROSIVE
	Phenyl phosphorus thiodichloride	8	UN2799	II	8	B2, B15, IB2, T7, TP2	154	202	242	Forbidden	30 L	B	40	137	CORROSIVE
	Phenyl urea pesticides, liquid, toxic	6.1	UN3002	I	6.1	T14, TP2, TP27	None	201	243	1 L	30 L	B	40	151	POISON
				II	6.1	T7, TP2	None	202	243	5 L	60 L	B	40	151	POISON
				III	6.1	T4, TP1	153	203	241	60 L	220 L	A	40	151	POISON
	Phenylacetonitrile, liquid	6.1	UN2470	III	6.1	IB3, T4, TP1	153	203	241	60 L	220 L	A	26	152	POISON
	Phenylacetyl chloride	8	UN2577	II	8	B2, IB2, T7, TP2	154	202	242	1 L	30 L	C	40	156	CORROSIVE
	Phenylcarbylamine chloride	6.1	UN1672	I	6.1	2, B9, B14, B32, T20, TP2, TP13, TP38, TP45	None	227	244	Forbidden	Forbidden	D	40	151	POISON INHALATION HAZARD*
	m-Phenylene diaminediperchlorate (dry)	Forbidden													
+	**Phenylenediamines** *(o-; m-; p-;)*	6.1	UN1673	III	6.1	IB8, IP3, T1, TP33	153	213	240	100 kg	200 kg	A		153	POISON
	Phenylhydrazine	6.1	UN2572	II	6.1	IB2, T7, TP2	153	202	243	5 L	60 L	A	40	153	POISON
	Phenylmercuric acetate	6.1	UN1674	II	6.1	IB8, IP2, IP4, T3, TP33	153	212	242	25 kg	100 kg	A		151	POISON
G	**Phenylmercuric compounds, n.o.s.**	6.1	UN2026	I	6.1	IB7, IP1, T6, TP33	None	211	242	5 kg	50 kg	A		151	POISON
				II	6.1	IB8, IP2, IP4, T3, TP33	153	212	242	25 kg	100 kg	A		151	POISON
				III	6.1	IB8, IP3, T1, TP33	153	213	240	100 kg	200 kg	A		151	POISON
	Phenylmercuric hydroxide	6.1	UN1894	II	6.1	IB8, IP2, IP4, T3, TP33	153	212	242	25 kg	100 kg	A		151	POISON
	Phenylmercuric nitrate	6.1	UN1895	II	6.1	IB8, IP2, IP4, T3, TP33	153	212	242	25 kg	100 kg	A		151	POISON

Sym-bols (1)	Hazardous materials descriptions and proper shipping names (2)	Hazard class or Division (3)	Identifi-cation Numbers (4)	PG (5)	Label Codes (6)	Special Provisions (§172.102) (7)	(8) Packaging (§173.***) Excep-tions (8A)	Non-bulk (8B)	Bulk (8C)	(9) Quantity limitations (see §§173.27 and 175.75) Passenger aircraft/rail (9A)	Cargo aircraft only (9B)	(10) Vessel stowage Location (10A)	Other (10B)	ERG Guide No.	Placard Advisory Consult §172.504 & §172.505 to determine placard *Denotes placard for any quantity
	Phenyltrichlorosilane	8	UN1804	II	8	A7, B6, N34, T10, TP2, TP7, TP13	None	206	242	Forbidden	30 L	C	40	156	CORROSIVE
	Phosgene	2.3	UN1076		2.3, 8	1, B7, B46	None	192	314	Forbidden	Forbidden	D	40	125	POISON GAS*
	9-Phosphabicyclononanes *or* **Cyclooctadiene phosphines**	4.2	UN2940	II	4.2	A19, IB6, IP2, T3, TP33	None	212	241	15 kg	50 kg	A		135	SPONTANEOUSLY COMBUSTIBLE
	Phosphine	2.3	UN2199		2.3, 2.1	1	None	192	245	Forbidden	Forbidden	D	40	119	POISON GAS*
	Phosphoric acid solution	8	UN1805	III	8	A7, IB3, N34, T4, TP1	154	203	241	5 L	60 L	A		154	CORROSIVE
	Phosphoric acid, solid	8	UN3453	III	8	IB8, IP3, T1, TP33	154	213	240	25 kg	100 kg	A		154	CORROSIVE
	Phosphoric acid triethyleneimine, see **Tris-(1-aziridinyl)phosphine oxide, solution**														
	Phosphoric anhydride, see **Phosphorus pentoxide**														
	Phosphorous acid	8	UN2834	III	8	IB8, IP3, T1, TP33	154	213	240	25 kg	100 kg	A	48	154	CORROSIVE
	Phosphorus, amorphous	4.1	UN1338	III	4.1	A1, A19, B1, B9, B26, IB8, IP3, T1, TP33	None	213	243	25 kg	100 kg	A	74	133	FLAMMABLE SOLID
	Phosphorus bromide, see **Phosphorus tribromide**														
	Phosphorus chloride, see **Phosphorus trichloride**														
	Phosphorus heptasulfide, *free from yellow or white phosphorus*	4.1	UN1339	II	4.1	A20, IB4, N34, T3, TP33	None	212	240	15 kg	50 kg	B	74	139	FLAMMABLE SOLID
	Phosphorus oxybromide	8	UN1939	II	8	B8, IB8, IP2, IP4, N41, N43, T3, TP33	None	212	240	Forbidden	50 kg	C	12, 40	137	CORROSIVE
	Phosphorus oxybromide, molten	8	UN2576	II	8	B2, B8, IB1, N41, N43, T7, TP3, TP13	None	202	242	Forbidden	Forbidden	C	40	137	CORROSIVE
+	**Phosphorus oxychloride**	6.1	UN1810	I	6.1, 8	2, B9, B14, B32, B77, N34, T20, TP2, TP13, TP38, TP45	None	227	244	Forbidden	Forbidden	D	40	137	POISON INHALATION HAZARD*
	Phosphorus pentabromide	8	UN2691	II	8	A7, IB8, IP2, IP4, N34, T3, TP33	154	212	240	Forbidden	50 kg	B	12, 40, 53, 55	137	CORROSIVE
	Phosphorus pentachloride	8	UN1806	II	8	A7, IB8, IP2, IP4, N34, T3, TP33	None	212	240	Forbidden	50 kg	C	40, 44, 89, 100, 141	137	CORROSIVE
	Phosphorus Pentafluoride	2.3	UN2198		2.3, 8	2, B9, B14	None	302, 304	314, 315	Forbidden	Forbidden	D	40	125	POISON GAS*
	Phosphorus pentasulfide, *free from yellow or white phosphorus*	4.3	UN1340	II	4.3, 4.1	A20, B59, IB4, T3, TP33	151	212	242	15 kg	50 kg	B	74	139	DANGEROUS WHEN WET*
	Phosphorus pentoxide	8	UN1807	II	8	A7, IB8, IP2, IP4, N34, T3, TP33	154	212	240	15 kg	50 kg	A		137	CORROSIVE
	Phosphorus sesquisulfide, *free from yellow or white phosphorus*	4.1	UN1341	II	4.1	A20, IB4, N34, T3, TP33	None	212	240	15 kg	50 kg	B	74	139	FLAMMABLE SOLID
	Phosphorus tribromide	8	UN1808	II	8	A3, A6, A7, B2, B25, IB2, N34, N43, T7, TP2	None	202	242	Forbidden	30 L	C	40	137	CORROSIVE
	Phosphorus trichloride	6.1	UN1809	I	6.1, 8	2, B9, B14, B15, B32, B77, N34, T20, TP2, TP13, TP38, TP45	None	227	244	Forbidden	Forbidden	C	40	137	POISON INHALATION HAZARD*
	Phosphorus trioxide	8	UN2578	III	8	IB8, IP3, T1, TP33	154	213	240	25 kg	100 kg	A	12	157	CORROSIVE
	Phosphorus trisulfide, *free from yellow or white phosphorus*	4.1	UN1343	II	4.1	A20, IB4, N34, T3, TP33	None	212	240	15 kg	50 kg	B	74	139	FLAMMABLE SOLID
	Phosphorus, white dry *or* **Phosphorus, white, under water** *or* **Phosphorus white, in solution** *or* **Phosphorus, yellow dry** *or* **Phosphorus, yellow, under water** *or* **Phosphorus, yellow, in solution**	4.2	UN1381	I	4.2, 6.1	B9, B26, N34, T9, TP3, TP31	None	188	243	Forbidden	Forbidden	E		136	SPONTANEOUSLY COMBUSTIBLE
	Phosphorus white, molten	4.2	UN2447	I	4.2, 6.1	B9, B26, N34, T21, TP3, TP7, TP26	None	188	243	Forbidden	Forbidden	D		136	SPONTANEOUSLY COMBUSTIBLE
	Phosphorus (white or red) and a chlorate, mixtures of	Forbidden													
	Phosphoryl chloride, see **Phosphorus oxychloride**														
	Phthalic anhydride *with more than .05 percent maleic anhydride*	8	UN2214	III	8	IB8, IP3, T1, TP33	154	213	240	25 kg	100 kg	A		156	CORROSIVE

Symbols (1)	Hazardous materials descriptions and proper shipping names (2)	Hazard class or Division (3)	Identification Numbers (4)	PG (5)	Label Codes (6)	Special Provisions (§172.102) (7)	(8) Packaging (§173.***) Exceptions (8A)	Non-bulk (8B)	Bulk (8C)	(9) Quantity limitations (see §§173.27 and 175.75) Passenger aircraft/rail (9A)	Cargo aircraft only (9B)	(10) Vessel stowage Location (10A)	Other (10B)	ERG Guide No.	Placard Advisory Consult §172.504 & §172.505 to determine placard *Denotes placard for any quantity
	Picolines	3	UN2313	III	3	B1, IB3, T4, TP1	150	203	242	60 L	220 L	A	40	129	FLAMMABLE
	Picric acid, *see* **Trinitrophenol,** *etc.*														
	Picrite, *see* **Nitroguanidine,** *etc.*														
	Picryl chloride, *see* **Trinitrochlorobenzene**														
	Pine oil	3	UN1272	III	3	B1, IB3, T2, TP1	150	203	242	60 L	220 L	A		129	FLAMMABLE
	alpha-Pinene	3	UN2368	III	3	B1, IB3, T2, TP1	150	203	242	60 L	220 L	A		128	FLAMMABLE
	Piperazine	8	UN2579	III	8	IB8, IP3, T1, TP33	154	213	240	25 kg	100 kg	A	12, 52	153	CORROSIVE
	Piperidine	8	UN2401	I	8, 3	A10, T10, TP2	None	201	243	0.5 L	2.5 L	B	52	132	CORROSIVE
	Pivaloyl chloride, see **Trimethylacetyl chloride**														
	Plastic molding compound *in dough, sheet or extruded rope form evolving flammable vapor*	9	UN3314	III	9	32, IB8, IP3, IP7	155	221	221	100 kg	200 kg	E	19, 21, 25, 87, 144	171	CLASS 9
	Plastic solvent, n.o.s., see **Flammable liquids, n.o.s.**														
	Plastics, nitrocellulose-based, self-heating, n.o.s.	4.2	UN2006	III	4.2		None	213	None	Forbidden	Forbidden	C		135	SPONTANEOUSLY COMBUSTIBLE
	Poisonous gases, n.o.s., see **Compressed** *or* **liquefied gases, flammable** *or* **toxic, n.o.s.**														
	Polyalkylamines, n.o.s., see **Amines,** *etc.*														
	Polyamines, flammable, corrosive, n.o.s. *see* **Amines, flammable, corrosive, n.o.s.**														
	Polyamines, liquid, corrosive, n.o.s. *see* **Amines, liquid, corrosive, n.o.s.**														
	Polyamines, liquid, corrosive, flammable, n.o.s. *see* **Amines, liquid, corrosive, flammable, n.o.s.**														
	Polychlorinated biphenyls, liquid	9	UN2315	II	9	9, 81, 140, IB3, T4, TP1	155	202	241	100 L	220 L	A	95	171	CLASS 9
	Polychlorinated biphenyls, solid	9	UN3432	II	9	9, 81, 140, IB8, IP2, IP4, T3, TP33	155	212	240	100 kg	220 kg	A	95	171	CLASS 9
	Polyester resin kit	3	UN3269		3	40, 149	165	165	None	5 kg	5 kg	B		128	FLAMMABLE
	Polyhalogenated biphenyls, liquid *or* **Polyhalogenated terphenyls liquid**	9	UN3151	II	9	IB2	155	204	241	100 L	220 L	A	95	171	CLASS 9
	Polyhalogenated biphenyls, solid *or* **Polyhalogenated terphenyls, solid**	9	UN3152	II	9	IB8, IP2, IP4, T3, TP33	155	204	241	100 kg	200 kg	A	95	171	CLASS 9
	Polymeric beads expandable, *evolving flammable vapor*	9	UN2211	III	9	32, IB8, IP3, IP7, T1, TP33	155	221	221	100 kg	200 kg	E	19, 21, 25, 87, 144	133	CLASS 9
	Potassium	4.3	UN2257	I	4.3	A7, A19, A20, B27, IB4, IP1, N6, N34, T9, TP7, TP33	None	211	244	Forbidden	15 kg	D	52	138	DANGEROUS WHEN WET*
	Potassium arsenate	6.1	UN1677	II	6.1	IB8, IP2, IP4, T3, TP33	153	212	242	25 kg	100 kg	A		151	POISON
	Potassium arsenite	6.1	UN1678	II	6.1	IB8, IP2, IP4, T3, TP33	153	212	242	25 kg	100 kg	A		154	POISON
	Potassium bisulfite solution, see **Bisulfites, aqueous solutions, n.o.s.**														
	Potassium borohydride	4.3	UN1870	I	4.3	A19, N40	None	211	242	Forbidden	15 kg	E	52	138	DANGEROUS WHEN WET*
	Potassium bromate	5.1	UN1484	II	5.1	IB8, IP2, IP4, T3, TP33	152	212	242	5 kg	25 kg	A	56, 58	140	OXIDIZER
	Potassium carbonyl	Forbidden													
	Potassium chlorate	5.1	UN1485	II	5.1	A9, IB8, IP2, IP4, N34, T3, TP33	152	212	242	5 kg	25 kg	A	56, 58	140	OXIDIZER
	Potassium chlorate, aqueous solution	5.1	UN2427	II	5.1	A2, IB2, T4, TP1	152	202	241	1 L	5 L	B	56, 58, 133	140	OXIDIZER
				III	5.1	A2, IB2, T4, TP1	152	203	241	2.5 L	30 L	B	56, 58, 69, 133	140	OXIDIZER
	Potassium chlorate mixed with mineral oil, see **Explosive, blasting, type C**														
	Potassium cuprocyanide	6.1	UN1679	II	6.1	IB8, IP2, IP4, T3, TP33	153	212	242	25 kg	100 kg	A	52	157	POISON

Symbols (1)	Hazardous materials descriptions and proper shipping names (2)	Hazard class or Division (3)	Identification Numbers (4)	PG (5)	Label Codes (6)	Special Provisions (§172.102) (7)	(8) Packaging (§173.***) Exceptions (8A)	Non-bulk (8B)	Bulk (8C)	(9) Quantity limitations (see §§173.27 and 175.75) Passenger aircraft/rail (9A)	Cargo aircraft only (9B)	(10) Vessel stowage Location (10A)	Other (10B)	ERG Guide No.	Placard Advisory Consult §172.504 & §172.505 to determine placard *Denotes placard for any quantity
	Potassium cyanide, solid	6.1	UN1680	I	6.1	B69, B77, IB7, IP1, N74, N75, T6, TP33	None	211	242	5 kg	50 kg	B	52	157	POISON
	Potassium cyanide solution	6.1	UN3413	I	6.1	B69, B77, N74, N75, T14, TP2, TP13	None	201	243	1 L	30 L	B	52	157	POISON
				II	6.1	B69, B77, IB2, N74, N75, T11, TP2, TP13, TP27	153	202	243	5 L	60 L	B	52	157	POISON
				III	6.1	B69, B77, IB3, N74, N75, T7, TP2, TP13, TP28	153	203	241	60 L	220 L	A	52	157	POISON
	Potassium dichloro isocyanurate or Potassium dichloro-s-triazinetrione, see **Dichloroisocyanuric acid, dry** *or* **Dichloroisocyanuric acid salts** *etc.*														
	Potassium dithionite *or* **Potassium hydrosulfite**	4.2	UN1929	II	4.2	A8, A19, A20, IB6, IP2, T3, TP33	None	212	241	15 kg	50 kg	E	13	135	SPONTANEOUSLY COMBUSTIBLE
	Potassium fluoride, solid	6.1	UN1812	III	6.1	IB8, IP3, T1, TP33	153	213	240	100 kg	200 kg	A	52	154	POISON
	Potassium fluoride solution	6.1	UN3422	III	6.1	IB3, T4, TP1	153	203	241	60 L	220 L	A	52	154	POISON
	Potassium fluoroacetate	6.1	UN2628	I	6.1	IB7, IP1, T6, TP33	None	211	242	5 kg	50 kg	E		151	POISON
	Potassium fluorosilicate	6.1	UN2655	III	6.1	IB8, IP3, T1, TP33	153	213	240	100 kg	200 kg	A	52	151	POISON
	Potassium hydrate, see **Potassium hydroxide, solid**														
	Potassium hydrogen fluoride, see **Potassium hydrogen difluoride**														
	Potassium hydrogen fluoride solution, see **Corrosive liquids, n.o.s.**														
	Potassium hydrogen sulfate	8	UN2509	II	8	A7, IB8, IP2, IP4, N34, T3, TP33	154	212	240	15 kg	50 kg	A		154	CORROSIVE
	Potassium hydrogendifluoride solid	8	UN1811	II	8, 6.1	IB8, IP2, IP4, N3, N34, T3, TP33	154	212	240	15 kg	50 kg	A	25, 40, 52	154	CORROSIVE
	Potassium hydrogendifluoride solution	8	UN3421	II	8, 6.1	IB2, N3, N34, T7, TP2	154	202	243	1 L	30 L	A	25, 40, 52	154	CORROSIVE
				III	8, 6.1	IB3, N3, N34, T4, TP1	154	203	241	5 L	60 L	A	40, 52	154	CORROSIVE
	Potassium hydrosulfite, *see* **Potassium dithionite**														
	Potassium hydroxide, liquid, see **Potassium hydroxide, solution**														
	Potassium hydroxide, solid	8	UN1813	II	8	IB8, IP2, IP4, T3, TP33	154	212	240	15 kg	50 kg	A	52	154	CORROSIVE
	Potassium hydroxide, solution	8	UN1814	II	8	B2, IB2, T7, TP2	154	202	242	1 L	30 L	A	52	154	CORROSIVE
				III	8	IB3, T4, TP1	154	203	241	5 L	60 L	A	52	154	CORROSIVE
	Potassium hypochlorite, solution, see **Hypochlorite solutions,** *etc.*														
	Potassium, metal alloys, liquid	4.3	UN1420	I	4.3	A7, A19, A20, B27	None	201	244	Forbidden	1 L	E	40, 52	138	DANGEROUS WHEN WET*
	Potassium, metal alloys, solid	4.3	UN3403	I	4.3	A19, A20, B27, IB4, IP1, T9, TP7, TP33	None	211	244	Forbidden	15 kg	D		138	DANGEROUS WHEN WET*
	Potassium metavanadate	6.1	UN2864	II	6.1	IB8, IP2, IP4, T3, TP33	153	212	242	25 kg	100 kg	A		151	POISON
	Potassium monoxide	8	UN2033	II	8	IB8, IP2, IP4, T3, TP33	154	212	240	15 kg	50 kg	A	29, 52	154	CORROSIVE
	Potassium nitrate	5.1	UN1486	III	5.1	A1, A29, IB8, IP3, T1, TP33, W1	152	213	240	25 kg	100 kg	A		140	OXIDIZER
	Potassium nitrate and sodium nitrite mixtures	5.1	UN1487	II	5.1	B78, IB8, IP2, IP4, T3, TP33	152	212	240	5 kg	25 kg	A	56, 58	140	OXIDIZER
	Potassium nitrite	5.1	UN1488	II	5.1	IB8, IP2, IP4, T3, TP33	152	212	242	5 kg	25 kg	A	56, 58	140	OXIDIZER
	Potassium perchlorate	5.1	UN1489	II	5.1	IB6, IP2, T3, TP33	152	212	242	5 kg	25 kg	A	56, 58	140	OXIDIZER
	Potassium permanganate	5.1	UN1490	II	5.1	IB8, IP2, IP4, T3, TP33	152	212	240	5 kg	25 kg	D	56, 58, 138	140	OXIDIZER

Symbols (1)	Hazardous materials descriptions and proper shipping names (2)	Hazard class or Division (3)	Identification Numbers (4)	PG (5)	Label Codes (6)	Special Provisions (§172.102) (7)	(8) Packaging (§173.***) Exceptions (8A)	Non-bulk (8B)	Bulk (8C)	(9) Quantity limitations (see §§173.27 and 175.75) Passenger aircraft/rail (9A)	Cargo aircraft only (9B)	(10) Vessel stowage Location (10A)	Other (10B)	ERG Guide No.	Placard Advisory Consult §172.504 & §172.505 to determine placard *Denotes placard for any quantity
	Potassium peroxide	5.1	UN1491	I	5.1	A20, IB6, IP1, N34	None	211	None	Forbidden	15 kg	B	13, 52, 66, 75,	144	OXIDIZER
	Potassium persulfate	5.1	UN1492	III	5.1	A1, A29, IB8, IP3, T1, TP33	152	213	240	25 kg	100 kg	A	58, 145	140	OXIDIZER
	Potassium phosphide	4.3	UN2012	I	4.3, 6.1	A19, N40	None	211	None	Forbidden	15 kg	E	40, 52, 85	139	DANGEROUS WHEN WET*
	Potassium selenate, see **Selenates** *or* **Selenites**														
	Potassium selenite, see **Selenates** *or* **Selenites**														
	Potassium sodium alloys, liquid	4.3	UN1422	I	4.3	A7, A19, B27, N34, N40, T9, TP3, TP7, TP31	None	201	244	Forbidden	1 L	E	40, 52	138	DANGEROUS WHEN WET*
	Potassium sodium alloys, solid	4.3	UN3404	I	4.3	A19, B27, N34, N40, T9, TP7, TP33	None	211	244	Forbidden	15 kg	D	52	138	DANGEROUS WHEN WET*
	Potassium sulfide, anhydrous *or* **Potassium sulfide** *with less than 30 percent water of crystallization*	4.2	UN1382	II	4.2	A19, A20, B16, IB6, IP2, N34, T3, TP33	None	212	241	15 kg	50 kg	A	52	135	SPONTANEOUSLY COMBUSTIBLE
	Potassium sulfide, hydrated *with not less than 30 percent water of crystallization*	8	UN1847	II	8	IB8, IP2, IP4, T3, TP33	154	212	240	15 kg	50 kg	A	52	153	CORROSIVE
	Potassium superoxide	5.1	UN2466	I	5.1	A20, IB6, IP1	None	211	None	Forbidden	15 kg	B	13, 52, 66, 75	143	OXIDIZER
	Powder cake, wetted *or* **Powder paste, wetted** *with not less than 17 percent alcohol by mass*	1.1C	UN0433	II	1.1C		None	62	None	Forbidden	Forbidden	10		112	EXPLOSIVES 1.1*
	Powder cake, wetted *or* **Powder paste, wetted** *with not less than 25 percent water, by mass*	1.3C	UN0159	II	1.3C		None	62	None	Forbidden	Forbidden	10		112	EXPLOSIVES 1.3*
	Powder paste, *see* **Powder cake,** *etc.*														
	Powder, smokeless	1.1C	UN0160	II	1.1C		None	62	None	Forbidden	Forbidden		26E	112	EXPLOSIVES 1.1*
	Powder, smokeless	1.3C	UN0161	II	1.3C		None	62	None	Forbidden	Forbidden		26E	112	EXPLOSIVES 1.3*
	Powder, smokeless	1.4C	UN0509		1.4C		None	62	None	Forbidden	75 kg	06		114	EXPLOSIVES 1.4
	Power device, explosive, see **Cartridges, power device**														
	Primers, cap type	1.4S	UN0044	II	None		None	62	None	25 kg	100 kg	05		114	NONE
	Primers, cap type	1.1B	UN0377	II	1.1B		None	62	None	Forbidden	Forbidden	11		112	EXPLOSIVES 1.1*
	Primers, cap type	1.4B	UN0378	II	1.4B		None	62	None	Forbidden	75 kg	06		114	EXPLOSIVES 1.4
	Primers, small arms, see **Primers, cap type**														
	Primers, tubular	1.3G	UN0319	II	1.3G		None	62	None	Forbidden	Forbidden	07		112	EXPLOSIVES 1.3*
	Primers, tubular	1.4G	UN0320	II	1.4G		None	62	None	Forbidden	75 kg	06		114	EXPLOSIVES 1.4
	Primers, tubular	1.4S	UN0376	II	None		None	62	None	25 kg	100 kg	05		114	NONE
	Printing ink, *flammable or* **Printing ink related material** *(including printing ink thinning or reducing compound), flammable*	3	UN1210	I	3	T11, TP1, TP8	150	173	243	1 L	30 L	E		129	FLAMMABLE
				II	3	149, IB2, T4, TP1, TP8	150	173	242	5 L	60 L	B		129	FLAMMABLE
				III	3	B1, IB3, T2, TP1	150	173	242	60 L	220 L	A		129	FLAMMABLE
	Projectiles, illuminating, see **Ammunition, illuminating,** *etc.*														
	Projectiles, *inert with tracer*	1.4S	UN0345	II	1.4S			62	62	25 kg	100 kg	01		114	EXPLOSIVES 1.4
	Projectiles, *inert, with tracer*	1.3G	UN0424	II	1.3G			62	62	Forbidden	Forbidden	03		112	EXPLOSIVES 1.3*
	Projectiles, *inert, with tracer*	1.4G	UN0425	II	1.4G			62	62	Forbidden	75 kg	02		114	EXPLOSIVES 1.4
	Projectiles, *with burster or expelling charge*	1.2D	UN0346	II	1.2D			62	62	Forbidden	Forbidden	03		112	EXPLOSIVES 1.2*
	Projectiles, *with burster or expelling charge*	1.4D	UN0347	II	1.4D			62	62	Forbidden	75 kg	02		114	EXPLOSIVES 1.4
	Projectiles, *with burster or expelling charge*	1.2F	UN0426	II	1.2F			62	None	Forbidden	Forbidden	08		112	EXPLOSIVES 1.2*
	Projectiles, *with burster or expelling charge*	1.4F	UN0427	II	1.4F			62	None	Forbidden	Forbidden	08		114	EXPLOSIVES 1.4
	Projectiles, *with burster or expelling charge*	1.2G	UN0434	II	1.2G			62	62	Forbidden	Forbidden	03		112	EXPLOSIVES 1.2*
	Projectiles, *with burster or expelling charge*	1.4G	UN0435	II	1.4G			62	62	Forbidden	75 kg	02		114	EXPLOSIVES 1.4
	Projectiles, *with bursting charge*	1.1F	UN0167	II	1.1F			62	None	Forbidden	Forbidden	08		112	EXPLOSIVES 1.1*
	Projectiles, *with bursting charge*	1.1D	UN0168	II	1.1D			62	62	Forbidden	Forbidden	03		112	EXPLOSIVES 1.1*
	Projectiles, *with bursting charge*	1.2D	UN0169	II	1.2D			62	62	Forbidden	Forbidden	03		112	EXPLOSIVES 1.2*
	Projectiles, *with bursting charge*	1.2F	UN0324	II	1.2F			62	None	Forbidden	Forbidden	08		112	EXPLOSIVES 1.2*
	Projectiles, *with bursting charge*	1.4D	UN0344	II	1.4D			62	62	Forbidden	75 kg	02		114	EXPLOSIVES 1.4

Symbols (1)	Hazardous materials descriptions and proper shipping names (2)	Hazard class or Division (3)	Identification Numbers (4)	PG (5)	Label Codes (6)	Special Provisions (§172.102) (7)	(8) Packaging (§173.***) Exceptions (8A)	Non-bulk (8B)	Bulk (8C)	(9) Quantity limitations (see §§173.27 and 175.75) Passenger aircraft/rail (9A)	Cargo aircraft only (9B)	(10) Vessel stowage Location (10A)	Other (10B)	ERG Guide No.	Placard Advisory Consult §172.504 & §172.505 to determine placard *Denotes placard for any quantity
	Propadiene, stabilized	2.1	UN2200		2.1		None	304	314, 315	Forbidden	150 kg	B	40	116P	FLAMMABLE GAS
	Propadiene mixed with methyl acetylene, see **Methyl acetylene and propadiene mixtures, stabilized**														
	Propane, *see* **Petroleum gases, liquefied**	2.1	UN1978		2.1	19, T50	306	304	314, 315	Forbidden	150 kg	E	40	115	FLAMMABLE GAS
	Propanethiols	3	UN2402	II	3	A6, IB2, T4, TP1, TP13	150	202	242	5 L	60 L	E	95, 102	130	FLAMMABLE
	n-Propanol *or* **Propyl alcohol, normal**	3	UN1274	II	3	B1, IB2, T4, TP1	150	202	242	5 L	60 L	B		129	FLAMMABLE
				III	3	B1, IB3, T2, TP1	150	203	242	60 L	220 L	A		129	FLAMMABLE
	Propellant, liquid	1.3C	UN0495	II	1.3C	37	None	62	None	Forbidden	Forbidden	10		112	EXPLOSIVES 1.3*
	Propellant, liquid	1.1C	UN0497	II	1.1C	37	None	62	None	Forbidden	Forbidden	10		112	EXPLOSIVES 1.1*
	Propellant, solid	1.1C	UN0498	II	1.1C		None	62	None	Forbidden	Forbidden		26E	112	EXPLOSIVES 1.1*
	Propellant, solid	1.3C	UN0499	II	1.3C		None	62	None	Forbidden	Forbidden		26E	112	EXPLOSIVES 1.3*
	Propellant, solid	1.4C	UN0501		1.4C		None	62	None	Forbidden	Forbidden	A	24E	114	EXPLOSIVES 1.4
	Propionaldehyde	3	UN1275	II	3	IB2, T7, TP1	150	202	242	5 L	60 L	E		129	FLAMMABLE
	Propionic acid *with not less than 90% acid by mass*	8	UN3463	II	8, 3	IB2, T7, TP2	154	202	243	1 L	30 L	A		132	CORROSIVE
	Propionic acid *with not less than 10% and less than 90% acid by mass*	8	UN1848	III	8	IB3, T4, TP1	154	203	241	5 L	60 L	A		132	CORROSIVE
	Propionic anhydride	8	UN2496	III	8	IB3, T4, TP1	154	203	241	5 L	60 L	A		156	CORROSIVE
	Propionitrile	3	UN2404	II	3, 6.1	IB2, T7, TP1, TP13	None	202	243	Forbidden	60 L	E	40	131	FLAMMABLE
	Propionyl chloride	3	UN1815	II	3, 8	IB1, T7, TP1	150	202	243	1 L	5 L	B	40	132	FLAMMABLE
	n-Propyl acetate	3	UN1276	II	3	IB2, T4, TP1	150	202	242	5 L	60 L	B		129	FLAMMABLE
	Propyl alcohol, *see* **Propanol**														
	n-Propyl benzene	3	UN2364	III	3	B1, IB3, T2, TP1	150	203	242	60 L	220 L	A		128	FLAMMABLE
	Propyl chloride see, **1-Chloropropane**														
	n-Propyl chloroformate	6.1	UN2740	I	6.1, 3, 8	2, B9, B14, B32, B77, N34, T20, TP2, TP13, TP38, TP44	None	227	244	Forbidden	Forbidden	B	21, 40, 100	155	POISON INHALATION HAZARD*
	Propyl formates	3	UN1281	II	3	IB2, T4, TP1	150	202	242	5 L	60 L	B		129	FLAMMABLE
	n-Propyl isocyanate	6.1	UN2482	I	6.1, 3	1, B9, B14, B30, T20, TP2, TP13, TP38, TP44	None	226	244	Forbidden	Forbidden	D	40	155	POISON INHALATION HAZARD*
	Propyl mercaptan, see **Propanethiols**														
	n-Propyl nitrate	3	UN1865	II	3	IB9	150	202	None	5 L	60 L	D	44, 89, 90, 100	131	FLAMMABLE
	Propylamine	3	UN1277	II	3, 8	A7, IB2, N34, T7, TP1	150	202	243	1 L	5 L	E	40	132	FLAMMABLE
	Propylene, *see* **Petroleum gases, liquefied**	2.1	UN1077		2.1	19, T50	306	304	314, 315	Forbidden	150 kg	E	40	115	FLAMMABLE GAS
	Propylene chlorohydrin	6.1	UN2611	II	6.1, 3	IB2, T7, TP2, TP13	153	202	243	5 L	60 L	A	12, 40, 48	131	POISON
	Propylene oxide	3	UN1280	I	3	A3, N34, T11, TP2, TP7	None	201	243	1 L	30 L	E	40	127P	FLAMMABLE
	Propylene tetramer	3	UN2850	III	3	B1, IB3, T2, TP1	150	203	242	60 L	220 L	A		128	FLAMMABLE
	1,2-Propylenediamine	8	UN2258	II	8, 3	A3, A6, IB2, N34, T7, TP2	None	202	243	1 L	30 L	A	40	132	CORROSIVE
	Propyleneimine, stabilized	3	UN1921	I	3, 6.1	A3, N34, T14, TP2, TP13	None	201	243	1 L	30 L	B	40	131P	FLAMMABLE
	Propyltrichlorosilane	8	UN1816	II	8, 3	A7, B2, B6, N34, T10, TP2, TP7, TP13	None	206	243	Forbidden	30 L	C	40	155	CORROSIVE
	Prussic acid, see **Hydrogen cyanide**														
	Pyrethroid pesticide, liquid, flammable, toxic, *flash point less than 23 degrees C*	3	UN3350	I	3, 6.1	T14, TP2, TP13, TP27	None	201	243	Forbidden	30 L	B	40	131	FLAMMABLE
				II	3, 6.1	IB2, T11, TP2, TP13, TP27	150	202	243	1 L	60 L	B	40	131	FLAMMABLE
	Pyrethroid pesticide, liquid, toxic	6.1	UN3352	I	6.1	T14, TP2, TP13, TP27	None	201	243	1 L	30 L	B	40	151	POISON

Symbols (1)	Hazardous materials descriptions and proper shipping names (2)	Hazard class or Division (3)	Identification Numbers (4)	PG (5)	Label Codes (6)	Special Provisions (§172.102) (7)	(8) Packaging (§173.***) Exceptions (8A)	Non-bulk (8B)	Bulk (8C)	(9) Quantity limitations (see §§173.27 and 175.75) Passenger aircraft/rail (9A)	Cargo aircraft only (9B)	(10) Vessel stowage Location (10A)	Other (10B)	ERG Guide No.	Placard Advisory Consult §172.504 & §172.505 to determine placard *Denotes placard for any quantity
				II	6.1	IB2, T11, TP2, TP27	153	202	243	5 L	60 L	B	40	151	POISON
				III	6.1	IB3, T7, TP2, TP28	153	203	241	60 L	220 L	A	40	151	POISON
	Pyrethroid pesticide, liquid, toxic, flammable, *flash point not less than 23 degrees C*	6.1	UN3351	I	6.1, 3	T14, TP2, TP13, TP27	None	201	243	1 L	30 L	B	40	131	POISON
				II	6.1, 3	IB2, T11, TP2, TP13, TP27	153	202	243	5 L	60 L	B	40	131	POISON
				III	6.1, 3	IB3, T7, TP2, TP28	153	203	241	60 L	220 L	B	40	131	POISON
	Pyrethroid pesticide, solid, toxic	6.1	UN3349	I	6.1	IB7, IP1, T6, TP33	None	211	242	5 kg	50 kg	A	40	151	POISON
				II	6.1	IB8, IP2, IP4, T3, TP33	153	212	242	25 kg	100 kg	A	40	151	POISON
				III	6.1	IB8, IP3, T1, TP33	153	213	240	100 kg	200 kg	A	40	151	POISON
	Pyridine	3	UN1282	II	3	IB2, T4, TP2	None	202	242	5 L	60 L	B	21, 100	129	FLAMMABLE
	Pyridine perchlorate	Forbidden													
G	**Pyrophoric liquid, inorganic, n.o.s.**	4.2	UN3194	I	4.2		None	181	244	Forbidden	Forbidden	D	78	135	SPONTANEOUSLY COMBUSTIBLE
G	**Pyrophoric liquids, organic, n.o.s.**	4.2	UN2845	I	4.2	B11, T22, TP2, TP7	None	181	244	Forbidden	Forbidden	D	78	135	SPONTANEOUSLY COMBUSTIBLE
G	**Pyrophoric metals, n.o.s.,** *or* **Pyrophoric alloys, n.o.s.**	4.2	UN1383	I	4.2	B11, T21, TP7, TP33	None	187	242	Forbidden	Forbidden	D		135	SPONTANEOUSLY COMBUSTIBLE
G	**Pyrophoric solid, inorganic, n.o.s.**	4.2	UN3200	I	4.2	T21, TP7, TP33	None	187	242	Forbidden	Forbidden	D		135	SPONTANEOUSLY COMBUSTIBLE
G	**Pyrophoric solids, organic, n.o.s.**	4.2	UN2846	I	4.2		None	187	242	Forbidden	Forbidden	D		135	SPONTANEOUSLY COMBUSTIBLE
	Pyrosulfuryl chloride	8	UN1817	II	8	B2, IB2, T8, TP2, TP12	154	202	242	1 L	30 L	C	40	137	CORROSIVE
	Pyroxylin solution or solvent, see **Nitrocellulose**														
	Pyrrolidine	3	UN1922	II	3, 8	IB2, T7, TP1	150	202	243	1 L	5 L	B	40, 52	132	FLAMMABLE
	Quebrachitol pentanitrate	Forbidden													
	Quicklime, see **Calcium oxide**														
	Quinoline	6.1	UN2656	III	6.1	IB3, T4, TP1	153	203	241	60 L	220 L	A	12	154	POISON
	R 12, see **Dichlorodifluoromethane**														
	R 12B1, see **Chlorodifluorobromomethane**														
	R 13, see **Chlorotrifluoromethane**														
	R 13B1, see **Bromotrifluoromethane**														
	R 14, see **Tetrafluoromethane**														
	R 21, see **Dichlorofluoromethane**														
	R 22, see **Chlorodifluoromethane**														
	R 114, see **Dichlorotetrafluoroethane**														
	R 115, see **Chloropentafluoroethane**														
	R 116, see **Hexafluoroethane**														
	R 124, see **Chlorotetrafluoroethane**														
	R 133a, see **Chlorotrifluoroethane**														
	R 152a, see **Diflouroethane**														
	R 500, see **Dichlorodifluoromethane and difluorethane,** *etc.*														
	R 502, see **Chlorodifluoromethane and chloropentafluoroethane mixture,** *etc.*														
	R 503, see **Chlorotrifluoromethane and trifluoromethane,** *etc.*														
	Radioactive material, excepted package-articles manufactured from natural uranium *or* **depleted uranium** *or* **natural thorium**	7	UN2909		None		422, 426	422, 426	422, 426			A		161	NONE
	Radioactive material, excepted package-empty packaging	7	UN2908		Empty		422, 428	422, 428	422, 428			A		161	NONE
	Radioactive material, excepted package-instruments *or* **articles**	7	UN2911		None		422, 424	422, 424				A		161	NONE

HAZARDOUS MATERIALS TABLE

Symbols (1)	Hazardous materials descriptions and proper shipping names (2)	Hazard class or Division (3)	Identification Numbers (4)	PG (5)	Label Codes (6)	Special Provisions (§172.102) (7)	(8) Packaging (§173.***) Exceptions (8A)	Non-bulk (8B)	Bulk (8C)	(9) Quantity limitations (see §§173.27 and 175.75) Passenger aircraft/rail (9A)	Cargo aircraft only (9B)	(10) Vessel stowage Location (10A)	Other (10B)	ERG Guide No.	Placard Advisory Consult §172.504 & §172.505 to determine placard *Denotes placard for any quantity
	Radioactive material, excepted package-limited quantity of material	7	UN2910		None		421, 422	421, 422	421, 422			A		161	NONE
	Radioactive material, low specific activity (LSA-I) *non fissile or fissile-excepted*	7	UN2912		7	A56, T5, TP4, W7	421, 422, 428	427	427			A	95, 129	162	RADIOACTIVE* (YELLOW III LABEL OR EXCLUSIVE USE SHIPMENTS)
	Radioactive material, low specific activity (LSA-II) *non fissile or fissile-excepted*	7	UN3321		7	A56, T5, TP4, W7	421, 422, 428	427	427			A	95, 129	162	RADIOACTIVE* (YELLOW III LABEL OR EXCLUSIVE USE SHIPMENTS)
	Radioactive material, low specific activity (LSA-III) *non fissile or fissile excepted*	7	UN3322		7	A56, T5, TP4, W7	421, 422, 428	427	427			A	95, 129	162	RADIOACTIVE* (YELLOW III LABEL OR EXCLUSIVE USE SHIPMENTS)
	Radioactive material, surface contaminated objects (SCO-I *or* **SCO-II)** *non fissile or fissile-excepted*	7	UN2913		7	A56	421, 422, 428	427	427			A	95	162	RADIOACTIVE* (YELLOW III LABEL OR EXCLUSIVE USE SHIPMENTS)
	Radioactive material, transported under special arrangement, *non fissile or fissile excepted*	7	UN2919		7	A56, 139						A	95, 105	163	RADIOACTIVE* (YELLOW III LABEL OR EXCLUSIVE USE SHIPMENTS)
	Radioactive material, transported under special arrangement, fissile	7	UN3331		7	A56, 139						A	95, 105	165	RADIOACTIVE* (YELLOW III LABEL OR EXCLUSIVE USE SHIPMENTS)
	Radioactive material, Type A package, fissile *non-special form*	7	UN3327		7	A56, W7, W8	453	417	417			A	95, 105, 131	165	RADIOACTIVE* (YELLOW III LABEL OR EXCLUSIVE USE SHIPMENTS)
	Radioactive material, Type A package *non-special form, non fissile or fissile-excepted*	7	UN2915		7	A56, W7, W8	None	415, 418, 419	415, 418, 419			A	95, 130	163	RADIOACTIVE* (YELLOW III LABEL OR EXCLUSIVE USE SHIPMENTS)
	Radioactive material, Type A package, special form *non fissile or fissile-excepted*	7	UN3332		7	A56, W7, W8		415, 476	415, 476			A	95	164	RADIOACTIVE* (YELLOW III LABEL OR EXCLUSIVE USE SHIPMENTS)
	Radioactive material, Type A package, special form, fissile	7	UN3333		7	A56, W7, W8	453	417, 476	417, 476			A	95, 105	165	RADIOACTIVE* (YELLOW III LABEL OR EXCLUSIVE USE SHIPMENTS)
	Radioactive material, Type B(M) package, fissile	7	UN3329		7	A56	453	417	417			A	95, 105	165	RADIOACTIVE* (YELLOW III LABEL OR EXCLUSIVE USE SHIPMENTS)
	Radioactive material, Type B(M) package *non fissile or fissile-excepted*	7	UN2917		7	A56		416	416			A	95, 105	163	RADIOACTIVE* (YELLOW III LABEL OR EXCLUSIVE USE SHIPMENTS)
	Radioactive material, Type B(U) package, fissile	7	UN3328		7	A56	453	417	417			A	95, 105	165	RADIOACTIVE* (YELLOW III LABEL OR EXCLUSIVE USE SHIPMENTS)
	Radioactive material, Type B(U) package *non fissile or fissile-excepted*	7	UN2916		7	A56		416	416			A	95, 105	163	RADIOACTIVE* (YELLOW III LABEL OR EXCLUSIVE USE SHIPMENTS)
	Radioactive material, uranium hexafluoride *non fissile or fissile-excepted*	7	UN2978		7, 8		423	420, 427	420, 427			A	95, 132	166	RADIOACTIVE* (YELLOW III LABEL OR EXCLUSIVE USE SHIPMENTS)
	Radioactive material, uranium hexafluoride, fissile	7	UN2977		7, 8		453	417, 420	417, 420			A	95, 132	166	RADIOACTIVE* (YELLOW III LABEL OR EXCLUSIVE USE SHIPMENTS)

Symbols (1)	Hazardous materials descriptions and proper shipping names (2)	Hazard class or Division (3)	Identification Numbers (4)	PG (5)	Label Codes (6)	Special Provisions (§172.102) (7)	(8) Packaging (§173.***) Exceptions (8A)	Non-bulk (8B)	Bulk (8C)	(9) Quantity limitations (see §§173.27 and 175.75) Passenger aircraft/rail (9A)	Cargo aircraft only (9B)	(10) Vessel stowage Location (10A)	Other (10B)	ERG Guide No.	Placard Advisory Consult §172.504 & §172.505 to determine placard *Denotes placard for any quantity
A W	**Rags, oily**	4.2	UN1856	III	4.2		151	213	240	Forbidden	Forbidden	A		133	SPONTANEOUSLY COMBUSTIBLE
	Railway torpedo, see **Signals, railway track, explosive**														
	RC 318, see **Octafluorocyclobutane**														
	RDX and cyclotetramethylenetetranitramine, wetted *or* **desensitized** *see* **RDX and HMX mixtures, wetted** *or* **desensitized**														
	RDX and HMX mixtures, wetted *with not less than 15 percent water by mass or* **RDX and HMX mixtures, desensitized** *with not less than 10 percent phlegmatizer by mass*	1.1D	UN0391	II	1.1D		None	62	None	Forbidden	Forbidden	10		112	EXPLOSIVES 1.1*
	RDX and Octogen mixtures, wetted *or* **desensitized** *see* **RDX and HMX mixtures, wetted** *or* **desensitized** *etc.*														
	RDX, see **Cyclotrimethylenetrinitramine,** *etc.*														
	RDX, see **Cyclotrimethylenetrinitramine,** *etc.*														
	Receptacles, small, containing gas *or* **gas cartridges** *(flammable) without release device, not refillable and not exceeding 1 L capacity*	2.1	UN2037		2.1		306	304	None	1 kg	15 kg	B	40	115	FLAMMABLE GAS
	Receptacles, small, containing gas *or* **gas cartridges** *(non-flammable) without release device, not refillable and not exceeding 1 L capacity*	2.2	UN2037		2.2		306	304	None	1 kg	15 kg	B	40	115	NONFLAMMABLE GAS
	Receptacles, small, containing gas *or* **gas cartridges** *(oxidizing) without release device, not refillable and not exceeding 1 L capacity*	2.2	UN2037		2.2, 5.1	A14	306	304	None	1 kg	15 kg	B	40	115	NONFLAMMABLE GAS
	Red phosphorus, see **Phosphorus, amorphous**														
	Refrigerant gas R 404A	2.2	UN3337		2.2	T50	306	304	314, 315	75 kg	150 kg	A		126	NONFLAMMABLE GAS
	Refrigerant gas R 407A	2.2	UN3338		2.2	T50	306	304	314, 315	75 kg	150 kg	A		126	NONFLAMMABLE GAS
	Refrigerant gas R 407B	2.2	UN3339		2.2	T50	306	304	314, 315	75 kg	150 kg	A		126	NONFLAMMABLE GAS
	Refrigerant gas R 407C	2.2	UN3340		2.2	T50	306	304	314, 315	75 kg	150 kg	A		126	NONFLAMMABLE GAS
G	**Refrigerant gases, n.o.s.**	2.2	UN1078		2.2	T50	306	304	314, 315	75 kg	150 kg	A		126	NONFLAMMABLE GAS
D	**Refrigerant gases, n.o.s.** *or* **Dispersant gases, n.o.s.**	2.1	NA1954		2.1	T50	306	304	314, 315	Forbidden	150 kg	D	40	115	FLAMMABLE GAS
	Refrigerating machines, *containing flammable, non-toxic, liquefied gas*	2.1	UN3358		2.1		306, 307	306	306	Forbidden	Forbidden	D	40	115	FLAMMABLE GAS
	Refrigerating machines, *containing non-flammable, non-toxic gases, or ammonia solutions (UN2672)*	2.2	UN2857		2.2	A53	306, 307	306	306, 307	450 kg	450 kg	A		126	NONFLAMMABLE GAS
	Regulated medical waste, n.o.s. *or* **Clinical waste, unspecified, n.o.s.** *or* **(BIO) Medical waste, n.o.s.** *or* **Biomedical waste, n.o.s.** *or* **Medical waste, n.o.s.**	6.2	UN3291	II	6.2	41, A13	134	197	197	No limit	No limit	B	40	158	NONE
	Release devices, explosive	1.4S	UN0173	II	1.4S		None	62	62	25 kg	100 kg	05		114	EXPLOSIVES 1.4
	Resin solution, *flammable*	3	UN1866	I	3	B52, T11, TP1, TP8, TP28	150	201	243	1 L	30 L	E		127	FLAMMABLE
				II	3	149, B52, IB2, T4, TP1, TP8	150	173	242	5 L	60 L	B		127	FLAMMABLE
				III	3	B1, B52, IB3, T2, TP1	150	173	242	60 L	220 L	A		127	FLAMMABLE
	Resorcinol	6.1	UN2876	III	6.1	IB8, IP3, T1, TP33	153	213	240	100 kg	200 kg	A		153	POISON
	Rifle grenade, see **Grenades,** *hand or rifle, etc.*														
	Rifle powder, see **Powder, smokeless** *(UN 0160)*														
	Rivets, explosive	1.4S	UN0174	II	1.4S		None	62	62	25 kg	100 kg	05		114	EXPLOSIVES 1.4
	Road asphalt or tar liquid, see **Tars, liquid,***etc.*														
	Rocket motors	1.3C	UN0186	II	1.3C	109	None	62	62	Forbidden	220 kg	03		112	EXPLOSIVES 1.3*
	Rocket motors	1.1C	UN0280	II	1.1C	109	None	62	62	Forbidden	Forbidden	03		112	EXPLOSIVES 1.1*
	Rocket motors	1.2C	UN0281	II	1.2C	109	None	62	62	Forbidden	Forbidden	03		112	EXPLOSIVES 1.2*
	Rocket motors, liquid fueled	1.2J	UN0395	II	1.2J	109	None	62	None	Forbidden	Forbidden	04	23E	112	EXPLOSIVES 1.2*

Symbols (1)	Hazardous materials descriptions and proper shipping names (2)	Hazard class or Division (3)	Identification Numbers (4)	PG (5)	Label Codes (6)	Special Provisions (§172.102) (7)	(8) Packaging (§173.***) Exceptions (8A)	Non-bulk (8B)	Bulk (8C)	(9) Quantity limitations (see §§173.27 and 175.75) Passenger aircraft/rail (9A)	Cargo aircraft only (9B)	(10) Vessel stowage Location (10A)	Other (10B)	ERG Guide No.	Placard Advisory Consult §172.504 & §172.505 to determine placard *Denotes placard for any quantity
	Rocket motors, liquid fueled	1.3J	UN0396	II	1.3J	109	None	62	None	Forbidden	Forbidden	04	23E	112	EXPLOSIVES 1.3*
	Rocket motors with hypergolic liquids *with or without an expelling charge*	1.3L	UN0250	II	1.3L	109	None	62	None	Forbidden	Forbidden	08	8E, 14E, 15E	112	EXPLOSIVES 1.3*
	Rocket motors with hypergolic liquids *with or without an expelling charge*	1.2L	UN0322	II	1.2L	109	None	62	None	Forbidden	Forbidden	08	8E, 14E, 15E	112	EXPLOSIVES 1.2*
	Rockets, line-throwing	1.2G	UN0238	II	1.2G		None	62	None	Forbidden	Forbidden	07		112	EXPLOSIVES 1.2*
	Rockets, line-throwing	1.3G	UN0240	II	1.3G		None	62	None	Forbidden	75 kg	07		112	EXPLOSIVES 1.3*
	Rockets, line-throwing	1.4G	UN0453	II	1.4G		None	62	None	Forbidden	75 kg	06		114	EXPLOSIVES 1.4
	Rockets, liquid fueled *with bursting charge*	1.1J	UN0397	II	1.1J		None	62	None	Forbidden	Forbidden	04	23E	112	EXPLOSIVES 1.1*
	Rockets, liquid fueled *with bursting charge*	1.2J	UN0398	II	1.2J		None	62	None	Forbidden	Forbidden	04	23E	112	EXPLOSIVES 1.2*
	Rockets, *with bursting charge*	1.1F	UN0180	II	1.1F		None	62	None	Forbidden	Forbidden	08		112	EXPLOSIVES 1.1*
	Rockets,*with bursting charge*	1.1E	UN0181	II	1.1E		None	62	62	Forbidden	Forbidden	03		112	EXPLOSIVES 1.1*
	Rockets, *with bursting charge*	1.2E	UN0182	II	1.2E		None	62	62	Forbidden	Forbidden	03		112	EXPLOSIVES 1.2*
	Rockets,*with bursting charge*	1.2F	UN0295	II	1.2F		None	62	None	Forbidden	Forbidden	08		112	EXPLOSIVES 1.2*
	Rockets,*with expelling charge*	1.2C	UN0436	II	1.2C		None	62	62	Forbidden	Forbidden	03		112	EXPLOSIVES 1.2*
	Rockets, *with expelling charge*	1.3C	UN0437	II	1.3C		None	62	62	Forbidden	Forbidden	03		112	EXPLOSIVES 1.3*
	Rockets, *with expelling charge*	1.4C	UN0438	II	1.4C		None	62	62	Forbidden	75 kg	02		114	EXPLOSIVES 1.4
	Rockets, *with inert head*	1.3C	UN0183	II	1.3C		None	62	62	Forbidden	Forbidden	03		112	EXPLOSIVES 1.3*
	Rockets, *with inert head*	1.2C	UN0502		1.2C		None	62	62	Forbidden	Forbidden	B	1E, 5E	112	EXPLOSIVES 1.2*
	Rosin oil	3	UN1286	II	3	IB2, T4, TP1	150	202	242	5 L	60 L	B		127	FLAMMABLE
				III	3	B1, IB3, T2, TP1	150	203	242	60 L	220 L	A		127	FLAMMABLE
	Rubber scrap *or* **shoddy,** *powdered or granulated, not exceeding 840 microns and rubber content exceeding 45%*	4.1	UN1345	II	4.1	IB8, IP2, IP4, T3, TP33	151	212	240	15 kg	50 kg	A		133	FLAMMABLE SOLID
	Rubber solution	3	UN1287	II	3	149, IB2, T4, TP1, TP8	150	202	242	5 L	60 L	B		127	FLAMMABLE
				III	3	B1, IB3, T2, TP1	150	203	242	60 L	220 L	A		127	FLAMMABLE
	Rubidium	4.3	UN1423	I	4.3	22, A7, A19, IB4, IP1, N34, N40, N45	None	211	242	Forbidden	15 kg	D	52	138	DANGEROUS WHEN WET*
	Rubidium hydroxide	8	UN2678	II	8	IB8, IP2, IP4, T3, TP33	154	212	240	15 kg	50 kg	A	29, 52	154	CORROSIVE
	Rubidium hydroxide solution	8	UN2677	II	8	B2, IB2, T7, TP2	154	202	242	1 L	30 L	A	29, 52	154	CORROSIVE
				III	8	IB3, T4, TP1	154	203	241	5 L	60 L	A	29, 52	154	CORROSIVE
	Safety fuse, see **Fuse, safety**														
G	**Samples, explosive,** *other than initiating explosives*		UN0190	II		113	None	62	None	Forbidden	Forbidden	14		—	—
	Sand acid, see **Fluorosilicic acid**														
	Seed cake, *containing vegetable oil solvent extractions and expelled seeds, with not more than 10 percent of oil and when the amount of moisture is higher than 11 percent, with not more than 20 percent of oil and moisture combined.*	4.2	UN1386	III	None	IB8, IP3, IP6, N7	None	213	241	Forbidden	Forbidden	A	13	135	SPONTANEOUSLY COMBUSTIBLE
I	**Seed cake** *with more than 1.5 percent oil and not more than 11 percent moisture*	4.2	UN1386	III	None	IB8, IP3, IP6, N7	None	213	241	Forbidden	Forbidden	E	13	135	SPONTANEOUSLY COMBUSTIBLE
I	**Seed cake** *with not more than 1.5 percent oil and not more than 11 percent moisture*	4.2	UN2217	III	None	IB8, IP3, IP6, N7	None	213	241	Forbidden	Forbidden	A	13	135	SPONTANEOUSLY COMBUSTIBLE
G	**Selenates** *or* **Selenites**	6.1	UN2630	I	6.1	IB7, IP1, T6, TP33	None	211	242	5 kg	50 kg	E		151	POISON
	Selenic acid	8	UN1905	I	8	IB7, IP1, N34, T6, TP33	None	211	242	Forbidden	25 kg	A		154	CORROSIVE
G	**Selenium compound, liquid, n.o.s.**	6.1	UN3440	I	6.1	T14, TP2, TP27	None	201	243	1 L	30 L	B		151	POISON
				II	6.1	IB2, T11, TP2, TP27	153	202	243	5 L	60 L	B		151	POISON
				III	6.1	IB3, T7, TP1, TP28	153	203	241	60 L	220 L	A		151	POISON
G	**Selenium compound, solid, n.o.s.**	6.1	UN3283	I	6.1	IB7, IP1, T6, TP33	None	211	242	5 kg	50 kg	B		151	POISON
				II	6.1	IB8, IP2, IP4, T3, TP33	153	212	242	25 kg	100 kg	B		151	POISON
				III	6.1	IB8, IP3, T1, TP33	153	213	240	100 kg	200 kg	A		151	POISON
	Selenium disulfide	6.1	UN2657	II	6.1	IB8, IP2, IP4, T3, TP33	153	212	242	25 kg	100 kg	A		153	POISON

Sym-bols (1)	Hazardous materials descriptions and proper shipping names (2)	Hazard class or Division (3)	Identifi-cation Numbers (4)	PG (5)	Label Codes (6)	Special Provisions (§172.102) (7)	(8) Packaging (§173.***) Excep-tions (8A)	Non-bulk (8B)	Bulk (8C)	(9) Quantity limitations (see §§173.27 and 175.75) Passenger aircraft/rail (9A)	Cargo aircraft only (9B)	(10) Vessel stowage Location (10A)	Other (10B)	ERG Guide No.	Placard Advisory Consult §172.504 & §172.505 to determine placard *Denotes placard for any quantity
	Selenium hexafluoride	2.3	UN2194		2.3, 8	1	None	302	None	Forbidden	Forbidden	D	40	125	POISON GAS*
	Selenium nitride	Forbidden													
	Selenium oxychloride	8	UN2879	I	8, 6.1	A3, A6, A7, N34, T10, TP2, TP13	None	201	243	0.5 L	2.5 L	E	40	157	CORROSIVE
	Self-defense spray, aerosol, see **Aerosols,** *etc.*														
+ A D	**Self-defense spray, non-pressurized**	9	NA3334	III	9	A37	155	203	None	No limit	No limit	A		171	CLASS 9
G	**Self-heating liquid, corrosive, inorganic, n.o.s.**	4.2	UN3188	II	4.2, 8	IB2	None	202	243	1 L	5 L	C		136	SPONTANEOUSLY COMBUSTIBLE
				III	4.2, 8	IB2	None	203	241	5 L	60 L	C		136	SPONTANEOUSLY COMBUSTIBLE
G	**Self-heating liquid, corrosive, organic, n.o.s.**	4.2	UN3185	II	4.2, 8	IB2	None	202	243	1 L	5 L	C		136	SPONTANEOUSLY COMBUSTIBLE
				III	4.2, 8	IB2	None	203	241	5 L	60 L	C		136	SPONTANEOUSLY COMBUSTIBLE
G	**Self-heating liquid, inorganic, n.o.s.**	4.2	UN3186	II	4.2	IB2	None	202	242	1 L	5 L	C		135	SPONTANEOUSLY COMBUSTIBLE
				III	4.2	IB2	None	203	241	5 L	60 L	C		135	SPONTANEOUSLY COMBUSTIBLE
G	**Self-heating liquid, organic, n.o.s.**	4.2	UN3183	II	4.2	IB2	None	202	242	1 L	5 L	C		135	SPONTANEOUSLY COMBUSTIBLE
				III	4.2	IB2	None	203	241	5 L	60 L	C		135	SPONTANEOUSLY COMBUSTIBLE
G	**Self-heating liquid, toxic, inorganic, n.o.s.**	4.2	UN3187	II	4.2, 6.1	IB2	None	202	243	1 L	5 L	C		136	SPONTANEOUSLY COMBUSTIBLE
				III	4.2, 6.1	IB2	None	203	241	5 L	60 L	C		136	SPONTANEOUSLY COMBUSTIBLE
G	**Self-heating liquid, toxic, organic, n.o.s.**	4.2	UN3184	II	4.2, 6.1	IB2	None	202	243	1 L	5 L	C		136	SPONTANEOUSLY COMBUSTIBLE
				III	4.2, 6.1	IB2	None	203	241	5 L	60 L	C		136	SPONTANEOUSLY COMBUSTIBLE
G	**Self-heating solid, corrosive, inorganic, n.o.s.**	4.2	UN3192	II	4.2, 8	IB5, IP2, T3, TP33	None	212	242	15 kg	50 kg	C		136	SPONTANEOUSLY COMBUSTIBLE
				III	4.2, 8	IB8, IP3, T1, TP33	None	213	242	25 kg	100 kg	C		136	SPONTANEOUSLY COMBUSTIBLE
G	**Self-heating solid, corrosive, organic, n.o.s.**	4.2	UN3126	II	4.2, 8	IB5, IP2, T3, TP33	None	212	242	15 kg	50 kg	C		136	SPONTANEOUSLY COMBUSTIBLE
				III	4.2, 8	IB8, IP3, T1, TP33	None	213	242	25 kg	100 kg	C		136	SPONTANEOUSLY COMBUSTIBLE
G	**Self-heating solid, inorganic, n.o.s.**	4.2	UN3190	II	4.2	IB6, IP2, T3, TP33	None	212	241	15 kg	50 kg	C		135	SPONTANEOUSLY COMBUSTIBLE
				III	4.2	IB8, IP3, T1, TP33	None	213	241	25 kg	100 kg	C		135	SPONTANEOUSLY COMBUSTIBLE
G	**Self-heating solid, organic, n.o.s.**	4.2	UN3088	II	4.2	IB6, IP2, T3, TP33	None	212	241	15 kg	50 kg	C		135	SPONTANEOUSLY COMBUSTIBLE
				III	4.2	IB8, IP3, T1, TP33	None	213	241	25 kg	100 kg	C		135	SPONTANEOUSLY COMBUSTIBLE
G	**Self-heating solid, oxidizing, n.o.s.**	4.2	UN3127		4.2, 5.1		None	214	214	Forbidden	Forbidden			135	SPONTANEOUSLY COMBUSTIBLE
G	**Self-heating solid, toxic, inorganic, n.o.s.**	4.2	UN3191	II	4.2, 6.1	IB5, IP2, T3, TP33	None	212	242	15 kg	50 kg	C		136	SPONTANEOUSLY COMBUSTIBLE
				III	4.2, 6.1	IB8, IP3, T1, TP33	None	213	242	25 kg	100 kg	C		136	SPONTANEOUSLY COMBUSTIBLE
G	**Self-heating solid, toxic, organic, n.o.s.**	4.2	UN3128	II	4.2, 6.1	IB5, IP2, T3, TP33	None	212	242	15 kg	50 kg	C		136	SPONTANEOUSLY COMBUSTIBLE
				III	4.2, 6.1	IB8, IP3, T1, TP33	None	213	242	25 kg	100 kg	C		136	SPONTANEOUSLY COMBUSTIBLE
	Self-propelled vehicle, see **Engines** *or* **Batteries** *etc.*														
G	**Self-reactive liquid type B**	4.1	UN3221	II	4.1	53	None	224	None	Forbidden	Forbidden	D	52, 53	149	FLAMMABLE SOLID
G	**Self-reactive liquid type B, temperature controlled**	4.1	UN3231	II	4.1	53	None	224	None	Forbidden	Forbidden	D	2, 52, 53	150	FLAMMABLE SOLID

Symbols (1)	Hazardous materials descriptions and proper shipping names (2)	Hazard class or Division (3)	Identification Numbers (4)	PG (5)	Label Codes (6)	Special Provisions (§172.102) (7)	(8) Packaging (§173.***) Exceptions (8A)	Non-bulk (8B)	Bulk (8C)	(9) Quantity limitations (see §§173.27 and 175.75) Passenger aircraft/rail (9A)	Cargo aircraft only (9B)	(10) Vessel stowage Location (10A)	Other (10B)	ERG Guide No.	Placard Advisory Consult §172.504 & §172.505 to determine placard *Denotes placard for any quantity
G	**Self-reactive liquid type C**	4.1	UN3223	II	4.1		None	224	None	5 L	10 L	D	52, 53	149	FLAMMABLE SOLID
G	**Self-reactive liquid type C, temperature controlled**	4.1	UN3233	II	4.1		None	224	None	Forbidden	Forbidden	D	2, 52, 53	150	FLAMMABLE SOLID
G	**Self-reactive liquid type D**	4.1	UN3225	II	4.1		None	224	None	5 L	10 L	D	52, 53	149	FLAMMABLE SOLID
G	**Self-reactive liquid type D, temperature controlled**	4.1	UN3235	II	4.1		None	224	None	Forbidden	Forbidden	D	2, 52, 53	150	FLAMMABLE SOLID
G	**Self-reactive liquid type E**	4.1	UN3227	II	4.1		None	224	None	10 L	25 L	D	52, 53	149	FLAMMABLE SOLID
G	**Self-reactive liquid type E, temperature controlled**	4.1	UN3237	II	4.1		None	224	None	Forbidden	Forbidden	D	2, 52, 53	150	FLAMMABLE SOLID
G	**Self-reactive liquid type F**	4.1	UN3229	II	4.1		None	224	None	10 L	25 L	D	52, 53	149	FLAMMABLE SOLID
G	**Self-reactive liquid type F, temperature controlled**	4.1	UN3239	II	4.1		None	224	None	Forbidden	Forbidden	D	2, 52, 53	150	FLAMMABLE SOLID
G	**Self-reactive solid type B**	4.1	UN3222	II	4.1	53	None	224	None	Forbidden	Forbidden	D	52, 53	149	FLAMMABLE SOLID
G	**Self-reactive solid type B, temperature controlled**	4.1	UN3232	II	4.1	53	None	224	None	Forbidden	Forbidden	D	2, 52, 53	150	FLAMMABLE SOLID
G	**Self-reactive solid type C**	4.1	UN3224	II	4.1		None	224	None	5 kg	10 kg	D	52, 53	149	FLAMMABLE SOLID
G	**Self-reactive solid type C, temperature controlled**	4.1	UN3234	II	4.1		None	224	None	Forbidden	Forbidden	D	2, 52, 53	150	FLAMMABLE SOLID
G	**Self-reactive solid type D**	4.1	UN3226	II	4.1		None	224	None	5 kg	10 kg	D	52, 53	149	FLAMMABLE SOLID
G	**Self-reactive solid type D, temperature controlled**	4.1	UN3236	II	4.1		None	224	None	Forbidden	Forbidden	D	2, 52, 53	150	FLAMMABLE SOLID
G	**Self-reactive solid type E**	4.1	UN3228	II	4.1		None	224	None	10 kg	25 kg	D	52, 53	149	FLAMMABLE SOLID
G	**Self-reactive solid type E, temperature controlled**	4.1	UN3238	II	4.1		None	224	None	Forbidden	Forbidden	D	2, 52, 53	150	FLAMMABLE SOLID
G	**Self-reactive solid type F**	4.1	UN3230	II	4.1		None	224	None	10 kg	25 kg	D	52, 53	149	FLAMMABLE SOLID
G	**Self-reactive solid type F, temperature controlled**	4.1	UN3240	II	4.1		None	224	None	Forbidden	Forbidden	D	2, 52, 53	150	FLAMMABLE SOLID
	Shale oil	3	UN1288	I	3	T11, TP1, TP8, TP27	None	201	243	1 L	30 L	B		128	FLAMMABLE
				II	3	IB2, T4, TP1, TP8	150	202	242	5 L	60 L	B		128	FLAMMABLE
				III	3	B1, IB3, T2, TP1	150	203	242	60 L	220 L	A		128	FLAMMABLE
	Shaped charges, see **Charges, shaped,** *etc.*														
	Signal devices, hand	1.4G	UN0191	II	1.4G		None	62	None	Forbidden	75 kg	06		114	EXPLOSIVES 1.4
	Signal devices, hand	1.4S	UN0373	II	1.4S		None	62	None	25 kg	100 kg	05		114	EXPLOSIVES 1.4
	Signals, distress, *ship*	1.1G	UN0194	II	1.1G		None	62	None	Forbidden	Forbidden	07		112	EXPLOSIVES 1.1*
	Signals, distress, *ship*	1.3G	UN0195	II	1.3G		None	62	None	Forbidden	75 kg	07		112	EXPLOSIVES 1.3*
	Signals, distress, *ship*	1.4G	UN0505		1.4G		None	62	None	Forbidden	75 kg	06		114	EXPLOSIVES 1.4
	Signals, distress, *ship*	1.4S	UN0506		1.4S		None	62	None	25 kg	100 kg	05		114	EXPLOSIVES 1.4
	Signals, highway, see **Signal devices, hand**														
	Signals, railway track, explosive	1.1G	UN0192	II	1.1G		None	62	None	Forbidden	Forbidden	07		112	EXPLOSIVES 1.1*
	Signals, railway track, explosive	1.4S	UN0193	II	1.4S		None	62	None	25 kg	100 kg	05		114	EXPLOSIVES 1.4
	Signals, railway track, explosive	1.3G	UN0492		1.3G		None	62	None	Forbidden	Forbidden	07		112	EXPLOSIVES 1.3*
	Signals, railway track, explosive	1.4G	UN0493		1.4G		None	62	None	Forbidden	75 kg	06		114	EXPLOSIVES 1.4
	Signals, ship distress, water-activated, see **Contrivances, water-activated,** *etc.*														
	Signals, smoke	1.1G	UN0196	II	1.1G		None	62	None	Forbidden	Forbidden	07		112	EXPLOSIVES 1.1*
	Signals, smoke	1.4G	UN0197	II	1.4G		None	62	None	Forbidden	75 kg	06		114	EXPLOSIVES 1.4
	Signals, smoke	1.2G	UN0313	II	1.2G		None	62	None	Forbidden	Forbidden	07		112	EXPLOSIVES 1.2*
	Signals, smoke	1.4S	UN0507		1.4S		None	62	None	25 kg	100 kg	05		114	EXPLOSIVES 1.4
	Signals, smoke	1.3G	UN0487	II	1.3G		None	62	None	Forbidden	Forbidden	07		112	EXPLOSIVES 1.3*
	Silane	2.1	UN2203		2.1		None	302	None	Forbidden	Forbidden	E	40, 57, 104	116	FLAMMABLE GAS
	Silicofluoric acid, see **Fluorosilicic acid**														
	Silicon chloride, see **Silicon tetrachloride**														
	Silicon powder, amorphous	4.1	UN1346	III	4.1	A1, IB8, IP3, T1, TP33	None	213	240	25 kg	100 kg	A	74	170	FLAMMABLE SOLID
	Silicon tetrachloride	8	UN1818	II	8	A3, A6, B2, B6, T10, TP2, TP7, TP13	None	202	242	Forbidden	30 L	C	40	157	CORROSIVE
	Silicon tetrafluoride	2.3	UN1859		2.3, 8	2	None	302	None	Forbidden	Forbidden	D	40	125	POISON GAS*
	Silver acetylide (dry)	Forbidden													
	Silver arsenite	6.1	UN1683	II	6.1	IB8, IP2, IP4, T3, TP33	153	212	242	25 kg	100 kg	A		151	POISON
	Silver azide (dry)	Forbidden													

Symbols (1)	Hazardous materials descriptions and proper shipping names (2)	Hazard class or Division (3)	Identification Numbers (4)	PG (5)	Label Codes (6)	Special Provisions (§172.102) (7)	(8) Packaging (§173.***) Exceptions (8A)	Non-bulk (8B)	Bulk (8C)	(9) Quantity limitations (see §§173.27 and 175.75) Passenger aircraft/rail (9A)	Cargo aircraft only (9B)	(10) Vessel stowage Location (10A)	Other (10B)	ERG Guide No.	Placard Advisory Consult §172.504 & §172.505 to determine placard *Denotes placard for any quantity
	Silver chlorite (dry)	Forbidden													
	Silver cyanide	6.1	UN1684	II	6.1	IB8, IP2, IP4, T3, TP33	153	212	242	25 kg	100 kg	A	40, 52	151	POISON
	Silver fulminate (dry)	Forbidden													
	Silver nitrate	5.1	UN1493	II	5.1	IB8, IP2, IP4, T3, TP33	152	212	242	5 kg	25 kg	A		140	OXIDIZER
	Silver oxalate (dry)	Forbidden													
	Silver picrate (dry)	Forbidden													
	Silver picrate, wetted *with not less than 30 percent water, by mass*	4.1	UN1347	I	4.1	23	None	211	None	Forbidden	Forbidden	D	28, 36	113	FLAMMABLE SOLID
	Sludge, acid	8	UN1906	II	8	A3, A7, B2, IB2, N34, T8, TP2, TP28	None	202	242	Forbidden	30 L	C	14	153	CORROSIVE
D	**Smokeless powder for small arms** *(100 pounds or less)*	4.1	NA3178	I	4.1	16	None	171	None	Forbidden	7.3 kg	A		133	FLAMMABLE SOLID
	Soda lime *with more than 4 percent sodium hydroxide*	8	UN1907	III	8	IB8, IP3, T1, TP33	154	213	240	25 kg	100 kg	A	52	154	CORROSIVE
	Sodium	4.3	UN1428	I	4.3	A7, A8, A19, A20, B9, B48, B68, IB4, IP1, N34, T9, TP7, TP33, TP46	None	211	244	Forbidden	15 kg	D	52	138	DANGEROUS WHEN WET*
A	**Sodium aluminate, solid**	8	UN2812	III	8	IB8, IP3, T1, TP33	154	213	240	25 kg	100 kg	A		154	CORROSIVE
	Sodium aluminate, solution	8	UN1819	II	8	B2, IB2, T7, TP2	154	202	242	1 L	30 L	A	52	154	CORROSIVE
				III	8	IB3, T4, TP1	154	203	241	5 L	60 L	A	52	154	CORROSIVE
	Sodium aluminum hydride	4.3	UN2835	II	4.3	A8, A19, A20, IB4, T3, TP33	151	212	242	Forbidden	50 kg	E	52	138	DANGEROUS WHEN WET*
	Sodium ammonium vanadate	6.1	UN2863	II	6.1	IB8, IP2, IP4, T3, TP33	153	212	242	25 kg	100 kg	A		154	POISON
	Sodium arsanilate	6.1	UN2473	III	6.1	IB8, IP3, T1, TP33	153	213	240	100 kg	200 kg	A		154	POISON
	Sodium arsenate	6.1	UN1685	II	6.1	IB8, IP2, IP4, T3, TP33	153	212	242	25 kg	100 kg	A		151	POISON
	Sodium arsenite, aqueous solutions	6.1	UN1686	II	6.1	IB2, T7, TP2	153	202	243	5 L	60 L	A		154	POISON
				III	6.1	IB3, T4, TP2	153	203	241	60 L	220 L	A		154	POISON
	Sodium arsenite, solid	6.1	UN2027	II	6.1	IB8, IP2, IP4, T3, TP33	153	212	242	25 kg	100 kg	A		151	POISON
	Sodium azide	6.1	UN1687	II	6.1	IB8, IP2, IP4	153	212	242	25 kg	100 kg	A	36, 52, 91	153	POISON
	Sodium bifluoride, see **Sodium hydrogendifluoride**														
	Sodium bisulfite, solution, see **Bisulfites, aqueous solutions, n.o.s.**														
	Sodium borohydride	4.3	UN1426	I	4.3	N40	None	211	242	Forbidden	15 kg	E	52	138	DANGEROUS WHEN WET*
	Sodium borohydride and sodium hydroxide solution, *with not more than 12 percent sodium borohydride and not more than 40 percent sodium hydroxide by mass*	8	UN3320	II	8	B2, IB2, N34, T7, TP2	154	202	242	1 L	30 L	A	52	157	CORROSIVE
				III	8	B2, IB3, N34, T4, TP2	154	203	241	5 L	60 L	A	52	157	CORROSIVE
	Sodium bromate	5.1	UN1494	II	5.1	IB8, IP2, IP4, T3, TP33	152	212	242	5 kg	25 kg	A	56, 58	141	OXIDIZER
	Sodium cacodylate	6.1	UN1688	II	6.1	IB8, IP2, IP4, T3, TP33	153	212	242	25 kg	100 kg	A	52	152	POISON
	Sodium carbonate peroxyhydrate	5.1	UN3378	II	5.1	IB8, IP2, IP4, T3, TP33	152	212	240	5 kg	25 kg	A	13, 48, 75	140	OXIDIZER
				III	5.1	IB8, IP3, T1, TP33	152	213	240	25 kg	100 kg	A	13, 48, 75	140	OXIDIZER
	Sodium chlorate	5.1	UN1495	II	5.1	A9, IB8, IP2, IP4, N34, T3, TP33	152	212	240	5 kg	25 kg	A	56, 58	140	OXIDIZER
	Sodium chlorate, aqueous solution	5.1	UN2428	II	5.1	A2, IB2, T4, TP1	152	202	241	1 L	5 L	B	56, 58, 133	140	OXIDIZER
				III	5.1	A2, IB2, T4, TP1	152	203	241	2.5 L	30 L	B	56, 58, 69, 133	140	OXIDIZER
	Sodium chlorate mixed with dinitrotoluene, see **Explosive blasting, type C**														

Symbols (1)	Hazardous materials descriptions and proper shipping names (2)	Hazard class or Division (3)	Identification Numbers (4)	PG (5)	Label Codes (6)	Special Provisions (§172.102) (7)	(8) Packaging (§173.***) Exceptions (8A)	Non-bulk (8B)	Bulk (8C)	(9) Quantity limitations (see §§173.27 and 175.75) Passenger aircraft/rail (9A)	Cargo aircraft only (9B)	(10) Vessel stowage Location (10A)	Other (10B)	ERG Guide No.	Placard Advisory Consult §172.504 & §172.505 to determine placard *Denotes placard for any quantity
	Sodium chlorite	5.1	UN1496	II	5.1	A9, IB8, IP2, IP4, N34, T3, TP33	None	212	242	5 kg	25 kg	A	56, 58	143	OXIDIZER
	Sodium chloroacetate	6.1	UN2659	III	6.1	IB8, IP3, T1, TP33	153	213	240	100 kg	200 kg	A		151	POISON
	Sodium cuprocyanide, solid	6.1	UN2316	I	6.1	IB7, IP1, T6, TP33	None	211	242	5 kg	50 kg	A	52	157	POISON
	Sodium cuprocyanide, solution	6.1	UN2317	I	6.1	T14, TP2, TP13	None	201	243	1 L	30 L	B	40, 52	157	POISON
	Sodium cyanide, solid	6.1	UN1689	I	6.1	B69, B77, IB7, N74, N75, T6, TP33	None	211	242	5 kg	50 kg	B	52	157	POISON
	Sodium cyanide solution	6.1	UN3414	I	6.1	B69, B77, N74, N75, T14, TP2, TP13	None	201	243	1 L	30 L	B	52	157	POISON
				II	6.1	B69, B77, IB2, N74, N75, T11, TP2, TP13, TP27	153	202	243	5 L	60 L	B	52	157	POISON
				III	6.1	B69, B77, IB3, N74, N75, T7, TP2, TP13, TP28	153	203	241	60 L	220 L	A	52	157	POISON
	Sodium dichloroisocyanurate or Sodium dichloro-s-triazinetrione, see **Dichloroisocyanuric acid** *etc.*														
	Sodium dinitro-o-cresolate, *dry or wetted with less than 15 percent water, by mass*	1.3C	UN0234	II	1.3C		None	62	None	Forbidden	Forbidden	10	5E	112	EXPLOSIVES 1.3*
	Sodium dinitro-o-cresolate, wetted, *with not less than 10% water, by mass*	4.1	UN3369	I	4.1	162, A8, A19, N41, N84	None	211	None	0.5 kg	0.5 kg	E	36	113	FLAMMABLE SOLID
	Sodium dinitro-o-cresolate, wetted *with not less than 15 percent water, by mass*	4.1	UN1348	I	4.1, 6.1	23, A8, A19, A20, N41	None	211	None	1 kg	15 kg	E	28, 36	113	FLAMMABLE SOLID
	Sodium dithionite *or* **Sodium hydrosulfite**	4.2	UN1384	II	4.2	A19, A20, IB6, IP2, T3, TP33	None	212	241	15 kg	50 kg	E	13	135	SPONTANEOUSLY COMBUSTIBLE
	Sodium fluoride, solid	6.1	UN1690	III	6.1	IB8, IP3, T1, TP33	153	213	240	100 kg	200 kg	A	52	154	POISON
	Sodium fluoride solution	6.1	UN3415	III	6.1	IB3, T4, TP1	153	203	241	60 L	220 L	A	52	154	POISON
	Sodium fluoroacetate	6.1	UN2629	I	6.1	IB7, IP1, T6, TP33	None	211	242	5 kg	50 kg	E		151	POISON
	Sodium fluorosilicate	6.1	UN2674	III	6.1	IB8, IP3, T1, TP33	153	213	240	100 kg	200 kg	A	52	154	POISON
	Sodium hydrate, see **Sodium hydroxide, solid**														
	Sodium hydride	4.3	UN1427	I	4.3	A19, N40	None	211	242	Forbidden	15 kg	E	52	138	DANGEROUS WHEN WET*
	Sodium hydrogendifluoride	8	UN2439	II	8	IB8, IP2, IP4, N3, N34, T3, TP33	154	212	240	15 kg	50 kg	A	12, 25, 40, 52	154	CORROSIVE
	Sodium hydrosulfide, *with less than 25 percent water of crystallization*	4.2	UN2318	II	4.2	A7, A19, A20, IB6, IP2, T3, TP33	None	212	241	15 kg	50 kg	A	52	135	SPONTANEOUSLY COMBUSTIBLE
	Sodium hydrosulfide *with not less than 25 percent water of crystallization*	8	UN2949	II	8	A7, IB8, IP2, IP4, T7, TP2	154	212	240	15 kg	50 kg	A	52	154	CORROSIVE
	Sodium hydrosulfite, *see* **Sodium dithionite**														
	Sodium hydroxide, solid	8	UN1823	II	8	IB8, IP2, IP4, T3, TP33	154	212	240	15 kg	50 kg	A	52	154	CORROSIVE
	Sodium hydroxide solution	8	UN1824	II	8	B2, IB2, N34, T7, TP2	154	202	242	1 L	30 L	A	52	154	CORROSIVE
				III	8	IB3, N34, T4, TP1	154	203	241	5 L	60 L	A	52	154	CORROSIVE
	Sodium hypochlorite, solution, see **Hypochlorite solutions** *etc.*														
	Sodium metal, liquid alloy, see **Alkali metal alloys, liquid, n.o.s.**														
	Sodium methylate	4.2	UN1431	II	4.2, 8	A7, A19, IB5, IP2, T3, TP33	None	212	242	15 kg	50 kg	B		138	SPONTANEOUSLY COMBUSTIBLE
	Sodium methylate solutions *in alcohol*	3	UN1289	II	3, 8	IB2, T7, TP1, TP8	150	202	243	1 L	5 L	B		132	FLAMMABLE
				III	3, 8	B1, IB3, T4, TP1	150	203	242	5 L	60 L	A		132	FLAMMABLE
	Sodium monoxide	8	UN1825	II	8	IB8, IP2, IP4, T3, TP33	154	212	240	15 kg	50 kg	A	52	157	CORROSIVE
	Sodium nitrate	5.1	UN1498	III	5.1	A1, A29, IB8, IP3, T1, TP33, W1	152	213	240	25 kg	100 kg	A		140	OXIDIZER
	Sodium nitrate and potassium nitrate mixtures	5.1	UN1499	III	5.1	A1, A29, IB8, IP3, T1, TP33, W1	152	213	240	25 kg	100 kg	A		140	OXIDIZER

Symbols (1)	Hazardous materials descriptions and proper shipping names (2)	Hazard class or Division (3)	Identification Numbers (4)	PG (5)	Label Codes (6)	Special Provisions (§172.102) (7)	(8) Packaging (§173.***) Exceptions (8A)	Non-bulk (8B)	Bulk (8C)	(9) Quantity limitations (see §§173.27 and 175.75) Passenger aircraft/rail (9A)	Cargo aircraft only (9B)	(10) Vessel stowage Location (10A)	Other (10B)	ERG Guide No.	Placard Advisory Consult §172.504 & §172.505 to determine placard *Denotes placard for any quantity
	Sodium nitrite	5.1	UN1500	III	5.1, 6.1	A1, A29, IB8, IP3, T1, TP33	152	213	240	25 kg	100 kg	A	56, 58	140	OXIDIZER
	Sodium pentachlorophenate	6.1	UN2567	II	6.1	IB8, IP2, IP4, T3, TP33	153	212	242	25 kg	100 kg	A		154	POISON
	Sodium perborate monohydrate	5.1	UN3377	III	5.1	IB8, IP3, T1, TP33	152	213	240	25 kg	100 kg	A	13, 48, 75	140	OXIDIZER
	Sodium perchlorate	5.1	UN1502	II	5.1	IB6, IP2, T3, TP33	152	212	242	5 kg	25 kg	A	56, 58	140	OXIDIZER
	Sodium permanganate	5.1	UN1503	II	5.1	IB6, IP2, T3, TP33	152	212	242	5 kg	25 kg	D	56, 58, 138	140	OXIDIZER
	Sodium peroxide	5.1	UN1504	I	5.1	A20, IB5, IP1, N34	None	211	None	Forbidden	15 kg	B	13, 52, 66, 75	144	OXIDIZER
	Sodium peroxoborate, anhydrous	5.1	UN3247	II	5.1	IB8, IP2, IP4, T3, TP33	152	212	240	5 kg	25 kg	A	13, 25	140	OXIDIZER
	Sodium persulfate	5.1	UN1505	III	5.1	A1, IB8, IP3, T1, TP33	152	213	240	25 kg	100 kg	A	58, 145	140	OXIDIZER
	Sodium phosphide	4.3	UN1432	I	4.3, 6.1	A19, N40	None	211	None	Forbidden	15 kg	E	40, 52, 85	139	DANGEROUS WHEN WET*
	Sodium picramate, *dry or wetted with less than 20 percent water, by mass*	1.3C	UN0235	II	1.3C		None	62	None	Forbidden	Forbidden	10	5E	112	EXPLOSIVES 1.3*
	Sodium picramate, wetted *with not less than 20 percent water, by mass*	4.1	UN1349	I	4.1	23, A8, A19, N41	None	211	None	Forbidden	15 kg	E	28, 36	113	FLAMMABLE SOLID
	Sodium picryl peroxide	Forbidden													
	Sodium potassium alloys, see **Potassium sodium alloys**														
	Sodium selenate, see **Selenates** *or* **Selenites**														
	Sodium sulfide, anhydrous *or* **Sodium sulfide** *with less than 30 percent water of crystallization*	4.2	UN1385	II	4.2	A19, A20, IB6, IP2, N34, T3, TP33	None	212	241	15 kg	50 kg	A	52	135	SPONTANEOUSLY COMBUSTIBLE
	Sodium sulfide, hydrated *with not less than 30 percent water*	8	UN1849	II	8	IB8, IP2, IP4, T3, TP33	154	212	240	15 kg	50 kg	A	52	153	CORROSIVE
	Sodium superoxide	5.1	UN2547	I	5.1	A20, IB6, IP1, N34	None	211	None	Forbidden	15 kg	E	13, 52, 66, 75	143	OXIDIZER
	Sodium tetranitride	Forbidden													
G	**Solids containing corrosive liquid, n.o.s.**	8	UN3244	II	8	49, IB5, T3, TP33	154	212	240	15 kg	50 kg	B	40	154	CORROSIVE
G	**Solids containing flammable liquid, n.o.s.**	4.1	UN3175	II	4.1	47, IB6, IP2, T3, TP33	151	212	240	15 kg	50 kg	B		133	FLAMMABLE SOLID
G	**Solids containing toxic liquid, n.o.s.**	6.1	UN3243	II	6.1	48, IB2, T2, TP33	153	212	240	25 kg	100 kg	B	40	151	POISON
	Sounding devices, explosive	1.2F	UN0204	II	1.2F		None	62	62	Forbidden	Forbidden	08		112	EXPLOSIVES 1.2*
	Sounding devices, explosive	1.1F	UN0296	II	1.1F		None	62	62	Forbidden	Forbidden	08		112	EXPLOSIVES 1.1*
	Sounding devices, explosive	1.1D	UN0374	II	1.1D		None	62	62	Forbidden	Forbidden	07		112	EXPLOSIVES 1.1*
	Sounding devices, explosive	1.2D	UN0375	II	1.2D		None	62	62	Forbidden	Forbidden	07		112	EXPLOSIVES 1.2*
	Spirits of salt, see **Hydrochloric acid**														
	Squibs, see **Igniters** *etc.*														
	Stannic chloride, anhydrous	8	UN1827	II	8	B2, IB2, T7, TP2	154	202	242	1 L	30 L	C		137	CORROSIVE
	Stannic chloride pentahydrate	8	UN2440	III	8	IB8, IP3, T1, TP33	154	213	240	25 kg	100 kg	A		154	CORROSIVE
	Stannic phosphide	4.3	UN1433	I	4.3, 6.1	A19, N40	None	211	242	Forbidden	15 kg	E	40, 52, 85	139	DANGEROUS WHEN WET*
	Steel swarf, see **Ferrous metal borings,** *etc.*														
	Stibine	2.3	UN2676		2.3, 2.1	1	None	304	None	Forbidden	Forbidden	D	40	119	POISON GAS*
	Storage batteries, wet, see **Batteries,** *wet etc.*														
	Strontium arsenite	6.1	UN1691	II	6.1	IB8, IP2, IP4, T3, TP33	153	212	242	25 kg	100 kg	A		151	POISON
	Strontium chlorate	5.1	UN1506	II	5.1	A1, A9, IB8, IP2, IP4, N34, T3, TP33	152	212	242	5 kg	25 kg	A	56, 58	143	OXIDIZER
	Strontium nitrate	5.1	UN1507	III	5.1	A1, A29, IB8, IP3, T1, TP33	152	213	240	25 kg	100 kg	A		140	OXIDIZER
	Strontium perchlorate	5.1	UN1508	II	5.1	IB6, IP2, T3, TP33	152	212	242	5 kg	25 kg	A	56, 58	140	OXIDIZER
	Strontium peroxide	5.1	UN1509	II	5.1	IB6, IP2, T3, TP33	152	212	242	5 kg	25 kg	A	13, 52, 66, 75	143	OXIDIZER

Sym-bols (1)	Hazardous materials descriptions and proper shipping names (2)	Hazard class or Division (3)	Identifi-cation Numbers (4)	PG (5)	Label Codes (6)	Special Provisions (§172.102) (7)	(8) Packaging (§173.***) Excep-tions (8A)	Non-bulk (8B)	Bulk (8C)	(9) Quantity limitations (see §§173.27 and 175.75) Passenger aircraft/rail (9A)	Cargo aircraft only (9B)	(10) Vessel stowage Location (10A)	Other (10B)	ERG Guide No.	Placard Advisory Consult §172.504 & §172.505 to determine placard *Denotes placard for any quantity
	Strontium phosphide	4.3	UN2013	I	4.3, 6.1	A19, N40	None	211	None	Forbidden	15 kg	E	40, 52, 85	139	DANGEROUS WHEN WET*
	Strychnine *or* **Strychnine salts**	6.1	UN1692	I	6.1	IB7, IP1, T6, TP33	None	211	242	5 kg	50 kg	A	40	151	POISON
	Styphnic acid, *see* **Trinitroresorcinol,** *etc.*												14E		
	Styrene monomer, stabilized	3	UN2055	III	3	B1, IB3, T2, TP1	150	203	242	60 L	220 L	A		128P	FLAMMABLE
G	**Substances, explosive, n.o.s.**	1.1L	UN0357	II	1.1L		None	62	None	Forbidden	Forbidden		8E, 14E, 15E, 17E	112	EXPLOSIVES 1.1*
G	**Substances, explosive, n.o.s.**	1.2L	UN0358	II	1.2L		None	62	None	Forbidden	Forbidden		8E, 14E, 15E, 17E	112	EXPLOSIVES 1.2*
G	**Substances, explosive, n.o.s.**	1.3L	UN0359	II	1.3L		None	62	None	Forbidden	Forbidden		8E, 14E, 15E, 17E	112	EXPLOSIVES 1.3*
G	**Substances, explosive, n.o.s.**	1.1A	UN0473	II	1.1A	111	None	62	None	Forbidden	Forbidden	12		112	EXPLOSIVES 1.1*
G	**Substances, explosive, n.o.s.**	1.1C	UN0474	II	1.1C		None	62	None	Forbidden	Forbidden	10		112	EXPLOSIVES 1.1*
G	**Substances, explosive, n.o.s.**	1.1D	UN0475	II	1.1D		None	62	None	Forbidden	Forbidden	10		112	EXPLOSIVES 1.1*
G	**Substances, explosive, n.o.s.**	1.1G	UN0476	II	1.1G		None	62	None	Forbidden	Forbidden	08		112	EXPLOSIVES 1.1*
G	**Substances, explosive, n.o.s.**	1.3C	UN0477	II	1.3C		None	62	None	Forbidden	Forbidden	10		112	EXPLOSIVES 1.3*
G	**Substances, explosive, n.o.s.**	1.3G	UN0478	II	1.3G		None	62	None	Forbidden	Forbidden	08		112	EXPLOSIVES 1.3*
G	**Substances, explosive, n.o.s.**	1.4C	UN0479	II	1.4C		None	62	None	Forbidden	75 kg	09		114	EXPLOSIVES 1.4
G	**Substances, explosive, n.o.s.**	1.4D	UN0480	II	1.4D		None	62	None	Forbidden	75 kg	09		114	EXPLOSIVES 1.4
G	**Substances, explosive, n.o.s.**	1.4S	UN0481	II	1.4S		None	62	None	25 kg	75 kg	05		114	EXPLOSIVES 1.4
G	**Substances, explosive, n.o.s.**	1.4G	UN0485	II	1.4G		None	62	None	Forbidden	75 kg	08		114	EXPLOSIVES 1.4
G	**Substances, explosive, very insensitive, n.o.s.,** *or* **Substances, EVI, n.o.s**	1.5D	UN0482	II	1.5D		None	62	None	Forbidden	Forbidden	10		112	EXPLOSIVES 1.5
	Substituted nitrophenol pesticides, liquid, flammable, toxic, *flash point less than 23 degrees C*	3	UN2780	I	3, 6.1	T14, TP2, TP13, TP27	None	201	243	Forbidden	30 L	B	40	131	FLAMMABLE
				II	3, 6.1	IB2, T11, TP2, TP13, TP27	150	202	243	1 L	60 L	B	40	131	FLAMMABLE
	Substituted nitrophenol pesticides, liquid, toxic	6.1	UN3014	I	6.1	T14, TP2, TP13, TP27	None	201	243	1 L	30 L	B	40	153	POISON
				II	6.1	IB2, T11, TP2, TP13, TP27	153	202	243	5 L	60 L	B	40	153	POISON
				III	6.1	IB3, T7, TP2, TP28	153	203	241	60 L	220 L	A	40	153	POISON
	Substituted nitrophenol pesticides, liquid, toxic, flammable *flash point not less than 23 degrees C*	6.1	UN3013	I	6.1, 3	T14, TP2, TP13, TP27	None	201	243	1 L	30 L	B	40	131	POISON
				II	6.1, 3	IB2, T11, TP2, TP13, TP27	153	202	243	5 L	60 L	B	40	131	POISON
				III	6.1, 3	B1, IB3, T7, TP2, TP28	153	203	242	60 L	220 L	A	40	131	POISON
	Substituted nitrophenol pesticides, solid, toxic	6.1	UN2779	I	6.1	IB7, IP1, T6, TP33	None	211	242	5 kg	50 kg	A	40	153	POISON
				II	6.1	IB8, IP2, IP4, T3, TP33	153	212	242	25 kg	100 kg	A	40	153	POISON
				III	6.1	IB8, IP3, T1, TP33	153	213	240	100 kg	200 kg	A	40	153	POISON
	Sucrose octanitrate (dry)	Forbidden													
	Sulfamic acid	8	UN2967	III	8	IB8, IP3, T1, TP33	154	213	240	25 kg	100 kg	A		154	CORROSIVE
D	**Sulfur**	9	NA1350	III	9	30, IB8, IP2	None	None	240	No limit	No limit	A	19, 74	133	CLASS 9
I	**Sulfur**	4.1	UN1350	III	4.1	30, IB8, IP3, T1, TP33	None	None	240	25 kg	100 kg	A	19, 74	133	FLAMMABLE SOLID
	Sulfur and chlorate, loose mixtures of	Forbidden													
	Sulfur chlorides	8	UN1828	I	8	5, A3, A7, A10, B10, B77, N34, T20, TP2	None	201	243	Forbidden	2.5 L	C	40	137	CORROSIVE
	Sulfur dichloride, see **Sulfur chlorides**														
	Sulfur dioxide	2.3	UN1079		2.3, 8	3, B14, T50, TP19	None	304	314, 315	Forbidden	Forbidden	D	40	125	POISON GAS*
	Sulfur dioxide solution, see **Sulfurous acid**														
	Sulfur hexafluoride	2.2	UN1080		2.2		306	304	314, 315	75 kg	150 kg	A		126	NONFLAMMABLE GAS

Symbols (1)	Hazardous materials descriptions and proper shipping names (2)	Hazard class or Division (3)	Identification Numbers (4)	PG (5)	Label Codes (6)	Special Provisions (§172.102) (7)	(8) Packaging (§173.***) Exceptions (8A)	Non-bulk (8B)	Bulk (8C)	(9) Quantity limitations (see §§173.27 and 175.75) Passenger aircraft/rail (9A)	Cargo aircraft only (9B)	(10) Vessel stowage Location (10A)	Other (10B)	ERG Guide No.	Placard Advisory Consult §172.504 & §172.505 to determine placard *Denotes placard for any quantity
D	**Sulfur, molten**	9	NA2448	III	9	30, IB3, T1, TP3	None	213	247	Forbidden	Forbidden	C	61	133	CLASS 9
I	**Sulfur, molten**	4.1	UN2448	III	4.1	30, IB1, T1, TP3	None	213	247	Forbidden	Forbidden	C	74	133	FLAMMABLE SOLID
	Sulfur tetrafluoride	2.3	UN2418		2.3, 8	1	None	302	245	Forbidden	Forbidden	D	40	125	POISON GAS*
+	**Sulfur trioxide, stabilized**	8	UN1829	I	8, 6.1	2, B9, B14, B32, B49, B77, N34, T20, TP4, TP13, TP25, TP26, TP38, TP45	None	227	244	Forbidden	Forbidden	A	40	137	CORROSIVE, POISON INHALATION HAZARD*
	Sulfuretted hydrogen, see **Hydrogen sulfide**														
	Sulfuric acid, fuming *with less than 30 percent free sulfur trioxide*	8	UN1831	I	8	A3, A7, B84, N34, T20, TP2, TP13	None	201	243	Forbidden	2.5 L	C	14, 40	137	CORROSIVE
+	**Sulfuric acid, fuming** *with 30 percent or more free sulfur trioxide*	8	UN1831	I	8, 6.1	2, B9, B14, B32, B77, B84, N34, T20, TP2, TP13	None	227	244	Forbidden	Forbidden	C	14, 40	137	CORROSIVE, POISON INHALATION HAZARD*
	Sulfuric acid, spent	8	UN1832	II	8	A3, A7, B2, B83, B84, IB2, N34, T8, TP2	None	202	242	Forbidden	30 L	C	14	137	CORROSIVE
	Sulfuric acid *with more than 51 percent acid*	8	UN1830	II	8	A3, A7, B3, B83, B84, IB2, N34, T8, TP2	154	202	242	1 L	30 L	C	14	137	CORROSIVE
	Sulfuric acid *with not more than 51% acid*	8	UN2796	II	8	A3, A7, B2, B15, IB2, N6, N34, T8, TP2	154	202	242	1 L	30 L	B		157	CORROSIVE
	Sulfuric and hydrofluoric acid mixtures, *see* **Hydrofluoric and sulfuric acid mixtures**														
	Sulfuric anhydride, see **Sulfur trioxide, stabilized**														
	Sulfurous acid	8	UN1833	II	8	B3, IB2, T7, TP2	154	202	242	1 L	30 L	B	40	154	CORROSIVE
+	**Sulfuryl chloride**	6.1	UN1834	I	6.1, 8	1, B6, B9, B10, B14, B30, B77, N34, T22, TP2, TP13, TP38, TP44	None	226	244	Forbidden	Forbidden	D	40	137	POISON INHALATION HAZARD*
	Sulfuryl fluoride	2.3	UN2191		2.3	4	None	304	314, 315	Forbidden	Forbidden	D	40	123	POISON GAS*
	Tars, liquid *including road oils and cutback bitumens*	3	UN1999	II	3	149, B13, IB2, T3, TP3, TP29	150	202	242	5 L	60 L	B		130	FLAMMABLE
				III	3	B1, B13, IB3, T1, TP3	150	203	242	60 L	220 L	A		130	FLAMMABLE
	Tear gas candles	6.1	UN1700	II	6.1, 4.1		None	340	None	Forbidden	50 kg	D	40	159	POISON
	Tear gas cartridges, see **Ammunition, tear-producing,** *etc.*														
D	**Tear gas devices** *with more than 2 percent tear gas substances, by mass*	6.1	NA1693	I	6.1		None	340	None	Forbidden	Forbidden	D	40	159	POISON
				II	6.1		None	340	None	Forbidden	Forbidden	D	40	159	POISON
	Tear gas devices, with not more than 2 percent tear gas substances, by mass, see **Aerosols,** *etc.*														
	Tear gas grenades, see **Tear gas candles**														
G	**Tear gas substances, liquid, n.o.s.**	6.1	UN1693	I	6.1		None	201	None	Forbidden	Forbidden	D	40	159	POISON
				II	6.1	IB2	None	202	None	Forbidden	5 L	D	40	159	POISON
G	**Tear gas substance, solid, n.o.s.**	6.1	UN3448	I	6.1	T6, TP33	None	211	242	Forbidden	Forbidden	D	40	159	POISON
				II	6.1	IB8, IP2, IP4, T3, TP33	None	212	242	Forbidden	25 kg	D	40	159	POISON
G	**Tellurium compound, solid, n.o.s.**	6.1	UN3284	I	6.1	IB7, IP1, T6, TP33	None	211	242	5 kg	50 kg	B		151	POISON
				II	6.1	IB8, IP2, IP4, T3, TP33	153	212	242	25 kg	100 kg	B		151	POISON
				III	6.1	IB8, IP3, T1, TP33	153	213	240	100 kg	200 kg	A		151	POISON
	Tellurium hexafluoride	2.3	UN2195		2.3, 8	1	None	302	None	Forbidden	Forbidden	D	40	125	POISON GAS*
	Terpene hydrocarbons, n.o.s.	3	UN2319	III	3	B1, IB3, T4, TP1, TP29	150	203	242	60 L	220 L	A		128	FLAMMABLE
	Terpinolene	3	UN2541	III	3	B1, IB3, T2, TP1	150	203	242	60 L	220 L	A		128	FLAMMABLE
	Tetraazido benzene quinone	Forbidden													

HAZARDOUS MATERIALS TABLE

Symbols (1)	Hazardous materials descriptions and proper shipping names (2)	Hazard class or Division (3)	Identification Numbers (4)	PG (5)	Label Codes (6)	Special Provisions (§172.102) (7)	(8) Packaging (§173.***) Exceptions (8A)	Non-bulk (8B)	Bulk (8C)	(9) Quantity limitations (see §§173.27 and 175.75) Passenger aircraft/rail (9A)	Cargo aircraft only (9B)	(10) Vessel stowage Location (10A)	Other (10B)	ERG Guide No.	Placard Advisory Consult §172.504 & §172.505 to determine placard *Denotes placard for any quantity
	Tetrabromoethane	6.1	UN2504	III	6.1	IB3, T4, TP1	153	203	241	60 L	220 L	A		159	POISON
	1,1,2,2-Tetrachloroethane	6.1	UN1702	II	6.1	IB2, N36, T7, TP2	153	202	243	5 L	60 L	A	40	151	POISON
	Tetrachloroethylene	6.1	UN1897	III	6.1	IB3, N36, T4, TP1	153	203	241	60 L	220 L	A	40	160	POISON
	Tetraethyl dithiopyrophosphate	6.1	UN1704	II	6.1	IB2, T7, TP2	153	212	242	25 kg	100 kg	D	40	153	POISON
	Tetraethyl silicate	3	UN1292	III	3	B1, IB3, T2, TP1	150	203	242	60 L	220 L	A		129	FLAMMABLE
	Tetraethylammonium perchlorate (dry)	Forbidden													
	Tetraethylenepentamine	8	UN2320	III	8	IB3, T4, TP1	154	203	241	5 L	60 L	A	52	153	CORROSIVE
	1,1,1,2-Tetrafluoroethane *or* **Refrigerant gas R 134a**	2.2	UN3159		2.2	T50	306	304	314, 315	75 kg	150 kg	A		126	NONFLAMMABLE GAS
	Tetrafluoroethylene, stabilized	2.1	UN1081		2.1		306	304	None	Forbidden	150 kg	E	40	116P	FLAMMABLE GAS
	Tetrafluoromethane, *or* **Refrigerant gas R 14**	2.2	UN1982		2.2		None	302	None	75 kg	150 kg	A		126	NONFLAMMABLE GAS
	1,2,3,6-Tetrahydrobenzaldehyde	3	UN2498	III	3	B1, IB3, T2, TP1	150	203	242	60 L	220 L	A		129	FLAMMABLE
	Tetrahydrofuran	3	UN2056	II	3	IB2, T4, TP1	None	202	242	5 L	60 L	B		127	FLAMMABLE
	Tetrahydrofurfurylamine	3	UN2943	III	3	B1, IB3, T2, TP1	150	203	242	60 L	220 L	A		129	FLAMMABLE
	Tetrahydrophthalic anhydrides *with more than 0.05 percent of maleic anhydride*	8	UN2698	III	8	IB8, IP3, T1, TP33	154	213	240	25 kg	100 kg	A		156	CORROSIVE
	1,2,3,6-Tetrahydropyridine	3	UN2410	II	3	IB2, T4, TP1	150	202	242	5 L	60 L	B		129	FLAMMABLE
	Tetrahydrothiophene	3	UN2412	II	3	IB2, T4, TP1	150	202	242	5 L	60 L	B		130	FLAMMABLE
	Tetramethylammonium hydroxide, solid	8	UN3423	II	8	B2, IB8, IP2, IP4, T3, TP33	154	213	240	15 kg	50 kg	A	52	153	CORROSIVE
	Tetramethylammonium hydroxide solution	8	UN1835	II	8	B2, IB2, T7, TP2	154	202	242	1 L	30 L	A	52	153	CORROSIVE
				III	8	B2, IB3, T7, TP2	154	203	241	5 L	60 L	A	52	153	CORROSIVE
	Tetramethylene diperoxide dicarbamide	Forbidden													
	Tetramethylsilane	3	UN2749	I	3	A7, T14, TP2	None	201	243	Forbidden	30 L	D		130	FLAMMABLE
	Tetranitro diglycerin	Forbidden													
	Tetranitroaniline	1.1D	UN0207	II	1.1D		None	62	None	Forbidden	Forbidden	10		112	EXPLOSIVES 1.1*
+	**Tetranitromethane**	6.1	UN1510	I	6.1, 5.1	2, B32, T20, TP2, TP13, TP38, TP44	None	227	None	Forbidden	Forbidden	D	40, 66	143	POISON INHALATION HAZARD*
	2,3,4,6-Tetranitrophenol	Forbidden													
	2,3,4,6-Tetranitrophenyl methyl nitramine	Forbidden													
	2,3,4,6-Tetranitrophenylnitramine	Forbidden													
	Tetranitroresorcinol (dry)	Forbidden													
	2,3,5,6-Tetranitroso-1,4-dinitrobenzene	Forbidden													
	2,3,5,6-Tetranitroso nitrobenzene (dry)	Forbidden													
	Tetrapropylorthotitanate	3	UN2413	III	3	B1, IB3, T4, TP1	150	203	242	60 L	220 L	A		128	FLAMMABLE
	Tetrazene, *see* **Guanyl nitrosaminoguanyltetrazene**														
	Tetrazine (dry)	Forbidden													
	Tetrazol-1-acetic acid	1.4C	UN0407	II	1.4C		None	62	None	Forbidden	75 kg	09		114	EXPLOSIVES 1.4
	1H-Tetrazole	1.1D	UN0504		1.1D		None	62	None	Forbidden	Forbidden	B	1E, 5E	112	EXPLOSIVES 1.1*
	Tetrazolyl azide (dry)	Forbidden													
	Tetryl, *see* **Trinitrophenylmethylnitramine**														
A I W	**Textile waste, wet**	4.2	UN1857	III	4.2		151	213	240	Forbidden	Forbidden	A		133	SPONTANEOUSLY COMBUSTIBLE
	Thallium chlorate	5.1	UN2573	II	5.1, 6.1	IB6, IP2, T3, TP33	152	212	242	5 kg	25 kg	A	56, 58	141	OXIDIZER
	Thallium compounds, n.o.s.	6.1	UN1707	II	6.1	IB8, IP2, IP4, T3, TP33	153	212	242	25 kg	100 kg	A		151	POISON
	Thallium nitrate	6.1	UN2727	II	6.1, 5.1	IB6, IP2, T3, TP33	153	212	242	5 kg	25 kg	A		141	POISON
	4-Thiapentanal	6.1	UN2785	III	6.1	IB3, T4, TP1	153	203	241	60 L	220 L	D	25, 49	152	POISON
	Thioacetic acid	3	UN2436	II	3	IB2, T4, TP1	150	202	242	5 L	60 L	B		129	FLAMMABLE
	Thiocarbamate pesticide, liquid, flammable, toxic, *flash point less than 23 degrees C*	3	UN2772	I	3, 6.1	T14, TP2, TP13, TP27	None	201	243	Forbidden	30 L	B	40	131	FLAMMABLE
				II	3, 6.1	IB2, T11, TP13, TP27	150	202	243	1 L	60 L	B	40	131	FLAMMABLE

Symbols (1)	Hazardous materials descriptions and proper shipping names (2)	Hazard class or Division (3)	Identification Numbers (4)	PG (5)	Label Codes (6)	Special Provisions (§172.102) (7)	(8) Packaging (§173.***) Exceptions (8A)	Non-bulk (8B)	Bulk (8C)	(9) Quantity limitations (see §§173.27 and 175.75) Passenger aircraft/rail (9A)	Cargo aircraft only (9B)	(10) Vessel stowage Location (10A)	Other (10B)	ERG Guide No.	Placard Advisory Consult §172.504 & §172.505 to determine placard *Denotes placard for any quantity
	Thiocarbamate pesticide, liquid, toxic, flammable, *flash point not less than 23 degrees C*	6.1	UN3005	I	6.1, 3	T14, TP2, TP13	None	201	243	1 L	30 L	B	40	131	POISON
				II	6.1, 3	IB2, T11, TP2, TP13, TP27	153	202	243	5 L	60 L	B	40	131	POISON
				III	6.1, 3	IB3, T7, TP2, TP28	153	203	242	60 L	220 L	A	40	131	POISON
	Thiocarbamate pesticide, liquid, toxic	6.1	UN3006	I	6.1	T14, TP2, TP13	None	201	243	1 L	30 L	B	40	151	POISON
				II	6.1	IB2, T11, TP2, TP13, TP27	153	202	243	5 L	60 L	B	40	151	POISON
				III	6.1	IB3, T7, TP2, TP28	153	203	241	60 L	220 L	A	40	151	POISON
	Thiocarbamate pesticides, solid, toxic	6.1	UN2771	I	6.1	IB7, IP1, T6, TP33	None	211	242	5 kg	50 kg	A	40	151	POISON
				II	6.1	IB8, IP2, IP4, T3, TP33	153	212	242	25 kg	100 kg	A	40	151	POISON
				III	6.1	IB8, IP3, T1, TP33	153	213	240	100 kg	200 kg	A	40	151	POISON
	Thiocarbonylchloride, see **Thiophosgene**														
	Thioglycol	6.1	UN2966	II	6.1	IB2, T7, TP2	153	202	243	5 L	60 L	A		153	POISON
	Thioglycolic acid	8	UN1940	II	8	A7, B2, IB2, N34, T7, TP2	154	202	242	1 L	30 L	A		153	CORROSIVE
	Thiolactic acid	6.1	UN2936	II	6.1	IB2, T7, TP2	153	202	243	5 L	60 L	A		153	POISON
	Thionyl chloride	8	UN1836	I	8	B6, B10, N34, T10, TP2, TP13	None	201	243	Forbidden	Forbidden	C	40	137	CORROSIVE
	Thiophene	3	UN2414	II	3	IB2, T4, TP1	150	202	242	5 L	60 L	B	40	130	FLAMMABLE
+	**Thiophosgene**	6.1	UN2474	I	6.1	2, B9, B14, B32, N33, N34, T20, TP2, TP13, TP38, TP45	None	227	244	Forbidden	Forbidden	D	40, 52	157	POISON INHALATION HAZARD*
	Thiophosphoryl chloride	8	UN1837	II	8	A3, A7, B2, B8, B25, IB2, N34, T7, TP2	None	202	242	Forbidden	30 L	C	40	157	CORROSIVE
	Thiourea dioxide	4.2	UN3341	II	4.2	IB6, IP2, T3, TP33	None	212	241	15 kg	50 kg	D		135	SPONTANEOUSLY COMBUSTIBLE
				III	4.2	IB8, IP3, T1, TP33	None	213	241	25 kg	100 kg	D		135	SPONTANEOUSLY COMBUSTIBLE
	Tin chloride, fuming, see **Stannic chloride, anhydrous**														
	Tin perchloride or Tin tetrachloride, see **Stannic chloride, anhydrous**														
	Tinctures, medicinal	3	UN1293	II	3	IB2, T4, TP1, TP8	150	202	242	5 L	60 L	B		127	FLAMMABLE
				III	3	B1, IB3, T2, TP1	150	203	242	60 L	220 L	A		127	FLAMMABLE
	Tinning flux, see **Zinc chloride**														
	Tires and tire assemblies, *see* **Air, compressed** *or* **Nitrogen, compressed**														
	Titanium disulphide	4.2	UN3174	III	4.2	IB8, IP3, T1, TP33	None	213	241	25 kg	100 kg	A		135	SPONTANEOUSLY COMBUSTIBLE
	Titanium hydride	4.1	UN1871	II	4.1	A19, A20, IB4, N34, T3, TP33	None	212	241	15 kg	50 kg	E		170	FLAMMABLE SOLID
	Titanium powder, dry	4.2	UN2546	I	4.2		None	211	242	Forbidden	Forbidden	D		135	SPONTANEOUSLY COMBUSTIBLE
				II	4.2	A19, A20, IB6, IP2, N5, N34, T3, TP33	None	212	241	15 kg	50 kg	D		135	SPONTANEOUSLY COMBUSTIBLE
				III	4.2	IB8, IP3, T1, TP33	None	213	241	25 kg	100 kg	D		135	SPONTANEOUSLY COMBUSTIBLE
	Titanium powder, wetted *with not less than 25 percent water (a visible excess of water must be present) (a) mechanically produced, particle size less than 53 microns; (b) chemically produced, particle size less than 840 microns*	4.1	UN1352	II	4.1	A19, A20, IB6, IP2, N34, T3, TP33	None	212	240	15 kg	50 kg	E	74	170	FLAMMABLE SOLID
	Titanium sponge granules *or* **Titanium sponge powders**	4.1	UN2878	III	4.1	A1, IB8, IP3, T1, TP33	None	213	240	25 kg	100 kg	D	74	170	FLAMMABLE SOLID
+	**Titanium tetrachloride**	6.1	UN1838	I	6.1, 8	2, B7, B9, B14, B32, B77, T20, TP2, TP13, TP38, TP45	None	227	244	Forbidden	Forbidden	D	40	137	POISON INHALATION HAZARD*
	Titanium trichloride mixtures	8	UN2869	II	8	A7, IB8, IP2, IP4, N34, T3, TP33	154	212	240	15 kg	50 kg	A	40	157	CORROSIVE

Symbols (1)	Hazardous materials descriptions and proper shipping names (2)	Hazard class or Division (3)	Identification Numbers (4)	PG (5)	Label Codes (6)	Special Provisions (§172.102) (7)	(8) Packaging (§173.***) Exceptions (8A)	Non-bulk (8B)	Bulk (8C)	(9) Quantity limitations (see §§173.27 and 175.75) Passenger aircraft/rail (9A)	Cargo aircraft only (9B)	(10) Vessel stowage Location (10A)	Other (10B)	ERG Guide No.	Placard Advisory Consult §172.504 & §172.505 to determine placard *Denotes placard for any quantity
				III	8	A7, IB8, IP3, N34, T1, TP33	154	213	240	25 kg	100 kg	A	40	157	CORROSIVE
	Titanium trichloride, pyrophoric *or* **Titanium trichloride mixtures, pyrophoric**	4.2	UN2441	I	4.2, 8	N34	None	181	244	Forbidden	Forbidden	D	40	135	SPONTANEOUSLY COMBUSTIBLE
	TNT mixed with aluminum, see **Tritonal**														
	TNT, *see* **Trinitrotoluene,** *etc.*														
	Toluene	3	UN1294	II	3	IB2, T4, TP1	150	202	242	5 L	60 L	B		130	FLAMMABLE
+	**Toluene diisocyanate**	6.1	UN2078	II	6.1	IB2, T7, TP2, TP13	153	202	243	5 L	60 L	D	25, 40	156	POISON
	Toluene sulfonic acid, see **Alkyl,** *or* **Aryl sulfonic acid** *etc.*														
+	**Toluidines, liquid**	6.1	UN1708	II	6.1	IB2, T7, TP2	153	202	243	5 L	60 L	A		153	POISON
	Toluidines, solid	6.1	UN3451	II	6.1	IB8, IP2, IP4, T3, TP33	153	212	242	25 kg	100 kg	A		153	POISON
	2,4-Toluylenediamine, solid *or* **2,4-Toluenediamine, solid**	6.1	UN1709	III	6.1	IB8, IP3, T1, TP33	153	213	240	100 kg	200 kg	A		151	POISON
	2,4-Toluylenediamine solution *or* **2,4-Toluenediamine solution**	6.1	UN3418	III	6.1	IB3, T4, TP1	153	203	241	60 L	220 L	A		151	POISON
	Torpedoes, liquid fueled, *with inert head*	1.3J	UN0450	II	1.3J			62	None	Forbidden	Forbidden	04	23E	112	EXPLOSIVES 1.3*
	Torpedoes, liquid fueled, *with or without bursting charge*	1.1J	UN0449	II	1.1J			62	None	Forbidden	Forbidden	04	23E	112	EXPLOSIVES 1.1*
	Torpedoes *with bursting charge*	1.1E	UN0329	II	1.1E			62	62	Forbidden	Forbidden	03		112	EXPLOSIVES 1.1*
	Torpedoes *with bursting charge*	1.1F	UN0330	II	1.1F			62	None	Forbidden	Forbidden	08		112	EXPLOSIVES 1.1*
	Torpedoes *with bursting charge*	1.1D	UN0451	II	1.1D			62	62	Forbidden	Forbidden	03		112	EXPLOSIVES 1.1*
G	**Toxic by inhalation liquid, corrosive, flammable, n.o.s.** *with an inhalation toxicity lower than or equal to 200 ml/m3 and saturated vapor concentration greater than or equal to 500 LC50*	6.1	UN3492	I	6.1, 8, 3	1, B9, B14, B30, B72, T22, TP2, TP13, TP27, TP38, TP44	None	226	244	Forbidden	Forbidden	D	40, 125	—	POISON INHALATION HAZARD*
G	**Toxic by inhalation liquid, corrosive, flammable, n.o.s.** *with an inhalation toxicity lower than or equal to 1000 ml/m3 and saturated vapor concentration greater than or equal to 10 LC50*	6.1	UN3493	I	6.1, 8, 3	2, B9, B14, B32, B74, T20, TP2, TP13, TP27, TP38, TP45	None	227	244	Forbidden	Forbidden	D	40, 125	—	POISON INHALATION HAZARD*
G	**Toxic by inhalation liquid, flammable, corrosive, n.o.s.** *with an inhalation toxicity lower than or equal to 200 ml/m3 and saturated vapor concentration greater than or equal to 500 LC50*	6.1	UN3488	I	6.1, 3, 8	1, B9, B14, B30, B72, T22, TP2, TP13, TP27, TP38, TP44	None	226	244	Forbidden	Forbidden	D	40, 125	—	POISON INHALATION HAZARD*
G	**Toxic by inhalation liquid, flammable, corrosive, n.o.s.** *with an inhalation toxicity lower than or equal to 1000 ml/m3 and saturated vapor concentration greater than or equal to 10 LC50*	6.1	UN3489	I	6.1, 3, 8	2, B9, B14, B32, B74, T20, TP2, TP13, TP27, TP38, TP45	None	227	244	Forbidden	Forbidden	D	40, 125	—	POISON INHALATION HAZARD*
G	**Toxic by inhalation liquid, n.o.s.** *with an inhalation toxicity lower than or equal to 200 ml/m3 and saturated vapor concentration great than or equal to 500 LC50*	6.1	UN3381	I	6.1	1, B9, B14, B30, T22, TP2, TP13, TP27, TP38, TP44	None	226	244	Forbidden	Forbidden	D	40	151	POISON INHALATION HAZARD*
G	**Toxic by inhalation liquid, n.o.s.** *with an inhalation toxicity lower than or equal to 1000 ml/m3 and saturated vapor concentration greater than or equal to 10 LC50*	6.1	UN3382	I	6.1	2, B9, B14, B32, T20, TP2, TP13, TP27, TP38, TP45	None	227	244	Forbidden	Forbidden	D	40	151	POISON INHALATION HAZARD*
G	**Toxic by inhalation liquid, flammable, n.o.s.** *with an inhalation toxicity lower than or equal to 200 ml/m3 and saturated vapor concentration greater than or equal to 500 LC50*	6.1	UN3383	I	6.1, 3	1, B9, B14, B30, T22, TP2, TP13, TP27, TP38, TP44	None	226	244	Forbidden	Forbidden	D	40	131	POISON INHALATION HAZARD*
G	**Toxic by inhalation liquid, flammable, n.o.s.** *with an inhalation toxicity lower than or equal to 1000 ml/m3 and saturated vapor concentration greater than or equal to 10 LC50*	6.1	UN3384	I	6.1, 3	2, B9, B14, B32, T20, TP2, TP13, TP27, TP38, TP45	None	227	244	Forbidden	Forbidden	D	40	131	POISON INHALATION HAZARD*
G	**Toxic by inhalation liquid, water-reactive, n.o.s.** *with a inhalation toxicity lower than or equal to 200 ml/m3 and saturated vapor concentration greater than or equal to 500 LC50*	6.1	UN3385	I	6.1, 4.3	1, B9, B14, B30, T22, TP2, TP13, TP38, TP44	None	226	244	Forbidden	Forbidden	D	40	139	POISON INHALATION HAZARD*
G	**Toxic by inhalation liquid, water-reactive, n.o.s.** *with an inhalation toxicity lower than or equal to 1000 ml/m3 and saturated vapor concentration greater than or equal to 10 LC50*	6.1	UN3386	I	6.1, 4.3	2, B9, B14, B32, T20, TP2, TP13, TP38, TP44	None	227	244	Forbidden	Forbidden	D	40	139	POISON INHALATION HAZARD*
G	**Toxic by inhalation liquid, water-reactive, flammable, n.o.s.** *with an inhalation toxicity lower than or equal to 200 ml/m3 and saturated vapor concentration greater than or equal to 500 LC50*	6.1	UN3490	I	6.1, 4.3, 3	1, B9, B14, B30, B72, T22, TP2, TP13, TP27, TP38, TP44	None	226	244	Forbidden	Forbidden	D	21, 28, 40, 49	—	POISON INHALATION HAZARD*
G	**Toxic by inhalation liquid, water-reactive, flammable, n.o.s.** *with an inhalation toxicity lower than or equal to 1000 ml/m3 and saturated vapor concentration greater than or equal to 10 LC50*	6.1	UN3491	I	6.1, 4.3, 3	2, B9, B14, B32, B74, T20, TP2, TP13, TP27, TP38, TP45	None	227	244	Forbidden	Forbidden	D	21, 28, 40, 49	—	POISON INHALATION HAZARD*

Symbols (1)	Hazardous materials descriptions and proper shipping names (2)	Hazard class or Division (3)	Identification Numbers (4)	PG (5)	Label Codes (6)	Special Provisions (§172.102) (7)	(8) Packaging (§173.***) Exceptions (8A)	Non-bulk (8B)	Bulk (8C)	(9) Quantity limitations (see §§173.27 and 175.75) Passenger aircraft/rail (9A)	Cargo aircraft only (9B)	(10) Vessel stowage Location (10A)	Other (10B)	ERG Guide No.	Placard Advisory Consult §172.504 & §172.505 to determine placard *Denotes placard for any quantity
G	**Toxic by inhalation liquid, oxidizing, n.o.s.** *with an inhalation toxicity lower than or equal to 200 ml/m3 and saturated vapor concentration greater than or equal to 500 LC50*	6.1	UN3387	I	6.1, 5.1	1, B9, B14, B30, T22, TP2, TP13, TP38, TP44	None	226	244	Forbidden	Forbidden	D	40	142	POISON INHALATION HAZARD*
G	**Toxic by inhalation liquid, oxidizing, n.o.s.** *with an inhalation toxicity lower than or equal to 1000 ml/m3 and saturated vapor concentration greater than or equal to 10 LC50*	6.1	UN3388	I	6.1, 5.1	2, B9, B14, B32, T20, TP2, TP13, TP38, TP44	None	227	244	Forbidden	Forbidden	D	40	142	POISON INHALATION HAZARD*
G	**Toxic by inhalation liquid, corrosive, n.o.s.** *with an inhalation toxicity lower than or equal to 200 ml/m3 and saturated vapor concentration greater than or equal to 500 LC50*	6.1	UN3389	I	6.1, 8	1, B9, B14, B30, T22, TP2, TP13, TP27, TP38, TP44	None	226	244	Forbidden	Forbidden	D	40	154	POISON INHALATION HAZARD*
G	**Toxic inhalation liquid, corrosive, n.o.s.** *with an inhalation toxicity lower than or equal to 1000 ml/m3 and saturated vapor concentration greater than or equal to 10 LC50*	6.1	UN3390	I	6.1, 8	2, B9, B14, B32, T20, TP2, TP13, TP27, TP38, TP45	None	227	244	Forbidden	Forbidden	D	40	154	POISON INHALATION HAZARD*
G	**Toxic liquid, corrosive, inorganic, n.o.s.**	6.1	UN3289	I	6.1, 8	T14, TP2, TP13, TP27	None	201	243	0.5 L	2.5 L	A		154	POISON
				II	6.1, 8	IB2, T11, TP2, TP27	153	202	243	1 L	30 L	A		154	POISON
G	**Toxic liquid, inorganic, n.o.s.**	6.1	UN3287	I	6.1	T14, TP2, TP13, TP27	None	201	243	1 L	30 L	A		151	POISON
				II	6.1	IB2, T11, TP2, TP27	153	202	243	5 L	60 L	A		151	POISON
				III	6.1	IB3, T7, TP1, TP28	153	203	241	60 L	220 L	A		151	POISON
G	**Toxic liquids, corrosive, organic, n.o.s.**	6.1	UN2927	I	6.1, 8	T14, TP2, TP13, TP27	None	201	243	0.5 L	2.5 L	B	40	154	POISON
				II	6.1, 8	IB2, T11, TP2, TP27	153	202	243	1 L	30 L	B	40	154	POISON
G	**Toxic liquids, flammable, organic, n.o.s.**	6.1	UN2929	I	6.1, 3	T14, TP2, TP13, TP27	None	201	243	1 L	30 L	B	40	131	POISON
				II	6.1, 3	IB2, T11, TP2, TP13, TP27	153	202	243	5 L	60 L	B	40	131	POISON
G	**Toxic, liquids, organic, n.o.s.**	6.1	UN2810	I	6.1	T14, TP2, TP13, TP27	None	201	243	1 L	30 L	B	40	153	POISON
				II	6.1	IB2, T11, TP2, TP13, TP27	153	202	243	5 L	60 L	B	40	153	POISON
				III	6.1	IB3, T7, TP1, TP28	153	203	241	60 L	220 L	A	40	153	POISON
G	**Toxic liquids, oxidizing, n.o.s.**	6.1	UN3122	I	6.1, 5.1	A4	None	201	243	Forbidden	2.5 L	C		142	POISON
				II	6.1, 5.1	IB2	153	202	243	1 L	5 L	C		142	POISON
G	**Toxic liquids, water-reactive, n.o.s.**	6.1	UN3123	I	6.1, 4.3	A4	None	201	243	Forbidden	1 L	E	40	139	POISON, DANGEROUS WHEN WET*
				II	6.1, 4.3	IB2	None	202	243	1 L	5 L	E	40	139	POISON, DANGEROUS WHEN WET*
G	**Toxic solid, corrosive, inorganic, n.o.s.**	6.1	UN3290	I	6.1, 8	IB7, T6, TP33	None	211	242	1 kg	25 kg	A		154	POISON
				II	6.1, 8	IB6, IP2, T3, TP33	153	212	242	15 kg	50 kg	A		154	POISON
G	**Toxic solid, inorganic, n.o.s.**	6.1	UN3288	I	6.1	IB7, T6, TP33	None	211	242	5 kg	50 kg	A		151	POISON
				II	6.1	IB8, IP2, IP4, T3, TP33	153	212	242	25 kg	100 kg	A		151	POISON
				III	6.1	IB8, IP3, T1, TP33	153	213	240	100 kg	200 kg	A		151	POISON
G	**Toxic solids, corrosive, organic, n.o.s.**	6.1	UN2928	I	6.1, 8	IB7, T6, TP33	None	211	242	1 kg	25 kg	B	40	154	POISON
				II	6.1, 8	IB6, IP2, T3, TP33	153	212	242	15 kg	50 kg	B	40	154	POISON
G	**Toxic solids, flammable, organic, n.o.s.**	6.1	UN2930	I	6.1, 4.1	IB6, T6, TP33	None	211	242	1 kg	15 kg	B		134	POISON
				II	6.1, 4.1	IB8, IP2, IP4, T3, TP33	153	212	242	15 kg	50 kg	B		134	POISON
G	**Toxic solids, organic, n.o.s.**	6.1	UN2811	I	6.1	IB7, T6, TP33	None	211	242	5 kg	50 kg	B		154	POISON
				II	6.1	IB8, IP2, IP4, T3, TP33	153	212	242	25 kg	100 kg	B		154	POISON
				III	6.1	IB8, IP3, T1, TP33	153	213	240	100 kg	200 kg	A		154	POISON
G	**Toxic solids, oxidizing, n.o.s.**	6.1	UN3086	I	6.1, 5.1	T6, TP33	None	211	242	1 kg	15 kg	C		141	POISON

Sym-bols (1)	Hazardous materials descriptions and proper shipping names (2)	Hazard class or Division (3)	Identifi-cation Numbers (4)	PG (5)	Label Codes (6)	Special Provisions (§172.102) (7)	(8) Packaging (§173.***) Excep-tions (8A)	Non-bulk (8B)	Bulk (8C)	(9) Quantity limitations (see §§173.27 and 175.75) Passenger aircraft/rail (9A)	Cargo aircraft only (9B)	(10) Vessel stowage Location (10A)	Other (10B)	ERG Guide No.	Placard Advisory Consult §172.504 & §172.505 to determine placard *Denotes placard for any quantity
				II	6.1, 5.1	IB6, IP2, T3, TP33	153	212	242	15 kg	50 kg	C		141	POISON
G	**Toxic solids, self-heating, n.o.s.**	6.1	UN3124	I	6.1, 4.2	A5, T6, TP33	None	211	242	5 kg	15 kg	D	40	136	POISON
				II	6.1, 4.2	IB6, IP2, T3, TP33	None	212	242	15 kg	50 kg	D	40	136	POISON
G	**Toxic solids, water-reactive, n.o.s.**	6.1	UN3125	I	6.1, 4.3	A5, T6, TP33	None	211	242	5 kg	15 kg	D	40	139	POISON, DANGEROUS WHEN WET*
				II	6.1, 4.3	IB6, IP2, T3, TP33	153	212	242	15 kg	50 kg	D	40	139	POISON, DANGEROUS WHEN WET*
G	**Toxins extracted from living sources, liquid, n.o.s.**	6.1	UN3172	I	6.1	141	None	201	243	1 L	30 L	B	40	153	POISON
				II	6.1	141, IB2	None	202	243	5 L	60 L	B	40	153	POISON
				III	6.1	141, IB3	153	203	241	60 L	220 L	A	40	153	POISON
G	**Toxins extracted from living sources, solid, n.o.s.**	6.1	UN3462	I	6.1	141, IB7, IP1, T6, TP33	None	211	243	5 kg	50 kg	B		153	POISON
				II	6.1	141, IB8, IP2, IP4, T3, TP33	None	212	243	25 kg	100 kg	B		153	POISON
				III	6.1	141, IB8, IP3, T1, TP33	153	213	241	100 kg	200 kg	A		153	POISON
D	**Toy Caps**	1.4S	NA0337	II	1.4S		None	62	None	25 kg	100 kg	05		114	EXPLOSIVES 1.4
	Tracers for ammunition	1.3G	UN0212	II	1.3G		None	62	None	Forbidden	Forbidden	07		112	EXPLOSIVES 1.3*
	Tracers for ammunition	1.4G	UN0306	II	1.4G		None	62	None	Forbidden	75 kg	06		114	EXPLOSIVES 1.4
	Tractors, see **Vehicle**, *etc.*														
	Tri-(b-nitroxyethyl) ammonium nitrate	Forbidden													
	Triallyl borate	6.1	UN2609	III	6.1	IB3	153	203	241	60 L	220 L	A	13	156	POISON
	Triallylamine	3	UN2610	III	3, 8	B1, IB3, T4, TP1	None	203	242	5 L	60 L	A	40	132	FLAMMABLE
	Triazine pesticides, liquid, flammable, toxic, *flash point less than 23 degrees C*	3	UN2764	I	3, 6.1	T14, TP2, TP13, TP27	None	201	243	Forbidden	30 L	B	40	131	FLAMMABLE
				II	3, 6.1	IB2, T11, TP2, TP13, TP27	150	202	243	1 L	60 L	B	40	131	FLAMMABLE
	Triazine pesticides, liquid, toxic	6.1	UN2998	I	6.1	T14, TP2, TP13, TP27	None	201	243	1 L	30 L	B	40	151	POISON
				II	6.1	IB2, T11, TP2, TP13, TP27	153	202	243	5 L	60 L	B	40	151	POISON
				III	6.1	IB3, T7, TP2, TP28	153	203	241	60 L	220 L	A	40	151	POISON
	Triazine pesticides, liquid, toxic, flammable, *flash point not less than 23 degrees C*	6.1	UN2997	I	6.1, 3	T14, TP2, TP13, TP27	None	201	243	1 L	30 L	B	40	131	POISON
				II	6.1, 3	IB2, T11, TP2, TP13, TP27	153	202	243	5 L	60 L	B	40	131	POISON
				III	6.1, 3	IB3, T7, TP2, TP28	153	203	242	60 L	220 L	A	40	131	POISON
	Triazine pesticides, solid, toxic	6.1	UN2763	I	6.1	IB7, IP1, T6, TP33	None	211	242	5 kg	50 kg	A	40	151	POISON
				II	6.1	IB8, IP2, IP4, T3, TP33	153	212	242	25 kg	100 kg	A	40	151	POISON
				III	6.1	IB8, IP3, T1, TP33	153	213	240	100 kg	200 kg	A	40	151	POISON
	Tributylamine	6.1	UN2542	II	6.1	IB2, T7, TP2	153	202	243	5 L	60 L	A		153	POISON
	Tributylphosphane	4.2	UN3254	I	4.2	T21, TP7, TP33	None	211	242	Forbidden	Forbidden	D	136	135	SPONTANEOUSLY COMBUSTIBLE
	Trichloro-s-triazinetrione dry, with more than 39 percent available chlorine, see **Trichloroisocyanuric acid, dry**														
	Trichloroacetic acid	8	UN1839	II	8	A7, IB8, IP2, IP4, N34, T3, TP33	154	212	240	15 kg	50 kg	A		153	CORROSIVE
	Trichloroacetic acid, solution	8	UN2564	II	8	A3, A6, A7, B2, IB2, N34, T7, TP2	154	202	242	1 L	30 L	B		153	CORROSIVE
				III	8	A3, A6, A7, IB3, N34, T4, TP1	154	203	241	5 L	60 L	B	8	153	CORROSIVE

Symbols (1)	Hazardous materials descriptions and proper shipping names (2)	Hazard class or Division (3)	Identification Numbers (4)	PG (5)	Label Codes (6)	Special Provisions (§172.102) (7)	(8) Packaging (§173.***) Exceptions (8A)	Non-bulk (8B)	Bulk (8C)	(9) Quantity limitations (see §§173.27 and 175.75) Passenger aircraft/rail (9A)	Cargo aircraft only (9B)	(10) Vessel stowage Location (10A)	Other (10B)	ERG Guide No.	Placard Advisory Consult §172.504 & §172.505 to determine placard *Denotes placard for any quantity
+	**Trichloroacetyl chloride**	8	UN2442	II	8, 6.1	2, B9, B14, B32, N34, T20, TP2, TP38, TP45	None	227	244	Forbidden	Forbidden	D	40	156	CORROSIVE, POISON INHALATION HAZARD*
	Trichlorobenzenes, liquid	6.1	UN2321	III	6.1	IB3, T4, TP1	153	203	241	60 L	220 L	A		153	POISON
	Trichlorobutene	6.1	UN2322	II	6.1	IB2, T7, TP2	153	202	243	5 L	60 L	A	25, 40	152	POISON
	1,1,1-Trichloroethane	6.1	UN2831	III	6.1	IB3, N36, T4, TP1	153	203	241	60 L	220 L	A	40	160	POISON
	Trichloroethylene	6.1	UN1710	III	6.1	IB3, N36, T4, TP1	153	203	241	60 L	220 L	A	40	160	POISON
	Trichloroisocyanuric acid, dry	5.1	UN2468	II	5.1	IB8, IP2, IP4, T3, TP33	152	212	240	5 kg	25 kg	A	13	140	OXIDIZER
	Trichloromethyl perchlorate	Forbidden													
	Trichlorosilane	4.3	UN1295	I	4.3, 3, 8	N34, T14, TP2, TP7, TP13	None	201	244	Forbidden	Forbidden	D	21, 28, 40, 49, 100	139	DANGEROUS WHEN WET*
	Tricresyl phosphate *with more than 3 percent ortho isomer*	6.1	UN2574	II	6.1	A3, IB2, N33, N34, T7, TP2	153	202	243	5 L	60 L	A		151	POISON
	Triethyl phosphite	3	UN2323	III	3	B1, IB3, T2, TP1	150	203	242	60 L	220 L	A		130	FLAMMABLE
	Triethylamine	3	UN1296	II	3, 8	IB2, T7, TP1	None	202	243	1 L	5 L	B	40	132	FLAMMABLE
	Triethylenetetramine	8	UN2259	II	8	B2, IB2, T7, TP2	154	202	242	1 L	30 L	B	40	153	CORROSIVE
	Trifluoroacetic acid	8	UN2699	I	8	A3, A6, A7, B4, N3, N34, N36, T10, TP2	None	201	243	0.5 L	2.5 L	B	12, 40	154	CORROSIVE
	Trifluoroacetyl chloride	2.3	UN3057		2.3, 8	2, B7, B9, B14, T50, TP21	None	304	314, 315	Forbidden	Forbidden	D	40	125	POISON GAS*
	Trifluorochloroethylene, stabilized	2.3	UN1082		2.3, 2.1	3, B14, T50	None	304	314, 315	Forbidden	Forbidden	D	40	119P	POISON GAS*
	1,1,1-Trifluoroethane *or* **Refrigerant gas R 143a**	2.1	UN2035		2.1	T50	306	304	314, 315	Forbidden	150 kg	B	40	115	FLAMMABLE GAS
	Trifluoromethane *or* **Refrigerant gas R 23**	2.2	UN1984		2.2		306	304	314, 315	75 kg	150 kg	A		126	NONFLAMMABLE GAS
	Trifluoromethane, refrigerated liquid	2.2	UN3136		2.2	T75, TP5	306	None	314, 315	50 kg	500 kg	D		120	NONFLAMMABLE GAS
	2-Trifluoromethylaniline	6.1	UN2942	III	6.1	IB3	153	203	241	60 L	220 L	A		153	POISON
	3-Trifluoromethylaniline	6.1	UN2948	II	6.1	IB2, T7, TP2	153	202	243	5 L	60 L	A	40	153	POISON
	Triformoxime trinitrate	Forbidden													
	Triisobutylene	3	UN2324	III	3	B1, IB3, T4, TP1	150	203	242	60 L	220 L	A		128	FLAMMABLE
	Triisopropyl borate	3	UN2616	II	3	IB2, T4, TP1	150	202	242	5 L	60 L	A		129	FLAMMABLE
				III	3	B1, IB3, T2, TP1	150	203	242	60 L	220 L	A		129	FLAMMABLE
D	**Trimethoxysilane**	6.1	NA9269	I	6.1, 3	2, B9, B14, B32, T20, TP4, TP13, TP38, TP45	None	227	244	Forbidden	Forbidden	E	40	132	POISON INHALATION HAZARD*
	Trimethyl borate	3	UN2416	II	3	IB2, T7, TP1	150	202	242	5 L	60 L	B		129	FLAMMABLE
	Trimethyl phosphite	3	UN2329	III	3	B1, IB3, T2, TP1	150	203	242	60 L	220 L	A		130	FLAMMABLE
	1,3,5-Trimethyl-2,4,6-trinitrobenzene	Forbidden													
	Trimethylacetyl chloride	6.1	UN2438	I	6.1, 8, 3	2, B3, B9, B14, B32, N34, T20, TP2, TP13, TP38, TP45	None	227	244	Forbidden	Forbidden	D	25, 40	132	POISON INHALATION HAZARD*
	Trimethylamine, anhydrous	2.1	UN1083		2.1	N87, T50	306	304	314, 315	Forbidden	150 kg	B	40	118	FLAMMABLE GAS
	Trimethylamine, aqueous solutions *with not more than 50 percent trimethylamine by mass*	3	UN1297	I	3, 8	T11, TP1	None	201	243	0.5 L	2.5 L	D	40, 135	132	FLAMMABLE
				II	3, 8	B1, IB2, T7, TP1	150	202	243	1 L	5 L	B	40, 41	132	FLAMMABLE
				III	3, 8	B1, IB3, T7, TP1	150	203	242	5 L	60 L	A	40, 41	132	FLAMMABLE
	1,3,5-Trimethylbenzene	3	UN2325	III	3	B1, IB3, T2, TP1	150	203	242	60 L	220 L	A		129	FLAMMABLE
	Trimethylchlorosilane	3	UN1298	II	3, 8	A3, A7, B77, N34, T10, TP2, TP7, TP13	None	206	243	1 L	5 L	E	40	155	FLAMMABLE
	Trimethylcyclohexylamine	8	UN2326	III	8	IB3, T4, TP1	154	203	241	5 L	60 L	A		153	CORROSIVE
	Trimethylene glycol diperchlorate	Forbidden													
	Trimethylhexamethylene diisocyanate	6.1	UN2328	III	6.1	IB3, T4, TP2, TP13	153	203	241	60 L	220 L	B		156	POISON

HAZARDOUS MATERIALS TABLE

Symbols	Hazardous materials descriptions and proper shipping names	Hazard class or Division	Identification Numbers	PG	Label Codes	Special Provisions (§172.102)	(8) Packaging (§173.***)			(9) Quantity limitations (see §§173.27 and 175.75)		(10) Vessel stowage		ERG Guide No.	Placard Advisory Consult §172.504 & §172.505 to determine placard *Denotes placard for any quantity
							Exceptions	Non-bulk	Bulk	Passenger aircraft/rail	Cargo aircraft only	Location	Other		
(1)	(2)	(3)	(4)	(5)	(6)	(7)	(8A)	(8B)	(8C)	(9A)	(9B)	(10A)	(10B)		
	Trimethylhexamethylenediamines	8	UN2327	III	8	IB3, T4, TP1	154	203	241	5 L	60 L	A		153	CORROSIVE
	Trimethylol nitromethane trinitrate	Forbidden													
	Trinitro-m-cresol	1.1D	UN0216	II	1.1D		None	62	None	Forbidden	Forbidden	10	5E	112	EXPLOSIVES 1.1*
	2,4,6-Trinitro-1,3-diazobenzene	Forbidden													
	2,4,6-Trinitro-1,3,5-triazido benzene (dry)	Forbidden													
	Trinitroacetic acid	Forbidden													
	Trinitroacetonitrile	Forbidden													
	Trinitroamine cobalt	Forbidden													
	Trinitroaniline *or* **Picramide**	1.1D	UN0153	II	1.1D		None	62	None	Forbidden	Forbidden	10		112	EXPLOSIVES 1.1*
	Trinitroanisole	1.1D	UN0213	II	1.1D		None	62	None	Forbidden	Forbidden	10		112	EXPLOSIVES 1.1*
	Trinitrobenzene, *dry or wetted with less than 30 percent water, by mass*	1.1D	UN0214	II	1.1D		None	62	None	Forbidden	Forbidden	10		112	EXPLOSIVES 1.1*
	Trinitrobenzene, wetted, *with not less than 10% water by mass*	4.1	UN3367	I	4.1	162, A8, A19, N41, N84	None	211	None	0.5 kg	0.5 kg	E	36	113	FLAMMABLE SOLID
	Trinitrobenzene, wetted *with not less than 30 percent water, by mass*	4.1	UN1354	I	4.1	23, A2, A8, A19, N41	None	211	None	0.5 kg	0.5 kg	E	28	113	FLAMMABLE SOLID
	Trinitrobenzenesulfonic acid	1.1D	UN0386	II	1.1D		None	62	None	Forbidden	Forbidden	10	5E	112	EXPLOSIVES 1.1*
	Trinitrobenzoic acid, *dry or wetted with less than 30 percent water, by mass*	1.1D	UN0215	II	1.1D		None	62	None	Forbidden	Forbidden	10		112	EXPLOSIVES 1.1*
	Trinitrobenzoic acid, wetted, *with not less than 10% water by mass*	4.1	UN3368	I	4.1	162, A8, A19, N41, N84	None	211	None	0.5 kg	0.5 kg	E	36	113	FLAMMABLE SOLID
	Trinitrobenzoic acid, wetted *with not less than 30 percent water, by mass*	4.1	UN1355	I	4.1	23, A2, A8, A19, N41	None	211	None	0.5 kg	0.5 kg	E	28	113	FLAMMABLE SOLID
	Trinitrochlorobenzene *or* **Picryl chloride**	1.1D	UN0155	II	1.1D		None	62	None	Forbidden	Forbidden	10		112	EXPLOSIVES 1.1*
	Trinitrochlorobenzene (picryl chloride), wetted, *with not less than 10% water by mass*	4.1	UN3365	I	4.1	162, A8, A19, N41, N84	None	211	None	0.5 kg	0.5 kg	E	36	113	FLAMMABLE SOLID
	Trinitroethanol	Forbidden													
	Trinitroethylnitrate	Forbidden													
	Trinitrofluorenone	1.1D	UN0387	II	1.1D		None	62	None	Forbidden	Forbidden	10		112	EXPLOSIVES 1.1*
	Trinitromethane	Forbidden													
	1,3,5-Trinitronaphthalene	Forbidden													
	Trinitronaphthalene	1.1D	UN0217	II	1.1D		None	62	None	Forbidden	Forbidden	10		112	EXPLOSIVES 1.1*
	Trinitrophenetole	1.1D	UN0218	II	1.1D		None	62	None	Forbidden	Forbidden	10		112	EXPLOSIVES 1.1*
	Trinitrophenol *or* **Picric acid,** *dry or wetted with less than 30 percent water, by mass*	1.1D	UN0154	II	1.1D		None	62	None	Forbidden	Forbidden	10	5E	112	EXPLOSIVES 1.1*
	Trinitrophenol (picric acid), wetted, *with not less than 10 percent water by mass*	4.1	UN3364	I	4.1	162, A8, A19, N41, N84	None	211	None	0.5 kg	0.5 kg	E	36	113	FLAMMABLE SOLID
	Trinitrophenol, wetted *or* **Picric acid, wetted,** *with not less than 30 percent water by mass*	4.1	UN1344	I	4.1	23, A8, A19, N41	None	211	None	1 kg	15 kg	E	28, 36	113	FLAMMABLE SOLID
	2,4,6-Trinitrophenyl guanidine (dry)	Forbidden													
	2,4,6-Trinitrophenyl nitramine	Forbidden													
	2,4,6-Trinitrophenyl trimethylol methyl nitramine trinitrate (dry)	Forbidden													
	Trinitrophenylmethylnitramine *or* **Tetryl**	1.1D	UN0208	II	1.1D		None	62	None	Forbidden	Forbidden	10		112	EXPLOSIVES 1.1*
	Trinitroresorcinol *or* **Styphnic acid,** *dry or wetted with less than 20 percent water, or mixture of alcohol and water, by mass*	1.1D	UN0219	II	1.1D		None	62	None	Forbidden	Forbidden	10	5E	112	EXPLOSIVES 1.1*
	Trinitroresorcinol, wetted *or* **Styphnic acid, wetted** *with not less than 20 percent water, or mixture of alcohol and water by mass*	1.1D	UN0394	II	1.1D		None	62	None	Forbidden	Forbidden	10	5E	112	EXPLOSIVES 1.1*
	2,4,6-Trinitroso-3-methyl nitraminoanisole	Forbidden													
	Trinitrotetramine cobalt nitrate	Forbidden													
	Trinitrotoluene and Trinitrobenzene mixtures *or* **TNT and trinitrobenzene mixtures** *or* **TNT and hexanitrostilbene mixtures** *or* **Trinitrotoluene and hexanitrostilnene mixtures**	1.1D	UN0388	II	1.1D		None	62	None	Forbidden	Forbidden	10		112	EXPLOSIVES 1.1*

Symbols	Hazardous materials descriptions and proper shipping names	Hazard class or Division	Identification Numbers	PG	Label Codes	Special Provisions (§172.102)	(8) Packaging (§173.***)			(9) Quantity limitations (see §§173.27 and 175.75)		(10) Vessel stowage		ERG Guide No.	Placard Advisory Consult §172.504 & §172.505 to determine placard *Denotes placard for any quantity
							Exceptions	Non-bulk	Bulk	Passenger aircraft/rail	Cargo aircraft only	Location	Other		
(1)	(2)	(3)	(4)	(5)	(6)	(7)	(8A)	(8B)	(8C)	(9A)	(9B)	(10A)	(10B)		
	Trinitrotoluene mixtures containing Trinitrobenzene and Hexanitrostilbene *or* **TNT mixtures containing trinitrobenzene and hexanitrostilbene**	1.1D	UN0389	II	1.1D		None	62	None	Forbidden	Forbidden	10		112	EXPLOSIVES 1.1*
	Trinitrotoluene *or* **TNT,** *dry or wetted with less than 30 percent water, by mass*	1.1D	UN0209	II	1.1D		None	62	None	Forbidden	Forbidden	10		112	EXPLOSIVES 1.1*
	Trinitrotoluene (TNT), wetted, *with not less than 10 percent water by mass*	4.1	UN3366	I	4.1	162, A8, A19, N41, N84	None	211	None	0.5 kg	0.5 kg	E	36	113	FLAMMABLE SOLID
	Trinitrotoluene, wetted *or* **TNT, wetted,** with not less than 30 percent water by mass	4.1	UN1356	I	4.1	23, A2, A8, A19, N41	None	211	None	0.5 kg	0.5 kg	E	28, 36	113	FLAMMABLE SOLID
	Tripropylamine	3	UN2260	III	3, 8	B1, IB3, T4, TP1	150	203	242	5 L	60 L	A	40	132	FLAMMABLE
	Tripropylene	3	UN2057	II	3	IB2, T4, TP1	150	202	242	5 L	60 L	B		128	FLAMMABLE
				III	3	B1, IB3, T2, TP1	150	203	242	60 L	220 L	A		128	FLAMMABLE
	Tris-(1-aziridinyl)phosphine oxide, solution	6.1	UN2501	II	6.1	IB2, T7, TP2	153	202	243	5 L	60 L	A		152	POISON
				III	6.1	IB3, T4, TP1	153	203	241	60 L	220 L	A		152	POISON
	Tris, bis-bifluoroamino diethoxy propane (TVOPA)	Forbidden													
	Tritonal	1.1D	UN0390	II	1.1D		None	62	None	Forbidden	Forbidden	10		112	EXPLOSIVES 1.1*
	Tungsten hexafluoride	2.3	UN2196		2.3, 8	2, N86	None	338	None	Forbidden	Forbidden	D	40	125	POISON GAS*
	Turpentine	3	UN1299	III	3	B1, IB3, T2, TP1	150	203	242	60 L	220 L	A		128	FLAMMABLE
	Turpentine substitute	3	UN1300	I	3	T11, TP1, TP8, TP27	None	201	243	1 L	30 L	B		128	FLAMMABLE
				II	3	IB2, T4, TP1	150	202	242	5 L	60 L	B		128	FLAMMABLE
				III	3	B1, IB3, T2, TP1	150	203	242	60 L	220 L	A		128	FLAMMABLE
	Undecane	3	UN2330	III	3	B1, IB3, T2, TP1	150	203	242	60 L	220 L	A		128	FLAMMABLE
	Urea hydrogen peroxide	5.1	UN1511	III	5.1, 8	A1, A7, A29, IB8, IP3, T1, TP33	152	213	240	25 kg	100 kg	A	13	140	OXIDIZER
	Urea nitrate, *dry or wetted with less than 20 percent water, by mass*	1.1D	UN0220	II	1.1D	119	None	62	None	Forbidden	Forbidden	10		112	EXPLOSIVES 1.1*
	Urea nitrate, wetted, *with not less than 10 percent water by mass*	4.1	UN3370	I	4.1	162, A8, A19, N41, N83	None	211	None	0.5 kg	0.5 kg	E	36	113	FLAMMABLE SOLID
	Urea nitrate, wetted *with not less than 20 percent water, by mass*	4.1	UN1357	I	4.1	23, 39, A8, A19, N41	None	211	None	1 kg	15 kg	E	28, 36	113	FLAMMABLE SOLID
	Urea peroxide, see **Urea hydrogen peroxide**														
	Valeraldehyde	3	UN2058	II	3	IB2, T4, TP1	150	202	242	5 L	60 L	B		129	FLAMMABLE
	Valeric acid, see **Corrosive liquids, n.o.s.**														
	Valeryl chloride	8	UN2502	II	8, 3	A3, A6, A7, B2, IB2, N34, T7, TP2	154	202	243	1 L	30 L	C	40	132	CORROSIVE
G	**Vanadium compound, n.o.s.**	6.1	UN3285	I	6.1	IB7, IP1, T6, TP33	None	211	242	5 kg	50 kg	B		151	POISON
				II	6.1	IB8, IP2, IP4, T3, TP33	153	212	242	25 kg	100 kg	B		151	POISON
				III	6.1	IB8, IP3, T1, TP33	153	213	240	100 kg	200 kg	A		151	POISON
	Vanadium oxytrichloride	8	UN2443	II	8	A3, A6, A7, B2, B16, IB2, N34, T7, TP2	154	202	242	Forbidden	30 L	C	40	137	CORROSIVE
	Vanadium pentoxide, *non-fused form*	6.1	UN2862	III	6.1	IB8, IP3, T1, TP33	153	213	240	100 kg	200 kg	A	40	151	POISON
	Vanadium tetrachloride	8	UN2444	I	8	A3, A6, A7, B4, N34, T10, TP2	None	201	243	Forbidden	2.5 L	C	40	137	CORROSIVE
	Vanadium trichloride	8	UN2475	III	8	IB8, IP3, T1, TP33	154	213	240	25 kg	100 kg	A	40	157	CORROSIVE
	Vanadyl sulfate	6.1	UN2931	II	6.1	IB8, IP2, IP4, T3, TP33	153	212	242	25 kg	100 kg	A		151	POISON
	Vehicle, flammable gas powered *or* **Vehicle, fuel cell, flammable gas powered**	9	UN3166		9	135	220	220	220	Forbidden	No limit	A		128	CLASS 9
	Vehicle, flammable liquid powered *or* **Vehicle, fuel cell, flammable liquid powered**	9	UN3166		9	135	220	220	220	No limit	No limit	A		128	CLASS 9
	Very signal cartridge, see **Cartridges, signal**														
	Vinyl acetate, stabilized	3	UN1301	II	3	IB2, T4, TP1	150	202	242	5 L	60 L	B		129P	FLAMMABLE
	Vinyl bromide, stabilized	2.1	UN1085		2.1	N86, T50	306	304	314, 315	Forbidden	150 kg	B	40	116P	FLAMMABLE GAS
	Vinyl butyrate, stabilized	3	UN2838	II	3	IB2, T4, TP1	150	202	242	5 L	60 L	B		129P	FLAMMABLE

HAZARDOUS MATERIALS TABLE

Sym-bols (1)	Hazardous materials descriptions and proper shipping names (2)	Hazard class or Division (3)	Identification Numbers (4)	PG (5)	Label Codes (6)	Special Provisions (§172.102) (7)	(8) Packaging (§173.***) Exceptions (8A)	Non-bulk (8B)	Bulk (8C)	(9) Quantity limitations (see §§173.27 and 175.75) Passenger aircraft/rail (9A)	Cargo aircraft only (9B)	(10) Vessel stowage Location (10A)	Other (10B)	ERG Guide No.	Placard Advisory Consult §172.504 & §172.505 to determine placard *Denotes placard for any quantity
	Vinyl chloride, stabilized	2.1	UN1086		2.1	21, B44, N86, T50	306	304	314, 315	Forbidden	150 kg	B	40	116P	FLAMMABLE GAS
	Vinyl chloroacetate	6.1	UN2589	II	6.1, 3	IB2, T7, TP2	153	202	243	5 L	60 L	A		155	POISON
	Vinyl ethyl ether, stabilized	3	UN1302	I	3	A3, T11, TP2	None	201	243	1 L	30 L	D		127P	FLAMMABLE
	Vinyl fluoride, stabilized	2.1	UN1860		2.1	N86	306	304	314, 315	Forbidden	150 kg	E	40	116P	FLAMMABLE GAS
	Vinyl isobutyl ether, stabilized	3	UN1304	II	3	IB2, T4, TP1	150	202	242	5 L	60 L	B		127P	FLAMMABLE
	Vinyl methyl ether, stabilized	2.1	UN1087		2.1	B44, T50	306	304	314, 315	Forbidden	150 kg	B	40	116P	FLAMMABLE GAS
	Vinyl nitrate polymer	Forbidden													
	Vinylidene chloride, stabilized	3	UN1303	I	3	T12, TP2, TP7	150	201	243	1 L	30 L	E	40	130P	FLAMMABLE
	Vinylpyridines, stabilized	6.1	UN3073	II	6.1, 3, 8	IB1, T7, TP2, TP13	153	202	243	1 L	30 L	B	21, 40, 52	131P	POISON
	Vinyltoluenes, stabilized	3	UN2618	III	3	B1, IB3, T2, TP1	150	203	242	60 L	220 L	A		130P	FLAMMABLE
	Vinyltrichlorosilane, stabilized	3	UN1305	II	3, 8	A3, A7, B6, N34, T10, TP2, TP7, TP13	None	206	243	1 L	5 L	B	40	155P	FLAMMABLE
	Warheads, rocket *with burster or expelling charge*	1.4D	UN0370	II	1.4D		None	62	62	Forbidden	75 kg	02		114	EXPLOSIVES 1.4
	Warheads, rocket *with burster or expelling charge*	1.4F	UN0371	II	1.4F		None	62	None	Forbidden	Forbidden	08		114	EXPLOSIVES 1.4
	Warheads, rocket *with bursting charge*	1.1D	UN0286	II	1.1D		None	62	62	Forbidden	Forbidden	03		112	EXPLOSIVES 1.1*
	Warheads, rocket *with bursting charge*	1.2D	UN0287	II	1.2D		None	62	62	Forbidden	Forbidden	03		112	EXPLOSIVES 1.2*
	Warheads, rocket *with bursting charge*	1.1F	UN0369	II	1.1F		None	62	None	Forbidden	Forbidden	08		112	EXPLOSIVES 1.1*
	Warheads, torpedo *with bursting charge*	1.1D	UN0221	II	1.1D		None	62	62	Forbidden	Forbidden	03		112	EXPLOSIVES 1.1*
G	**Water-reactive liquid, corrosive, n.o.s.**	4.3	UN3129	I	4.3, 8	T14, TP2, TP7	None	201	243	Forbidden	1 L	D		138	DANGEROUS WHEN WET*
				II	4.3, 8	IB1, T11, TP2	None	202	243	1 L	5 L	E	85	138	DANGEROUS WHEN WET*
				III	4.3, 8	IB2, T7, TP1	None	203	242	5 L	60 L	E		138	DANGEROUS WHEN WET*
G	**Water-reactive liquid, n.o.s.**	4.3	UN3148	I	4.3	T9, TP2, TP7	None	201	244	Forbidden	1L	E	40	138	DANGEROUS WHEN WET*
				II	4.3	IB1, T7, TP2	None	202	243	1 L	5 L	E	40	138	DANGEROUS WHEN WET*
				III	4.3	IB2, T7, TP1	None	203	242	5 L	60 L	E	40	138	DANGEROUS WHEN WET*
G	**Water-reactive liquid, toxic, n.o.s.**	4.3	UN3130	I	4.3, 6.1	A4	None	201	243	Forbidden	1 L	D		139	DANGEROUS WHEN WET*
				II	4.3, 6.1	IB1	None	202	243	1 L	5 L	E	85	139	DANGEROUS WHEN WET*
				III	4.3, 6.1	IB2	None	203	242	5 L	60 L	E	85	139	DANGEROUS WHEN WET*
G	**Water-reactive solid, corrosive, n.o.s.**	4.3	UN3131	I	4.3, 8	IB4, IP1, N40, T9, TP7, TP33	None	211	242	Forbidden	15 kg	D		138	DANGEROUS WHEN WET*
				II	4.3, 8	IB6, IP2, T3, TP33	151	212	242	15 kg	50 kg	E	85	138	DANGEROUS WHEN WET*
				III	4.3, 8	IB8, IP4, T1, TP33	151	213	241	25 kg	100 kg	E	85	138	DANGEROUS WHEN WET*
G	**Water-reactive solid, flammable, n.o.s.**	4.3	UN3132	I	4.3, 4.1	IB4, N40	None	211	242	Forbidden	15 kg	D		138	DANGEROUS WHEN WET*
				II	4.3, 4.1	IB4, T3, TP33	151	212	242	15 kg	50 kg	E		138	DANGEROUS WHEN WET*
				III	4.3, 4.1	IB6, T1, TP33	151	213	241	25 kg	100 kg	E		138	DANGEROUS WHEN WET*
G	**Water-reactive solid, n.o.s.**	4.3	UN2813	I	4.3	IB4, N40, T9, TP7, TP33	None	211	242	Forbidden	15 kg	E	40	138	DANGEROUS WHEN WET*
				II	4.3	IB7, IP2, T3, TP33	151	212	242	15 kg	50 kg	E	40	138	DANGEROUS WHEN WET*
				III	4.3	IB8, IP4, T1, TP33	151	213	241	25 kg	100 kg	E	40	138	DANGEROUS WHEN WET*

Symbols (1)	Hazardous materials descriptions and proper shipping names (2)	Hazard class or Division (3)	Identification Numbers (4)	PG (5)	Label Codes (6)	Special Provisions (§172.102) (7)	(8) Packaging (§173.***) Exceptions (8A)	Non-bulk (8B)	Bulk (8C)	(9) Quantity limitations (see §§173.27 and 175.75) Passenger aircraft/rail (9A)	Cargo aircraft only (9B)	(10) Vessel stowage Location (10A)	Other (10B)	ERG Guide No.	Placard Advisory Consult §172.504 & §172.505 to determine placard *Denotes placard for any quantity
G	**Water-reactive solid, oxidizing, n.o.s.**	4.3	UN3133	II	4.3, 5.1		None	214	214	Forbidden	Forbidden	E	40	138	DANGEROUS WHEN WET*
				III	4.3, 5.1		None	214	214	Forbidden	Forbidden	E	40	138	DANGEROUS WHEN WET*
G	**Water-reactive solid, self-heating, n.o.s.**	4.3	UN3135	I	4.3, 4.2	N40	None	211	242	Forbidden	15 kg	E		138	DANGEROUS WHEN WET*
				II	4.3, 4.2	IB5, IP2, T3, TP33	None	212	242	15 kg	50 kg	E		138	DANGEROUS WHEN WET*
				III	4.3, 4.2	IB8, IP4, T1, TP33	None	213	241	25 kg	100 kg	E		138	DANGEROUS WHEN WET*
G	**Water-reactive solid, toxic, n.o.s.**	4.3	UN3134	I	4.3, 6.1	A8, IB4, IP1, N40	None	211	242	Forbidden	15 kg	D		139	DANGEROUS WHEN WET*
				II	4.3, 6.1	IB5, IP2, T3, TP33	151	212	242	15 kg	50 kg	E	85	139	DANGEROUS WHEN WET*
				III	4.3, 6.1	IB8, IP4, T1, TP33	151	213	241	25 kg	100 kg	E	85	139	DANGEROUS WHEN WET*
	Wheel chair, electric, see **Battery powered vehicle** *or* **Battery powered equipment**														
	White acid, see **Hydrofluoric acid**														
I	**White asbestos** *(chrysotile, actinolite, anthophyllite, tremolite)*	9	UN2590	III	9	156, IB8, IP2, IP3, T1, TP33	155	216	240	200 kg	200 kg	A	34, 40	171	CLASS 9
	Wood preservatives, liquid	3	UN1306	II	3	149, IB2, T4, TP1, TP8	150	202	242	5 L	60 L	B		129	FLAMMABLE
				III	3	B1, IB3, T2, TP1	150	203	242	60 L	220 L	A	40	129	FLAMMABLE
A I W	**Wool waste, wet**	4.2	UN1387	III	4.2		151	213	240	Forbidden	Forbidden	A		133	SPONTANEOUSLY COMBUSTIBLE
	Xanthates	4.2	UN3342	II	4.2	IB6, IP2, T3, TP33	None	212	241	15 kg	50 kg	D	40	135	SPONTANEOUSLY COMBUSTIBLE
				III	4.2	IB8, IP3, T1, TP33	None	213	241	25 kg	100 kg	D	40	135	SPONTANEOUSLY COMBUSTIBLE
	Xenon, compressed	2.2	UN2036		2.2		306, 307	302	None	75 kg	150 kg	A		121	NONFLAMMABLE GAS
	Xenon, refrigerated liquid *(cryogenic liquids)*	2.2	UN2591		2.2	T75, TP5	320	None	None	50 kg	500 kg	D		120	NONFLAMMABLE GAS
	Xylenes	3	UN1307	II	3	IB2, T4, TP1	150	202	242	5 L	60 L	B		130	FLAMMABLE
				III	3	B1, IB3, T2, TP1	150	203	242	60 L	220 L	A		130	FLAMMABLE
	Xylenols, solid	6.1	UN2261	II	6.1	IB8, IP2, IP4, T7, T3, TP33	153	212	242	25 kg	100 kg	A		153	POISON
	Xylenols, liquid	6.1	UN3430	II	6.1	IB2, T7, TP2	153	202	243	5 L	60 L	A		153	POISON
	Xylidines, liquid	6.1	UN1711	II	6.1	IB2, T7, TP2	153	202	243	5 L	60 L	A		153	POISON
	Xylidines, solid	6.1	UN3452	II	6.1	IB8, IP2, IP4, T7, T3, TP33	153	212	242	25 kg	100 kg	A		153	POISON
	Xylyl bromide, liquid	6.1	UN1701	II	6.1	A3, A6, A7, IB2, N33, T7, TP2, TP13	None	340	None	Forbidden	60 L	D	40	152	POISON
	Xylyl bromide, solid	6.1	UN3417	II	6.1	A3, A6, A7, IB8, IP2, IP4, N33, T3, TP33	None	340	None	25 kg	100 kg	B	40	152	POISON
	p-Xylyl diazide	Forbidden													
	Zinc ammonium nitrite	5.1	UN1512	II	5.1	IB8, IP4, T3, TP33	None	212	242	5 kg	25 kg	E		140	OXIDIZER
	Zinc arsenate *or* **Zinc arsenite** *or* **Zinc arsenate and zinc arsenite mixtures.**	6.1	UN1712	II	6.1	IB8, IP2, IP4, T3, TP33	153	212	242	25 kg	100 kg	A		151	POISON
	Zinc ashes	4.3	UN1435	III	4.3	A1, A19, IB8, IP4, T1, TP33	151	213	241	25 kg	100 kg	A		138	DANGEROUS WHEN WET*
	Zinc bisulfite solution, see **Bisulfites, aqueous solutions, n.o.s.**														
	Zinc bromate	5.1	UN2469	III	5.1	A1, A29, IB8, IP3, T1, TP33	152	213	240	25 kg	100 kg	A	56, 58	140	OXIDIZER
	Zinc chlorate	5.1	UN1513	II	5.1	A9, IB8, IP2, IP4, N34, T3, TP33	152	212	242	5 kg	25 kg	A	56, 58	140	OXIDIZER
	Zinc chloride, anhydrous	8	UN2331	III	8	IB8, IP3, T1, TP33	None	213	240	25 kg	100 kg	A		154	CORROSIVE

Symbols (1)	Hazardous materials descriptions and proper shipping names (2)	Hazard class or Division (3)	Identification Numbers (4)	PG (5)	Label Codes (6)	Special Provisions (§172.102) (7)	(8) Packaging (§173.***) Exceptions (8A)	Non-bulk (8B)	Bulk (8C)	(9) Quantity limitations (see §§173.27 and 175.75) Passenger aircraft/rail (9A)	Cargo aircraft only (9B)	(10) Vessel stowage Location (10A)	Other (10B)	ERG Guide No.	Placard Advisory Consult §172.504 & §172.505 to determine placard *Denotes placard for any quantity
	Zinc chloride, solution	8	UN1840	III	8	IB3, T4, TP1	154	203	241	5 L	60 L	A		154	CORROSIVE
	Zinc cyanide	6.1	UN1713	I	6.1	IB7, IP1, T6, TP33	None	211	242	5 kg	50 kg	A	52	151	POISON
	Zinc dithionite *or* **Zinc hydrosulfite**	9	UN1931	III	None	IB8, IP3, T1, TP33	155	204	240	100 kg	200 kg	A	49	171	CLASS 9
	Zinc ethyl, see **Diethylzinc**														
	Zinc fluorosilicate	6.1	UN2855	III	6.1	IB8, IP3, T1, TP33	153	213	240	100 kg	200 kg	A	52	151	POISON
	Zinc hydrosulfite, *see* **Zinc dithionite**														
	Zinc muriate solution, see **Zinc chloride, solution**														
	Zinc nitrate	5.1	UN1514	II	5.1	IB8, IP2, IP4, T3, TP33	152	212	240	5 kg	25 kg	A		140	OXIDIZER
	Zinc permanganate	5.1	UN1515	II	5.1	IB6, IP2, T3, TP33	152	212	242	5 kg	25 kg	D	56, 58, 138	140	OXIDIZER
	Zinc peroxide	5.1	UN1516	II	5.1	IB6, IP2, T3, TP33	152	212	242	5 kg	25 kg	A	13, 52, 66, 75	143	OXIDIZER
	Zinc phosphide	4.3	UN1714	I	4.3, 6.1	A19, N40	None	211	None	Forbidden	15 kg	E	40, 52, 85	139	DANGEROUS WHEN WET*
	Zinc powder *or* **Zinc dust**	4.3	UN1436	I	4.3, 4.2	A19, IB4, IP1, N40	None	211	242	Forbidden	15 kg	A	52, 53	138	DANGEROUS WHEN WET*
				II	4.3, 4.2	A19, IB7, IP2, T3, TP33	None	212	242	15 kg	50 kg	A	52, 53	138	DANGEROUS WHEN WET*
				III	4.3, 4.2	IB8, IP4, T1, TP33	None	213	242	25 kg	100 kg	A	52, 53	138	DANGEROUS WHEN WET*
	Zinc resinate	4.1	UN2714	III	4.1	A1, IB6, T1, TP33	151	213	240	25 kg	100 kg	A		133	FLAMMABLE SOLID
	Zinc selenate, see **Selenates** *or* **Selenites**														
	Zinc selenite, see **Selenates** *or* **Selenites**														
	Zinc silicofluoride, see **Zinc fluorosilicate**														
	Zirconium, dry, *coiled wire, finished metal sheets, strip (thinner than 254 microns but not thinner than 18 microns)*	4.1	UN2858	III	4.1	A1	151	213	240	25 kg	100 kg	A		170	FLAMMABLE SOLID
	Zirconium, dry, *finished sheets, strip or coiled wire*	4.2	UN2009	III	4.2	A1, A19	None	213	240	25 kg	100 kg	D		135	SPONTANEOUSLY COMBUSTIBLE
	Zirconium hydride	4.1	UN1437	II	4.1	A19, A20, IB4, N34, T3, TP33	None	212	240	15 kg	50 kg	E		138	FLAMMABLE SOLID
	Zirconium nitrate	5.1	UN2728	III	5.1	A1, A29, IB8, IP3, T1, TP33	152	213	240	25 kg	100 kg	A		140	OXIDIZER
	Zirconium picramate, *dry or wetted with less than 20 percent water, by mass*	1.3C	UN0236	II	1.3C		None	62	None	Forbidden	Forbidden	10	5E	112	EXPLOSIVES 1.3*
	Zirconium picramate, wetted *with not less than 20 percent water, by mass*	4.1	UN1517	I	4.1	23, N41	None	211	None	1 kg	15 kg	D	28, 36	113	FLAMMABLE SOLID
	Zirconium powder, dry	4.2	UN2008	I	4.2	T21, TP7, TP33	None	211	242	Forbidden	Forbidden	D		135	SPONTANEOUSLY COMBUSTIBLE
				II	4.2	A19, A20, IB6, IP2, N5, N34, T3, TP33	None	212	241	15 kg	50 kg	D		135	SPONTANEOUSLY COMBUSTIBLE
				III	4.2	IB8, IP3, T1, TP33	None	213	241	25 kg	100 kg	D		135	SPONTANEOUSLY COMBUSTIBLE
	Zirconium powder, wetted *with not less than 25 percent water (a visible excess of water must be present) (a) mechanically produced, particle size less than 53 microns; (b) chemically produced, particle size less than 840 microns*	4.1	UN1358	II	4.1	A19, A20, IB6, IP2, N34, T3, TP33	None	212	241	15 kg	50 kg	E	74	170	FLAMMABLE SOLID
	Zirconium scrap	4.2	UN1932	III	4.2	IB8, IP3, N34, T1, TP33	None	213	240	Forbidden	Forbidden	D		135	SPONTANEOUSLY COMBUSTIBLE
	Zirconium suspended in a liquid	3	UN1308	I	3		None	201	243	Forbidden	Forbidden	B		170	FLAMMABLE
				II	3	IB2	None	202	242	5 L	60 L	B		170	FLAMMABLE
				III	3	B1, IB2	150	203	242	60 L	220 L	B		170	FLAMMABLE
	Zirconium tetrachloride	8	UN2503	III	8	IB8, IP3, T1, TP33	154	213	240	25 kg	100 kg	A		137	CORROSIVE

Appendix A to §172.101 — List of Hazardous Substances and Reportable Quanties

1. This Appendix lists materials and their corresponding reportable quantities (RQs) that are listed or designated as "hazardous substances" under section 101(14) of the Comprehensive Environmental Response, Compensation, and Liability Act, 42 U.S.C. 9601(14) (CERCLA; 42 U.S.C. 9601 et seq). This listing fulfills the requirement of CERCLA, 42 U.S.C. 9656 (a), that all "hazardous substances," as defined in 42 U.S.C. 9601 (14), be listed and regulated as hazardous materials under 49 U.S.C. 5101-5127. That definition includes substances listed under sections 311(b)(2)(A) and 307(a) of the Federal Water Pollution Control Act, 33 U.S.C. 1321(b)(2)(A) and 1317(a), section 3001 of the Solid Waste Disposal Act, 42 U.S.C. 6921, and Section 112 of the Clean Air Act, 42 U.S.C. 7412. In addition, this list contains materials that the Administrator of the Environmental Protection Agency has determined to be hazardous substances in accordance with section 102 of CERCLA, 42 U.S.C. 9602. It should be noted that 42 U.S.C. 9656(b) provides that common and contract carriers may be held liable under laws other than CERCLA for the release of a hazardous substance as defined in that Act, during transportation that commenced before the effective date of the listing and regulating of that substance as a hazardous material under 49 U.S.C. 5101-5127.

2. This Appendix is divided into two TABLES which are entitled "TABLE 1—HAZARDOUS SUBSTANCES OTHER THAN RADIONUCLIDES" and "TABLE 2—RADIONUCLIDES." A material listed in this Appendix is regulated as a hazardous material and a hazardous substance under this subchapter if it meets the definition of a hazardous substance in §171.8 of this subchapter.

3. The procedure for selecting a proper shipping name for a hazardous substance is set forth in §172.101(c).

4. Column 1 of TABLE 1, entitled *"Hazardous substance"*, contains the names of those elements and compounds that are hazardous substances. Following the listing of elements and compounds is a listing of waste streams. These waste streams appear on the list in numerical sequence and are referenced by the appropriate "D", "F", or "K" numbers. Column 2 of TABLE 1, entitled "Reportable quantity (RQ)", contains the reportable quantity (RQ), in pounds and kilograms, for each hazardous substance listed in Column 1 of TABLE 1.

5. A series of notes is used throughout TABLE 1 and TABLE 2 to provide additional information concerning certain hazardous substances. These notes are explained at the end of each TABLE.

6. TABLE 2 lists radionuclides that are hazardous substances and their corresponding RQ's. The RQ's in TABLE 2 for radionuclides are expressed in units of curies and terabecquerels, whereas those in TABLE 1 are expressed in units of pounds and kilograms. If a material is listed in both TABLE 1 and TABLE 2, the lower RQ shall apply. Radionuclides are listed in alphabetical order. The RQs for radionuclides are given in the radiological unit of measure of curie, abbreviated "Ci", followed, in parentheses, by an equivalent unit measured in terabecquerels, abbreviated "TBq".

7. For mixtures of radionuclides, the following requirements shall be used in determining if a package contains an RQ of a hazardous substance: (i) if the identity and quantity (in curies or terabecquerels) of each radionuclide in a mixture or solution is known, the ratio between the quantity per package (in curies or terabecquerels) and the RQ for the radionuclide must be determined for each radionuclide. A package contains an RQ of a hazardous substance when the sum of the ratios for the radionuclides in the mixture or solution is equal to or greater than one; (ii) if the identity of each radionuclide in a mixture or solution is known but the quantity per package (in curies or terabecquerels) of one or more of the radionuclides is unknown, an RQ of a hazardous substance is present in a package when the total quantity (in curies or terabecquerels) of the mixture or solution is equal to or greater than the lowest RQ of any individual radionuclide in the mixture or solution; and (iii) if the identity of one or more radionuclides in a mixture or solution is unknown (or if the identity of a radionuclide by itself is unknown), an RQ of a hazardous substance is present when the total quantity (in curies or terabecquerels) in a package is equal to or greater than either one curie or the lowest RQ of any known individual radionuclide in the mixture or solution, whichever is lower.

TABLE 1—HAZARDOUS SUBSTANCES OTHER THAN RADIONUCLIDES

Hazardous substance	Reportable quantity (RQ) pounds (kilograms)
A2213	5000 (2270)
Acenaphthene	100 (45.4)
Acenaphthylene	5000 (2270)
Acetaldehyde	1000 (454)
Acetaldehyde, chloro-	1000 (454)
Acetaldehyde, trichloro-	5000 (2270)
Acetamide	100 (45.4)
Acetamide, N-(aminothioxomethyl)-	1000 (454)
Acetamide, N-(4-ethoxyphenyl)-	100 (45.4)
Acetamide, N-9H-fluoren-2-yl-	1 (0.454)
Acetamide 2-fluoro-	100 (45.4)
Acetic acid	5000 (2270)
Acetic acid, (2,4-dichlorophenoxy)-, salts & esters	100 (45.4)
Acetic acid, ethyl ester	5000 (2270)
Acetic acid, fluoro-, sodium salt	10 (4.54)
Acetic acid, lead(2+) salt	10 (4.54)
Acetic acid, thallium(1+) salt	100 (45.4)
Acetic acid, (2,4,5-trichlorophenoxy)-	1000 (454)
Acetic anhydride	5000 (2270)
Acetone	5000 (2270)
Acetone cyanohydrin	10 (4.54)
Acetonitrile	5000 (2270)
Acetophenone	5000 (2270)
2-Acetylaminofluorene	1 (0.454)
Acetyl bromide	5000 (2270)
Acetyl chloride	5000 (2270)
1-Acetyl-2-thiourea	1000 (454)
Acrolein	1 (0.454)
Acrylamide	5000 (2270)
Acrylic acid	5000 (2270)
Acrylonitrile	100 (45.4)
Adipic acid	5000 (2270)
Aldicarb	1 (0.454)
Aldicarb sulfone	100 (45.4)
Aldrin	1 (0.454)
Allyl alcohol	100 (45.4)
Allyl chloride	1000 (454)
Aluminum phosphide	100 (45.4)
Aluminum sulfate	5000 (2270)
4-Aminobiphenyl	1 (0.454)
5-(Aminomethyl)-3-isoxazolol	1000 (454)
4-Aminopyridine	1000 (454)
Amitrole	10 (4.54)
Ammonia	100 (45.4)
Ammonium acetate	5000 (2270)
Ammonium benzoate	5000 (2270)
Ammonium bicarbonate	5000 (2270)
Ammonium bichromate	10 (4.54)
Ammonium bifluoride	100 (45.4)
Ammonium bisulfite	5000 (2270)
Ammonium carbamate	5000 (2270)
Ammonium carbonate	5000 (2270)
Ammonium chloride	5000 (2270)
Ammonium chromate	10 (4.54)
Ammonium citrate, dibasic	5000 (2270)
Ammonium dichromate @	10 (4.54)

TABLE 1—HAZARDOUS SUBSTANCES OTHER THAN RADIONUCLIDES, Continued

Hazardous substance	Reportable quantity (RQ) pounds (kilograms)
Ammonium fluoborate	5000 (2270)
Ammonium fluoride	100 (45.4)
Ammonium hydroxide	1000 (454)
Ammonium oxalate	5000 (2270)
Ammonium picrate	10 (4.54)
Ammonium silicofluoride	1000 (454)
Ammonium sulfamate	5000 (2270)
Ammonium sulfide	100 (45.4)
Ammonium sulfite	5000 (2270)
Ammonium tartrate	5000 (2270)
Ammonium thiocyanate	5000 (2270)
Ammonium vanadate	1000 (454)
Amyl acetate	5000 (2270)
iso-Amyl acetate	
sec-Amyl acetate	
tert-Amyl acetate	
Aniline	5000 (2270)
o-Anisidine	100 (45.4)
Anthracene	5000 (2270)
Antimony [c]	5000 (2270)
Antimony pentachloride	1000 (454)
Antimony potassium tartrate	100 (45.4)
Antimony tribromide	1000 (454)
Antimony trichloride	1000 (454)
Antimony trifluoride	1000 (454)
Antimony trioxide	1000 (454)
Argentate(1-), bis(cyano-C)-, potassium	1 (0.454)
Aroclor 1016	1 (0.454)
Aroclor 1221	1 (0.454)
Aroclor 1232	1 (0.454)
Aroclor 1242	1 (0.454)
Aroclor 1248	1 (0.454)
Aroclor 1254	1 (0.454)
Aroclor 1260	1 (0.454)
Aroclors	1 (0.454)
Arsenic [c]	1 (0.454)
Arsenic acid H_3AsO_4	1 (0.454)
Arsenic disulfide	1 (0.454)
Arsenic oxide As_2O_3	1 (0.454)
Arsenic oxide As_2O_5	1 (0.454)
Arsenic pentoxide	1 (0.454)
Arsenic trichloride	1 (0.454)
Arsenic trioxide	1 (0.454)
Arsenic trisulfide	1 (0.454)
Arsine, diethyl-	1 (0.454)
Arsinic acid, dimethyl-	1 (0.454)
Arsonous dichloride, phenyl-	1 (0.454)
Asbestos [cc]	1 (0.454)
Auramine	100 (45.4)
Azaserine	1 (0.454)
Aziridine	1 (0.454)
Aziridine, 2-methyl-	1 (0.454)
Azirino[2',3':3,4]pyrrolo[1,2-a]indole-4,7-dione, 6-amino-8-[[(aminocarbonyl)oxy]methyl]-1,1a,2,8,8a,8b-hexahydro-8a-methoxy-5-methyl-, [1aS-(1aalpha,8beta,8aalpha, 8balpha)]-	10 (4.54)
Barban	10 (4.54)
Barium cyanide	10 (4.54)
Bendiocarb	100 (45.4)
Bendiocarb phenol	1000 (454)
Benomyl	10 (4.54)
Benz[j]aceanthrylene, 1,2-dihydro-3-methyl-	10 (4.54)
Benz[c]acridine	100 (45.4)
Benzal chloride	5000 (2270)
Benzamide, 3,5-dichloro-N-(1,1-dimethyl-2-propynyl)-	5000 (2270)
Benz[a]anthracene	10 (4.54)
1,2-Benzanthracene	10 (4.54)
Benz[a]anthracene, 7,12-dimethyl-	1 (0.454)
Benzenamine	5000 (2270)
Benzenamine, 4,4'-carbonimidoylbis (N,N dimethyl-	100 (45.4)
Benzenamine, 4-chloro-	1000 (454)
Benzenamine, 4-chloro-2-methyl-, hydrochloride	100 (45.4)
Benzenamine, N,N-dimethyl-4-(phenylazo)-	10 (4.54)
Benzenamine, 2-methyl-	100 (45.4)
Benzenamine, 4-methyl-	100 (45.4)
Benzenamine, 4,4'-methylenebis[2-chloro-	10 (4.54)
Benzenamine, 2-methyl-, hydrochloride	100 (45.4)
Benzenamine, 2-methyl-5-nitro-	100 (45.4)
Benzenamine, 4-nitro-	5000 (2270)
Benzene	10 (4.54)
Benzeneacetic acid, 4-chloro-α-(4-chlorophenyl)-α-hydroxy-, ethyl ester	10 (4.54)
Benzene, 1-bromo-4-phenoxy-	100 (45.4)
Benzenebutanoic acid, 4-[bis(2-chloroethyl)amino]-	10 (4.54)
Benzene, chloro-	100 (45.4)
Benzene, (chloromethyl)-	100 (45.4)
Benzenediamine, ar-methyl-	10 (4.54)
1,2-Benzenedicarboxylic acid, bis(2-ethylhexyl) ester	100 (45.4)
1,2-Benzenedicarboxylic acid, dibutyl ester	10 (4.54)
1,2-Benzenedicarboxylic acid, diethyl ester	1000 (454)
1,2-Benzenedicarboxylic acid, dimethyl ester	5000 (2270)
1,2-Benzenedicarboxylic acid, dioctyl ester	5000 (2270)
Benzene, 1,2-dichloro-	100 (45.4)
Benzene, 1,3-dichloro-	100 (45.4)
Benzene, 1,4-dichloro-	100 (45.4)
Benzene, 1,1'-(2,2-dichloroethylidene) bis[4-chloro-	1 (0.454)
Benzene, (dichloromethyl)-	5000 (2270)
Benzene, 1,3-diisocyanatomethyl-	100 (45.4)
Benzene, dimethyl-	100 (45.4)
1,3-Benzenediol	5000 (2270)
1,2-Benzenediol,4-[1-hydroxy-2-(methylamino) ethyl]-	1000 (454)
Benzeneethanamine, alpha,alpha-dimethyl-	5000 (2270)
Benzene, hexachloro-	10 (4.54)
Benzene, hexahydro-	1000 (454)
Benzene, methyl-	1000 (454)
Benzene, 1-methyl-2,4-dinitro-	10 (4.54)
Benzene, 2-methyl-1,3-dinitro-	100 (45.4)
Benzene, (1-methylethyl)-	5000 (2270)
Benzene, nitro-	1000 (454)
Benzene, pentachloro-	10 (4.54)
Benzene, pentachloronitro-	100 (45.4)
Benzenesulfonic acid chloride	100 (45.4)
Benzenesulfonyl chloride	100 (45.4)
Benzene,1,2,4,5-tetrachloro-	5000 (2270)
Benzenethiol	100 (45.4)
Benzene,1,1'-(2,2,2-trichloroethylidene) bis[4-chloro-	1 (0.454)
Benzene,1,1'-(2,2,2-trichloroethylidene) bis[4-methoxy-	1 (0.454)
Benzene, (trichloromethyl)-	10 (4.54)
Benzene, 1,3,5-trinitro-	10 (4.54)
Benzidine	1 (0.454)
Benzo[a]anthracene	10 (4.54)
1,3-Benzodioxole, 5-(1-propenyl)-1	100 (45.4)
1,3-Benzodioxole, 5-(2-propenyl)-	100 (45.4)
1,3-Benzodioxole, 5-propyl-	10 (4.54)
1,3-Benzodioxol-4-ol, 2,2-dimethyl-	1000 (454)
1,3-Benzodioxol-4-ol, 2,2-dimethyl-, methyl carbamate	100 (45.4)
Benzo[b]fluoranthene	1 (0.454)
Benzo(k)fluoranthene	5000 (2270)
7-Benzofuranol, 2,3-dihydro-2,2-dimethyl-	10 (4.54)
7-Benzofuranol, 2,3-dihydro-2,2-dimethyl-, methylcarbamate	10 (4.54)
Benzoic acid	5000 (2270)
Benzoic acid, 2-hydroxy-, compd. With (3aS-cis)-1,2,3,3a,8,8a-hexahydro-1,3a,8-trimethylpyrrolo [2,3-b]indol-5-yl methylcarbamate ester (1:1)	100 (45.4)
Benzonitrile	5000 (2270)
Benzo[rst]pentaphene	10 (4.54)
Benzo[ghi]perylene	5000 (2270)
2H-1-Benzopyran-2-one, 4-hydroxy-3-(3-oxo-1-phenylbutyl)-, & salts	100 (45.4)
Benzo[a]pyrene	1 (0.454)
3,4-Benzopyrene	1 (0.454)
p-Benzoquinone	10 (4.54)
Benzotrichloride	10 (4.54)

TABLE 1—HAZARDOUS SUBSTANCES OTHER THAN RADIONUCLIDES, Continued

Hazardous substance	Reportable quantity (RQ) pounds (kilograms)
Benzoyl chloride	1000 (454)
Benzyl chloride	100 (45.4)
Beryllium ¢	10 (4.54)
Beryllium chloride	1 (0.454)
Beryllium fluoride	1 (0.454)
Beryllium nitrate	1 (0.454)
Beryllium powder ¢	10 (4.54)
alpha-BHC	10 (4.54)
beta-BHC	1 (0.454)
delta-BHC	1 (0.454)
gamma-BHC	1 (0.454)
2,2'-Bioxirane	10 (4.54)
Biphenyl	100 (45.4)
[1,1'-Biphenyl]-4,4'-diamine	1 (0.454)
[1,1'-Biphenyl]-4,4'-diamine,3,3'-dichloro-	1 (0.454)
[1,1'-Biphenyl]-4,4'-diamine,3,3'-dimethoxy-	100 (45.4)
[1,1'-Biphenyl]-4,4'-diamine,3,3'-dimethyl-	10 (4.54)
Bis(2-chloroethoxy) methane	1000 (454)
Bis(2-chloroethyl) ether	10 (4.54)
Bis(chloromethyl) ether	10 (4.54)
Bis(2-ethylhexyl) phthalate	100 (45.4)
Bromoacetone	1000 (454)
Bromoform	100 (45.4)
Bromomethane	1000 (454)
4-Bromophenyl phenyl ether	100 (45.4)
Brucine	100 (45.4)
1,3-Butadiene	10 (4.54)
1,3-Butadiene, 1,1,2,3,4,4-hexachloro-	1 (0.454)
1-Butanamine, N-butyl-N-nitroso-	10 (4.54)
1-Butanol	5000 (2270)
2-Butanone	5000 (2270)
2-Butanone, 3,3-dimethyl-1(methylthio)-, O [(methylamino) carbonyl] oxime	100 (45.4)
2-Butanone peroxide	10 (4.54)
2-Butenal	100 (45.4)
2-Butene, 1,4-dichloro-	1 (0.454)
2-Butenoic acid, 2-methyl-, 7-[[2,3-dihydroxy-2-(1-methoxyethyl)-3-methyl-1-oxobutoxy] methyl]-2,3,5,7a-tetrahydro-1H-pyrrolizin-1-yl ester, [1S-[1alpha(Z), 7(2S*,3R*),7aalpha]]-	10 (4.54)
Butyl acetate	5000 (2270)
iso-Butyl acetate	....
sec-Butyl acetate	....
tert-Butyl acetate	....
n-Butyl alcohol	5000 (2270)
Butylamine	1000 (454)
iso-Butylamine	....
sec-Butylamine	....
tert-Butylamine	....
Butyl benzyl phthalate	100 (45.4)
n-Butyl phthalate	10 (4.54)
Butyric acid	5000 (2270)
iso-Butyric acid	....
Cacodylic acid	1 (0.454)
Cadmium ¢	10 (4.54)
Cadmium acetate	10 (4.54)
Cadmium bromide	10 (4.54)
Cadmium chloride	10 (4.54)
Calcium arsenate	1 (0.454)
Calcium arsenite	1 (0.454)
Calcium carbide	10 (4.54)
Calcium chromate	10 (4.54)
Calcium cyanamide	1000 (454)
Calcium cyanide $Ca(CN)_2$	10 (4.54)
Calcium dodecylbenzenesulfonate	1000 (454)
Calcium hypochlorite	10 (4.54)
Captan	10 (4.54)
Carbamic acid, 1H-benzimidazol-2-yl, methyl ester	10 (4.54)
Carbamic acid, [1-[(butylamino)carbonyl]-1H-benzimidazol-2-yl]-, methyl ester	10 (4.54)
Carbamic acid, (3-chlorophenyl)-, 4-chloro-2-butynyl ester	10 (4.54)
Carbamic acid, [(dibutylamino)-thio]methyl-, 2,3-dihydro-2,2-dimethyl-7-benzofuranyl ester	1000 (454)
Carbamic acid, dimethyl-,1-[(dimethyl-amino)carbonyl]-5-methyl-1H-pyrazol-3-yl ester	1 (0.454)
Carbamic acid, dimethyl-, 3-methyl-1-(1-methylethyl)-1H-pyrazol-5-yl ester	100 (45.4)
Carbamic acid, ethyl ester	100 (45.4)
Carbamic acid, methyl-, 3-methylphenyl ester	1000 (454)
Carbamic acid, methylnitroso-, ethyl ester	1 (0.454)
Carbamic acid, [1,2-phenylenebis(iminocarbonothioyl)] bis-, dimethyl ester	10 (4.54)
Carbamic acid, phenyl-, 1-methylethyl ester	1000 (454)
Carbamic chloride, dimethyl-	1 (0.454)
Carbamodithioic acid, 1,2-ethanediylbis-, salts & esters	5000 (2270)
Carbamothioic acid, bis(1-methylethyl)-, S-(2,3-dichloro-2-propenyl) ester	100 (45.4)
Carbamothioic acid, bis(1-methylethyl)-, S-(2,3,3-trichloro-2-propenyl) ester	100 (45.4)
Carbamothioic acid, dipropyl-, S-(phenylmethyl) ester	5000 (2270)
Carbaryl	100 (45.4)
Carbendazim	10 (4.54)
Carbofuran	10 (4.54)
Carbofuran phenol	10 (4.54)
Carbon disulfide	100 (45.4)
Carbonic acid, dithallium(1+) salt	100 (45.4)
Carbonic dichloride	10 (4.54)
Carbonic difluoride	1000 (454)
Carbonochloridic acid, methyl ester	1000 (454)
Carbon oxyfluoride	1000 (454)
Carbon tetrachloride	10 (4.54)
Carbonyl sulfide	100 (45.4)
Carbosulfan	1000 (454)
Catechol	100 (45.4)
Chloral	5000 (2270)
Chloramben	100 (45.4)
Chlorambucil	10 (4.54)
Chlordane	1 (0.454)
Chlordane, alpha & gamma isomers	1 (0.454)
CHLORDANE (TECHNICAL MIXTURE AND METABOLITES)	1 (0.454)
Chlorinated camphene	1 (0.454)
Chlorine	10 (4.54)
Chlornaphazine	100 (45.4)
Chloroacetaldehyde	1000 (454)
Chloroacetic acid	100 (45.4)
2-Chloroacetophenone	100 (45.4)
p-Chloroaniline	1000 (454)
Chlorobenzene	100 (45.4)
Chlorobenzilate	10 (4.54)
p-Chloro-m-cresol	5000 (2270)
Chlorodibromomethane	100 (45.4)
1-Chloro-2,3-epoxypropane	100 (45.4)
Chloroethane	100 (45.4)
2-Chloroethyl vinyl ether	1000 (454)
Chloroform	10 (4.54)
Chloromethane	100 (45.4)
Chloromethyl methyl ether	10 (4.54)
beta-Chloronaphthalene	5000 (2270)
2-Chloronaphthalene	5000 (2270)
2-Chlorophenol	100 (45.4)
o-Chlorophenol	100 (45.4)
4-Chlorophenyl phenyl ether	5000 (2270)
1-(o-Chlorophenyl)thiourea	100 (45.4)
Chloroprene	100 (45.4)
3-Chloropropionitrile	1000 (454)
Chlorosulfonic acid	1000 (454)
4-Chloro-o-toluidine, hydrochloride	100 (45.4)
Chlorpyrifos	1 (0.454)
Chromic acetate	1000 (454)
Chromic acid	10 (4.54)
Chromic acid H_2CrO_4, calcium salt	10 (4.54)
Chromic sulfate	1000 (454)

HAZARDOUS SUBSTANCES RQ

TABLE 1—HAZARDOUS SUBSTANCES OTHER THAN RADIONUCLIDES, Continued

Hazardous substance	Reportable quantity (RQ) pounds (kilograms)
Chromium [¢]	5000 (2270)
Chromous chloride	1000 (454)
Chrysene	100 (45.4)
Cobaltous bromide	1000 (454)
Cobaltous formate	1000 (454)
Cobaltous sulfamate	1000 (454)
Coke Oven Emissions	1 (0.454)
Copper [¢]	5000 (2270)
Copper chloride [@]	10 (4.54)
Copper cyanide Cu(CN)	10 (4.54)
Coumaphos	10 (4.54)
Creosote	1 (0.454)
Cresol (cresylic acid)	100 (45.4)
m-Cresol	100 (45.4)
o-Cresol	100 (45.4)
p-Cresol	100 (45.4)
Cresols (isomers and mixture)	100 (45.4)
Cresylic acid (isomers and mixture)	100 (45.4)
Crotonaldehyde	100 (45.4)
Cumene	5000 (2270)
m-Cumenyl methylcarbamate	10 (4.54)
Cupric acetate	100 (45.4)
Cupric acetoarsenite	1 (0.454)
Cupric chloride	10 (4.54)
Cupric nitrate	100 (45.4)
Cupric oxalate	100 (45.4)
Cupric sulfate	10 (4.54)
Cupric sulfate, ammoniated	100 (45.4)
Cupric tartrate	100 (45.4)
Cyanides (soluble salts and complexes) not otherwise specified	10 (4.54)
Cyanogen	100 (45.4)
Cyanogen bromide (CN)Br	1000 (454)
Cyanogen chloride (CN)Cl	10 (4.54)
2,5-Cyclohexadiene-1,4-dione	10 (4.54)
Cyclohexane	1000 (454)
Cyclohexane, 1,2,3,4,5,6-hexachloro-, (1α, 2α, 3β-, 4α, 5α, 6β)	1 (0.454)
Cyclohexanone	5000 (2270)
2-Cyclohexyl-4,6-dinitrophenol	100 (45.4)
1,3-Cyclopentadiene, 1,2,3,4,5,5-hexachloro-	10 (4.54)
Cyclophosphamide	10 (4.54)
2,4-D Acid	100 (45.4)
2,4-D Ester	100 (45.4)
2,4-D, salts and esters	100 (45.4)
Daunomycin	10 (4.54)
DDD	1 (0.454)
4,4'-DDD	1 (0.454)
DDE (72-55-9) [#]	1 (0.454)
DDE (3547-04-4) [#]	5000 (2270)
4,4'-DDE	1 (0.454)
DDT	1 (0.454)
4,4'-DDT	1 (0.454)
DEHP	100 (45.4)
Diallate	100 (45.4)
Diazinon	1 (0.454)
Diazomethane	100 (45.4)
Dibenz[a,h]anthracene	1 (0.454)
1,2:5,6-Dibenzanthracene	1 (0.454)
Dibenzo[a,h]anthracene	1 (0.454)
Dibenzofuran	100 (45.4)
Dibenzo[a,i]pyrene	10 (4.54)
1,2-Dibromo-3-chloropropane	1 (0.454)
Dibromoethane	1 (0.454)
Dibutyl phthalate	10 (4.54)
Di-n-butyl phthalate	10 (4.54)
Dicamba	1000 (454)
Dichlobenil	100 (45.4)
Dichlone	1 (0.454)
Dichlorobenzene	100 (45.4)
1,2-Dichlorobenzene	100 (45.4)

Hazardous substance	Reportable quantity (RQ) pounds (kilograms)
1,3-Dichlorobenzene	100 (45.4)
1,4-Dichlorobenzene	100 (45.4)
m-Dichlorobenzene	100 (45.4)
o-Dichlorobenzene	100 (45.4)
p-Dichlorobenzene	100 (45.4)
3,3'-Dichlorobenzidine	1 (0.454)
Dichlorobromomethane	5000 (2270)
1,4-Dichloro-2-butene	1 (0.454)
Dichlorodifluoromethane	5000 (2270)
1,1-Dichloroethane	1000 (454)
1,2-Dichloroethane	100 (45.4)
1,1-Dichloroethylene	100 (45.4)
1,2-Dichloroethylene	1000 (454)
Dichloroethyl ether	10 (4.54)
Dichloroisopropyl ether	1000 (454)
Dichloromethane	1000 (454)
Dichloromethoxyethane	1000 (454)
Dichloromethyl ether	10 (4.54)
2,4-Dichlorophenol	100 (45.4)
2,6-Dichlorophenol	100 (45.4)
Dichlorophenylarsine	1 (0.454)
Dichloropropane	1000 (454)
1,1-Dichloropropane	....
1,3-Dichloropropane	....
1,2-Dichloropropane	1000 (454)
Dichloropropane-Dichloropropene (mixture)	100 (45.4)
Dichloropropene	100 (45.4)
2,3-Dichloropropene	....
1,3-Dichloropropene	100 (45.4)
2,2-Dichloropropionic acid	5000 (2270)
Dichlorvos	10 (4.54)
Dicofol	10 (4.54)
Dieldrin	1 (0.454)
1,2:3,4-Diepoxybutane	10 (4.54)
Diethanolamine	100 (45.4)
Diethylamine	100 (45.4)
N,N-Diethylaniline	1000 (454)
Diethylarsine	1 (0.454)
Diethylene glycol, dicarbamate	5000 (2270)
1,4-Diethyleneoxide	100 (45.4)
Diethylhexyl phthalate	100 (45.4)
N,N'-Diethylhydrazine	10 (4.54)
O,O-Diethyl S-methyl dithiophosphate	5000 (2270)
Diethyl-p-nitrophenyl phosphate	100 (45.4)
Diethyl phthalate	1000 (454)
O,O-Diethyl O-pyrazinyl phosphorothioate	100 (45.4)
Diethylstilbestrol	1 (0.454)
Diethyl sulfate	10 (4.54)
Dihydrosafrole	10 (4.54)
Diisopropylfluorophosphate (DFP)	100 (45.4)
1,4:5,8-Dimethanonaphthalene, 1,2,3,4,10,10-hexachloro-1,4,4a,5,8,8a-hexahydro-, (1alpha, 4alpha, 4abeta, 5alpha, 8alpha, 8abeta)-	1 (0.454)
1,4:5,8-Dimethanonaphthalene, 1,2,3,4,10,10-hexachloro-1,4,4a,5,8,8a-hexahydro-, (1alpha, 4alpha, 4abeta, 5beta, 8beta, 8abeta)-	1 (0.454)
2,7:3,6-Dimethanonaphth[2,3-b]oxirene,3,4,5,6,9,9-hexachloro-1a,2,2a,3,6,6a,7,7a-octahydro-, (1aalpha, 2beta, 2aalpha, 3beta, 6beta, 6aalpha, 7beta, 7aalpha)-	1 (0.454)
2,7:3,6-Dimethanonaphth[2, 3-b]oxirene,3,4,5,6,9,9-hexachloro-1a,2,2a,3,6,6a,7,7a-octahydro-, (1aalpha, 2beta, 2abeta, 3alpha, 6alpha, 6abeta, 7beta, 7aalpha)-, & metabolites	1 (0.454)
Dimethoate	10 (4.54)
3,3'-Dimethoxybenzidine	100 (45.4)
Dimethylamine	1000 (454)
Dimethyl aminoazobenzene	10 (4.54)
p-Dimethylaminoazobenzene	10 (4.54)
N,N-Dimethylaniline	100 (45.4)
7,12-Dimethylbenz[a]anthracene	1 (0.454)
3,3'-Dimethylbenzidine	10 (4.54)
alpha,alpha-Dimethylbenzylhydroperoxide	10 (4.54)

TABLE 1—HAZARDOUS SUBSTANCES OTHER THAN RADIONUCLIDES, Continued

Hazardous substance	Reportable quantity (RQ) pounds (kilograms)
Dimethylcarbamoyl chloride	1 (0.454)
Dimethylformamide	100 (45.4)
1,1-Dimethylhydrazine	10 (4.54)
1,2-Dimethylhydrazine	1 (0.454)
Dimethylhydrazine, unsymmetrical @	10 (4.54)
alpha,alpha-Dimethylphenethylamine	5000 (2270)
2,4-Dimethylphenol	100 (45.4)
Dimethyl phthalate	5000 (2270)
Dimethyl sulfate	100 (45.4)
Dimetilan	1 (0.454)
Dinitrobenzene (mixed)	100 (45.4)
m-Dinitrobenzene	
o-Dinitrobenzene	
p-Dinitrobenzene	
4,6-Dinitro-o-cresol, and salts	10 (4.54)
Dinitrogen tetroxide @	10 (4.54)
Dinitrophenol	10 (4.54)
2,5-Dinitrophenol	
2,6-Dinitrophenol	
2,4-Dinitrophenol	10 (4.54)
Dinitrotoluene	10 (4.54)
3,4-Dinitrotoluene	
2,4-Dinitrotoluene	10 (4.54)
2,6-Dinitrotoluene	100 (45.4)
Dinoseb	1000 (454)
Di-n-octyl phthalate	5000 (2270)
1,4-Dioxane	100 (45.4)
1,2-Diphenylhydrazine	10 (4.54)
Diphosphoramide, octamethyl-	100 (45.4)
Diphosphoric acid, tetraethyl ester	10 (4.54)
Dipropylamine	5000 (2270)
Di-n-propylnitrosamine	10 (4.54)
Diquat	1000 (454)
Disulfoton	1 (0.454)
Dithiobiuret	100 (45.4)
1,3-Dithiolane-2-carboxaldehyde, 2,4-dimethyl-, O-[(methylamino)-carbonyl]oxime	100 (45.4)
Diuron	100 (45.4)
Dodecylbenzenesulfonic acid	1000 (454)
Endosulfan	1 (0.454)
alpha-Endosulfan	1 (0.454)
beta-Endosulfan	1 (0.454)
Endosulfan sulfate	1 (0.454)
Endothall	1000 (454)
Endrin	1 (0.454)
Endrin aldehyde	1 (0.454)
Endrin, & metabolites	1 (0.454)
Epichlorohydrin	100 (45.4)
Epinephrine	1000 (454)
1,2-Epoxybutane	100 (45.4)
Ethanal	1000 (454)
Ethanamine, N,N-diethyl-	5000 (2270)
Ethanamine, N-ethyl-N-nitroso-	1 (0.454)
1,2-Ethanediamine, N,N-dimethyl-N'-2-pyridinyl-N'-(2-thienylmethyl)-	5000 (2270)
Ethane, 1,2-dibromo-	1 (0.454)
Ethane, 1,1-dichloro-	1000 (454)
Ethane, 1,2-dichloro-	100 (45.4)
Ethanedinitrile	100 (45.4)
Ethane, hexachloro-	100 (45.4)
Ethane, 1,1'-[methylenebis(oxy)]bis[2-chloro-	1000 (454)
Ethane, 1,1'-oxybis-	100 (45.4)
Ethane, 1,1'-oxybis[2-chloro-	10 (4.54)
Ethane, pentachloro-	10 (4.54)
Ethane, 1,1,1,2-tetrachloro-	100 (45.4)
Ethane, 1,1,2,2-tetrachloro-	100 (45.4)
Ethanethioamide	10 (4.54)
Ethane, 1,1,1-trichloro-	1000 (454)
Ethane, 1,1,2-trichloro-	100 (45.4)
Ethanimidothioic acid, 2-(dimethylamino)-N-hydroxy-2-oxo-, methyl ester	5000 (2270)
Ethanimidothioic acid, 2-(dimethylamino)-N-[[(methylamino) carbonyl]oxy]-2-oxo-, methyl ester	100 (45.4)
Ethanimidothioic acid, N-[[(methylamino) carbonyl]oxy]-, methyl ester	100 (45.4)
Ethanimidothioic acid, N,N'[thiobis[(methylimino)carbonyloxy]] bis-, dimethyl ester	100 (45.4)
Ethanol, 2-ethoxy-	1000 (454)
Ethanol, 2,2'-(nitrosoimino)bis-	1 (0.454)
Ethanol, 2,2'-oxybis-, dicarbamate	5000 (2270)
Ethanone, 1-phenyl-	5000 (2270)
Ethene, chloro-	1 (0.454)
Ethene, (2-chloroethoxy)-	1000 (454)
Ethene, 1,1-dichloro-	100 (45.4)
Ethene, 1,2-dichloro-(E)	1000 (454)
Ethene, tetrachloro-	100 (45.4)
Ethene, trichloro-	100 (45.4)
Ethion	10 (4.54)
Ethyl acetate	5000 (2270)
Ethyl acrylate	1000 (454)
Ethylbenzene	1000 (454)
Ethyl carbamate	100 (45.4)
Ethyl chloride	100 (45.4)
Ethyl cyanide	10 (4.54)
Ethylenebisdithiocarbamic acid, salts & esters	5000 (2270)
Ethylenediamine	5000 (2270)
Ethylenediamine-tetraacetic acid (EDTA)	5000 (2270)
Ethylene dibromide	1 (0.454)
Ethylene dichloride	100 (45.4)
Ethylene glycol	5000 (2270)
Ethylene glycol monoethyl ether	1000 (454)
Ethylene oxide	10 (4.54)
Ethylenethiourea	10 (4.54)
Ethylenimine	1 (0.454)
Ethyl ether	100 (45.4)
Ethylidene dichloride	1000 (454)
Ethyl methacrylate	1000 (454)
Ethyl methanesulfonate	1 (0.454)
Ethyl methyl ketone @	5000 (2270)
Famphur	1000 (454)
Ferric ammonium citrate	1000 (454)
Ferric ammonium oxalate	1000 (454)
Ferric chloride	1000 (454)
Ferric fluoride	100 (45.4)
Ferric nitrate	1000 (454)
Ferric sulfate	1000 (454)
Ferrous ammonium sulfate	1000 (454)
Ferrous chloride	100 (45.4)
Ferrous sulfate	1000 (454)
Fluoranthene	100 (45.4)
Fluorene	5000 (2270)
Fluorine	10 (4.54)
Fluoroacetamide	100 (45.4)
Fluoroacetic acid, sodium salt	10 (4.54)
Formaldehyde	100 (45.4)
Formetanate hydrochloride	100 (45.4)
Formic acid	5000 (2270)
Formparanate	100 (45.4)
Fulminic acid, mercury(2+)salt	10 (4.54)
Fumaric acid	5000 (2270)
Furan	100 (45.4)
2-Furancarboxyaldehyde	5000 (2270)
2,5-Furandione	5000 (2270)
Furan, tetrahydro-	1000 (454)
Furfural	5000 (2270)
Furfuran	100 (45.4)
Glucopyranose, 2-deoxy-2-(3-methyl-3-nitrosoureido)-, D-	1 (0.454)
D-Glucose, 2-deoxy-2-[[(methylnitrosoamino)-carbonyl]amino]-	1 (0.454)
Glycidylaldehyde	10 (4.54)
Guanidine, N-methyl-N'-nitro-N-nitroso-	10 (4.54)

TABLE 1—HAZARDOUS SUBSTANCES OTHER THAN RADIONUCLIDES, Continued

Hazardous substance	Reportable quantity (RQ) pounds (kilograms)
Guthion	1 (0.454)
Heptachlor	1 (0.454)
Heptachlor epoxide	1 (0.454)
Hexachlorobenzene	10 (4.54)
Hexachlorobutadiene	1 (0.454)
Hexachlorocyclopentadiene	10 (4.54)
Hexachloroethane	100 (45.4)
Hexachlorophene	100 (45.4)
Hexachloropropene	1000 (454)
Hexaethyl tetraphosphate	100 (45.4)
Hexamethylene-1,6-diisocyanate	100 (45.4)
Hexamethylphosphoramide	1 (0.454)
Hexane	5000 (2270)
Hexone	5000 (2270)
Hydrazine	1 (0.454)
Hydrazinecarbothioamide	100 (45.4)
Hydrazine, 1,2-diethyl-	10 (4.54)
Hydrazine, 1,1-dimethyl-	10 (4.54)
Hydrazine, 1,2-dimethyl-	1 (0.454)
Hydrazine, 1,2-diphenyl-	10 (4.54)
Hydrazine, methyl-	10 (4.54)
Hydrochloric acid	5000 (2270)
Hydrocyanic acid	10 (4.54)
Hydrofluoric acid	100 (45.4)
Hydrogen chloride	5000 (2270)
Hydrogen cyanide	10 (4.54)
Hydrogen fluoride	100 (45.4)
Hydrogen phosphide	100 (45.4)
Hydrogen sulfide H2S	100 (45.4)
Hydroperoxide, 1-methyl-1-phenylethyl-	10 (4.54)
Hydroquinone	100 (45.4)
2-Imidazolidinethione	10 (4.54)
Indeno(1,2,3-cd)pyrene	100 (45.4)
Iodomethane	100 (45.4)
1,3-Isobenzofurandione	5000 (2270)
Isobutyl alcohol	5000 (2270)
Isodrin	1 (0.454)
Isolan	100 (45.4)
Isophorone	5000 (2270)
Isoprene	100 (45.4)
Isopropanolamine dodecylbenzenesulfonate	1000 (454)
3-Isopropylphenyl N-methylcarbamate	10 (4.54)
Isosafrole	100 (45.4)
3(2H)-Isoxazolone, 5-(aminomethyl)-	1000 (454)
Kepone	1 (0.454)
Lasiocarpine	10 (4.54)
Lead [¢]	10 (4.54)
Lead acetate	10 (4.54)
Lead arsenate	1 (0.454)
Lead, bis(acetato-O)tetrahydroxytri-	10 (4.54)
Lead chloride	10 (4.54)
Lead fluoborate	10 (4.54)
Lead fluoride	10 (4.54)
Lead iodide	10 (4.54)
Lead nitrate	10 (4.54)
Lead phosphate	10 (4.54)
Lead stearate	10 (4.54)
Lead subacetate	10 (4.54)
Lead sulfate	10 (4.54)
Lead sulfide	10 (4.54)
Lead thiocyanate	10 (4.54)
Lindane	1 (0.454)
Lindane (all isomers)	1 (0.454)
Lithium chromate	10 (4.54)
Malathion	100 (45.4)
Maleic acid	5000 (2270)
Maleic anhydride	5000 (2270)
Maleic hydrazide	5000 (2270)
Malononitrile	1000 (454)
Manganese, bis(dimethylcarbamodithioato-S,S')-	10 (4.54)
Manganese dimethyldithiocarbamate	10 (4.54)
MDI	5000 (2270)
MEK	5000 (2270)
Melphalan	1 (0.454)
Mercaptodimethur	10 (4.54)
Mercuric cyanide	1 (0.454)
Mercuric nitrate	10 (4.54)
Mercuric sulfate	10 (4.54)
Mercuric thiocyanate	10 (4.54)
Mercurous nitrate	10 (4.54)
Mercury	1 (0.454)
Mercury, (acetato-O)phenyl-	100 (45.4)
Mercury fulminate	10 (4.54)
Methacrylonitrile	1000 (454)
Methanamine, N-methyl-	1000 (454)
Methanamine, N-methyl-N-nitroso-	10 (4.54)
Methane, bromo-	1000 (454)
Methane, chloro-	100 (45.4)
Methane, chloromethoxy-	10 (4.54)
Methane, dibromo-	1000 (454)
Methane, dichloro-	1000 (454)
Methane, dichlorodifluoro-	5000 (2270)
Methane, iodo-	100 (45.4)
Methane, isocyanato-	10 (4.54)
Methane, oxybis(chloro-	10 (4.54)
Methanesulfenyl chloride, trichloro-	100 (45.4)
Methanesulfonic acid, ethyl ester	1 (0.454)
Methane, tetrachloro-	10 (4.54)
Methane, tetranitro-	10 (4.54)
Methanethiol	100 (45.4)
Methane, tribromo-	100 (45.4)
Methane, trichloro-	10 (4.54)
Methane, trichlorofluoro-	5000 (2270)
Methanimidamide, N,N-dimethyl-N'-[3-[[(methylamino) carbonyl] oxy] phenyl]-, monohydrochloride	100 (45.4)
Methanimidamide, N,N-dimethyl-N'-[2-methyl-4-[[(methylamino)carbonyl] oxy]phenyl]-	100 (45.4)
6,9-Methano-2,4,3-benzodioxathiepin,6,7,8,9,10,10-hexachloro-1,5,5a,6,9,9a-hexahydro-, 3-oxide	1 (0.454)
4,7-Methano-1H-indene, 1,4,5,6,7,8,8-heptachloro-3a,4,7,7a-tetrahydro-	1 (0.454)
4,7-Methano-1H-indene, 1,2,4,5,6,7,8,8-octachloro-2,3,3a,4,7,7a-hexahydro-	1 (0.454)
Methanol	5000 (2270)
Methapyrilene	5000 (2270)
1,3,4-Metheno-2H-cyclobuta[cd]pentalen-2-one, 1,1a,3,3a,4,5,5,5a,5b,6-decachlorooctahydro-	1 (0.454)
Methiocarb	10 (4.54)
Methomyl	100 (45.4)
Methoxychlor	1 (0.454)
Methyl alcohol	5000 (2270)
Methylamine [@]	100 (45.4)
2-Methyl aziridine	1 (0.454)
Methyl bromide	1000 (454)
1-Methylbutadiene	100 (45.4)
Methyl chloride	100 (45.4)
Methyl chlorocarbonate	1000 (454)
Methyl chloroform	1000 (454)
Methyl chloroformate [@]	1000 (454)
Methyl chloromethyl ether [@]	10 (4.54)
3-Methylcholanthrene	10 (4.54)
4,4'-Methylenebis(2-chloroaniline)	10 (4.54)
Methylene bromide	1000 (454)
Methylene chloride	1000 (454)
4,4'-Methylenedianiline	10 (4.54)
Methylene diphenyl diisocyanate	5000 (2270)
Methyl ethyl ketone	5000 (2270)
Methyl ethyl ketone peroxide	10 (4.54)
Methyl hydrazine	10 (4.54)
Methyl iodide	100 (45.4)
Methyl isobutyl ketone	5000 (2270)
Methyl isocyanate	10 (4.54)
2-Methyllactonitrile	10 (4.54)

TABLE 1—HAZARDOUS SUBSTANCES OTHER THAN RADIONUCLIDES, Continued

Hazardous substance	Reportable quantity (RQ) pounds (kilograms)
Methyl mercaptan	100 (45.4)
Methyl methacrylate	1000 (454)
Methyl parathion	100 (45.4)
4-Methyl-2-pentanone	5000 (2270)
Methyl tert-butyl ether	1000 (454)
Methylthiouracil	10 (4.54)
Metolcarb	1000 (454)
Mevinphos	10 (4.54)
Mexacarbate	1000 (454)
Mitomycin C	10 (4.54)
MNNG	10 (4.54)
Monoethylamine	100 (45.4)
Monomethylamine	100 (45.4)
Naled	10 (4.54)
5,12-Naphthacenedione, 8-acetyl-10-[(3-amino-2,3,6-trideoxy-alpha-L-lyxo-hexopyranosyl)oxy]-7,8,9,10-tetrahydro-6,8,11-trihydroxy-1-methoxy-, (8S-cis)-	10 (4.54)
1-Naphthalenamine	100 (45.4)
2-Naphthalenamine	10 (4.54)
Naphthalenamine, N,N'-bis(2-chloroethyl)-	100 (45.4)
Naphthalene	100 (45.4)
Naphthalene, 2-chloro-	5000 (2270)
1,4-Naphthalenedione	5000 (2270)
2,7-Naphthalenedisulfonic acid, 3,3'-[(3,3'-dimethyl-(1,1'-biphenyl)-4,4'-diyl)-bis(azo)]bis(5-amino-4-hydroxy)-tetrasodium salt	10 (4.54)
1-Naphthalenol, methylcarbamate	100 (45.4)
Naphthenic acid	100 (45.4)
1,4-Naphthoquinone	5000 (2270)
alpha-Naphthylamine	100 (45.4)
beta-Naphthylamine	10 (4.54)
alpha-Naphthylthiourea	100 (45.4)
Nickel ¢	100 (45.4)
Nickel ammonium sulfate	100 (45.4)
Nickel carbonyl Ni(CO)4, (T-4)-	10 (4.54)
Nickel chloride	100 (45.4)
Nickel cyanide $Ni(CN)_2$	10 (4.54)
Nickel hydroxide	10 (4.54)
Nickel nitrate	100 (45.4)
Nickel sulfate	100 (45.4)
Nicotine, & salts	100 (45.4)
Nitric acid	1000 (454)
Nitric acid, thallium (1+) salt	100 (45.4)
Nitric oxide	10 (4.54)
p-Nitroaniline	5000 (2270)
Nitrobenzene	1000 (454)
4-Nitrobiphenyl	10 (4.54)
Nitrogen dioxide	10 (4.54)
Nitrogen oxide NO	10 (4.54)
Nitrogen oxide NO_2	10 (4.54)
Nitroglycerine	10 (4.54)
Nitrophenol (mixed)	100 (45.4)
m-Nitrophenol	....
o-Nitrophenol	100 (45.4)
p-Nitrophenol	100 (45.4)
2-Nitrophenol	100 (45.4)
4-Nitrophenol	100 (45.4)
2-Nitropropane	10 (4.54)
N-Nitrosodi-n-butylamine	10 (4.54)
N-Nitrosodiethanolamine	1 (0.454)
N-Nitrosodiethylamine	1 (0.454)
N-Nitrosodimethylamine	10 (4.54)
N-Nitrosodiphenylamine	100 (45.4)
N-Nitroso-N-ethylurea	1 (0.454)
N-Nitroso-N-methylurea	1 (0.454)
N-Nitroso-N-methylurethane	1 (0.454)
N-Nitrosomethylvinylamine	10 (4.54)
N-Nitrosomorpholine	1 (0.454)
N-Nitrosopiperidine	10 (4.54)
N-Nitrosopyrrolidine	1 (0.454)
Nitrotoluene	1000 (454)
m-Nitrotoluene	....
o-Nitrotoluene	....
p-Nitrotoluene	....
5-Nitro-o-toluidine	100 (45.4)
Octamethylpyrophosphoramide	100 (45.4)
Osmium oxide OsO_4, (T-4)-	1000 (454)
Osmium tetroxide	1000 (454)
7-Oxabicyclo[2.2.1]heptane-2,3-dicarboxylic acid	1000 (454)
Oxamyl	100 (45.4)
1,2-Oxathiolane, 2,2-dioxide	10 (4.54)
2H-1,3,2-Oxazaphosphorin-2-amine, N,N-bis(2-chloroethyl) tetrahydro-, 2-oxide	10 (4.54)
Oxirane	10 (4.54)
Oxiranecarboxyaldehyde	10 (4.54)
Oxirane, (chloromethyl)-	100 (45.4)
Paraformaldehyde	1000 (454)
Paraldehyde	1000 (454)
Parathion	10 (4.54)
PCBs	1 (0.454)
PCNB	100 (45.4)
Pentachlorobenzene	10 (4.54)
Pentachloroethane	10 (4.54)
Pentachloronitrobenzene	100 (45.4)
Pentachlorophenol	10 (4.54)
1,3-Pentadiene	100 (45.4)
Perchloroethylene	100 (45.4)
Perchloromethyl mercaptan @	100 (45.4)
Phenacetin	100 (45.4)
Phenanthrene	5000 (2270)
Phenol	1000 (454)
Phenol, 2-chloro-	100 (45.4)
Phenol, 4-chloro-3-methyl-	5000 (2270)
Phenol, 2-cyclohexyl-4,6-dinitro-	100 (45.4)
Phenol, 2,4-dichloro-	100 (45.4)
Phenol, 2,6-dichloro-	100 (45.4)
Phenol, 4,4'-(1,2-diethyl-1,2-ethenediyl)bis-, (E)	1 (0.454)
Phenol, 2,4-dimethyl-	100 (45.4)
Phenol, 4-(dimethylamino)-3,5-dimethyl-, methylcarbamate (ester)	1000 (454)
Phenol, (3,5-dimethyl-4-(methylthio)-, methylcarbamate	10 (4.54)
Phenol, 2,4-dinitro-	10 (4.54)
Phenol, methyl-	100 (45.4)
Phenol, 2-methyl-4,6-dinitro-, & salts	10 (4.54)
Phenol, 2,2'-methylenebis[3,4,6-trichloro-	100 (45.4)
Phenol, 2-(1-methylethoxy)-, methylcarbamate	100 (45.4)
Phenol, 3-(1-methylethyl)-, methyl carbamate	10 (4.54)
Phenol, 3-methyl-5-(1-methylethyl)-, methyl carbamate	1000 (454)
Phenol, 2-(1-methylpropyl)-4,6-dinitro-	1000 (454)
Phenol, 4-nitro-	100 (45.4)
Phenol, pentachloro-	10 (4.54)
Phenol, 2,3,4,6-tetrachloro-	10 (4.54)
Phenol, 2,4,5-trichloro-	10 (4.54)
Phenol, 2,4,6-trichloro-	10 (4.54)
Phenol, 2,4,6-trinitro-, ammonium salt	10 (4.54)
L-Phenylalanine, 4-[bis(2-chloroethyl)amino]-	1 (0.454)
p-Phenylenediamine	5000 (2270)
Phenyl mercaptan @	100 (45.4)
Phenylmercury acetate	100 (45.4)
Phenylthiourea	100 (45.4)
Phorate	10 (4.54)
Phosgene	10 (4.54)
Phosphine	100 (45.4)
Phosphoric acid	5000 (2270)
Phosphoric acid, diethyl 4-nitrophenyl ester	100 (45.4)
Phosphoric acid, lead(2+) salt (2:3)	10 (4.54)
Phosphorodithioic acid, O,O-diethyl S-[2-(ethylthio)ethyl] ester	1 (0.454)
Phosphorodithioic acid, O,O-diethyl S-[(ethylthio)methyl] ester	10 (4.54)
Phosphorodithioic acid, O,O-diethyl S-methyl ester	5000 (2270)
Phosphorodithioic acid, O,O-dimethyl S-[2-(methylamino)-2-oxoethyl] ester	10 (4.54)
Phosphorofluoridic acid, bis(1-methylethyl) ester	100 (45.4)

TABLE 1—HAZARDOUS SUBSTANCES OTHER THAN RADIONUCLIDES, Continued

Hazardous substance	Reportable quantity (RQ) pounds (kilograms)
Phosphorothioic acid, O,O-diethyl O-(4-nitrophenyl) ester	10 (4.54)
Phosphorothioic acid, O,O-diethyl O-pyrazinyl ester	100 (45.4)
Phosphorothioic acid, O-[4-[(dimethylamino) sulfonyl]phenyl] O,O-dimethyl ester	1000 (454)
Phosphorothioic acid, O,O-dimethyl O-(4-nitrophenyl) ester	100 (45.4)
Phosphorus	1 (0.454)
Phosphorus oxychloride	1000 (454)
Phosphorus pentasulfide	100 (45.4)
Phosphorus sulfide	100 (45.4)
Phosphorus trichloride	1000 (454)
Phthalic anhydride	5000 (2270)
Physostigmine	100 (45.4)
Physostigmine salicylate	100 (45.4)
2-Picoline	5000 (2270)
Piperidine, 1-nitroso-	10 (4.54)
Plumbane, tetraethyl-	10 (4.54)
POLYCHLORINATED BIPHENYLS	1 (0.454)
Potassium arsenate	1 (0.454)
Potassium arsenite	1 (0.454)
Potassium bichromate	10 (4.54)
Potassium chromate	10 (4.54)
Potassium cyanide K(CN)	10 (4.54)
Potassium hydroxide	1000 (454)
Potassium permanganate	100 (45.4)
Potassium silver cyanide	1 (0.454)
Promecarb	1000 (454)
Pronamide	5000 (2270)
Propanal, 2-methyl-2-(methyl-sulfonyl)-, O-[(methylamino)carbonyl] oxime	100 (45.4)
Propanal, 2-methyl-2-(methylthio)-, O-[(methylamino)carbonyl] oxime	1 (0.454)
1-Propanamine	5000 (2270)
1-Propanamine, N-propyl-	5000 (2270)
1-Propanamine, N-nitroso-N-propyl-	10 (4.54)
Propane, 1,2-dibromo-3-chloro-	1 (0.454)
Propane, 1,2-dichloro-	1000 (454)
Propanedinitrile	1000 (454)
Propanenitrile	10 (4.54)
Propanenitrile, 3-chloro-	1000 (454)
Propanenitrile, 2-hydroxy-2-methyl-	10 (4.54)
Propane, 2-nitro-	10 (4.54)
Propane, 2,2'-oxybis[2-chloro-	1000 (454)
1,3-Propane sultone	10 (4.54)
1,2,3-Propanetriol, trinitrate	10 (4.54)
Propanoic acid, 2-(2,4,5-trichlorophenoxy)-	100 (45.4)
1-Propanol, 2,3-dibromo-, phosphate (3:1)	10 (4.54)
1-Propanol, 2-methyl-	5000 (2270)
2-Propanone	5000 (2270)
2-Propanone, 1-bromo-	1000 (454)
Propargite	10 (4.54)
Propargyl alcohol	1000 (454)
2-Propenal	1 (0.454)
2-Propenamide	5000 (2270)
1-Propene, 1,3-dichloro-	100 (45.4)
1-Propene, 1,1,2,3,3,3-hexachloro-	1000 (454)
2-Propenenitrile	100 (45.4)
2-Propenenitrile, 2-methyl-	1000 (454)
2-Propenoic acid	5000 (2270)
2-Propenoic acid, ethyl ester	1000 (454)
2-Propenoic acid, 2-methyl-, ethyl ester	1000 (454)
2-Propenoic acid, 2-methyl-, methyl ester	1000 (454)
2-Propen-1-ol	100 (45.4)
Propham	1000 (454)
beta-Propiolactone	10 (4.54)
Propionaldehyde	1000 (454)
Propionic acid	5000 (2270)
Propionic anhydride	5000 (2270)
Propoxur (Baygon)	100 (45.4)
n-Propylamine	5000 (2270)
Propylene dichloride	1000 (454)
Propylene oxide	100 (45.4)
1,2-Propylenimine	1 (0.454)
2-Propyn-1-ol	1000 (454)
Prosulfocarb	5000 (2270)
Pyrene	5000 (2270)
Pyrethrins	1 (0.454)
3,6-Pyridazinedione, 1,2-dihydro-	5000 (2270)
4-Pyridinamine	1000 (454)
Pyridine	1000 (454)
Pyridine, 2-methyl-	5000 (2270)
Pyridine, 3-(1-methyl-2-pyrrolidinyl)-, (S)-, & salts	100 (45.4)
2,4-(1H,3H)-Pyrimidinedione, 5-[bis(2-chloroethyl)amino]-	10 (4.54)
4(1H)-Pyrimidinone, 2,3-dihydro-6-methyl-2-thioxo-	10 (4.54)
Pyrrolidine, 1-nitroso-	1 (0.454)
Pyrrolo[2,3-b] indol-5-ol,1,2,3,3a,8,8a-hexahydro-1,3a,8-trimethyl-, methylcarbamate (ester), (3aS-cis)-	100 (45.4)
Quinoline	5000 (2270)
Quinone	10 (4.54)
Quintobenzene	100 (45.4)
RADIONUCLIDES	See Table 2
Reserpine	5000 (2270)
Resorcinol	5000 (2270)
Safrole	100 (45.4)
Selenious acid	10 (4.54)
Selenious acid, dithallium (1+) salt	1000 (454)
Selenium [c]	100 (45.4)
Selenium dioxide	10 (4.54)
Selenium oxide	10 (4.54)
Selenium sulfide SeS2	10 (4.54)
Selenourea	1000 (454)
L-Serine, diazoacetate (ester)	1 (0.454)
Silver [c]	1000 (454)
Silver cyanide Ag(CN)	1 (0.454)
Silver nitrate	1 (0.454)
Silvex (2,4,5-TP)	100 (45.4)
Sodium	10 (4.54)
Sodium arsenate	1 (0.454)
Sodium arsenite	1 (0.454)
Sodium azide	1000 (454)
Sodium bichromate	10 (4.54)
Sodium bifluoride	100 (45.4)
Sodium bisulfite	5000 (2270)
Sodium chromate	10 (4.54)
Sodium cyanide Na(CN)	10 (4.54)
Sodium dodecylbenzenesulfonate	1000 (454)
Sodium fluoride	1000 (454)
Sodium hydrosulfide	5000 (2270)
Sodium hydroxide	1000 (454)
Sodium hypochlorite	100 (45.4)
Sodium methylate	1000 (454)
Sodium nitrite	100 (45.4)
Sodium phosphate, dibasic	5000 (2270)
Sodium phosphate, tribasic	5000 (2270)
Sodium selenite	100 (45.4)
Streptozotocin	1 (0.454)
Strontium chromate	10 (4.54)
Strychnidin-10-one, & salts	10 (4.54)
Strychnidin-10-one, 2,3-dimethoxy-	100 (45.4)
Strychnine, & salts	10 (4.54)
Styrene	1000 (454)
Styrene oxide	100 (45.4)
Sulfur chlorides [@]	1000 (454)
Sulfuric acid	1000 (454)
Sulfuric acid, dimethyl ester	100 (45.4)
Sulfuric acid, dithallium (1+) salt	100 (45.4)
Sulfur monochloride	1000 (454)
Sulfur phosphide	100 (45.4)
2,4,5-T	1000 (454)
2,4,5-T acid	1000 (454)
2,4,5-T amines	5000 (2270)

TABLE 1—HAZARDOUS SUBSTANCES OTHER THAN RADIONUCLIDES, Continued

Hazardous substance	Reportable quantity (RQ) pounds (kilograms)
2,4,5-T esters	1000 (454)
2,4,5-T salts	1000 (454)
TCDD	1 (0.454)
TDE	1 (0.454)
1,2,4,5-Tetrachlorobenzene	5000 (2270)
2,3,7,8-Tetrachlorodibenzo-p-dioxin	1 (0.454)
1,1,1,2-Tetrachloroethane	100 (45.4)
1,1,2,2-Tetrachloroethane	100 (45.4)
Tetrachloroethylene	100 (45.4)
2,3,4,6-Tetrachlorophenol	10 (4.54)
Tetraethyl pyrophosphate	10 (4.54)
Tetraethyl lead	10 (4.54)
Tetraethyldithiopyrophosphate	100 (45.4)
Tetrahydrofuran	1000 (454)
Tetranitromethane	10 (4.54)
Tetraphosphoric acid, hexaethyl ester	100 (45.4)
Thallic oxide	100 (45.4)
Thallium [c]	1000 (454)
Thallium (I) acetate	100 (45.4)
Thallium (I) carbonate	100 (45.4)
Thallium chloride TlCl	100 (45.4)
Thallium (I) nitrate	100 (45.4)
Thallium oxide Tl_2O_3	100 (45.4)
Thallium (I) selenite	1000 (454)
Thallium (I) sulfate	100 (45.4)
Thioacetamide	10 (4.54)
Thiodicarb	100 (45.4)
Thiodiphosphoric acid, tetraethyl ester	100 (45.4)
Thiofanox	100 (45.4)
Thioimidodicarbonic diamide $[(H_2N)C(S)]_2NH$	100 (45.4)
Thiomethanol	100 (45.4)
Thioperoxydicarbonic diamide $[(H_2N)C(S)]_2S_2$, tetramethyl-	10 (4.54)
Thiophanate-methyl	10 (4.54)
Thiophenol	100 (45.4)
Thiosemicarbazide	100 (45.4)
Thiourea	10 (4.54)
Thiourea, (2-chlorophenyl)-	100 (45.4)
Thiourea, 1-naphthalenyl-	100 (45.4)
Thiourea, phenyl-	100 (45.4)
Thiram	10 (4.54)
Tirpate	100 (45.4)
Titanium tetrachloride	1000 (454)
Toluene	1000 (454)
Toluenediamine	10 (4.54)
2,4-Toluene diamine	10 (4.54)
Toluene diisocyanate	100 (45.4)
2,4-Toluene diisocyanate	100 (45.4)
o-Toluidine	100 (45.4)
p-Toluidine	100 (45.4)
o-Toluidine hydrochloride	100 (45.4)
Toxaphene	1 (0.454)
2,4,5-TP acid	100 (45.4)
2,4,5-TP esters	100 (45.4)
Triallate	100 (45.4)
1H-1,2,4-Triazol-3-amine	10 (4.54)
Trichlorfon	100 (45.4)
1,2,4-Trichlorobenzene	100 (45.4)
1,1,1-Trichloroethane	1000 (454)
1,1,2-Trichloroethane	100 (45.4)
Trichloroethylene	100 (45.4)
Trichloromethanesulfenyl chloride	100 (45.4)
Trichloromonofluoromethane	5000 (2270)
Trichlorophenol	10 (4.54)
2,3,4-Trichlorophenol	
2,3,5-Trichlorophenol	
2,3,6-Trichlorophenol	
3,4,5-Trichlorophenol	
2,4,5-Trichlorophenol	10 (4.54)
2,4,6-Trichlorophenol	10 (4.54)
Triethanolamine dodecylbenzenesulfonate	1000 (454)
Triethylamine	5000 (2270)
Trifluralin	10 (4.54)
Trimethylamine	100 (45.4)
2,2,4-Trimethylpentane	1000 (454)
1,3,5-Trinitrobenzene	10 (4.54)
1,3,5-Trioxane, 2,4,6-trimethyl-	1000 (454)
Tris(2,3-dibromopropyl) phosphate	10 (4.54)
Trypan blue	10 (4.54)
D002 Unlisted Hazardous Wastes Characteristic of Corrosivity	100 (45.4)
D001 Unlisted Hazardous Wastes Characteristic of Ignitability	100 (45.4)
D003 Unlisted Hazardous Wastes Characteristic of Reactivity	100 (45.4)
D004–D043 Unlisted Hazardous Wastes Characteristic of Toxicity:	
Arsenic (D004)	1 (0.454)
Barium (D005)	1000 (454)
Benzene (D018)	10 (4.54)
Cadmium (D006)	10 (4.54)
Carbon tetrachloride (D019)	10 (4.54)
Chlordane (D020)	1 (0.454)
Chlorobenzene (D021)	100 (45.4)
Chloroform (D022)	10 (4.54)
Chromium (D007)	10 (4.54)
o-Cresol (D023)	100 (45.4)
m-Cresol (D024)	100 (45.4)
p-Cresol (D025)	100 (45.4)
Cresol (D026)	100 (45.4)
2,4-D (D016)	100 (45.4)
1,4-Dichlorobenzene (D027)	100 (45.4)
1,2-Dichloroethane (D028)	100 (45.4)
1,1-Dichloroethylene (D029)	100 (45.4)
2,4-Dinitrotoluene (D030)	10 (4.54)
Endrin (D012)	1 (0.454)
Heptachlor (and epoxide) (D031)	1 (0.454)
Hexachlorobenzene (D032)	10 (4.54)
Hexachlorobutadiene (D033)	1 (0.454)
Hexachloroethane (D034)	100 (45.4)
Lead (D008)	10 (4.54)
Lindane (D013)	1 (0.454)
Mercury (D009)	1 (0.454)
Methoxychlor (D014)	1 (0.454)
Methyl ethyl ketone (D035)	5000 (2270)
Nitrobenzene (D036)	1000 (454)
Pentachlorophenol (D037)	10 (4.54)
Pyridine (D038)	1000 (454)
Selenium (D010)	10 (4.54)
Silver (D011)	1 (0.454)
Tetrachloroethylene (D039)	100 (45.4)
Toxaphene (D015)	1 (0.454)
Trichloroethylene (D040)	100 (45.4)
2,4,5-Trichlorophenol (D041)	10 (4.54)
2,4,6-Trichlorophenol (D042)	10 (4.54)
2,4,5-TP (D017)	100 (45.4)
Vinyl chloride (D043)	1 (0.454)
Uracil mustard	10 (4.54)
Uranyl acetate	100 (45.4)
Uranyl nitrate	100 (45.4)
Urea, N-ethyl-N-nitroso-	1 (0.454)
Urea, N-methyl-N-nitroso-	1 (0.454)
Urethane	100 (45.4)
Vanadic acid, ammonium salt	1000 (454)
Vanadium oxide V_2O_5	1000 (454)
Vanadium pentoxide	1000 (454)
Vanadyl sulfate	1000 (454)
Vinyl acetate	5000 (2270)
Vinyl acetate monomer	5000 (2270)
Vinylamine, N-methyl-N-nitroso-	10 (4.54)
Vinyl bromide	100 (45.4)
Vinyl chloride	1 (0.454)
Vinylidene chloride	100 (45.4)

TABLE 1—HAZARDOUS SUBSTANCES OTHER THAN RADIONUCLIDES, Continued

Hazardous substance	Reportable quantity (RQ) pounds (kilograms)
Warfarin, & salts	100 (45.4)
Xylene	100 (45.4)
m-Xylene	1000 (454)
o-Xylene	1000 (454)
p-Xylene	100 (45.4)
Xylene (mixed)	100 (45.4)
Xylenes (isomers and mixture)	100 (45.4)
Xylenol	1000 (454)
Yohimban-16-carboxylic acid,11,17-dimethoxy-18-[(3,4,5-trimethoxybenzoyl)oxy]-, methyl ester (3beta,16beta,17alpha,18beta, 20alpha)	5000 (2270)
Zinc [c]	1000 (454)
Zinc acetate	1000 (454)
Zinc ammonium chloride	1000 (454)
Zinc, bis(dimethylcarbamodithioato-S,S')-	10 (4.54)
Zinc borate	1000 (454)
Zinc bromide	1000 (454)
Zinc carbonate	1000 (454)
Zinc chloride	1000 (454)
Zinc cyanide $Zn(CN)_2$	10 (4.54)
Zinc fluoride	1000 (454)
Zinc formate	1000 (454)
Zinc hydrosulfite	1000 (454)
Zinc nitrate	1000 (454)
Zinc phenolsulfonate	5000 (2270)
Zinc phosphide Zn_3P_2	100 (45.4)
Zinc silicofluoride	5000 (2270)
Zinc sulfate	1000 (454)
Ziram	10 (4.54)
Zirconium nitrate	5000 (2270)
Zirconium potassium fluoride	1000 (454)
Zirconium sulfate	5000 (2270)
Zirconium tetrachloride	5000 (2270)
F001	10 (4.54)
(a) Tetrachloroethylene	100 (45.4)
(b) Trichloroethylene	100 (45.4)
(c) Methylene chloride	1000 (454)
(d) 1,1,1-Trichloroethane	1000 (454)
(e) Carbon tetrachloride	10 (4.54)
(f) Chlorinated fluorocarbons	5000 (2270)
F002	10 (4.54)
(a) Tetrachloroethylene	100 (45.4)
(b) Methylene chloride	1000 (454)
(c) Trichloroethylene	100 (45.4)
(d) 1,1,1-Trichloroethane	1000 (454)
(e) Chlorobenzene	100 (45.4)
(f) 1,1,2-Trichloro-1,2,2-trifluoroethane	5000 (2270)
(g) o-Dichlorobenzene	100 (45.4)
(h) Trichlorofluoromethane	5000 (2270)
(i) 1,1,2-Trichloroethane	100 (45.4)
F003	100 (45.4)
(a) Xylene	1000 (454)
(b) Acetone	5000 (2270)
(c) Ethyl acetate	5000 (2270)
(d) Ethylbenzene	1000 (454)
(e) Ethyl ether	100 (45.4)
(f) Methyl isobutyl ketone	5000 (2270)
(g) n-Butyl alcohol	5000 (2270)
(h) Cyclohexanone	5000 (2270)
(i) Methanol	5000 (2270)
F004	100 (45.4)
(a) Cresols/Cresylic acid	100 (45.4)
(b) Nitrobenzene	1000 (454)
F005	100 (45.4)
(a) Toluene	1000 (454)
(b) Methyl ethyl ketone	5000 (2270)
(c) Carbon disulfide	100 (45.4)
(d) Isobutanol	5000 (2270)
(e) Pyridine	1000 (454)
F006	10 (4.54)
F007	10 (4.54)
F008	10 (4.54)
F009	10 (4.54)
F010	10 (4.54)
F011	10 (4.54)
F012	10 (4.54)
F019	10 (4.54)
F020	1 (0.454)
F021	1 (0.454)
F022	1 (0.454)
F023	1 (0.454)
F024	1 (0.454)
F025	1 (0.454)
F026	1 (0.454)
F027	1 (0.454)
F028	1 (0.454)
F032	1 (0.454)
F034	1 (0.454)
F035	1 (0.454)
F037	1 (0.454)
F038	1 (0.454)
F039	1 (0.454)
K001	1 (0.454)
K002	10 (4.54)
K003	10 (4.54)
K004	10 (4.54)
K005	10 (4.54)
K006	10 (4.54)
K007	10 (4.54)
K008	10 (4.54)
K009	10 (4.54)
K010	10 (4.54)
K011	10 (4.54)
K013	10 (4.54)
K014	5000 (2270)
K015	10 (4.54)
K016	1 (0.454)
K017	10 (4.54)
K018	1 (0.454)
K019	1 (0.454)
K020	1 (0.454)
K021	10 (4.54)
K022	1 (0.454)
K023	5000 (2270)
K024	5000 (2270)
K025	10 (4.54)
K026	1000 (454)
K027	10 (4.54)
K028	1 (0.454)
K029	1 (0.454)
K030	1 (0.454)
K031	1 (0.454)
K032	10 (4.54)
K033	10 (4.54)
K034	10 (4.54)
K035	1 (0.454)
K036	1 (0.454)
K037	1 (0.454)
K038	10 (4.54)
K039	10 (4.54)
K040	10 (4.54)
K041	1 (0.454)
K042	10 (4.54)
K043	10 (4.54)
K044	10 (4.54)
K045	10 (4.54)
K046	10 (4.54)
K047	10 (4.54)
K048	10 (4.54)
K049	10 (4.54)
K050	10 (4.54)
K051	10 (4.54)

TABLE 1—HAZARDOUS SUBSTANCES OTHER THAN RADIONUCLIDES, Continued

Hazardous substance	Reportable quantity (RQ) pounds (kilograms)
K052	10 (4.54)
K060	1 (0.454)
K061	10 (4.54)
K062	10 (4.54)
K064	10 (4.54)
K065	10 (4.54)
K066	10 (4.54)
K069	10 (4.54)
K071	1 (0.454)
K073	10 (4.54)
K083	100 (45.4)
K084	1 (0.454)
K085	10 (4.54)
K086	10 (4.54)
K087	100 (45.4)
K088	10 (4.54)
K090	10 (4.54)
K091	10 (4.54)
K093	5000 (2270)
K094	5000 (2270)
K095	100 (45.4)
K096	100 (45.4)
K097	1 (0.454)
K098	1 (0.454)
K099	10 (4.54)
K100	10 (4.54)
K101	1 (0.454)
K102	1 (0.454)
K103	100 (45.4)
K104	10 (4.54)
K105	10 (4.54)
K106	1 (0.454)
K107	10 (4.54)
K108	10 (4.54)
K109	10 (4.54)
K110	10 (4.54)
K111	10 (4.54)
K112	10 (4.54)
K113	10 (4.54)
K114	10 (4.54)
K115	10 (4.54)
K116	10 (4.54)
K117	1 (0.454)
K118	1 (0.454)
K123	10 (4.54)
K124	10 (4.54)
K125	10 (4.54)
K126	10 (4.54)
K131	100 (45.4)
K132	1000 (454)
K136	1 (0.454)
K141	1 (0.454)
K142	1 (0.454)
K143	1 (0.454)
K144	1 (0.454)
K145	1 (0.454)
K147	1 (0.454)
K148	1 (0.454)
K149	10 (4.54)
K150	10 (4.54)
K151	10 (4.54)
K156	10 (4.54)
K157	10 (4.54)
K158	10 (4.54)
K159	10 (4.54)
K161	1 (0.454)
K169	10 (4.54)
K170	1 (0.454)
K171	1 (0.454)
K172	1 (0.454)
K174	1 (0.454)
K175	1 (0.454)
K176	1 (0.454)
K177	5000 (2270)
K178	1000 (454)
K181	1 (0.454)

¢ The RQ for these hazardous substances is limited to those pieces of the metal having a diameter smaller than 100 micrometers (0.004 inches).

¢¢ The RQ for asbestos is limited to friable forms only.

@ Indicates that the name was added by PHMSA because (1) the name is a synonym for a specific hazardous substance and (2) the name appears in the Hazardous Materials Table as a proper shipping name.

To provide consistency with EPA regulations, two entries with different CAS numbers are provided. Refer to the EPA Table 302.4—List of Hazardous Substances and Reportable Quantities for an explanation of the two entries.

Reserved

TABLE 2—RADIONUCLIDES

(1)—Radionuclide	(2)— Atomic Number	(3)—Reportable Quantity (RQ) Ci(TBq)
Actinium-224	89	100 (3.7)
Actinium-225	89	1 (.037)
Actinium-226	89	10 (.37)
Actinium-227	89	0.001 (.000037)
Actinium-228	89	10 (.37)
Aluminum-26	13	10 (.37)
Americium-237	95	1000 (37)
Americium-238	95	100 (3.7)
Americium-239	95	100 (3.7)
Americium-240	95	10 (.37)
Americium-241	95	0.01 (.00037)
Americium-242	95	100 (3.7)
Americium-242m	95	0.01 (.00037)
Americium-243	95	0. 01 (.00037)
Americium-244	95	10 (.37)
Americium-244m	95	1000 (37)
Americium-245	95	1000 (37)
Americium-246	95	1000 (37)
Americium-246m	95	1000 (37)
Antimony-115	51	1000 (37)
Antimony-116	51	1000 (37)
Antimony-116m	51	100 (3.7)
Antimony-117	51	1000 (37)
Antimony-118m	51	10 (.37)
Antimony-119	51	1000 (37)
Antimony-120 (16 min)	51	1000 (37)
Antimony-120 (5.76 day)	51	10 (.37)
Antimony-122	51	10 (.37)
Antimony-124	51	10 (.37)
Antimony-124m	51	1000 (37)
Antimony-125	51	10 (.37)
Antimony-126	51	10 (.37)
Antimony-126m	51	1000 (37)
Antimony-127	51	10 (.37)
Antimony-128 (10.4 min)	51	1000 (37)
Antimony-128 (9.01 hr)	51	10 (.37)
Antimony-129	51	100 (3.7)
Antimony-130	51	100 (3.7)
Antimony-131	51	1000 (37)
Argon-39	18	1000 (37)
Argon-41	18	10 (.37)
Arsenic-69	33	1000 (37)
Arsenic-70	33	100 (3.7)
Arsenic-71	33	100 (3.7)
Arsenic-72	33	10 (.37)
Arsenic-73	33	100 (3.7)
Arsenic-74	33	10 (.37)
Arsenic-76	33	100 (3.7)
Arsenic-77	33	1000 (37)
Arsenic-78	33	100 (3.7)
Astatine-207	85	100 (3.7)
Astatine-211	85	100 (3.7)
Barium-126	56	1000 (37)
Barium-128	56	10 (.37)
Barium-131	56	10 (.37)
Barium-131m	56	1000 (37)
Barium-133	56	10 (.37)
Barium-133m	56	100 (3.7)
Barium-135m	56	1000 (37)
Barium-139	56	1000 (37)
Barium-140	56	10 (.37)
Barium-141	56	1000 (37)
Barium-142	56	1000 (37)
Berkelium-245	97	100 (3.7)
Berkelium-246	97	10 (.37)
Berkelium-247	97	0. 01 (.00037)
Berkelium-249	97	1 (.037)
Berkelium-250	97	100 (3.7)
Beryllium-7	4	100 (3.7)
Beryllium-10	4	1 (.037)
Bismuth-200	83	100 (3.7)
Bismuth-201	83	100 (3.7)
Bismuth-202	83	1000 (37)
Bismuth-203	83	10 (.37)
Bismuth-205	83	10 (.37)
Bismuth-206	83	10 (.37)
Bismuth-207	83	10 (.37)
Bismuth-210	83	10 (.37)
Bismuth-210m	83	0.1 (.0037)
Bismuth-212	83	100 (3.7)
Bismuth-213	83	100 (3.7)
Bismuth-214	83	100 (3.7)
Bromine-74	35	100 (3.7)
Bromine-74m	35	100 (3.7)
Bromine-75	35	100 (3.7)
Bromine-76	35	10 (.37)
Bromine-77	35	100 (3.7)
Bromine-80m	35	1000 (37)
Bromine-80	35	1000 (37)
Bromine-82	35	10 (.37)
Bromine-83	35	1000 (37)
Bromine-84	35	100 (3.7)
Cadmium-104	48	1000 (37)
Cadmium-107	48	1000 (37)
Cadmium-109	48	1 (.037)
Cadmium-113	48	0.1 (.0037)
Cadmium-113m	48	0.1 (.0037)
Cadmium-115	48	100 (3.7)
Cadmium-115m	48	10 (.37)
Cadmium-117	48	100 (3.7)
Cadmium-117m	48	10 (.37)
Calcium-41	20	10 (.37)
Calcium-45	20	10 (.37)
Calcium-47	20	10 (.37)
Californium-244	98	1000 (37)
Californium-246	98	10 (.37)
Californium-248	98	0.1 (.0037)
Californium-249	98	0.01 (.00037)
Californium-250	98	0.01 (.00037)
Californium-251	98	0.01 (.00037)
Californium-252	98	0.1 (.0037)
Californium-253	98	10 (.37)
Californium-254	98	0.1 (.0037)
Carbon-11	6	1000 (37)

TABLE 2—RADIONUCLIDES, Continued

(1)—Radionuclide	(2)— Atomic Number	(3)—Reportable Quantity (RQ) Ci(TBq)
Carbon-14	6	10 (.37)
Cerium-134	58	10 (.37)
Cerium-135	58	10 (.37)
Cerium-137	58	1000 (37)
Cerium-137m	58	100 (3.7)
Cerium-139	58	100 (3.7)
Cerium-141	58	10 (.37)
Cerium-143	58	100 (3.7)
Cerium-144	58	1 (.037)
Cesium-125	55	1000 (37)
Cesium-127	55	100 (3.7)
Cesium-129	55	100 (3.7)
Cesium-130	55	1000 (37)
Cesium-131	55	1000 (37)
Cesium-132	55	10 (.37)
Cesium-134	55	1 (.037)
Cesium-134m	55	1000 (37)
Cesium-135	55	10 (.37)
Cesium-135m	55	100 (3.7)
Cesium-136	55	10 (.37)
Cesium-137	55	1 (.037)
Cesium-138	55	100 (3.7)
Chlorine-36	17	10 (.37)
Chlorine-38	17	100 (3.7)
Chlorine-39	17	100 (3.7)
Chromium-48	24	100 (3.7)
Chromium-49	24	1000 (37)
Chromium-49	24	1000 (37)
Chromium-51	24	1000 (37)
Cobalt-55	27	10 (.37)
Cobalt-56	27	10 (.37)
Cobalt-57	27	100 (3.7)
Cobalt-58	27	10 (.37)
Cobalt-58m	27	1000 (37)
Cobalt-60	7	10 (.37)
Cobalt-60m	27	1000 (37)
Cobalt-61	27	1000 (37)
Cobalt-62m	27	1000 (37)
Copper-60	29	100 (3.7)
Copper-61	29	100 (3.7)
Copper-64	29	1000 (37)
Copper-67	29	100 (3.7)
Curium-238	96	1000 (37)
Curium-240	96	1 (.037)
Curium-241	96	10 (.37)
Curium-242	96	1 (.037)
Curium-243	96	0.01 (.00037)
Curium-244	96	0.01 (.00037)
Curium-245	96	0.01 (.00037)
Curium-246	96	0.01 (.00037)
Curium-247	96	0.01 (.00037)
Curium-248	96	0.001 (.000037)
Curium-249	96	1000 (37)
Dysprosium-155	66	100 (3.7)
Dysprosium-157	66	100 (3.7)
Dysprosium-159	66	100 (3.7)
Dysprosium-165	66	1000 (37)
Dysprosium-166	66	10 (.37)
Einsteinium-250	99	10 (.37)
Einsteinium-251	99	1000 (37)
Einsteinium-253	99	10 (.37)
Einsteinium-254	99	0.1 (.0037)
Einsteinium-254m	99	1 (.037)
Erbium-161	68	100 (3.7)
Erbium-165	68	1000 (37)
Erbium-169	68	100 (3.7)
Erbium-171	68	100 (3.7)
Erbium-172	68	10 (.37)
Europium-145	63	10 (.37)
Europium-146	63	10 (.37)
Europium-147	63	10 (.37)
Europium-148	63	10 (.37)
Europium-149	63	100 (3.7)
Europium-150 (12.6 hr)	63	1000 (37)
Europium-150 (34.2 yr)	63	10 (.37)
Europium-152	63	10 (.37)
Europium-152m	63	100 (3.7)
Europium-154	63	10 (.37)
Europium-155	63	10 (.37)
Europium-156	63	10 (.37)
Europium-157	63	10 (.37)
Europium-158	63	1000 (37)
Fermium-252	100	10 (.37)
Fermium-253	100	10 (.37)
Fermium-254	100	100 (3.7)
Fermium-255	100	100 (3.7)
Fermium-257	100	1 (.037)
Fluorine-18	9	1000 (37)
Francium-222	87	100 (3.7)
Francium-223	87	100 (3.7)
Gadolinium-145	64	100 (3.7)
Gadolinium-146	64	10 (.37)
Gadolinium-147	64	10 (.37)
Gadolinium-148	64	0.001 (.000037)
Gadolinium-149	64	100 (3.7)
Gadolinium-151	64	100 (3.7)
Gadolinium-152	64	0.001 (.000037)
Gadolinium-153	64	10 (.37)
Gadolinium-159	64	1000 (37)
Gallium-65	31	1000 (37)
Gallium-66	31	10 (.37)
Gallium-67	31	100 (3.7)
Gallium-68	31	1000 (37)
Gallium-70	31	1000 (37)
Gallium-72	31	10 (.37)
Gallium-73	31	100 (3.7)
Germanium-66	32	100 (3.7)
Germanium-67	32	1000 (37)
Germanium-68	32	10 (.37)
Germanium-69	32	10 (.37)
Germanium-71	32	1000 (37)
Germanium-75	32	1000 (37)
Germanium-77	32	10 (.37)
Germanium-78	32	1000 (37)
Gold-193	79	100 (3.7)
Gold-194	79	10 (.37)

TABLE 2—RADIONUCLIDES, Continued

(1)—Radionuclide	(2)—Atomic Number	(3)—Reportable Quantity (RQ) Ci(TBq)
Gold-195	79	100 (3.7)
Gold-198	79	100 (3.7)
Gold-198m	79	10 (.37)
Gold-199	79	100 (3.7)
Gold-200	79	1000 (37)
Gold-200m	79	10 (.37)
Gold-201	79	1000 (37)
Hafnium-170	72	100 (3.7)
Hafnium-172	72	1 (.037)
Hafnium-173	72	100 (3.7)
Hafnium-175	72	100 (3.7)
Hafnium-177m	72	1000 (37)
Hafnium-178m	72	0.1 (.0037)
Hafnium-179m	72	100 (3.7)
Hafnium-180m	72	100 (3.7)
Hafnium-181	72	10 (.37)
Hafnium-182	72	0.1 (.0037)
Hafnium-182m	72	100 (3.7)
Hafnium-183	72	100 (3.7)
Hafnium-184	72	100 (3.7)
Holmium-155	67	1000 (37)
Holmium-157	67	1000 (37)
Holmium-159	67	1000 (37)
Holmium-161	67	1000 (37)
Holmium-162	67	1000 (37)
Holmium-162m	67	1000 (37)
Holmium-164	67	1000 (37)
Holmium-164m	67	1000 (37)
Holmium-166	67	100 (3.7)
Holmium-166m	67	1 (.037)
Holmium-167	67	100 (3.7)
Hydrogen-3	1	100 (3.7)
Indium-109	49	100 (3.7)
Indium-110 (69.1 min)	49	100 (3.7)
Indium-110 (4.9 hr)	49	10 (.37)
Indium-111	49	100 (3.7)
Indium-112	49	1000 (37)
Indium-113m	49	1000 (37)
Indium-114m	49	10 (.37)
Indium-115	49	0.1 (.0037)
Indium-115m	49	100 (3.7)
Indium-116m	49	100 (3.7)
Indium-117	49	1000 (37)
Indium-117m	49	100 (3.7)
Indium-119m	49	1000 (37)
Iodine-120	53	10 (.37)
Iodine-120m	53	100 (3.7)
Iodine-121	53	100 (3.7)
Iodine-123	53	10 (.37)
Iodine-124	53	0.1 (.0037)
Iodine-125	53	0.01 (.00037)
Iodine-126	53	0.01 (.00037)
Iodine-128	53	1000 (37)
Iodine-129	53	0.001 (.000037)
Iodine-130	53	1 (.037)
Iodine-131	53	0.01 (.00037)
Iodine-132	53	10 (.37)
Iodine-132m	53	10 (.37)
Iodine-133	53	0.1 (.0037)
Iodine-134	53	100 (3.7)
Iodine-135	3	10 (.37)
Iridium-182	7	1000 (37)
Iridium-184	77	100 (3.7)
Iridium-185	77	100 (3.7)
Iridium-186	77	10 (.37)
Iridium-187	77	100 (3.7)
Iridium-188	77	10 (.37)
Iridium-189	77	100 (3.7)
Iridium-190	77	10 (.37)
Iridium-190m	77	1000 (37)
Iridium-192	77	10 (.37)
Iridium-192m	77	100 (3.7)
Iridium-194	77	100 (3.7)
Iridium-194m	77	10 (.37)
Iridium-195	77	1000 (37)
Iridium-195m	77	100 (3.7)
Iron-52	26	100 (3.7)
Iron-55	26	100 (3.7)
Iron-59	26	10 (.37)
Iron-60	26	0.1 (.0037)
Krypton-74	36	10 (.37)
Krypton-76	36	10 (.37)
Krypton-77	36	10 (.37)
Krypton-79	36	100 (3.7)
Krypton-81	36	1000 (37)
Krypton-83m	36	1000 (37)
Krypton-85	36	1000 (37)
Krypton-85m	36	100 (3.7)
Krypton-87	36	10 (.37)
Krypton-88	36	10 (.37)
Lanthanum-131	57	1000 (37)
Lanthanum-132	57	100 (3.7)
Lanthanum-135	57	1000 (37)
Lanthanum-137	57	10 (.37)
Lanthanum-138	57	1 (.037)
Lanthanum-140	57	10 (.37)
Lanthanum-141	57	1000 (37)
Lanthanum-142	57	100 (3.7)
Lanthanum-143	57	1000 (37)
Lead-195m	82	1000 (37)
Lead-198	82	100 (3.7)
Lead-199	82	100 (3.7)
Lead-200	82	100 (3.7)
Lead-201	82	100 (3.7)
Lead-202	82	1 (.037)
Lead-202m	82	10 (.37)
Lead-203	82	100 (3.7)
Lead-205	82	100 (3.7)
Lead-209	82	1000 (37)
Lead-210	82	0.01 (.00037)
Lead-211	82	100 (3.7)
Lead-212	82	10 (.37)
Lead-214	82	100 (3.7)
Lutetium-169	71	10 (.37)
Lutetium-170	71	10 (.37)
Lutetium-171	71	10 (.37)

TABLE 2—RADIONUCLIDES, Continued

(1)—Radionuclide	(2)— Atomic Number	(3)—Reportable Quantity (RQ) Ci(TBq)
Lutetium-172	71	10 (.37)
Lutetium-173	71	100 (3.7)
Lutetium-174	71	10 (.37)
Lutetium-174m	71	10 (.37)
Lutetium-176	71	1 (.037)
Lutetium-176m	71	1000 (37)
Lutetium-177	71	100 (3.7)
Lutetium-177m	71	10 (.37)
Lutetium-178	71	1000 (37)
Lutetium-178m	71	1000 (37)
Lutetium-179	71	1000 (37)
Magnesium-28	12	10 (.37)
Manganese-51	25	1000 (37)
Manganese-52	25	10 (.37)
Manganese-52m	25	1000 (37)
Manganese-53	25	1000 (37)
Manganese-54	25	10 (.37)
Manganese-56	25	100 (3.7)
Mendelevium-257	101	100 (3.7)
Mendelevium-258	101	1 (.037)
Mercury-193	80	100 (3.7)
Mercury-193m	80	10 (.37)
Mercury-194	80	0.1 (.0037)
Mercury-195	80	100 (3.7)
Mercury-195m	80	100 (3.7)
Mercury-197	80	1000 (37)
Mercury-197m	80	1000 (37)
Mercury-199m	80	1000 (37)
Mercury-203	80	10 (.37)
Molybdenum-90	42	100 (3.7)
Molybdenum-93	42	100 (3.7)
Molybdenum-93m	42	10 (.37)
Molybdenum-99	42	100 (3.7)
Molybdenum-101	42	1000 (37)
Neodymium-136	60	1000 (37)
Neodymium-138	60	1000 (37)
Neodymium-139	60	1000 (37)
Neodymium-139m	60	100 (3.7)
Neodymium-141	60	1000 (37)
Neodymium-147	60	10 (.37)
Neodymium-149	60	100 (3.7)
Neodymium-151	60	1000 (37)
Neptunium-232	93	1000 (37)
Neptunium-233	93	1000 (37)
Neptunium-234	93	10 (.37)
Neptunium-235	93	1000 (37)
Neptunium-236 (1.2 E 5 yr)	93	0.1 (.0037)
Neptunium-236 (22.5 hr)	93	100 (3.7)
Neptunium-237	93	0.01 (.00037)
Neptunium-238	93	10 (.37)
Neptunium-239	93	100 (3.7)
Neptunium-240	93	100 (3.7)
Nickel-56	28	10 (.37)
Nickel-57	28	10 (.37)
Nickel-59	28	100 (3.7)
Nickel-63	28	100 (3.7)
Nickel-65	28	100 (3.7)
Nickel-66	28	10 (.37)
Niobium-88	41	100 (3.7)
Niobium-89 (66 min)	41	100 (3.7)
Niobium-89 (122 min)	41	100 (3.7)
Niobium-90	41	10 (.37)
Niobium-93m	41	100 (3.7)
Niobium-94	41	10 (.37)
Niobium-95	41	10 (.37)
Niobium-95m	41	100 (3.7)
Niobium-96	41	10 (.37)
Niobium-97	41	100 (3.7)
Niobium-98	41	1000 (37)
Osmium-180	76	1000 (37)
Osmium-181	76	100 (3.7)
Osmium-182	76	100 (3.7)
Osmium-185	76	10 (.37)
Osmium-189m	76	1000 (37)
Osmium-191	76	100 (3.7)
Osmium-191m	76	1000 (37)
Osmium-193	76	100 (3.7)
Osmium-194	76	1 (.037)
Palladium-100	46	100 (3.7)
Palladium-101	46	100 (3.7)
Palladium-103	46	100 (3.7)
Palladium-107	46	100 (3.7)
Palladium-109	46	1000 (37)
Phosphorus-32	15	0.1 (.0037)
Phosphorus-33	15	1 (.037)
Platinum-186	78	100 (3.7)
Platinum-188	78	100 (3.7)
Platinum-189	78	100 (3.7)
Platinum-191	78	100 (3.7)
Platinum-193	78	1000 (37)
Platinum-193m	78	100 (3.7)
Platinum-195m	78	100 (3.7)
Platinum-197	78	1000 (37)
Platinum-197m	78	1000 (37)
Platinum-199	78	1000 (37)
Platinum-200	78	100 (3.7)
Plutonium-234	94	1000 (37)
Plutonium-235	94	1000 (37)
Plutonium-236	94	0.1 (.0037)
Plutonium-237	94	1000 (37)
Plutonium-238	94	0.01 (.00037)
Plutonium-239	94	0.01 (.00037)
Plutonium-240	94	0.01 (.00037)
Plutonium-241	94	1 (.037)
Plutonium-242	94	0.01 (.00037)
Plutonium-243	94	1000 (37)
Plutonium-244	94	0.01 (.00037)
Plutonium-245	94	100 (3.7)
Polonium-203	84	100 (3.7)
Polonium-205	84	100 (3.7)
Polonium-207	84	10 (.37)
Polonium-210	84	0.01 (.00037)
Potassium-40	19	1 (.037)
Potassium-42	19	100 (3.7)
Potassium-43	19	10 (.37)
Potassium-44	19	100 (3.7)

TABLE 2—RADIONUCLIDES, Continued

(1)—Radionuclide	(2)—Atomic Number	(3)—Reportable Quantity (RQ) Ci(TBq)
Potassium-45	19	1000 (37)
Praseodymium-136	59	1000 (37)
Praseodymium-137	59	1000 (37)
Praseodymium-138m	59	100 (3.7)
Praseodymium-139	59	1000 (37)
Praseodymium-142	59	100 (3.7)
Praseodymium-142m	59	1000 (37)
Praseodymium-143	59	10 (.37)
Praseodymium-144	59	1000 (37)
Praseodymium-145	59	1000 (37)
Praseodymium-147	59	1000 (37)
Promethium-141	61	1000 (37)
Promethium-143	61	100 (3.7)
Promethium-144	61	10 (.37)
Promethium-145	61	100 (3.7)
Promethium-146	61	10 (.37)
Promethium-147	61	10 (.37)
Promethium-148	61	10 (.37)
Promethium-148m	61	10 (.37)
Promethium-149	61	100 (3.7)
Promethium-150	61	100 (3.7)
Promethium-151	61	100 (3.7)
Protactinium-227	91	100 (3.7)
Protactinium-228	91	10 (.37)
Protactinium-230	91	10 (.37)
Protactinium-231	91	0.01 (.00037)
Protactinium-232	91	10 (.37)
Protactinium-233	91	100 (3.7)
Protactinium-234	91	10 (.37)
RADIONUCLIDES§†		1 (.037)
Radium-223	88	1 (.037)
Radium-224	88	10 (.37)
Radium-225	88	1 (.037)
Radium-226**	88	0.1 (.0037)
Radium-227	88	1000 (37)
Radium-228	88	0.1 (.0037)
Radon-220	86	0.1 (.0037)
Radon-222	86	0.1 (.0037)
Rhenium-177	75	1000 (37)
Rhenium-178	75	1000 (37)
Rhenium-181	75	100 (3.7)
Rhenium-182 (12.7 hr)	75	10 (.37)
Rhenium-182 (64.0 hr)	75	10 (.37)
Rhenium-184	75	10 (.37)
Rhenium-184m	75	10 (.37)
Rhenium-186	75	100 (3.7)
Rhenium-186m	75	10 (.37)
Rhenium-187	75	1000 (37)
Rhenium-188	75	1000 (37)
Rhenium-188m	75	1000 (37)
Rhenium-189	75	1000 (37)
Rhodium-99	45	10 (.37)
Rhodium-99m	45	100 (3.7)
Rhodium-100	45	10 (.37)
Rhodium-101	45	10 (.37)
Rhodium-101m	45	100 (3.7)
Rhodium-102	45	10 (.37)
Rhodium-102m	45	10 (.37)
Rhodium-103m	45	1000 (37)
Rhodium-105	45	100 (3.7)
Rhodium-106m	45	10 (.37)
Rhodium-107	45	1000 (37)
Rubidium-79	37	1000 (37)
Rubidium-81	37	100 (3.7)
Rubidium-81m	37	1000 (37)
Rubidium-82m	37	10 (.37)
Rubidium-83	37	10 (.37)
Rubidium-84	37	10 (.37)
Rubidium-86	37	10 (.37)
Rubidium-88	37	1000 (37)
Rubidium-89	37	1000 (37)
Rubidium-87	37	10 (.37)
Ruthenium-94	44	1000 (37)
Ruthenium-97	44	100 (3.7)
Ruthenium-103	44	10 (.37)
Ruthenium-105	44	100 (3.7)
Ruthenium-106	44	1 (.037)
Samarium-141	62	1000 (37)
Samarium-141m	62	1000 (37)
Samarium-142	62	1000 (37)
Samarium-145	62	100 (3.7)
Samarium-146	62	0.01 (.00037)
Samarium-147	62	0.01 (.00037)
Samarium-151	62	10 (.37)
Samarium-153	62	100 (3.7)
Samarium-155	62	1000 (37)
Samarium-156	62	100 (3.7)
Scandium-43	21	1000 (37)
Scandium-44	21	100 (3.7)
Scandium-44m	21	10 (.37)
Scandium-46	21	10 (.37)
Scandium-47	21	100 (3.7)
Scandium-48	21	10 (.37)
Scandium-49	21	1000 (37)
Selenium-70	34	1000 (37)
Selenium-73	34	10 (.37)
Selenium-73m	34	100 (3.7)
Selenium-75	34	10 (.37)
Selenium-79	34	10 (.37)
Selenium-81	34	1000 (37)
Selenium-81m	34	1000 (37)
Selenium-83	34	1000 (37)
Silicon-31	14	1000 (37)
Silicon-32	14	1 (.037)
Silver-102	47	100 (3.7)
Silver-103	47	1000 (37)
Silver-104	47	1000 (37)
Silver-104m	47	1000 (37)
Silver-105	47	10 (.37)
Silver-106	47	1000 (37)
Silver-106m	47	10 (.37)
Silver-108m	47	10 (.37)
Silver-110m	47	10 (.37)
Silver-111	47	10 (.37)
Silver-112	47	100 (3.7)
Silver-115	47	1000 (37)

TABLE 2—RADIONUCLIDES, Continued

(1)—Radionuclide	(2)—Atomic Number	(3)—Reportable Quantity (RQ) Ci(TBq)
Sodium-22	11	10 (.37)
Sodium-24	11	10 (.37)
Strontium-80	38	100 (3.7)
Strontium-81	38	1000 (37)
Strontium-83	38	100 (3.7)
Strontium-85	38	10 (.37)
Strontium-85m	38	1000 (37)
Strontium-87m	38	100 (3.7)
Strontium-89	38	10 (.37)
Strontium-90	38	0.1 (.0037)
Strontium-91	38	10 (.37)
Strontium-92	38	100 (3.7)
Sulfur-35	16	1 (.037)
Tantalum-172	73	100 (3.7)
Tantalum-173	73	100 (3.7)
Tantalum-174	73	100 (3.7)
Tantalum-175	73	100 (3.7)
Tantalum-176	73	10 (.37)
Tantalum-177	73	1000 (37)
Tantalum-178	73	1000 (37)
Tantalum-179	73	1000 (37)
Tantalum-180	73	100 (3.7)
Tantalum-180m	73	1000 (37)
Tantalum-182	73	10 (.37)
Tantalum-182m	73	1000 (37)
Tantalum-183	73	100 (3.7)
Tantalum-184	73	10 (.37)
Tantalum-185	73	1000 (37)
Tantalum-186	73	1000 (37)
Technetium-93	43	100 (3.7)
Technetium-93m	43	1000 (37)
Technetium-94	43	10 (.37)
Technetium-94m	43	100 (3.7)
Technetium-96	43	10 (.37)
Technetium-96m	43	1000 (37)
Technetium-97	43	100 (3.7)
Technetium-97m	43	100 (3.7)
Technetium-98	43	10 (.37)
Technetium-99	43	10 (.37)
Technetium-99m	43	100 (3.7)
Technetium-101	43	1000 (37)
Technetium-104	43	1000 (37)
Tellurium-116	52	1000 (37)
Tellurium-121	52	10 (.37)
Tellurium-121m	52	10 (.37)
Tellurium-123	52	10 (.37)
Tellurium-123m	52	10 (.37)
Tellurium-125m	52	10 (.37)
Tellurium-127	52	1000 (37)
Tellurium-127m	52	10 (.37)
Tellurium-129	52	1000 (37)
Tellurium-129m	52	10 (.37)
Tellurium-131	52	1000 (37)
Tellurium-131m	52	10 (.37)
Tellurium-132	52	10 (.37)
Tellurium-133	52	1000 (37)
Tellurium-133m	52	1000 (37)
Tellurium-134	52	1000 (37)
Terbium-147	65	100 (3.7)
Terbium-149	65	100 (3.7)
Terbium-150	65	100 (3.7)
Terbium-151	65	10 (.37)
Terbium-153	65	100 (3.7)
Terbium-154	65	10 (.37)
Terbium-155	65	100 (3.7)
Terbium-156m (5.0 hr)	65	1000 (37)
Terbium-156m (24.4 hr)	65	1000 (37)
Terbium-156	65	10 (.37)
Terbium-157	65	100 (3.7)
Terbium-158	65	10 (.37)
Terbium-160	65	10 (.37)
Terbium-161	65	100 (3.7)
Thallium-194	81	1000 (37)
Thallium-194m	81	100 (3.7)
Thallium-195	81	100 (3.7)
Thallium-197	81	100 (3.7)
Thallium-198	81	10 (.37)
Thallium-198m	81	100 (3.7)
Thallium-199	81	100 (3.7)
Thallium-200	81	10 (.37)
Thallium-201	81	1000 (37)
Thallium-202	81	10 (.37)
Thallium-204	81	10 (.37)
Thorium (Irradiated)	90	***
Thorium (Natural)	90	**
Thorium-226	90	100 (3.7)
Thorium-227	90	1 (.037)
Thorium-228	90	0.01 (.00037)
Thorium-229	90	0.001 (.000037)
Thorium-230	90	0.01 (.00037)
Thorium-231	90	100 (3.7)
Thorium-232**	90	0.001 (.000037)
Thorium-234	90	100 (3.7)
Thulium-162	69	1000 (37)
Thulium-166	69	10 (.37)
Thulium-167	69	100 (3.7)
Thulium-170	69	10 (.37)
Thulium-171	69	100 (3.7)
Thulium-172	69	100 (3.7)
Thulium-173	69	100 (3.7)
Thulium-175	69	1000 (37)
Tin-110	50	100 (3.7)
Tin-111	50	1000 (37)
Tin-113	50	10 (.37)
Tin-117m	50	100 (3.7)
Tin-119m	50	10 (.37)
Tin-121	50	1000 (37)
Tin-121m	50	10 (.37)
Tin-123	50	10 (.37)
Tin-123m	50	1000 (37)
Tin-125	50	10 (.37)
Tin-126	50	1 (.037)
Tin-127	50	100 (3.7)
Tin-128	50	1000 (37)
Titanium-44	22	1 (.037)
Titanium-45	22	1000 (37)

TABLE 2—RADIONUCLIDES, Continued

(1)—Radionuclide	(2)—Atomic Number	(3)—Reportable Quantity (RQ) Ci(TBq)
Tungsten-176	74	1000 (37)
Tungsten-177	74	100 (3.7)
Tungsten-178	74	100 (3.7)
Tungsten-179	74	1000 (37)
Tungsten-181	74	100 (3.7)
Tungsten-185	74	10 (.37)
Tungsten-187	74	100 (3.7)
Tungsten-188	74	10 (.37)
Uranium (Depleted)	92	***
Uranium (Irradiated)	92	***
Uranium (Natural)	92	**
Uranium Enriched 20% or greater	92	***
Uranium Enriched less than 20%	92	***
Uranium-230	92	1 (.037)
Uranium-231	92	1000 (37)
Uranium-232	92	0.01 (.00037)
Uranium-233	92	0.1 (.0037)
Uranium-234**	92	0.1 (.0037)
Uranium-235**	92	0.1 (.0037)
Uranium-236	92	0.1 (.0037)
Uranium-237	92	100 (3.7)
Uranium-238**	92	0.1 (.0037)
Uranium-239	92	1000 (37)
Uranium-240	92	1000 (37)
Vanadium-47	23	1000 (37)
Vanadium-48	23	10 (.37)
Vanadium-49	23	1000 (37)
Xenon-120	54	100 (3.7)
Xenon-121	54	10 (.37)
Xenon-122	54	100 (3.7)
Xenon-123	54	10 (.37)
Xenon-125	54	100 (3.7)
Xenon-127	54	100 (3.7)
Xenon-129m	54	1000 (37)
Xenon-131m	54	1000 (37)
Xenon-133	54	1000 (37)
Xenon-133m	54	1000 (37)
Xenon-135	54	100 (3.7)
Xenon-135m	54	10 (.37)
Xenon-138	54	10 (.37)
Ytterbium-162	70	1000 (37)
Ytterbium-166	70	10 (.37)
Ytterbium-167	70	1000 (37)
Ytterbium-169	70	10 (.37)
Ytterbium-175	70	100 (3.7)
Ytterbium-177	70	1000 (37)
Ytterbium-178	70	1000 (37)
Yttrium-86	39	10 (.37)
Yttrium-86m	39	1000 (37)
Yttrium-87	39	10 (.37)
Yttrium-88	39	10 (.37)
Yttrium-90	39	10 (.37)
Yttrium-90m	39	100 (3.7)
Yttrium-91	39	10 (.37)
Yttrium-91m	39	1000 (37)
Yttrium-92	39	100 (3.7)
Yttrium-93	39	100 (3.7)
Yttrium-94	39	1000 (37)
Yttrium-95	39	1000 (37)
Zinc-62	30	100 (3.7)
Zinc-63	30	1000 (37)
Zinc-65	30	10 (.37)
Zinc-69	30	1000 (37)
Zinc-69m	30	100 (3.7)
Zinc-71m	30	100 (3.7)
Zinc-72	30	100 (3.7)
Zirconium-86	40	100 (3.7)
Zirconium-88	40	10 (.37)
Zirconium-89	40	100 (3.7)
Zirconium-93	40	1 (.037)
Zirconium-95	40	10 (.37)
Zirconium-97	40	10 (.37)

§ The RQs for all radionuclides apply to chemical compounds containing the radionuclides and elemental forms regardless of the diameter of pieces of solid material.

† The RQ of one curie applies to all radionuclides not otherwise listed. Whenever the RQs in Table 1—HAZARDOUS SUBSTANCES OTHER THAN RADIONUCLIDES and this table conflict, the lowest RQ shall apply. For example, uranyl acetate and uranyl nitrate have RQs shown in TABLE 1 of 100 pounds, equivalent to about one-tenth the RQ level for unanium-238 in this table.

** The method to determine RQs for mixtures or solutions of radionuclides can be found in paragraph 7 of the note preceding TABLE 1 of this appendix. RQs for the following four common radionuclide mixtures are provided: radium-226 in secular equilibrium with its daughters (0.053 curie); natural uranium (0.1 curie); natural uranium in secular equilibrium with its daughters (0.052 curie); and natural thorium in secular equilibrium with its daughters (0.011 curie).

*** Indicates that the name was added by PHMSA because it appears in the list of radionuclides in 49 CFR 173.435. The reportable quantity (RQ), if not specifically listed elsewhere in this appendix, shall be determined in accordance with the procedures in paragraph 7 of this appendix.

Reserved

Appendix B to §172.101 — List of Marine Pollutants

1. See §171.4 of this subchapter for applicability to marine pollutants. This appendix lists potential marine pollutants as defined in §171.8 of this subchapter.

2. Marine pollutants listed in this appendix are not necessarily listed by name in the §172.101 Table. If a marine pollutant not listed by name or by synonym in the §172.101 Table meets the definition of any hazard Class 1 through 8, then you must determine the class and division of the material in accordance with §173.2a of this subchapter. You must also select the most appropriate hazardous material description and proper shipping name. If a marine pollutant not listed by name or by synonym in the §172.101 Table does not meet the definition of any Class 1 through 8, then you must offer it for transportation under the most appropriate of the following two Class 9 entries: "Environmentally hazardous substances, liquid, n.o.s." UN 3082, or "Environmentally hazardous substances, solid, n.o.s." UN3077.

3. This appendix contains two columns. The first column, entitled "S.M.P." (for severe marine pollutants), identifies whether a material is a severe marine pollutant. If the letters "PP" appear in this column for a material, the material is a severe marine pollutant, otherwise it is not. The second column, entitled "Marine Pollutant", lists the marine pollutants.

4. If a material is not listed in this appendix and meets the criteria for a marine pollutant as provided in Chapter 2.9 of the IMDG Code, (incorporated by reference; see §171.7 of this subchapter), the material may be transported as a marine pollutant in accordance with the applicable requirements of this subchapter.

5. If a material or a solution meeting the definition of a marine pollutant in §171.8 of this subchapter does not meet the criteria for a marine pollutant as provided in section 2.9.3.3 and 2.9.3.4 of the IMDG Code, (incorporated by reference; see §171.7 of this subchapter), it may be excepted from the requirements of this subchapter as a marine pollutant if that exception is approved by the Associate Administrator.

LIST OF MARINE POLLUTANTS

S.M.P. (1)	Marine Pollutant (2)
	Acetone cyanohydrin, stabilized
	Acetylene tetrabromide
	Acetylene tetrachloride
	Acraldehyde, inhibited
	Acrolein, inhibited
	Acrolein stabilized
	Acrylic aldehyde, inhibited
	Alcohol C-12 - C-16 poly(1-6) ethoxylate
	Alcohol C-6 C-17 (secondary)poly(3-6) ethoxylate
	Aldicarb
PP	Aldrin
	Alkybenzenesulphonates, branched and straight chain (excluding C11-C13 straight chain or branched chain homologues)
	Alkyl (C12-C14) dimethylamine
	Alkyl (C7-C9) nitrates
	Allyl bromide
	ortho-Aminoanisole
	Aminocarb
	Ammonium dinitro-o-cresolate
	n-Amylbenzene
PP	Azinphos-ethyl
PP	Azinphos-methyl
	Barium cyanide
	Bendiocarb
	Benomyl
	Benquinox
	Benzyl chlorocarbonate
	Benzyl chloroformate
PP	Binapacryl
	N,N-Bis (2-hydroxyethyl) oleamide (LOA)
PP	Brodifacoum
	Bromine cyanide
	Bromoacetone
	Bromoallylene
	Bromobenzene
	ortho-Bromobenzyl cyanide
	Bromocyane
	Bromoform
PP	Bromophos-ethyl
	3-Bromopropene
	Bromoxynil
	Butanedione
	2-Butenal, stabilized
	Butyl benzyl phthalate
	N-tert-butyl-*N*-cyclopropyl-6-methylthio-1,3,5-triazine- 2,4-diamine
	para-tertiary-butyltoluene
PP	Cadmium compounds
	Cadmium sulphide
	Calcium arsenate
	Calcium arsenate and calcium arsenite, mixtures, solid
	Calcium cyanide
PP	Camphechlor
	Carbaryl
	Carbendazim
	Carbofuran
	Carbon tetrabromide
	Carbon tetrachloride
PP	Carbophenothion
	Cartap hydrochloride
PP	Chlordane
	Chlorfenvinphos
PP	Chlorinated paraffins (C-10 - C-13)
PP	Chlorinated paraffins (C14 - C17), with more than 1% shorter chain length
	Chlorine
	Chlorine cyanide, inhibited
	Chlomephos
	Chloroacetone, stabilized

LIST OF MARINE POLLUTANTS, Continued

S.M.P. (1)	Marine Pollutant (2)
	1-Chloro-2,3-Epoxypropane
	2-Chloro-6-nitrotoluene
	4-Chloro-2-nitrotoluene
	Chloro-ortho-nitrotoluene
	2-Chloro-5-trifluoromethylnitrobenzene
	para-Chlorobenzyl chloride, liquid or solid
	Chlorodinitrobenzenes, liquid or solid
	1-Chloroheptane
	1-Chlorohexane
	Chloronitroanilines
	Chloronitrotoluenes, *liquid*
	Chloronitrotoluenes, *solid*
	1-Chlorooctane
PP	Chlorophenolates, liquid
PP	Chlorophenolates, solid
	Chlorophenyltrichlorosilane
	Chloropicrin
	Chlorotoluenes (meta-; para-)
PP	Chlorpyriphos
PP	Chlorthiophos
	Cocculus
	Coconitrile
	Copper acetoarsenite
	Copper arsenite
PP	Copper chloride
PP	Copper chloride (solution)
PP	Copper cyanide
PP	Copper metal powder
PP	Copper sulphate, anhydrous, hydrates
	Coumachlor
PP	Coumaphos
PP	Cresyl diphenyl phosphate
	Crotonaldehyde, stabilized
	Crotonic aldehyde, stabilized
	Crotoxyphos
	Cupric arsenite
PP	Cupric chloride
PP	Cupric cyanide
PP	Cupric sulfate
	Cupriethylenediamine solution
PP	Cuprous chloride
	Cyanide mixtures
	Cyanide solutions
	Cyanides, inorganic, n.o.s.
	Cyanogen bromide
	Cyanogen chloride, inhibited
	Cyanogen chloride, stabilized
	Cyanophos
PP	1,5,9-Cyclododecatriene
PP	Cyhexatin
PP	Cymenes (o-;m-;p-)
PP	Cypermethrin
PP	DDT
	Decycloxytetrahydrothiophene dioxide
	Decyl acrylate
	DEF
	Desmedipham
	Di-allate
	Di-n-Butyl phthalate
	1,4-Di-tert-butylbenzene
PP	Dialifos
	4,4′-Diaminodiphenylmethane
PP	Diazinon
	1,3-Dibromobenzene
PP	Dichlofenthion
	Dichloroanilines
	1,3-Dichlorobenzene
	1,4-Dichlorobenzene
	Dichlorobenzene (meta-; para-)
	2,2-Dichlorodiethyl ether
	Dichlorodimethyl ether, symmetrical
	Di-(2-chloroethyl) ether
	1,1-Dichloroethylene, inhibited
	1,6-Dichlorohexane
	Dichlorophenyltrichlorosilane
PP	Dichlorvos
PP	Dichlofop-methyl
	Dicrotophos
PP	Dieldrin
	Diisopropylbenzenes
	Diisopropylnaphthalenes, mixed isomers
PP	Dimethoate
PP	N,N-Dimethyldodecylamine
	Dimethylhydrazine, symmetrical
	Dimethylhydrazine, unsymmetrical
	Dinitro-o-cresol, *solid*
	Dinitro-o-cresol, *solution*
	Dinitrochlorobenzenes, liquid or solid
	Dinitrophenol, *dry or wetted with less than 15 percent water, by mass*
	Dinitrophenol solutions
	Dinitrophenol, wetted *with not less than 15 percent water, by mass*
	Dinitrophenolates *alkali metals, dry or wetted with less than 15 per cent water, by mass*
	Dinitrophenolates, wetted *with not less than 15 percent water, by mass*
	Dinobuton
	Dinoseb
	Dinoseb acetate
	Dioxacarb
	Dioxathion
	Dipentene
	Diphacinone
	Diphenyl
PP	Diphenylamine chloroarsine
PP	Diphenylchloroarsine, solid *or* liquid
	2,4-Di-tert-butylphenol
PP	2,6-Di-tert-Butylphenol
	Disulfoton
	DNOC
	DNOC (pesticide)
	Dodecyl diphenyl oxide disulphonate
PP	Dodecyl hydroxypropyl sulfide
	1-Dodecylamine
PP	Dodecylphenol
	Drazoxolon
	Edifenphos
PP	Endosulfan

LIST OF MARINE POLLUTANTS, Continued

S.M.P. (1)	Marine Pollutant (2)
PP	Endrin
	Epibromohydrin
	Epichlorohydrin
PP	EPN
PP	Esfenvalerate
PP	Ethion
	Ethoprophos
	Ethyl fluid
	Ethyl mercaptan
	2-Ethyl-3-propylacrolein
	Ethyl tetraphosphate
	Ethyldichloroarsine
	2-Ethylhexaldehyde
	Ethylene dibromide and methyl bromide mixtures, liquid
	2-Ethylhexyl nitrate
	Fenamiphos
PP	Fenbutatin oxide
PP	Fenchlorazole-ethyl
PP	Fenitrothion
PP	Fenoxapro-ethyl
PP	Fenoxaprop-P-ethyl
PP	Fenpropathrin
	Fensulfothion
PP	Fenthion
PP	Fentin acetate
PP	Fentin hydroxide
	Ferric arsenate
	Ferric arsenite
	Ferrous arsenate
PP	Fonofos
	Formetanate
PP	Furathiocarb (ISO)
PP	gamma-BHC
	Gasoline, leaded
PP	Heptachlor
	n-Heptaldehyde
	Heptenophos
	normal-Heptyl chloride
	n-Heptylbenzene
PP	Hexachlorobutadiene
PP	1,3-Hexachlorobutadiene
	Hexaethyl tetraphosphate, *liquid*
	Hexaethyl tetraphosphate, *solid*
	normal-Hexyl chloride
	n-Hexylbenzene
	Hydrocyanic acid, anhydrous, stabilized containing less than 3% water
	Hydrocyanic acid, anhydrous, stabilized, containing less than 3% water and absorbed in a porous inert material
	Hydrocyanic acid, aqueous solutions *not more than 20% hydrocyanic acid*
	Hydrogen cyanide solution in alcohol, *with not more than 45% hydrogen cyanide*
	Hydrogen cyanide, stabilized *with less than 3% water*
	Hydrogen cyanide, stabilized *with less than 3% water and absorbed in a porous inert material*
	Hydroxydimethylbenzenes, liquid or solid
	Ioxynil
	Isobenzan
	Isobutyl butyrate
	Isobutylbenzene
	Isodecyl acrylate
	Isodecyl diphenyl phosphate
	Isofenphos
	Isooctyl nitrate
	Isoprocarb
	Isotetramethylbenzene
PP	Isoxathion
	Lead acetate
	Lead arsenates
	Lead arsenites
	Lead compounds, soluble, n.o.s.
	Lead cyanide
	Lead nitrate
	Lead perchlorate, solid or solution
	Lead tetraethyl
	Lead tetramethyl
PP	Lindane
	Linuron
	London Purple
	Magnesium arsenate
	Malathion
	Mancozeb (ISO)
	Maneb
	Maneb preparation, stabilized against self-heating
	Maneb preparations *with not less than 60% maneb*
	Maneb stabilized or Maneb preparations, stabilized *against self-heating*
	Manganese ethylene-1,2-bisdithiocarbamate
	Manganeseethylene-1,2-bisdithiocarbamate, stabilized against self-heating
	Mecarbam
	Mephosfolan
	Mercaptodimethur
PP	Mercuric acetate
PP	Mercuric ammonium chloride
PP	Mercuric arsenate
PP	Mercuric benzoate
PP	Mercuric bisulphate
PP	Mercuric bromide
PP	Mercuric chloride
PP	Mercuric cyanide
PP	Mercuric gluconate
	Mercuric iodide
PP	Mercuric nitrate
PP	Mercuric oleate
PP	Mercuric oxide
PP	Mercuric oxycyanide, desensitized
PP	Mercuric potassium cyanide
PP	Mercuric Sulphate
PP	Mercuric thiocyanate
PP	Mercurol
PP	Mercurous acetate
PP	Mercurous bisulphate
PP	Mercurous bromide
PP	Mercurous chloride

LIST OF MARINE POLLUTANTS, Continued

S.M.P. (1)	Marine Pollutant (2)
PP	Mercurous nitrate
PP	Mercurous salicylate
PP	Mercurous sulphate
PP	Mercury acetates
PP	Mercury ammonium chloride
PP	Mercury based pesticide, liquid, flammable, toxic
PP	Mercury based pesticide, liquid, toxic, flammable
PP	Mercury based pesticide, liquid, toxic
PP	Mercury based pesticide, solid, toxic
PP	Mercury benzoate
PP	Mercury bichloride
PP	Mercury bisulphates
PP	Mercury bromides
PP	Mercury compounds, liquid, n.o.s.
PP	Mercury compounds, solid, n.o.s.
PP	Mercury cyanide
PP	Mercury gluconate
PP	NMercury (I) (mercurous) compounds (pesticides)
PP	NMercury (II) (mercuric) compounds (pesticides)
	Mercury iodide
PP	Mercury nucleate
PP	Mercury oleate
PP	Mercury oxide
PP	Mercury oxycyanide, desensitized
PP	Mercury potassium cyanide
PP	Mercury potassium iodide
PP	Mercury salicylate
PP	Mercury sulfates
PP	Mercury thiocyanate
	Metam-sodium
	Methamidophos
	Methanethiol
	Methidathion
	Methomyl
	ortho-Methoxyaniline
	Methyl bromide and ethylene dibromide mixtures, liquid
	1-Methyl-2-ethylbenzene
	Methyl mercaptan
	3-Methylacrolein, stabilized
	Methylchlorobenzenes
	Methylnitrophenols
	alpha-Methylstyrene
	Methyltrithion
	Methylvinylbenzenes, inhibited
PP	Mevinphos
	Mexacarbate
	Mirex
	Monocratophos
	Motor fuel anti-knock mixtures
	Motor fuel anti-knock mixtures or compunds
	Nabam
	Naled
PP	Nickel carbonyl
PP	Nickel cyanide
PP	Nickel tetracarbonyl
	3-Nitro-4-chlorobenzotrifluoride

S.M.P. (1)	Marine Pollutant (2)
	Nitrobenzene
	Nitrobenzotrifluorides, liquid or solid
	Nitroxylenes, liquid or solid
	Nonylphenol
	normal-Octaldehyde
	Oleylamine
PP	Organotin compounds, liquid, n.o.s.
PP	Organotin compounds, (pesticides)
PP	Organotin compounds, solid, n.o.s.
PP	Organotin pesticides, liquid, flammable, toxic, n.o.s. *flash point less than 23 deg.C*
PP	Organotin pesticides, liquid, toxic, flam-mable, n.o.s.
PP	Organotin pesticides, liquid, toxic, n.o.s.
PP	Organotin pesticides, solid, toxic, n.o.s.
	Orthoarsenic acid
PP	Osmium tetroxide
	Oxamyl
	Oxydisulfoton
	Paraoxon
PP	Parathion
PP	Parathion-methyl
PP	PCBs
	Pentachloroethane
PP	Pentachlorophenol
	Pentalin
	n-Pentylbenzene
	Perchloroethylene
	Perchloromethylmercaptan
	Petrol, leaded
PP	Phenarsazine chloride
	d-Phenothrin
PP	Phenthoate
	1-Phenylbutane
	2-Phenylbutane
	Phenylcyclohexane
PP	Phenylmercuric acetate
PP	Phenylmercuric compounds, n.o.s.
PP	Phenylmercuric hydroxide
PP	Phenylmercuric nitrate
PP	Phorate
PP	Phosaione
	Phosmet
PP	Phosphamidon
PP	Phosphorus, white, molten
PP	Phosphorus, white *or* yellow dry *or* under water *or* in solution
PP	Phosphorus white, or yellow, molten
PP	Phosphorus, yellow, molten
	Pindone (and salts of)
	Pirimicarb
PP	Pirimiphos-ethyl
PP	Polychlorinated biphenyls
PP	Polyhalogenated biphenyls, liquid *or* Ter-phenyls liquid
PP	Polyhalogenated biphenyls, solid *or* Ter-phenyls solid
PP	Potassium cuprocyanide
	Potassium cyanide, solid
	Potassium cyanide, solution

LIST OF MARINE POLLUTANTS, Continued

S.M.P. (1)	Marine Pollutant (2)
PP	Potassium cyanocuprate (I)
PP	Potassium cyanomercurate
PP	Potassium mercuric iodide
	Promecarb
	Propachlor
	Propaphos
	Propenal, inhibited
	Propoxur
	Prothoate
	Prussic acid, anhydrous, stabilized
	Prussic acid, anhydrous, stabilized, absorbed in a porous inert material
PP	Pyrazophos
	Quinalphos
PP	Quizalofop
PP	Quizalofop-p-ethyl
	Rotenone
	Salithion
PP	Silafluofen
	Silver arsenite
	Silver cyanide
	Silver orthoarsenite
PP	Sodium copper cyanide, solid
PP	Sodium copper cyanide solution
PP	Sodium cuprocyanide, solid
PP	Sodium cuprocyanide, solution
	Sodium cyanide, solid
	Sodium cyanide, solution
	Sodium dinitro-o-cresolate, *dry or wetted* with less than 15 percent water, by mass
	Sodium dinitro-ortho-cresolate, wetted *with not less than 15 percent water, by mass*
PP	Sodium pentachlorophenate
	Strychnine *or* Strychnine salts
	Sulfotep
PP	Sulprophos
	Tallow nitrile
	Temephos
	TEPP
PP	Terbufos
	Tetrabromoethane
	Tetrabromomethane
	1,1,2,2-Tetrachloroethane
	Tetrachloroethylene
	Tetrachloromethane
	Tetraethyl dithiopyrophosphate
PP	Tetraethyl lead, liquid
	Tetramethrin
	Tetramethyllead
	Thallium chlorate
	Thallium compounds, n.o.s.
	Thallium compounds (pesticides)
	Thallium nitrate
	Thallium sulfate
	Thallous chlorate
	Thiocarbonyl tetrachloride
	Triaryl phosphates, isopropylated
PP	Triaryl phosphates, n.o.s.
	Triazophos
	Tribromomethane
PP	Tributyltin compounds
	Trichlorfon
PP	1,2,3-Trichlorobenzene
	Trichlorobenzenes, liquid
	Trichlorobutene
	Trichlorobutylene
	Trichloromethane sulphuryl chloride
	Trichloromethyl sulphochloride
	Trichloronat
	Tricresyl phosphate (less than 1% ortho-isomer)
PP	Tricresyl phosphate, not less than 1% ortho-isomer but not more than 3% orthoisomer
PP	Tricresyl phosphate *with more than 3 per cent ortho isomer*
	Triethylbenzene
	Triisopropylated phenyl phosphates
	Trimethylene dichloride
	Triphenyl phosphate/tert-butylated triphenyl phosphates mixtures containing 5% to 10% tri-phenyl phosphates
PP	Triphenyl phosphate/tert-butylated triphenyl phosphates mixtures containing 10% to 48% triphenyl phosphates
PP	Triphenylphosphate
PP	Triphenyltin compounds
	Tritolyl phosphate (less than 1% ortho-isomer)
PP	Tritolyl phosphate (not less than 1% ortho-isomer)
	Trixylenyl phosphate
	Vinylidene chloride, inhibited
	Vinylidene chloride, stabilized
	Warfarin (and salts of)
PP	White phosphorus, dry
PP	White phosphorus, wet
	White spirit, low (15-20%) aromatic
PP	Yellow phosphorus, dry
PP	Yellow phosphorus, wet
	Zinc bromide
	Zinc cyanide

Reserved

§172.102 Special provisions.

(a) *General.* When Column 7 of the §172.101 Table refers to a special provision for a hazardous material, the meaning and requirements of that provision are as set forth in this section. When a special provision specifies packaging or packaging requirements—

(1) The special provision is in addition to the standard requirements for all packagings prescribed in §173.24 of this subchapter and any other applicable packaging requirements in subparts A and B of part 173 of this subchapter; and

(2) To the extent a special provision imposes limitations or additional requirements on the packaging provisions set forth in Column 8 of the §172.101 Table, packagings must conform to the requirements of the special provision.

(b) *Description of codes for special provisions.* Special provisions contain packaging provisions, prohibitions, exceptions from requirements for particular quantities or forms of materials and requirements or prohibitions applicable to specific modes of transportation, as follows:

(1) A code consisting only of numbers (for example, "11") is multi-modal in application and may apply to bulk and non-bulk packagings.

(2) A code containing the letter "A" refers to a special provision which applies only to transportation by aircraft.

(3) A code containing the letter "B" refers to a special provision that applies only to bulk packaging requirements. Unless otherwise provided in this subchapter, these special provisions do not apply to UN, IM Specification portable tanks or IBCs.

(4) A code containing the letters "IB" or "IP" refers to a special provision that applies only to transportation in IBCs.

(5) A code containing the letter "N" refers to a special provision which applies only to non-bulk packaging requirements.

(6) A code containing the letter "R" refers to a special provision which applies only to transportation by rail.

(7) A code containing the letter "T" refers to a special provision which applies only to transportation in UN or IM Specification portable tanks.

(8) A code containing the letters "TP" refers to a portable tank special provision for UN or IM Specification portable tanks that is in addition to those provided by the portable tank instructions or the requirements in part 178 of this subchapter.

(9) A code containing the letter "W" refers to a special provision that applies only to transportation by water.

(c) *Tables of special provisions.* The following tables list, and set forth the requirements of, the special provisions referred to in Column 7 of the §172.101 Table.

(1) *Numeric provisions.* These provisions are multimodal and apply to bulk and non-bulk packagings:

Code / Special Provisions

1 This material is poisonous by inhalation (see §171.8 of this subchapter) in Hazard Zone A (see §173.116(a) or §173.133(a) of this subchapter), and must be described as an inhalation hazard under the provisions of this subchapter.

2 This material is poisonous by inhalation (see §171.8 of this subchapter) in Hazard Zone B (see §173.116(a) or §173.133(a) of this subchapter), and must be described as an inhalation hazard under the provisions of this subchapter.

3 This material is poisonous by inhalation (see §171.8 of this subchapter) in Hazard Zone C (see §173.116(a) of this subchapter), and must be described as an inhalation hazard under the provisions of this subchapter.

4 This material is poisonous by inhalation (see §171.8 of this subchapter) in Hazard Zone D (see §173.116(a) of this subchapter), and must be described as an inhalation hazard under the provisions of this subchapter.

5 If this material meets the definition for a material poisonous by inhalation (see §171.8 of this subchapter), a shipping name must be selected which identifies the inhalation hazard, in Division 2.3 or Division 6.1, as appropriate.

6 This material is poisonous-by-inhalation and must be described as an inhalation hazard under the provisions of this subchapter.

8 A hazardous substance that is not a hazardous waste may be shipped under the shipping description "Other regulated substances, liquid or solid, n.o.s.", as appropriate. In addition, for solid materials, special provision B54 applies.

9 Packaging for certain PCBs for disposal and storage is prescribed by EPA in 40 CFR 761.60 and 761.65.

11 The hazardous material must be packaged as either a liquid or a solid, as appropriate, depending on its physical form at 55°C (131°F) at atmospheric pressure.

12 In concentrations greater than 40 percent, this material has strong oxidizing properties and is capable of starting fires in contact with combustible materials. If appropriate, a package containing this material must conform to the additional labeling requirements of §172.402 of this subchapter.

13 The words "Inhalation Hazard" shall be entered on each shipping paper in association with the shipping description, shall be marked on each non-bulk package in association with the proper shipping name and identification number, and shall be marked on two opposing sides of each bulk package. Size of marking on bulk package must conform to §172.302(b) of this subchapter. The requirements of §§172.203(m) and 172.505 of this subchapter do not apply.

14 Motor fuel antiknock mixtures are:

a. Mixtures of one or more organic lead mixtures (such as tetraethyl lead, triethylmethyl lead, diethyldimethyl lead, ethyltrimethyl lead, and tetramethyl lead) with one or more halogen compounds (such as ethylene dibromide and ethylene dichloride), hydrocarbon solvents or other equally efficient stabilizers; or

b. tetraethyl lead.

15 This entry applies to "Chemical kits" and "First aid kits" containing one or more compatible items of hazardous materials in boxes, cases, etc. that, for example, are used for medical, analytical, diagnostic, testing, or repair purposes. Kits that are carried on board transport vehicles for first aid or operating purposes are not subject to the requirements of this subchapter.

16 This description applies to smokeless powder and other solid propellants that are used as powder for small arms and have been classed as Division 1.3 and 4.1 in accordance with §173.56 of this subchapter.

18 This description is authorized only for fire extinguishers listed in §173.309(b) of this subchapter meeting the following conditions:

a. Each fire extinguisher may only have extinguishing contents that are nonflammable, non-poisonous, non-corrosive and commercially free from corroding components.

b. Each fire extinguisher must be charged with a nonflammable, non-poisonous, dry gas that has a dew-point at or below minus 46.7°C (minus 52°F) at 101kPa (1 atmosphere) and is free of corroding components, to not more than the service pressure of the cylinder.

c. A fire extinguisher may not contain more than 30% carbon dioxide by volume or any other corrosive extinguishing agent.

d. Each fire extinguisher must be protected externally by suitable corrosion-resisting coating.

19 For domestic transportation only, the identification number "UN1075" may be used in place of the identification number specified in Column (4) of the §172.101 Table. The identification number used must be consistent on package markings, shipping papers and emergency response information.

21 This material must be stabilized by appropriate means (e.g., addition of chemical inhibitor, purging to remove oxygen) to prevent dangerous polymerization (see §173.21(f) of this subchapter).

22 If the hazardous material is in dispersion in organic liquid, the organic liquid must have a flash point above 50°C (122°F).

23 This material may be transported under the provisions of Division 4.1 only if it is so packed that the percentage of diluent will not fall below that stated in the shipping description at any time during transport. Quantities of not more than 500 g per package with not less than 10 percent water by mass may also be classed in Division 4.1, provided a negative test result is obtained when tested in accordance with test series 6(c) of the UN Manual of Tests and Criteria (IBR, see §171.7 of this subchapter).

24 Alcoholic beverages containing more than 70 percent alcohol by volume must be transported as materials in Packing Group II. Alcoholic beverages containing more than 24 percent but not more than 70 percent alcohol by volume must be transported as materials in Packing Group III.

26 This entry does not include ammonium permanganate, the transport of which is prohibited except when approved by the Associate Administrator.

28 The dihydrated sodium salt of dichloroisocyanuric acid is not subject to the requirements of this subchapter.

29 For transportation by motor vehicle, rail car or vessel, production runs (exceptions for prototypes can be found in §173.185(e)) of not more than 100 lithium cells or batteries are excepted from the testing requirements of §173.185(a)(1) if—

a. For a lithium metal cell or battery, the lithium content is not more than 1.0 g per cell and the aggregate lithium content is not more than 2.0 g per battery, and, for a lithium-ion cell or battery, the equivalent lithium content is not more than 1.5 g per cell and the aggregate equivalent lithium content is not more than 8 g per battery;

b. The cells and batteries are transported in an outer packaging that is a metal, plastic or plywood drum or metal, plastic or wooden box that meets the criteria for Packing Group I packagings; and

c. Each cell and battery is individually packed in an inner packaging inside an outer packaging and is surrounded by cushioning material that is non-combustible, and non-conductive.

30 Sulfur is not subject to the requirements of this subchapter if transported in a non-bulk packaging or if formed to a specific shape (for example, prills, granules, pellets, pastilles, or flakes). A bulk packaging containing sulfur is not subject to the placarding requirements of subpart F of this part, if it is marked with the appropriate identification number as required by subpart D of this part. Molten sulfur must be marked as required by §172.325 of this subchapter.

31 Materials which have undergone sufficient heat treatment to render them non-hazardous are not subject to the requirements of this subchapter.

32 Polymeric beads and molding compounds may be made from polystyrene, poly(methyl methacrylate) or other polymeric material.

33 Ammonium nitrates and mixtures of an inorganic nitrite with an ammonium salt are prohibited.

34 The commercial grade of calcium nitrate fertilizer, when consisting mainly of a double salt (calcium nitrate and ammonium nitrate) containing not more than 10 percent ammonium nitrate and at least 12 percent water of crystallization, is not subject to the requirements of this subchapter.

35 Antimony sulphides and oxides which do not contain more than 0.5 percent of arsenic calculated on the total mass do not meet the definition of Division 6.1.

37 Unless it can be demonstrated by testing that the sensitivity of the substance in its frozen state is no greater than in its liquid state, the substance must remain liquid during normal transport conditions. It must not freeze at temperatures above -15°C (5°F).

38 If this material shows a violent effect in laboratory tests involving heating under confinement, the labeling requirements of Special Provision 53 apply, and the material must be packaged in accordance with packing method OP6 in §173.225 of this subchapter. If the SADT of the technically pure substance is higher than 75°C, the technically pure substance and formulations derived from it are not self-reactive materials and, if not meeting any other hazard class, are not subject to the requirements of this subchapter.

39 This substance may be carried under provisions other than those of Class 1 only if it is so packed that the percentage of water will not fall below that stated at any time during transport. When phlegmatized with water and inorganic inert material, the content of urea nitrate must not exceed 75 percent by mass and the mixture should not be capable of being detonated by test 1(a)(i) or test 1(a) (ii) in the UN Manual of Tests and Criteria (IBR, see §171.7 of this subchapter).

40 Polyester resin kits consist of two components: A base material (Class 3, Packing Group II or III) and an activator (organic peroxide), each separately packed in an inner packaging. The organic peroxide must be type D, E, or F, not requiring temperature control. The components may be placed in the same outer packaging provided they will not interact dangerously in the event of leakage. The Packing Group assigned will be II or III, according to the classification criteria for Class 3, applied to the base material. Additionally, unless otherwise excepted in this subchapter, polyester resin kits must be packaged in specification combination packagings based on the performance level of the base material contained within the kit.

41 This material at the Packing Group II hazard criteria level may be transported in Large Packagings.

43 The membrane filters, including paper separators and coating or backing materials, that are present in transport, must not be able to propagate a detonation as tested by one of the tests de scribed in the UN Manual of Tests and Criteria, Part I, Test series 1(a) (IBR, see §171.7 of this subchapter). On the basis of the results of suitable burning rate tests, and taking into account the standard tests in the UN Manual of Tests and Criteria, Part III, subsection 33.2.1 (IBR, see §171.7 of this subchapter), nitrocellulose membrane filters in the form in which they are to be transported that do not meet the criteria for a Division 4.1 material are not subject to the requirements of this subchapter. Packagings must be so constructed that explosion is not possible by reason of increased internal pressure. Nitrocellulose membrane filters covered by this entry, each with a mass not exceeding 0.5 g, are not subject to the requirements of this subchapter when contained individually in an article or a sealed packet.

44 The formulation must be prepared so that it remains homogeneous and does not separate during transport. Formulations with low nitrocellulose contents and neither showing dangerous properties when tested for their ability to detonate, deflagrate or explode when heated under defined confinement by the appropriate test methods and criteria in the UN Manual of Tests and Criteria (IBR, see §171.7 of this subchapter), nor classed as a Division 4.1 (flammable solid) when tested in accordance with the procedures specified in §173.124 of this subchapter (chips, if necessary, crushed and sieved to a particle size of less than 1.25 mm), are not subject to the requirements of this subchapter.

45 Temperature should be maintained between 18°C (64.4°F) and 40°C (104°F). Tanks containing solidified methacrylic acid must not be reheated during transport.

46 This material must be packed in accordance with packing method OP6 (see §173.225 of this subchapter). During transport, it must be protected from direct sunshine and stored (or kept) in a cool and well-ventilated place, away from all sources of heat.

47 Mixtures of solids that are not subject to this subchapter and flammable liquids may be transported under this entry without first applying the classification criteria of Division 4.1, provided there is no free liquid visible at the time the material is loaded or at the time the packaging or transport unit is closed. Except when the liquids are fully absorbed in solid material contained in sealed bags, each packaging must correspond to a design type that has passed a leakproofness test at the Packing Group II level. Small inner packagings consisting of sealed packets and articles containing less than 10 mL of a Class 3 liquid in Packing Group II or III absorbed onto a solid material are not subject to this subchapter provided there is no free liquid in the packet or article.

48 Mixtures of solids which are not subject to this subchapter and toxic liquids may be transported under this entry without first applying the classification criteria of Division 6.1, provided there is no free liquid visible at the time the material is loaded or at the time the packaging or transport unit is closed. Each packaging must correspond to a design type that has passed a leakproofness test at the Packing Group II level. This entry may not be used for solids containing a Packing Group I liquid.

49 Mixtures of solids which are not subject to this subchapter and corrosive liquids may be transported under this entry without first applying the classification criteria of Class 8, provided there is no free liquid visible at the time the material is loaded or at the time the packaging or transport unit is closed. Each packaging must correspond to a design type that has passed a leakproofness test at the Packing Group II level.

50 Cases, cartridge, empty with primer which are made of metallic or plastic casings and meeting the classification criteria of Division 1.4 are not regulated for domestic transportation.

51 This description applies to items previously described as "Toy propellant devices, Class C" and includes reloadable kits. Model rocket motors containing 30 grams or less propellant are classed as Division 1.4S and items containing more than 30 grams of propellant but not more than 62.5 grams of propellant are classed as Division 1.4C.

52 This entry may only be used for substances that do not exhibit explosive properties of Class 1 (explosive) when tested in accordance with Test Series 1 and 2 of Class 1 (explosive) in the UN Manual of Tests and Criteria, Part I (incorporated by reference; see §171.7 of this subchapter).

53 Packages of these materials must bear the subsidiary risk label, "EXPLOSIVE", and the subsidiary hazard class/division must be entered in parentheses immediately following the primary hazard class in the shipping description, unless otherwise provided in this subchapter or through an approval issued by the Associate Administrator, or the competent authority of the country of origin. A copy of the approval shall accompany the shipping papers.

54 Maneb or maneb preparations not meeting the definition of Division 4.3 or any other hazard class are not subject to the requirements of this subchapter when transported by motor vehicle, rail car, or aircraft.

55 This device must be approved in accordance with §173.56 of this subchapter by the Associate Administrator.

56 A means to interrupt and prevent detonation of the detonator from initiating the detonating cord must be installed between each electric detonator and the detonating cord ends of the jet perforating guns before the charged jet perforating guns are offered for transportation.

57 Maneb or Maneb preparations stabilized against self-heating need not be classified in Division 4.2 when it can be demonstrated by testing that a volume of 1 m^3 of substance does not self-ignite and that the temperature at the center of the sample does not exceed 200°C, when the sample is maintained at a temperature of not less than 75°C ± 2°C for a period of 24 hours, in accordance with procedures set forth for testing self-heating materials in the UN Manual of Tests and Criteria (IBR, see §171.7 of this subchapter).

58 Aqueous solutions of Division 5.1 inorganic solid nitrate substances are considered as not meeting the criteria of Division 5.1 if the concentration of the substances in solution at the minimum temperature encountered in transport is not greater than 80% of the saturation limit.

59 Ferrocerium, stabilized against corrosion, with a minimum iron content of 10 percent is not subject to the requirements of this subchapter.

61 A chemical oxygen generator is spent if its means of ignition and all or a part of its chemical contents have been expended.

62 Oxygen generators (see §171.8 of this subchapter) are not authorized for transportation under this entry.

64 The group of alkali metals includes lithium, sodium, potassium, rubidium, and caesium.

65 The group of alkaline earth metals includes magnesium, calcium, strontium, and barium.

66 Formulations of these substances containing not less than 30 percent non-volatile, non-flammable phlegmatizer are not subject to this subchapter.

70 Black powder that has been classed in accordance with the requirements of §173.56 of this subchapter may be reclassed and offered for domestic transportation as a Division 4.1 material if it is offered for transportation and transported in accordance with the limitations and packaging requirements of §173.170 of this subchapter.

74 During transport, this material must be protected from direct sunshine and stored or kept in a cool and well-ventilated place, away from all sources of heat.

78 This entry may not be used to describe compressed air which contains more than 23.5 percent oxygen. Compressed air containing greater than 23.5 percent oxygen must be shipped using the description "Compressed gas, oxidizing, n.o.s., UN3156."

79 This entry may not be used for mixtures that meet the definition for oxidizing gas.

81 Polychlorinated biphenyl items, as defined in 40 CFR 761.3, for which specification packagings are impractical, may be packaged in non-specification packagings meeting the general packaging requirements of subparts A and B of part 173 of this subchapter. Alternatively, the item itself may be used as a packaging if it meets the general packaging requirements of subparts A and B of part 173 of this subchapter.

102 The ends of the detonating cord must be tied fast so that the explosive cannot escape. The articles may be transported as in Division 1.4 Compatibility Group D (1.4D) if all of the conditions specified in §173.63(a) of this subchapter are met.

103 Detonators which will not mass detonate and undergo only limited propagation in the shipping package may be assigned to 1.4B classification code. Mass detonate means that more than 90 percent of the devices tested in a package explode practically simultaneously. Limited propagation means that if one detonator near the center of a shipping package is exploded, the aggregate weight of explosives, excluding ignition and delay charges, in this and all additional detonators in the outside packaging that explode may not exceed 25 grams.

105 The word "Agents" may be used instead of "Explosives" when approved by the Associate Administrator.

106 The recognized name of the particular explosive may be specified in addition to the type.

107 The classification of the substance is expected to vary especially with the particle size and packaging but the border lines have not been experimentally determined; appropriate classifications should be verified following the test procedures in §§173.57 and 173.58 of this subchapter.

108 Fireworks must be so constructed and packaged that loose pyrotechnic composition will not be present in packages during transportation.

109 Rocket motors must be nonpropulsive in transportation unless approved in accordance with §173.56 of this subchapter. A rocket motor to be considered "nonpropulsive" must be capable of unrestrained burning and must not appreciably move in any direction when ignited by any means.

110 Fire extinguishers transported under UN1044 and oxygen cylinders transported for emergency use under UN1072 may include installed actuating cartridges (cartridges, power device of Division 1.4C or 1.4S), without changing the classification of Division 2.2, provided the aggregate quantity of deflagrating (propellant) explosives does not exceed 3.2 grams per cylinder. Oxygen cylinders with installed actuating cartridges as prepared for transportation must have an effective means of preventing inadvertent activation.

111 Explosive substances of Division 1.1 Compatibility Group A (1.1A) are forbidden for transportation if dry or not desensitized, unless incorporated in a device.

113 The sample must be given a tentative approval by an agency or laboratory in accordance with §173.56 of this subchapter.

114 Jet perforating guns, charged, oil well, without detonator may be reclassed to Division 1.4 Compatibility Group D (1.4D) if the following conditions are met:

a. The total weight of the explosive contents of the shaped charges assembled in the guns does not exceed 90.5 kg (200 pounds) per vehicle; and

b. The guns are packaged in accordance with Packing Method US1 as specified in §173.62 of this subchapter.

115 Boosters with detonator, detonator assemblies and boosters with detonators in which the total explosive charge per unit does not exceed 25 g, and which will not mass detonate and undergo only limited propagation in the shipping package may be assigned to 1.4B classification code. Mass detonate means more than 90 percent of the devices tested in a package explode practically simultaneously. Limited propagation means that if one booster near the center of the package is exploded, the aggregate weight of explosives, excluding ignition and delay charges, in this and all additional boosters in the outside packaging that explode may not exceed 25 g.

116 Fuzes, detonating may be classed in Division 1.4 if the fuzes do not contain more than 25 g of explosive per fuze and are made and packaged so that they will not cause functioning of other fuzes, explosives or other explosive devices if one of the fuzes detonates in a shipping packaging or in adjacent packages.

117 If shipment of the explosive substance is to take place at a time that freezing weather is anticipated, the water contained in the explosive substance must be mixed with denatured alcohol so that freezing will not occur.

118 This substance may not be transported under the provisions of Division 4.1 unless specifically authorized by the Associate Administrator.

119 This substance, when in quantities of not more than 11.5 kg (25.3 pounds), with not less than 10 percent water, by mass, also may be classed as Division 4.1, provided a negative test result is obtained when tested in accordance with test series 6(c) of the UN Manual of Tests and Criteria (IBR, see §171.7 of this subchapter).

120 The phlegmatized substance must be significantly less sensitive than dry PETN.

121 This substance, when containing less alcohol, water or phlegmatizer than specified, may not be transported unless approved by the Associate Administrator.

123 Any explosives, blasting, type C containing chlorates must be segregated from explosives containing ammonium nitrate or other ammonium salts.

125 Lactose or glucose or similar materials may be used as a phlegmatizer provided that the substance contains not less than 90%, by mass, of phlegmatizer. These mixtures may be classified in Division 4.1 when tested in accordance with test series 6(c) of the UN Manual of Tests and Criteria (IBR, see §171.7 of this subchapter) and approved by the Associate Administrator. Testing must be conducted on at least three packages as prepared for transport. Mixtures containing at least 98%, by mass, of phlegmatizer are not subject to the requirements of this subchapter. Packages containing mixtures with not less than 90% by mass, of phlegmatizer need not bear a POISON subsidiary risk label.

127 Mixtures containing oxidizing and organic materials transported under this entry may not meet the definition and criteria of a Class 1 material. (See §173.50 of this subchapter.)

128 Regardless of the provisions of §172.101(c)(12), aluminum smelting by-products and aluminum remelting by-products described under this entry, meeting the definition of Class 8, Packing Group II and III may be classed as a Division 4.3 material and transported under this entry. The presence of a Class 8 hazard must be communicated as required by this Part for subsidiary hazards.

129 These materials may not be classified and transported unless authorized by the Associate Administrator on the basis of results from Series 2 Test and a Series 6(c) Test from the UN Manual of Tests and Criteria (IBR, see §171.7 of this subchapter) on packages as prepared for transport. The packing group assignment and packaging must be approved by the Associate Administrator on the basis of the criteria in §173.21 of this subchapter and the package type used for the Series 6(c) test.

130 "Batteries, dry, sealed, n.o.s.," commonly referred to as dry batteries, are hermetically sealed and generally utilize metals (other than lead) and/or carbon as electrodes. These batteries are typically used for portable power applications. The rechargeable (and some non-rechargeable) types have gelled alkaline electrolytes (rather than acidic) making it difficult for them to generate hydrogen or oxygen when overcharged and therefore, differentiating them from non-spillable batteries. Dry batteries specifically covered by another entry in the §172.101 Table must be transported in accordance with the requirements applicable to that entry. For example, nickel-metal hydride batteries transported by vessel in certain quantities are covered by another entry (*see* Batteries, nickel-metal hydride, UN3496). Dry batteries not specifically covered by another entry in the §172.101 Table are covered by this entry (*i.e.*, Batteries, dry, sealed, n.o.s.) and are not subject to requirements of this subchapter except for the following:

(a) *Incident reporting*. For transportation by aircraft, a telephone report in accordance with §171.15(a) is required if a fire, violent rupture, explosion or dangerous evolution of heat (*i.e.*, an amount of heat sufficient to be dangerous to packaging or personal safety to include charring of packaging, melting of packaging, scorching of packaging, or other evidence) occurs as a direct result of a dry battery. For all modes of transportation, a written report submitted, retained, and updated in accordance with §171.16 is required if a fire, violent rupture, explosion or dangerous evolution of heat occurs as a direct result of a dry battery or batterypowered device.

(b) *Preparation for transport*. Batteries and battery-powered device(s) containing batteries must be prepared and packaged for transport in a manner to prevent:

(1) A dangerous evolution of heat;

(2) Short circuits, including but not limited to the following methods:

(i) Packaging each battery or each battery-powered device when practicable, in fully enclosed inner packagings made of non-conductive material;

(ii) Separating or packaging batteries in a manner to prevent contact with other batteries, devices or conductive materials (*e.g.*, metal) in the packagings; or

(iii) Ensuring exposed terminals or connectors are protected with non-conductive caps, non-conductive tape, or by other appropriate means; and

(3) Damage to terminals. If not impact resistant, the outer packaging should not be used as the sole means of protecting the battery terminals from damage or short circuiting. Batteries must be securely cushioned and packed to prevent shifting which could loosen terminal caps or reorient the terminals to produce short circuits. Batteries contained in devices must be securely installed. Terminal protection methods include but are not limited to the following:

(i) Securely attaching covers of sufficient strength to protect the terminals;

(ii) Packaging the battery in a rigid plastic packaging; or

SPECIAL PROVISIONS

(iii) Constructing the battery with terminals that are recessed or otherwise protected so that the terminals will not be subjected to damage if the package is dropped.

(c) *Additional air transport requirements.* For a battery whose voltage (electrical potential) exceeds 9 volts—

(1) When contained in a device, the device must be packaged in a manner that prevents unintentional activation or must have an independent means of preventing unintentional activation (*e.g.*, packaging restricts access to activation switch, switch caps or locks, recessed switches, trigger locks, temperature sensitive circuit breakers, *etc.*); and

(2) An indication of compliance with this special provision must be provided by marking each package with the words "not restricted" or by including the words "not restricted" on a transport document such as an air waybill accompanying the shipment.

(d) *Used or spent battery exception.* Used or spent dry batteries of both nonrechargeable and rechargeable designs, with a marked rating up to 9-volt that are combined in the same package and transported by highway or rail for recycling, reconditioning, or disposal are not subject to this special provision or any other requirement of the HMR. Note that batteries utilizing different chemistries (*i.e.*, those battery chemistries specifically covered by another entry in the §172.101 Table) as well as dry batteries with a marked rating greater than 9-volt may not be combined with used or spent batteries in the same package. Note also that this exception does not apply to batteries that have been reconditioned for reuse.

131 This material may not be offered for transportation unless approved by the Associate Administrator.

132 This entry may only be used for uniform, ammonium nitrate based fertilizer mixtures, containing nitrogen, phosphate or potash, meeting the following criteria: (1) Contains not more than 70% ammonium nitrate and not more than 0.4% total combustible, organic material calculated as carbon or (2) Contains not more than 45% ammonium nitrate and unrestricted combustible material.

134 This entry only applies to vehicles, machinery and equipment powered by wet batteries, sodium batteries, or lithium batteries that are transported with these batteries installed. Examples of such items are electrically-powered cars, lawn mowers, wheelchairs, and other mobility aids. Self-propelled vehicles or equipment that also contain an internal combustion engine must be consigned under the entry "Engine, internal combustion, flammable gas powered" or "Engine, internal combustion, flammable liquid powered" or "Vehicle, flammable gas powered" or "Vehicle, flammable liquid powered," as appropriate. These entries include hybrid electric vehicles powered by both an internal combustion engine and batteries. Additionally, selfpropelled vehicles or equipment that contain a fuel cell engine must be consigned under the entries "Engine, fuel cell, flammable gas powered" or "Engine, fuel cell, flammable liquid powered" or "Vehicle, fuel cell, flammable gas powered" or "Vehicle, fuel cell, flammable liquid powered," as appropriate. These entries include hybrid electric vehicles powered by a fuel cell engine, an internal combustion engine, and batteries.

135 Internal combustion engines installed in a vehicle must be consigned under the entries "Vehicle, flammable gas powered" or "Vehicle, flammable liquid powered," as appropriate. These entries include hybrid electric vehicles powered by both an internal combustion engine and wet, sodium or lithium batteries installed. If a fuel cell engine is installed in a vehicle, the vehicle must be consigned using the entries "Vehicle, fuel cell, flammable gas powered" or "Vehicle, fuel cell, flammable liquid powered," as appropriate. These entries include hybrid electric vehicles powered by a fuel cell, an internal combustion engine, and wet, sodium or lithium batteries installed.

136 This entry only applies to machinery and apparatus containing hazardous materials as in integral element of the machinery or apparatus. It may not be used to describe machinery or apparatus for which a proper shipping name exists in the §172.101 Table. Except when approved by the Associate Administrator, machinery or apparatus may only contain hazardous materials for which exceptions are referenced in Column (8) of the §172.101 Table and are provided in part 173, subpart D, of this subchapter. Hazardous materials shipped under this entry are excepted from the labeling requirements of this subchapter unless offered for transportation or transported by aircraft and are not subject to the placarding requirements of part 172, subpart F, of this subchapter. Orientation markings as described in §172.312(a)(2) are required when liquid hazardous materials may escape due to incorrect orientation. The machinery or apparatus, if unpackaged, or the packaging in which it is contained shall be marked "Dangerous goods in machinery" or "Dangerous goods in apparatus", as appropriate, with the identification number UN3363. For transportation by aircraft, machinery or apparatus may not contain any material forbidden for transportation by passenger or cargo aircraft. The Associate Administrator may except from the requirements of this subchapter, equipment, machinery and apparatus provided:

a. It is shown that it does not pose a significant risk in transportation;

b. The quantities of hazardous materials do not exceed those specified in §173.4a of this subchapter; and

c. The equipment, machinery or apparatus conforms with §173.222 of this subchapter.

137 Cotton, dry; flax, dry; sisal, dry; and tampico fiber, dry are not subject to the requirements of this subchapter when they are baled in accordance with ISO 8115, "Cotton Bales—Dimensions and Density" (IBR, see §171.7 of this subchapter) to a density of not less than 360 kg/m^3 (22.1 lb/ft^3) for cotton, 400 kg/m^3 (24.97 lb/ft^3) for flax, 620 kg/m^3 (38.71 lb/ft^3) for sisal and 360 kg/m^3 (22.1 lb/ft^3) for tampico fiber and transported in a freight container or closed transport vehicle.

138 Lead compounds which, when mixed in a ratio of 1:1,000 with 0.07 M (Molar concentration) hydrochloric acid and stirred for one hour at a temperature of 23 °C ± 2 °C, exhibit a solubility of 5% or less are considered insoluble and are not subject to the requirements of this subchapter unless they meet criteria as another hazard class or division.

139 Use of the "special arrangement" proper shipping names for international shipments must be made under an IAEA Certificate of Competent Authority issued by the Associate Administrator in accordance with the requirements in §173.471, §173.472, or §173.473 of this subchapter. Use of these proper shipping names for domestic shipments may be made only under a DOT special permit, as defined in, and in accordance with the requirements of subpart B of part 107 of this subchapter.

140 This material is regulated only when it meets the defining criteria for a hazardous substance or a marine pollutant. In addition, the column 5 reference is modified to read "III" on those occasions when this material is offered for transportation or transported by highway or rail.

141 A toxin obtained from a plant, animal, or bacterial source containing an infectious substance, or a toxin contained in an infectious substance, must be classed as Division 6.2, described as an infectious substance, and assigned to UN 2814 or UN 2900, as appropriate.

142 These hazardous materials may not be classified and transported unless authorized by the Associate Administrator. The Associate Administrator will base the authorization on results from Series 2 tests and a Series 6(c) test from the UN Manual of Tests and Criteria (IBR, see §171.7 of this subchapter) on packages as prepared for transport in accordance with the requirements of this subchapter.

144 If transported as a residue in an underground storage tank (UST), as defined in 40 CFR 280.12, that has been cleaned and purged or rendered inert according to the American Petroleum Institute (API) Standard 1604 (IBR, see §171.7 of this subchapter), then the tank and this material are not subject to any other requirements of this subchapter. However, sediments remaining in the tank that meet the definition for a hazardous material are subject to the applicable regulations of this subchapter.

145 This entry applies to formulations that neither detonate in the cavitated state nor deflagrate in laboratory testing, show no effect when heated under confinement, exhibit no explosive power, and are thermally stable (self-accelerating decomposition temperature (SADT) at 60°C (140°F) or higher for a 50 kg (110.2 lbs.) package). Formulations not meeting these criteria must be transported under the provisions applicable to the appropriate entry in the Organic Peroxide Table in §173.225 of this subchapter.

146 This description may be used for a material that poses a hazard to the environment but does not meet the definition for a hazardous waste or a hazardous substance, as defined in §171.8 of this subchapter, or any hazard class, as defined in part 173 of this subchapter, if it is designated as environmentally hazardous by another Competent Authority. This provision may be used for both domestic and international shipments.

147 This entry applies to non-sensitized emulsions, suspensions, and gels consisting primarily of a mixture of ammonium nitrate and fuel, intended to produce a Type E blasting explosive only after further processing prior to use. The mixture for emulsions typically has the following composition: 60-85% ammonium nitrate; 5-30% water; 2-8% fuel; 0.5-4% emulsifier or thickening agent; 0-10% soluble flame suppressants; and trace additives. Other inorganic nitrate salts may replace part of the ammonium nitrate. The mixture for suspensions and gels typically has the following composition: 60–85% ammonium nitrate; 0–5% sodium or potassium perchlorate; 0–17% hexamine nitrate or monomethylamine nitrate; 5-30% water; 2-15% fuel; 0.5-4% thickening agent; 0-10% soulble flame suppressants; and trace additives. Other inorganic nitrate salts may replace part of the ammonium nitrate. These substances must satisfactorily pass Test Series 8 of the UN Manual of Tests and Criteria, Part I, Section 18 (IBR, see §171.7 of this subchapter), and may not be classified and transported unless approved by the Associate Administrator.

149 Except for transportation by aircraft, when transported as a limited quantity or a consumer commodity, the maximum net capacity specified in §173.150(b)(2) of this subchapter for inner packagings may be increased to 5 L (1.3 gallons).

150 This description may be used only for uniform mixtures of fertilizers containing ammonium nitrate as the main ingredient within the following composition limits:

a. Not less than 90% ammonium nitrate with not more than 0.2% total combustible, organic material calculated as carbon, and with added matter, if any, that is inorganic and inert when in contact with ammonium nitrate; or

b. Less than 90% but more than 70% ammonium nitrate with other inorganic materials, or more than 80% but less than 90% ammonium nitrate mixed with calcium carbonate and/or dolomite and/or mineral calcium sulphate, and not more than 0.4% total combustible, organic material calculated as carbon; or

c. Ammonium nitrate-based fertilizers containing mixtures of ammonium nitrate and ammonium sulphate with more than 45% but less than 70% ammonium nitrate, and not more than 0.4% total combustible, organic material calculated as carbon such that the sum of the percentage of compositions of ammonium nitrate and ammonium sulphate exceeds 70%.

151 If this material meets the definition of a flammable liquid in §173.120 of this subchapter, a FLAMMABLE LIQUID label is also required and the basic description on the shipping paper must indicate the Class 3 subsidiary hazard.

155 Fish meal or fish scrap may not be transported if the temperature at the time of loading either exceeds 35°C (95°F), or exceeds 5°C (41°F) above the ambient temperature, whichever is higher.

156 Asbestos that is immersed or fixed in a natural or artificial binder material, such as cement, plastic, asphalt, resins or mineral ore, or contained in manufactured products is not subject to the requirements of this subchapter.

159 This material must be protected from direct sunshine and kept in a cool, well-ventilated place away from sources of heat.

160 This entry applies to articles that are used as life-saving vehicle air bag inflators, air bag modules or seat-belt pretensioners containing Class 1 (explosive) materials or materials of other hazard classes. Air bag inflators and modules must be tested in accordance with Test series 6(c) of Part I of the UN Manual of Tests and Criteria (incorporated by reference; see §171.7 of this subchapter), with no explosion of the device, no fragmentation of device casing or pressure vessel, and no projection hazard or thermal effect that would significantly hinder fire-fighting or other emergency response efforts in the immediate vicinity. If the air bag inflator unit satisfactorily passes the series 6(c) test, it is not necessary to repeat the test on the air bag module.

161 For domestic transport, air bag inflators, air bag modules or seat belt pretensioners that meet the criteria for a Division 1.4G explosive must be transported using the description, "Articles, pyrotechnic for technical purposes," UN0431.

162 This material may be transported under the provisions of Division 4.1 only if it is packed so that at no time during transport will the percentage of diluent fall below the percentage that is stated in the shipping description.

163 Substances must satisfactorily pass Test Series 8 of the UN Manual of Tests and Criteria, Part I, Section 18 (IBR, see §171.7 of this subchapter).

164 Substances must not be transported under this entry unless approved by the Associate Administrator on the basis of the results of appropriate tests according to Part I of the UN Manual of Tests and Criteria (IBR, see §171.7 of this subchapter). The material must be packaged so that the percentage of diluent does not fall below that stated in the approval at any time during transportation.

165 These substances are susceptible to exothermic decomposition at elevated temperatures. Decomposition can be initiated by heat, moisture or by impurities (e.g., powdered metals (iron, manganese, cobalt, magnesium)). During the course of transportation, these substances must be shaded from direct sunlight and all sources of heat and be placed in adequately ventilated areas.

166 When transported in non-friable tablet form, calcium hypochlorite, dry, may be transported as a Packing Group III material.

167 These storage systems must always be considered as containing hydrogen. A metal hydride storage system installed in or intended to be installed in a vehicle or equipment or in vehicle or equipment components must be approved for transport by the Associate Administrator. A copy of the approval must accompany each shipment.

168 For lighters containing a Division 2.1 gas (see §171.8 of this subchapter), representative samples of each new lighter design must be examined and successfully tested as specified in §173.308(b)(3). For criteria in determining what is a new lighter design, see §173.308(b)(1). For transportation of new lighter design samples for examination and testing, see §173.308(b)(2). The examination and testing of each lighter design must be performed by a person authorized by the Associate Administrator under the provisions of subpart E of part 107 of this chapter, as specified in §173.308(a)(4). For continued use of approvals dated prior to January 1, 2012, see §173.308(b)(5).

For non-pressurized lighters containing a Class 3 (flammable liquid) material, its design, description, and packaging must be approved by the Associate Administrator prior to being offered for transportation or transported in commerce. In addition, a lighter design intended to contain a non-pressurized Class 3 material is excepted from the examination and testing criteria specified in §173.308(b)(3). An unused lighter or a lighter that is cleaned of residue and purged of vapors is not subject to the requirements of this subchapter.

169 This entry applies to lighter refills (see §171.8 of this subchapter) that contain a Division 2.1 (flammable) gas but do not contain an ignition device. Lighter refills offered for transportation under this entry may not exceed 4 fluid ounces capacity (7.22 cubic inches) or contain more than 65 grams of fuel. A lighter refill exceeding 4 fluid ounces capacity (7.22 cubic inches) or containing more than 65 grams of fuel must be classed as a Division 2.1 material, described with the proper shipping name appropriate for the material, and packaged in the packaging specified in part 173 of this subchapter for the flammable gas contained therein. In addition, a container exceeding 4 fluid ounces volumetric capacity (7.22 cubic inches) or containing more than 65 grams of fuel may not be connected or manifolded to a lighter or similar device and must also be described and packaged according to the fuel contained therein. For transportation by passenger-carrying aircraft, the net mass of lighter refills may not exceed 1 kg per package, and, for cargo-only aircraft, the net mass of lighter refills may not exceed 15 kg per package. See §173.306(h) of this subchapter.

170 Air must be eliminated from the vapor space by nitrogen or other means.

171 This entry may only be used when the material is transported in non-friable tablet form or for granular or powered mixtures that have been shown to meet the PG III criteria in §173.127.

172 This entry includes alcohol mixtures containing up to 5% petroleum products.

173 For adhesives, printing inks, printing ink-related materials, paints, paint-related materials, and resin solutions which are assigned to UN3082, and do not meet the definition of another hazard class, metal or plastic packaging for substances of packing groups II and III in quantities of 5 L (1.3 gallons) or less per packaging are not required to meet the UN performance package testing when transported:

a. Except for transportation by aircraft, in palletized loads, a pallet box or unit load device (*e.g.* individual packaging placed or stacked and secured by strapping, shrink or stretch-wrapping or other suitable means to a pallet). For vessel transport, the palletized loads, pallet boxes or unit load devices must be firmly packed and secured in closed cargo transport units; or

b. Except for transportation by aircraft, as an inner packaging of a combination packaging with a maximum net mass of 40 kg (88 pounds). For transportation by aircraft, as an inner packaging of a combination packaging with a maximum gross mass of 30 kg when packaged as a limited quantity in accordance with §173.27(f).

175 This substance must be stabilized when in concentrations of not more than 99%.

176 This entry must be used for formaldehyde solutions containing methanol as a stabilizer. Formaldehyde solutions not containing methanol and not meeting the Class 3 flammable liquid criteria must be described using a different proper shipping name.

177 Gasoline, or, ethanol and gasoline mixtures, for use in internal combustion engines (*e.g.*, in automobiles, stationary engines and other engines) must be assigned to Packing Group II regardless of variations in volatility.

188 *Small lithium cells and batteries.* Lithium cells or batteries, including cells or batteries packed with or contained in equipment, are not subject to any other requirements of this subchapter if they meet all of the following:

a. *Primary lithium batteries and cells.* (1) Primary lithium batteries and cells are forbidden for transport aboard passenger-carrying aircraft. The outside of each package that contains primary (nonrechargeable) lithium batteries or cells must be marked "PRIMARY LITHIUM BATTERIES—FORBIDDEN FOR TRANSPORT ABOARD PASSENGER AIRCRAFT" or "LITHIUM METAL BATTERIES—FORBIDDEN FOR TRANSPORT ABOARD PASSENGER AIRCRAFT" on a background of contrasting color. The letters in the marking must be:

(i) At least 12 mm (0.5 inch) in height on packages having a gross weight of more than 30 kg (66 pounds); or

(ii) At least 6 mm (0.25 inch) on packages having a gross weight of 30 kg (66 pounds) or less, except that smaller font may be used as necessary to fit package dimensions; and

(2) The provisions of paragraph (a)(1) do not apply to packages that contain 5 kg (11 pounds) net weight or less of primary lithium batteries or cells that are contained in or packed with equipment and the package contains no more than the number of lithium batteries or cells necessary to power the piece of equipment;

b. For a lithium metal or lithium alloy cell, the lithium content is not more than 1.0 g. For a lithium-ion cell, the equivalent lithium content is not more than 1.5 g;

c. For a lithium metal or lithium alloy battery, the aggregate lithium content is not more than 2.0 g. For a lithium-ion battery, the aggregate equivalent lithium content is not more than 8 g;

d. Effective October 1, 2009, the cell or battery must be of a type proven to meet the requirements of each test in the UN Manual of Tests and Criteria (IBR; see §171.7 of this subchapter);

e. Cells or batteries are separated or packaged in a manner to prevent short circuits and are packed in a strong outer packaging or are contained in equipment;

f. Effective October 1, 2008, except when contained in equipment, each package containing more than 24 lithium cells or 12 lithium batteries must be:

(1) Marked to indicate that it contains lithium batteries, and special procedures should be followed if the package is damaged;

(2) Accompanied by a document indicating that the package contains lithium batteries and special procedures should be followed if the package is damaged;

(3) Capable of withstanding a 1.2 meter drop test in any orientation without damage to cells or batteries contained in the package, without shifting of the contents that would allow short circuiting and without release of package contents; and

(4) Gross weight of the package may not exceed 30 kg (66 pounds). This requirement does not apply to lithium cells or batteries packed with equipment;

g. Electrical devices must conform to §173.21;

h. For transportation by aircraft, a telephone report in accordance with §171.15(a) is required if a fire, violent rupture, explosion or dangerous evolution of heat (*i.e.*, an amount of heat sufficient to be dangerous to packaging or personal safety to include charring of packaging, melting of packaging, scorching of packaging, or other evidence) occurs as a direct result of a lithium battery. For all modes of transportation, a written report submitted, retained, and updated in accordance with §171.16 is required if a fire, violent rupture, explosion or dangerous evolution of heat occurs as a direct result of a lithium battery or battery-powered device; and

i. Lithium batteries or cells are not authorized aboard an aircraft in checked or carry-on luggage except as provided in §175.10.

189 *Medium lithium cells and batteries.* Effective October 1, 2008, when transported by motor vehicle or rail car, lithium cells or batteries, including cells or batteries packed with or contained in equipment, are not subject to any other requirements of this subchapter if they meet all of the following:

a. The lithium content anode of each cell, when fully charged, is not more than 5 grams.

b. The aggregate lithium content of the anode of each battery, when fully charged, is not more than 25 grams.

c. The cells or batteries are of a type proven to meet the requirements of each test in the UN Manual of Tests and Criteria (IBR; see §171.7 of this subchapter). A cell or battery and equipment containing a cell or battery that was first transported prior to January 1, 2006 and is of a type proven to meet the criteria of Class 9 by testing in accordance with the tests in the UN Manual of Tests and Criteria, Third revised edition, 1999, need not be retested.

d. Cells or batteries are separated or packaged in a manner to prevent short circuits and are packed in a strong outer packaging or are contained in equipment.

e. The outside of each package must be marked "LITHIUM BATTERIES— FORBIDDEN FOR TRANSPORT ABOARD AIRCRAFT AND VESSEL" on a background of contrasting color, in letters:

(1) At least 12 mm (0.5 inch) in height on packages having a gross weight of more than 30 kg (66 pounds); or

(2) At least 6 mm (0.25 inch) on packages having a gross weight of 30 kg (66 pounds) or less, except that smaller font may be used as necessary to fit package dimensions.

f. Except when contained in equipment, each package containing more than 24 lithium cells or 12 lithium batteries must be:

(1) Marked to indicate that it contains lithium batteries, and special procedures should be followed if the package is damaged;

(2) Accompanied by a document indicating that the package contains lithium batteries and special procedures should be followed if the package is damaged;

(3) Capable of withstanding a 1.2 meter drop test in any orientation without damage to cells or batteries contained in the package, without shifting of the contents that would allow short circuiting and without release of package contents; and

(4) Gross weight of the package may not exceed 30 kg (66 pounds). This requirement does not apply to lithium cells or batteries packed with equipment.

g. Electrical devices must conform to §173.21 of this subchapter; and

h. A written report submitted, retained, and updated in accordance with §171.16 is required if a fire, violent rupture, explosion or dangerous evolution of heat (*i.e.*, an amount of heat sufficient to be dangerous to packaging or personal safety to include charring of packaging, melting of packaging, scorching of packaging, or other evidence) occurs as a direct result of a lithium battery or battery-powered device.

190 Until the effective date of the standards set forth in Special Provision 189, medium lithium cells or batteries, including cells or batteries packed with or contained in equipment, are not subject to any other requirements of this subchapter if they meet all of the following:

a. *Primary lithium batteries and cells.* (1) Primary lithium batteries and cells are forbidden for transport aboard passenger-carrying aircraft. The outside of each package that contains primary (nonrechargeable) lithium batteries or cells must be marked "PRIMARY LITHIUM BATTERIES—FORBIDDEN FOR TRANSPORT ABOARD PASSENGER AIRCRAFT" or "LITHIUM METAL BATTERIES—FORBIDDEN FOR TRANSPORT ABOARD PASSENGER AIRCRAFT" on a background of contrasting color. The letters in the marking must be:

(i) At least 12 mm (0.5 inch) in height on packages having a gross weight of more than 30 kg (66 pounds); or

(ii) At least 6 mm (0.25 inch) on packages having a gross weight of 30 kg (66 pounds) or less, except that smaller font may be used as necessary to fit package dimensions; and

(2) The provisions of paragraph (a)(1) do not apply to packages that contain 5 kg (11 pounds) net weight or less of primary lithium batteries or cells that are contained in or packed with equipment and the package contains no more than the number of lithium batteries or cells necessary to power the piece of equipment.

b. The lithium content of each cell, when fully charged, is not more than 5 grams.

c. The aggregate lithium content of each battery, when fully charged, is not more than 25 grams.

d. The cells or batteries are of a type proven to meet the requirements of each test in the UN Manual of Tests and Criteria (IBR; see §171.7 of this subchapter). A cell or battery and equipment containing a cell or battery that was first transported prior to January 1, 2006 and is of a type proven to meet the criteria of Class 9 by testing in accordance with the tests in the UN Manual of Tests and Criteria, Third Revised Edition, 1999, need not be retested.

e. Cells or batteries are separated so as to prevent short circuits and are packed in a strong outer packaging or are contained in equipment.

f. Electrical devices must conform to §173.21 of this subchapter.

198 Nitrocellulose solutions containing not more than 20% nitrocellulose may be transported as paint, perfumery products, or printing ink, as applicable, provided the nitrocellulose contains no more 12.6% nitrogen (by dry mass). *See* UN1210, UN1263, UN1266, UN3066, UN3469, and UN3470.

237 "Batteries, dry, containing potassium hydroxide solid, *electric storage*" must be prepared and packaged in accordance with the requirements of §173.159(a), (b), and (c). For transportation by aircraft, the provisions of §173.159(b)(2) are applicable.

332 Magnesium nitrate hexahydrate is not subject to the requirements of this subchapter.

335 Mixtures of solids that are not subject to this subchapter and environmentally hazardous liquids or solids may be classified as "Environmentally hazardous substances, solid, n.o.s," UN3077 and may be transported under this entry, provided there is no free liquid visible at the time the material is loaded or at the time the packaging or transport unit is closed. Each transport unit must be leakproof when used as bulk packaging.

340 This entry applies only to the vessel transportation of nickel-metal hydride batteries as cargo. Nickel-metal hydride button cells or nickel-metal hydride cells or batteries packed with or contained in battery-powered devices transported by vessel are not subject to the requirements of this special provision. *See* "Batteries, dry, sealed, n.o.s." in the §172.101 Hazardous Materials Table (HMT) of this part for transportation requirements for nickelmetal hydride batteries transported by other modes and for nickel-metal hydride button cells or nickel-metal hydride cells or batteries packed with or contained in battery-powered devices transported by vessel. Nickel-metal hydride batteries subject to this special provision are subject only to the following requirements: (1) The batteries must be prepared and packaged for transport in a manner to prevent a dangerous evolution of heat, short circuits, and damage to terminals; and are subject to the incident reporting in accordance with §171.16 of this subchapter if a fire, violent rupture, explosion or dangerous evolution of heat (*i.e.*, an amount of heat sufficient to be dangerous to packaging or personal safety to include charring of packaging, melting of packaging, scorching of packaging, or other evidence) occurs as a direct result of a nickel metal hydride battery; and (2) when loaded in a cargo transport unit in a total quantity of 100 kg gross mass or more, the shipping paper requirements of Subpart C of this part, the manifest requirements of §176.30 of this subchapter, and the vessel stowage requirements assigned to this entry in Column (10) of the §172.101 Hazardous Materials Table.

342 Glass inner packagings (such as ampoules or capsules) intended only for use in sterilization devices, when containing less than 30 mL of ethylene oxide per inner packaging with not more than 300 mL per outer packaging, may be transported in accordance with §173.4a of this subchapter, irrespective of the restriction of §173.4a(b) provided that:

a. After filling, each glass inner packaging must be determined to be leak-tight by placing the glass inner packaging in a hot water bath at a temperature and for a period of time sufficient to ensure that an internal pressure equal to the vapor pressure of ethylene oxide at 55 °C is achieved. Any glass inner packaging showing evidence of leakage, distortion or other defect under this test must not be transported under the terms of this special provision;

b. In addition to the packaging required in §173.4a, each glass inner packaging must be placed in a sealed plastic bag compatible with ethylene oxide and capable of containing the contents in the event of breakage or leakage of the glass inner packaging; and

c. Each glass inner packaging is protected by a means of preventing puncture of the plastic bag (*e.g.*, sleeves or cushioning) in the event of damage to the packaging (*e.g.*, by crushing).

343 A bulk packaging that emits hydrogen sulfide in sufficient concentration that vapors evolved from the crude oil can present an inhalation hazard must be marked as specified in §172.327 of this part.

345 "Nitrogen, refrigerated liquid (*cryogenic liquid*), UN1977" transported in open cryogenic receptacles with a maximum capacity of 1 L are not subject to the requirements of this subchapter. The receptacles must be constructed with glass double walls having the space between the walls vacuum insulated and each receptacle must be transported in an outer packaging with sufficient cushioning and absorbent materials to protect the receptacle from damage.

346 "Nitrogen, refrigerated liquid (*cryogenic liquid*), UN1977" transported in accordance with the requirements for open cryogenic receptacles in §173.320 and this special provision are not subject to any other requirements of this subchapter. The receptacle must contain no hazardous materials other than the liquid nitrogen which must be fully absorbed in a porous material in the receptacle.

347 Effective July 1, 2011, for transportation by aircraft, this entry may only be used if the results of Test series 6(d) of Part I of the UN Manual of Tests and Criteria (IBR, *see* §171.7 of this subchapter) have demonstrated that any hazardous effects from accidental functioning are confined to within the package. Effective January 1, 2012, for transportation by vessel, this entry may only be used if the results of Test Series 6(d) of Part I of the UN Manual of Tests and Criteria (IBR, *see* §171.7 of this subchapter) have demonstrated that any hazardous effects from accidental functioning are confined to within the package. Effective January 1, 2014, for transportation domestically by highway or rail, this entry may only be used if the results of Test Series 6(d) of Part I of the UN Manual of Tests and Criteria (IBR, *see* §171.7 of this subchapter) have demonstrated that any hazardous effects from accidental functioning are confined to within the package. Testing must be performed or witnessed by a person who is approved by the Associate Administrator (see §173.56(b) of this subchapter). All successfully conducted tests or reassignment to another compatibility group require the issuance of a new or revised approval by the Associate Administrator prior to transportation on or after the dates specified for each authorized mode of transport in this special provision.

349 Mixtures of hypochlorite with an ammonium salt are forbidden for transport. A hypochlorite solution, UN1791, is a Class 8 corrosive material.

350 Ammonium bromate, ammonium bromate aqueous solutions, and mixtures of a bromate with an ammonium salt are forbidden for transport.

351 Ammonium chlorate, ammonium chlorate aqueous solutions, and mixtures of a chlorate with an ammonium salt are forbidden for transport.

352 Ammonium chlorite, ammonium chlorite aqueous solutions, and mixtures of a chlorite with an ammonium salt are forbidden for transport.

353 Ammonium permanganate, ammonium permanganate aqueous solutions, and mixtures of a permanganate with an ammonium salt are forbidden for transport.

357 A bulk packaging that emits hydrogen sulfide in sufficient concentration that vapors evolved from the crude oil can present an inhalation hazard must be marked as specified in §172.327 of this part.

(2) *"A" codes.* These provisions apply only to transportation by aircraft:

Code / Special Provisions

A1 Single packagings are not permitted on passenger aircraft.

A2 Single packagings are not permitted on aircraft.

A3 For combination packagings, if glass inner packagings (including ampoules) are used, they must be packed with absorbent material in tightly closed metal receptacles before packing in outer packagings.

A4 Liquids having an inhalation toxicity of Packing Group I are not permitted on aircraft.

A5 Solids having an inhalation toxicity of Packing Group I are not permitted on passenger aircraft and may not exceed a maximum net quantity per package of 15 kg (33 pounds) on cargo aircraft.

A6 For combination packagings, if plastic inner packagings are used, they must be packed in tightly closed metal receptacles before packing in outer packagings.

A7 Steel packagings must be corrosion-resistant or have protection against corrosion.

A8 For combination packagings, if glass inner packagings (including ampoules) are used, they must be packed with cushioning material in tightly closed metal receptacles before packing in outer packagings.

A9 For combination packagings, if plastic bags are used, they must be packed in tightly closed metal receptacles before packing in outer packagings.

A10 When aluminum or aluminum alloy construction materials are used, they must be resistant to corrosion.

A11 For combination packagings, when metal inner packagings are permitted, only specification cylinders constructed of metals which are compatible with the hazardous material may be used.

A13 Bulk packagings are not authorized for transportation by aircraft.

A14 This material is not authorized to be transported as a limited quantity or consumer commodity in accordance with §173.306 of this subchapter when transported aboard an aircraft.

A19 Combination packagings consisting of outer fiber drums or plywood drums, with inner plastic packagings, are not authorized for transportation by aircraft.

A20 Plastic bags as inner receptacles of combination packagings are not authorized for transportation by aircraft.

A29 Combination packagings consisting of outer expanded plastic boxes with inner plastic bags are not authorized for transportation by aircraft.

A30 Ammonium permanganate is not authorized for transportation on aircraft.

A34 Aerosols containing a corrosive liquid in Packing Group II charged with a gas are not permitted for transportation by aircraft.

A35 This includes any material which is not covered by any of the other classes but which has an anesthetic, narcotic, noxious or other similar properties such that, in the event of spillage or leakage on an aircraft, extreme annoyance or discomfort could be caused to crew members so as to prevent the correct performance of assigned duties.

A37 This entry applies only to a material meeting the definition in §171.8 of this subchapter for self-defense spray.

A53 Refrigerating machines and refrigerating machine components are not subject to the requirements of this subchapter when containing less than 12 kg (26.4 pounds) of a non-flammable gas or when containing 12 L (3 gallons) or less of ammonia solution (UN2672) (see §173.307 of this subchapter).

A54 Lithium batteries or lithium batteries contained or packed with equipment that exceed the maximum gross weight allowed by Column (9B) of the §172.101 Table may only be transported on cargo aircraft if approved by the Associate Administrator.

A55 Prototype lithium batteries and cells that are packed with not more than 24 cells or 12 batteries per packaging that have not completed the test requirements in Sub-section 38.3 of the UN Manual of Tests and Criteria (incorporated by reference; see §171.7 of this subchapter) may be transported by cargo aircraft if approved by the Associate Administrator and provided the following requirements are met:

a. The cells and batteries must be transported in rigid outer packagings that conform to the requirements of Part 178 of this subchapter at the Packing Group I performance level; and

b. Each cell and battery must be protected against short circuiting, must be surrounded by cushioning material that is non-combustible and non-conductive, and must be individually packed in an inner packaging that is placed inside an outer specification packaging.

A56 Radioactive material with a subsidiary hazard of Division 4.2, Packing Group I, must be transported in Type B packages when offered for transportation by aircraft. Radioactive material with a subsidiary hazard of Division 2.1 is forbidden from transport on passenger aircraft.

A60 Sterilization devices, when containing less than 30 mL per inner packaging with not more than 150 mL per outer packaging, may be transported in accordance with the provisions in §173.4a, irrespective of §173.4a(b), provided such packagings were first subjected to comparative fire testing. Comparative fire testing must show no difference in burning rate between a package as prepared for transport (including the substance to be transported) and an identical package filled with water.

A82 The quantity limits in columns (9A) and (9B) do not apply to human or animal body parts, whole organs or whole bodies known to contain or suspected of containing an infectious substance.

A100 Primary (non-rechargeable) lithium batteries and cells are forbidden for transport aboard passenger carrying aircraft. Secondary (rechargeable) lithium batteries and cells are authorized aboard passenger carrying aircraft in packages that do not exceed a gross weight of 5 kg.

A101 A primary lithium battery or cell packed with or contained in equipment is forbidden for transport aboard a passenger carrying aircraft unless the equipment and the battery conform to the following provisions and the package contains no more than the number of lithium batteries or cells necessary to power the intended piece of equipment:

(1) The lithium content of each cell, when fully charged, is not more than 5 grams.

(2) The aggregate lithium content of the anode of each battery, when fully charged, is not more than 25 grams.

(3) The net weight of lithium batteries does not exceed 5 kg (11 pounds).

A103 Equipment is authorized aboard passenger carrying aircraft if the gross weight of the inner package of secondary lithium batteries or cells packed with the equipment does not exceed 5 kg (11 pounds).

A104 The net weight of secondary lithium batteries or cells contained in equipment may not exceed 5 kg (11 pounds) in packages that are authorized aboard passenger carrying aircraft.

A105 The total net quantity of dangerous goods contained in one package, excluding magnetic material, must not exceed the following:

a. 1 kg (2.2 pounds) in the case of solids;

b. 0.5 L (0.1 gallons) in the case of liquids;

c. 0.5 kg (1.1 pounds) in the case of Division 2.2 gases; or

d. any combination thereof.

A112 Notwithstanding the quantity limits shown in Column (9A) and (9B) for this entry, the following IBCs are authorized for transportation aboard passenger and cargo-only aircraft. Each IBC may not exceed a maximum net quantity of 1,000 kg:

a. Metal: 11A, 11B, 11N, 21A, 21B and 21N

b. Rigid plastics: 11H1, 11H2, 21H1 and 21H2

c. Composite with plastic inner receptacle: 11HZ1, 11HZ2, 21HZ1 and 21HZ2

d. Fiberboard: 11G

e. Wooden: 11C, 11D and 11F (with inner liners)

f. Flexible: 13H2, 13H3, 13H4, 13H5, 13L2, 13L3, 13L4, 13M1 and 13M2 (flexible IBCs must be sift-proof and water resistant or must be fitted with a sift-proof and water resistant liner).

(3) *"B" codes.* These provisions apply only to bulk packagings. Except as otherwise provided in this subchapter, these special provisions do not apply to UN portable tanks or IBCs:

Code / Special Provisions

B1 If the material has a flash point at or above 38°C (100°F) and below 93°C (200°F), then the bulk packaging requirements of §173.241 of this subchapter are applicable. If the material has a flash point of less than 38°C (100°F), then the bulk packaging requirements of §173.242 of this subchapter are applicable.

B2 MC 300, MC 301, MC 302, MC 303, MC 305, and MC 306, and DOT 406 cargo tanks are not authorized.

B3 MC 300, MC 301, MC 302, MC 303, MC 305, and MC 306, and DOT 406 cargo tanks and DOT 57 portable tanks are not authorized.

B4 MC 300, MC 301, MC 302, MC 303, MC 305, and MC 306, and DOT 406 cargo tanks are not authorized.

B5 Only ammonium nitrate solutions with 35 percent or less water that will remain completely in solution under all conditions of transport at a maximum lading temperature of 116°C (240°F) are authorized for transport in the following bulk packagings: MC 307, MC 312, DOT 407 and DOT 412 cargo tanks with at least 172 kPa (25 psig) design pressure. The packaging shall be designed for a working temperature of at least 121°C (250°F). Only Specifications MC 304, MC 307 or DOT 407 cargo tank motor vehicles are authorized for transportation by vessel.

B6 Packagings shall be made of steel.

B7 Safety relief devices are not authorized on multi-unit tank car tanks. Openings for safety relief devices on multi-unit tank car tanks shall be plugged or blank flanged.

B8 Packagings shall be made of nickel, stainless steel, or steel with nickel, stainless steel, lead or other suitable corrosion resistant metallic lining.

B9 Bottom outlets are not authorized.

B10 MC 300, MC 301, MC 302, MC 303, MC 305 and MC 306 and DOT 406 cargo tanks, and DOT 57 portable tanks are not authorized.

B11 Tank car tanks must have a test pressure of at least 2,068.5 kPa (300 psig). Cargo and portable tanks must have a design pressure of at least 1,207 kPa (175 psig).

B13 A nonspecification cargo tank motor vehicle authorized in §173.247 of this subchapter must be at least equivalent in design and in construction to a DOT 406 cargo tank or MC 306 cargo tank (if constructed before August 31, 1995), except as follows:

a. Packagings equivalent to MC 306 cargo tanks are excepted from the certification, venting, and emergency flow requirements of the MC 306 specification.

b. Packagings equivalent to DOT 406 cargo tanks are excepted from §§178.345-7(d)(5), circumferential reinforcements; 178.345-10, pressure relief; 178.345-11, outlets; 178.345-14, marking, and 178.345-15, certification.

c. Packagings are excepted from the design stress limits at elevated temperatures, as described in Section VIII of the ASME Code (IBR, see §171.7 of this subchapter). However, the design stress limits may not exceed 25 percent of the stress for 0 temper at the maximum design temperature of the cargo tank, as specified in the Aluminum Association's "Aluminum Standards and Data" (IBR, see §171.7 of this subchapter).

B14 Each bulk packaging, except a tank car or a multi-unit-tank car tank, must be insulated with an insulating material so that the overall thermal conductance at 15.5°C (60°F) is no more than 1.5333 kilojoules per hour per square meter per degree Celsius (0.075 Btu per hour per square foot per degree Fahrenheit) temperature differential. Insulating materials must not promote corrosion to steel when wet.

B15 Packagings must be protected with non-metallic linings impervious to the lading or have a suitable corrosion allowance.

B16 The lading must be completely covered with nitrogen, inert gas or other inert materials.

B18 Open steel hoppers or bins are authorized.

B23 Tanks must be made of steel that is rubber lined or unlined. Unlined tanks must be passivated before being placed in service. If unlined tanks are washed out with water, they must be repassivated prior to return to service. Lading in unlined tanks must be inhibited so that the corrosive effect on steel is not greater than that of hydrofluoric acid of 65 percent concentration.

B25 Packagings must be made from monel or nickel or monel-lined or nickel-lined steel.

B26 Tanks must be insulated. Insulation must be at least 100 mm (3.9 inches) except that the insulation thickness may be reduced to 51 mm (2 inches) over the exterior heater coils. Interior heating coils are not authorized. The packaging may not be loaded with a material outside of the packaging's design temperature range. In addition, the material also must be covered with an inert gas or the container must be filled with water to the tank's capacity. After unloading, the residual material also must be covered with an inert gas or the container must be filled with water to the tank's capacity.

B27 Tanks must have a service pressure of 1,034 kPa (150 psig). Tank car tanks must have a test pressure rating of 1,379 kPa (200 psig). Lading must be blanketed at all times with a dry inert gas at a pressure not to exceed 103 kPa (15 psig).

B28 Packagings must be made of stainless steel.

B30 MC 312, MC 330, MC 331 and DOT 412 cargo tanks and DOT 51 portable tanks must be made of stainless steel, except that steel other than stainless steel may be used in accordance with the provisions of §173.24b(b) of this subchapter. Thickness of stainless steel for tank shell and heads for cargo tanks and portable tanks must be the greater of 7.62 mm (0.300 inch) or the thickness required for a tank with a design pressure at least equal to 1.5 times the vapor pressure of the lading at 46°C (115°F). In addition, MC 312 and DOT 412 cargo tank motor vehicles must:

a. Be ASME Code (U) stamped for 100% radiography of all pressure-retaining welds;

b. Have accident damage protection which conforms with §178.345-8 of this subchapter;

c. Have a MAWP or design pressure of at least 87 psig: and

d. Have a bolted manway cover.

B32 MC 312, MC 330, MC 331, DOT 412 cargo tanks and DOT 51 portable tanks must be made of stainless steel, except that steel other than stainless steel may be used in accordance with the provisions of §173.24b(b) of this subchapter. Thickness of stainless steel for tank shell and heads for cargo tanks and portable tanks must be the greater of 6.35 mm (0.250 inch) or the thickness required for a tank with a design pressure at least equal to 1.3 times the vapor pressure of the lading at 46°C (115°F). In addition, MC 312 and DOT 412 cargo tank motor vehicles must:

a. Be ASME Code (U) stamped for 100% radiography of all pressure-retaining welds;

b. Have accident damage protection which conforms with §178.345-8 of this subchapter;

c. Have a MAWP or design pressure of at least 87 psig; and

d. Have a bolted manway cover.

B33 MC 300, MC 301, MC 302, MC 303, MC 305, MC 306, and DOT 406 cargo tanks equipped with a 1 psig normal vent used to transport gasoline must conform to Table I of this Special Provision. Based on the volatility class determined by using ASTM D439 and the Reid vapor pressure (RVP) of the particular gasoline, the maximum lading pressure and maximum ambient temperature permitted during the loading of gasoline may not exceed that listed in Table I.

TABLE I—MAXIMUM AMBIENT TEMPERATURE—GASOLINE

ASTM D439 volatility class	Maximum lading and ambient temperature (see note 1)
A (RVP<=9.0 psia)	131°F
B (RVP<=10.0 psia)	124°F
C (RVP<=11.5 psia)	116°F
D (RVP<=13.5 psia)	107°F
E (RVP<=15.0 psia)	100°F

Note 1: Based on maximum lading pressure of 1 psig at top of cargo tank.

B35 Tank cars containing hydrogen cyanide may be alternatively marked "Hydrocyanic acid, liquefied" if otherwise conforming to marking requirements in subpart D of this part. Tank cars marked "HYDROCYANIC ACID" prior to October 1, 1991 do not need to be remarked.

B37 The amount of nitric oxide charged into any tank car tank may not exceed 1,379 kPa (200 psig) at 21°C (70°F).

B42 Tank cars constructed before March 16, 2009, must have a test pressure of 34.47 Bar (500 psig) or greater and conform to Class 105J. Each tank car must have a reclosing pressure relief device having a start-to-discharge pressure of 10.34 Bar (150 psig). The tank car specification may be marked to indicate a test pressure of 13.79 Bar (200 psig).

B44 All parts of valves and safety relief devices in contact with lading must be of a material which will not cause formation of acetylides.

B45 Each tank must have a reclosing combination pressure relief device equipped with stainless steel or platinum rupture discs approved by the AAR Tank Car Committee.

B46 The detachable protective housing for the loading and unloading valves of multi-unit tank car tanks must withstand tank test pressure and must be approved by the Associate Administrator.

B47 Each tank may have a reclosing pressure relief device having a start-to-discharge pressure setting of 310 kPa (45 psig).

B48 Portable tanks in sodium metal service may be visually inspected at least once every 5 years instead of being retested hydrostatically. Date of the visual inspection must be stenciled on the tank near the other required markings.

B49 Tanks equipped with interior heater coils are not authorized. Single unit tank car tanks must have a reclosing pressure relief device having a start-to-discharge pressure set at no more than 1551 kPa (225 psig).

B50 Each valve outlet of a multi-unit tank car tank must be sealed by a threaded solid plug or a threaded cap with inert luting or gasket material. Valves must be of stainless steel and the caps, plugs, and valve seats must be of a material that will not deteriorate as a result of contact with the lading.

B52 Notwithstanding the provisions of §173.24b of this subchapter, non-reclosing pressure relief devices are authorized on DOT 57 portable tanks.

B53 Packagings must be made of either aluminum or steel.

B54 Open-top, sift-proof rail cars are also authorized.

B55 Water-tight, sift-proof, closed-top, metal-covered hopper cars, equipped with a venting arrangement (including flame arrestors) approved by the Associate Administrator are also authorized.

B56 Water-tight, sift-proof, closed-top, metal-covered hopper cars also authorized if the particle size of the hazardous material is not less than 149 microns.

B57 Class 115A tank car tanks used to transport chloroprene must be equipped with a non-reclosing pressure relief device of a diameter not less than 305 mm (12 inches) with a maximum rupture disc burst pressure of 310 kPa (45 psig).

B59 Water-tight, sift-proof, closed-top, metal-covered hopper cars are also authorized provided that the lading is covered with a nitrogen blanket.

B60 DOT Specification 106A500X multi-unit tank car tanks that are not equipped with a pressure relief device of any type are authorized. For the transportation of phosgene, the outage must be sufficient to prevent tanks from becoming liquid full at 55°C (130°F).

B61 Written procedures covering details of tank car appurtenances, dome fittings, safety devices, and marking, loading, handling, inspection, and testing practices must be approved by the Associate Administrator before any single unit tank car tank is offered for transportation.

B65 Tank cars constructed before March 16, 2009, must have a test pressure of 34.47 Bar (500 psig) or greater and conform to Class 105A. Each tank car must have a reclosing pressure relief device having a start-to-discharge pressure of 15.51 Bar (225 psig). The tank car specification may be marked to indicate a test pressure of 20.68 Bar (300 psig).

B66 Each tank must be equipped with gas tight valve protection caps. Outage must be sufficient to prevent tanks from becoming liquid full at 55°C (130°F). Specification 110A500W tanks must be stainless steel.

B67 All valves and fittings must be protected by a securely attached cover made of metal not subject to deterioration by the lading, and all valve openings, except safety valve, must be fitted with screw plugs or caps to prevent leakage in the event of valve failure.

B68 Sodium must be in a molten condition when loaded and allowed to solidify before shipment. Outage must be at least 5 percent at 98°C (208°F). Bulk packagings must have exterior heating coils fusion welded to the tank shell which have been properly stress relieved. The only tank car tanks authorized are Class DOT 105 tank cars having a test pressure of 2,069 kPa (300 psig) or greater.

B69 Dry sodium cyanide or potassium cyanide may be shipped in the following sift-proof and weather-resistant packagings: metal covered hopper cars, covered motor vehicles, portable tanks, or non-specification bins.

B70 If DOT 103ANW tank car tank is used: All cast metal in contact with the lading must have 96.7 percent nickel content; and the lading must be anhydrous and free from any impurities.

B76 Tank cars constructed before March 16, 2009, must have a test pressure of 20.68 Bar (300 psig) or greater and conform to Class 105S, 112J, 114J or 120S. Each tank car must have a reclosing pressure relief device having a start-to-discharge pressure of 10.34 Bar (150 psig). The tank car specification may be marked to indicate a test pressure of 13.79 Bar (200 psig).

B77 Other packaging are authorized when approved by the Associate Administrator.

B78 Tank cars must have a test pressure of 4.14 Bar (60 psig) or greater and conform to Class 103, 104, 105, 109, 111, 112, 114 or 120. Heater pipes must be of welded construction designed for a test pressure of 500 pounds per square inch. A 25 mm (1 inch) woven lining of asbestos or other approved material must be placed between the bolster slabbing and the bottom of the tank. If a tank car tank is equipped with a non-reclosing pressure relief device, the rupture disc must be perforated with a 3.2 mm (0.13 inch) diameter hole. If a tank car tank is equipped with a reclosing pressure relief valve, the tank must also be equipped with a vacuum relief valve.

B80 Each cargo tank must have a minimum design pressure of 276 kPa (40 psig).

B81 Venting and pressure relief devices for tank car tanks and cargo tanks must be approved by the Associate Administrator.

B82 Cargo tanks and portable tanks are not authorized.

B83 Bottom outlets are prohibited on tank car tanks transporting sulfuric acid in concentrations over 65.25 percent.

B84 Packagings must be protected with non-metallic linings impervious to the lading or have a suitable corrosion allowance for sulfuric acid or spent sulfuric acid in concentration up to 65.25 percent.

B85 Cargo tanks must be marked with the name of the lading accordance with the requirements of §172.302(b).

B90 Steel tanks conforming or equivalent to ASME specifications which contain solid or semisolid residual motor fuel antiknock mixture (including rust, scale, or other contaminants) may be shipped by rail freight or highway. The tank must have been designed and constructed to be capable of withstanding full vacuum. All openings must be closed with gasketed blank flanges or vapor tight threaded closures.

B115 Rail cars, highway trailers, roll-on/roll-off bins, or other non-specification bulk packagings are authorized. Packagings must be sift-proof, prevent liquid water from reaching the hazardous material, and be provided with sufficient venting to preclude dangerous accumulation of flammable, corrosive, or toxic gaseous emissions such as methane, hydrogen, and ammonia. The material must be loaded dry.

(4) *IB Codes and IP Codes.* These provisions apply only to transportation in IBCs and Large Packagings. Table 1 authorizes IBCs for specific proper shipping names through the use of IB Codes assigned in the §172.101 table of this subchapter. Table 2 defines IP Codes on the use of IBCs that are assigned to specific commodities in the §172.101 Table of this subchapter. Table 3 authorizes Large Packagings for specific proper shipping names through the use of IB Codes assigned in the §172.101 table of this subchapter. Large Packagings are authorized for the Packing Group III entries of specific proper shipping names when either Special Provision IB3 or IB8 is assigned to that entry in the §172.101 Table. When no IB code is assigned in the §172.101 Table for a specific proper shipping name, or in §173.225(e) Organic Peroxide Table for Type F organic peroxides, use of an IBC or Large Packaging for the material may be authorized when approved by the Associate Administrator. The letter "Z" shown in the marking code for composite IBCs must be replaced with a capital code letter designation found in §178.702(a)(2) of this subchapter to specify the material used for the other packaging. Tables 1, 2, and 3 follow:

SPECIAL PROVISIONS

TABLE 1—IB CODES (IBC CODES)

IBC code	Authorized IBCs
IB1	*Authorized IBCs:* Metal (31A, 31B and 31N). *Additional Requirement:* Only liquids with a vapor pressure less than or equal to 110 kPa at 50°C (1.1 bar at 122 °F), or 130 kPa at 55°C (1.3 bar at 131°F) are authorized.
IB2	*Authorized IBCs*: Metal (31A, 31B and 31N); Rigid plastics (31H1 and 31H2); Composite (31HZ1). *Additional Requirement:* Only liquids with a vapor pressure less than or equal to 110 kPa at 50°C (1.1 bar at 122 °F), or 130 kPa at 55°C (1.3 bar at 131°F) are authorized.
IB3	*Authorized IBCs:* Metal (31A, 31B and 31N); Rigid plastics (31H1 and 31H2); Composite (31HZ1 and 31HA2, 31HB2, 31HN2, 31HD2 and 31HH2). *Additional Requirement:* Only liquids with a vapor pressure less than or equal to 110 kPa at 50°C (1.1 bar at 12 °F), or 130 kPa at 55°C (1.3 bar at 131°F) are authorized, except for UN2672 (also *see* Special Provision IP8 in Table 2 for UN2672).
IB4	*Authorized IBCs:* Metal (11A, 11B, 11N, 21A, 21B and 21N).
IB5	*Authorized IBCs:* Metal (11A, 11B, 11N, 21A, 21B and 21N); Rigid plastics (11H1, 11H2, 21H1, 21H2, 31H1 and 31H2); Composite (11HZ1, 21HZ1 and 31HZ1).
IB6	*Authorized IBCs:* Metal (11A, 11B, 11N, 21A, 21B and 21N); Rigid plastics (11H1, 11H2, 21H1, 21H2, 31H1 and 31H2); Composite (11HZ1, 11HZ2, 21HZ1, 21HZ2, 31HZ1 and 31HZ2). *Additional Requirement:* Composite IBCs 11HZ2 and 21HZ2 may not be used when the hazardous materials being transported may become liquid during transport.
IB7	*Authorized IBCs:* Metal (11A, 11B, 11N, 21A, 21B and 21N); Rigid plastics (11H1, 11H2, 21H1, 21H2, 31H1 and 31H2); Composite (11HZ1, 11HZ2, 21HZ1, 21HZ2, 31HZ1 and 31HZ2); Wooden (11C, 11D and 11F). *Additional Requirement:* Liners of wooden IBCs must be sift-proof.
IB8	*Authorized IBCs:* Metal (11A, 11B, 11N, 21A, 21B and 21N); Rigid plastics (11H1, 11H2, 21H1, 21H2, 31H1 and 31H2); Composite (11HZ1, 11HZ2, 21HZ1, 21HZ2, 31HZ1 and 31HZ2); Fiberboard (11G); Wooden (11C, 11D and 11F); Flexible (13H1, 13H2, 13H3, 13H4, 13H5, 13L1, 13L2, 13L3, 13L4, 13M1 or 13M2).
IB9	IBCs are only authorized if approved by the Associate Administrator.

TABLE 2—IP CODES

IP Code	
IP1	IBCs must be packed in closed freight containers or a closed transport vehicle.
IP2	When IBCs other than metal or rigid plastics IBCs are used, they must be offered for transportation in a closed freight container or a closed transport vehicle.
IP3	Flexible IBCs must be sift-proof and water-resistant or must be fitted with a sift-proof and water-resistant liner.
IP4	Flexible, fiberboard or wooden IBCs must be sift-proof and water-resistant or be fitted with a sift-proof and water-resistant liner.
IP5	IBCs must have a device to allow venting. The inlet to the venting device must be located in the vapor space of the IBC under maximum filling conditions.
IP6	Non-specification bulk bins are authorized.
IP7	For UN identification numbers 1327, 1363, 1364, 1365, 1386, 1841, 2211, 2217, 2793 and 3314, IBCs are not required to meet the IBC performance tests specified in part 178, subpart N of this subchapter.
IP8	Ammonia solutions may be transported in rigid or composite plastic IBCs (31H1, 31H2 and 31HZ1) that have successfully passed, without leakage or permanent deformation, the hydrostatic test specified in §178.814 of this subchapter at a test pressure that is not less than 1.5 times the vapor pressure of the contents at 55 °C (131 °F).
IP13	Transportation by vessel in IBCs is prohibited.
IP14	Air must be eliminated from the vapor space by nitrogen or other means.
IP15	For UN2031 with more than 55% nitric acid, rigid plastic IBCs and composite IBCs with a rigid plastic inner receptacle are authorized for two years from the date of IBC manufacture.
IP20	Dry sodium cyanide or potassium cyanide is also permitted in siftproof, water-resistant, fiberboard IBCs when transported in closed freight containers or transport vehicles.

TABLE 3—IB CODES

[Large packaging authorizations]

IB3	Authorized Large Packagings (LIQUIDS) (PG III materials only)[2]
Inner packagings:	Large outer packagings:
Glass 10 liter	steel (50A).
Plastics 30 liter	aluminum (50B).
Metal 40 liter	metal other than steel or aluminum (50N).
	rigid plastics (50H).
	natural wood (50C).
	plywood (50D).
	reconstituted wood (50F).
	rigid fiverboard (50G).

TABLE 3—IB CODES, Continued

[Large packaging authorizations]

IB8	Authorized Large Packagings (SOLIDS) (PG III materials only)[2]
Inner packagings:	Large outer packagings:
Glass 10 kg	steel (50A).
Plastics 50 kg	aluminum (50B).
Metal 50 kg	metal other than steel or aluminum (50N).
Paper 50 kg	flexible plastics (51H)[1].
Fiber 50 kg	rigid plastics (50H).
	natural wood (50C).
	plywood (50D).
	reconstituted wood (50F).
	rigid fiberboard (50G).

[1]Flexible plastic (51H) Large Packagings are only authorized for use with flexible inner packagings.

[2]Except when authorized under Special Provision 41.

(5) *"N" codes.* These provisions apply only to non-bulk packagings:

Code / Special Provisions

N3 Glass inner packagings are permitted in combination or composite packagings only if the hazardous material is free from hydrofluoric acid.

N4 For combination or composite packagings, glass inner packagings, other than ampoules, are not permitted.

N5 Glass materials of construction are not authorized for any part of a packaging which is normally in contact with the hazardous material.

N6 Battery fluid packaged with electric storage batteries, wet or dry, must conform to the packaging provisions of §173.159(g) or (h) of this subchapter.

N7 The hazard class or division number of the material must be marked on the package in accordance with §172.302 of this subchapter. However, the hazard label corresponding to the hazard class or division may be substituted for the marking.

N8 Nitroglycerin solution in alcohol may be transported under this entry only when the solution is packed in metal cans of not more than 1 L capacity each, overpacked in a wooden box containing not more than 5 L. Metal cans must be completely surrounded with absorbent cushioning material. Wooden boxes must be completely lined with a suitable material impervious to water and nitroglycerin.

N11 This material is excepted for the specification packaging requirements of this subchapter if the material is packaged in strong, tight non-bulk packaging meeting the requirements of subparts A and B of part 173 of this subchapter.

N12 Plastic packagings are not authorized.

N20 A 5M1 multi-wall paper bag is authorized if transported in a closed transport vehicle.

N25 Steel single packagings are not authorized.

N32 Aluminum materials of construction are not authorized for single packagings.

N33 Aluminum drums are not authorized.

N34 Aluminum construction materials are not authorized for any part of a packaging which is normally in contact with the hazardous material.

N36 Aluminum or aluminum alloy construction materials are permitted only for halogenated hydrocarbons that will not react with aluminum.

N37 This material may be shipped in an integrally-lined fiber drum (1G) which meets the general packaging requirements of subpart B of part 173 of this subchapter, the requirements of part 178 of this subchapter at the packing group assigned for the material and to any other special provisions of column 7 of the §172.101 table.

N40 This material is not authorized in the following packagings:

a. A combination packaging consisting of a 4G fiberboard box with inner receptacles of glass or earthenware;

b. A single packaging of a 4C2 sift-proof, natural wood box; or

c. A composite packaging 6PG2 (glass, porcelain or stoneware receptacles within a fiberboard box).

N41 Metal construction materials are not authorized for any part of a packaging which is normally in contact with the hazardous material.

N42 1A1 drums made of carbon steel with thickness of body and heads of not less than 1.3 mm (0.050 inch) and with a corrosion-resistant phenolic lining are authorized for stabilized benzyl chloride if tested and certified to the Packing Group I performance level at a specific gravity of not less than 1.8.

N43 Metal drums are permitted as single packagings only if constructed of nickel or monel.

N45 Copper cartridges are authorized as inner packagings if the hazardous material is not in dispersion.

N65 Outage must be sufficient to prevent cylinders or spheres from becoming liquid full at 55°C (130°F). The vacant space (outage) may be charged with a nonflammable nonliquefied compressed gas if the pressure in the cylinder or sphere at 55°C (130°F) does not exceed 125 percent of the marked service pressure.

N72 Packagings must be examined by the Bureau of Explosives and approved by the Associate Administrator.

N73 Packagings consisting of outer wooden or fiberboard boxes with inner glass, metal or other strong containers; metal or fiber drums; kegs or barrels; or strong metal cans are authorized and need not conform to the requirements of part 178 of this subchapter.

N74 Packages consisting of tightly closed inner containers of glass, earthenware, metal or polyethylene, capacity not over 0.5 kg (1.1 pounds) securely cushioned and packed in outer wooden barrels or wooden or fiberboard boxes, not over 15 kg (33 pounds) net weight, are authorized and need not conform to the requirements of part 178 of this subchapter.

N75 Packages consisting of tightly closed inner packagings of glass, earthenware or metal, securely cushioned and packed in outer wooden barrels or wooden or fiberboard boxes, capacity not over 2.5 kg (5.5 pounds) net weight, are authorized and need not conform to the requirements of part 178 of this subchapter.

N76 For materials of not more than 25 percent active ingredient by weight, packages consisting of inner metal packagings not greater than 250 mL (8 ounces) capacity each, packed in strong outer packagings together with sufficient absorbent material to completely absorb the liquid contents are authorized and need not conform to the requirements of part 178 of this subchapter.

N77 For materials of not more than two percent active ingredients by weight, packagings need not conform to the requirements of part 178 of this subchapter, if liquid contents are absorbed in an inert material.

N78 Packages consisting of inner glass, earthenware, or polyethylene or other nonfragile plastic bottles or jars not over 0.5 kg (1.1 pounds) capacity each, or metal cans not over five pounds capacity each, packed in outer wooden boxes, barrels or kegs, or fiberboard boxes are authorized and need not conform to the requirements of part 178 of this subchapter. Net weight of contents in fiberboard boxes may not exceed 29 kg (64 pounds). Net weight of contents in wooden boxes, barrels or kegs may not exceed 45 kg (99 pounds).

N79 Packages consisting of tightly closed metal inner packagings not over 0.5 kg (1.1 pounds) capacity each, packed in outer wooden or fiberboard boxes, or wooden barrels, are authorized and need not conform to the requirements of part 178 of this subchapter. Net weight of contents may not exceed 15 kg (33 pounds).

N80 Packages consisting of one inner metal can, not over 2.5 kg (5.5 pounds) capacity, packed in an outer wooden or fiberboard box, or a wooden barrel, are authorized and need not conform to the requirements of part 178 of this subchapter.

N82 See §173.115 of this subchapter for classification criteria for flammable aerosols.

N83 This material may not be transported in quantities of more than 11.5 kg (25.4 lbs) per package.

N84 The maximum quantity per package is 500 g (1.1 lbs.).

N85 Packagings certified at the Packing Group I performance level may not be used.

N86 UN pressure receptacles made of aluminum alloy are not authorized.

N87 The use of copper valves on UN pressure receptacles is prohibited.

N88 Any metal part of a UN pressure receptacle in contact with the contents may not contain more than 65% copper, with a tolerance of 1%.

N89 When steel UN pressure receptacles are used, only those bearing the "H" mark are authorized.

N90 Metal packagings are not authorized.

(6) *"R" codes.* These provisions apply only to transportation by rail. [Reserved]

(7) *"T" codes.* (i) These provisions apply to the transportation of hazardous materials in UN portable tanks. Portable tank instructions specify the requirements applicable to a portable tank when used for the transportation of a specific hazardous material. These requirements must be met in addition to the design and construction specifications in part 178 of this subchapter. Portable tank instructions T1 through T22 specify the applicable minimum test pressure, the mini mum shell thickness (in reference steel), bottom opening requirements and pressure relief requirements. Liquefied compressed gases are assigned to portable tank instruction T50. Refrigerated liquefied gases that are authorized to be transported in portable tanks are specified in tank instruction T75.

(ii) The following table specifies the portable tank requirements applicable to "T" Codes T1 through T22. Column 1 specifies the "T" Code. Column 2 specifies the minimum test pressure, in bar (1 bar = 14.5 psig), at which the periodic hydrostatic testing required by §180.605 of this subchapter must be conducted. Column 3 specifies the section reference for minimum shell thickness or, alternatively, the minimum shell thickness value. Column 4 specifies the applicability of §178.275(g)(3) of this subchapter for the pressure relief devices. When the word "Normal" is indicated, §178.275(g)(3)) of this subchapter does not apply. Column 5 references applicable requirements for bottom openings in part 178 of this subchapter. "Prohibited" means bottom openings are prohibited, and "Prohibited for liquids" means bottom openings are authorized for solid material only. The table follows:

TABLE OF PORTABLE TANK T CODES T1—T22

[Portable tank codes T1—T22 apply to liquid and solid hazardous materials of Classes 3 through 9 which are transported in portable tanks.]

Portable tank instruction (1)	Minimum test pressure (bar) (2)	Minimum shell thickness (in mm-reference steel) (*See* §178.274(d)) (3)	Pressure-relief requirements (*See* §178.275(g)) (4)	Bottom opening requirements (*See* §178.275(d)) (5)
T1	1.5	§178.274(d)(2)	Normal	§178.275(d)(2).
T2	1.5	§178.274(d)(2)	Normal	§178.275(d)(3).
T3	2.65	§178.274(d)(2)	Normal	§178.275(d)(2).
T4	2.65	§178.274(d)(2)	Normal	§178.275(d)(3).
T5	2.65	§178.274(d)(2)	§178.275(g)(3)	Prohibited.
T6	4	§178.274(d)(2)	Normal	§178.275(d)(2).
T7	4	§178.274(d)(2)	Normal	§178.275(d)(3).
T8	4	§178.274(d)(2)	Normal	Prohibited.
T9	4	6 mm	Normal	Prohibited for liquids. §178.275(d)(2).
T10	4	6 mm	§178.275(g)(3)	Prohibited.
T11	6	§178.274(d)(2)	Normal	§178.275(d)(3).
T12	6	§178.274(d)(2)	§178.275(g)(3)	§178.275(d)(3).
T13	6	6 mm	Normal	Prohibited.
T14	6	6 mm	§178.275(g)(3)	Prohibited.
T15	10	§178.274(d)(2)	Normal	§178.275(d)(3).
T16	10	§178.274(d)(2)	§178.275(g)(3)	§178.275(d)(3).
T17	10	6 mm	Normal	§178.275(d)(3).
T18	10	6 mm	§178.275(g)(3)	§178.275(d)(3).
T19	10	6 mm	§178.275(g)(3)	Prohibited.
T20	10	8 mm	§178.275(g)(3)	Prohibited.
T21	10	10 mm	Normal	Prohibited for liquids. §178.275(d)(2).
T22	10	10 mm	§178.275(g)(3)	Prohibited.

(iii) T50. When portable tank instruction T50 is referenced in Column (7) of the §172.101 Table, the applicable liquefied compressed gases are authorized to be transported in portable tanks in accordance with the requirements of §173.313 of this subchapter.

(iv) T75. When portable tank instruction T75 is referenced in Column (7) of the §172.101 Table, the applicable refrigerated liquefied gases are authorized to be transported in portable tanks in accordance with the requirements of §178.277 of this subchapter.

(v) *UN and IM portable tank codes / special provisions.* When a specific portable tank instruction is specified by a "T" Code in Column (7) of the §172.101 Table for a specific hazardous material, a specification portable tank conforming to an alternative tank instruction may be used if:

(A) The alternative portable tank has a higher or equivalent test pressure (for example, 4 bar when 2.65 bar is specified);

(B) The alternative portable tank has greater or equivalent wall thickness (for example, 10 mm when 6 mm is specified);

(C) The alternative portable tank has a pressure relief device as specified in the "T" Code. If a frangible disc is required in series with the reclosing pressure relief device for the specified portable tank, the alternative portable tank must be fitted with a frangible disc in series with the reclosing pressure relief device; and

(D) With regard to bottom openings—

(1) When two effective means are specified, the alternative portable tank is fitted with bottom openings having two or three effective means of closure or no bottom openings; or

(2) When three effective means are specified, the portable tank has no bottom openings or three effective means of closure; or

(3) When no bottom openings are authorized, the alternative portable tank must not have bottom openings.

(vi) Except when an organic peroxide is authorized under §173.225(g), if a hazardous material is not assigned a portable tank "T" Code, the hazardous material may not be transported in a portable tank unless approved by the Associate Administrator.

(8) *"TP" codes.* (i) These provisions apply to the transportation of hazardous materials in IM and UN Specification portable tanks. Portable tank special provisions are assigned to certain hazardous materials to specify requirements that are in addition to those provided by the portable tank instructions or the requirements in part 178 of this subchapter. Portable tank special provisions are designated with the abbreviation TP (tank provision) and are assigned to specific hazardous materials in Column (7) of the §172.101 Table.

(ii) The following is a list of the portable tank special provisions:

Code / Special Provisions

TP1 The maximum degree of filling must not exceed the degree of filling determined by the following:

$$\left(\text{Degree of filling} = \frac{97}{1 + \alpha(t_r - t_f)}\right).$$

Where:

t_r is the maximum mean bulk temperature during transport, and tf is the temperature in degrees celsius of the liquid during filling.

TP2 a. The maximum degree of filling must not exceed the degree of filling determined by the following:

$$\left(\text{Degree of filling} = \frac{95}{1 + \alpha(t_r - t_f)}\right).$$

Where:

t_r is the maximum mean bulk temperature during transport,

t_f is the temperature in degrees celsius of the liquid during filling, and

α is the mean coefficient of cubical expansion of the liquid between the mean temperature of the liquid during filling (t_f) and the maximum mean bulk temperature during transportation (t_r) both in degrees celsius.

b. For liquids transported under ambient conditions α may be calculated using the formula:

$$\alpha = \frac{d_{15} - d_{50}}{35 d_{50}}$$

Where:

d_{15} and d_{50} are the densities (in units of mass per unit volume) of the liquid at 15°C (59°F) and 50°C (122°F), respectively.

TP3 The maximum degree of filling (in %) for solids transported above their melting points and for elevated temperature liquids shall be determined by the following:

$$\left(\text{Degree of filling} = 95\frac{d_r}{d_f}\right).$$

Where: d_f and d_r are the mean densities of the liquid at the mean temperature of the liquid during filling and the maximum mean bulk temperature during transport respectively.

TP4 The maximum degree of filling for portable tanks must not exceed 90%.

TP5 For a portable tank used for the transport of flammable refrigerated liquefied gases or refrigerated liquefied oxygen, the maximum rate at which the portable tank may be filled must not exceed the liquid flow capacity of the primary pressure relief system rated at a pressure not exceeding 120 percent of the portable tank's design pressure. For portable tanks used for the transport of refrigerated liquefied helium and refrigerated liquefied atmospheric gas (except oxygen), the maximum rate at which the tank is filled must not exceed the liquid flow capacity of the pressure relief device rated at 130 percent of the portable tank's design pressure. Except for a portable tank containing refrigerated liquefied helium, a portable tank shall have an outage of at least two percent below the inlet of the pressure relief device or pressure control valve, under conditions of incipient opening, with the portable tank in a level attitude. No outage is required for helium.

TP6 The tank must be equipped with a pressure release device which prevent a tank from bursting under fire engulfment conditions (the conditions prescribed in CGA pamphlet S-1.2 (see §171.7 of this subchapter) or alternative conditions approved by the Associate Administrator may be used to consider the fire engulfment condition), taking into account the properties of the hazardous material to be transported.

TP7 The vapor space must be purged of air by nitrogen or other means.

TP8 A portable tank having a minimum test pressure of 1.5 bar (150 kPa) may be used when the flash point of the hazardous material transported is greater than 0°C (32°F).

TP9 A hazardous material assigned to special provision TP9 in Column (7) of the §172.101 Table may only be transported in a portable tank if approved by the Associate Administrator.

TP10 The portable tank must be fitted with a lead lining at least 5 mm (0.2 inches) thick. The lead lining must be tested annually to ensure that it is intact and functional. Another suitable lining material may be used if approved by the Associate Administrator.

TP13 Self-contained breathing apparatus must be provided when this hazardous material is transported by sea.

TP16 The portable tank must be protected against over and under pressurization which may be experienced during transportation. The means of protection must be approved by the approval agency designated to approve the portable tank in accordance with the procedures in part 107, subpart E, of this subchapter. The pressure relief device must be preceded by a frangible disk in accordance with the requirements in §178.275(g)(3) of this subchapter to prevent crystallization of the product in the pressure relief device.

TP17 Only inorganic non-combustible materials may be used for thermal insulation of the tank.

TP18 The temperature of this material must be maintained between 18°C (64.4°F) and 40°C (104°F) while in transportation. Portable tanks containing solidified methacrylic acid must not be re heated during transportation.

TP19 The calculated wall thickness must be increased by 3 mm at the time of construction. Wall thickness must be verified ultrasonically at intervals midway between periodic hydraulic tests (every 2.5 years). The portable tank must not be used if the wall thickness is less than that prescribed by the applicable T code in Column (7) of the Table for this material.

TP20 This hazardous material must only be transported in insulated tanks under a nitrogen blanket.

TP21 The wall thickness must not be less than 8 mm. Portable tanks must be hydraulically tested and internally inspected at intervals not exceeding 2.5 years.

TP22 Lubricants for portable tank fittings (for example, gaskets, shut-off valves, flanges) must be oxygen compatible.

TP24 The portable tank may be fitted with a device to prevent the build up of excess pressure due to the slow decomposition of the hazardous material being transported. The device must be in the vapor space when the tank is filled under maximum filling conditions. This device must also prevent an unacceptable amount of leakage of liquid in the case of overturning.

TP25 Sulphur trioxide 99.95% pure and above may be transported in tanks without an inhibitor provided that it is maintained at a temperature equal to or above 32.5°C (90.5°F).

TP26 The heating device must be exterior to the shell. For UN3176, this requirement only applies when the hazardous material reacts dangerously with water.

TP27 A portable tank having a minimum test pressure of 4 bar (400 kPa) may be used provided the calculated test pressure is 4 bar or less based on the MAWP of the hazardous material, as defined in §178.275 of this subchapter, where the test pressure is 1.5 times the MAWP.

TP28 A portable tank having a minimum test pressure of 2.65 bar (265 kPa) may be used provided the calculated test pressure is 2.65 bar or less based on the MAWP of the hazardous material, as defined in §178.275 of this subchapter, where the test pressure is 1.5 times the MAWP.

TP29 A portable tank having a minimum test pressure of 1.5 bar (150.0 kPa) may be used provided the calculated test pressure is 1.5 bar or less based on the MAWP of the hazardous materials, as defined in §178.275 of this subchapter, where the test pressure is 1.5 times the MAWP.

TP30 This hazardous material may only be transported in insulated tanks.

TP31 This hazardous material may only be transported in tanks in the solid state.

TP32 Portable tanks may be used subject to the following conditions:

a. Each portable tank constructed of metal must be fitted with a pressure-relief device consisting of a reclosing spring loaded type, a frangible disc or a fusible element. The set to discharge for the spring loaded pressure relief device and the burst pressure for the frangible disc, as applicable, must not be greater than 2.65 bar for portable tanks with minimum test pressures greater than 4 bar;

b. The suitability for transport in tanks must be demonstrated using test 8(d) in Test Series 8 (see UN Manual of Tests and Criteria, Part 1, Subsection 18.7) (IBR, see §171.7 of this subchapter) or an alternative means approved by the Associate Administrator.

TP33 The portable tank instruction assigned for this substance applies for granular and powdered solids and for solids which are filled and discharged at temperatures above their melting point which are cooled and transported as a solid mass. Solid substances transported or offered for transport above their melting point are authorized for transportation in portable tanks conforming to the provisions of portable tank instruction T4 for solid substances of packing group III or T7 for solid substances of packing group II, unless a tank with more stringent requirements for minimum shell thickness, maximum allowable working pressure, pressure-relief devices or bottom outlets are assigned in which case the more stringent tank instruction and special provisions shall apply. Filling limits must be in accordance with portable tank special provision TP3. Solids meeting the definition of an elevated temperature material must be transported in accordance with the applicable requirements of this subchapter.

TP36 For material assigned this portable tank special provision, portable tanks used to transport such material may be equipped with fusible elements in the vapor space of the portable tank.

TP37 IM portable tanks are only authorized for the shipment of hydrogen peroxide solutions in water containing 72% or less hydrogen peroxide by weight. Pressure relief devices shall be designed to prevent the entry of foreign matter, the leak age of liquid and the development of any dangerous excess pressure. In addition, the portable tank must be designed so that internal surfaces may be effectively cleaned and passivated. Each tank must be equipped with pressure relief de vices conforming to the following requirements:

Concentration of hydrogen per peroxide solution	Total[1]
52% or less	11
Over 52%, but not greater than 60%	22
Over 60%, but not greater than 72%	32

[1]Total venting capacity in standard cubic feet hour (S.C.F.H.) per pound of hydrogen peroxide solution.

TP38 Each portable tank must be insulated with an insulating material so that the overall thermal conductance at 15.5°C (60°F) is no more than 1.5333 kilojoules per hour per square meter per degree Celsius (0.075 Btu per hour per square foot per degree Fahrenheit) temperature differential. Insulating materials may not promote corrosion to steel when wet.

TP44 Each portable tank must be made of stainless steel, except that steel other than stainless steel may be used in accordance with the provisions of §173.24b(b) of this subchapter. Thickness of stainless steel for tank shell and heads must be the greater of 7.62 mm (0.300 inch) or the thickness required for a portable tank with a design pressure at least equal to 1.5 times the vapor pressure of the hazardous material at 46°C (115°F).

TP45 Each portable tank must be made of stainless steel, except that steel other than stainless steel may be used in accordance with the provisions of 173.24b(b) of this subchapter. Thickness of stainless steel for portable tank shells and heads must be the greater of 6.35 mm (0.250 inch) or the thickness required for a portable tank with a design pressure at least equal to 1.3 times the vapor pressure of the hazardous material at 46°C (115°F).

TP46 Portable tanks in sodium metal service are not required to be hydrostatically retested.

(9) *"W"codes.* These provisions apply only to transportation by water:

Code / Special Provisions

W1 This substance in a non friable prill or granule form is not subject to the requirements of this subchapter when tested in accordance with the UN Manual of Test and Criteria (IBR, *see* §171.7 of this subchapter) and is found to not meet the definition or criteria for inclusion in Division 5.1.

W7 Vessel stowage category for uranyl nitrate hexahydrate solution is "D" as defined in §172.101(k)(4).

W8 Vessel stowage category for pyrophoric thorium metal or pyrophoric uranium metal is "D" as defined in §172.101(k)(4).

W9 When offered for transportation by water, the following Specification packagings are not authorized unless approved by the Associate Administrator: woven plastic bags, plastic film bags, textile bags, paper bags, IBCs and bulk packagings.

W41 When offered for transportation by water, this material must be packaged in bales and be securely and tightly bound with rope, wire or similar means.

PART 172

Subpart C—Shipping Papers

§172.200 Applicability.

(a) *Description of hazardous materials required.* Except as otherwise provided in this subpart, each person who offers a hazardous material for transportation shall describe the hazardous material on the shipping paper in the manner required by this subpart.

(b) This subpart does not apply to any material, other than a hazardous substance, hazardous waste or marine pollutant, that is —

(1) Identified by the letter "A" in Column 1 of the §172.101 Table, except when the material is offered or intended for transportation by air; or

(2) Identified by the letter "W" in Column 1 of the §172.101 Table, except when the material is offered or intended for transportation by water, or

(3) A limited quantity package unless the material is offered or intended for transportation by air or vessel and, until December 31, 2013, a package of ORM–D material authorized by this subchapter in effect on October 1, 2010 when offered for transportation by highway or rail.

(4) Category B infectious substances prepared in accordance with §173.199.

§172.201 Preparation and rentention of shipping papers.

(a) *Contents.* When a description of hazardous material is required to be included on a shipping paper, that description must conform to the following requirements:

(1) When a hazardous material and a material not subject to the requirements of this subchapter are described on the same shipping paper, the hazardous material description entries required by §172.202 and those additional entries that may be required by §172.203:

(i) Must be entered first, or

(ii) Must be entered in a color that clearly contrasts with any description on the shipping paper of a material not subject to the requirements of this subchapter, except that a description on a reproduction of a shipping paper may be highlighted, rather than printed, in a contrasting color (the provisions of this paragraph apply only to the basic description required by §172.202(a)(1), (2), (3), and (4)), or

(iii) Must be identified by the entry of an "X" placed before the basic shipping description required by §172.202 in a column captioned "HM." (The "X" may be replaced by "RQ," if appropriate.)

(2) The required shipping description on a shipping paper and all copies thereof used for transportation purposes, must be legible and printed (manually or mechanically) in English.

(3) Unless it is specifically authorized or required in this subchapter, the required shipping description may not contain any code or abbreviation.

(4) A shipping paper may contain additional information concerning the material provided the information is not inconsistent with the required description. Unless otherwise permitted or required by this subpart, additional information must be placed after the basic description required by §172.202(a).

(b) [Reserved]

(c) *Continuation page.* A shipping paper may consist of more than one page, if each page is consecutively numbered and the first page bears a notation specifying the total number of pages included in the shipping paper. For example, "Page 1 of 4 pages."

(d) *Emergency response telephone number.* Except as provided in §172.604(c), a shipping paper must contain an emergency response telephone number and, if utilizing an emergency response information telephone number service provider, identify the person (by name or contract number) who has a contractual agreement with the service provider, as prescribed in subpart G of this part.

(e) *Retention and Recordkeeping.* Each person who provides a shipping paper must retain a copy of the shipping paper required by §172.200(a), or an electronic image thereof, that is accessible at or through its principal place of business and must make the shipping paper available, upon request, to an authorized official of a Federal, State, or local government agency at reasonable times and locations. For a hazardous waste, the shipping paper copy must be retained for three years after the material is accepted by the initial carrier. For all other hazardous materials, the shipping paper must be retained for two years after the material is accepted by the initial carrier. Each shipping paper copy must include the date of acceptance by the initial carrier, except that, for rail, vessel, or air shipments, the date on the shipment waybill, airbill, or bill of lading may be used in place of the date of acceptance by the initial carrier. A motor carrier (as defined in §390.5 of subchapter B of chapter III of subtitle B) using a shipping paper without change for multiple shipments of one or more hazardous materials having the same shipping name and identification number may retain a single copy of the shipping paper, instead of a copy for each shipment made, if the carrier also retains a record of each shipment made, to include shipping name, identfication number, quantity transported, and date of shipment.

§172.202 Description of hazardous material on shipping papers.

(a) The shipping description of a hazardous material on the shipping paper must include:

(1) The identification number prescribed for the material as shown in Column (4) of the §172.101 table;

(2) The proper shipping name prescribed for the material in Column (2) of the §172.101 table;

(3) The hazard class or division number prescribed for the material, as shown in Column (3) of the §172.101 table. The subsidiary hazard class or division number is not required to be entered when a corresponding subsidiary hazard label is not required. Except for combustible liquids, the subsidiary hazard class(es) or

subsidiary division number(s) must be entered in parentheses immediately following the primary hazard class or division number. In addition—

(i) The words "Class" or "Division" may be included preceding the primary and subsidiary hazard class or division numbers.

(ii) The hazard class need not be included for the entry "Combustible liquid, n.o.s."

(iii) For domestic shipments, primary and subsidiary hazard class or division names may be entered following the numerical hazard class or division, or following the basic description.

(4) The packing group in Roman numerals, as designated for the hazardous material in Column (5) of the §172.101 table. Class 1 (explosives) materials; self-reactive substances; batteries other than those containing lithium, lithium ions, or sodium; Division 5.2 materials; and entries that are not assigned a packing group (*e.g.* , Class 7) are excepted from this requirement. The packing group may be preceded by the letters "PG" (for example, "PG II"); and

(5) Except for transportation by aircraft, the total quantity of hazardous materials covered by the description must be indicated (by mass or volume, or by activity for Class 7 materials) and must include an indication of the applicable unit of measurement, for example, "200 kg" (440 pounds) or "50 L" (13 gallons). The following provisions also apply:

(i) For Class 1 materials, the quantity must be the net explosive mass. For an explosive that is an article, such as Cartridges, small arms, the net explosive mass may be expressed in terms of the net mass of either the article or the explosive materials contained in the article.

(ii) For hazardous materials in salvage packaging, an estimate of the total quantity is acceptable.

(iii) The following are excepted from the requirements of paragraph (a)(5) of this section:

(A) Bulk packages, provided some indication of the total quantity is shown, for example, "1 cargo tank" or "2 IBCs."

(B) Cylinders, provided some indication of the total quantity is shown, for example, "10 cylinders."

(C) Packages containing only residue.

(6) For transportation by aircraft, the total net mass per package, must be shown unless a gross mass is indicated in Columns (9A) or (9B) of the §172.101 table in which case the total gross mass per package must be shown; or, for Class 7 materials, the quantity of radioactive material must be shown by activity. The following provisions also apply:

(i) For empty uncleaned packaging, only the number and type of packaging must be shown;

(ii) For chemical kits and first aid kits, the total net mass of hazardous materials must be shown. Where the kits contain only liquids, or solids and liquids, the net mass of liquids within the kits is to be calculated on a 1 to 1 basis, i.e., 1 L (0.3 gallons) equals 1 kg (2.2 pounds);

(iii) For dangerous goods in machinery or apparatus, the individual total quantities or an estimate of the individual total quantities of dangerous goods in solid, liquid or gaseous state, contained in the article must be shown;

(iv) For dangerous goods transported in a salvage packaging, an estimate of the quantity of dangerous goods per package must be shown;

(v) For cylinders, total quantity may be indicated by the number of cylinders, for example, "10 cylinders;"

(vi) For items where "No Limit" is shown in Column (9A) or (9B) of the §172.101 table, the quantity shown must be the net mass or volume of the material. For articles (*e.g.*, UN2800 and UN3166) the quantity must be the gross mass, followed by the letter "G"; and

(7) The number and type of packages must be indicated. The type of packages must be indicated by description of the package (for example, "12 drums"). Indication of the packaging specification number ("1H1") may be included in the description of the package (for example, "12 1H1 drums" or "12 drums (UN 1A1)"). Abbreviations may be used for indicating packaging types (for example, "cyl." for "cylinder") provided the abbreviations are commonly accepted and recognizable.

(b) Except as provided in this subpart, the basic description specified in paragraphs (a)(1), (2), (3), and (4) of this section must be shown in sequence with no additional information interspersed. For example, "UN2744, Cyclobutyl chloroformate, 6.1, (8, 3), PG II." The shipping description sequences in effect on December 31, 2006, may be used until January 1, 2013.

(c) The total quantity of the material covered by one description must appear before or after, or both before and after, the description required and authorized by this subpart. The type of packaging and destination marks may be entered in any appropriate manner before or after the basic description. Abbreviations may be used to express units of measurement and types of packagings.

(d) Technical and chemical group names may be entered in parentheses between the proper shipping name and hazard class or following the basic description. An appropriate modifier, such as "contains" or "containing," and/or the percentage of the technical constituent may also be used. For example: "Flammable liquids, n.o.s. (contains Xylene and Benzene), 3, UN 1993, II".

(e) Except for those materials in the UN Recommendations, the ICAO Technical Instructions, or the IMDG Code (IBR, see §171.7 of this subchapter), a material that is not a hazardous material according to this subchapter may not be offered for transportation or transported when its description on a shipping paper includes a hazard class or an identification number specified in the §172.101 Table.

§172.203 Additional description requirements.

(a) *Special permits.* Except as provided in §173.23 of this subchapter, each shipping paper issued in connection with a shipment made under a special permit must bear the notation "DOT–SP" followed by the special permit number assigned and located so that the notation is clearly associated with the description to which the special permit applies. Each shipping paper issued in connection with a shipment made under an exemption or special permit issued prior to October 1, 2007, may bear the notation "DOT–E" followed by the number assigned and so located that the notation is clearly associated with the description to which it applies.

(b) *Limited quantities.* When a shipping paper is required by this subchapter, the description for a material offered for transportation as "limited quantity," as authorized by this subchapter, must include the words "Limited Quantity" or "Ltd Qty" following the basic description.

(c) *Hazardous substances.* (1) Except for Class 7 (radioactive) materials described in accordance with paragraph (d) of this section, if the proper shipping name for a material that is a hazardous substance does not identify the hazardous substance by name, the name of the hazardous substance must be entered in parentheses in association with the basic description. If the material contains two or more hazardous substances, at least two hazardous substances, including the two with the lowest reportable quantities (RQs), must be identified. For a hazardous waste, the waste code (e.g., D001), if appropriate, may be used to identify the hazardous substance.

(2) The letters "RQ" must be entered on the shipping paper either before or after the basic description required by §172.202 for each hazardous substance (see definition in §171.8 of this subchapter). For example: "RQ, UN 1098, Allyl alcohol, 6.1, I, Toxic-inhalation hazard, Zone B"; or "UN 3077, Environmentally hazardous substances, solid, n.o.s., 9, III, RQ (Adipic acid)".

(d) *Radioactive material.* The description for a shipment of a Class 7 (radioactive) material must include the following additional entries as appropriate:

(1) The name of each radionuclide in the Class 7 (radioactive) material that is listed in §173.435 of this subchapter. For mixtures of radionuclides, the radionuclides required to be shown must be determined in accordance with §173.433(g) of this subchapter. Abbreviations, *e.g.*, "^{99}Mo," are authorized.

(2) A description of the physical and chemical form of the material, if the material is not in special form (generic chemical description is acceptable for chemical form).

(3) The activity contained in each package of the shipment in terms of the appropriate SI units (*e.g.*, Becquerels (Bq), Terabecquerels (TBq), etc.). The activity may also be stated in appropriate customary units (Curies (Ci), milliCuries (mCi), microCuries (uCi), etc.) in parentheses following the SI units. Abbreviations are authorized. Except for plutonium-239 and plutonium-241, the weight in grams or kilograms of fissile radionuclides may be inserted instead of activity units. For plutonium-239 and plutonium-241, the weight in grams of fissile radionuclides may be inserted in addition to the activity units.

(4) The category of label applied to each package in the shipment. For example: "RADIOACTIVE WHITE-I."

(5) The transport index assigned to each package in the shipment bearing RADIOACTIVE YELLOW-II or RADIOACTIVE YELLOW-III labels.

(6) For a package containing fissile Class 7 (radioactive) material:

(i) The words "Fissile Excepted" if the package is excepted pursuant to §173.453 of this subchapter; or otherwise

(ii) The criticality safety index for that package.

(7) For a package approved by the U.S. Department of Energy (DOE) or U.S. Nuclear Regulatory Commission (NRC), a notation of the package identification marking as prescribed in the applicable DOE or NRC approval (see §173.471 of the subchapter).

(8) For an export shipment or a shipment in a foreign made package, a notation of the package identification marking as prescribed in the applicable International Atomic Energy Agency (IAEA) Certificate of Competent Authority which has been issued for the package (see §173.473 of the subchapter).

(9) For a shipment required by this subchapter to be consigned as exclusive use:

(i) An indication that the shipment is consigned as exclusive use; or

(ii) If all the descriptions on the shipping paper are consigned as exclusive use, then the statement "Exclusive Use Shipment" may be entered only once on the shipping paper in a clearly visible location.

(10) For the shipment of a package containing a highway route controlled quantity of Class 7 (radioactive) materials (see §173.403 of this subchapter) the words "Highway route controlled quantity" or "HRCQ" must be entered in association with the basic description.

(e) *Empty packagings*

(1) The description on the shipping paper for a packaging containing the residue of a hazardous material may include the words "RESIDUE: Last Contained * * *" in association with the basic description of the hazardous material last contained in the packaging.

(2) The description on the shipping paper for a tank car containing the residue of a hazardous material must include the phrase, "RESIDUE: LAST CONTAINED * * *" before the basic description.

(f) *Transportation by air.* A statement indicating that the shipment is within the limitations prescribed for either passenger and cargo aircraft or cargo aircraft only must be entered on the shipping paper.

(g) *Transportation by rail.* (1) A shipping paper prepared by a rail carrier for a rail car, freight container, transport vehicle or portable tank that contains hazardous materials must include the reporting mark and number when displayed on the rail car, freight container, transport vehicle or portable tank.

(2) The shipping paper for each DOT-113 tank car containing a Division 2.1 material or its residue must contain an appropriate notation, such as "DOT 113", and the statement "Do not hump or cut off car while in motion."

(3) When shipments of elevated temperature materials are transported under the exception permitted in §173.247(h)(3) of this subchapter, the shipping paper must contain an appropriate notation, such as "Maximum operating speed 15 mph."

(h) *Transportation by highway.* Following the basic description for a hazardous material in a Specification MC 330 or MC 331 cargo tank, there must be entered for—

(1) *Anhydrous ammonia.*

(i) The words "0.2 PERCENT WATER" to indicate the suitability for shipping anhydrous ammonia in a cargo tank made of quenched and tempered steel as authorized by §173.315(a), Note 14 of this subchapter, or

(ii) The words "NOT FOR Q and T TANKS" when the anhydrous ammonia does not contain 0.2 percent or more water by weight.

(2) *Liquefied petroleum gas.* (i) The word "NONCORROSIVE" or "NONCOR" to indicate the suitability for shipping "NONCORROSIVE" liquefied petroleum gas in a cargo tank made of quenched and tempered steel as authorized by §173.315(a), Note 15 of this subchapter, or

(ii) The words "NOT FOR Q and T TANKS" for grades of liquefied petroleum gas other than "Noncorrosive".

(i) *Transportation by water.* Each shipment by water must have the following additional shipping paper entries:

(1) The name of the shipper.

(2) Minimum flashpoint if 60 °C (140 °F) or below (in °C closed cup (c.c.)) in association with the basic description. For lab packs packaged in conformance with §173.12(b) of this subchapter, an indication that the lowest flashpoint of all hazardous materials contained in the lab pack is below 23 °C or that the flash point is not less than 23 °C but not more than 60 °C must be identified on the shipping paper in lieu of the minimum flashpoint.

(3) For a hazardous material consigned under an "n.o.s." entry not included in the segregation groups listed in section 3.1.4 of the IMDG Code but belonging, in the opinion of the consignor, to one of these groups, the appropriate segregation group must be shown in association with the basic description (for example, IMDG Code segregation group–1 Acids). When no segregation group is applicable, there is no requirement to indicate that condition.

(j) [Reserved]

(k) *Technical names for "n.o.s." and other generic descriptions.* Unless otherwise excepted, if a material is described on a shipping paper by one of the proper shipping names identified by the letter "G" in Column (1) of the §172.101 Table, the technical name of the hazardous material must be entered in parentheses in association with the basic description. For example "Corrosive liquid, n.o.s. (Octanoyl chloride), 8, UN 1760, II", or "Corrosive liquid, n.o.s., 8, UN 1760, II (contains Octanoyl chloride)". The word "contains" may be used in association with the technical name, if appropriate. For organic peroxides which may qualify for more than one generic listing depending on concentration, the technical name must include the actual concentration being shipped or the concentration range for the appropriate generic listing. For example, "Organic peroxide type B, solid, 5.2, UN 3102 (dibenzoyl peroxide, 52–100%)" or "Organic peroxide type E, solid, 5.2, UN 3108, (dibenzoyl peroxide, paste, <52%)". Shipping descriptions for toxic materials that meet the criteria of Division 6.1, PG I or II (as specified in §173.132(a) of this subchapter) or Division 2.3 (as specified in §173.115(c) of this subchapter) and are identified by the letter "G" in column (1) of the §172.101 Table, must have the technical name of the toxic constituent entered in parentheses in association with the basic description. A material classed as Division 6.2 and assigned identification number UN 2814 or UN 2900 that is suspected to contain an unknown Category A infectious substance must have the words "suspected Category A infectious substance" entered in parentheses in place of the technical name as part of the proper shipping description. For additional technical name options, see the definition for "Technical name" in §171.8. A technical name should not be marked on the outer package of a Division 6.2 material (see §172.301(b)).

(1) If a hazardous material is a mixture or solution of two or more hazardous materials, the technical names of at least two components most predominantly contributing to the hazards of the mixture or solution must be entered on the shipping paper as required by paragraph (k) of this section. For example, "Flammable liquid, corrosive, n.o.s., 3, UN 2924, II (contains Methanol, Potassium hydroxide)".

(2) The provisions of this paragraph do not apply—

(i) To a material that is a hazardous waste and described using the proper shipping name "Hazardous waste, liquid *or* solid, n.o.s.", classed as a miscellaneous Class 9, provided the EPA hazardous waste number is included on the shipping paper in association with the basic description, or provided the material is described in accordance with the provisions of §172.203(c) of this part.

(ii) To a material for which the hazard class is to be determined by testing under the criteria in §172.101(c)(11).

(iii) If the n.o.s. description for the material (other than a mixture of hazardous materials of different classes meeting the definitions of more than one hazard class) contains the name of the chemical element or group which is primarily responsible for the material being included in the hazard class indicated.

(iv) If the n.o.s. description for the material (which is a mixture of hazardous materials of different classes meeting the definition of more than one hazard class) contains the name of the chemical element or group responsible for the material meeting the definition of one of these classes. In such cases, only the technical name of the component that is not appropriately identified in the n.o.s. description shall be entered in parentheses.

(l) *Marine pollutants.* (1) If the proper shipping name for a material which is a marine pollutant does not identify by name the component which makes the material a marine pollutant, the name of that component must appear in parentheses in association with the basic description. Where two or more components which make a material a marine pollutant are present, the names of at least two of the components most predominantly contributing to the marine pollutant designation must appear in parentheses in association with the basic description.

(2) The words "Marine Pollutant" shall be entered in association with the basic description for a material which is a marine pollutant.

(3) Except for transportation by vessel, marine pollutants subject to the provisions of 49 CFR 130.11 are excepted from the requirements of paragraph (l) of this section if a phrase indicating the material is an oil is placed in association with the basic description.

(4) Except when all or part of transportation is by vessel, marine pollutants in non-bulk packagings are not subject to the requirements of paragraphs (l)(1) and (l)(2) of this section (see §171.4 of this subchapter).

(m) *Poisonous Materials.* Notwithstanding the hazard class to which a material is assigned, for materials that are poisonous by inhalation (see §171.8 of this subchapter), the words "Poison-Inhalation Hazard" or "Toxic-Inhalation Hazard" and the words "Zone A", "Zone B", "Zone C", or "Zone D" for gases or "Zone A" or "Zone B" for liquids, as appropriate, shall be entered on the shipping paper immediately following the shipping description. The word "Poison" or "Toxic" need not be repeated if it otherwise appears in the shipping description.

(n) *Elevated temperature materials.* If a liquid material in a package meets the definition of an elevated temperature material in §171.8 of this subchapter, and the fact that it is an elevated temperature material is not disclosed in the proper shipping name (for example, when the words "Molten" or "Elevated temperature" are part of the proper shipping name), the word "HOT" must immediately precede the proper shipping name of the material on the shipping paper.

(o) *Organic peroxides and self-reactive materials.* The description on a shipping paper for a Division 4.1 (self-reactive) material or a Division 5.2 (organic peroxide) material must include the following additional information, as appropriate:

(1) If notification or competent authority approval is required, the shipping paper must contain a statement of approval of the classification and conditions of transport.

(2) For Division 4.1 (self-reactive) and Division 5.2 (organic peroxide) materials that require temperature control during transport, the control and emergency temperature must be included on the shipping paper.

(3) The word "SAMPLE" must be included in association with the basic description when a sample of a Division 4.1 (self-reactive) material (see §173.224(c)(3) of this subchapter) or Division 5.2 (organic peroxide) material (see §173.225(b)(2) of this subchapter) is offered for transportation.

(p) *Liquefied petroleum gas (LPG).* The word "non-odorized" must immediately precede the proper shipping name on a shipping paper when non-odorized liquefied petroleum gas is offered for transportation.

§172.204 Shipper's certification.

(a) *General.* Except as provided in paragraphs (b) and (c) of this section, each person who offers a hazardous material for transportation shall certify that the material is offered for transportation in accordance with this subchapter by printing (manually or mechanically) on the shipping paper containing the required shipping description the certification contained in paragraph (a)(1) of this section or the certification (declaration) containing the language contained in paragraph (a)(2) of this section.

(1) "This is to certify that the above-named materials are properly classified, described, packaged, marked and labeled, and are in proper condition for transportation according to the applicable regulations of the Department of Transportation."

NOTE: In line one of the certification the words "herein-named" may be substituted for the words "above-named".

(2) "I hereby declare that the contents of this consignment are fully and accurately described above by the proper shipping name, and are classified, packaged, marked and labelled/placarded, and are in all respects in proper condition for transport according to applicable international and national governmental regulations."

(b) *Exceptions.*

(1) Except for a hazardous waste, no certification is required for a hazardous material offered for transportation by motor vehicle and transported:

(i) In a cargo tank supplied by the carrier, or

(ii) By the shipper as a private carrier except for a hazardous material that is to be reshipped or transferred from one carrier to another.

(2) No certification is required for the return of an empty tank car which previously contained a hazardous material and which has not been cleaned or purged.

(c) *Transportation by air—*

(1) *General.* Certification containing the following language may be used in place of the certification required by paragraph (a) of this section:

I hereby certify that the contents of this consignment are fully and accurately described above by proper shipping name and are classified, packaged, marked and labeled, and in proper condition for carriage by air according to applicable national governmental regulations.

NOTE TO PARAGRAPH (c)(1): In the certification, the word "packed" may be used instead of the word "packaged" until October 1, 2010.

(2) *Certificate in duplicate.* Each person who offers a hazardous material to an aircraft operator for transportation by air shall provide two copies of the certification required in this section. (See §175.30 of this subchapter.)

(3) *Additional certification requirements.* Effective October 1, 2006, each person who offers a hazardous material for transportation by air must add to the certification required in this section the following statement:

"I declare that all of the applicable air transport requirements have been met."

(i) Each person who offers any package or overpack of hazardous materials for transport by air must ensure that:

(A) The articles or substances are not prohibited for transport by air (see the §172.101 Table);

(B) The articles or substances are properly classed, marked and labeled and otherwise in a condition for transport as required by this subchapter;

(C) The articles or substances are packaged in accordance with all the applicable air transport requirements, including appropriate types of packaging that conform to the packing requirements and the "A" Special Provisions in §172.102; inner packaging and maximum quantity per package limits; the compatibility requirements (see, for example, §173.24 of this subchapter); and requirements for closure for both inner and outer packagings, absorbent materials, and pressure differential in §173.27 of this subchapter. Other requirements may also apply. For example, single packagings may be prohibited, inner packaging may need to be packed in intermediate packagings, and certain materials may be required to be transported in packagings meeting a more stringent performance level.

(ii) [Reserved]

(4) *Radioactive material.* Each person who offers any radioactive material for transportation aboard a passenger-carrying aircraft shall sign (mechanically or manually) a printed certificate stating that the shipment contains radioactive material intended for use in, or incident to, research, or medical diagnosis or treatment.

(d) *Signature.* The certifications required by paragraph (a) or (c) of this section:

(1) Must be legibly signed by a principal, officer, partner, or employee of the shipper or his agent; and

(2) May be legibly signed manually, by typewriter, or by other mechanical means.

§172.205 Hazardous waste manifest.

(a) No person may offer, transport, transfer, or deliver a hazardous waste (waste) unless an EPA Form 8700-22 and 8700-22A (when necessary) hazardous waste manifest (manifest) is prepared in accordance with 40 CFR 262.20 and is signed, carried, and given as required of that person by this section.

(b) The shipper (generator) shall prepare the manifest in accordance with 40 CFR part 262.

(c) The original copy of the manifest must be dated by, and bear the handwritten signature of, the person representing:

(1) The shipper (generator) of the waste at the time it is offered for transportation, and

(2) The initial carrier accepting the waste for transportation.

(d) A copy of the manifest must be dated by, and bear the handwritten signature of the person representing:

(1) Each subsequent carrier accepting the waste for transportation, at the time of acceptance, and

(2) The designated facility receiving the waste, upon receipt.

(e) A copy of the manifest bearing all required dates and signatures must be:

(1) Given to a person representing each carrier accepting the waste for transportation,

(2) Carried during transportation in the same manner as required by this subchapter for shipping papers,

(3) Given to a person representing the designated facility receiving the waste,

(4) Returned to the shipper (generator) by the carrier that transported the waste from the United States to a foreign destination with a notation of the date of departure from the United States, and

(5) Retained by the shipper (generator) and by the initial and each subsequent carrier for three years from the date the waste was accepted by the initial carrier. Each retained copy must bear all required signatures and dates up to and including those entered by the next person who received the waste.

(f) *Transportation by rail.* Notwithstanding the requirements of paragraphs (d) and (e) of this section, the following requirements apply:

(1) When accepting hazardous waste from a non-rail transporter, the initial rail transporter must:

(i) Sign and date the manifest acknowledging acceptance of the hazardous waste;

(ii) Return a signed copy of the manifest to the non-rail transporter;

(iii) Forward at least three copies of the manifest to:

(A) The next non-rail transporter, if any;

(B) The designated facility, if the shipment is delivered to that facility by rail; or

(C) The last rail transporter designated to handle the waste in the United States; and

(iv) Retain one copy of the manifest and rail shipping paper in accordance with 40 CFR 263.22.

(2) Rail transporters must ensure that a shipping paper containing all the information required on the manifest (excluding the EPA identification numbers, generator certification and signatures) and, for exports, an EPA Acknowledgment of Consent accompanies the hazardous waste at all times. Intermediate rail transporters are not required to sign either the manifest or shipping paper.

(3) When delivering hazardous waste to the designated facility, a rail transporter must:

(i) Obtain the date of delivery and handwritten signature of the owner or operator of the designated facility on the manifest or the shipping paper (if the manifest has not been received by the facility); and

(ii) Retain a copy of the manifest or signed shipping paper in accordance with 40 CFR 263.22.

(4) When delivering hazardous waste to a non-rail transporter, a rail transporter must:

(i) Obtain the date of delivery and the handwritten signature of the next non-rail transporter on the manifest; and

(ii) Retain a copy of the manifest in accordance with 40 CFR 263.22.

(5) Before accepting hazardous waste from a rail transporter, a non-rail transporter must sign and date the manifest and provide a copy to the rail transporter.

(g) The person delivering a hazardous waste to an initial rail carrier shall send a copy of the manifest, dated and signed by a representative of the rail carrier, to the person representing the designated facility.

(h) A hazardous waste manifest required by 40 CFR part 262, containing all of the information required by this subpart, may be used as the shipping paper required by this subpart.

(i) The shipping description for a hazardous waste must be modified as required by §172.101(c)(9).

Subpart D—Marking

§172.300 Applicability.

(a) Each person who offers a hazardous material for transportation shall mark each package, freight container, and transport vehicle containing the hazardous material in the manner required by this subpart.

(b) When assigned the function by this subpart, each carrier that transports a hazardous material shall mark each package, freight container, and transport vehicle containing the hazardous material in the manner required by this subpart.

(c) Unless otherwise provided in a specific rule, stocks of preprinted packagings marked in accordance with this subpart prior to the effective date of a final rule may be continued in use, in the manner previously authorized, until depleted or for a one-year period subsequent to the compliance date of the marking amendment, whichever is less.

§172.301 General marking requirements for non-bulk packagings.

(a) *Proper shipping name and identification number.* (1) Except as otherwise provided by this subchapter, each person who offers a hazardous material for transportation in a non-bulk packaging must mark the package with the proper shipping name and identification number (preceded by "UN", "NA" or "ID," as appropriate) for the material as shown in the §172.101 Table.

(2) The proper shipping name for a hazardous waste (as defined in §171.8 of this subchapter) is not required to include the word "waste" if the package bears the EPA marking prescribed by 40 CFR 262.32.

(3) *Large quantities of a single hazardous material in non-bulk packages.* A transport vehicle or freight container containing only a single hazardous material in non-bulk packages must be marked, on each side and each end as specified in §172.332 or §172.336, with the identification number specified for the hazardous material in the §172.101 Table, subject to the following provisions and limitations:

(i) Each package is marked with the same proper shipping name and identification number;

(ii) The aggregate gross weight of the hazardous material is 4,000 kg (8,820 pounds) or more;

(iii) All of the hazardous material is loaded at one loading facility;

(iv) The transport vehicle or freight container contains no other material, hazardous or otherwise; and

(v) The identification number marking requirement of this paragraph (a)(3) does not apply to Class 1, Class 7, or to non-bulk packagings for which identification numbers are not required.

(b) *Technical names.* In addition to the marking required by paragraph (a) of this section, each non-bulk packaging containing a hazardous material subject to the provisions of §172.203(k) of this part, except for a Division 6.2 material, must be marked with the technical name in parentheses in association with the proper shipping name in accordance with the requirements and exceptions specified for display of technical descriptions on shipping papers in §172.203(k) of this part. A technical name should not be marked on the outer package of a Division 6.2 material.

(c) *Special permit packagings.* Except as provided in §173.23 of this subchapter, the outside of each package authorized by a special permit must be plainly and durably marked "DOT–SP" followed by the special permit number assigned. Packages authorized by an exemption issued prior to October 1, 2007, may be plainly and durably marked "DOT–E" in lieu of "DOT–SP" followed by the number assigned as specified in the most recent version of that exemption.

(d) *Consignee's or consignor's name and address.* Each person who offers for transportation a hazardous material in a non-bulk package shall mark that package with the name and address of the consignor or consignee except when the package is—

(1) Transported by highway only and will not be transferred from one motor carrier to another; or

(2) Part of a carload lot, truckload lot or freight container load, and the entire contents of the rail car, truck or freight container are shipped from one consignor to one consignee.

(e) *Previously marked packagings.* A package which has been previously marked as required for the material it contains and on which the marking remains legible, need not be remarked. (For empty packagings, see §173.29 of this subchapter.)

(f) *NON-ODORIZED marking on cylinders containing LPG.* No person may offer for transportation or transport a specification cylinder, except a Specification 2P or 2Q container or a Specification 39 cylinder, that contains an unodorized Liquefied petroleum gas (LPG) unless it is legibly marked NON-ODORIZED or NOT ODORIZED in letters not less than 6.3 mm (0.25 inches) in height near the marked proper shipping name required by paragraph (a) of this section.

§172.302 General marking requirements for bulk packagings.

(a) *Identification numbers.* Except as otherwise provided in this subpart, no person may offer for transportation or transport a hazardous material in a bulk packaging unless the packaging is marked as required by §172.332 with the identification number specified for the material in the §172.101 Table—

(1) On each side and each end, if the packaging has a capacity of 3,785 L (1,000 gallons) or more;

(2) On two opposing sides, if the packaging has a capacity of less than 3,785 L (1,000 gallons); or

(3) For cylinders permanently installed on a tube trailer motor vehicle, on each side and each end of the motor vehicle.

(b) *Size of markings.* Except as otherwise provided, markings required by this subpart on bulk packagings must—

(1) Have a width of at least 6.0 mm (0.24 inch) and a height of at least 100 mm (3.9 inches) for rail cars;

(2) Have a width of at least 4.0 mm (0.16 inch) and a height of at least 25 mm (one inch) for portable tanks with capacities of less than 3,785 L (1,000 gallons) and IBCs; and

(3) Have a width of at least 6.0 mm (0.24 inch) and a height of at least 50 mm (2.0 inches) for cargo tanks and other bulk packagings.

(c) *Special permit packagings*. Except as provided in §173.23 of this subchapter, the outside of each package used under the terms of a special permit must be plainly and durably marked "DOT–SP" followed by the special permit number assigned. Packages authorized by an exemption issued prior to October 1, 2007 may be plainly and durably marked "DOT–E" in lieu of "DOT-SP" followed by the number assigned as specified in the most recent version of that exemption.

(d) Each bulk packaging marked with a proper shipping name, common name or identification number as required by this subpart must remain marked when it is emptied unless it is—

(1) Sufficiently cleaned of residue and purged of vapors to remove any potential hazard; or

(2) Refilled, with a material requiring different markings or no markings, to such an extent that any residue remaining in the packaging is no longer hazardous.

(e) Additional requirements for marking portable tanks, cargo tanks, tank cars, multi-unit tank car tanks, and other bulk packagings are prescribed in §§172.326, 172.328, 172.330, and 172.331, respectively, of this subpart.

(f) A bulk packaging marked prior to October 1, 1991, in conformance to the regulations of this subchapter in effect on September 30, 1991, need not be remarked if the key words of the proper shipping name are identical to those currently specified in the §172.101 Table. For example, a tank car marked "NITRIC OXIDE" need not be remarked "NITRIC OXIDE, COMPRESSED".

(g) A rail car, freight container, truck body or trailer in which the lading has been fumigated with any hazardous material, or is undergoing fumigation, must be marked as specified in §173.9 of this subchapter.

§172.303 Prohibited markings.

(a) No person may offer for transportation or transport a package which is marked with the proper shipping name, the identification number of a hazardous material or any other markings indicating that the material is hazardous (e.g., RQ, INHALATION HAZARD) unless the package contains the identified hazardous material or its residue.

(b) This section does not apply to—

(1) Transportation of a package in a transport vehicle or freight container if the package is not visible during transportation and is loaded by the shipper and unloaded by the shipper or consignee.

(2) Markings on a package which are securely covered in transportation.

(3) The marking of a shipping name on a package when the name describes a material not regulated under this subchapter.

§172.304 Marking requirements.

(a) The marking required in this subpart —

(1) Must be durable, in English and printed on or affixed to the surface of a package or on a label, tag, or sign;

(2) Must be displayed on a background of sharply contrasting color;

(3) Must be unobscured by labels or attachments; and

(4) Must be located away from any other marking (such as advertising) that could substantially reduce its effectiveness.

§172.308 Authorized abbreviations.

(a) Abbreviations may not be used in a proper shipping name marking except as authorized in this section.

(b) The abbreviation "ORM" may be used in place of the words "Other Regulated Material."

(c) Abbreviations which appear as authorized descriptions in Column 2 of the §172.101 Table (e.g., "TNT" and "PCB") are authorized.

§172.310 Class 7 (radioactive) materials.

In addition to any other markings required by this subpart, each package containing Class 7 (radioactive) materials must be marked as follows:

(a) Each package with a gross mass greater than 50 kg (110 lb) must have its gross mass including the unit of measurement (which may be abbreviated) marked on the outside of the package.

(b) Each industrial, Type A, Type B(U), or Type B(M) package must be legibly and durably marked on the outside of the packaging, in letters at least 13 mm (0.5 in) high, with the words "TYPE IP–1," "TYPE IP–2," "TYPE IP–3," "TYPE A," "TYPE B(U)" or "TYPE B(M)," as appropriate. A package which does not conform to Type IP–1, Type IP–2, Type IP–3, Type A, Type B(U) or Type B(M) requirements may not be so marked.

(c) Each package which conforms to an IP–1, IP–2, IP–3 or a Type A package design must be legibly and durably marked on the outside of the packaging with the international vehicle registration code of the country of origin of the design. The international vehicle registration code for packages designed by a United States company or agency is the symbol "USA."

(d) Each package which conforms to a Type B(U) or Type B(M) package design must have the outside of the outermost receptacle, which is resistant to the effects of fire and water, plainly marked by embossing, stamping or other means resistant to the effects of fire and water with a radiation symbol that conforms to the requirements of Appendix B of this part.

(e) Each Type B(U), Type B(M) or fissile material package destined for export shipment must also be marked "USA" in conjunction with the specification marking, or other package certificate identification. (See §§173.471, 173.472, and 173.473 of this subchapter.)

§172.312 Liquid hazardous materials in non-bulk packagings.

(a) Except as provided in this section, each non-bulk combination package having inner packagings containing liquid hazardous materials, single packaging fitted with vents, or open cryogenic receptacle intended for the transport of refrigerated liquefied gases must be:

(1) Packed with closures upward, and

(2) Legibly marked with package orientation markings that are similar to the illustration shown in this paragraph, on two opposite vertical sides of the package with the arrows pointing in the correct upright direction. The arrows must be either black or red on white or other suitable contrasting background and commensurate with the size of the package. Depicting a rectangular border around the arrows is optional.

Package orientation

(b) Arrows for purposes other than indicating proper package orientation may not be displayed on a package containing a liquid hazardous material.

(c) The requirements of paragraph (a) of this section do not apply to—

(1) A non-bulk package with inner packagings which are cylinders.

(2) Except when offered or intended for transportation by aircraft, packages containing flammable liquids in inner packagings of 1L or less prepared in accordance with §173.150(b) or (c) of this subchapter.

(3) When offered or intended for transportation by aircraft, packages containing flammable liquids in inner packagings of 120 mL (4 fluid oz.) or less prepared in accordance with §173.150(b) or (c) of this subchapter when packed with sufficient absorption material between the inner and outer packagings to completely absorb the liquid contents.

(4) Liquids contained in manufactured articles (e.g., alcohol or mercury in thermometers) which are leak-tight in all orientations.

(5) A non-bulk package with hermetically sealed inner packagings not exceeding 500 mL each.

(6) Packages containing liquid infectious substances in primary receptacles not exceeding 50 mL (1.7 oz.).

(7) Class 7 radioactive material in Type A, IP-2, IP-3, Type B(U), or Type B(M) packages.

§172.313 Poisonous hazardous materials.

In addition to any other markings required by this subpart:

(a) A material poisonous by inhalation (see §171.8 of this subchapter) shall be marked "Inhalation Hazard" in association with the required labels or placards, as appropriate, and shipping name when required. The marking must be on two opposing sides of a bulk packaging. (See §172.302(b) of this subpart for size of markings on bulk packages.) When the words "Inhalation Hazard" appear on the label, as prescribed in §§172.416 and 172.429, or placard, as prescribed in §§172.540 and 172.555, the "Inhalation Hazard" marking is not required on the package.

(b) Each non-bulk plastic outer packaging used as a single or composite packaging for materials meeting the definition of Division 6.1 (in §173.132 of this subchapter) shall be permanently marked, by embossment or other durable means, with the word "POISON" in letters at least 6.3 mm (0.25 inch) in height. Additional text or symbols related to hazard warning may be included in the marking. The marking shall be located within 150 mm (6 inches) of the closure of the packaging.

(c) A transport vehicle or freight container containing a material poisonous by inhalation in non-bulk packages shall be marked, on each side and each end as specified in §172.332 or §172.336, with the identification number specified for the hazardous material in the §172.101 Table, subject to the following provisions and limitations:

(1) The material is in Hazard Zone A or B;

(2) The transport vehicle or freight container is loaded at one facility with 1,000 kg (2,205 pounds) or more aggregate gross weight of the material in non-bulk packages marked with the same proper shipping name and identification number; and

(3) If the transport vehicle or freight container contains more than one material meeting the provisions of this paragraph (c), it shall be marked with the identification number for one material, determined as follows:

(i) For different materials in the same hazard zone, with the identification number of the material having the greatest aggregate gross weight; and

(ii) For different materials in both Hazard Zones A and B, with the identification number for the Hazard Zone A material.

(d) For a packaging containing a Division 6.1 PG III material, "PG III" may be marked adjacent to the POISON label. (See §172.405(c).)

§172.315 Limited quantities.

(a) Except for transportation by aircraft or as otherwise provided in this subchapter, a package containing a limited quantity of hazardous material is not required to be marked with the proper shipping name and identification (ID) number when marked in accordance with the white square-on-point limited quantity marking as follows:

(1) The limited quantity marking must be durable, legible and of a size relative to the package that is readily visible. The marking must be applied on at least one side or one end of the outer packaging. The width of the border forming the square-on-point must be at least 2 mm and the minimum dimension of each side must be 100 mm unless the package size requires a reduced size marking that must be no less than 50 mm on each side. When intended for transportation by vessel, a cargo transport unit (*see* §176.2 of this subchapter) containing only limited quantity material must be suitably marked on one side or end of the exterior of the unit with an identical mark except that it must have minimum dimensions of 250 mm on each side.

(2) The top and bottom portions of the square-on-point and the border forming the square-on-point must be black and the center white or of a suitable contrasting background as follows:

(b) For transportation by aircraft, a limited quantity package conforming to Table 3 of §173.27(f) of this subchapter must be marked as follows:

(1) The marking must be durable, legible and of a size relative to the package as to be readily visible. The marking must be applied on at least one side or one end of the outer packaging. The width of the border forming the square-on-point must be at least 2 mm and the minimum dimension of each side must be 100 mm unless the package size requires a reduced size marking that must be no less than 50 mm on each side.

(2) The top and bottom portions of the square-on-point and the border forming the square-on-point must be black and the center white or of a suitable contrasting background and the symbol "Y" must be black and located in the center of the square-on-point and be clearly visible as follows:

(c) As applicable, package markings required by this subpart (*e.g.,* technical name, "RQ") must be in association with the marking required by paragraph (a) or (b) of this section.

(d) *Transitional exception.* Except for transportation by aircraft, until December 31, 2013, a package properly marked in accordance with §172.316 is not required to be marked with the limited quantity marking required by this section. For transportation by aircraft, until December 31, 2012, a package properly marked in accordance with §172.316 is not required to be marked with the limited quantity marking required by this section.

§172.316 Packagings containing materials classed as ORM–D.

(a) Each non-bulk packaging containing a material classed as ORM–D must be marked on at least one side or end with the ORM–D designation immediately following or below the proper shipping name of the material. The ORM designation must be placed within a rectangle that is approximately 6.3 mm (0.25 inches) larger on each side than the designation. The designation for ORM–D must be:

(1) Until December 31, 2012, ORM–D–AIR for an ORM–D that is prepared for air shipment and packaged in accordance with §§173.63, 173.150 through 173.155, 173.306 and the applicable requirements in §173.27.

(2) Until December 31, 2013, ORM–D for an ORM–D that is packaged in accordance with §§173.63, 173.150 through 173.155 and 173.306.

(b) When the ORM–D marking including the proper shipping name can not be affixed on the package surface, it may be on an attached tag.

(c) The marking ORM–D is the certification by the person offering the packaging for transportation that the material is properly described, classed, packaged, marked and labeled (when appropriate) and in proper condition for transportation according to the applicable regulations of this subchapter. This form of certification does not preclude the requirement for a certificate on a shipping paper when required by subpart C of this part.

§172.317 KEEP AWAY FROM HEAT handling mark.

(a) *General.* For transportation by aircraft, each package containing self-reactive substances of Division 4.1 or organic peroxides of Division 5.2 must be marked with the KEEP AWAY FROM HEAT handling mark specified in this section.

(b) *Location and design.* The marking must be a rectangle measuring at least 105 mm (4.1 inches) in height by 74 mm (2.9 inches) in width. Markings with not less than half this dimension are permissible where the dimensions of the package can only bear a smaller mark.

(c) *KEEP AWAY FROM HEAT handling mark.* The KEEP AWAY FROM HEAT handling mark must conform to the following:

(1) Except for size, the KEEP AWAY FROM HEAT handling mark must appear as follows:

(2) The symbol, letters and border must be black and the background white, except for the starburst which must be red.

(3) The KEEP AWAY FROM HEAT handling marking required by paragraph (a) of this section must be durable, legible and displayed on a background of contrasting color.

§172.320 Explosive hazardous materials.

(a) Except as otherwise provided in paragraphs (b), (c), (d) and (e) of this section, each package containing a Class 1 material must be marked with the EX-number for each substance, article or device contained therein.

(b) Except for fireworks approved in accordance with §173.56(j) of this subchapter, a package of Class 1 materials may be marked, in lieu of the EX-number required by paragraph (a) of this section, with a national stock number issued by the Department of Defense or identifying information, such as a product code required by regulations for commercial explosives specified in 27 CFR part 555, if the national stock number or identifying information can be specifically associated with the EX-number assigned.

(c) When more than five different Class 1 materials are packed in the same package, the package may be marked with only five of the EX-numbers, national stock numbers, product codes, or combination thereof.

(d) The requirements of this section do not apply if the EX-number, product code or national stock number of each explosive item described under a proper shipping description is shown in association with the shipping description required by §172.202(a) of this part. Product codes and national stock numbers must be traceable to the specific EX-number assigned by the Associate Administrator.

(e) The requirements of this section do not apply to the following Class 1 materials:

(1) Those being shipped to a testing agency in accordance with §173.56(d) of this subchapter;

(2) Those being shipped in accordance with §173.56(e) of this subchapter, for the purposes of developmental testing;

(3) Those which meet the requirements of §173.56(h) of this subchapter and therefore are not subject to the approval process of §173.56 of this subchapter;

(4) [Reserved];

(5) Those that are transported in accordance with §173.56(c)(2) of this subchapter and, therefore, are covered by a national security classification currently in effect.

§172.322 Marine pollutants.

(a) For vessel transportation of each non-bulk packaging that contains a marine pollutant—

(1) If the proper shipping name for a material which is a marine pollutant does not identify by name the component which makes the material a marine pollutant, the name of that component must be marked on the package in parentheses in association with the marked proper shipping name. Where two or more components which make a material a marine pollutant are present, the names of at least two of the components most predominantly contributing to the marine pollutant designation must appear in parentheses in association with the marked proper shipping name; and

(2) The MARINE POLLUTANT mark shall be placed in association with the hazard warning labels required by subpart E of this part or, in the absence of any labels, in association with the marked proper shipping name.

(b) A bulk packaging that contains a marine pollutant must—

(1) Be marked with the MARINE POLLUTANT mark on at least two opposing sides or two ends other than the bottom if the packaging has a capacity of less than 3,785 L (1,000 gallons). The mark must be visible from the direction it faces. The mark may be displayed in black lettering on a square-on-point configuration having the same outside dimensions as a placard; or

(2) Be marked on each end and each side with the MARINE POLLUTANT mark if the packaging has a capacity of 3,785 L (1,000 gallons) or more. The mark must be visible from the direction it faces. The mark may be displayed in black lettering on a square-on-point configuration having the same outside dimensions as a placard.

(c) A transport vehicle or freight container that contains a package subject to the marking requirements of paragraph (a) or (b) of this section must be marked with the MARINE POLLUTANT mark. The mark must appear on each side and each end of the transport vehicle or freight container, and must be visible from the direction it faces. This requirement may be met by the marking displayed on a freight container or portable tank loaded on a motor vehicle or rail car. This mark may be displayed in black lettering on a white square-on-point configuration having the same outside dimensions as a placard.

(d) The MARINE POLLUTANT mark is not required—

(1) On single packagings or combination packagings where each single package or each inner packaging of combination packagings has:

(i) A net quantity of 5 L (1.3 gallons) or less for liquids; or

(ii) A net mass of 5 kg (11 pounds) or less for solids.

(2) On a combination packaging containing a marine pollutant, other than a severe marine pollutant, in inner packagings each of which contains:

(i) 5 L (1.3 gallons) or less net capacity for liquids; or

(ii) 5 kg (11 pounds) or less net capacity for solids.

(3) Except for transportation by vessel, on a bulk packaging, freight container or transport vehicle that bears a label or placard specified in Subparts E or F of this part.

(4) On a package of limited quantity material marked in accordance with §172.315 of this part.

(e) *MARINE POLLUTANT mark.* Effective January 14, 2010 the MARINE POLLUTANT mark must conform to the following:

(1) Except for size, the MARINE POLLUTANT mark must appear as follows:

Symbol (fish and tree): Black on white or suitable contrasting background.

MARKING

(2) The symbol and border must be black and the background white, or the symbol, border and background must be of contrasting color to the surface to which the mark is to be affixed. Each side of the mark must be—

(i) At least 100 mm (3.9 inches) for marks applied to:

(A) Non-bulk packages, except in the case of packages which, because of their size, can only bear smaller marks;

(B) Bulk packages with a capacity of less than 3,785 L (1,000 gallons); or

(ii) At least 250 mm (9.8 inches) for marks applied to all other bulk packages.

(f) *Exceptions*. See §171.4(c).

§172.323 Infectious substances.

(a) In addition to other requirements of this subpart, a bulk packaging containing a regulated medical waste, as defined in §173.134(a)(5) of this subchapter, must be marked with a BIOHAZARD marking conforming to 29 CFR 1910.1030(g)(1)(i)—

(1) On two opposing sides or two ends other than the bottom if the packaging has a capacity of less than 3,785 L (1,000 gallons). The BIOHAZARD marking must measure at least 152.4 mm (6 inches) on each side and must be visible from the direction it faces.

(2) On each end and each side if the packaging has a capacity of 3,785 L (1,000 gallons) or more. The BIOHAZARD marking must measure at least 152.4 mm (6 inches) on each side and must be visible from the direction it faces.

(b) For a bulk packaging contained in or on a transport vehicle or freight container, if the BIOHAZARD marking on the bulk packaging is not visible, the transport vehicle or freight container must be marked as required by paragraph (a) of this section on each side and each end.

(c) The background color for the BIOHAZARD marking required by paragraph (a) of this section must be orange and the symbol and letters must be black. Except for size the BIOHAZARD marking must appear as follows:

(d) The BIOHAZARD marking required by paragraph (a) of this section must be displayed on a background of contrasting color. It may be displayed on a plain white square-on-point configuration having the same outside dimensions as a placard, as specified in §172.519(c) of this part.

§172.324 Hazardous substances in non-bulk packagings.

For each non-bulk package that contains a hazardous substance—

(a) Except for packages of radioactive material labeled in accordance with §172.403, if the proper shipping name of a material that is a hazardous substance does not identify the hazardous substance by name, the name of the hazardous substance must be marked on the package, in parentheses, in association with the proper shipping name. If the material contains two or more hazardous substances, at least two hazardous substances, including the two with the lowest reportable quantities (RQs), must be identified. For a hazardous waste, the waste code (e.g., D001), if appropriate, may be used to identify the hazardous substance.

(b) The letters "RQ" must be marked on the package in association with the proper shipping name.

(c) A package of limited quantity material marked in accordance with §172.315 must also be marked in accordance with the applicable requirements of this section.

§172.325 Elevated temperature materials.

(a) Except as provided in paragraph (b) of this section, a bulk packaging containing an elevated temperature material must be marked on two opposing sides with the word "HOT" in black or white Gothic lettering on a contrasting background. The marking must be displayed on the packaging itself or in black lettering on a plain white square-on-point configuration having the same outside dimensions as a placard. (See §172.302(b) for size of markings on bulk packagings.)

(b) Bulk packagings containing molten aluminum or molten sulfur must be marked "MOLTEN ALUMINUM" or "MOLTEN SULFUR", respectively, in the same manner as prescribed in paragraph (a) of this section.

(c) If the identification number is displayed on a white-square-on-point display configuration, as prescribed in §172.336(b), the word "HOT" may be displayed in the upper corner of the same white-square-on-point display configuration. The word "HOT" must be in black letters having a height of at least 50 mm (2.0 inches). Except for size, these markings shall be as illustrated for an Elevated temperature material, liquid, n.o.s.:

§172.326 Portable tanks.

(a) *Shipping name.* No person may offer for transportation or transport a portable tank containing a hazardous material unless it is legibly marked on two opposing sides with the proper shipping name specified for the material in the §172.101 Table. For transportation by vessel, the minimum height for a proper shipping name marked on a portable tank is 65 mm (2.5 inches).

(b) *Owner's name.* The name of the owner or of the lessee, if applicable, must be displayed on a portable tank that contains a hazardous material.

(c) *Identification numbers.* (1) If the identification number markings required by §172.302(a) are not visible, a transport vehicle or freight container used to transport a portable tank containing a hazardous material must be marked on each side and each end as required by §172.332 with the identification number specified for the material in the §172.101 Table.

(2) Each person who offers a portable tank containing a hazardous material to a motor carrier, for transportation in a transport vehicle or freight container, shall provide the motor carrier with the required identification numbers on placards, orange panels, or the white square-on-point configuration, as appropriate, for each side and each end of the transport vehicle or freight container from which identification numbers on the portable tank are not visible.

(d) *NON-ODORIZED marking on portable tanks containing LPG.* After September 30, 2006, no person may offer for transportation or transport a portable tank containing liquefied petroleum gas (LPG) that is unodorized as authorized in §173.315(b)(1) unless it is legibly marked NON-ODORIZED or NOT ODORIZED on two opposing sides near the marked proper shipping name required by paragraph (a) of this section, or near the placards.

§172.327 Petroleum sour crude oil in bulk packaging.

A Bulk packaging used to transport petroleum crude oil containing hydrogen sulfide (*i.e.,* sour crude oil) in sufficient concentration that vapors evolved from the crude oil may present an inhalation hazard must include a marking, label, tag, or sign to warn of the toxic hazard as follows:

(a) The marking must be durable, legible and of a size relative to the package as to be readily visible and similar to the illustration shown in this paragraph with the minimum dimension of each side of the marking at least 100 mm (3.9 inches). The width of the border forming the square-on-point marking must be at least 5 mm. The marking must be displayed at each location (*e.g.,* manhole, loading head) where exposure to hydrogen sulfide vapors may occur.

(b) The border of the square-on-point must be black or red on a white or other suitable contrasting background. The symbol must be black and located in the center of the square-on-point and be clearly visible as follows:

(c) As an alternative to the marking required in (a) and (b) of this section, a label, tag, or sign may be displayed at each location (*e.g.,* manhole, loading head) where exposure to hydrogen sulfide vapors may occur. The label, tag, or sign must be durable, in English, and printed legibly and of a size relative to the package with a warning statement such as "Danger, Possible Hydrogen Sulfide Inhalation Hazard" to communicate the possible risk of exposure to harmful concentrations of hydrogen sulfide gas.

§172.328 Cargo tanks.

(a) *Providing and affixing identification numbers.* Unless a cargo tank is already marked with the identification numbers required by this subpart, the identification numbers must be provided or affixed as follows:

(1) A person who offers a hazardous material to a motor carrier for transportation in a cargo tank shall provide the motor carrier the identification numbers on placards or shall affix orange panels containing the required identification numbers, prior to or at the time the material is offered for transportation.

(2) A person who offers a cargo tank containing a hazardous material for transportation shall affix the required identification numbers on panels or placards prior to or at the time the cargo tank is offered for transportation.

(3) For a cargo tank transported on or in a transport vehicle or freight container, if the identification number marking on the cargo tank required by §172.302(a) would not normally be visible during transportation—

(i) The transport vehicle or freight container must be marked as required by §172.332 on each side and each end with the identification number specified for the material in the §172.101 Table; and

(ii) When the cargo tank is permanently installed within an enclosed cargo body of the transport vehicle or freight container, the identification number marking required by §172.302(a) need only be displayed on each side and end of a cargo tank that is visible when the cargo tank is accessed.

(b) *Required markings: Gases.* Except for certain nurse tanks which must be marked as specified in §173.315(m) of this subchapter, each cargo tank transporting a Class 2 material subject to this subchapter must be marked, in lettering no less than 50 mm (2.0 inches), on each side and each end with—

(1) The proper shipping name specified for the gas in the §172.101 Table; or

(2) An appropriate common name for the material (e.g., "Refrigerant Gas").

(c) *QT/NQT markings.* Each MC 330 and MC 331 cargo tank must be marked near the specification plate, in letters no less than 50 mm (2.0 inches) in height, with—

(1) "QT", if the cargo tank is constructed of quenched and tempered steel; or

(2) "NQT", if the cargo tank is constructed of other than quenched and tempered steel.

(d) After October 3, 2005, each on-vehicle manually-activated remote shutoff device for closure of the internal self-closing stop valve must be identified by marking "Emergency Shutoff" in letters at least 0.75 inches in height, in a color that contrasts with its background, and located in an area immediately adjacent to the means of closure.

(e) *NON-ODORIZED marking on cargo tanks containing LPG.* After September 30, 2006, no person may offer for transportation or transport a cargo tank containing liquefied petroleum gas (LPG) that is unodorized as authorized in §173.315(b)(1) unless it is legibly marked NON-ODORIZED or NOT ODORIZED on two opposing sides near the marked proper shipping name as specified in paragraph (b)(1) of this section, or near the placards.

§172.330 Tank cars and multi-unit tank car tanks.

(a) *Shipping name and identification number.* No person may offer for transportation or transport a hazardous material—

(1) In a tank car unless the following conditions are met:

(i) The tank car must be marked on each side and each end as required by §172.302 with the identification number specified for the material in the §172.101 Table; and

(ii) A tank car containing any of the following materials must be marked on each side with the key words of the proper shipping name specified for the material in the §172.101 Table, or with a common name authorized for the material in this subchapter (e.g., "Refrigerant Gas"):

Acrolein, stabilized
Ammonia, anhydrous, liquefied
Ammonia solutions (more than 50% ammonia)
Bromine *or* Bromine solutions
Bromine chloride
Chloroprene, stabilized
Dispersant gas *or* Refrigerant gas (as defined in §173.115 of this subchapter)
Division 2.1 materials
Division 2.2 materials (in Class DOT 107 tank cars only)
Division 2.3 materials
Formic acid
Hydrocyanic acid, aqueous solutions
Hydrofluoric acid, solution
Hydrogen cyanide, stabilized (less than 3% water)
Hydrogen fluoride, anhydrous
Hydrogen peroxide, aqueous solutions (greater than 20% hydrogen peroxide)
Hydrogen peroxide, stabilized
Hydrogen peroxide and peroxyacetic acid mixtures
Nitric acid (other than red fuming)
Phosphorus, amorphous
Phosphorus, white dry *or* Phosphorus, white, under water *or* Phosphorus white, in solution, *or* Phosphorus, yellow dry *or* Phosphorus, yellow, under water *or* Phosphorus, yellow, in solution
Phosphorus white, molten
Potassium nitrate and sodium nitrate mixtures
Potassium permanganate
Sulfur trioxide, stabilized
Sulfur trioxide, uninhibited

(2) In a multi-unit tank car tank, unless the tank is marked on two opposing sides, in letters and numerals no less than 50 mm (2.0 inches) high—

(i) With the proper shipping name specified for the material in the §172.101 Table or with a common name authorized for the material in this subchapter (e.g., "Refrigerant Gas"); and

(ii) With the identification number specified for the material in the §172.101 Table, unless marked in accordance with §§172.302(a) and 172.332 of this subpart.

(b) A motor vehicle or rail car used to transport a multi-unit tank car tank containing a hazardous material must be marked on each side and each end, as required by §172.332, with the identification number specified for the material in the §172.101 Table.

(c) After September 30, 2006, no person may offer for transportation or transport a tank car or multi-unit tank car tank containing liquefied petroleum gas (LPG) that is unodorized unless it is legibly marked NON-ODORIZED or NOT ODORIZED on two opposing sides near the marked proper shipping name required by paragraphs (a)(1) and (a)(2) of this section, or near the placards. The NON-ODORIZED or NOT ODORIZED marking may appear on a tank car or multi-unit tank car tank used for both unodorized and odorized LPG.

§172.331 Bulk packagings other than portable tanks, cargo tanks, tank cars and multi-unit tank car tanks.

(a) Each person who offers a hazardous material to a motor carrier for transportation in a bulk packaging shall provide the motor carrier with the required identification numbers on placards or plain white square-on-point display configurations, as authorized, or shall affix orange panels containing the required identification numbers to the packaging prior to or at the time the material is offered for transportation, unless the packaging is already marked with the identification number as required by this subchapter.

(b) Each person who offers a bulk packaging containing a hazardous material for transportation shall affix to the packaging the required identification numbers on orange panels, square-on-point configurations or placards, as appropriate, prior to, or at the time the packaging is offered for transportation unless it is already marked with identification numbers as required by this subchapter.

(c) For a bulk packaging contained in or on a transport vehicle or freight container, if the identification number marking on the bulk packaging (e.g., an IBC) required by §172.302(a) is not visible, the transport vehicle or freight container must be marked as required by §172.332 on each side and each end with the identification number specified for the material in the §172.101 Table.

§172.332 Identification number markings.

(a) *General.* When required by §172.301, §172.302, §172.313, §172.326, §172.328, §172.330, or §172.331, identification number markings must be displayed on orange panels or placards as specified in this section, or on white square-on-point configurations as prescribed in §172.336(b).

(b) *Orange panels.* Display of an identification number on an orange panel shall be in conformance with the following:

(1) The orange panel must be 160 mm (6.3 inches) high by 400 mm (15.7 inches) wide with a 15 mm (0.6 inches) black outer border. The identification number shall be displayed in 100 mm (3.9 inches) black Helvetica Medium numerals on the orange panel. Measurements may vary from those specified plus or minus 5 mm (0.2 inches).

(2) The orange panel may be made of any durable material prescribed for placards in §172.519, and shall be of the orange color specified for labels or placards in appendix A to this part.

(3) The name and hazard class of a material may be shown in the upper left border of the orange panel in letters not more than 18 points (0.25 in.) high.

(4) Except for size and color, the orange panel and identification numbers shall be as illustrated for Liquefied petroleum gas:

1075

(c) *Placards*: Display of an identification number on a hazard warning placard shall be in conformance with the following:

(1) The identification number shall be displayed across the center area of the placard in 88 mm (3.5 inches) black Alpine Gothic or Alternate Gothic No. 3 numerals on a white background 100 mm (3.9 inches) high and approximately 215 mm (8.5 inches) wide and may be outlined with a solid or dotted line border.

(2) The top of the 100 mm (3.9 inches) high white background shall be approximately 40 mm (1.6 inches) above the placard horizontal centerline.

(3) An identification number may be displayed only on a placard corresponding to the primary hazard class of the hazardous material.

(4) For a COMBUSTIBLE placard used to display an identification number, the entire background below the white background for the identification number must be white during transportation by rail and may be white during transportation by highway.

(5) The name of the hazardous material and the hazard class may be shown in letters not more than 18 points high immediately within the upper border of the space on the placard bearing the identification number of the material.

(6) If an identification number is placed over the word(s) on a placard, the word(s) should be substantially covered to maximize the effectiveness of the identification number.

(d) Except for size and color, the display of an identification number on a placard shall be as illustrated for Acetone:

§172.334 Identification numbers; prohibited display.

(a) No person may display an identification number on a RADIOACTIVE, EXPLOSIVES 1.1, 1.2, 1.3, 1.4, 1.5 or 1.6, DANGEROUS, or subsidiary hazard placard.

(b) No person may display an identification number on a placard, orange panel or white square-on-point display configuration unless—

(1) The identification number is specified for the material in §172.101;

(2) The identification number is displayed on the placard, orange panel or white square-on-point configuration authorized by §172.332 or §172.336(b), as appropriate, and any placard used for display of the identification number corresponds to the hazard class of the material specified in §172.504;

(3) Except as provided under §172.336(c)(4) or (c)(5), the package, freight container, or transport vehicle on which the number is displayed contains the hazardous material associated with that identification number in §172.101.

(c) Except as required by §172.332(c)(4) for a combustible liquid, the identification number of a material may be displayed only on the placards required by the tables in §172.504.

(d) Except as provided in §172.336, a placard bearing an identification number may not be used to meet the requirements of Subpart F of the Part unless it is the correct identification number for all hazardous materials of the same class in the transport vehicle or freight container on which it is displayed.

(e) Except as specified in §172.338, an identification number may not be displayed on an orange panel on a cargo tank unless affixed to the cargo tank by the person offering the hazardous material for transportation in the cargo tank.

(f) If a placard is required by §172.504, an identification number may not be displayed on an orange panel unless it is displayed in proximity to the placard.

(g) No person shall add any color, number, letter, symbol, or word other than as specified in this subchapter, to any identification number marking display which is required or authorized by this subchapter.

§172.336 Identification numbers; special provisions.

(a) When not required or prohibited by this subpart, identification numbers may be displayed on a transport vehicle or a freight container in the manner prescribed by this subpart.

(b) Identification numbers, when required, must be displayed on either orange panels (see §172.332(b)) or on a plain white square-on-point display configuration having the same outside dimensions as a placard. In addition, for materials in hazard classes for which placards are specified and identification number displays are required, but for which identification numbers may not be displayed on the placards authorized for the material (see §172.334(a)), identification numbers must be displayed on orange panels or on the plain square-on-point display configuration in association with the required placards. An identification number displayed on a white square-on-point display configuration is not considered to be a placard.

(1) The 100 mm (3.9 inch) by 215 mm (8.5 inches) area containing the identification number shall be located as prescribed by §172.332(c)(1) and (c)(2) and may be outlined with a solid or dotted line border.

(2) [Reserved]

(c) Identification numbers are not required:

(1) On the ends of a portable tank, cargo tank or tank car having more than one compartment if hazardous materials having different identification numbers are being transported therein. In such a circumstance, the identification numbers on the sides of the tank shall be displayed in the same sequence as the compartments containing the materials they identify.

(2) On a cargo tank containing only gasoline, if the cargo tank is marked "Gasoline" on each side and rear in letters no less than 50 mm (2 inches) high, or is placarded in accordance with §172.542(c).

(3) On a cargo tank containing only fuel oil, if the cargo tank is marked "Fuel Oil" on each side and rear in letters no less the 50 mm (2 inches) high, or is placarded in accordance with §172.544(c).

(4) For each of the different liquid petroleum distillate fuels, including gasoline and gasohol, in a compartmented cargo tank or tank car, if the identification number is displayed for the distillate fuel having the lowest flash point. After October 1, 2010, if a compartmented cargo tank or tank car contains such fuels together with a gasoline and alcohol fuel blend containing more than ten percent ethanol, the identification number "3475" or "1987" must also be displayed as appropriate in addition to the identification number for the liquid petroleum distillate fuel having the lowest flash point.

(5) For each of the different liquid petroleum distillate fuels, including gasoline and gasohol transported in a cargo tank, if the identification number is displayed for the liquid petroleum distillate fuel having the lowest flash point.

(6) For each of the different liquid petroleum distillate fuels, including gasoline and gasohol, transported in a cargo tank, if the identification number is displayed for the liquid petroleum distillate fuel having the lowest flash point. After October 1, 2010, if a cargo tank is used to transport a gasoline and alcohol fuel blend containing more than ten percent ethanol, the identification number "3475" must also be displayed in addition to the identification number for the liquid petroleum distillate fuel having the lowest flash point.

(7) On nurse tanks meeting the provisions of §173.315(m) of this subchapter.

(d) When a bulk packaging is labeled instead of placarded in accordance with §172.514(c) of this subchapter, identification number markings may be displayed on the package in accordance with the marking requirements of §172.301(a)(1) of this subchapter.

§172.338 Replacement of identification numbers.

If more than one of the identification number markings on placards, orange panels, or white square-on-point display configurations that are required to be displayed are lost, damaged or destroyed during transportation, the carrier shall replace all the missing or damaged identification numbers as soon as practicable. However, in such a case, the numbers may be entered by hand on the appropriate placard, orange panel or white square-on-point display configuration providing the correct identification numbers are entered legibly using an indelible marking material. When entered by hand, the identification numbers must be located in the white display area specified in §172.332. This section does not preclude required compliance with the placarding requirements of Subpart F of the subchapter.

Subpart E—Labeling

§172.400 General labeling requirements.

(a) Except as specified in §172.400a, each person who offers for transportation or transports a hazardous material in any of the following packages or containment devices, shall label the package or containment device with labels specified for the material in the §172.101 Table and in this subpart:

(1) A non-bulk package;

(2) A bulk packaging, other than a cargo tank, portable tank, or tank car, with a volumetric capacity of less than 18 m^3 (640 cubic feet), unless placarded in accordance with subpart F of this part;

(3) A portable tank of less than 3785 L (1000 gallons) capacity, unless placarded in accordance with subpart F of this part;

(4) A DOT Specification 106 or 110 multi-unit tank car tank, unless placarded in accordance with subpart F of this part; and

(5) An overpack, freight container or unit load device, of less than 18 m^3 (640 cubic feet), which contains a package for which labels are required, unless placarded or marked in accordance with §172.512 of this part.

(b) Labeling is required for a hazardous material which meets one or more hazard class definitions, in accordance with Column 6 of the §172.101 Table and the following table:

Hazard class or division	Label name	Label design or section reference
1.1	EXPLOSIVES 1.1	172.411
1.2	EXPLOSIVES 1.2	172.411
1.3	EXPLOSIVES 1.3	172.411
1.4	EXPLOSIVES 1.4	172.411
1.5	EXPLOSIVES 1.5	172.411
1.6	EXPLOSIVES 1.6	172.411
2.1	FLAMMABLE GAS	172.417
2.2	NONFLAMMABLE GAS	172.415
2.3	POISON GAS	172.416
3 (flammable liquid)	FLAMMABLE LIQUID	172.419
Combustible liquid	(none)	
4.1	FLAMMABLE SOLID	172.420
4.2	SPONTANEOUSLY COMBUSTIBLE	172.422
4.3	DANGEROUS WHEN WET	172.423
5.1	OXIDIZER	172.426
5.2	ORGANIC PEROXIDE	172.427
6.1 (material poisonous by inhalation (see §171.8 of this subchapter))	POISON INHALATION HAZARD	172.429
6.1 (other than material poisonous by inhalation)	POISON	172.430
6.2	INFECTIOUS SUBSTANCE[1]	172.432
7 (see §172.403)	RADIOACTIVE WHITE-I	172.436
7	RADIOACTIVE YELLOW-II	172.438
7	RADIOACTIVE YELLOW-III	172.440
7 (fissile radioactive material; see §172.402)	FISSILE	172.441
7 (empty packages, see §173.428 of this subchapter)	EMPTY	172.450
8	CORROSIVE	172.442
9	CLASS 9	172.446

[1]The ETIOLOGIC AGENT label specified in regulations of the Department of Health and Human Services at 42 CFR 72.3 may apply to packages of infectious substances.

§172.400a Exceptions from labeling.

(a) Notwithstanding the provisions of §172.400, a label is not required on—

(1) A Dewar flask meeting the requirements in §173.320 of this subchapter or a cylinder containing a Division 2.1, 2.2, or 2.3 material that is—

(i) Not overpacked; and

(ii) Durably and legibly marked in accordance with CGA C–7, Appendix A (IBR; see §171.7 of this subchapter).

(2) A package or unit of military explosives (including ammunition) shipped by or on behalf of the DOD when in—

(i) Freight containerload, carload or truckload shipments, if loaded and unloaded by the shipper or DOD; or

(ii) Unitized or palletized break-bulk shipments by cargo vessel under charter to DOD if at least one required label is displayed on each unitized or palletized load.

(3) A package containing a hazardous material other than ammunition that is—

(i) Loaded and unloaded under the supervision of DOD personnel, and

(ii) Escorted by DOD personnel in a separate vehicle.

(4) A compressed gas cylinder permanently mounted in or on a transport vehicle.

(5) A freight container, aircraft unit load device or portable tank, which—

(i) Is placarded in accordance with subpart F of this part, or

(ii) Conforms to paragraph (a)(3) or (b)(3) of §172.512.

(6) An overpack or unit load device in or on which labels representative of each hazardous material in the overpack or unit load device are visible.

(7) A package of low specific activity radioactive material and surface contaminated objects, when transported under §173.427(a)(6)(vi) of this subchapter.

(b) Certain exceptions to labeling requirements are provided for small quantities and limited quantities in applicable sections in part 173 of this subchapter.

(c) Notwithstanding the provisions of §172.402(a), a Division 6.1 subsidiary hazard label is not required on a package containing a Class 8 (corrosive) material which has a subsidiary hazard of Division 6.1 (poisonous) if the toxicity of the material is based solely on the corrosive destruction of tissue rather than systemic poisoning. In addition, a Division 4.1 subsidiary hazard label is not required on a package bearing a Division 4.2 label.

(d) A package containing a material poisonous by inhalation (see §171.8 of this subchapter) in a closed transport vehicle or freight container may be excepted from the POISON INHALATION HAZARD or POISON GAS label or placard, under the conditions set forth in §171.23(b)(10) of this subchapter.

§172.401 Prohibited labeling.

(a) Except as otherwise provided in this section, no person may offer for transportation and no carrier may transport a package bearing a label specified in this subpart unless:

(1) The package contains a material that is a hazardous material, and

(2) The label represents a hazard of the hazardous material in the package.

(b) No person may offer for transportation and no carrier may transport a package bearing any marking or label which by its color, design, or shape could be confused with or conflict with a label prescribed by this part.

(c) The restrictions in paragraphs (a) and (b) of this section, do not apply to packages labeled in conformance with:

(1) The UN Recommendations (IBR, see §171.7 of this subchapter);

(2) The IMDG Code (IBR, see §171.7 of this subchapter);

(3) The ICAO Technical Instructions (IBR, see §171.7 of this subchapter);

(4) The TDG Regulations (IBR, see §171.7 of this subchapter).

(5) The Globally Harmonized System of Classification and Labelling of Chemicals (GHS) (IBR, see §171.7 of this subchapter).

(d) The provisions of paragraph (a) of this section do not apply to a packaging bearing a label if that packaging is:

(1) Unused or cleaned and purged of all residue;

(2) Transported in a transport vehicle or freight container in such a manner that the packaging is not visible during transportation; and

(3) Loaded by the shipper and unloaded by the shipper or consignee.

§172.402 Additional Labeling requirements.

(a) *Subsidiary hazard labels.* Each package containing a hazardous material—

(1) Shall be labeled with primary and subsidiary hazard labels as specified in Column 6 of the §172.101 Table (unless excepted in paragraph (a)(2) of this section); and

(2) For other than Class 1 or Class 2 materials (for subsidiary labeling requirements for Class 1 or Class 2 materials see paragraph (e) or paragraphs (f) and (g), respectively, of this section), if not already labeled under paragraph (a)(1) of this section, shall be labeled with subsidiary hazard labels in accordance with the following table:

SUBSIDIARY HAZARD LABELS

Subsidiary hazard level (packing group)	Subsidiary Hazard (Class or Division)						
	3	4.1	4.2	4.3	5.1	6.1	8
I	X	***	***	X	X	X	X
II	X	X	X	X	X	X	X
III	*	X	X	X	X	X	X

X—Required for all modes.

*—Required for all modes, except for a material with a flash point at or above 38° C (100° F) transported by rail or highway.

**—Reserved

***—Impossible as subsidiary hazard.

(b) *Display of hazard class on labels.* The appropriate hazard class or division number must be displayed in the lower corner of a primary hazard label and a subsidiary hazard label.

(c) *Cargo Aircraft Only label* . Each person who offers for transportation or transports by aircraft a package containing a hazardous material which is authorized on cargo aircraft only shall label the package with a CARGO AIRCRAFT ONLY label specified in §172.448 of this subpart.

(d) *Class 7 (Radioactive) Materials.* Except as otherwise provided in this paragraph, each package containing a Class 7 material that also meets the definition of one or more additional hazard classes must be labeled as a Class 7 material as required by §172.403 and for each additional hazard.

(1) For a package containing a Class 7 material that also meets the definition of one or more additional hazard classes, whether or not the material satisfies §173.4a(b)(7) of this subchapter, a subsidiary label is not required on the package if the material conforms to the remaining criteria in §173.4a of this subchapter.

(2) Each package or overpack containing fissile material, other than fissile-excepted material (see §173.453 of this subchapter) must bear two FISSILE labels, affixed to opposite sides of the package or overpack, which conforms to the figure shown in §172.441; such labels, where applicable, must be affixed adjacent to the labels for radioactive materials.

(e) *Class 1 (explosive) Materials.* In addition to the label specified in Column 6 of the §172.101 Table, each package of Class 1 material that also meets the definition for:

(1) Division 6.1, Packing Groups I or II, shall be labeled POISON or POISON INHALATION HAZARD, as appropriate.

(2) Class 7, shall be labeled in accordance with §172.403 of this subpart.

(f) *Division 2.2 materials.* In addition to the label specified in Column 6 of the §172.101 Table , each package of Division 2.2 material that also meets the definition for an oxidizing gas (see §171.8 of this subchapter) must be labeled OXIDIZER.

(g) *Division 2.3 materials.* In addition to the label specified in Column 6 of the §172.101 Table, each package of Division 2.3 material that also meets the definition for:

(1) Division 2.1, must be labeled Flammable Gas;

(2) Division 5.1, must be labeled Oxidizer; and

(3) Class 8, must be labeled Corrosive.

§172.403 Class 7 (radioactive) material.

(a) Unless excepted from labeling by §§173.421 through 173.427 of this subchapter, each package of radioactive material must be labeled as provided in this section.

(b) The proper label to affix to a package of Class 7 (radioactive) material is based on the radiation level at the surface of the package and the transport index. The proper category of label must be determined in accordance with paragraph (c) of this section. The label to be applied must be the highest category required for any of the two determining conditions for the package. RADIOACTIVE WHITE–I is the lowest category and RADIOACTIVE YELLOW–III is the highest. For example, a package with a transport index of 0.8 and a maximum surface radiation level of 0.6 millisievert (60 millirems) per hour must bear a RADIOACTIVE YELLOW–III label.

(c) Category of label to be applied to Class 7 (radioactive) materials packages:

Transport index	Maximum radiation level at any point on the external surface	Label category[1]
0[2]	Less than or equal to 0.005 mSv/h (0.5 mrem/h).	WHITE–I.
More than 0 but not more than 1	Greater than 0.005 mSv/h (0.5 mrem/h) but less than or equal to 0.5 mSv/h (50 mrem/h).	YELLOW–II.
More than 1 but not more than 10	Greater than 0.5 mSv/h (50 mrem/h) but less than or equal to 2 mSv/h (200 mrem/h).	YELLOW–III.
More than 10	Greater than 2 mSv/h (200 mrem/h) but less than or equal to 10mSv/h (1,000 mrem/h).	YELLOW–III (Must be shipped under exclusive use provisions; see 173.441(b) of this subchapter).

[1]Any package containing a "highway route controlled quantity" (§173.403 of this subchapter) must be labelled as RADIOACTIVE YELLOW–III.

[2]If the measured TI is not greater than 0.05, the value may be considered to be zero.

(d) *EMPTY* label. See §173.428(d) of this subchapter for EMPTY labeling requirements.

(e) *FISSILE label*. For packages required in §172.402 to bear a FISSILE label, each such label must be completed with the criticality safety index (CSI) assigned in the NRC or DOE package design approval, or in the certificate of approval for special arrangement or the certificate of approval for the package design issued by the Competent Authority for import and export shipments. For overpacks and freight containers required in §172.402 to bear a FISSILE label, the CSI on the label must be the sum of the CSIs for all of the packages contained in the overpack or freight container.

(f) Each package required by this section to be labeled with a RADIOACTIVE label must have two of these labels, affixed to opposite sides of the package. (See §172.406(e)(3) for freight container label requirements).

(g) The following applicable items of information must be entered in the blank spaces on the RADIOACTIVE label by legible printing (manual or mechanical), using durable weather resistant means of marking:

(1) *Contents.* Except for LSA–I material, the names of the radionuclides as taken from the listing of radionuclides in §173.435 of this subchapter (symbols which conform to established radiation protection terminology are authorized, *i.e.*, ^{99}Mo, ^{60}Co, *etc.*). For mixtures of radionuclides, with consideration of space available on the label, the radionuclides that must be shown must be determined in accordance with §173.433(g) of this subchapter. For LSA–I material, the term "LSA–I" may be used in place of the names of the radionuclides.

(2) *Activity.* The activity in the package must be expressed in appropriate SI units (*e.g.*, Becquerels (Bq), Terabecquerels (TBq), *etc.*). The activity may also be stated in appropriate customary units (Curies (Ci), milli-Curies (mCi), microCuries (uCi), *etc.*) in parentheses following the SI units. Abbreviations are authorized. Except for plutonium-239 and plutonium-241, the weight in grams or kilograms of fissile radionuclides may be inserted instead of activity units. For plutonium-239 and plutonium-241, the weight in grams of fissile radionu-clides may be inserted in addition to the activity units.

(3) *Transport index*. (see §173.403 of this subchapter.)

(h) When one or more packages of Class 7 (radioactive) material are placed within an overpack, the overpack must be labeled as prescribed in this section, except as follows:

(1) The "contents" entry on the label may state "mixed" in place of the names of the radionuclides unless each inside package contains the same radionuclide(s).

(2) The "activity" entry on the label must be determined by adding together the number of becquerels of the Class 7 (radioactive) materials packages contained therein.

(3) For an overpack, the transport index (TI) must be determined by adding together the transport indices of the Class 7 (radioactive) materials packages contained therein, except that for a rigid overpack, the transport index (TI) may alternatively be determined by direct measurement as prescribed in §173.403 of this subchapter under the definition for "transport index," taken by the person initially offering the packages contained within the overpack for shipment.

(4) The category of Class 7 label for the overpack must be determined from the table in §172.403(c) using the TI derived according to paragraph (h)(3) of this section, and the maximum radiation level on the surface of the overpack.

(5) The category of the Class 7 label of the overpack, and not that of any of the packages contained therein, must be used in accordance with Table 1 of §172.504(e) to determine when the transport vehicle must be placarded.

(6) For fissile material, the criticality safety index which must be entered on the overpack FISSILE label is the sum of the criticality safety indices of the individual

packages in the overpack, as stated in the certificate of approval for the package design issued by the NRC or the U.S. Competent Authority.

§172.404 Labels for mixed and consolidated packaging.

(a) *Mixed packaging.* When compatible hazardous materials having different hazard classes are packed within the same packaging, or within the same outside container or overpack as described in §173.25, the packaging, outside container or overpack must be labeled as required for each class of hazardous material contained therein.

(b) *Consolidated packaging.* When two or more packages containing compatible hazardous materials are placed within the same outside container or overpack, the outside container or overpack must be labeled as required for each class of hazardous material contained therein, unless labels representative of each hazardous material in the outside container or overpack are visible.

(c) *Consolidation bins used by a single motor carrier.* Notwithstanding the provisions of paragraph (b) of this section, labeling of a consolidation bin is not required under the following conditions:

(1) The consolidation bin must be reusable, made of materials such as plastic, wood, or metal and must have a capacity of 64 cubic feet or less;

(2) Hazardous material packages placed in the consolidation bin must be properly labeled in accordance with this subpart;

(3) Packages must be compatible as specified in §177.848 of this subchapter;

(4) Packages may only be placed within the consolidation bin and the bin be loaded on a motor vehicle by an employee of a single motor carrier;

(5) Packages must be secured within the consolidation bin by other packages or by other suitable means in such a manner as to prevent shifting of, or significant relative motion between, the packages that would likely compromise the integrity of any package;

(6) The consolidation bin must be clearly and legibly marked on a tag or fixed display device with an indication of each hazard class or division contained within the bin;

(7) The consolidation bin must be properly blocked and braced within the transport vehicle; and

(8) Consolidation bins may only be transported by a single motor carrier, or on railcars transporting such vehicles.

§172.405 Authorized label modifications.

(a) For Classes 1, 2, 3, 4, 5, 6, and 8, text indicating a hazard (for example FLAMMABLE LIQUID) is not required on a primary or subsidiary label.

(b) For a package containing Oxygen, compressed, or Oxygen, refrigerated liquid, the OXIDIZER label specified in §172.426 of this subpart, modified to display the word "OXYGEN" instead of "OXIDIZER", and the class number "2" instead of "5.1", may be used in place of the NON-FLAMMABLE GAS and OXIDIZER labels. Notwithstanding the provisions of paragraph (a) of this section, the word "OXYGEN" must appear on the label.

(c) For a package containing a Division 6.1, Packing Group III material, the POISON label specified in §172.430 may be modified to display the text "PG III" instead of "POISON" or "TOXIC" below the mid line of the label. Also see §172.313(d).

§172.406 Placement of labels.

(a) *General.* (1) Except as provided in paragraphs (b) and (e) of this section, each label required by this subpart must—

(i) Be printed on or affixed to a surface (other than the bottom) of the package or containment device containing the hazardous material; and

(ii) Be located on the same surface of the package and near the proper shipping name marking, if the package dimensions are adequate.

(2) Except as provided in paragraph (e) of this section, duplicate labeling is not required on a package or containment device (such as to satisfy redundant labeling requirements).

(b) *Exceptions.* A label may be printed on or placed on a securely affixed tag, or may be affixed by other suitable means to:

(1) A package that contains no radioactive material and which has dimensions less than those of the required label;

(2) A cylinder; and

(3) A package which has such an irregular surface that a label cannot be satisfactorily affixed.

(c) *Placement of multiple labels.* When primary and subsidiary hazard labels are required, they must be displayed next to each other. Placement conforms to this requirement if labels are within 150 mm (6 inches) of one another.

(d) *Contrast with background.* Each label must be printed on or affixed to a background of contrasting color, or must have a dotted or solid line outer border.

(e) *Duplicate labeling.* Generally, only one of each different required label must be displayed on a package. However, duplicate labels must be displayed on at least two sides or two ends (other than the bottom) of—

(1) Each package or overpack having a volume of 1.8 m^3 (64 cubic feet) or more;

(2) Each non-bulk package containing a radioactive material;

(3) Each DOT 106 or 110 multi-unit tank car tank. Labels must be displayed on each end;

(4) Each portable tank of less than 3,785 L (1000 gallons) capacity;

(5) Each freight container or aircraft unit load device having a volume of 1.8 m^3 (64 cubic feet) or more, but less than 18 m^3 (640 cubic feet). One of each required label must be displayed on or near the closure; and

(6) An IBC having a volume of 1.8 m^3 (64 cubic feet) or more.

(f) *Visibility.* A label must be clearly visible and may not be obscured by markings or attachments.

§172.411 EXPLOSIVE 1.1, 1.2, 1.3, 1.4, 1.5 and 1.6 labels, and EXPLOSIVE Subsidiary label.

(a) Except for size and color, the EXPLOSIVE 1.1, EXPLOSIVE 1.2 and EXPLOSIVE 1.3 labels must be as follows:

(b) In addition to complying with §172.407, the background color on the EXPLOSIVE 1.1, EXPLOSIVE 1.2 and EXPLOSIVE 1.3 labels must be orange. The "**" must be replaced with the appropriate division number and compatibility group letter. The compatibility group letter must be the same size as the division number and must be shown as a capitalized Roman letter.

(c) Except for size and color, the EXPLOSIVE 1.4, EXPLOSIVE 1.5 and EXPLOSIVE 1.6 labels must be as follows:

EXPLOSIVE 1.4:

EXPLOSIVE 1.5:

EXPLOSIVE 1.6:

(d) In addition to complying with §172.407, the background color on the EXPLOSIVE 1.4, EXPLOSIVE 1.5 and EXPLOSIVE 1.6 label must be orange. The "*" must be replaced with the appropriate compatibility group. The compatibility group letter must be shown as a capitalized Roman letter. Division numbers must measure at least 30 mm (1.2 inches) in height and at least 5 mm (0.2 inches) in width.

(e) An EXPLOSIVE subsidiary label is required for materials identified in Column (6) of the HMT as having an explosive subsidiary hazard. The division number or compability group letter may be displayed on the subsidiary hazard label. Except for size and color, the EXPLOSIVE subsidiary label must be as follows:

(f) The EXPLOSIVE subsidiary label must comply with §172.407.

§172.415 NON-FLAMMABLE GAS label.

(a) Except for size and color, the NON-FLAMMABLE GAS label must be as follows:

(b) In addition to complying with §172.407, the background color on the NON-FLAMMABLE GAS label must be green.

§172.416 POISON GAS label.

(a) Except for size and color, the POISON GAS label must be as follows:

(b) In addition to complying with §172.407, the background on the POISON GAS label and the symbol must be white. The background of the upper diamond must be black and the lower point of the upper diamond must be 14 mm (0.54 inches) above the horizontal center line.

§172.417 FLAMMABLE GAS label.

(a) Except for size and color, the FLAMMABLE GAS label must be as follows:

(b) In addition to complying with §172.407, the background color on the FLAMMABLE GAS label must be red.

§172.419 FLAMMABLE LIQUID label.

(a) Except for size and color the FLAMMABLE LIQUID label must be as follows:

(b) In addition to complying with §172.407, the background color on the FLAMMABLE LIQUID label must be red.

§172.420 FLAMMABLE SOLID label.

(a) Except for size and color, the FLAMMABLE SOLID label must be as follows:

(b) In addition to complying with §172.407, the background on the FLAMMABLE SOLID label must be white with vertical red stripes equally spaced on each side of a red stripe placed in the center of the label. The red vertical stripes must be spaced so that, visually, they appear equal in width to the white spaces between them. The symbol (flame) and text (when used) must be overprinted. The text "FLAMMABLE SOLID" may be placed in a white rectangle.

§172.422 SPONTANEOUSLY COMBUSTIBLE label.

(a) Except for size and color, the SPONTANEOUSLY COMBUSTIBLE label must be as follows:

(b) In addition to complying with §172.407, the background color on the lower half of the SPONTANEOUSLY COMBUSTIBLE label must be red and the upper half must be white.

§172.423 DANGEROUS WHEN WET label.

(a) Except for size and color, the DANGEROUS WHEN WET label must be as follows:

(b) In addition to complying with §172.407, the background color on the DANGEROUS WHEN WET label must be blue.

§172.426 OXIDIZER label.

(a) Except for size and color, the OXIDIZER label must be as follows:

(b) In addition to complying with §172.407, the background color on the OXIDIZER label must be yellow.

§172.427 ORGANIC PEROXIDE label.

(a) Except for size and color, the ORGANIC PEROXIDE label must be as follows:

(b) In addition to complying with §172.407, the background on the ORGANIC PEROXIDE label must be red in the top half and yellow in the lower half.

§172.429 POISON INHALATION HAZARD label.

(a) Except for size and color, the POISON INHALATION HAZARD label must be as follows:

(b) In addition to complying with §172.407, the background on the POISON INHALATION HAZARD label and the symbol must be white. The background of the upper diamond must be black and the lower point of the upper diamond must be 14 mm (0.54 inches) above the horizontal center line.

§172.430 POISON label.

(a) Except for size and color, the POISON label must be as follows:

LABELING

(b) In addition to complying with §172.407, the background on the POISON label must be white. The word "TOXIC" may be used in lieu of the word "POISON".

§172.431 [Reserved]

§172.432 INFECTIOUS SUBSTANCE label.

(a) Except for size and color, the INFECTIOUS SUBSTANCE label must be as follows:

(b) In addition to complying with §172.407, the background on the INFECTIOUS SUBSTANCE label must be white.

(c) Labels conforming to requirements in place on August 18, 2011 may continue to be used until October 1, 2014.

§172.436 RADIOACTIVE WHITE–I label.

(a) Except for size and color, the RADIOACTIVE WHITE–I label must be as follows:

(b) In addition to complying with §172.407, the background on the RADIOACTIVE WHITE–I label must be white. The printing and symbol must be black, except for the "I" which must be red.

§172.438 RADIOACTIVE YELLOW–II label.

(a) Except for size and color, the RADIOACTIVE YELLOW–II must be as follows:

(b) In addition to complying with §172.407, the background color on the RADIOACTIVE YELLOW–II label must be yellow in the top half and white in the lower half. The printing and symbol must be black, except for the "II" which must be red.

§172.440 RADIOACTIVE YELLOW–III label.

(a) Except for size and color, the RADIOACTIVE YELLOW–III label must be as follows:

(b) In addition to complying with §172.407, the background color on the RADIOACTIVE YELLOW–III label must be yellow in the top half and white in the lower half. The printing and symbol must be black, except for the "III" which must be red.

§172.441 FISSILE label.

(a) Except for size and color, the FISSILE label must be as follows:

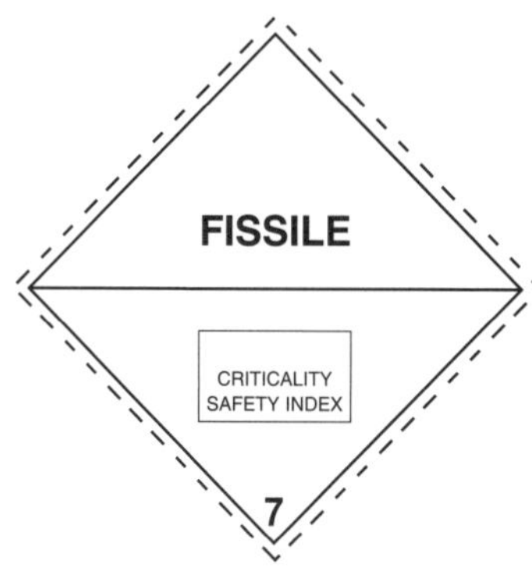

(b) In addition to complying with §172.407, the background color on the FISSILE label must be white.

§172.442 CORROSIVE label.

(a) Except for size and color, the CORROSIVE label must be as follows:

(b) In addition to complying with §172.407, the background on the CORROSIVE label must be white in the top half and black in the lower half.

§172.446 CLASS 9 label.

(a) Except for size and color, the "CLASS 9" (miscellaneous hazardous materials) label must be as follows:

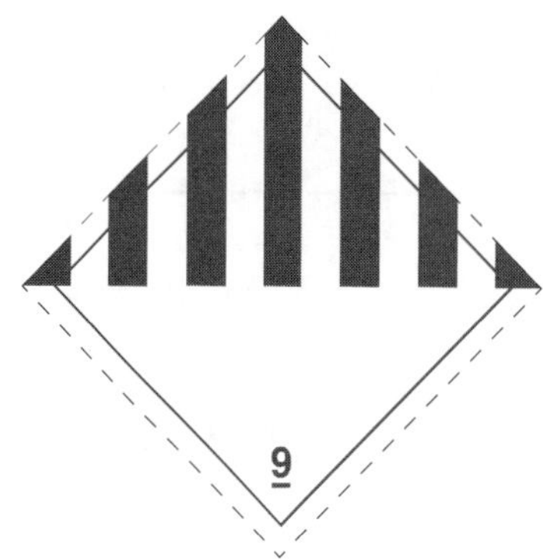

(b) In addition to complying with §172.407, the background on the CLASS 9 label must be white with seven black vertical stripes on the top half. The black vertical stripes must be spaced, so that, visually, they appear equal in width to the six white spaces between them. The lower half of the label must be white with the class number "9" underlined and centered at the bottom. The solid horizontal line dividing the lower and upper half of the label is optional.

(c) Labels conforming to requirements in place on August 18, 2011 may continue to be used until October 1, 2014.

§172.448 CARGO AIRCRAFT ONLY label.

(a) Except for size and color, the CARGO AIRCRAFT ONLY label must be as follows:

(b) The CARGO AIRCRAFT ONLY label must be black on an orange background.

(c) A CARGO AIRCRAFT ONLY label conforming to the specifications in this section and in §172.407(c)(2) in effect on October 1, 2008, may be used until January 1, 2013.

§172.450 EMPTY label.

(a) Each EMPTY label, except for size, must be as follows:

(1) Each side must be at least 6 inches (152 mm) with each letter at least 1 inch (25.4 mm) in height.

(2) The label must be white with black printing.

(b) [Reserved]

Subpart F—Placarding

§172.500 Applicability of placarding requirements.

(a) Each person who offers for transportation or transports any hazardous material subject to this subchapter shall comply with the applicable placarding requirements of this subpart.

(b) This subpart does not apply to—

(1) Infectious substances;

(2) Hazardous materials classed as ORM–D;

(3) Hazardous materials authorized by this subchapter to be offered for transportation as a limited quantity when identified as such on a shipping paper in accordance with §172.203(b) or when marked as such in accordance with §172.315;

(4) Hazardous materials prepared in accordance with §173.13 of this subchapter;

(5) Hazardous materials which are packaged as small quantities under the provisions of §§173.4, 173.4a, 173.4b of this subchapter; and

(6) Combustible liquids in non-bulk packagings.

§172.502 Prohibited and permissive placarding.

(a) *Prohibited placarding.* Except as provided in paragraph (b) of this section, no person may affix or display on a packaging, freight container, unit load device, motor vehicle or rail car—

(1) Any placard described in this subpart unless—

(i) The material being offered or transported is a hazardous material;

(ii) The placard represents a hazard of the hazardous material being offered or transported; and

(iii) Any placarding conforms to the requirements of this subpart.

(2) Any sign, advertisement, slogan (such as "Drive Safely"), or device that, by its color, design, shape or content, could be confused with any placard prescribed in this subpart.

(b) *Exceptions.* (1) The restrictions in paragraph (a) of this section do not apply to a bulk packaging, freight container, unit load device, transport vehicle or rail car which is placarded in conformance with TDG Regulations, the IMDG Code or the UN Recommendations (IBR, see §171.7 of this subchapter).

(2) The restrictions of paragraph (a) of this section do not apply to the display of a BIOHAZARD marking, a "HOT" marking, a sour crude oil hazard marking, or an identification number on a white square-on-point configuration in accordance with §§172.323(c), 172.325(c), 172.327(a), or 172.336(b) of this part, respectively.

(3) The restrictions in paragraph (a)(2) of this section do not apply until October 1, 2001 to a safety sign or safety slogan (e.g., "Drive Safely" or "Drive Carefully"), which was permanently marked on a transport vehicle, bulk packaging, or freight container on or before August 21, 1997.

(c) *Permissive placarding.* Placards may be displayed for a hazardous material, even when not required, if the placarding otherwise conforms to the requirements of this subpart.

§172.503 Identification number display on placards.

For procedures and limitations pertaining to the display of identification numbers on placards, see §172.334.

§172.504 General placarding requirements.

(a) *General.* Except as otherwise provided in this subchapter, each bulk packaging, freight container, unit load device, transport vehicle or rail car containing any quantity of a hazardous material must be placarded on each side and each end with the type of placards specified in Tables 1 and 2 of this section and in accordance with other placarding requirements of this subpart, including the specifications for the placards named in the tables and described in detail in §§172.519 through 172.560.

(b) *DANGEROUS placard* . A freight container, unit load device, transport vehicle, or rail car which contains non-bulk packages with two or more categories of hazardous materials that require different placards specified in Table 2 of paragraph (e) of this section may be placarded with a DANGEROUS placard instead of the separate placarding specified for each of the materials in Table 2 of paragraph (e) of this section. However, when 1,000 kg (2,205 pounds) aggregate gross weight or more of one category of material is loaded therein at one loading facility on a freight container, unit load device, transport vehicle, or rail car, the placard specified in Table 2 of paragraph (e) of this section for that category must be applied.

(c) *Exception for less than 454 kg (1,001 pounds).* Except for bulk packagings and hazardous materials subject to §172.505, when hazardous materials covered by Table 2 of this section are transported by highway or rail, placards are not required on —

(1) A transport vehicle or freight container which contains less than 454 kg (1001 pounds) aggregate gross weight of hazardous materials covered by Table 2 of paragraph (e) of this section; or

(2) A rail car loaded with transport vehicles or freight containers, none of which is required to be placarded.

The exceptions provided in paragraph (c) of this section do not prohibit the display of placards in the manner prescribed in this subpart, if not otherwise prohibited (see §172.502), on transport vehicles or freight containers which are not required to be placarded.

(d) *Exception for empty non-bulk packages.* Except for hazardous materials subject to §172.505, a non-bulk packaging that contains only the residue of a hazardous material covered by Table 2 of paragraph (e) of this section need not be included in determining placarding requirements.

(e) *Placarding tables.* Placards are specified for hazardous materials in accordance with the following tables:

TABLE 1

Category of material (Hazard class or division number and additional description, as appropriate)	Placard name	Placard design section reference (§)
1.1	EXPLOSIVES 1.1	172.522
1.2	EXPLOSIVES 1.2	172.522
1.3	EXPLOSIVES 1.3	172.522
2.3	POISON GAS	172.540
4.3	DANGEROUS WHEN WET	172.548
5.2 (Organic peroxide, Type B, liquid *or* solid, temperature controlled).	ORGANIC PEROXIDE	172.552
6.1 (material poisonous by inhalation (see §171.8 of this subchapter))	POISON INHALATION HAZARD	172.555
7 (Radioactive Yellow III label only)	RADIOACTIVE[1]	172.556

[1]RADIOACTIVE placard also required for exclusive use shipments of low specific activity material and surface contaminated objects transported in accordance with §173.427(b)(4) and (5) or (c) of this subchapter.

TABLE 2

Category of material (Hazard class or division number and additional description, as appropriate)	Placard name	Placard design section reference (§)
1.4	EXPLOSIVES 1.4	172.523
1.5	EXPLOSIVES 1.5	172.524
1.6	EXPLOSIVES 1.6	172.525
2.1	FLAMMABLE GAS	172.532
2.2	NON-FLAMMABLE GAS	172.528
3	FLAMMABLE	172.542
Combustible liquid	COMBUSTIBLE	172.544
4.1	FLAMMABLE SOLID	172.546
4.2	SPONTANEOUSLY COMBUSTIBLE	172.547
5.1	OXIDIZER	172.550
5.2 (Other than organic peroxide, Type B, liquid or solid, temperature controlled)	ORGANIC PEROXIDE	172.552
6.1 (other than material poisonous by inhalation)	POISON	172.554
6.2	(None)	
8	CORROSIVE	172.558
9	CLASS 9 (see §172.504(f)(9))	172.560
ORM-D	(None)	

(f) *Additional placarding exceptions.* (1) When more than one division placard is required for Class 1 materials on a transport vehicle, rail car, freight container or unit load device, only the placard representing the lowest division number must be displayed.

(2) A FLAMMABLE placard may be used in place of a COMBUSTIBLE placard on—

(i) A cargo tank or portable tank.

(ii) A compartmented tank car which contains both flammable and combustible liquids.

(3) A NON-FLAMMABLE GAS placard is not required on a transport vehicle which contains non-flammable gas if the transport vehicle also contains flammable gas or oxygen and it is placarded with FLAMMABLE GAS or OXYGEN placards, as required.

(4) OXIDIZER placards are not required for Division 5.1 materials on freight containers, unit load devices, transport vehicles or rail cars which also contain Division 1.1 or 1.2 materials and which are placarded with EXPLOSIVES 1.1 or 1.2 placards, as required.

(5) For transportation by transport vehicle or rail car only, an OXIDIZER placard is not required for Division 5.1 materials on a transport vehicle, rail car or freight container which also contains Division 1.5 explosives and is placarded with EXPLOSIVES 1.5 placards, as required.

(6) The EXPLOSIVE 1.4 placard is not required for those Division 1.4 Compatibility Group S (1.4S) materials that are not required to be labeled 1.4S.

(7) For domestic transportation of oxygen, compressed or oxygen, refrigerated liquid, the OXYGEN placard in §172.530 of this subpart may be used in place of a NON-FLAMMABLE GAS placard.

(8) For domestic transportation, a POISON INHALATION HAZARD placard is not required on a transport vehicle or freight container that is already placarded with the POISON GAS placard.

(9) For Class 9, a CLASS 9 placard is not required for domestic transportation, including that portion of international transportation, defined in §171.8 of this subchapter, which occurs within the United States. However, a bulk packaging must be marked with the appropriate identification number on a CLASS 9 placard, an orange panel, or a white square-on-point display configuration as required by subpart D of this part.

(10) For Division 6.1, PG III materials, a POISON placard may be modified to display the text "PG III" below the mid line of the placard.

(11) For domestic transportation, a POISON placard is not required on a transport vehicle or freight container required to display a POISON INHALATION HAZARD or POISON GAS placard.

(g) For shipments of Class 1 (explosive materials) by aircraft or vessel, the applicable compatibility group letter must be displayed on the placards, or labels when applicable, required by this section. When more than one compatibility group placard is required for Class 1 materials, only one placard is required to be displayed, as provided in paragraphs (g)(1) through (g)(4) of this section. For the purposes of paragraphs (g)(1) through (g)(4), there is a distinction between the phrases *explosive articles* and *explosive substances. Explosive article* means an article containing an explosive substance; examples include a detonator, flare, primer or fuse. *Explosive substance* means a substance contained in a packaging that is not contained in an article; examples include black powder and smokeless powder.

(1) Explosive articles of compatibility groups C, D or E may be placarded displaying compatibility group E.

(2) Explosive articles of compatibility groups C, D, or E, when transported with those in compatibility group N, may be placarded displaying compatibility group D.

(3) Explosive substances of compatibility groups C and D may be placarded displaying compatibility group D.

(4) Explosive articles of compatibility groups C, D, E or G, except for fireworks, may be placarded displaying compatibility group E.

§172.505 Placarding for subsidiary hazards.

(a) Each transport vehicle, freight container, portable tank, unit load device, or rail car that contains a poisonous material subject to the "Poison Inhalation Hazard" shipping description of §172.203(m) must be placarded with a POISON INHALATION HAZARD or POISON GAS placard, as appropriate, on each side and each end, in addition to any other placard required for that material in §172.504. Duplication of the POISON INHALATION HAZARD or POISON GAS placard is not required.

(b) In addition to the RADIOACTIVE placard which may be required by §172.504(e) of this subpart, each transport vehicle, portable tank or freight container that contains 454 kg (1001 pounds) or more gross weight of fissile or low specific activity uranium hexafluoride shall be placarded with a CORROSIVE placard on each side and each end.

(c) Each transport vehicle, portable tank, freight container or unit load device that contains a material which has a subsidiary hazard of being dangerous when wet, as defined in §173.124 of this subchapter, shall be placarded with DANGEROUS WHEN WET placards, on each side and each end, in addition to the placards required by §172.504.

(d) Hazardous materials that possess secondary hazards may exhibit subsidiary placards that correspond to the placards described in this part, even when not required by this part (see also §172.519(b)(4) of this subpart).

§172.506 Providing and affixing placards: highway.

(a) Each person offering a motor carrier a hazardous material for transportation by highway shall provide to the motor carrier the required placards for the material being offered prior to or at the same time the material is offered for transportation, unless the carrier's motor vehicle is already placarded for the material as required by this subpart.

(1) No motor carrier may transport a hazardous material in a motor vehicle unless the placards required for the hazardous material are affixed thereto as required by this subpart.

(2) [Reserved]

(b) [Reserved]

§172.507 Special placarding provisions: highway.

(a) Each motor vehicle used to transport a package of highway route controlled quantity Class 7 (radioactive) materials (see §173.403 of this subchapter) must have the required RADIOACTIVE warning placard placed on a square background as described in §172.527.

(b) A nurse tank, meeting the provisions of §173.315(m) of this subchapter, is not required to be placarded on an end containing valves, fittings, regulators or gauges when those appurtenances prevent the markings and placard from being properly placed and visible.

§172.508 Providing and affixing placards: rail.

(a) Each person offering a hazardous material for transportation by rail shall affix to the rail car containing the material, the placards specified by this subpart. Placards displayed on motor vehicles transport containers, or portable tanks may be used to satisfy this requirement, if the placards otherwise conform to the provisions of this subpart.

(b) No rail carrier may accept a rail car containing a hazardous material for transportation unless the placards for the hazardous material are affixed thereto as required by this subpart.

§172.510 Special placarding provisions: rail.

(a) *White square background.* The following must have the specified placards placed on a white square background, as described in §172.527:

(1) Division 1.1 and 1.2 (explosive) materials which require EXPLOSIVES 1.1 or EXPLOSIVES 1.2 placards affixed to the rail car;

(2) Materials classed in Division 2.3 Hazard Zone A or 6.1 Packing Group I Hazard Zone A which require POISON GAS or POISON placards affixed to the rail car, including tank cars containing only a residue of the material; and

(3) Class DOT 113 tank cars used to transport a Division 2.1 (flammable gas) material, including tank cars containing only a residue of the material.

(b) *Chemical ammunition.* Each rail car containing Division 1.1 or 1.2 (explosive) ammunition which also meets the definition of a material poisonous by inhalation (see §171.8 of this subchapter) must be placarded EXPLOSIVES 1.1 or EXPLOSIVES 1.2 and POISON GAS or POISON INHALATION HAZARD.

§172.512 Freight containers and aircraft unit load devices.

(a) *Capacity of 640 cubic feet or more.* Each person who offers for transportation, and each person who loads and transports, a hazardous material in a freight container or aircraft unit load device having a capacity of 640 cubic feet or more shall affix to the freight container or aircraft unit load device the placards specified for the material in accordance with §172.504. However:

(1) The placarding exception provided in §172.504(c) applies to motor vehicles transporting freight containers and aircraft unit load devices,

(2) The placarding exception provided in §172.504(c), applies to each freight container and aircraft unit load device being transported for delivery to a consignee immediately following an air or water shipment, and,

(3) Placarding is not required on a freight container or aircraft unit load device if it is only transported by air and is identified as containing a hazardous material in the manner provided in part 7, chapter 2, section 2.7, of the ICAO Technical Instructions (IBR, see §171.7 of this subchapter).

(b) *Capacity less than 18 m^3 (640 cubic feet).*

(1) Each person who offers for transportation by air, and each person who loads and transports by air, a hazardous material in a freight container or aircraft unit load device having a capacity of less than 18 m3 (640 cubic feet) shall affix one placard of the type specified by paragraph (a) of this section unless the freight container or aircraft unit load device:

(i) Is labeled in accordance with subpart E of this part, including §172.406(e);

(ii) Contains radioactive materials requiring the Radioactive Yellow III label and is placarded with one Radioactive placard and is labeled in accordance with subpart E of this part, including §172.406(e); or,

(iii) Is identified as containing a hazardous material in the manner provided in part 7, chapter 2, section 2.7, of the ICAO Technical Instructions.

(2) When hazardous materials are offered for transportation, not involving air transportation, in a freight container having a capacity of less than 640 cubic feet the freight container need not be placarded. However, if not placarded, it must be labeled in accordance with subpart E of this part.

(c) Notwithstanding paragraphs (a) and (b) of this section, packages containing hazardous materials, other than ORM–D, offered for transportation by air in freight containers are subject to the inspection requirements of §175.30 of this chapter.

§172.514 Bulk packagings.

(a) Except as provided in paragraph (c) of this section, each person who offers for transportation a bulk packaging which contains a hazardous material, shall affix the placards specified for the material in §§172.504 and 172.505.

(b) Each bulk packaging that is required to be placarded when it contains a hazardous material, must remain placarded when it is emptied, unless it—

(1) Is sufficiently cleaned of residue and purged of vapors to remove any potential hazard;

(2) Is refilled, with a material requiring different placards or no placards, to such an extent that any residue remaining in the packaging is no longer hazardous; or

(3) Contains the residue of a hazardous substance in Class 9 in a quantity less than the reportable quantity, and conforms to §173.29(b)(1) of this subchapter.

(c) Exceptions. The following packagings may be placarded on only two opposite sides or, alternatively, may be labeled instead of placarded in accordance with subpart E of this part:

(1) A portable tank having a capacity of less than 3,785 L (1000 gallons);

(2) A DOT 106 or 110 multi-unit tank car tank;

(3) A bulk packaging other than a portable tank, cargo tank, or tank car (*e.g.*, a bulk bag or box) with a volumetric capacity of less than 18 cubic meters (640 cubic feet);

(4) *An IBC.* For an IBC labeled in accordance with subpart E of this part instead of placarded, the IBC may display the proper shipping name and UN identification number in accordance with the size requirements of §172.302(b)(2) in place of the UN number on an orange panel or placard; and

(5) A Large Packaging as defined in §171.8 of this subchapter.

§172.516 Visibility and display of placards.

(a) Each placard on a motor vehicle and each placard on a rail car must be clearly visible from the direction it faces, except from the direction of another transport vehicle or rail car to which the motor vehicle or rail car is coupled. This requirement may be met by the placards displayed on the freight containers or portable tanks loaded on a motor vehicle or rail car.

(b) The required placarding of the front of a motor vehicle may be on the front of a truck-tractor instead of or in addition to the placarding on the front of the cargo body to which a truck-tractor is attached.

(c) Each placard on a transport vehicle, bulk packaging, freight container or aircraft unit load device must—

(1) Be securely attached or affixed thereto or placed in a holder thereon. (See appendix C to this part.)

(2) Be located clear of appurtenances and devices such as ladders, pipes, doors, and tarpaulins;

(3) So far as practicable, be located so that dirt or water is not directed to it from the wheels of the transport vehicle;

(4) Be located away from any marking (such as advertising) that could substantially reduce its effectiveness, and in any case at least 3 inches (76.0 mm.) away from such marking;

(5) Have the words or identification number (when authorized) printed on it displayed horizontally, reading from left to right;

(6) Be maintained by the carrier in a condition so that the format, legibility, color, and visibility of the placard will not be substantially reduced due to damage, deterioration, or obscurement by dirt or other matter;

(7) Be affixed to a background of contrasting color, or must have a dotted or solid line outer border which contrasts with the background color.

(d) Recommended specifications for a placard holder are set forth in appendix C of this part. Except for a placard holder similar to that contained in appendix C to this part, the means used to attach a placard may not obscure any part of its surface other than the borders.

(e) A placard or placard holder may be hinged provided the required format, color, and legibility of the placard are maintained.

§172.519 General specifications for placards.

(a) *Strength and durability*. Placards must conform to the following:

(1) A placard may be made of any plastic, metal or other material capable of withstanding, without deterioration or a substantial reduction in effectiveness, a 30-day exposure to open weather conditions.

(2) A placard made of tagboard must be at least equal to that designated commercially as white tagboard. Tagboard must have a weight of at least 80 kg (176 pounds) per ream of 610 by 910 mm (24 by 36-inch) sheets, waterproofing materials included. In addition, each placard made of tagboard must be able to pass a 414 kPa (60 p.s.i.) Mullen test.

(3) Reflective or retroreflective materials may be used on a placard if the prescribed colors, strength and durability are maintained.

(b) *Design.* (1) Except as provided in §172.332 of this part, each placard must be as described in this subpart, and except for size and color, the printing, inner border and symbol must be as shown in §§172.521 through 172.560 of this subpart, as appropriate.

(2) The dotted line border shown on each placard is not part of the placard specification. However, a dotted or solid line outer border may be used when needed to indicate the full size of a placard that is part of a larger format or is on a background of a noncontrasting color.

(3) For other than Class 7 or the DANGEROUS placard, text indicating a hazard (for example, "FLAMMABLE") is not required. Text may be omitted from the OXYGEN placard only if the specific identification number is displayed on the placard.

(4) For a placard corresponding to the primary or subsidiary hazard class of a material, the hazard class or division number must be displayed in the lower corner of the placard. However, a permanently affixed subsidiary placard meeting the specifications of this section which were in effect on October 1, 2001, (such as, a placard without the hazard class or division number displayed in the lower corner of the placard) and which was installed prior to September 30, 2001, may continue to be used as a subsidiary placard in domestic transportation by rail or highway, provided the color tolerances are maintained and are in accordance with the display requirements in this subchapter.

(c) *Size.* (1) Each placard prescribed in this subpart must measure at least 250 mm (9.84 inches) on each side and must have a solid line inner border approximately 12.7 mm (0.5 inches) from each edge.

(2) Except as otherwise provided in this subpart, the hazard class or division number, as appropriate, must be shown in numerals measuring at least 41 mm (1.6 inches) in height.

(3) Except as otherwise provided in this subpart, when text indicating a hazard is displayed on a placard, the printing must be in letters measuring at least 41 mm (1.6 inches) in height.

(d) *Color.* (1) The background color, symbol, text, numerals and inner border on a placard must be as specified in §§172.521 through 172.560 of this subpart, as appropriate.

(2) Black and any color on a placard must be able to withstand, without substantial change—

(i) A 72-hour fadeometer test (for a description of equipment designed for this purpose, see ASTM G 23–69 or ASTM G 26-70); and

(ii) A 30-day exposure to open weather.

(3) Upon visual examination, a color on a placard must fall within the color tolerances displayed on the appropriate Hazardous Materials Label and Placard Color Tolerance Chart (see §172.407(d)(4)). As an alternative, the PANTONE® formula guide coated/uncoated as specified for colors in §172.407(d)(5) may be used.

(4) The placard color must extend to the inner border and may extend to the edge of the placard in the area designated on each placard except the color on the CORROSIVE and RADIOACTIVE placards (black and yellow, respectively) must extend only to the inner border.

(e) *Form identifications.* A placard may contain form identification information, including the name of its maker, provided that information is printed outside of the solid line inner border in no larger than 10-point type.

(f) *Exceptions.* When hazardous materials are offered for transportation or transported under the provisions of subpart C of part 171 of this subchapter, a placard conforming to the specifications in the ICAO Technical Instructions, the IMDG Code, or the Transport Canada TDG Regulations (IBR, see §171.7 of this subchapter) may be used in place of a corresponding placard conforming to the requirements of this subpart. However, a bulk packaging, transport vehicle, or freight container containing a material poisonous by inhalation (see §171.8 of this subchapter) must be placarded in accordance with this subpart (see §171.23(b)(10) of this subchapter).

(g) *Trefoil symbol.* The trefoil symbol on the RADIOACTIVE placard must meet the appropriate specification in Appendix B of this part.

§172.521 DANGEROUS placard.

(a) Except for size and color, the DANGEROUS placard must be as follows:

(b) In addition to meeting the requirements of §172.519, and appendix B to this part, the DANGEROUS placard must have a red upper and lower triangle. The placard center area and ½-inch (12.7 mm.) border must be white. The inscription must be black with the ⅛-inch (3.2 mm.) border marker in the white area at each end of the inscription red.

§172.522 EXPLOSIVES 1.1, EXPLOSIVES 1.2 and EXPLOSIVES 1.3 placards.

(a) Except for size and color, the EXPLOSIVES 1.1, EXPLOSIVES 1.2 and EXPLOSIVES 1.3 placards must be as follows:

(b) In addition to complying with §172.519 of this subpart, the background color on the EXPLOSIVES 1.1, EXPLOSIVES 1.2, and EXPLOSIVES 1.3 placards must be orange. The "*" shall be replaced with the appropriate division number and, when required, appropriate compatibility group letter. The symbol, text, numerals and inner border must be black.

§172.523 EXPLOSIVES 1.4 placard.

(a) Except for size and color, the EXPLOSIVES 1.4 placard must be as follows:

(b) In addition to complying with §172.519 of this subpart, the background color on the EXPLOSIVES 1.4 placard must be orange. The "*" shall be replaced, when required, with the appropriate compatibility group letter. The division numeral, 1.4, must measure at least 64 mm (2.5 inches) in height. The text, numerals and inner border must be black.

§172.524 EXPLOSIVES 1.5 placard.

(a) Except for size and color, the EXPLOSIVES 1.5 placard must be as follows:

(b) In addition to complying with the §172.519 of this subpart, the background color on the EXPLOSIVES 1.5 placard must be orange. The "*" shall be replaced, when required, with the appropriate compatibility group letter. The division numeral, 1.5, must measure at least 64 mm (2.5 inches) in height. The text, numerals and inner border must be black.

§172.525 EXPLOSIVES 1.6 placard.

(a) Except for size and color the EXPLOSIVES 1.6 placard must be as follows:

(b) In addition to complying with §172.519 of this subpart, the background color on the EXPLOSIVES 1.6 placard must be orange. The "*" shall be replaced, when required, with the appropriate compatibility group letter. The division numeral, 1.6, must measure at least 64 mm (2.5 inches) in height. The text, numerals and inner border must be black.

§172.527 Background requirements for certain placards.

(a) Except for size and color, the square background required by §172.510(a) for certain placards on rail cars, and §172.507 for placards on motor vehicles containing a package of highway route controlled quantity radioactive materials, must be as follows:

(b) In addition to meeting the requirements of §172.519 for minimum durability and strength, the square background must consist of a white square measuring 14¼ inches (362.0 mm.) on each side surrounded by a black border extending to 15¼ inches (387.0 mm.) on each side.

§172.528 NON-FLAMMABLE GAS placard.

(a) Except for size and color, the NON-FLAMMABLE GAS placard must be as follows:

(b) In addition to complying with §172.519, the background color on the NON-FLAMMABLE GAS placard must be green. The letters in both words must be at least 38 mm (1.5 inches) high. The symbol, text, class number and inner border must be white.

§172.530 OXYGEN placard.

(a) Except for size and color, the OXYGEN placard must be as follows:

(b) In addition to complying with §172.519, the background color on the OXYGEN placard must be yellow. The symbol, text, class number and inner border must be black.

§172.532 FLAMMABLE GAS placard.

(a) Except for size and color, the FLAMMABLE GAS placard must be as follows:

(b) In addition to complying with §172.519, the background color on the FLAMMABLE GAS placard must be red. The symbol, text, class number and inner border must be white.

§172.540 POISON GAS placard.

(a) Except for size and color, the POISON GAS placard must be as follows:

(b) In addition to complying with §172.519, the background on the POISON GAS placard and the symbol must be white. The background of the upper diamond must be black and the lower point of the upper diamond must be 65 mm (2⅝ inches) above the horizontal center line. The text, class number, and inner border must be black.

§172.542 FLAMMABLE placard.

(a) Except for size and color, the FLAMMABLE placard must be as follows:

(b) In addition to complying with §172.519, the background color on the FLAMMABLE placard must be red. The symbol, text, class number and inner border must be white.

(c) The word "GASOLINE" may be used in place of the word "FLAMMABLE" on a placard that is displayed on a cargo tank or a portable tank being used to transport gasoline by highway. The word "GASOLINE" must be shown in white.

§172.544 COMBUSTIBLE placard.

(a) Except for size and color, the COMBUSTIBLE placard must be as follows:

(b) In addition to complying with §172.519, the background color on the COMBUSTIBLE placard must be red. The symbol, text, class number and inner border must be white. On a COMBUSTIBLE placard with a white bottom as prescribed by §172.332(c)(4), the class number must be red or black.

(c) The words "FUEL OIL" may be used in place of the word "COMBUSTIBLE" on a placard that is displayed on a cargo tank or portable tank being used to transport by highway fuel oil that is not classed as a flammable liquid. The words "FUEL OIL" must be shown in white.

§172.546 FLAMMABLE SOLID placard.

(a) Except for size and color, the FLAMMABLE SOLID placard must be as follows:

(b) In addition to complying with §172.519, the background on the FLAMMABLE SOLID placard must be white with seven vertical red stripes. The stripes must be equally spaced, with one red stripe placed in the center of the label. Each red stripe and each white space between two red stripes must be 25 mm (1.0 inches) wide. The letters in the word "SOLID" must be at least 38.1 mm (1.5 inches) high. The symbol, text, class number and inner border must be black.

§172.547 SPONTANEOUSLY COMBUSTIBLE placard.

(a) Except for size and color, the SPONTANEOUSLY COMBUSTIBLE placard must be as follows:

(b) In addition to complying with §172.519, the background color on the SPONTANEOUSLY COMBUSTIBLE placard must be red in the lower half and white in upper half. The letters in the word "SPONTANEOUSLY" must be at least 12 mm (0.5 inch) high. The symbol, text, class number and inner border must be black.

§172.548 DANGEROUS WHEN WET placard.

(a) Except for size and color, the DANGEROUS WHEN WET placard must be as follows:

(b) In addition to complying with §172.519, the background color on the DANGEROUS WHEN WET placard must be blue. The letters in the words "WHEN WET" must be at least 25 mm (1.0 inches) high. The symbol, text, class number and inner border must be white.

§172.550 OXIDIZER placard.

(a) Except for size and color, the OXIDIZER placard must be as follows:

(b) In addition to complying with §172.519, the background color on the OXIDIZER placard must be yellow. The symbol, text, division number and inner border must be black.

§172.552 ORGANIC PEROXIDE placard.

(a) Except for size and color, the ORGANIC PEROXIDE placard must be as follows:

(b) In addition to complying with §172.519, the background on the ORGANIC PEROXIDE placard must be red in the top half and yellow in the lower half. The text, division number and inner border must be black; the symbol may be either black or white.

(c) For transportation by highway, a Division 5.2 placard conforming to the specifications in this section in effect on December 31, 2006 may continue to be used until January 1, 2014.

§172.553 [Reserved]

§172.554 POISON placard.

(a) Except for size and color, the POISON placard must be as follows:

(b) In addition to complying with §172.519, the background on the POISON placard must be white. The symbol, text, class number and inner border must be black. The word “TOXIC” may be used in lieu of the word “POISON”.

§172.555 POISON INHALATION HAZARD placard.

(a) Except for size and color, the POISON INHALATION HAZARD placard must be as follows:

(b) In addition to complying with §172.519, the background on the POISON INHALATION HAZARD placard and the symbol must be white. The background of the upper diamond must be black and the lower point of the upper diamond must be 65 mm (2⅝ inches) above the horizontal center line. The text, class number, and inner border must be black.

§172.556 RADIOACTIVE placard.

(a) Except for size and color, the RADIOACTIVE placard must be as follows:

(b) In addition to complying with §172.519, the background color on the RADIOACTIVE placard must be white in the lower portion with a yellow triangle in the upper portion. The base of the yellow triangle must be 29 mm ± 5 mm (1.1 inches ± 0.2 inches) above the placard horizontal center line. The symbol, text, class number and inner border must be black.

§172.558 CORROSIVE placard.

(a) Except for size and color, the CORROSIVE placard must be as follows:

(b) In addition to complying with §172.519, the background color on the CORROSIVE placard must be black in the lower portion with a white triangle in the upper portion. The base of the white triangle must be 38 mm ± 5 mm (1.5 inches ± 0.2 inches) above the placard horizontal center line. The text and class number must be white. The symbol and inner border must be black.

§172.560 CLASS 9 placard.

(a) Except for size and color the CLASS 9 (miscellaneous hazardous materials) placard must be as follows:

(b) In addition to conformance with §172.519, the background on the CLASS 9 placard must be white with seven black vertical stripes on the top half extending from the top of the placard to one inch above the horizontal centerline. The black vertical stripes must be spaced so that, visually, they appear equal in width to the six white spaces between them. The space below the vertical lines must be white with the class number 9 underlined and centered at the bottom.

Reserved

Subpart G—Emergency Response Information

§172.600 Applicability and general requirements.

(a) *Scope.* Except as provided in paragraph (d) of this section, this subpart prescribes requirements for providing and maintaining emergency response information during transportation and at facilities where hazardous materials are loaded for transportation, stored incidental to transportation or otherwise handled during any phase of transportation.

(b) *Applicability.* This subpart applies to persons who offer for transportation, accept for transportation, transfer or otherwise handle hazardous materials during transportation.

(c) *General requirements.* No person to whom this subpart applies may offer for transportation, accept for transportation, transfer, store or otherwise handle during transportation a hazardous material unless:

(1) Emergency response information conforming to this subpart is immediately available for use at all times the hazardous material is present; and

(2) Emergency response information, including the emergency response telephone number, required by this subpart is immediately available to any person who, as a representative of a Federal, State or local government agency, responds to an incident involving a hazardous material, or is conducting an investigation which involves a hazardous material.

(d) *Exceptions.* The requirements of this subpart do not apply to hazardous material which is excepted from the shipping paper requirements of this subchapter or a material properly classified as an ORM–D.

§172.602 Emergency response information.

(a) *Information required.* For purposes of this subpart, the term "emergency response information" means information that can be used in the mitigation of an incident involving hazardous materials and, as a minimum, must contain the following information:

(1) The basic description and technical name of the hazardous material as required by §§172.202 and 172.203(k), the ICAO Technical Instructions, the IMDG Code, or the TDG Regulations, as appropriate (IBR, see §171.7 of this subchapter);

(2) Immediate hazards to health;

(3) Risks of fire or explosion;

(4) Immediate precautions to be taken in the event of an accident or incident;

(5) Immediate methods for handling fires;

(6) Initial methods for handling spills or leaks in the absence of fire; and

(7) Preliminary first aid measures.

(b) *Form of information.* The information required for a hazardous material by paragraph (a) of this section must be:

(1) Printed legibly in English;

(2) Available for use away from the package containing the hazardous material; and

(3) Presented—

(i) On a shipping paper;

(ii) In a document, other than a shipping paper, that includes both the basic description and technical name of the hazardous material as required by §§172.202 and 172.203(k), the ICAO Technical Instructions, the IMDG Code or the TDG Regulations, as appropriate, and the emergency response information required by this subpart (e.g., a material safety data sheet); or

(iii) Related to the information on a shipping paper, a written notification to pilot-in-command, or a dangerous cargo manifest, in a separate document (e.g., an emergency response guidance document), in a manner that cross-references the description of the hazardous material on the shipping paper with the emergency response information contained in the document. Aboard aircraft, the ICAO "Emergency Response Guidance for Aircraft Incidents Involving Dangerous Goods" and aboard vessels, the IMO "Emergency Procedures for Ships Carrying Dangerous Goods", or equivalent documents, may be used to satisfy the requirements of this section for a separate document.

(c) *Maintenance of information.* Emergency response information shall be maintained as follows:

(1) *Carriers.* Each carrier who transports a hazardous material shall maintain the information specified in paragraph (a) of this section and §172.606 of this part in the same manner as prescribed for shipping papers, except that the information must be maintained in the same manner aboard aircraft as the notification of pilot-in-command, and aboard vessels in the same manner as the dangerous cargo manifest. This information must be immediately accessible to train crew personnel, drivers of motor vehicles, flight crew members, and bridge personnel on vessels for use in the event of incidents involving hazardous materials.

(2) *Facility operators.* Each operator of a facility where a hazardous material is received, stored or handled during transportation, shall maintain the information required by paragraph (a) of this section whenever the hazardous material is present. This information must be in a location that is immediately accessible to facility personnel in the event of an incident involving the hazardous material.

§172.604 Emergency response telephone number.

(a) A person who offers a hazardous material for transportation must provide an emergency response telephone number, including the area code, for use in the event of an emergency involving the hazardous material. For telephone numbers outside the United States, the international access code or the "+" (plus) sign, country code, and city code, as appropriate, must be included. The telephone number must be—

(1) Monitored at all times the hazardous material is in transportation, including storage incidental to transportation;

(2) The telephone number of a person who is either knowledgeable of the hazardous material being shipped and has comprehensive emergency response and incident mitigation information for that material, or has immediate access to a person who possesses such knowledge and information. A telephone number that requires

a call back (such as an answering service, answering machine, or beeper device) does not meet the requirements of paragraph (a) of this section; and

(3) Entered on a shipping paper, as follows:

(i) Immediately following the description of the hazardous material required by subpart C of this part; or

(ii) Entered once on the shipping paper in a prominent, readily identifiable, and clearly visible manner that allows the information to be easily and quickly found, such as by highlighting, use of a larger font or a font that is a different color from other text and information, or otherwise setting the information apart to provide for quick and easy recognition. This provision may be used only if the telephone number applies to each hazardous material entered on the shipping paper, and if it is indicated that the telephone number is for emergency response information (for example: "EMERGENCY CONTACT: * * *").

(b) The telephone number required by paragraph (a) of this section must be —

(1) The number of the person offering the hazardous material for transportation when that person is also the emergency response information provider (ERI provider). The name of the person, or contract number or other unique identifier assigned by an ERI provider, identified with the emergency response telephone number must be entered on the shipping paper immediately before, after, above, or below the emergency response telephone number unless the name is entered elsewhere on the shipping paper in a prominent, readily identifiable, and clearly visible manner that allows the information to be easily and quickly found; or

(2) The number of an agency or organization capable of, and accepting responsibility for, providing the detailed information required by paragraph (a)(2) of this section. The person who is registered with the ERI provider must ensure that the agency or organization has received current information on the material before it is offered for transportation. The person who is registered with the ERI provider must be identified by name, or contract number or other unique identifier assigned by the ERI provider, on the shipping paper immediately before, after, above, or below the emergency response telephone number in a prominent, readily identifiable, and clearly visible manner that allows the information to be easily and quickly found, unless the name or identifier is entered elsewhere in a prominent manner as provided in paragraph (b)(1) of this section.

(c) A person preparing shipping papers for continued transportation in commerce must include the information required by this section. If the person preparing shipping papers for continued transportation in commerce elects to assume responsibility for providing the emergency response telephone number required by this section, the person must ensure that all the requirements of this section are met.

(d) The requirements of this section do not apply to—

(1) Hazardous materials that are offered for transportation under the provisions applicable to limited quantities; and

(2) Materials properly described under the following shipping names:

Battery powered equipment.
Battery powered vehicle.
Carbon dioxide, solid.
Castor bean.
Castor flake.
Castor meal.
Castor pomace.
Consumer commodity.
Dry ice.
Engines, internal combustion.
Fish meal, stabilized.
Fish scrap, stabilized.
Refrigerating machine.
Vehicle, flammable gas powered.
Vehicle, flammable liquid powered.
Wheelchair, electric.

(3) Transportation vehicles or freight containers containing lading that has been fumigated and displaying the FUMIGANT marking (*see* §172.302(g)) as required by §173.9 of this subchapter, unless other hazardous materials are present in the cargo transport unit.

§172.606 Carrier information contact.

(a) Each carrier who transports or accepts for transportation a hazardous material for which a shipping paper is required shall instruct the operator of a motor vehicle, train, aircraft, or vessel to contact the carrier (e.g., by telephone or mobile radio) in the event of an incident involving the hazardous material.

(b) For transportation by highway, if a transport vehicle, (e.g., a semi-trailer or freight container-on-chassis) contains hazardous material for which a shipping paper is required and the vehicle is separated from its motive power and parked at a location other than a facility operated by the consignor or consignee or a facility (e.g., a carrier's terminal or a marine terminal) subject to the provisions of §172.602(c)(2), the carrier shall—

(1) Mark the transport vehicle with the telephone number of the motor carrier on the front exterior near the brake hose and electrical connections or on a label, tag, or sign attached to the vehicle at the brake hose or electrical connection; or

(2) Have the shipping paper and emergency response information readily available on the transport vehicle.

(c) The requirements specified in paragraph (b) of this section do not apply to an unattended motor vehicle separated from its motive power when the motor vehicle is marked on an orange panel, a placard, or a plain white square-on-point configuration with the identification number of each hazardous material loaded therein, and the marking or placard is visible on the outside of the motor vehicle.

Subpart H—Training

§172.700 Purpose and scope.

(a) *Purpose*. This subpart prescribes requirements for training hazmat employees.

(b) *Scope*. Training as used in this subpart means a systematic program that ensures a hazmat employee has familiarity with the general provisions of this subchapter, is able to recognize and identify hazardous materials, has knowledge of specific requirements of this subchapter applicable to functions performed by the employee, and has knowledge of emergency response information, self-protection measures and accident prevention methods and procedures (see §172.704).

(c) *Modal-specific training requirements*. Additional training requirements for the individual modes of transportation are prescribed in parts 174, 175, 176, and 177 of this subchapter.

§172.701 Federal-State relationship.

This subpart and the parts referenced in §172.700(c) prescribe minimum training requirements for the transportation of hazardous materials. For motor vehicle drivers, however, a State may impose more stringent training requirements only if those requirements—

(a) Do not conflict with the training requirements in this subpart and in part 177 of this subchapter; and

(b) Apply only to drivers domiciled in that State.

§172.702 Applicability and responsibility for training and testing.

(a) A hazmat employer shall ensure that each of its hazmat employees is trained in accordance with the requirements prescribed in this subpart.

(b) Except as provided in §172.704(c)(1), a hazmat employee who performs any function subject to the requirements of this subchapter may not perform that function unless instructed in the requirements of this subchapter that apply to that function. It is the duty of each hazmat employer to comply with the applicable requirements of this subchapter and to thoroughly instruct each hazmat employee in relation thereto.

(c) Training may be provided by the hazmat employer or other public or private sources.

(d) A hazmat employer shall ensure that each of its hazmat employees is tested by appropriate means on the training subjects covered in §172.704.

§172.704 Training requirements.

(a) Hazmat employee training must include the following:

(1) *General awareness/familiarization training*. Each hazmat employee shall be provided general awareness/familiarization training designed to provide familiarity with the requirements of this subchapter, and to enable the employee to recognize and identify hazardous materials consistent with the hazard communication standards of this subchapter.

(2) *Function-specific training*. (i) Each hazmat employee must be provided function-specific training concerning requirements of this subchapter, or exemptions or special permits issued under subchapter A of this chapter, that are specifically applicable to the functions the employee performs.

(ii) As an alternative to function-specific training on the requirements of this subchapter, training relating to the requirements of the ICAO Technical Instructions and the IMDG Code may be provided to the extent such training addresses functions authorized by subpart C of part 171 of this subchapter.

(3) *Safety training*. Each hazmat employee shall receive safety training concerning—

(i) Emergency response information required by subpart G of part 172;

(ii) Measures to protect the employee from the hazards associated with hazardous materials to which they may be exposed in the work place, including specific measures the hazmat employer has implemented to protect employees from exposure; and

(iii) Methods and procedures for avoiding accidents, such as the proper procedures for handling packages containing hazardous materials.

(4) *Security awareness training*. Each hazmat employee must receive training that provides an awareness of security risks associated with hazardous materials transportation and methods designed to enhance transportation security. This training must also include a component covering how to recognize and respond to possible security threats. New hazmat employees must receive the security awareness training required by this paragraph within 90 days after employment.

(5) *In-depth security training*. Each hazmat employee of a person required to have a security plan in accordance with subpart I of this part who handles hazardous materials covered by the plan, performs a regulated function related to the hazardous materials covered by the plan, or is responsible for implementing the plan must be trained concerning the security plan and its implementation. Security training must include company security objectives, organizational security structure, specific security procedures, specific security duties and responsibilities for each employee, and specific actions to be taken by each employee in the event of a security breach.

(b) *OSHA, EPA, and other training*. Training conducted by employers to comply with the hazard communication programs required by the Occupational Safety and Health Administration of the Department of Labor (29 CFR 1910.120 or 1910.1200) or the Environmental Protection Agency (40 CFR 311.1), or training conducted by employers to comply with security training programs required by other Federal or international agencies, may be used to satisfy the training requirements in paragraph (a) of this section to the extent that such training addresses the training components specified in paragraph (a) of this section.

(c) *Initial and recurrent training*—(1) *Initial training*. A new hazmat employee, or a hazmat employee who changes job functions may perform those functions prior to the completion of training provided—

(i) The employee performs those functions under the direct supervision of a properly trained and knowledgeable hazmat employee; and

(ii) The training is completed within 90 days after employment or a change in job function.

(2) *Recurrent training.* A hazmat employee must receive the training required by this subpart at least once every three years. For in-depth security training required under paragraph (a)(5) of this section, a hazmat employee must be trained at least once every three years or, if the security plan for which training is required is revised during the three-year recurrent training cycle, within 90 days of implementation of the revised plan.

(3) *Relevant Training.* Relevant training received from a previous employer or other source may be used to satisfy the requirements of this subpart provided a current record of training is obtained from hazmat employees' previous employer.

(4) *Compliance.* Each hazmat employer is responsible for compliance with the requirements of this subchapter regardless of whether the training required by this subpart has been completed.

(d) *Recordkeeping.* A record of current training, inclusive of the preceding three years, in accordance with this section shall be created and retained by each hazmat employer for as long as that employee is employed by that employer as a hazmat employee and for 90 days thereafter. The record shall include:

(1) The hazmat employee's name;

(2) The most recent training completion date of the hazmat employee's training;

(3) A description, copy, or the location of the training materials used to meet the requirements in paragraph (a) of this section;

(4) The name and address of the person providing the training; and

(5) Certification that the hazmat employee has been trained and tested, as required by this subpart.

(e) *Limitations.* The following limitations apply:

(1) A hazmat employee who repairs, modifies, reconditions, or tests packagings, as qualified for use in the transportation of hazardous materials, and who does not perform any other function subject to the requirements of this subchapter, is not subject to the training requirement of paragraph (a)(3) of this section.

(2) A railroad maintenance-of-way employee or railroad signalman, who does not perform any function subject to the requirements of this subchapter, is not subject to the training requirements of paragraphs (a)(2), (a)(4), or (a)(5) of this section. Initial training for a railroad maintenance-of-way employee or railroad signalman in accordance with this section must be completed by October 1, 2006.

Subpart I—Safety and Security Plans

§172.800 Purpose and applicability.

(a) *Purpose*. This subpart prescribes requirements for development and implementation of plans to address security risks related to the transportation of hazardous materials in commerce.

(b) *Applicability*. Each person who offers for transportation in commerce or transports in commerce one or more of the following hazardous materials must develop and adhere to a transportation security plan for hazardous materials that conforms to the requirements of this subpart. As used in this section, "large bulk quantity" refers to a quantity greater than 3,000 kg (6,614 pounds) for solids or 3,000 liters (792 gallons) for liquids and gases in a single packaging such as a cargo tank motor vehicle, portable tank, tank car, or other bulk container.

(1) Any quantity of a Division 1.1, 1.2, or 1.3 material;

(2) A quantity of a Division 1.4, 1.5, or 1.6 material requiring placarding in accordance with subpart F of this part;

(3) A large bulk quantity of Division 2.1 material;

(4) A large bulk quantity of Division 2.2 material with a subsidiary hazard of 5.1;

(5) Any quantity of a material poisonous by inhalation, as defined in §171.8 of this subchapter;

(6) A large bulk quantity of a Class 3 material meeting the criteria for Packing Group I or II;

(7) A quantity of desensitized explosives meeting the definition of Division 4.1 or Class 3 material requiring placarding in accordance with subpart F of this part;

(8) A large bulk quantity of a Division 4.2 material meeting the criteria for Packing Group I or II;

(9) A quantity of a Division 4.3 material requiring placarding in accordance with subpart F of this part;

(10) A large bulk quantity of a Division 5.1 material in Packing Groups I and II; perchlorates; or ammonium nitrate, ammonium nitrate fertilizers, or ammonium nitrate emulsions, suspensions, or gels;

(11) Any quantity of organic peroxide, Type B, liquid or solid, temperature controlled;

(12) A large bulk quantity of Division 6.1 material (for a material poisonous by inhalation see paragraph (5) above);

(13) A select agent or toxin regulated by the Centers for Disease Control and Prevention under 42 CFR part 73 or the United States Department of Agriculture under 9 CFR part 121;

(14) A quantity of uranium hexafluoride requiring placarding under §172.505(b);

(15) International Atomic Energy Agency (IAEA) Code of Conduct Category 1 and 2 materials including Highway Route Controlled quantities as defined in 49 CFR 173.403 or known radionuclides in forms listed as RAM-QC by the Nuclear Regulatory Commission;

(16) A large bulk quantity of Class 8 material meeting the criteria for Packing Group I.

(c) *Exceptions*. Transportation activities of a farmer, who generates less than $500,000 annually in gross receipts from the sale of agricultural commodities or products, are not subject to this subpart if such activities are:

(1) Conducted by highway or rail;

(2) In direct support of their farming operations; and

(3) Conducted within a 150-mile radius of those operations.

§172.802 Components of a security plan.

(a) The security plan must include an assessment of transportation security risks for shipments of the hazardous materials listed in §172.800, including site-specific or location-specific risks associated with facilities at which the hazardous materials listed in §172.800 are prepared for transportation, stored, or unloaded incidental to movement, and appropriate measures to address the assessed risks. Specific measures put into place by the plan may vary commensurate with the level of threat at a particular time. At a minimum, a security plan must include the following elements:

(1) *Personnel security*. Measures to confirm information provided by job applicants hired for positions that involve access to and handling of the hazardous materials covered by the security plan. Such confirmation system must be consistent with applicable Federal and State laws and requirements concerning employment practices and individual privacy.

(2) *Unauthorized access*. Measures to address the assessed risk that unauthorized persons may gain access to the hazardous materials covered by the security plan or transport conveyances being prepared for transportation of the hazardous materials covered by the security plan.

(3) *En route security*. Measures to address the assessed security risks of shipments of hazardous materials covered by the security plan en route from origin to destination, including shipments stored incidental to movement.

(b) The security plan must also include the following:

(1) Identification by job title of the senior management official responsible for overall development and implementation of the security plan;

(2) Security duties for each position or department that is responsible for implementing the plan or a portion of the plan and the process of notifying employees when specific elements of the security plan must be implemented; and

(3) A plan for training hazmat employees in accordance with §172.704 (a)(4) and (a)(5) of this part.

(c) The security plan, including the transportation security risk assessment developed in accordance with paragraph (a) of this section, must be in writing and must be retained for as long as it remains in effect. The security plan must be reviewed at least annually and revised and/or updated as necessary to reflect changing circumstances. The most recent version of the security plan, or portions thereof, must be available to the employees who are responsible for implementing it, consistent with personnel security clearance or background investigation restrictions and a demonstrated need to know. When the security plan is updated or revised, all employees responsible for implementing it must be notified and all copies of the plan must be maintained as of the date of the most recent revision.

(d) Each person required to develop and implement a security plan in accordance with this subpart must maintain a copy of the security plan (or an electronic file thereof) that is accessible at, or through, its principal place of business and must make the security plan available upon request, at a reasonable time and location, to an authorized official of the Department of Transportation or the Department of Homeland Security.

§172.804 Relationship to other Federal requirements.

To avoid unnecessary duplication of security requirements, security plans that conform to regulations, standards, protocols, or guidelines issued by other Federal agencies, international organizations, or industry organizations may be used to satisfy the requirements in this subpart, provided such security plans address the requirements specified in this subpart.

§172.820 Additional planning requirements for transportation by rail.

(a) *General.* Each rail carrier transporting in commerce one or more of the following materials is subject to the additional safety and security planning requirements of this section:

(1) More than 2,268 kg (5,000 lbs) in a single carload of a Division 1.1, 1.2 or 1.3 explosive;

(2) A quantity of a material poisonous by inhalation in a single bulk packaging; or

(3) A highway route-controlled quantity of a Class 7 (radioactive) material, as defined in §173.403 of this subchapter.

(b) Not later than 90 days after the end of each calendar year, a rail carrier must compile commodity data for the previous calendar year for the materials listed in paragraph (a) of this section. The following stipulations apply to data collected:

(1) Commodity data must be collected by route, a line segment or series of line segments as aggregated by the rail carrier. Within the rail carrier selected route, the commodity data must identify the geographic location of the route and the total number of shipments by UN identification number for the materials specified in paragraph (a) of this section.

(2) A carrier may compile commodity data, by UN number, for all Class 7 materials transported (instead of only highway route controlled quantities of Class 7 materials) and for all Division 6.1 materials transported (instead of only Division 6.1 poison inhalation hazard materials).

(c) *Rail transportation route analysis.* For each calendar year, a rail carrier must analyze the safety and security risks for the transportation route(s), identified in the commodity data collected as required by paragraph (b) of this section. The route analysis must be in writing and include the factors contained in Appendix D to this part, as applicable.

(1) The safety and security risks present must be analyzed for the route and railroad facilities along the route. For purposes of this section, railroad facilities are railroad property including, but not limited to, classification and switching yards, storage facilities, and non-private sidings. This term does not include an offeror's facility, private track, private siding, or consignee's facility.

(2) In performing the analysis required by this paragraph, the rail carrier must seek relevant information from state, local, and tribal officials, as appropriate, regarding security risks to high-consequence targets along or in proximity to the route(s) utilized. If a rail carrier is unable to acquire relevant information from state, local, or tribal officials, then it must document that in its analysis. For purposes of this section, a high-consequence target means a property, natural resource, location, area, or other target designated by the Secretary of Homeland Security that is a viable terrorist target of national significance, the attack of which by railroad could result in catastrophic loss of life, significant damage to national security or defense capabilities, or national economic harm.

(d) *Alternative route analysis.* (1) For each calendar year, a rail carrier must identify practicable alternative routes over which it has authority to operate, if an alternative exists, as an alternative route for each of the transportation routes analyzed in accordance with paragraph (c) of this section. The carrier must perform a safety and security risk assessment of the alternative routes for comparison to the route analysis prescribed in paragraph (c) of this section. The alternative route analysis must be in writing and include the criteria in Appendix D of this part. When determining practicable alternative routes, the rail carrier must consider the use of interchange agreements with other rail carriers. The written alternative route analysis must also consider:

(i) Safety and security risks presented by use of the alternative route(s);

(ii) Comparison of the safety and security risks of the alternative(s) to the primary rail transportation route, including the risk of a catastrophic release from a shipment traveling along each route;

(iii) Any remediation or mitigation measures implemented on the primary or alternative route(s); and

(iv) Potential economic effects of using the alternative route(s), including but not limited to the economics of the commodity, route, and customer relationship.

(2) In performing the analysis required by this paragraph, the rail carrier should seek relevant information from state, local, and tribal officials, as appropriate, regarding security risks to high-consequence targets along or in proximity to the alternative routes. If a rail carrier determines that it is not appropriate to seek such relevant information, then it must explain its reasoning for that determination in its analysis.

(e) *Route Selection.* A carrier must use the analysis performed as required by paragraphs (c) and (d) of this section to select the route to be used in moving the materials covered by paragraph (a) of this section. The carrier must consider any remediation measures implemented on a route. Using this process, the carrier must at least annually review and select the practicable route posing the least overall safety and security risk. The rail carrier must retain in writing all route review and selection decision documentation and restrict the distribution, disclosure, and availability of information contained in the route analysis to covered persons with a need-to-know, as described in parts 15 and 1520 of this title. This documentation should include, but is not limited to, comparative analyses, charts, graphics or rail system maps.

(f) *Completion of route analyses.* (1) The rail transportation route analysis, alternative route analysis, and route selection process required under paragraphs (c), (d), and (e) of this section must be completed no later than the end of the calendar year following the year to which the analyses apply.

(2) The initial analysis and route selection determinations required under paragraphs (c), (d), and (e) of this section must include a comprehensive review of the entire system. Subsequent analyses and route selection determinations required under paragraphs (c), (d), and (e) of this section must include a comprehensive, system-wide review of all operational changes, infrastructure modifications, traffic adjustments, changes in the nature of high-consequence targets located along, or in proximity to, the route, and any other changes affecting the safety or security of the movements of the materials specified in paragraph (a) of this section that were implemented during the calendar year.

(3) A rail carrier need not perform a rail transportation route analysis, alternative route analysis, or route selection process for any hazardous material other than the materials specified in paragraph (a) of this section.

(g) *Rail carrier point of contact on routing issues.* Each rail carrier must identify a point of contact (including the name, title, phone number and e-mail address) on routing issues involving the movement of materials covered by this section in its security plan and provide this information to:

(1) State and/or regional Fusion Centers that have been established to coordinate with state, local and tribal officials on security issues and which are located within the area encompassed by the rail carrier's rail system; and

(2) State, local, and tribal officials in jurisdictions that may be affected by a rail carrier's routing decisions and who directly contact the railroad to discuss routing decisions.

(h) *Storage, delays in transit, and notification.* With respect to the materials specified in paragraph (a) of this section, each rail carrier must ensure the safety and security plan it develops and implements under this subpart includes all of the following:

(1) A procedure under which the rail carrier must consult with offerors and consignees in order to develop measures for minimizing, to the extent practicable, the duration of any storage of the material incidental to movement (see §171.8 of this subchapter).

(2) Measures to prevent unauthorized access to the materials during storage or delays in transit.

(3) Measures to mitigate risk to population centers associated with in-transit storage.

(4) Measures to be taken in the event of an escalating threat level for materials stored in transit.

(5) Procedures for notifying the consignee in the event of a significant delay during transportation; such notification must be completed within 48 hours after the carrier has identified the delay and must include a revised delivery schedule. A significant delay is one that compromises the safety or security of the hazardous material or delays the shipment beyond its normal expected or planned shipping time. Notification should be made by a method acceptable to both the rail carrier and consignee.

(i) *Recordkeeping.* (1) Each rail carrier must maintain a copy of the information specified in paragraphs (b), (c), (d), (e), and (f) of this section (or an electronic image thereof) that is accessible at, or through, its principal place of business and must make the record available upon request, at a reasonable time and location, to an authorized official of the Department of Transportation or the Department of Homeland Security. Records must be retained for a minimum of two years.

(2) Each rail carrier must restrict the distribution, disclosure, and availability of information collected or developed in accordance with paragraphs (c), (d), (e), and (f) of this section to covered persons with a need-to-know, as described in parts 15 and 1520 of this title.

(j) *Compliance and enforcement.* If the carrier's route selection documentation and underlying analyses are found to be deficient, the carrier may be required to revise the analyses or make changes in route selection. If DOT finds that a chosen route is not the safest and most secure practicable route available, the FRA Associate Administrator for Safety, in consultation with TSA, may require the use of an alternative route. Prior to making such a determination, FRA and TSA will consult with the Surface Transportation Board (STB) regarding whether the contemplated alternative route(s) would be economically practicable.

§172.822 Limitation on actions by states, local governments, and Indian tribes.

A law, order, or other directive of a state, political subdivision of a state, or an Indian tribe that designates, limits, or prohibits the use of a rail line (other than a rail line owned by a state, political subdivision of a state, or an Indian tribe) for the transportation of hazardous materials, including, but not limited to, the materials specified in §172.820(a), is preempted. 49 U.S.C. 5125, 20106.

Reserved

PART 173

Subpart A—General

§173.2 Hazardous materials classes and index to hazard class definitions.

The hazard class of a hazardous material is indicated either by its class (or division) number, its class name, or by the letters "ORM–D". The following table lists class numbers, division numbers, class or division names and those sections of this subchapter which contain definitions for classifying hazardous materials, including forbidden materials.

Class No.	Division No. (if any)	Name of class or division	49 CFR reference for definitions
None	....	Forbidden materials	173.21
None	....	Forbidden explosives	173.54
1	1.1	Explosives (with a mass explosion hazard)	173.50
1	1.2	Explosives (with a projection hazard)	173.50
1	1.3	Explosives (with predominantly a fire hazard)	173.50
1	1.4	Explosives (with no significant blast hazard)	173.50
1	1.5	Very insensitive explosives; blasting agents	173.50
1	1.6	Extremely insensitive detonating substances	173.50
2	2.1	Flammable gas	173.115
2	2.2	Non-flammable compressed gas	173.115
2	2.3	Poisonous gas	173.115
3	....	Flammable and combustible liquid	173.120
4	4.1	Flammable solid	173.124
4	4.2	Spontaneously combustible material	173.124
4	4.3	Dangerous when wet material	173.124
5	5.1	Oxidizer	173.127
5	5.2	Organic peroxide	173.128
6	6.1	Poisonous materials	173.132
6	6.2	Infectious substance (Etiologic agent)	173.134
7	....	Radioactive material	173.403
8	....	Corrosive material	173.136
9	....	Miscellaneous hazardous material	173.140
None	....	Other regulated material: ORM-D	173.144

§173.2a Classification of a material having more than one hazard.

(a) *Classification of a material having more than one hazard.* Except as provided in paragraph (c) of this section, a material not specifically listed in the §172.101 Table that meets the definition of more than one hazard class or division as defined in this part, shall be classed according to the highest applicable hazard class of the following hazard classes, which are listed in descending order of hazard:

(1) Class 7 (radioactive materials, other than limited quantities).

(2) Division 2.3 (poisonous gases).

(3) Division 2.1 (flammable gases).

(4) Division 2.2 (nonflammable gases).

(5) Division 6.1 (poisonous liquids), Packing Group I, poisonous-by-inhalation only.

(6) A material that meets the definition of a pyrophoric material in §173.124(b)(1) of this subchapter (Division 4.2).

(7) A material that meets the definition of a self-reactive material in §173.124(a)(2) of this subchapter (Division 4.1).

(8) Class 3 (flammable liquids), Class 8 (corrosive materials), Division 4.1 (flammable solids), Division 4.2 (spontaneously combustible materials), Division 4.3 (dangerous when wet materials), Division 5.1 (oxidizers) or Division 6.1 (poisonous liquids or solids other than Packing Group I, poisonous-by-inhalation). The hazard class and packing group for a material meeting more than one of these hazards shall be determined using the precedence table in paragraph (b) of this section.

(9) Combustible liquids.

(10) Class 9 (miscellaneous hazardous materials).

(b) *Precedence of hazard table for Classes 3 and 8 and Divisions 4.1, 4.2, 4.3, 5.1 and 6.1.* The following table ranks those materials that meet the definition of Classes 3 and 8 and Divisions 4.1, 4.2, 4.3, 5.1 and 6.1:

PRECEDENCE OF HAZARD TABLE

[Hazard class or division and packing group]

	4.2	4.3	5.1 I[1]	5.1 II[1]	5.1 III[1]	6.1, I dermal	6.1, I oral	6.1, II	6.1 III	8, I liquid	8, I solid	8, II liquid	8, II solid	8, III liquid	8, III solid
3 I[2]	.....	4.3	.....	.....	.....	3	3	3	3	3	([3])	3	([3])	3	([3])
3 II[2]	.....	4.3	.....	.....	.....	3	3	3	3	8	([3])	3	([3])	3	([3])
3 III[2]	.....	4.3	.....	.....	.....	6.1	6.1	6.1	3[4]	8	([3])	8	([3])	3	([3])
4.1 II[2]	4.2	4.3	5.1	4.1	4.1	6.1	6.1	4.1	4.1	([3])	8	([3])	4.1	([3])	4.1
4.1 III[2]	4.2	4.3	5.1	4.1	4.1	6.1	6.1	6.1	4.1	([3])	8	([3])	8	([3])	4.1
4.2 II	.....	4.3	5.1	4.2	4.2	6.1	6.1	4.2	4.2	8	8	4.2	4.2	4.2	4.2
4.2 III	.....	4.3	5.1	5.1	4.2	6.1	6.1	6.1	4.2	8	8	8	8	4.2	4.2
4.3 I	.....	.....	5.1	4.3	4.3	6.1	4.3	4.3	4.3	4.3	4.3	4.3	4.3	4.3	4.3
4.3 II	.....	.....	5.1	4.3	4.3	6.1	4.3	4.3	4.3	8	8	4.3	4.3	4.3	4.3
4.3 III	.....	.....	5.1	5.1	4.3	6.1	6.1	6.1	4.3	8	8	8	8	4.3	4.3
5.1 I[1]	.....	.....	.....	.....	.....	5.1	5.1	5.1	5.1	5.1	5.1	5.1	5.1	5.1	5.1
5.1 II[1]	.....	.....	.....	.....	.....	6.1	5.1	5.1	5.1	8	8	5.1	5.1	5.1	5.1
5.1 III[1]	.....	.....	.....	.....	.....	6.1	6.1	6.1	5.1	8	8	8	8	5.1	5.1
6.1 I, Dermal	.....	.....	.....	.....	.....	.....	.....	.....	.....	8	6.1	6.1	6.1	6.1	6.1
6.1 I, Oral	.....	.....	.....	.....	.....	.....	.....	.....	.....	8	6.1	6.1	6.1	6.1	6.1
6.1 II, Inhalation	.....	.....	.....	.....	.....	.....	.....	.....	.....	8	6.1	6.1	6.1	6.1	6.1
6.1 II, Dermal	.....	.....	.....	.....	.....	.....	.....	.....	.....	8	6.1	8	6.1	6.1	6.1
6.1 II, Oral	.....	.....	.....	.....	.....	.....	.....	.....	.....	8	8	8	6.1	6.1	6.1
6.1 III	.....	.....	.....	.....	.....	.....	.....	.....	.....	8	8	8	8	8	8

[1]See §173.127.

[2]Materials of Division 4.1 other than self-reactive substances and solid desensitized explosives, and materials of Class 3 other than liquid desensitized explosives.

[3]Denotes an impossible combination.

[4]For pesticides only, where a material has the hazards of Class 3, Packing Group III, and Division 6.1, Packing Group III, the primary hazard is Division 6.1, Packing Group III.

Note 1: The most stringent packing group assigned to a hazard of the material takes precedence over other packing groups; for example, a material meeting Class 3 PG II and Division 6.1 PG I (oral toxicity) is classified as Class 3 PG I.

Note 2: A material which meets the definition of Class 8 and has an inhalation toxicity by dusts and mists which meets criteria for Packing Group I specified in §173.133(a)(1) must be classed as Division 6.1 if the oral or dermal toxicity meets criteria for Packing Group I or II. If the oral or dermal toxicity meets criteria for Packing Group III or less, the material must be classed as Class 8.

(c) The following materials are not subject to the provisions of paragraph (a) of this section because of their unique properties:

(1) A Class 1 (explosive) material that meets any other hazard class or division as defined in this part shall be assigned a division in Class 1. Class 1 materials shall be classed and approved in accordance with §173.56 of this part;

(2) A Division 5.2 (organic peroxide) material that meets the definition of any other hazard class or division as defined in this part, shall be classed as Division 5.2;

(3) A Division 6.2 (infectious substance) material that also meets the definition of another hazard class or division, other than Class 7, or that also is a limited quantity Class 7 material, shall be classed as Division 6.2;

(4) A material that meets the definition of a wetted explosive in §173.124(a)(1) of this subchapter (Division 4.1). Wetted explosives are either specifically listed in the §172.101 Table or are approved by the Associate Administrator (see §173.124(a)(1) of this subchapter); and

(5) A limited quantity of a Class 7 (radioactive) material that meets the definition for more than one hazard class or division shall be classed in accordance with §173.423.

§173.4 Small quantities for highway and rail.

(a) When transported domestically by highway or rail in conformance with this section, quantities of Division 2.2 (except aerosols with no subsidiary hazard), Class 3, Division 4.1, Division 4.2 (PG II and III), Division 4.3 (PG II and III), Division 5.1, Division 5.2, Division 6.1, Class 7, Class 8, and Class 9 materials are not subject to any other requirements when—

(1) The maximum quantity of material per inner receptacle or article is limited to—

(i) Thirty (30) mL (1 ounce) for authorized liquids, other than Division 6.1, Packing Group I, Hazard Zone A or B materials;

(ii) Thirty (30) g (1 ounce) for authorized solid materials;

(iii) One (1) g (0.04 ounce) for authorized materials meeting the definition of a Division 6.1, Packing Group I, Hazard Zone A or B material; and

(iv) An activity level not exceeding that specified in §§173.421, 173.424, 173.425 or 173.426, as appropriate, for a package containing a Class 7 (radioactive) material.

(v) Thirty (30) mL water capacity (1.8 cubic inches) for authorized Division 2.2 materials.

(2) With the exception of temperature sensing devices, each inner receptacle:

(i) Is not liquid-full at 55°C (131°F), and

(ii) Is constructed of plastic having a minimum thickness of no less than 0.2 mm (0.008 inch), or earthenware, glass, or metal;

(3) Each inner receptacle with a removable closure has its closure held securely in place with wire, tape, or other positive means;

(4) Unless equivalent cushioning and absorbent material surrounds the inside packaging, each inner receptacle is securely packed in an inside packaging with cushioning and absorbent material that:

(i) Will not react chemically with the material, and

(ii) Is capable of absorbing the entire contents (if a liquid) of the receptacle;

(5) The inside packaging is securely packed in a strong outer packaging;

(6) The completed package, as demonstrated by prototype testing, is capable of sustaining—

(i) Each of the following free drops made from a height of 1.8 m (5.9 feet) directly onto a solid unyielding surface without breakage or leakage from any inner receptacle and without a substantial reduction in the effectiveness of the package:

(A) One drop flat on bottom;

(B) One drop flat on top;

(C) One drop flat on the long side;

(D) One drop flat on the short side; and

(E) One drop on a corner at the junction of three intersecting edges; and

(ii) A compressive load as specified in §178.606(c) of this subchapter.

Note to Paragraph (a)(6): Each of the tests in paragraph (a)(6) of this section may be performed on a different but identical package; *i.e.*, all tests need not be performed on the same package.

(7) Placement of the material in the package or packing different materials in the package does not result in a violation of §173.21;

(8) The gross mass of the completed package does not exceed 29 kg (64 pounds);

(9) The package is not opened or otherwise altered until it is no longer in commerce; and

(10)The shipper certifies conformance with this section by marking the outside of the package with the statement "This package conforms to 49 CFR 173.4 for domestic highway or rail transport only."

(b) A package containing a Class 7 (radioactive) material also must conform to the requirements of §173.421(a)(1) through (a)(5) or §173.424(a) through (g), as appropriate.

(c) Packages which contain a Class 2 (other than those authorized in paragraph (a) of this section), Division 4.2 (PG I), or Division 4.3 (PG I) material conforming to paragraphs (a)(1) through (10) of this section may be offered for transportation or transported if approved by the Associate Administrator.

(d) Lithium batteries and cells are not eligible for the exceptions provided in this section.

§173.4a Excepted quantities.

(a) Excepted quantities of materials other than articles transported in accordance with this section are not subject to any additional requirements of this subchapter except for:

(1) The shipper's responsibilities to properly class their material in accordance with §173.22 of this subchapter;

(2) Sections 171.15 and 171.16 of this subchapter pertaining to the reporting of incidents; and

(3) For a Class 7 (Radioactive) material the requirements for an excepted package.

(4) Packagings for which retention of liquid is a basic function must be capable of withstanding without leakage the pressure differential specified in §173.27(c) of this part.

(b) *Authorized materials.* Only materials authorized for transport aboard passenger aircraft and appropriately classed within one of the following hazard classes or divisions may be transported in accordance with this section:

(1) Division 2.2 material with no subsidiary hazard. An aerosol is not included as authorized Division 2.2 material;

(2) Class 3 materials;

(3) Class 4 (PG II and III) materials except for self-reactive materials;

(4) Division 5.1 (PG II and III);

(5) Division 5.2 materials only when contained in a chemical kit, first aid kit or a polyester resin kit;

(6) Division 6.1, other than PG I, Hazard Zone A or B material;

(7) Class 7, Radioactive material in excepted packages;

(8) Class 8 (PG II and III), except for UN2803 (Gallium) and UN2809 (Mercury); and

(9) Class 9, except for UN1845 (Carbon dioxide, solid or Dry ice), and lithium batteries and cells.

(c) *Inner packaging limits*. The maximum quantity of hazardous materials in each inner packaging is limited to:

(1) For toxic material with a Division 6.1 primary or subsidiary hazard, PG I or II—

(i) 1 g (0.04 ounce) for solids; or

(ii) 1 mL (0.03 ounce) for liquids;

(2) 30 g (1 ounce) or 30 mL (1 ounce) for solids or liquids other than those covered in paragraph (c)(1) of this section; and

(3) For gases a water capacity of 30 mL (1.8 cubic inches) or less.

(d) *Outer packaging aggregate quantity limits*. The maximum aggregate quantity of hazardous material contained in each outer packaging must not exceed the limits provided in the following paragraphs. For outer packagings containing more than one hazardous material, the aggregate quantity of hazardous material must not exceed the lowest permitted maximum aggregate quantity. The limits are as follows:

(1) For other than a Division 2.2 or Division 5.2 material:

(i) Packing Group I—300 g (0.66 pounds) for solids or 300 mL (0.08 gallons) for liquids;

(ii) Packing Group II—500 g (1.1 pounds) for solids or 500 mL (0.1 gallons) for liquids;

(iii) Packing Group III—1 kg (2.2 pounds) for solids or 1 L (0.2 gallons) for liquids;

(2) For Division 2.2 material, 1 L (61 cubic inches); or

(3) For Division 5.2 material, 500 g (1.1 pounds) for solids or 500 mL (0.1 gallons) for liquids.

(e) *Packaging materials*. Packagings used for the transport of excepted quantities must meet the following:

(1) Each inner receptacle must be constructed of plastic, or of glass, porcelain, stoneware, earthenware or metal. When used for liquid hazardous materials, plastic inner packagings must have a thickness of not less than 0.2 mm (0.008 inch).

(2) Each inner packaging with a removable closure must have its closure held securely in place with wire, tape or other positive means. Each inner receptacle having a neck with molded screw threads must have a leak proof, threaded type cap. The closure must not react chemically with the material.

(3) Each inner packaging must be securely packed in an intermediate packaging with cushioning material in such a way that, under normal conditions of transport, it cannot break, be punctured or leak its contents. The intermediate packaging must completely contain the contents in case of breakage or leakage, regardless of package orientation. For liquid hazardous materials, the intermediate packaging must contain sufficient absorbent material that:

(i) Will absorb the entire contents of the inner packaging.

(ii) Will not react dangerously with the material or reduce the integrity or function of the packaging materials.

(iii) The absorbent material may be the cushioning material.

(4) The intermediate packaging must be securely packed in a strong, rigid outer packaging.

(5) Placement of the material in the package or packing different materials in the package must not result in a violation of §173.21.

(6) Each package must be of such a size that there is adequate space to apply all necessary markings.

(7) The package is not opened or otherwise altered until it is no longer in commerce.

(8) Overpacks may be used and may also contain packages of hazardous material or other materials not subject to the HMR subject to the requirements of §173.25.

(f) *Package tests*. The completed package as prepared for transport, with inner packagings filled to not less than 95% of their capacity for solids or 98% for liquids, must be capable of withstanding, as demonstrated by testing which is appropriately documented, without breakage or leakage of any inner packaging and without significant reduction in effectiveness:

(1) Drops onto a solid unyielding surface from a height of 1.8 m (5.9 feet):

(i) Where the sample is in the shape of a box, it must be dropped in each of the following orientations:

(A) One drop flat on the bottom;

(B) One drop flat on the top;

(C) One drop flat on the longest side;

(D) One drop flat on the shortest side; and

(E) One drop on a corner at the junction of three intersecting edges.

(ii) Where the sample is in the shape of a drum, it must be dropped in each of the following orientations:

(A) One drop diagonally on the top chime, with the center of gravity directly above the point of impact;

(B) One drop diagonally on the base chime; and

(C) One drop flat on the side.

(2) A compressive load as specified in §178.606(c) of this subchapter. Each of the tests in this paragraph (f) of this section may be performed on a different but identical package; that is, all tests need not be performed on the same package.

(g) *Marking*. Excepted quantities of hazardous materials packaged, marked, and otherwise offered and transported in accordance with this section must be durably and legibly marked with the following marking:

(1) The "*" must be replaced by the primary hazard class, or when assigned, the division of each of the hazardous materials contained in the package. The "**" must be replaced by the name of the shipper or consignee if not shown elsewhere on the package.

(2) The marking must not be less than 100 mm (3.9 inches) by 100 mm (3.9 inches), and must be durable and clearly visible.

(3) When packages of excepted quantities are contained in an overpack, and the package marking required by this section is not visible inside the overpack, the excepted quantities marking must also be placed on the overpack. Additionally, an overpack containing packages of excepted quantities is not required to be marked with the word ″OVERPACK.″

(h) *Documentation.* (1) For transportation by highway or rail, no shipping paper is required.

(2) For transport by air, a shipping paper is not required, except that, if a document such as an air waybill accompanies a shipment, the document must include the statement "Dangerous Goods in Excepted Quantities" and indicate the number of packages.

(3) For transport by vessel, a shipping paper is required and must include the statement "Dangerous Goods in Excepted Quantities" and indicate the number of packages.

(i) *Training.* Each person who offers or transports excepted quantities of hazardous materials must know about the requirements of this section.

(j) *Restrictions.* Hazardous material packaged in accordance with this section may not be carried in checked or carry-on baggage.

§173.4b De minimis exceptions.

(a) Packing Group II and III materials in Class 3, Division 4.1, Division 4.2, Division 4.3, Division 5.1, Division 6.1, Class 8, and Class 9 do not meet the definition of a hazardous material in §171.8 of this subchapter when packaged in accordance with this section and, therefore, are not subject to the requirements of this subchapter.

(1) The maximum quantity of material per inner receptacle or article is limited to—

(i) One (1) mL (0.03 ounce) for authorized liquids; and

(ii) One (1) g (0.04 ounce) for authorized solid materials;

(2) Each inner receptacle with a removable closure has its closure held securely in place with wire, tape, or other positive means;

(3) Unless equivalent cushioning and absorbent material surrounds the inside packaging, each inner receptacle is securely packed in an inside packaging with cushioning and absorbent material that:

(i) Will not react chemically with the material, and

(ii) Is capable of absorbing the entire contents (if a liquid) of the receptacle;

(4) The inside packaging is securely packed in a strong outer packaging;

(5) The completed package is capable of sustaining—

(i) Each of the following free drops made from a height of 1.8 m (5.9 feet) directly onto a solid unyielding surface without breakage or leakage from any inner receptacle and without a substantial reduction in the effectiveness of the package:

(A) One drop flat on bottom;

(B) One drop flat on top;

(C) One drop flat on the long side;

(D) One drop flat on the short side; and

(E) One drop on a corner at the junction of three intersecting edges; and

(ii) A compressive load as specified in §178.606(c) of this subchapter. Each of the tests in this paragraph (a)(5) may be performed on a different but identical package; that is, all tests need not be performed on the same package.

(6) Placement of the material in the package or packing different materials in the package does not result in a violation of §173.21;

(7) The aggregate quantity of hazardous material per package does not exceed 100 g (0.22 pounds) for solids or 100 mL (3.38 ounces) for liquids;

(8) The gross mass of the completed package does not exceed 29 kg (64 pounds);

(9) The package is not opened or otherwise altered until it is no longer in commerce; and

(10) For transportation by aircraft:

(i) The hazardous material is authorized to be carried aboard passenger-carrying aircraft in Column 9A of the §172.101 Hazardous Materials Table; and

(ii) Material packed in accordance with this section may not be carried in checked or carry-on baggage.

(b) Non-infectious specimens, such as specimens of mammals, birds, amphibians, reptiles, fish, insects and other invertebrates containing small quantities of Ethanol (UN1170), Formaldehyde solution, flammable (UN1198), Alcohols, n.o.s. (UN1987) and Isopropanol (UN1219) are not subject to the requirements of this subchapter provided the following packaging, marking and documentation provisions, as applicable, are met:

(1) The specimens are:

(i) Wrapped in a paper towel or cheesecloth moistened with alcohol or an alcohol solution and placed in a plastic bag that is heat-sealed. Any free liquid in the bag must not exceed 30 mL; or

(ii) Placed in vials or other rigid containers with no more than 30 mL of alcohol or alcohol solution. The containers are placed in a plastic bag that is heat-sealed;

(2) The bagged specimens are placed in another plastic bag with sufficient absorbent material to absorb the entire liquid contents inside the primary receptacle. The outer plastic bag is then heat-sealed;

(3) The completed bag is placed in a strong outer packaging with sufficient cushioning material that conforms to subpart B of part 173;

(4) The aggregate net quantity of flammable liquid in one outer packaging may not exceed 1 L; and

(5) The outer package must be legibly marked "Scientific research specimens, 49 CFR 173.4b applies."

(6) *Documentation.* (i) For transportation by highway or rail, no shipping paper is required.

(ii) For transport by air, a shipping paper is not required, except that, if a document such as an air waybill accompanies a shipment of specimens containing hazardous materials excepted under the terms of this section, the document must include the statement "Scientific research specimens, 49 CFR 173.4b applies" and the number of packages indicated.

(iii) For transport by vessel, a shipping paper is not required; however, the Dangerous Cargo Manifest must include the statement "Scientific research specimens, 49

CFR 173.4b applies" and the number of packages indicated. Vessel stowage is the same as for hazardous materials in excepted quantities.

(7) *Training.* Each person who offers or transports excepted quantities of hazardous materials must know about the requirements of this section.

(8) *Restrictions.* For transportation by aircraft, hazardous material packaged in accordance with this section may not be carried in checked or carry-on baggage by a passenger or crew member.

§173.5 Agricultural operations.

(a) For other than a Class 2 material, the transportation of an agricultural product over local roads between fields of the same farm is excepted from the requirements of this subchapter. A Class 2 material transported over local roads between fields of the same farm is excepted from subparts G and H of part 172 of this sub-chapter. In either instance, transportation of the hazardous material is subject to the following conditions:

(1) It is transported by a farmer who is an intrastate private motor carrier; and

(2) The movement of the agricultural product conforms to requirements of the State in which it is transported and is specifically authorized by a State statute or regulation in effect before October 1, 1998.

(b) The transportation of an agricultural product to or from a farm, within 150 miles of the farm, is excepted from the requirements in subparts G and H of part 172 of this subchapter and from the specific packaging requirements of this subchapter when:

(1) It is transported by a farmer who is an intrastate private motor carrier;

(2)The total amount of agricultural product being transported on a single motor vehicle does not exceed:

(i) 7,300 kg (16,094 lbs.) of ammonium nitrate fertilizer properly classed as Division 5.1, PG III, in a bulk packaging, or

(ii) 1900 L (502 gallons) for liquids or gases, or 2,300 kg (5,070 lbs.) for solids, of any other agricultural product;

(3) The movement and packaging of the agricultural product conform to the requirements of the State in which it is transported and are specifically authorized by a State statute or regulation in effect before October 1, 1998; and

(4) Each person having any responsibility for transporting the agricultural product or preparing the agricultural product for shipment has been instructed in the applicable requirements of this subchapter.

(c) Formulated liquid agricultural products in specification packagings of 220 L (58 gallons) capacity, or less, with closures manifolded to a closed mixing system and equipped with positive dry disconnect devices may be transported by a private motor carrier between a final distribution point and an ultimate point of application or for loading aboard an airplane for aerial application.

(d) *Moveable fuel storage tenders.* A non-DOT specification cargo tank motor vehicle may be used to transport Liquefied petroleum gas, UN1075, including Propane, UN1978, as moveable fuel storage tender used exclusively for agricultural purposes when operated by a private carrier under the following conditions:

(1) The cargo tank must have a minimum design pressure of 250 psig.

(2) The cargo tank must meet the requirements of the HMR in effect at the time of its manufacture and must be marked accordingly. For questions regarding these requirements, contact PHMSA by either:

(i) Telephone (800) 467-4922 or (202) 366-4488 (local); or

(ii) By electronic mail (e-mail) to: *infocntr@dot.gov.*

(3) The cargo tank must have a water capacity of 1,200 gallons or less.

(4) The cargo tank must conform to applicable requirements in National Fire Protection Association (NFPA) 58, Liquefied Petroleum Gas Code (IBR, see §171.7 of this subchapter).

(5) The cargo tank must be securely mounted on a motor vehicle.

(6) The cargo tank must be filled in accordance with §173.315(b) for liquefied petroleum gas.

(7) The cargo tank must be painted white, aluminum, or other light-reflecting color.

(8) Transportation of the filled moveable fuel storage tender is limited to movements over local roads between fields using the shortest practical distance.

(9) Transportation of the moveable fuel storage tender between its point of use and a liquefied petroleum gas distribution facility is authorized only if the cargo tank contains no more than five percent of its water capacity. A movable fuel storage tender may only be filled at the consumer's premises or point of use.

(e) *Liquid soil pesticide fumigants.* MC 306 and DOT 406 cargo tank motor vehicles and DOT 57 portable tanks may be used to transport liquid soil pesticide fumigants, Pesticides, liquid, toxic, flammable, n.o.s., *flash point not less than 23 degrees C*, 6.1, UN2903, PG II, exclusively for agricultural operations by a private motor carrier between a bulk loading facility and a farm (including between farms). However, transportation is not to exceed 150 miles between the loading facility and the farm, and not more than five days are permitted for intermediate stops for temporary storage. Additionally, transport is permitted only under the following conditions:

(1) *Cargo tanks.* MC 306 and DOT 406 cargo tank motor vehicles must:

(i) Meet qualification and maintenance requirements (including periodic testing and inspection) in accordance with Subpart E of Part 180 of this subchapter;

(ii) Conform to the pressure relief system requirements specified in §173.243(b)(1);

(iii) For MC 306 cargo tanks, be equipped with stop-valves capable of being remotely closed by manual and mechanical means; and

(iv) For DOT 406 cargo tanks, conform to the bottom outlet requirements specified in §173.243(b)(2).

(2) *Portable tanks.* DOT 57 portable tanks must—

(i) Be constructed of stainless steel; and

(ii) Meet qualification and maintenance requirements of Subpart G of Part 180 of this subchapter.

(f) See §173.315(m) pertaining to nurse tanks of anhydrous ammonia.

(g) See §173.6 pertaining to materials of trade.

(h) See §172.800(b) pertaining to security plans.

§173.6 Materials of trade exceptions.

When transported by motor vehicle in conformance with this section, a material of trade (see §171.8 of this subchapter) is not subject to any other requirements of this subchapter besides those set forth or referenced in this section.

(a) *Materials and amounts.* A material of trade is limited to the following:

(1) A Class 3, 8, 9, Division 4.1, 5.1, 5.2, 6.1, or ORM–D material contained in a packaging having a gross mass or capacity not over—

(i) 0.5 kg (1 pound) or 0.5 L (1 pint) for a Packing Group I material;

(ii) 30 kg (66 pounds) or 30 L (8 gallons) for a Packing Group II, Packing Group III, or ORM–D material;

(iii) 1500 L (400 gallons) for a diluted mixture, not to exceed 2 percent concentration, of a Class 9 material.

(2) A Division 2.1 or 2.2 material in a cylinder with a gross weight not over 100 kg (220 pounds), or a permanently mounted tank manufactured to the ASME Code of not more than 70 gallon water capacity for a non-liquefied Division 2.2 material with no subsidiary hazard.

(3) A Division 4.3 material in Packing Group II or III contained in a packaging having a gross capacity not exceeding 30 mL (1 ounce).

(4) A Division 6.2 material, other than a Category A infectious substance, contained in human or animal samples (including, but not limited to, secreta, excreta, blood and its components, tissue and tissue fluids, and body parts) being transported for research, diagnosis, investigational activities, or disease treatment or prevention, or is a biological product or regulated medical waste. The material must be contained in a combination packaging. For liquids, the inner packaging must be leakproof, and the outer packaging must contain sufficient absorbent material to absorb the entire contents of the inner packaging. For sharps, the inner packaging (sharps container) must be constructed of a rigid material resistant to punctures and securely closed to prevent leaks or punctures, and the outer packaging must be securely closed to prevent leaks or punctures. For solids, liquids, and sharps, the outer packaging must be a strong, tight packaging securely closed and secured against shifting, including relative motion between packages, within the vehicle on which it is being transported.

(i) For other than a regulated medical waste, the amount of Division 6.2 material in a combination packaging must conform to the following limitations:

(A) One or more inner packagings, each of which may not contain more than 0.5 kg (1.1 lbs) or 0.5 L (17 ounces), and an outer packaging containing not more than 4 kg (8.8 lbs) or 4 L (1 gallon); or

(B) A single inner packaging containing not more than 16 kg (35.2 lbs) or 16 L (4.2 gallons) in a single outer packaging.

(ii) For a regulated medical waste, a combination packaging must consist of one or more inner packagings, each of which may not contain more than 4 kg (8.8 lbs) or 4 L (1 gallon), and an outer packaging containing not more than 16 kg (35.2 lbs) or 16 L (4.2 gallons).

(5) This section does not apply to a hazardous material that is self-reactive (see §173.124), poisonous by inhalation (see §173.133), or a hazardous waste.

(b) *Packaging.* (1) Packagings must be leak tight for liquids and gases, sift proof for solids, and be securely closed, secured against shifting, and protected against damage.

(2) Each material must be packaged in the manufacturer's original packaging, or a packaging of equal or greater strength and integrity.

(3) Outer packagings are not required for receptacles (e.g., cans and bottles) that are secured against shifting in cages, carts, bins, boxes or compartments.

(4) For gasoline, a packaging must be made of metal or plastic and conform to the requirements of this subchapter or to the requirements of the Occupational Safety and Health Administration of the Department of Labor contained in 29 CFR 1910.106(d)(2) or 1926.152(a)(1).

(5) A cylinder or other pressure vessel containing a Division 2.1 or 2.2 material must conform to packaging, qualification, maintenance, and use requirements of this subchapter, except that outer packagings are not required. Manifolding of cylinders is authorized provided all valves are tightly closed.

(c) *Hazard communication.* (1) A non-bulk packaging other than a cylinder (including a receptacle transported without an outer packaging) must be marked with a common name or proper shipping name to identify the material it contains, including the letters "RQ" if it contains a reportable quantity of a hazardous substance.

(2) A bulk packaging containing a diluted mixture of a Class 9 material must be marked on two opposing sides with the four-digit identification number of the material. The identification number must be displayed on placards, orange panels or, alternatively, a white square-on-point configuration having the same outside dimensions as a placard (at least 273 mm (10.8 inches) on a side), in the manner specified in §172.332(b) and(c) of this subchapter.

(3) A DOT specification cylinder (except DOT specification 39) must be marked and labeled as prescribed in this subchapter. Each DOT–39 cylinder must display the markings specified in §178.65(i).

(4) The operator of a motor vehicle that contains a material of trade must be informed of the presence of the hazardous material (including whether the package contains a reportable quantity) and must be informed of the requirements of this section.

(d) *Aggregate gross weight.* Except for a material of trade authorized by paragraph (a)(1)(iii) of this section, the aggregate gross weight of all materials of trade on a motor vehicle may not exceed 200 kg (440 pounds).

(e) *Other exceptions.* A material of trade may be transported on a motor vehicle under the provisions of this section with other hazardous materials without affecting its eligibility for exceptions provided by this section.

§173.8 Exceptions for non-specification packagings used in intrastate transportation.

(a) *Non-specification bulk packagings.* Notwithstanding requirements for specification packagings in subpart F of this part and parts 178 and 180 of this subchapter, a non-specification bulk packaging may be used for transportation of a hazardous material by an intrastate motor carrier until July 1, 2000, in accordance with the provisions of paragraph (d) of this section.

(b) *Non-specification cargo tanks for petroleum products.* Notwithstanding requirements for specification packagings in subpart F of this part and parts 178 and 180 of this subchapter, a non-specification cargo tank motor vehicle having a capacity of less than 13,250 L (3,500 gallons) may be used by an intrastate motor carrier for transportation of a flammable liquid petroleum product in accordance with the provisions of paragraph (d) of this section.

(c) *Permanently secured non-bulk tanks for petroleum products.* Notwithstanding requirements for specification packagings in subpart F of this part 173 and parts 178 and 180 of this subchapter, a non-specification metal tank permanently secured to a transport vehicle and protected against leakage or damage in the event of a turnover, having a capacity of less than 450 L (119 gallons), may be used by an intrastate motor carrier for transportation of a flammable liquid petroleum product in accordance with the provisions of paragraph (d) of this section.

(d) *Additional requirements.* A packaging used under the provisions of paragraphs (a), (b) or (c) of this section must—

(1) Be operated by an intrastate motor carrier and in use as a packaging for hazardous material before October 1, 1998;

(2) Be operated in conformance with the requirements of the State in which it is authorized;

(3) Be specifically authorized by a State statute or regulation in effect before October 1, 1998, for use as a packaging for the hazardous material being transported;

(4) Be offered for transportation and transported in conformance with all other applicable requirements of this subchapter;

(5) Not be used to transport a flammable cryogenic liquid, hazardous substance, hazardous waste, or marine pollutant (except for gasoline); and

(6) For a tank authorized under paragraph (b) or (c) of this section, conform to all requirements in part 180 (except for §180.405(g)) of this subchapter in the same manner as required for a DOT specification MC 306 cargo tank motor vehicle.

§173.12 Exceptions for shipment of waste materials.

(a) *Open head drums.* If a hazardous material that is a hazardous waste is required by this subchapter to be shipped in a closed head drum (i.e., a drum with a 7.0 cm (3 inches) or less bung opening) and the hazardous waste contains solids or semisolids that make its placement in a closed head drum impracticable, an equivalent (except for closure) open head drum may be used for the hazardous waste.

(b) *Lab packs.* (1) Waste materials prohibited by paragraph (b)(3) of this section are not authorized for transport in packages authorized by this paragraph (b). Waste materials classed as Class or Division 3, 4.1, 4.2, 4.3, 5.1, 5.2, 6.1, 8, or 9 are excepted from the specification packaging requirements of this subchapter for combination packagings if packaged in accordance with this paragraph (b) and transported for disposal or recovery by highway, rail or cargo vessel. In addition, a generic description from the §172.101 Hazardous Materials Table may be used in place of specific chemical names, when two or more chemically compatible waste materials in the same hazard class are packaged in the same outside packaging.

(2) Combination packaging requirements:

(i) *Inner packagings.* The inner packagings must be either glass, not exceeding 4 L (1 gallon) rated capacity, or metal or plastic, not exceeding 20 L (5.3 gallons) rated capacity. Inner packagings containing liquid must be surrounded by a chemically compatible absorbent material in sufficient quantity to absorb the total liquid contents.

(ii) *Outer packaging.* Each outer packaging may contain only one class of waste material. The following outer packagings are authorized except that Division 4.2 Packing Group I materials must be packaged using UN standard steel or plastic drums tested and marked to the Packing Group I performance level for liquids or solids; and bromine pentafluoride and bromine trifluoride may not be packaged using UN 4G fiberboard boxes:

(A) A UN 1A2 or UN 1B2 metal drum, a UN 1D plywood drum, a UN 1G fiber drum, or a UN 1H2 plastic drum, tested and marked to at least the Packing Group III performance level for liquids or solids;

(B) At a minimum, a double-walled UN 4G fiberboard box made out of 500 pound burst-strength fiberboard fitted with a polyethylene liner at least 3 mils (0.12 inches) thick and when filled during testing to 95 percent capacity with a solid material, successfully passes the tests prescribed in §§78.603 (drop) and 178.606 (stacking), and is capable of passing the tests prescribed in §178.608 (vibration) to at least the Packing Group II performance level for liquids or solids; or

(C) A UN 11G fiberboard intermediate bulk container (IBC) or a UN 11HH2 composite IBC, fitted with a polyethylene liner at least 6 mils (0.24 inches) thick, that successfully passes the tests prescribed in Subpart O of Part 178 and § 178.603 to at least the Packing Group II performance level for liquids or solids; a UN 11HH2 is composed of multiple layers of encapsulated corrugated fiberboard between inner and outer layers of woven coated polypropylene.

(iii) The gross weight of each completed combination package may not exceed 205 kg (452 lbs).

(3) *Prohibited materials.* The following waste materials may not be packaged or described under the provisions of this paragraph (b): a material poisonous-by-inhalation, a Division 6.1 Packing Group I material, chloric acid, and oleum (fuming sulfuric acid).

(c) *Reuse of packagings*. A previously used packaging may be reused for the shipment of waste material transported for disposal or recovery, not subject to the reconditioning and reuse provisions contained in §173.28 and part 178 of this subchapter, under the following conditions:

(1) Except as authorized by this paragraph, the waste must be packaged in accordance with this part and offered for transportation in accordance with the requirements of this subchapter.

(2) Transportation is performed by highway only.

(3) A package is not offered for transportation less than 24 hours after it is finally closed for transportation, and each package is inspected for leakage and is found to be free from leaks immediately prior to being offered for transportation.

(4) Each package is loaded by the shipper and unloaded by the consignee, unless the motor carrier is a private or contract carrier.

(5) The packaging may be used only once under this paragraph and may not be used again for shipment of hazardous materials except in accordance with §173.28.

(d) *Technical names for n.o.s. descriptions*. The requirements for the inclusion of technical names for n.o.s. descriptions on shipping papers and package markings, §§172.203 and 172.301 of this subchapter, respectively, do not apply to packagings prepared in accordance with paragraph (b) of this section, except that packages containing materials meeting the definition of a hazardous substance must be described as required in §172.203 of this subchapter and marked as required in §172.324 of this subchapter.

(e) *Segregation requirements*. Waste materials packaged according to paragraph (b) of this section and transported in conformance with this paragraph (e) are not subject to the segregation requirements in §§ 174.81(d), 176.83(b), and 177.848(d) if blocked and braced in such a manner that they are separated from incompatible materials by a minimum horizontal distance of 1.2 m (4 feet) and the packages are loaded at least 100 mm (4 inches) off the floor of the freight container, unit load device, transport vehicle, or rail car. The following conditions specific to incompatible materials also apply:

(1) *General restrictions*. The freight container, unit load device, transport vehicle, or rail car may not contain any Class 1 explosives, Class 7 radioactive material, or uncontainerized hazardous materials;

(2) *Waste cyanides and waste acids*. For waste cyanides stored, loaded, and transported with waste acids:

(i) The cyanide or a cyanide mixture may not exceed 2 kg (4.4 pounds) net weight per inner packaging and may not exceed 10 kg (22 pounds) net weight per outer packaging; a cyanide solution may not exceed 2 L (0.6 gallon) per inner packaging and may not exceed 10 L (3.0 gallons) per outer packaging; and

(ii) The acids must be packaged in lab packs in accordance paragraph (b) of this section or in single packagings authorized for the acid in Column (8B) of the § 172.101 Hazardous Materials Table of this subchapter not to exceed 208 L (55 gallons) capacity.

(3) *Waste Division 4.2 materials and waste Class 8 liquids*. For waste Division 4.2 materials stored, loaded, and transported with waste Class 8 liquids:

(i) The Division 4.2 material may not exceed 2 kg (4.4 pounds) net weight per inner packaging and may not exceed 10 kg (22 pounds) net weight per outer packaging; and

(ii) The Class 8 liquid must be packaged in lab packs in accordance with paragraph (b) of this section or in single packagings authorized for the material in Column (8B) of the §172.101 Hazardous Materials Table of this subchapter not to exceed 208 L (55 gallons) capacity.

(4) *Waste Division 6.1 Packing Group I, Hazard Zone A material and waste Class 3, Class 8 liquids, or Division 4.1, 4.2, 4.3, 5.1 and 5.2 materials*. For waste Division 6.1 Packing Group I, Hazard Zone A material stored, loaded, and transported with waste Class 8 liquids, or Division 4.2, 4.3, 5.1 and 5.2 materials:

(i) The Division 6.1 Packing Group I, Hazard Zone A material must be packaged in accordance with §173.226(c) of this subchapter and overpacked in a UN standard steel or plastic drum meeting the Packing Group I performance level;

(ii) The Class 8 liquid must be packaged in lab packs in accordance with paragraph (b) of this section or in single packagings authorized for the material in Column (8B) of the §172.101 Hazardous Materials Table of this subchapter not to exceed 208 L (55 gallons) capacity.

(iii) The Division 4.2 material may not exceed 2 kg (4.4 pounds) net weight per inner packaging and may not exceed 10 kg (22 pounds) net weight per outer packaging;

(iv) The Division 5.1 materials may not exceed 2 kg (4.4 pounds) net weight per inner packaging and may not exceed 10 kg (22 pounds) net weight per outer packaging. The aggregate net weight per freight container, unit load device, transport vehicle, or rail car may not exceed 100 kg (220 pounds);

(v) The Division 5.2 material may not exceed 1 kg (2.2 pounds) net weight per inner packaging and may not exceed 5 kg (11 pounds) net weight per outer packaging. Organic Peroxide, Type B material may not exceed 0.5 kg (1.1 pounds) net weight per inner packaging and may not exceed 2.5 kg (5.5 pounds) net weight per outer packaging. The aggregate net weight per freight container, unit load device, transport vehicle, or rail car may not exceed 50 kg (110 pounds).

(f) *Additional exceptions*. Lab packs conforming to the requirements of this section are not subject to the following:

(1) The overpack marking and labeling requirements in §173.25(a)(2) of this subchapter when secured to a pallet with shrink-wrap or stretch-wrap except that labels representative of each Hazard Class or Division in the overpack must be visibly displayed on two opposing sides.

(2) The restrictions for overpacks containing Class 8, Packing Group I material and Division 5.1, Packing Group I material in §173.25(a)(5) of this subchapter. These waste materials may be overpacked with other materials.

(g) *Household waste.* Household waste, as defined in §171.8 of this subchapter, is not subject to the requirements of this subchapter when transported in accordance with applicable state, local, or tribal requirements.

§173.13 Exceptions for class 3, divisions 4.1, 4.2, 4.3, 5.1, 6.1, and classes 8 and 9 materials.

(a) A Class 3, 8 or 9, or Division 4.1, 4.2, 4.3, 5.1, or 6.1 material is excepted from the labeling (except for the CARGO AIRCRAFT ONLY label), placarding and segregation requirements of this subchapter if prepared for transportation in accordance with the requirements of this section. A material that meets the definition of a material poisonous by inhalation may not be offered for transportation or transported under provisions of this section.

(b) A hazardous material conforming to requirements of this section may be transported by motor vehicle and rail car. In addition, packages prepared in accordance with this section may be transported by aircraft under the following conditions:

(1) *Cargo-only aircraft.* Only hazardous materials permitted to be transported aboard either a passenger or cargo-only aircraft by column (9A) or (9B) of the Hazardous Materials Table in §172.101 of this subchapter are authorized aboard cargo-only aircraft.

(2) *Passenger carrying aircraft.* Only hazardous materials permitted to be transported aboard a passenger aircraft by column (9A) of the Hazardous Materials Table in §172.101 of this subchapter are authorized aboard passenger aircraft. The completed package, assembled as for transportation, must be successfully tested in accordance with part 178 of this subchapter at the Packing Group I level. A hazardous material which meets the definition of a Division 5.1 (oxidizer) at the Packing Group I level in accordance with §173.127(b)(1)(i) of this sub-chapter may not be transported aboard a passenger aircraft.

(3) Packages offered for transportation aboard either passenger or cargo-only aircraft must meet the requirements for transportation by aircraft specified in §173.27 of this subchapter.

(c) A hazardous material permitted by paragraph (a) of this section must be packaged as follows:

(1) For liquids:

(i) The hazardous material must be placed in a tightly closed glass, plastic or metal inner packaging with a maximum capacity not exceeding 1.2 L. Sufficient outage must be provided such that the inner packaging will not become liquid full at 55 °C (130 °F). The net quantity (measured at 20 °C (68 °F)) of liquid in any inner packaging may not exceed 1 L. For transportation by aircraft, the net quantity in one package may not exceed the quantity specified in columns (9A) or (9B), as appropriate.

(ii) The inner packaging must be placed in a hermetically sealed barrier bag which is impervious to the lading, and then wrapped in a non-reactive absorbent material in sufficient quantity to completely absorb the contents of the inner packaging. Alternatively, the inner packaging may first be wrapped in a non-reactive absorbent material and then placed in the hermetically sealed barrier bag. The combination of inner packaging, absorbent material, and bag must be placed in a snugly fitting metal can.

(iii) The metal can must be securely closed. For liquids that are in Division 4.2 or 4.3, the metal can must be hermetically sealed. For Division 4.2 materials in Packing Group I, the metal can must be tested in accordance with part 178 of this subchapter at the Packing Group I performance level.

(iv) The metal can must be placed in a fiberboard box that is placed in a hermetically sealed barrier bag which is impervious to the lading.

(v) The intermediate packaging must be placed inside a securely closed, outer packaging conforming to §173.201.

(vi) Not more than four intermediate packagings are permitted in an outer packaging.

(2) For solids:

(i) The hazardous material must be placed in a tightly closed glass, plastic or metal inner packaging. The net quantity of material in any inner packaging may not exceed 2.85 kg (6.25 pounds). For transportation by aircraft, the net quantity in one package may not exceed the quantity specified in columns (9A) or (9B), as appropriate.

(ii) The inner packaging must be placed in a hermetically sealed barrier bag which is impervious to the lading.

(iii) The barrier bag and its contents must be placed in a fiberboard box that is placed in a hermetically sealed barrier bag which is impervious to the lading.

(iv) The intermediate packaging must be placed inside an outer packaging conforming to §173.211.

(v) Not more than four intermediate packagings are permitted in an outer packaging.

(d) The outside of the package must be marked, in association with the proper shipping name, with the statement: "This package conforms to 49 CFR 173.13."

Subpart B—Preparation of Hazardous Materials for Transportation

§173.22 Shipper's responsibility.

(a) Except as otherwise provided in this part, a person may offer a hazardous material for transportation in a packaging or container required by this part only in accordance with the following:

(1) The person shall class and describe the hazardous material in accordance with parts 172 and 173 of this subchapter, and

(2) The person shall determine that the packaging or container is an authorized packaging, including part 173 requirements, and that it has been manufactured, assembled, and marked in accordance with:

(i) Section 173.7(a) and parts 173, 178, or 179 of this subchapter;

(ii) A specification of the Department in effect at the date of manufacture of the packaging or container;

(iii) National or international regulations based on the UN Recommendations (IBR, see §171.7 or this subchapter), as authorized in §173.24(d)(2);

(iv) An approval issued under this subchapter; or

(v) An exemption or special permit issued under subchapter A of this chapter.

(3) In making the determination under paragraph (a)(2) of this section, the person may accept:

(i) Except for the marking on the bottom of a metal or plastic drum with a capacity over 100 L which has been reconditioned, remanufactured or otherwise converted, the manufacturer's certification, specification, approval, or exemption or special permit marking (see §§178.2 and 179.1 of this subchapter); or

(ii) With respect to cargo tanks provided by a carrier, the manufacturer's identification plate or a written certification of specification or exemption or special permit provided by the carrier.

(4) For a DOT Specification or UN standard packaging subject to the requirements of part 178 of this subchapter, a person must perform all functions necessary to bring the package into compliance with parts 173 and 178 of this subchapter, as identified by the packaging manufacturer or subsequent distributor (for example, applying closures consistent with the manufacturer's closure instructions) in accordance with §178.2 of this subchapter. A person must maintain a copy of the manufacturer's notification, including closure instructions (*see* §178.2(c) of this subchapter) unless permanently embossed or printed on the packaging. When applicable, a person must maintain a copy of any supporting documentation for an equivalent level of performance under the selective testing variation in §178.601(g)(1) of this subchapter. A copy of the notification, unless permanently embossed or printed on the packaging, and supporting documentation, when applicable, must be made available for inspection by a representative of the Department upon request for the time period of the packaging's periodic retest date, *i.e.*, every 12 months for single or composite packagings and every 24 months for combination packagings.

(b) No person may offer a motor carrier any hazardous material specified in 49 CFR 385.403 unless that motor carrier holds a safety permit issued by the Federal Motor Carrier Safety Administration.

(c) Prior to each shipment of fissile radioactive materials, and Type B or highway route controlled quantity packages of radioactive materials (see §173.403(l)), the shipper shall notify the consignee of the dates of shipment and expected arrival. The shipper shall also notify each consignee of any special loading/unloading instructions prior to his first shipment. For any shipment of irradiated reactor fuel, the shipper shall provide physical protection in compliance with a plan established under:

(1) Requirements prescribed by the U.S. Nuclear Regulatory Commission, or

(2) Equivalent requirements approved by the Associate Administrator.

§173.22a Use of packagings authorized under special permits.

(a) Except as provided in paragraph (b) of this section, no person may offer a hazardous material for transportation in a packaging the use of which is dependent upon an exemption or special permit issued under subpart B of part 107 of this title, unless that person is the holder of or a party to the exemption or special permit.

(b) If an exemption or special permit authorizes the use of a packaging for the transportation of a hazardous material by any person or class of persons other than or in addition to the holder of the exemption or special permit, that person or a member of that class of persons may use the packaging for the purposes authorized in the exemption or special permit subject to the terms specified therein. Copies of exemptions and special permits may be obtained by accessing the Hazardous Materials Safety Web site at *http://www.phmsa.dot.gov/hazmat/regs/sp-a* or by writing to the Associate Administrator for Hazardous Materials Safety, U.S. Department of Transportation, East Building, 1200 New Jersey Avenue, SE., Washington, DC 20590-0001, Attention: Records Center.

(c) When an exemption or special permit issued to a person who offers a hazardous material contains requirements that apply to a carrier of the hazardous material, the offeror shall furnish a copy of the current exemption or special permit to the carrier before or at the time a shipment is tendered.

§173.24 General requirements for packagings and packages.

(a) Applicability. Except as otherwise provided in this subchapter, the provisions of this section apply to—

(1) Bulk and non-bulk packagings;

(2) New packagings and packagings which are reused; and

(3) Specification and non-specification packagings.

(b) Each package used for the shipment of hazardous materials under this subchapter shall be designed, constructed, maintained, filled, its contents so limited, and closed, so that under conditions normally incident to transportation—

(1) Except as otherwise provided in this subchapter, there will be no identifiable (without the use of instruments) release of hazardous materials to the environment;

(2) The effectiveness of the package will not be substantially reduced; for example, impact resistance, strength, packaging compatibility, etc. must be maintained for the minimum and maximum temperatures, changes in humidity and pressure, and shocks, loadings and vibrations, normally encountered during transportation;

(3) There will be no mixture of gases or vapors in the package which could, through any credible spontaneous increase of heat or pressure, significantly reduce the effectiveness of the packaging;

(4) There will be no hazardous material residue adhering to the outside of the package during transport.

(c) Authorized packagings. A packaging is authorized for a hazardous material only if—

(1) The packaging is prescribed or permitted for the hazardous material in a packaging section specified for that material in Column 8 of the §172.101 Table and conforms to applicable requirements in the special provisions of Column 7 of the §172.101 Table and, for specification packagings (but not including UN standard packagings manufactured outside the United States), the specification requirements in parts 178 and 179 of this subchapter; or

(2) The packaging is permitted under, and conforms to, provisions contained in subparts B or C of part 171 of this subchapter or §§173.3, 173.4, 173.4a, 173.4b, 173.5, 173.5a, 173.6, 173.7, 173.8, 173.27, or §176.11 of this subchapter.

(d) *Specification packagings and UN standard packagings manufactured outside the U.S.*— (1) *Specification packagings*. A specification packaging, including a UN standard packaging manufactured in the United States, must conform in all details to the applicable specification or standard in part 178 or part 179 of this subchapter.

(2) UN standard packagings manufactured outside the United States. A UN standard packaging manufactured outside the United States, in accordance with national or international regulations based on the UN Recommendations (IBR, see §171.7 of this subchapter), may be imported and used and is considered to be an authorized packaging under the provisions of paragraph (c)(1) of this section, subject to the following conditions and limitations:

(i) The packaging fully conforms to applicable provisions in the UN Recommendations and the requirements of this subpart, including reuse provisions;

(ii) The packaging is capable of passing the prescribed tests in part 178 of this subchapter applicable to that standard; and

(iii) The competent authority of the country of manufacture provides reciprocal treatment for UN standard packagings manufactured in the U.S.

(e) Compatibility. (1) Even though certain packagings are specified in this part, it is, nevertheless, the responsibility of the person offering a hazardous material for transportation to ensure that such packagings are compatible with their lading. This particularly applies to corrosivity, permeability, softening, premature aging and embrittlement.

(2) Packaging materials and contents must be such that there will be no significant chemical or galvanic reaction between the materials and contents of the package.

(3) Plastic packagings and receptacles. (i) Plastic used in packagings and receptacles must be of a type compatible with the lading and may not be permeable to an extent that a hazardous condition is likely to occur during transportation, handling or refilling.

(ii) Each plastic packaging or receptacle which is used for liquid hazardous materials must be capable of withstanding without failure the procedure specified in Appendix B of this part ("Procedure for Testing Chemical Compatibility and Rate of Permeation in Plastic Packagings and Receptacles"). The procedure specified in appendix B of this part must be performed on each plastic packaging or receptacle used for Packing Group I materials. The maximum rate of permeation of hazardous lading through or into the plastic packaging or receptacles may not exceed 0.5 percent for materials meeting the definition of a Division 6.1 material according to §173.132 and 2.0 percent for other hazardous materials, when subjected to a temperature no lower than—

(A) 18°C (64°F) for 180 days in accordance with Test Method 1 in appendix B of this part;

(B) 50°C (122°F) for 28 days in accordance with Test Method 2 in appendix B of this part; or

(C) 60.5°C (141°F) for 14 days in accordance with Test Method 3 in appendix B of this part.

(iii) Alternative procedures or rates of permeation are permitted if they yield a level of safety equivalent to or greater than that provided by paragraph (e)(3)(ii) of this section and are specifically approved by the Associate Administrator.

(4) Mixed contents. Hazardous materials may not be packed or mixed together in the same outer packaging with other hazardous or nonhazardous materials if such materials are capable of reacting dangerously with each other and causing—

(i) Combustion or dangerous evolution of heat;

(ii) Evolution of flammable, poisonous, or asphyxiant gases; or

(iii) Formation of unstable or corrosive materials.

(5) Packagings used for solids, which may become liquid at temperatures likely to be encountered during transportation, must be capable of containing the hazardous material in the liquid state.

(f) Closures. (1) Closures on packagings shall be so designed and closed that under conditions (including the effects of temperature, pressure and vibration) normally incident to transportation—

(i) Except as provided in paragraph (g) of this section, there is no identifiable release of hazardous materials to the environment from the opening to which the closure is applied; and

(ii) The closure is leakproof and secured against loosening. For air transport, stoppers, corks or other such friction closures must be held in place by positive means.

(2) Except as otherwise provided in this subchapter, a closure (including gaskets or other closure components, if any) used on a specification packaging must conform to all applicable requirements of the specification and must be closed in accordance with information, as applicable, provided by the manufacturer's notification required by §178.2 of this subchapter.

(g) *Venting.* Venting of packagings, to reduce internal pressure which may develop by the evolution of gas from the contents, is permitted only when—

(1) Except for shipments of cryogenic liquids as specified in §173.320(c) and of carbon dioxide, solid (dry ice), transportation by aircraft is not involved;

(2) Except as otherwise provided in this subchapter, the evolved gases are not poisonous, likely to create a flammable mixture with air or be an asphyxiant under normal conditions of transportation;

(3) The packaging is designed so as to preclude an unintentional release of hazardous materials from the receptacle;

(4) For bulk packagings, other than IBCs, venting is authorized for the specific hazardous material by a special provision in the §172.101 table or by the applicable bulk packaging specification in part 178 of this subchapter; and

(5) Intermediate bulk packagings (IBCs) may be vented when required to reduce internal pressure that may develop by the evolution of gas subject to the requirements of paragraphs (g)(1) through (g)(3) of this section. The IBC must be of a type that has successfully passed (with the vent in place) the applicable design qualification tests with no release of hazardous material.

(h) Outage and filling limits—(1) *General.* When filling packagings and receptacles for liquids, sufficient ullage (outage) must be left to ensure that neither leakage nor permanent distortion of the packaging or receptacle will occur as a result of an expansion of the liquid caused by temperatures likely to be encountered during transportation. Requirements for outage and filling limits for non-bulk and bulk packagings are specified in §§173.24a(d) and 173.24b(a), respectively.

(2) *Compressed gases and cryogenic liquids.* Filling limits for compressed gases and cryogenic liquids are specified in §§173.301 through 173.306 for cylinders and §§173.314 through 173.319 for bulk packagings.

(i) *Air transportation.* Except as provided in subpart C of part 171 of this subchapter, packages offered or intended for transportation by aircraft must conform to the general requirements for transportation by aircraft in §173.27.

§173.24a Additional general requirements for non-bulk packagings and packages.

(a) *Packaging design.* Except as provided in §172.312 of this subchapter:

(1) *Inner packaging closures.* A combination packaging containing liquid hazardous materials must be packed so that closures on inner packagings are upright.

(2) *Friction.* The nature and thickness of the outer packaging must be such that friction during transportation is not likely to generate an amount of heat sufficient to alter dangerously the chemical stability of the contents.

(3) *Securing and cushioning.* Inner packagings of combination packagings must be so packed, secured and cushioned to prevent their breakage or leakage and to control their shifting within the outer packaging under conditions normally incident to transportation. Cushioning material must not be capable of reacting dangerously with the contents of the inner packagings or having its protective properties significantly weakened in the event of leakage.

(4) *Metallic devices.* Nails, staples and other metallic devices shall not protrude into the interior of the outer packaging in such a manner as to be likely to damage inner packagings or receptacles.

(5) *Vibration.* Each non-bulk package must be capable of withstanding, without rupture or leakage, the vibration test procedure specified in §178.608 of this subchapter.

(b) *Non-bulk packaging filling limits.* (1) A single or composite non-bulk packaging may be filled with a liquid hazardous material only when the specific gravity of the material does not exceed that marked on the packaging, or a specific gravity of 1.2 if not marked, except as follows:

(i) A Packing Group I packaging may be used for a Packing Group II material with a specific gravity not exceeding the greater of 1.8, or 1.5 times the specific gravity marked on the packaging, provided all the performance criteria can still be met with the higher specific gravity material;

(ii) A Packing Group I packaging may be used for a Packing Group III material with a specific gravity not exceeding the greater of 2.7, or 2.25 times the specific gravity marked on the packaging, provided all the performance criteria can still be met with the higher specific gravity material; and

(iii) A Packing Group II packaging may be used for a Packing Group III material with a specific gravity not exceeding the greater of 1.8, or 1.5 times the specific gravity marked on the packaging, provided all the performance criteria can still be met with the higher specific gravity material.

(2) Except as otherwise provided in this section, a non-bulk packaging may not be filled with a hazardous material to a gross mass greater than the maximum gross mass marked on the packaging.

(3) A single or composite non-bulk packaging which is tested and marked for liquid hazardous materials may be filled with a hazardous material to a gross mass, in kilograms, not exceeding the rated capacity of the packaging in liters, multiplied by the specific gravity marked on the packaging, or 1.2 if not marked. In addition:

(i) A single or composite non-bulk packaging which is tested and marked for Packing Group I liquid hazardous materials may be filled with a solid Packing Group II hazardous material to a gross mass, in kilograms, not exceeding the rated capacity of the packaging in liters, multiplied by 1.5, multiplied by the specific gravity marked on the packaging, or 1.2 if not marked.

(ii) A single or composite non-bulk packaging which is tested and marked for Packing Group I liquid hazardous materials may be filled with a solid Packing Group III hazardous material to a gross mass, in kilograms, not exceeding the rated capacity of the packaging in liters, multiplied by 2.25, multiplied by the specific gravity marked on the packaging, or 1.2 if not marked.

(iii) A single or composite non-bulk packaging which is tested and marked for Packing Group II liquid hazardous materials may be filled with a solid Packing Group III hazardous material to a gross mass, in kilograms, not exceeding the rated capacity of the packaging in liters, multiplied by 1.5, multiplied by the specific gravity marked on the packaging, or 1.2 if not marked.

(4) Packagings tested as prescribed in §178.605 of this subchapter and marked with the hydrostatic test pressure as prescribed in §178.503(a)(5) of this sub-chapter may be used for liquids only when the vapor pressure of the liquid conforms to one of the following:

(i) The vapor pressure must be such that the total pressure in the packaging (i.e., the vapor pressure of the liquid plus the partial pressure of air or other inert gases, less 100 kPa (15 psia)) at 55°C (131°F), determined on the basis of a maximum degree of filling in accordance with paragraph (d) of this section and a filling temperature of 15°C (59°F)), will not exceed two-thirds of the marked test pressure;

(ii) The vapor pressure at 50°C (122°F) must be less than four-sevenths of the sum of the marked test pressure plus 100 kPa (15 psia); or

(iii) The vapor pressure at 55°C (131°F) must be less than two-thirds of the sum of the marked test pressure plus 100 kPa (15 psia).

(5) No hazardous material may remain on the outside of a package after filling.

(c) *Mixed contents.* (1) An outer non-bulk packaging may contain more than one hazardous material only when—

(i) The inner and outer packagings used for each hazardous material conform to the relevant packaging sections of this part applicable to that hazardous material;

(ii) The package as prepared for shipment meets the performance tests prescribed in part 178 of this subchapter for the packing group indicating the highest order of hazard for the hazardous materials contained in the package;

(iii) Corrosive materials (except ORM–D) in bottles are further packed in securely closed inner receptacles before packing in outer packagings; and

(iv) For transportation by aircraft, the total net quantity does not exceed the lowest permitted maximum net quantity per package as shown in Column 9a or 9b, as appropriate, of the §172.101 Table. The permitted maximum net quantity must be calculated in kilograms if a package contains both a liquid and a solid.

(2) A packaging containing inner packagings of Division 6.2 materials may not contain other hazardous materials except—

(i) Refrigerants, such as dry ice or liquid nitrogen, as authorized under the HMR;

(ii) Anticoagulants used to stabilize blood or plasma; or

(iii) Small quantities of Class 3, Class 8, Class 9, or other materials in Packing Groups II or III used to stabilize or prevent degradation of the sample, provided the quantity of such materials does not exceed 30 mL (1 ounce) or 30 g (1 ounce) in each inner packaging. The maximum quantity in an outer package, including a hazardous material used to preserve or stabilize a sample, may not exceed 4 L (1 gallon) or 4 kg (8.8 pounds). Such preservatives are not subject to the requirements of this subchapter.

(d) Liquids must not completely fill a receptacle at a temperature of 55°C (131°F) or less.

§173.24b Additional general requirements for bulk packagings.

(a) *Outage and filling limits.* (1) Except as otherwise provided in this subchapter, liquids and liquefied gases must be so loaded that the outage is at least five percent for materials poisonous by inhalation, or at least one percent for all other materials, of the total capacity of a cargo tank, portable tank, tank car (including dome capacity), multi-unit tank car tank, or any compartment thereof, at the following reference temperatures—

(i) 46°C (115°F) for a noninsulated tank;

(ii) 43°C (110°F) for a tank car having a thermal protection system, incorporating a metal jacket that provides an overall thermal conductance at 15.5°C (60°F) of no more than 10.22 kilojoules per hour per square meter per degree Celsius (0.5 Btu per hour/per square foot/ per degree F) temperature differential; or

(iii) 41°C (105°F) for an insulated tank.

(2) Hazardous materials may not be loaded into the dome of a tank car. If the dome of the tank car does not provide sufficient outage, vacant space must be left in the shell to provide the required outage.

(b) *Equivalent steel.* For the purposes of this section, the reference stainless steel is stainless steel with a guaranteed minimum tensile strength of 51.7 deka newtons per square millimeter (75,000 psi) and a guaranteed elongation of 40 percent or greater. Where the regulations permit steel other than stainless steel to be used in place of a specified stainless steel (for example, as in §172.102 of this subchapter, special provision B30), the minimum thickness for the steel must be obtained from one of the following formulas, as appropriate:

Formula for metric units:

$$e_1 = (12.74e_0)/(Rm_1\ A_1)13$$

Formula for non-metric units:

$$e_1 = (144.2e_0)/(Rm_1\ A_1)13$$

where:

e_0 = Required thickness of the reference stainless steel in millimeters or inches respectively;

e_1 = Equivalent thickness of the steel used in millimeters or inches respectively;

Rm_1 = Specified minimum tensile strength of the steel used in deka-newtons per square millimeter or pounds per square inch respectively; and

A_1 = Specified minimum percentage elongation of the steel used multiplied by 100 (for example, 20 percent times 100 equals 20). Elongation values used must be determined from a 50 mm or 2 inch test specimen.

(c) Air pressure in excess of ambient atmospheric pressure may not be used to load or unload any lading which may create an air-enriched mixture within the flammability range of the lading in the vapor space of the tank.

(d) A bulk packaging may not be loaded with a hazardous material that:

(1) Is at a temperature outside of the packaging's design temperature range; or

(2) Except as otherwise provided in this subchapter, exceeds the maximum weight of lading marked on the specification plate.

(e) *Stacking of IBCs and Large Packagings.* (1) IBCs and Large Packagings not designed and tested to be stacked. No packages or freight (hazardous or otherwise) may be stacked upon an IBC or a Large Packaging that was not designed and tested to be stacked upon.

(2) IBCs and Large Packagings designed and tested to be stacked. The superimposed weight placed upon an IBC or a Large Packaging designed to be stacked may not exceed the maximum permissible stacking test mass marked on the packaging.

(f) *UN portable tanks.* (1) A UN portable tank manufactured in the United States must conform in all details to the applicable requirements in parts 172, 173, 178 and 180 of this subchapter.

(2) *UN portable tanks manufactured outside the United States.* A UN portable tank manufactured outside the United States, in accordance with national or international regulations based on the UN Recommendations (IBR, see §171.7 of this subchapter), which is an authorized packaging under §173.24 of this subchapter, may be filled, offered and transported in the United States, if the §172.101 Table of this subchapter authorizes the hazardous material for transportation in the UN portable tank and it conforms to the applicable T codes, and tank provision codes, or other special provisions assigned to the hazardous material in Column (7) of the Table. In addition, the portable tank must—

(i) Conform to applicable provisions in the UN Recommendations (IBR, see §171.7 of this subchapter) and the requirements of this subpart;

(ii) Be capable of passing the prescribed tests and inspections in part 180 of this subchapter applicable to the UN portable tank specification;

(iii) Be designed and manufactured according to the ASME Code (IBR, see §171.7 of this subchapter) or a pressure vessel design code approved by the Associate Administrator;

(iv) Be approved by the Associate Administrator when the portable tank is designed and constructed under the provisions of an alternative arrangement (see §178.274(a)(2) of this subchapter); and

(v) The competent authority of the country of manufacture must provide reciprocal treatment for UN portable tanks manufactured in the United States.

§173.25 Authorized packagings and overpacks.

(a) Authorized packages containing hazardous materials may be offered for transportation in an overpack as defined in §171.8 of this subchapter, if all of the following conditions are met:

(1) The package meets the requirements of §§173.21 and 173.24 of this subchapter.

(2) The overpack is marked with the proper shipping name and identification number, when applicable, and is labeled as required by this subchapter for each hazardous material contained therein, unless marking and labels representative of each hazardous material in the overpack are visible.

(3) Each package subject to the orientation marking requirements of §172.312 of this subchapter is packed in the overpack with its filing holes up and the overpack is marked with package orientation marking arrows on two opposite vertical sides of the overpack with the arrows pointing in the correct direction or orientation.

(4) The overpack is marked with the word "OVERPACK" when specification packagings are required, unless specification markings on the inside packages are visible.

(5) Packages containing Class 8 (corrosive) materials in Packing Group I or Division 5.1 (oxidizing) materials in Packing Group I may not be overpacked with any other materials.

(6) Where packages of limited quantity materials are overpacked and, until December 31, 2012 or December 31, 2013, packages bearing the ORM-D AIR or ORM-D marking, respectively, must be marked "OVERPACK" unless all marking required by this section are visible. Where packages of excepted quantities (*see* §173.4a of this part) are overpacked and all required markings are not visible through the overpack, they must be repeated on the overpack. An overpack containing packages of excepted quantities is not required to be marked "OVERPACK."

(b) Shrink-wrapped or stretch-wrapped trays may be used as outer packagings for inner packagings prepared in accordance with the limited quantity provisions or consumer commodity provisions of this subchapter, provided that—

(1) Inner packagings are not fragile, liable to break or be easily punctured, such as those made of glass, porcelain, stoneware or certain plastics; and

(2) Each complete package does not exceed 20 kg (44 lbs) gross weight.

(c) Hazardous materials which are required to be labeled POISON may be transported in the same motor vehicle with material that is marked or known to be foodstuffs, feed or any edible material intended for consumption by humans or animals provided the hazardous material is marked, labeled, and packaged in accordance with this subchapter, conforms to the requirements of paragraph (a) of this section and is overpacked as specified in §177.841(e) of this subchapter or in an overpack which is a UN 1A2, 1B2, or 1N2 drum tested and marked for a Packing Group II or higher performance level.

§173.28 Reuse, reconditioning and remanufacture of packagings.

(a) *General.* Packagings and receptacles used more than once must be in such condition, including closure devices and cushioning materials, that they conform in all respects to the prescribed requirements of this subchapter. Before reuse, each packaging must be inspected and may not be reused unless free from incompatible residue, rupture, or other damage which reduces its structural integrity. Packagings not meeting the minimum thickness requirements prescribed in paragraph (b)(4)(i) of this section may not be reused or reconditioned for reuse.

(b) *Reuse of non-bulk packaging.* A non-bulk packaging used more than once must conform to the following provisions and limitations:

(1) A non-bulk packaging which, upon inspection, shows evidence of a reduction in integrity may not be reused unless it is reconditioned in accordance with paragraph (c) of this section.

(2) Before reuse, packagings subject to the leakproofness test with air prescribed in §178.604 of this sub-chapter shall be—

(i) Retested without failure in accordance with §178.604 of this subchapter using an internal air pressure (gauge) of at least 48 kPa (7.0 psig) for Packing Group I and 20 kPa (3.0 psig) for Packing Group II and Packing Group III; and

(ii) Marked with the letter "L", with the name and address or symbol of the person conducting the test, and the last two digits of the year the test was conducted. Symbols, if used, must be registered with the Associate Administrator.

(3) Packagings made of paper (other than fiber-board), plastic film, or textile are not authorized for reuse;

(4) Metal and plastic drums and jerricans used as single packagings or the outer packagings of composite packagings are authorized for reuse only when they are marked in a permanent manner (e.g., embossed) in millimeters with the nominal (for metal packagings) or minimum (for plastic packagings) thickness of the packaging material, as required by §178.503(a)(9) of this subchapter, and—

(i) Except as provided in paragraph (b)(4)(ii) of this section, conform to the following minimum thickness criteria:

Maximum capacity not over	Minimum thickness of packaging material	
	Metal drum or jerrican	Plastic drum or jerrican
20 L	0.63 mm (0.025 inch)	1.1 mm (0.043 inch)
30 L	0.73 mm (0.029 inch)	1.1 mm (0.043 inch)
40 L	0.73 mm (0.029 inch)	1.8 mm (0.071 inch)
60 L	0.92 mm (0.036 inch)	1.8 mm (0.071 inch)
120 L	0.92 mm (0.036 inch)	2.2 mm (0.087 inch)
220 L	0.92 mm (0.036 inch)[1]	2.2 mm (0.087 inch)
450 L	1.77 mm (0.070 inch)	5.0 mm (0.197 inch)

[1]Metal drums or jerricans with a minimum thickness of 0.82 mm body and 1.09 mm heads which are manufactured and marked prior to January 1, 1997 may be reused. Metal drums or jerricans manufactured and marked on or after January 1, 1997, and intended for reuse, must be constructed with a minimum thickness of 0.82 mm body and 1.11 mm heads.

(ii) For stainless steel drums and jerricans, conform to a minimum wall thickness as determined by the following equivalence formula:

Formula for metric units

$$e_1 = \frac{21.4 \times e_0}{\sqrt[3]{Rm_1 \times A_1}}$$

Formula for U.S. standard units

$$e_1 = \frac{21.4 \times e_0}{\sqrt[3]{(Rm_1 \times A_1)/145}}$$

where:

e_1 = required equivalent wall thickness of the metal to be used (in mm or, for U.S. Standard units, use inches).

e_0 = required minimum wall thickness for the reference steel (in mm or, for U.S. Standard units, use inches).

Rm_1 = guaranteed minimum tensile strength of the metal to be used (in N/mm^2 or for U.S. Standard units, use psi).

A_1 = quaranteed minimum elongation (as a percentage) of the metal to be used on fracture under tensile stress (see paragraph (c) (1) of this section).

(5) Plastic inner receptacles of composite packagings must have a minimum thickness of 1.0 mm (0.039 inch).

(6) A previously used non-bulk packaging may be reused for the shipment of hazardous waste, not subject to the reconditioning and reuse provisions of this section, in accordance with §173.12(c).

(7) Notwithstanding the provisions of paragraph (b)(2) of this section, a packaging otherwise authorized for reuse may be reused without being leakproofness tested with air provided the packaging—

(i) Is refilled with a material which is compatible with the previous lading;

(ii) Is refilled and offered for transportation by the original filler;

(iii) Is transported in a transport vehicle or freight container under the exclusive use of the refiller of the packaging; and

(iv) Is constructed of—

(A) Stainless steel, monel or nickel with a thickness not less than one and one-half times the minimum thickness prescribed in paragraph (b)(4) of this section;

(B) Plastic, provided the packaging is not refilled for reuse on a date more than five years from the date of manufacture marked on the packaging in accordance with §178.503(a)(6) of this subchapter; or

(C) Another material or thickness when approved under the conditions established by the Associate Administrator for reuse without retesting.

(c) *Reconditioning of non-bulk packaging.* (1) For the purpose of this subchapter, reconditioning of metal drums is:

(i) Cleaning to base material of construction, with all former contents, internal and external corrosion, and any external coatings and labels removed;

(ii) Restoring to original shape and contour, with chimes (if any) straightened and sealed, and all non-integral gaskets replaced: and

(iii) Inspecting after cleaning but before painting, Packagings that have visible pitting, significant reduction in material thickness, metal fatigue, damaged threads or closures, or other significant defects, must be rejected.

(2) For the purpose of this subchapter, reconditioning of a non-bulk packaging other than a metal drum includes:

(i) Removal of all former contents, external coatings and labels, and cleaning to the original materials of construction;

(ii) Inspection after cleaning with rejection of packagings with visible damage such as tears, creases or cracks, or damaged threads or closures, or other significant defects;

(iii) Replacement of all non-integral gaskets and closure devices with new or refurbished parts, and cushioning and cushioning materials; and components including gaskets, closure devices and cushioning and cushioning material. (For a UN 1H1 plastic drum, replacing a removable gasket or closure device with another of the same design and material that provides equivalent performance does not constitute reconditioning); and

(iv) Ensuring that the packagings are restored to a condition that conforms in all respects with the prescribed requirements of this subchapter.

(3) A person who reconditions a packaging manufactured and marked under the provisions of subpart L of part 178 of this subchapter, shall mark that packaging as required by §178.503(c) and (d) of this subchapter. The marking is the certification of the reconditioner that the packaging conforms to the standard for which it is marked and that all functions performed by the recondi-tioner which are prescribed by this subchapter have been performed in compliance with this subchapter.

(4) The markings applied by the reconditioner may be different from those applied by the manufacturer at the time of original manufacture, but may not identify a greater performance capability than that for which the original design type had been tested (for example, the reconditioner may mark a drum which was originally marked as 1A1/Y1.8 as 1A1/Y1.2 or 1A1/Z2.0).

(5) Packagings which have significant defects which cannot be repaired may not be reused.

(d) *Remanufacture of non-bulk packagings.* For the purpose of this subchapter, remanufacture is the conversion of a non-specification, non-bulk packaging to a DOT specification or U.N. standard, the conversion of a packaging meeting one specification or standard to another specification or standard (for example, conversion of 1A1 non-removable head drums to 1A2 removable head drums) or the replacement of integral structural packaging components (such as non-removable heads on drums). A person who remanufactures a non-bulk packaging to conform to a specification or standard in part 178 of this subchapter is subject to the requirements of part 178 of this subchapter as a manufacturer.

(e) *Non-reusable containers.* A packaging marked as NRC according to the DOT specification or UN standard requirements of part 178 of this subchapter may be reused for the shipment of any material not required by this subchapter to be shipped in a DOT specification or UN standard packaging.

(f) A Division 6.2 packaging to be reused must be disinfected prior to reuse by any means effective for neutralizing the infectious substance the packaging previously contained. A secondary packaging or outer packaging conforming to the requirements of §173.196 or §173.199 need not be disinfected prior to reuse if no leakage from the primary receptacle has occurred. Drums or jerricans not meeting the minimum thickness requirements prescribed in paragraph (b)(4)(i) of this section may not be reused or reconditioned for reuse.

§173.29 Empty packagings.

(a) General. Except as otherwise provided in this section, an empty packaging containing only the residue of a hazardous material shall be offered for transportation and transported in the same manner as when it previously contained a greater quantity of that hazardous material.

(b) Notwithstanding the requirements of paragraph (a) of this section, an empty packaging is not subject to any other requirements of this subchapter if it conforms to the following provisions:

(1) Any hazardous material shipping name and identification number markings, any hazard warning labels or placards, and any other markings indicating that the material is hazardous (*e.g.*, RQ, INHALATION HAZARD) are removed, obliterated, or securely covered in transportation. This provision does not apply to transportation in a transport vehicle or a freight container if the packaging is not visible in transportation and the packaging is loaded by the shipper and unloaded by the shipper or consignee;

(2) The packaging—

(i) Is unused;

(ii) Is sufficiently cleaned of residue and purged of vapors to remove any potential hazard;

(iii) Is refilled with a material which is not hazardous to such an extent that any residue remaining in the packaging no longer poses a hazard; or

(iv) Contains only the residue of—

(A) An ORM-D material; or

(B) A Division 2.2 non-flammable gas, other than ammonia, anhydrous, and with no subsidiary hazard, at a gauge pressure less than 200 kPa (29.0 psig); at 20°C (68°F); and

(3) Any material contained in the packaging does not meet the definitions in §171.8 of this subchapter for a hazardous substance, a hazardous waste, or a marine pollutant.

(c) A non-bulk packaging containing only the residue of a hazardous material covered by Table 2 of §172.504 of this subchapter that is not a material poisonous by inhalation or its residue shipped under the subsidiary placarding provisions of §172.505—

(1) Does not have to be included in determining the applicability of the placarding requirements of subpart F of part 172 of this subchapter; and

(2) Is not subject to the shipping paper requirements of this subchapter when collected and transported by a contract or private carrier for reconditioning, remanufacture or reuse.

(d) Notwithstanding the stowage requirements in Column 10a of the §172.101 Table for transportation by vessel, an empty drum or cylinder may be stowed on deck or under deck.

(e) Specific provisions for describing an empty packaging on a shipping paper appear in §172.203(e) of this subchapter.

(f) [Reserved]

(g) A package which contains a residue of an elevated temperature material may remain marked in the same manner as when it contained a greater quantity of the material even though it no longer meets the definition in §171.8 of this subchapter for an elevated temperature material.

(h) A package that contains a residue of a hazardous substance, Class 9, listed in the §172.101 Table, Appendix A, Table I, that does not meet the definition of another hazard class and is not a hazardous waste or marine pollutant, may remain marked, labeled and, if applicable, placarded in the same manner as when it contained a greater quantity of the material even though it no longer meets the definition in §171.8 of this subchapter for a hazardous substance.

§173.30 Loading and unloading of transport vehicles.

A person who is subject to the loading and unloading regulations in this subchapter must load or unload hazardous materials into or from a transport vehicle or vessel in conformance with the applicable loading and unloading requirements of parts 174, 175, 176, and 177 of this subchapter.

Subpart C—Definitions, Classifcation and Packaging for Class I

§173.50 Class 1—definitions.

(a) *Explosive*. For the purposes of this subchapter, an **explosive** means any substance or article, including a device, which is designed to function by explosion (*i.e.*, an extremely rapid release of gas and heat) or which, by chemical reaction within itself, is able to function in a similar manner even if not designed to function by explosion, unless the substance or article is otherwise classed under the provisions of this subchapter. The term includes a pyrotechnic substance or article, unless the substance or article is otherwise classed under the provisions of this subchapter.

(b) Explosives in Class 1 are divided into six divisions as follows:

(1) **Division 1.1** consists of explosives that have a mass explosion hazard. A mass explosion is one which affects almost the entire load instantaneously.

(2) **Division 1.2** consists of explosives that have a projection hazard but not a mass explosion hazard.

(3) **Division 1.3** consists of explosives that have a fire hazard and either a minor blast hazard or a minor projection hazard or both, but not a mass explosion hazard.

(4) **Division 1.4** consists of explosives that present a minor explosion hazard. The explosive effects are largely confined to the package and no projection of fragments of appreciable size or range is to be expected. An external fire must not cause virtually instantaneous explosion of almost the entire contents of the package.

(5) **Division 1.5**[1]consists of very insensitive explosives. This division is comprised of substances which have a mass explosion hazard but are so insensitive that there is very little probability of initiation or of transition from burning to detonation under normal conditions of transport.

(6) **Division 1.6**[2]consists of extremely insensitive articles which do not have a mass explosive hazard. This division is comprised of articles which contain only extremely insensitive detonating substances and which demonstrate a negligible probability of accidental initiation or propagation.

[1] The probability of transition from burning to detonation is greater when large quantities are transported in a vessel.

[2] The risk from articles of Division 1.6 is limited to the explosion of a single article.

Subpart D—Definitions, Classification, Packing Group Assignments and Exceptions for Hazardous Material Other Than Class 1 and Class 7

§173.115 Class 2, Divisions 2.1, 2.2, and 2.3 — Definitions.

(a) *Division 2.1 (Flammable gas)*. For the purpose of this subchapter, a **flammable gas (Division 2.1)** means any material which is a gas at 20°C (68°F) or less and 101.3 kPa (14.7 psia) of pressure (a material which has a boiling point of 20°C (68°F) or less at 101.3 kPa (14.7 psia)) which—

(1) Is ignitable at 101.3 kPa (14.7 psia) when in a mixture of 13 percent or less by volume with air; or

(2) Has a flammable range at 101.3 kPa (14.7 psia) with air of at least 12 percent regardless of the lower limit. Except for aerosols, the limits specified in paragraphs (a)(1) and (a)(2) of this section shall be determined at 101.3 kPa (14.7 psia) of pressure and a temperature of 20 °C (68 °F) in accordance with the ASTM E681-85, Standard Test Method for Concentration Limits of Flammability of Chemicals or other equivalent method approved by the Associate Administrator. The flammability of aerosols is determined by the tests specified in paragraph (l) of this section.

(b) *Division 2.2 (non-flammable, nonpoisonous compressed gas— including compressed gas, liquefied gas, pressurized cryogenic gas, compressed gas in solution, asphyxiant gas and oxidizing gas)*. For the purpose of this subchapter, a **non-flammable, nonpoisonous compressed gas (Division 2.2)** means any material (or mixture) which—

(1) Exerts in the packaging a gauge pressure of 200 kPa (29.0 psig/43.8 psia) or greater at 20°C (68°F), is a liquefied gas or is a cryogenic liquid, and

(2) Does not meet the definition of Division 2.1 or 2.3.

(c) *Division 2.3 (Gas poisonous by inhalation)*. For the purpose of this subchapter, a **gas poisonous by inhalation (Division 2.3)** means a material which is a gas at 20°C (68°F) or less and a pressure of 101.3 kPa (14.7 psia) (a material which has a boiling point of 20°C (68°F) or less at 101.3 kPa (14.7 psia)) and which—

(1) Is known to be so toxic to humans as to pose a hazard to health during transportation, or

(2) In the absence of adequate data on human toxicity, is presumed to be toxic to humans because when tested on laboratory animals it has an LC50 value of not more than 5000 mL/m^3 (see §173.116(a) of this subpart for assignment of Hazard Zones A, B, C or D). LC_{50} values for mixtures may be determined using the formula in §173.133(b)(1)(i) or CGA P–20 (IBR, see §171.7 of this subchapter).

(d) *Non-liquefied compressed gas*. A gas, which when packaged under pressure for transportation is entirely gaseous at –50°C (–58°F) with a critical temperature less than or equal to –50°C (–58°F), is considered to be a non-liquefied compressed gas.

(e) *Liquefied compressed gas.* A gas, which when packaged under pressure for transportation is partially liquid at temperatures above −50°C (−58°F), is considered to be a liquefied compressed gas. A liquefied compressed gas is further categorized as follows:

(1) *High pressure liquefied gas* which is a gas with a critical temperature between −50°C (−58°F) and +65°C (149°F), and

(2) *Low pressure liquefied gas* which is a gas with a critical temperature above + 65°C (149°F).

(f) *Compressed gas in solution.* A **compressed gas in solution** is a non-liquefied compressed gas which is dissolved in a solvent.

(g) *Cryogenic liquid.* A **cryogenic liquid** means a refrigerated liquefied gas having a boiling point colder than −90°C (−130°F) at 101.3 kPa (14.7 psia). A material meeting this definition is subject to requirements of this subchapter without regard to whether it meets the definition of a non-flammable, non-poisonous compressed gas in paragraph (b) of this section.

(h) *Flammable range.* The term **flammable range** means the difference between the minimum and maximum volume percentages of the material in air that forms a flammable mixture.

(i) *Service pressure.* The term **service pressure** means the authorized pressure marking on the packaging. For example, for a cylinder marked "DOT 3A1800", the service pressure is 12410 kPa (1800 psig).

(j) *Refrigerant gas or Dispersant gas.* The terms **Refrigerant gas** and **Dispersant gas** apply to all nonpoisonous refrigerant gases; dispersant gases (fluorocarbons) listed in §172.101 of this subchapter and §§173.304, 173.314(c), 173.315(a), and 173.315(h) and mixtures thereof; and any other compressed gas having a vapor pressure not exceeding 260 psia at 54°C (130°F), used only as a refrigerant, dispersant, or blowing agent.

(k) For Division 2.2 gases, the oxidizing ability shall be determined by tests or by calculation in accordance with ISO 10156:1996 and ISO 10156–2:2005 (IBR, see §171.7 of this subchapter).

(l) The following applies to aerosols (see §171.8 of this subchapter):

(1) An aerosol must be assigned to Division 2.1 if the contents include 85% by mass or more flammable components and the chemical heat of combustion is 30 kJ/g or more;

(2) An aerosol must be assigned to Division 2.2 if the contents contain 1% by mass or less flammable components and the heat of combustion is less than 20 kJ/g.

(3) Aerosols not meeting the provisions of paragraphs (l)(1) or (l)(2) of this section must be classed in accordance with the appropriate tests of the UN Manual of Tests and Criteria (IBR, see §171.7 of this subchapter). An aerosol which was tested in accordance with the requirements of this subchapter in effect on December 31, 2005, is not required to be retested.

(4) Division 2.3 gases may not be transported in an aerosol container.

(5) When the contents are classified as Division 6.1, PG III or Class 8, PG II or III, the aerosol must be assigned a subsidiary hazard of Division 6.1 or Class 8, as appropriate.

(6) Substances of Division 6.1, PG I or II, and substances of Class 8, PG I are forbidden from transportation in an aerosol container.

(7) Flammable components are Class 3 flammable liquids, Class 4.1 flammable solids, or Division 2.1 flammable gases. The chemical heat of combustion must be determined in accordance with the UN Manual of Tests and Criteria (IBR, see §171.7 of this subchapter).

§173.116 Class 2—Assignment of hazard zone.

(a) The hazard zone of a Class 2, Division 2.3 material is assigned in Column 7 of the §172.101 Table. There are no hazard zones for Divisions 2.1 and 2.2. When the §172.101 Table provides more than one hazard zone for a Division 2.3 material, or indicates that the hazard zone be determined on the basis of the grouping criteria for Division 2.3, the hazard zone shall be determined by applying the following criteria:

Hazard zone	Inhalation toxicity
A................	LC_{50} less than or equal to 200 ppm.
B................	LC_{50} greater than 200 ppm and less than or equal to 1000 ppm.
C................	LC_{50} greater than 1000 ppm and less than or equal to 3000 ppm.
D................	LC_{50} greater than 3000 ppm or less than or equal to 5000 ppm.

(b) The criteria specified in paragraph (a) of this section are represented graphically in §173.133, Figure 1.

§173.120 Class 3—Definitions.

(a) *Flammable liquid.* For the purpose of this subchapter, a **flammable liquid (Class 3)** means a liquid having a flash point of not more than 60°C (140°F), or any material in a liquid phase with a flash point at or above 37.8°C (100°F) that is intentionally heated and offered for transportation or transported at or above its flash point in a bulk packaging, with the following exceptions:

(1) Any liquid meeting one of the definitions specified in §173.115.

(2) Any mixture having one or more components with a flash point of 60°C (140°F) or higher, that make up at least 99 percent of the total volume of the mixture, if the mixture is not offered for transportation or transported at or above its flash point.

(3) Any liquid with a flash point greater than 35°C (95°F) that does not sustain combustion according to ASTM D 4206 (IBR, see §171.7 of this subchapter) or the procedure in appendix H of this part.

(4) Any liquid with a flash point greater than 35°C (95°F) and with a fire point greater than 100°C (212°F) according to ISO 2592 (IBR, see §171.7 of this subchapter).

(5) Any liquid with a flash point greater than 35°C (95°F) which is in a water-miscible solution with a water content of more than 90 percent by mass.

(b) Combustible liquid. (1) For the purpose of this subchapter, a **combustible liquid** means any liquid that does not meet the definition of any other hazard class specified in this subchapter, and has a flash point above 60°C (140°F) and below 93°C (200°F).

(2) A flammable liquid with a flash point at or above 38°C (100°F) that does not meet the definition of any other hazard class, may be reclassed as a combustible liquid. This provision does not apply to transportation by vessel or aircraft, except where other means of transportation is impracticable. An elevated temperature material that meets the definition of a Class 3 material because it is intentionally heated and offered for transportation or transported at or above its flash point may not be reclassed as a combustible liquid.

(3) A combustible liquid that does not sustain combustion is not subject to the requirements of this subchapter as a combustible liquid. Either the test method specified in ASTM D 4206 or the procedure in appendix H of this part may be used to determine if a material sustains combustion when heated under test conditions and exposed to an external source of flame.

(c) Flash point. (1) **Flash point** means the minimum temperature at which a liquid gives off vapor within a test vessel in sufficient concentration to form an ignitable mixture with air near the surface of the liquid. It shall be determined as follows:

(i) For a homogeneous, single-phase, liquid having a viscosity less than 45 S.U.S. at 38 °C (100 °F) that does not form a surface film while under test, one of the following test procedures shall be used:

(A) Standard Method of Test for Flash Point by Tag Closed Cup Tester, (ASTM D 56) (IBR; *see* §171.7 of this subchapter);

(B) Standard Test Methods for Flash Point of Liquids by Small Scale Closed-Cup Apparatus, (ASTM D 3278) (IBR; *see* §171.7 of this subchapter); or

(C) Standard Test Methods for Flash Point by Small Scale Closed Tester, (ASTM D 3828) (IBR; *see* §171.7 of this subchapter).

(ii) For a liquid other than one meeting all the criteria of paragraph (c)(1)(i) of this section, one of the following test procedures must be used:

(A) Standard Test Methods for Flash Point by Pensky-Martens Closed Cup Tester, (ASTM D 93) (IBR; *see* §171.7 of this subchapter). For cutback asphalt, use Method B of ASTM D 93 or alternative tests authorized in this standard;

(B) Standard Test Methods for Flash Point of Liquids by Small Scale Closed-Cup Apparatus (ASTM D 3278) (IBR; *see* §171.7 of this subchapter);

(C) Determination of Flash/No Flash-Closed Cup Equilibrium Method (ISO 1516) (IBR; *see* §171.7 of this subchapter);

(D) Determination of Flash point-Closed Cup Equilibrium Method (ISO 1523) (IBR; *see* §171.7 of this subchapter);

(E) Determination of Flash Point-Pensky-Martens Closed Cup Method (ISO 2719) (IBR; *see* §171.7 of this subchapter);

(F) Determination of Flash Point-Rapid Equilibrium Closed Cup Method (ISO 3679) (IBR; *see* §171.7 of this subchapter);

(G) Determination of Flash/No Flash-Rapid Equilibrium Closed Cup Method (ISO 3680) (IBR; *see* §171.7 of this subchapter); or

(H) Determination of Flash Point-Abel Closed-Cup Method (ISO 13736) (IBR; *see* §171.7 of this subchapter).

(2) For a liquid that is a mixture of compounds that have different volatility and flash points, its flash point shall be determined as specified in paragraph (c)(1) of this section, on the material in the form in which it is to be shipped. If it is determined by this test that the flash point is higher than −7°C (20° F) a second test shall be made as follows: a portion of the mixture shall be placed in an open beaker (or similar container) of such dimensions that the height of the liquid can be adjusted so that the ratio of the volume of the liquid to the exposed surface area is 6 to one. The liquid shall be allowed to evaporate under ambient pressure and temperature (20 to 25°C (68 to 77°F)) for a period of 4 hours or until 10 percent by volume has evaporated, whichever comes first. A flash point is then run on a portion of the liquid remaining in the evaporation container and the lower of the two flash points shall be the flash point of the material.

(3) For flash point determinations by Setaflash closed tester, the glass syringe specified need not be used as the method of measurement of the test sample if a minimum quantity of 2 mL (0.1 ounce) is assured in the test cup.

(d) If experience or other data indicate that the hazard of a material is greater or less than indicated by the criteria specified in paragraphs (a) and (b) of this section, the Associate Administrator may revise the classification or make the material subject or not subject to the requirements of parts 171 through 185 of this subchapter.

(e) *Transitional provisions.* The Class 3 classification criteria in effect on December 31, 2006, may continue to be used until January 1, 2012.

§173.121 Class 3—Assignment of packing group.

(a)(1) The packing group of a Class 3 material is as assigned in column 5 of the §172.101 Table. When the §172.101 Table provides more than one packing group for a hazardous material, the packing group must be determined by applying the following criteria:

Packing group	Flash point (closed-cup)	Initial boiling point
I	. . . .	≤35°C (95°F)
II.	<23°C (73°F)	>35°C (95°F)
III	≥23°C, ≤ 60°C (≥73°F, ≤140°F)	>35°C (95°F)

(2) The initial boiling point of a Class 3 material may be determined by using one of the following test methods:

(i) Standard Test Method for Distillation of Petroleum Products at Atmospheric Pressure (ASTM D 86) (IBR; *see* §171.7 of this subchapter);

(ii) Standard Test Method for Distillation Range of Volatile Organic Liquids (ASTM D 1078) (IBR; *see* §171.7 of this subchapter);

(iii) Petroleum Products—Determination of Distillation Characteristics at Atmospheric Pressure (ISO 3405) (IBR; *see* §171.7 of this subchapter);

CLASSIFICATION & EXCEPTIONS

(iv) Petroleum Products—Determination of Boiling Range Distribution—Gas Chromatography Method (ISO 3924) (IBR; *see* §171.7 of this subchapter); or

(v) Volatile Organic Liquids—Determination of Boiling Range of Organic Solvents Used as Raw Materials (ISO 4626) (IBR; *see* §171.7 of this subchapter).

(b) *Criteria for inclusion of viscous Class 3 materials in Packing Group III.*

(1) Viscous Class 3 materials in Packing Group II with a flash point of less than 23°C (73°F) may be grouped in Packing Group III provided that—

(i) Less than 3 percent of the clear solvent layer separates in the solvent separation test;

(ii) The mixture does not contain any substances with a primary or a subsidiary risk of Division 6.1 or Class 8;

(iii) The capacity of the packaging is not more than 30 L (7.9 gallons); and

(iv) The viscosity and flash point are in accordance with the following table:

Flow time t in seconds	Jet diameter in mm	Flash point c.c.
20<t≤60	4	above 17°C (62.6°F).
60<t≤100	4	above 10°C (50°F).
20<t≤32	6	above 5°C (41°F).
32<t≤44	6	above −1°C (31.2°F).
44<t≤100	6	above −5°C (23°F).
100<t	6	−5°C (23°F) and below.

(2) The methods by which the tests referred to in paragraph (b)(1) of this section shall be performed are as follows:

(i) *Viscosity Test.* The flow time in seconds is determined at 23°C (73.4°F) using the ISO standard cup with a 4 mm (0.16 inch) jet as set forth in ISO 2431 (IBR, see §171.7 of this subchapter). Where the flow time exceeds 100 seconds, a further test is carried out using the ISO standard cup with a 6 mm (0.24 inch) jet.

(ii) *Solvent Separation Test.* This test is carried out at 23°C (73°F) using a 100.0 mL (3 ounces) measuring cylinder of the stoppered type of approximately 25.0 cm (9.8 inches) total height and of a uniform internal diameter of approximately 30 mm (1.2 inches) over the calibrated section. The sample should be stirred to obtain a uniform consistency, and poured in up to the 100 mL (3 ounces) mark. The stopper should be inserted and the cylinder left standing undisturbed for 24 hours. After 24 hours, the height of the upper separated layer should be measured and the percentage of this layer as compared with the total height of the sample calculated.

(c) *Transitional provisions.* The criteria for packing group assignments in effect on December 31, 2006, may continue to be used until January 1, 2012.

§173.124 Class 4, Divisions 4.1, 4.2 and 4.3—Definitions.

(a) *Division 4.1 (Flammable Solid).* For the purposes of this subchapter, **flammable solid (Division 4.1)** means any of the following three types of materials:

(1) Desensitized explosives that—

(i) When dry are Explosives of Class 1 other than those of compatibility group A, which are wetted with sufficient water, alcohol, or plasticizer to suppress explosive properties; and

(ii) Are specifically authorized by name either in the §172.101 Table or have been assigned a shipping name and hazard class by the Associate Administrator under the provisions of—

(A) A special permit issued under subchapter A of this chapter; or

(B) An approval issued under §173.56(i) of this part.

(2)(i) Self-reactive materials are materials that are thermally unstable and that can undergo a strongly exothermic decomposition even without participation of oxygen (air). A material is excluded from this definition if any of the following applies:

(A) The material meets the definition of an explosive as prescribed in subpart C of this part, in which case it must be classed as an explosive;

(B) The material is forbidden from being offered for transportation according to §172.101 of this subchapter or §173.21;

(C) The material meets the definition of an oxidizer or organic peroxide as prescribed in subpart D of this part, in which case it must be so classed;

(D) The material meets one of the following conditions:

(*1*) Its heat of decomposition is less than 300 J/g; or

(*2*) Its self-accelerating decomposition temperature (SADT) is greater than 75°C (167°F) for a 50 kg package; or

(*3*) It is an oxidizing substance in Division 5.1 containing less than 5.0% combustible organic substances; or

(E) The Associate Administrator has determined that the material does not present a hazard which is associated with a Division 4.1 material.

(ii) *Generic types.* Division 4.1 self-reactive materials are assigned to a generic system consisting of seven types. A self-reactive substance identified by technical name in the Self-Reactive Materials Table in §173.224 is assigned to a generic type in accordance with that Table. Self-reactive materials not identified in the Self-Reactive Materials Table in §173.224 are assigned to generic types under the procedures of paragraph (a)(2)(iii) of this section.

(A) *Type A.* Self-reactive material type A is a self-reactive material which, as packaged for transportation, can detonate or deflagrate rapidly. Transportation of type A self-reactive material is forbidden.

(B) *Type B.* Self-reactive material type B is a self-reactive material which, as packaged for transportation, neither detonates nor deflagrates rapidly, but is liable to undergo a thermal explosion in a package.

(C) *Type C.* Self-reactive material type C is a self-reactive material which, as packaged for transportation, neither detonates nor deflagrates rapidly and cannot undergo a thermal explosion.

(D) *Type D.* Self-reactive material type D is a self-reactive material which

(*1*) Detonates partially, does not deflagrate rapidly and shows no violent effect when heated under confinement;

(*2*) Does not detonate at all, deflagrates slowly and shows no violent effect when heated under confinement; or

(*3*) Does not detonate or deflagrate at all and shows a medium effect when heated under confinement.

(E) *Type E.* Self-reactive material type E is a self-reactive material which, in laboratory testing, neither detonates nor deflagrates at all and shows only a low or no effect when heated under confinement.

(F) *Type F.* Self-reactive material type F is a self-reactive material which, in laboratory testing, neither detonates in the cavitated state nor deflagrates at all and shows only a low or no effect when heated under confinement as well as low or no explosive power.

(G) *Type G.* Self-reactive material type G is a self-reactive material which, in laboratory testing, does not detonate in the cavitated state, will not deflagrate at all, shows no effect when heated under confinement, nor shows any explosive power. A type G self-reactive material is not subject to the requirements of this subchapter for self-reactive material of Division 4.1 provided that it is thermally stable (self-accelerating decomposition temperature is 50°C (122°F) or higher for a 50 kg (110 pounds) package). A self-reactive material meeting all characteristics of type G except thermal stability is classed as a type F self-reactive, temperature control material.

(iii) *Procedures for assigning a self-reactive material to a generic type.* A self-reactive material must be assigned to a generic type based on—

(A) Its physical state (i.e. liquid or solid), in accordance with the definition of liquid and solid in §171.8 of this subchapter;

(B) A determination as to its control temperature and emergency temperature, if any, under the provisions of §173.21(f);

(C) Performance of the self-reactive material under the test procedures specified in the UN Manual of Tests and Criteria (IBR, see §171.7 of this subchapter) and the provisions of paragraph (a)(2)(iii) of this section; and

(D) Except for a self-reactive material which is identified by technical name in the Self-Reactive Materials Table in §173.224(b) or a self-reactive material which may be shipped as a sample under the provisions of §173.224, the self-reactive material is approved in writing by the Associate Administrator. The person requesting approval shall submit to the Associate Administrator the tentative shipping description and generic type and—

(*1*) All relevant data concerning physical state, temperature controls, and tests results; or

(*2*) An approval issued for the self-reactive material by the competent authority of a foreign government.

(iv) *Tests.* The generic type for a self-reactive material must be determined using the testing protocol from Figure 14.2 (Flow Chart for Assigning Self-Reactive Substances to Division 4.1) from the UN Manual of Tests and Criteria.

(3) Readily combustible solids are materials that—

(i) Are solids which may cause a fire through friction, such as matches;

(ii) Show a burning rate faster than 2.2 mm (0.087 inches) per second when tested in accordance with the UN Manual of Tests and Criteria (IBR, see §171.7 of this sub-chapter); or

(iii) Any metal powders that can be ignited and react over the whole length of a sample in 10 minutes or less, when tested in accordance with the UN Manual of Tests and Criteria.

(b) *Division 4.2 (Spontaneously Combustible Material).* For the purposes of this subchapter, **spontaneously combustible material (Division 4.2)** means—

(1) A pyrophoric material. A **pyrophoric material** is a liquid or solid that, even in small quantities and without an external ignition source, can ignite within five (5) minutes after coming in contact with air when tested according to the UN Manual of Tests and Criteria.

(2) A self-heating material. A **self-heating material** is a material that through a process where the gradual reaction of that substance with oxygen (in air) generates heat. If the rate of heat production exceeds the rate of heat loss, then the temperature of the substance will rise which, after an induction time, may lead to self-ignition and combustion. A material of this type which exhibits spontaneous ignition or if the temperature of the sample exceeds 200°C (392°F) during the 24-hour test period when tested in accordance with UN Manual of Tests and Criteria (IBR; *see* §171.7 of this subchapter), is classed as a Division 4.2 material.

(c) *Division 4.3 (Dangerous when wet material).* For the purposes of this chapter, **dangerous when wet material (Division 4.3)** means a material that, by contact with water, is liable to become spontaneously flammable or to give off flammable or toxic gas at a rate greater than 1 L per kilogram of the material, per hour, when tested in accordance with UN Manual of Tests and Criteria.

§173.125 Class 4 — Assignment of packing group.

(a) The packing group of a Class 4 material is assigned in column (5) of the §172.101 Table. When the §172.101 Table provides more than one packing group for a hazardous material, the packing group shall be determined on the basis of test results following test methods given in the UN Manual of Tests and Criteria (IBR, see §171.7 of this subchapter) and by applying the appropriate criteria given in this section.

(b) Packing group criteria for readily combustible materials of Division 4.1 are as follows:

(1) Powdered, granular or pasty materials must be classified in Division 4.1 when the time of burning of one or more of the test runs, in accordance with the UN Manual of Tests and Criteria, is less than 45 seconds or the rate of burning is more than 2.2 mm/s. Powders of metals or metal alloys must be classified in Division 4.1 when they can be ignited and the reaction spreads over the whole length of the sample in 10 minutes or less.

(2) Packing group criteria for readily combustible materials of Division 4.1 are assigned as follows:

(i) For readily combustible solids (other than metal powders), Packing Group II if the burning time is less than 45 seconds and the flame passes the wetted zone. Packing Group II must be assigned to powders of metal or metal alloys if the zone of reaction spreads over the whole length of the sample in 5 minutes or less.

(ii) For readily combustible solids (other than metal powders), Packing Group III must be assigned if the burning rate time is less than 45 seconds and the wetted zone stops the flame propagation for at least 4 minutes. Packing Group III must be assigned to metal powders if the reaction spreads over the whole length of the sample in more than 5 minutes but not more than 10 minutes.

(c) Packing group criteria for Division 4.2 materials is as follows:

(1) Pyrophoric liquids and solids of Division 4.2 are assigned to Packing Group I.

(2) A self-heating material is assigned to—

(i) Packing Group II, if the material gives a positive test result when tested with a 25 mm cube size sample at 140°C; or

(ii) Packing Group III, if—

(A) A positive test result is obtained in a test using a 100 mm sample cube at 140°C and a negative test result is obtained in a test using a 25 mm sample cube at 140°C and the substance is transported in packagings with a volume of more than 3 cubic meters; or

(B) A positive test result is obtained in a test using a 100 mm sample cube at 120°C and a negative result is obtained in a test using a 25 mm sample cube at 140°C and the substance is transported in packagings with a volume of more than 450 L; or

(C) A positive result is obtained in a test using a 100 mm sample cube at 100°C and a negative result is obtained in a test using a 25 mm sample cube at 140°C and the substance is transported in packagings with a volume of less than 450 L.

(d) A Division 4.3 dangerous when wet material is assigned to—

(1) Packing Group I, if the material reacts vigorously with water at ambient temperatures and demonstrates a tendency for the gas produced to ignite spontaneously, or which reacts readily with water at ambient temperatures such that the rate of evolution of flammable gases is equal or greater than 10 L per kilogram of material over any one minute;

(2) Packing Group II, if the material reacts readily with water at ambient temperatures such that the maximum rate of evolution of flammable gases is equal to or greater than 20 L per kilogram of material per hour, and which does not meet the criteria for Packing Group I; or

(3) Packing Group III, if the material reacts slowly with water at ambient temperatures such that the maximum rate of evolution of flammable gases is greater than 1 L per kilogram of material per hour, and which does not meet the criteria for Packing Group I or II.

§173.127 Class 5, Division 5.1 — Definition and assignment of packing groups.

(a) *Definition.* For the purpose of this subchapter, **oxidizer (Division 5.1)** means a material that may, generally by yielding oxygen, cause or enhance the combustion of other materials.

(1) A solid material is classed as a Division 5.1 material if, when tested in accordance with the UN Manual of Tests and Criteria (IBR, see §171.7 of this subchapter), its mean burning time is less than or equal to the burning time of a 3:7 potassium bromate/cellulose mixture.

(2) A liquid material is classed as a Division 5.1 material if, when tested in accordance with the UN Manual of Tests and Criteria, it spontaneously ignites or its mean time for a pressure rise from 690 kPa to 2070 kPa gauge is less then the time of a 1:1 nitric acid (65 percent)/cellulose mixture.

(b) *Assignment of packing groups.* (1) The packing group of a Division 5.1 material which is a solid shall be assigned using the following criteria:

(i) Packing Group I, for any material which, in either concentration tested, exhibits a mean burning time less than the mean burning time of a 3:2 potassium bromate/cellulose mixture.

(ii) Packing Group II, for any material which, in either concentration tested, exhibits a mean burning time less than or equal to the mean burning time of a 2:3 potassium bromate/cellulose mixture and the criteria for Packing Group I are not met.

(iii) Packing Group III for any material which, in either concentration tested, exhibits a mean burning time less than or equal to the mean burning time of a 3:7 potassium bromate/cellulose mixture and the criteria for Packing Group I and II are not met.

(2) The packing group of a Division 5.1 material which is a liquid shall be assigned using the following criteria:

(i) Packing Group I for:

(A) Any material which spontaneously ignites when mixed with cellulose in a 1:1 ratio; or

(B) Any material which exhibits a mean pressure rise time less than the pressure rise time of a 1:1 perchloric acid (50 percent)/cellulose mixture.

(ii) Packing Group II, any material which exhibits a mean pressure rise time less than or equal to the pressure rise time of a 1:1 aqueous sodium chlorate solution (40 percent)/cellulose mixture and the criteria for Packing Group I are not met.

(iii) Packing Group III, any material which exhibits a mean pressure rise time less than or equal to the pressure rise time of a 1:1 nitric acid (65 percent)/cellulose mixture and the criteria for Packing Group I and II are not met.

§173.128 Class 5, Division 5.2 — Definitions and types.

(a) *Definitions.* For the purposes of this subchapter, **organic peroxide (Division 5.2)** means any organic compound containing oxygen (O) in the bivalent -O-O- structure and which may be considered a derivative of hydrogen peroxide, where one or more of the hydrogen atoms have been replaced by organic radicals, unless any of the following paragraphs applies:

(1) The material meets the definition of an explosive as prescribed in subpart C of this part, in which case it must be classed as an explosive;

(2) The material is forbidden from being offered for transportation according to §172.101 of this subchapter or §173.21;

(3) The Associate Administrator has determined that the material does not present a hazard which is associated with a Division 5.2 material; or

(4) The material meets one of the following conditions:

(i) For materials containing no more than 1.0 percent hydrogen peroxide, the available oxygen, as calculated using the equation in paragraph (a)(4)(ii) of this section, is not more than 1.0 percent, or

(ii) For materials containing more than 1.0 percent but not more than 7.0 percent hydrogen peroxide, the available oxygen, content (O_a) is not more than 0.5 percent, when determined using the equation:

$$O_a = 16 \times \sum_{i=1}^{k} \frac{n_i c_i}{m_i}$$

where, for a material containing k species of organic peroxides:
n_i=number of -O-O- groups per molecule of the *i*th species
c_i=concentration (mass percent) of the *i*th species
m_i=molecular mass of the *i*th species

(b) *Generic types*. Division 5.2 organic peroxides are assigned to a generic system which consists of seven types. An organic peroxide identified by technical name in the Organic Peroxides Table in §173.225 is assigned to a generic type in accordance with that Table. Organic peroxides not identified in the Organic Peroxides Table are assigned to generic types under the procedures of paragraph (c) of this section.

(1) *Type A*. Organic peroxide type A is an organic peroxide which can detonate or deflagrate rapidly as packaged for transport. Transportation of type A organic peroxides is forbidden.

(2) *Type B*. Organic peroxide type B is an organic peroxide which, as packaged for transport, neither detonates nor deflagrates rapidly, but can undergo a thermal explosion.

(3) *Type C*. Organic peroxide type C is an organic peroxide which, as packaged for transport, neither detonates nor deflagrates rapidly and cannot undergo a thermal explosion.

(4) *Type D*. Organic peroxide type D is an organic peroxide which—

(i) Detonates only partially, but does not deflagrate rapidly and is not affected by heat when confined;

(ii) Does not detonate, deflagrates slowly, and shows no violent effect if heated when confined; or

(iii) Does not detonate or deflagrate, and shows a medium effect when heated under confinement.

(5) *Type E*. Organic peroxide type E is an organic peroxide which neither detonates nor deflagrates and shows low, or no, effect when heated under confinement.

(6) *Type F*. Organic peroxide type F is an organic peroxide which will not detonate in a cavitated state, does not deflagrate, shows only a low, or no, effect if heated when confined, and has low, or no, explosive power.

(7) *Type G*. Organic peroxide type G is an organic peroxide which will not detonate in a cavitated state, will not deflagrate at all, shows no effect when heated under confinement, and shows no explosive power. A type G organic peroxide is not subject to the requirements of this subchapter for organic peroxides of Division 5.2 provided that it is thermally stable (self-accelerating decomposition temperature is 50°C (122°F) or higher for a 50 kg (110 pounds) package). An organic peroxide meeting all characteristics of type G except thermal stability and requiring temperature control is classed as a type F, temperature control organic peroxide.

(c) *Procedure for assigning an organic peroxide to a generic type*. An organic peroxide shall be assigned to a generic type based on—

(1) Its physical state (i.e., liquid or solid), in accordance with the definitions for liquid and solid in §171.8 of this subchapter;

(2) A determination as to its control temperature and emergency temperature, if any, under the provisions of §173.21(f); and

(3) Performance of the organic peroxide under the test procedures specified in the UN Manual of Tests and Criteria (IBR, see §171.7 of this subchapter), and the provisions of paragraph (d) of this section.

(d) *Approvals*. (1) An organic peroxide must be approved, in writing, by the Associate Administrator, before being offered for transportation or transported, including assignment of a generic type and shipping description, except for—

(i) An organic peroxide which is identified by technical name in the Organic Peroxides Table in §173.225(c);

(ii) A mixture of organic peroxides prepared according to §173.225(b); or

(iii) An organic peroxide which may be shipped as a sample under the provisions of §173.225(b).

(2) A person applying for an approval must submit all relevant data concerning physical state, temperature controls, and tests results or an approval issued for the organic peroxide by the competent authority of a foreign government.

(e) *Tests*. The generic type for an organic peroxide shall be determined using the testing protocol from Figure 20.1(a) (Classification and Flow Chart Scheme for Organic Peroxides) from the UN Manual of Tests and Criteria (IBR, see §171.7 of this subchapter).

§173.129 Class 5, Division 5.2 — Assignment of packing group.

All Division 5.2 materials are assigned to Packing Group II in Column 5 of the §172.101 Table.

§173.132 Class 6, Division 6.1 — Definitions.

(a) For the purpose of this subchapter, **poisonous material (Division 6.1)** means a material, other than a gas, which is known to be so toxic to humans as to afford a hazard to health during transportation, or which, in the absence of adequate data on human toxicity:

(1) Is presumed to be toxic to humans because it falls within any one of the following categories when tested on laboratory animals (whenever possible, animal test data that has been reported in the chemical literature should be used):

(i) *Oral Toxicity.* A liquid or solid with an LD_{50} for acute oral toxicity of not more than 300 mg/kg.

(ii) *Dermal Toxicity.* A material with an LD_{50} for acute dermal toxicity of not more than 1000 mg/kg.

(iii) *Inhalation Toxicity.* (A) A dust or mist with an LC_{50} for acute toxicity on inhalation of not more than 4 mg/L; or

(B) A material with a saturated vapor concentration in air at 20°C (68°F) greater than or equal to one-fifth of the LC_{50} for acute toxicity on inhalation of vapors and with an LC_{50} for acute toxicity on inhalation of vapors of not more than 5000 mL/m^3; or

(2) Is an irritating material, with properties similar to tear gas, which causes extreme irritation, especially in confined spaces.

(b) For the purposes of this subchapter—

(1) LD_{50} (median lethal dose) for acute oral toxicity is the statistically derived single dose of a substance that can be expected to cause death within 14 days in 50% of young adult albino rats when administered by the oral route. The LD_{50} value is expressed in terms of mass of test substance per mass of test animal (mg/kg).

(2) LD_{50} for acute dermal toxicity means that dose of the material which, administered by continuous contact for 24 hours with the shaved intact skin (avoiding abrading) of an albino rabbit, causes death within 14 days in half of the animals tested. The number of animals tested must be sufficient to give statistically valid results and be in conformity with good pharmacological practices. The result is expressed in mg/kg body mass.

(3) LC_{50} for acute toxicity on inhalation means that concentration of vapor, mist, or dust which, administered by continuous inhalation for one hour to both male and female young adult albino rats, causes death within 14 days in half of the animals tested. If the material is administered to the animals as a dust or mist, more than 90 percent of the particles available for inhalation in the test must have a diameter of 10 microns or less if it is reasonably foreseeable that such concentrations could be encountered by a human during transport. The result is expressed in mg/L of air for dusts and mists or in mL/m^3 of air (parts per million) for vapors. See §173.133(b) for LC_{50} determination for mixtures and for limit tests.

(i) When provisions of this subchapter require the use of the LC_{50} for acute toxicity on inhalation of dusts and mists based on a one-hour exposure and such data is not available, the LC_{50} for acute toxicity on inhalation based on a four-hour exposure may be multiplied by four and the product substituted for the one-hour LC_{50} for acute toxicity on inhalation.

(ii) When the provisions of this subchapter require the use of the LC_{50} for acute toxicity on inhalation of vapors based on a one-hour exposure and such data is not available, the LC_{50} for acute toxicity on inhalation based on a four-hour exposure may be multiplied by two and the product substituted for the one-hour LC_{50} for acute toxicity on inhalation.

(iii) A solid substance should be tested if at least 10 percent of its total mass is likely to be dust in a respirable range, e.g. the aerodynamic diameter of that particle-fraction is 10 microns or less. A liquid substance should be tested if a mist is likely to be generated in a leakage of the transport containment. In carrying out the test both for solid and liquid substances, more than 90% (by mass) of a specimen prepared for inhalation toxicity testing must be in the respirable range as defined in this paragraph (b)(3)(iii).

(c) For purposes of classifying and assigning packing groups to mixtures possessing oral or dermal toxicity hazards according to the criteria in §173.133(a)(1), it is necessary to determine the acute LD_{50} of the mixture. If a mixture contains more than one active constituent, one of the following methods may be used to determine the oral or dermal LD_{50} of the mixture:

(1) Obtain reliable acute oral and dermal toxicity data on the actual mixture to be transported;

(2) If reliable, accurate data is not available, classify the formulation according to the most hazardous constituent of the mixture as if that constituent were present in the same concentration as the total concentration of all active constituents; or

(3) If reliable, accurate data is not available, apply the formula:

$$\frac{C_A}{T_A} + \frac{C_B}{T_B} + \frac{C_Z}{T_Z} = \frac{100}{T_M}$$

where:
C = the % concentration of constituent A, B ... Z in the mixture;
T = the oral LD_{50} values of constituent A, B ... Z;
T_M = the oral LD_{50} value of the mixture.

Note to formula in paragraph (c)(3): This formula also may be used for dermal toxicities provided that this information is available on the same species for all constituents. The use of this formula does not take into account any potentiation or protective phenomena.

(d) The foregoing categories shall not apply if the Associate Administrator has determined that the physical characteristics of the material or its probable hazards to humans as shown by documented experience indicate that the material will not cause serious sickness or death.

(e) *Transitional provisions.* The Division 6.1 classification criteria in effect on December 31, 2006, may continue to be used until January 1, 2012.

§173.133 Assignment of packing group and hazard zones for Division 6.1 materials.

(a) The packing group of Division 6.1 materials shall be as assigned in Column 5 of the §172.101 Table. When the §172.101 Table provides more than one packing group or hazard zone for a hazardous material, the packing group and hazard zone shall be determined by applying the following criteria:

(1) The packing group assignment for routes of administration other than inhalation of vapors shall be in accordance with the following table:

Packing Group	Oral toxicity LD_{50} (mg/kg)	Dermal toxicity LD_{50} (mg/kg)	Inhalation toxicity by dusts and mists LC_{50}(mg/L)
I	≤ 5.0	≤ 50	≤ 0.2
II	> 5.0 and ≤ 50	> 50 and ≤ 200	> 0.2 and ≤ 2.0
III	> 50 and ≤ 300	> 200 and ≤ 1000	> 2.0 and ≤ 4.0

(2)(i) The packing group and hazard zone assignments for liquids (see §173.115(c) of this subpart for gases) based on inhalation of vapors shall be in accordance with the following Table:

Packing Group	Vapor concentration and toxicity
I (Hazard Zone A)	$V \geq 500\ LC_{50}$ and $LC_{50} \leq 200$ mL/M³.
I (Hazard Zone B)	$V \geq 10\ LC_{50}$; $LC_{50} \leq 1000$ mL/m³; and the criteria for Packing Group I, Hazard Zone A are not met.
II	$V \geq LC_{50}$; $LC_{50} \leq 3000$ mL/m³; and the criteria for Packing Group I, are not met.
III	$V \geq .2\ LC_{50}$; $LC_{50} \leq 5000$ mL/m³; and the criteria for Packing Groups I and II, are not met.

Note 1: V is the saturated vapor concentration in air of the material in mL/m³ at 20 °C and standard atmospheric pressure.

Note 2: A liquid in Division 6.1 meeting criteria for Packing Group I, Hazard Zones A or B stated in paragraph (a)(2) of this section is a material poisonous by inhalation subject to the additional hazard communication requirements in §§172.203(m), §172.313 and table 1 of §172.504(e) of this subchapter.

(ii) These criteria are represented graphically in Figure 1:

Figure 1

Inhalation Toxicity: Packing Group and Hazard Zone Borderlines

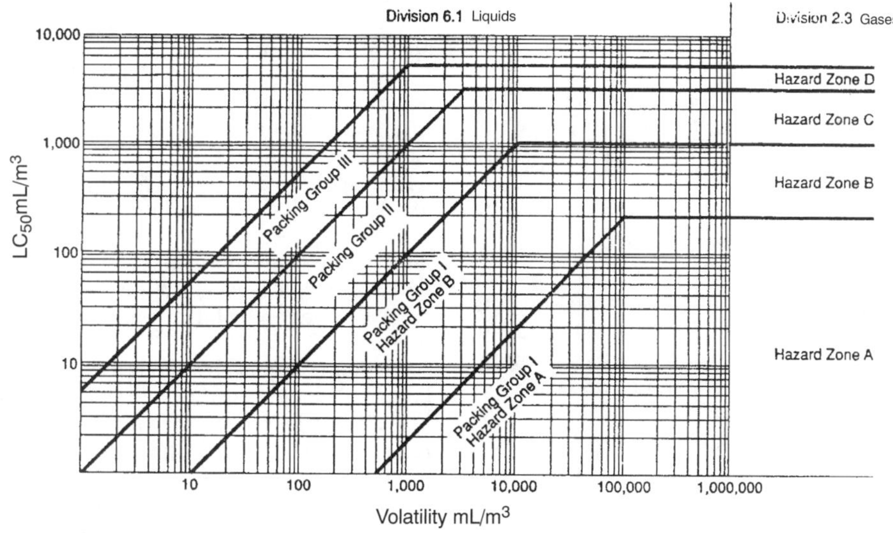

(3) When the packing group determined by applying these criteria is different for two or more (oral, dermal or inhalation) routes of administration, the packing group assigned to the material shall be that indicated for the highest degree of toxicity for any of the routes of administration.

(4) Notwithstanding the provisions of this paragraph, the packing group and hazard zone of a tear gas substance is as assigned in Column 5 of the §172.101 Table.

(b) The packing group and hazard zone for Division 6.1 mixtures that are poisonous (toxic) by inhalation may be determined by one of the following methods:

(1) Where LC_{50} data is available on each of the poisonous (toxic) substances comprising the mixture—

(i) The LC_{50} of the mixture is estimated using the formula:

$$LC_{50}(\text{mixture}) = \frac{1}{\sum_{i=1}^{n} \frac{f_i}{LC_{50i}}}$$

CLASSIFICATION & EXCEPTIONS

where f_i=mole fraction of the ith component substance of the liquid.

LC_{50i}=mean lethal concentration of the ith component substance in mL/m^3

(ii) The volatility of each component substance is estimated using the formula:

$$V_i = P_i \times \frac{10^6}{101.3} mL/m^3$$

where:

P_i=partial pressure of the *ith* component substance in kPa at 20 °C and one atmospheric pressure. Pi may be calculated according to Raoult's Law using appropriate activity coefficients. Where activity coefficients are not available, the coefficient may be assumed to be 1.0.

(iii) The ratio of the volatility to the LC_{50} is calculated using the formula:

$$R = \sum_{i=1}^{n} \frac{V_i}{LC_{50i}}$$

(iv) Using the calculated values LC_{50} (mixture) and R, the packing group for the mixture is determined as follows:

Packaging group (hazard zone)	Ratio of volatility and LC_{50}
I (Hazard Zone A)	R ≥500 and LC_{50} (mixture) ≤200 mL/m^3.
I (Hazard Zone B)	R ≥ 10 and LC_{50} (mixture) ≤ 1000 mL/m^3; and the criteria for Packing Group I, Hazard Zone A, are not met.
II	R ≥ 1 and LC_{50} (mixture) ≤ 3000 mL/m^3; and the criteria for Packing Group I, Hazard Zones A and B are not met.
III	R ≥ ⅕ and LC_{50} (mixture) ≤ 5000 mL/m^3; and the criteria for Packing Group I, Hazard Zones A and B and Packing Group II are not met.

(2) In the absence of LC_{50} data on the poisonous (toxic) constituent substances, the mixture may be assigned a packing group and hazard zone based on the following simplified threshold toxicity tests. When these threshold tests are used, the most restrictive packing group and hazard zone must be determined and used for the transportation of the mixture.

(i) A mixture is assigned to Packing Group I, Hazard Zone A only if both the following criteria are met:

(A) A sample of the liquid mixture is vaporized and diluted with air to create a test atmosphere of 200 mL/m^3 vaporized mixture in air. Ten albino rats (five male and five female) are exposed to the test atmosphere as determined by an analytical method appropriate for the material being classified for one hour and observed for fourteen days. If five or more of the animals die within the fourteen-day observation period, the mixture is presumed to have an LC_{50} equal to or less than 200 mL/m^3.

(B) A sample of the vapor in equilibrium with the liquid mixture is diluted with 499 equal volumes of air to form a test atmosphere. Ten albino rats (five male and five female) are exposed to the test atmosphere for one hour and observed for fourteen days. If five or more of the animals die within the fourteen-day observation period, the mixture is presumed to have a volatility equal to or greater than 500 times the mixture LC_{50}.

(ii) A mixture is assigned to Packing Group I, Hazard Zone B only if both the following criteria are met, and the mixture does not meet the criteria for Packing Group I, Hazard Zone A:

(A) A sample of the liquid mixture is vaporized and diluted with air to create a test atmosphere of 1000 mL/m^3 vaporized mixture in air. Ten albino rats (five male and five female) are exposed to the test atmosphere for one hour and observed for fourteen days. If five or more of the animals die within the fourteen-day observation period, the mixture is presumed to have an LC_{50} equal to or less than 1000 mL/m^3.

(B) A sample of the vapor in equilibrium with the liquid mixture is diluted with 9 equal volumes of air to form a test atmosphere. Ten albino rats (five male and five female) are exposed to the test atmosphere for one hour and observed for fourteen days. If five or more of the animals die within the fourteen-day observation period, the mixture is presumed to have a volatility equal to or greater than 10 times the mixture LC_{50}.

(iii) A mixture is assigned to Packing Group II only if both the following criteria are met, and the mixture does not meet the criteria for Packing Group I (Hazard Zones A or B):

(A) A sample of the liquid mixture is vaporized and diluted with air to create a test atmosphere of 3000 mL/m^3 vaporized mixture in air. Ten albino rats (five male and five female) are exposed to the test atmosphere for one hour and observed for fourteen days. If five or more of the animals die within the fourteen-day observation period, the mixture is presumed to have an LC_{50} equal to or less than 3000 mL/m^3.

(B) A sample of the vapor in equilibrium with the liquid mixture is used to form a test atmosphere. Ten albino rats (five male and five female) are exposed to the test atmosphere for one hour and observed for fourteen days. If five or more of the animals die within the fourteen-day observation period, the mixture is presumed to have a volatility equal to or greater than the mixture LC_{50}.

(iv) A mixture is assigned to Packing Group III only if both the following criteria are met, and the mixture does not meet the criteria for Packing Groups I (Hazard Zones A or B) or Packing Group II (Hazard Zone C):

(A) A sample of the liquid mixture is vaporized and diluted with air to create a test atmosphere of 5000 mL/m^3 vaporized mixture in air. Ten albino rats (five male and five female) are exposed to the test atmosphere for one hour and observed for fourteen days. If five or more of the animals die within the fourteen-day observation period, the mixture is presumed to have an LC_{50} equal to or less than 5000 mL/m^3.

(B) The vapor pressure of the liquid mixture is measured and if the vapor concentration is equal to or greater than 1000 mL/m^3, the mixture is presumed to have a volatility equal to or greater than ⅕ the mixture LC_{50}.

(c) *Transitional provisions.* The criteria for packing group assignments in effect on December 31, 2006, may continue to be used until January 1, 2012.

§173.134 Class 6, Division 6.2—Definitions and exceptions.

(a) *Definitions and classification criteria* For the purposes of this subchapter, the following definitions and classification criteria apply to Division 6.2 materials.

(1) **Division 6.2 (Infectious substance)** means a material known or reasonably expected to contain a pathogen. A pathogen is a microorganism (including bacteria, viruses, rickettsiae, parasites, fungi) or other agent, such as a proteinaceous infectious particle (prion), that can cause disease in humans or animals. An infectious substance must be assigned the identification number UN 2814, UN 2900, UN 3373, or UN 3291 as appropriate, and must be assigned to one of the following categories:

(i) **Category A:** An infectious substance in a form capable of causing permanent disability or life-threatening or fatal disease in otherwise healthy humans or animals when exposure to it occurs. An exposure occurs when an infectious substance is released outside of its protective packaging, resulting in physical contact with humans or animals. A Category A infectious substance must be assigned to identification number UN 2814 or UN 2900, as appropriate. Assignment to UN 2814 or UN 2900 must be based on the known medical history or symptoms of the source patient or animal, endemic local conditions, or professional judgment concerning the individual circumstances of the source human or animal.

(ii) **Category B:** An infectious substance that is not in a form generally capable of causing permanent disability or life-threatening or fatal disease in otherwise healthy humans or animals when exposure to it occurs. This includes Category B infectious substances transported for diagnostic or investigational purposes. A Category B infectious substance must be described as "Biological substance, Category B" and assigned identification number UN 3373. This does not include regulated medical waste, which must be assigned identification number UN 3291.

(2) **Biological product** means a virus, therapeutic serum, toxin, antitoxin, vaccine, blood, blood component or derivative, allergenic product, or analogous product, or arsphenamine or derivative of arsphenamine (or any other trivalent arsenic compound) applicable to the prevention, treatment, or cure of a disease or condition of human beings or animals. A *biological product* includes a material subject to regulation under 42 U.S.C. 262 or 21 U.S.C. 151–159. Unless otherwise excepted, a *biological product* known or reasonably expected to contain a pathogen that meets the definition of a Category A or B infectious substance must be assigned the identification number UN 2814, UN 2900, or UN 3373, as appropriate.

(3) **Culture** means an infectious substance containing a pathogen that is intentionally propagated. *Culture* does not include a human or animal patient specimen as defined in paragraph (a)(4) of this section.

(4) **Patient specimen** means human or animal material collected directly from humans or animals and transported for research, diagnosis, investigational activities, or disease treatment or prevention. *Patient specimen* includes excreta, secreta, blood and its components, tissue and tissue swabs, body parts, and specimens in transport media (*e.g.*, transwabs, culture media, and blood culture bottles).

(5) **Regulated medical waste or clinical waste or (bio) medical waste** means a waste or reusable material derived from the medical treatment of an animal or human, which includes diagnosis and immunization, or from biomedical research, which includes the production and testing of biological products. Regulated medical waste or clinical waste or (bio) medical waste containing a Category A infectious substance must be classed as an infectious substance, and assigned to UN 2814 or UN 2900, as appropriate.

(6) **Sharps** means any object contaminated with a pathogen or that may become contaminated with a pathogen through handling or during transportation and also capable of cutting or penetrating skin or a packaging material. *Sharps* includes needles, syringes, scalpels, broken glass, culture slides, culture dishes, broken capillary tubes, broken rigid plastic, and exposed ends of dental wires.

(7) **Toxin** means a Division 6.1 material from a plant, animal, or bacterial source. A *toxin* containing an infectious substance or a *toxin* contained in an infectious substance must be classed as Division 6.2, described as an infectious substance, and assigned to UN 2814 or UN 2900, as appropriate.

(8) **Used health care product** means a medical, diagnostic, or research device or piece of equipment, or a personal care product used by consumers, medical professionals, or pharmaceutical providers that does not meet the definition of a patient specimen, biological product, or regulated medical waste, is contaminated with potentially infectious body fluids or materials, and is not decontaminated or disinfected to remove or mitigate the infectious hazard prior to transportation.

(b) *Exceptions.* The following are not subject to the requirements of this subchapter as Division 6.2 materials:

(1) A material that does not contain an infectious substance or that is unlikely to cause disease in humans or animals.

(2) Non-infectious biological materials from humans, animals, or plants. Examples include non-infectious cells, tissue cultures, blood or plasma from individuals not suspected of having an infectious disease, DNA, RNA or other non-infectious genetic elements.

(3) A material containing micro-organisms that are non-pathogenic to humans or animals.

(4) A material containing pathogens that have been neutralized or inactivated such that they no longer pose a health risk.

(5) A material with a low probability of containing an infectious substance, or where the concentration of the infectious substance is at a level naturally occurring in the environment so it cannot cause disease when exposure to it occurs. Examples of these materials include: Foodstuffs; environmental samples, such as water or a sample of dust or mold; and substances that have been treated so that the pathogens have been neutralized or deactivated, such as a material treated by steam sterilization, chemical disinfection, or other appropriate method, so it no longer meets the definition of an infectious substance.

(6) A biological product, including an experimental or investigational product or component of a product, subject to Federal approval, permit, review, or licensing requirements, such as those required by the Food and Drug Administration of the U.S. Department of Health and Human Services or the U.S. Department of Agriculture.

(7) Blood collected for the purpose of blood transfusion or the preparation of blood products; blood products; plasma; plasma derivatives; blood components; tissues or organs intended for use in transplant operations; and human cell, tissues, and cellular and tissue-based products regulated under authority of the Public Health Service Act (42 U.S.C. 264–272) and/or the Food, Drug, and Cosmetic Act (21 U.S.C. 332 *et seq.*).

(8) Blood, blood plasma, and blood components collected for the purpose of blood transfusion or the preparation of blood products and sent for testing as part of the collection process, except where the person collecting the blood has reason to believe it contains an infectious substance, in which case the test sample must be shipped as a Category A or Category B infectious substance in accordance with §173.196 or §173.199, as appropriate.

(9) Dried blood spots or specimens for fecal occult blood detection placed on absorbent filter paper or other material.

(10) A Division 6.2 material, other than a Category A infectious substance, contained in a patient sample being transported for research, diagnosis, investigational activities, or disease treatment or prevention, or a biological product, when such materials are transported by a private or contract carrier in a motor vehicle used exclusively to transport such materials. Medical or clinical equipment and laboratory products may be transported aboard the same vehicle provided they are properly packaged and secured against exposure or contamination. If the human or animal sample or biological product meets the definition of regulated medical waste in paragraph (a)(5) of this section, it must be offered for transportation and transported in conformance with the appropriate requirements for regulated medical waste.

(11) A human or animal sample (including, but not limited to, secreta, excreta, blood and its components, tissue and tissue fluids, and body parts) being transported for routine testing not related to the diagnosis of an infectious disease, such as for drug/alcohol testing, cholesterol testing, blood glucose level testing, prostate specific antibody testing, testing to monitor kidney or liver function, or pregnancy testing, or for tests for diagnosis of non-infectious diseases, such as cancer biopsies, and for which there is a low probability the sample is infectious.

(12) Laundry and medical equipment and used health care products, as follows:

(i) Laundry or medical equipment conforming to the regulations of the Occupational Safety and Health Administration of the Department of Labor in 29 CFR 1910.1030. This exception includes medical equipment intended for use, cleaning, or refurbishment, such as reusable surgical equipment, or equipment used for testing where the components within which the equipment is contained essentially function as packaging. This exception does not apply to medical equipment being transported for disposal.

(ii) Used health care products not conforming to the requirements in 29 CFR 1910.1030 and being returned to the manufacturer or the manufacturer's designee are excepted from the requirements of this subchapter when offered for transportation or transported in accordance with this paragraph (b)(12). For purposes of this paragraph, a health care product is used when it has been removed from its original packaging. Used health care products contaminated with or suspected of contamination with a Category A infectious substance may not be transported under the provisions of this paragraph.

(A) Each used health care product must be drained of free liquid to the extent practicable and placed in a watertight primary container designed and constructed to assure that it remains intact under conditions normally incident to transportation. For a used health care product capable of cutting or penetrating skin or packaging material, the primary container must be capable of retaining the product without puncture of the packaging under normal conditions of transport. Each primary container must be marked with a BIOHAZARD marking conforming to 29 CFR 1910.1030(g)(1)(i).

(B) Each primary container must be placed inside a watertight secondary container designed and constructed to assure that it remains intact under conditions normally incident to transportation. The secondary container must be marked with a BIOHAZARD marking conforming to 29 CFR 1910.1030(g)(1)(i).

(C) The secondary container must be placed inside an outer packaging with sufficient cushioning material to prevent movement between the secondary container and the outer packaging. An itemized list of the contents of the primary container and information concerning possible contamination with a Division 6.2 material, including its possible location on the product, must be placed between the secondary container and the outside packaging.

(D) Each person who offers or transports a used health care product under the provisions of this paragraph must know about the requirements of this paragraph.

(13) Any waste or recyclable material, other than regulated medical waste, including—

(i) Household waste as defined in §171.8, when transported in accordance with applicable state, local, or tribal requirements.

(ii) Sanitary waste or sewage;

(iii) Sewage sludge or compost;

(iv) Animal waste generated in animal husbandry or food production; or

(v) Medical waste generated from households and transported in accordance with applicable state, local, or tribal requirements.

(14) Corpses, remains, and anatomical parts intended for interment, cremation, or medical research at a college, hospital, or laboratory.

(15) Forensic material transported on behalf of a U.S. Government, state, local or Indian tribal government agency, except that—

(i) Forensic material known or suspected to contain a Category B infectious substance must be shipped in a packaging conforming to the provisions of §173.24.

(ii) Forensic material known or suspected to contain a Category A infectious substance or an infectious substance listed as a select agent in 42 CFR Part 73 must be transported in packaging capable of meeting the test standards in §178.609 of this subchapter. The secondary packaging must be marked with a BIOHAZARD symbol conforming to specifications in 29 CFR 1910.1030(g)(1)(i). An itemized list of contents must be enclosed between the secondary packaging and the outer packaging.

(16) Agricultural products and food as defined in the Federal Food, Drug, and Cosmetics Act (21 U.S.C. 332 *et seq.*).

(c) *Exceptions for regulated medical waste.* The following provisions apply to the transportation of regulated medical waste:

(1) A regulated medical waste transported by a private or contract carrier is excepted from—

(i) The requirement for an "INFECTIOUS SUBSTANCE" label if the outer packaging is marked with a "BIOHAZARD" marking in accordance with 29 CFR 1910.1030; and

(ii) The specific packaging requirements of §173.197, if packaged in a rigid non-bulk packaging conforming to the general packaging requirements of §§173.24 and 173.24a and packaging requirements specified in 29 CFR 1910.1030, provided the material does not include a waste concentrated stock culture of an infectious substance. Sharps containers must be securely closed to prevent leaks or punctures.

(2) The following materials may be offered for transportation and transported as a regulated medical waste when packaged in a rigid non-bulk packaging conforming to the general packaging requirements of §§173.24 and 173.24a and packaging requirements specified in 29 CFR 1910.1030 and transported by a private or contract carrier in a vehicle used exclusively to transport regulated medical waste:

(i) Waste stock or culture of a Category B infectious substance;

(ii) Plant and animal waste regulated by the Animal and Plant Health Inspection Service (APHIS);

(iii) Waste pharmaceutical materials;

(iv) Laboratory and recyclable wastes;

(v) Infectious substances that have been treated to eliminate or neutralize pathogens;

(vi) Forensic materials being transported for final destruction;

(vii) Rejected or recalled health care products;

(viii) Documents intended for destruction in accordance with the Health Insurance Portability and Accountability Act of 1996 (HIPAA) requirements; and

(ix) Medical or clinical equipment and laboratory products provided they are properly packaged and secured against exposure or contamination. Sharps containers must be securely closed to prevent leaks or punctures.

(d) If an item listed in paragraph (b) or (c) of this section meets the definition of another hazard class or if it is a hazardous substance, hazardous waste, or marine pollutant, it must be offered for transportation and transported in accordance with applicable requirements of this subchapter.

§173.136 Class 8—Definitions.

(a) For the purpose of this subchapter, **corrosive material (Class 8)** means a liquid or solid that causes full thickness destruction of human skin at the site of contact within a specified period of time. A liquid, or a solid which may become liquid during transportation, that has a severe corrosion rate on steel or aluminum based on the criteria in §173.137(c)(2) is also a corrosive material. Whenever practical, *in vitro* test methods authorized in §173.137 of this part or historical data authorized in paragraph (c) of this section should be used to determine whether a material is corrosive.

(b) If human experience or other data indicate that the hazard of a material is greater or less than indicated by the results of the tests specified in paragraph (a) of this section, PHMSA may revise its classification or make the determination that the material is not subject to the requirements of this subchapter.

(c) Skin corrosion test data produced no later than September 30, 1995, using the procedures of Part 173, Appendix A, in effect on September 30, 1995 (see 49 CFR Part 173, Appendix A, revised as of October 1, 1994) for appropriate exposure times may be used for classification and assignment of packing group for Class 8 materials corrosive to skin.

§173.137 Class 8—Assignment of packing group.

The packing group of a Class 8 material is indicated in Column 5 of the §172.101 Table. When the §172.101 Table provides more than one packing group for a Class 8 material, the packing group must be determined using data obtained from tests conducted in accordance with the OECD Guideline for the Testing of Chemicals, Number 435, "*In Vitro* Membrane Barrier Test Method for Skin Corrosion" (IBR, *see* §171.7 of this subchapter) or Number 404, "Acute Dermal Irritation/Corrosion" (IBR, *see* §171.7 of this subchapter). A material that is determined not to be corrosive in accordance with OECD Guideline for the Testing of Chemicals, Number 430, "*In Vitro* Skin Corrosion: Transcutaneous Electrical Resistance Test (TER)" (IBR, *see* §171.7 of this subchapter) or Number 431, "*In Vitro* Skin Corrosion: Human Skin Model Test" (IBR, *see* §171.7 of this subchapter) may be considered not to be corrosive to human skin for the purposes of this subchapter without further testing. However, a material determined to be corrosive in accordance with Number 430 or Number 431 must be further tested using Number 435 or Number 404. The packing group assignment using data obtained from tests conducted in accordance with OECD Guideline Number 404 or Number 435 must be as follows:

(a) *Packing Group I.* Materials that cause full thickness destruction of intact skin tissue within an observation period of up to 60 minutes starting after the exposure time of three minutes or less.

(b) *Packing Group II*. Materials other than those meeting Packing Group I criteria that cause full thickness destruction of intact skin tissue within an observation period of up to 14 days starting after the exposure time of more than three minutes but not more than 60 minutes.

(c) *Packing Group III*. Materials, other than those meeting Packing Group I or II criteria—

(1) That cause full thickness destruction of intact skin tissue within an observation period of up to 14 days starting after the exposure time of more than 60 minutes but not more than 4 hours; or

(2) That do not cause full thickness destruction of intact skin tissue but exhibit a corrosion on either steel or aluminum surfaces exceeding 6.25 mm (0.25 inch) a year at a test temperature of 55 °C (130 °F) when tested on both materials. The corrosion may be determined in accordance with the UN Manual of Tests and Criteria (IBR, see §171.7 of this subchapter) or other equivalent test methods.

Note to §173.137: When an initial test on either a steel or aluminum surface indicates the material being tested is corrosive, the follow up test on the other surface is not required.

§173.140 Class 9—Definitions.

For the purposes of this subchapter, **miscellaneous hazardous material (Class 9)** means a material which presents a hazard during transportation but which does not meet the definition of any other hazard class. This class includes:

(a) Any material which has an anesthetic, noxious or other similar property which could cause extreme annoyance or discomfort to a flight crew member so as to prevent the correct performance of assigned duties; or

(b) Any material that meets the definition in §171.8 of this subchapter for an elevated temperature material, a hazardous substance, a hazardous waste, or a marine pollutant.

§173.141 Class 9—Assignment of packing group.

The packing group of a Class 9 material is as indicated in Column 5 of the §172.101 Table.

§173.144 Other Regulated Materials (ORM)—Definitions.

Until December 31, 2013 and for the purposes of this subchapter, **ORM-D material** means a material such as a consumer commodity, cartridges, small arms or cartridges, power devices which, although otherwise subject to the regulations of this subchapter, presents a limited hazard during transportation due to its form, quantity and packaging. It must be a material for which exceptions are provided in Column (8A) of the §172.101 Hazardous Materials Table.

§173.145 Other regulated materials—assignment of packing group.

Packing groups are not assigned to ORM-D materials.

§173.150 Exceptions for Class 3 (flammable and combustible liquids).

(a) *General*. Exceptions for hazardous materials shipments in the following paragraphs are permitted only if this section is referenced for the specific hazardous material in the §172.101 Table of this subchapter.

(b) *Limited quantities*. Limited quantities of flammable liquids (Class 3) and combustible liquids are excepted from labeling requirements, unless the material is offered for transportation or transported by aircraft, and are excepted from the specification packaging requirements of this subchapter when packaged in combination packagings according to this paragraph. For transportation by aircraft, the package must also conform to applicable requirements of §173.27 of this part (*e.g.*, authorized materials, inner packaging quantity limits and closure securement) and only hazardous material authorized aboard passenger-carrying aircraft may be transported as a limited quantity. A limited quantity package that conforms to the provisions of this section is not subject to the shipping paper requirements of subpart C of part 172 of this subchapter, unless the material meets the definition of a hazardous substance, hazardous waste, marine pollutant, or is offered for transportation and transported by aircraft or vessel, and is eligible for the exceptions provided in §173.156 of this part. In addition, shipments of limited quantities are not subject to subpart F (Placarding) of part 172 of this subchapter. Each package must conform to the packaging requirements of subpart B of this part and may not exceed 30 kg (66 pounds) gross weight. Except for transportation by aircraft, the following combination packagings are authorized:

(1) For flammable liquids in Packing Group I, inner packagings not over 0.5 L (0.1 gallon) net capacity each, packed in a strong outer packaging;

(2) For flammable liquids in Packing Group II, inner packagings not over 1.0 L (0.3 gallons) net capacity each, packed in a strong outer packaging.

(3) For flammable liquids in Packing Group III and combustible liquids, inner packagings not over 5.0 L (1.3 gallons) net capacity each, packed in a strong outer packaging.

(c) *Consumer commodities*. Until December 31, 2013, a limited quantity package containing a "consumer commodity" as defined in §171.8 of this subchapter, may be renamed "Consumer commodity" and reclassed as ORM-D or, until December 31, 2012, ORM-D-AIR material and offered for transportation and transported in accordance with the applicable provisions of this subchapter in effect on October 1, 2010.

(d) *Alcoholic beverages*. An alcoholic beverage (wine and distilled spirits as defined in 27 CFR 4.10 and 5.11) is not subject to the requirements of this subchapter if it—

(1) Contains 24 percent or less alcohol by volume;

(2) Is in an inner packaging of 5 L (1.3 gallons) or less, and for transportation on passenger-carrying aircraft conforms to §175.10(a)(4) of this subchapter as checked or carry-on baggage; or

(3) Is a Packing Group III alcoholic beverage in a packaging of 250 L (66 gallons) or less, unless transported by air.

(e) *Aqueous solutions of alcohol*. An aqueous solution containing 24 percent or less alcohol by volume and no other hazardous material—

(1) May be reclassed as a combustible liquid.

(2) Is not subject to the requirements of this subchapter if it contains no less than 50 percent water.

(f) *Combustible liquids*. (1) A flammable liquid with a flash point at or above 38 °C (100 °F) that does not meet the definition of any other hazard class, may be reclassed as a combustible liquid. This provision does not apply to transportation by vessel or aircraft, except where other means of transportation is impracticable.

(2) The requirements in this subchapter do not apply to a material classed as a combustible liquid in a non-bulk packaging unless the combustible liquid is a hazardous substance, a hazardous waste, or a marine pollutant.

(3) A combustible liquid that is in a bulk packaging or a combustible liquid that is a hazardous substance, a hazardous waste, or a marine pollutant is not subject to the requirements of this subchapter except those pertaining to:

(i) Shipping papers, waybills, switching orders, and hazardous waste manifests;

(ii) Marking of packages;

(iii) Display of identification numbers on bulk packages;

(iv) For bulk packaging only, placarding requirements of subpart F of part 172 of this subchapter;

(v) Carriage aboard aircraft and vessels (for packaging requirements for transport by vessel, see §176.340 of this subchapter);

(vi) Reporting incidents as prescribed by §§171.15 and 171.16 of this subchapter;

(vii) Packaging requirements of subpart B of this part and, in addition, non-bulk packagings must conform with requirements of §173.203;

(viii) The requirements of §§173.1, 173.21, 173.24, 173.24a, 173.24b, 174.1, 177.804, 177.817, 177.834(j), and 177.837(d) of this subchapter;

(ix) The training requirements of subpart H of part 172 of this subchapter.

(x) Emergency response information requirements of subpart G of part 172.

(4) A combustible liquid that is not a hazardous substance, a hazardous waste, or marine pollutant is not subject to the requirements of this subchapter if it is a mixture of one or more components that—

(i) Has a flash point at or above 93 °C (200 °F),

(ii) Comprises at least 99 percent of the volume of the mixture, and

(iii) Is not offered for transportation or transported as a liquid at a temperature at or above its flash point.

§173.151 Exceptions for Class 4.

(a) *General*. Exceptions for hazardous materials shipments in the following paragraphs are permitted only if this section is referenced for the specific hazardous material in the §172.101 Table of this subchapter.

(b) *Limited quantities of Division 4.1*. (1) Limited quantities of flammable solids (Division 4.1) in Packing Groups II and III and, where authorized by this section, charcoal briquettes (Division 4.2) in Packing Group III, are excepted from labeling requirements, unless the material is offered for transportation or transported by aircraft, and are excepted from the specification packaging requirements of this subchapter when packaged in combination packagings according to this paragraph. For transportation by aircraft, the package must also conform to applicable requirements of §173.27 of this part (*e.g.,* authorized materials, inner packaging quantity limits and closure securement) and only hazardous material authorized aboard passenger-carrying aircraft may be transported as a limited quantity. A limited quantity package that conforms to the provisions of this section is not subject to the shipping paper requirements of subpart C of part 172 of this subchapter, unless the material meets the definition of a hazardous substance, hazardous waste, marine pollutant, or is offered for transportation and transported by aircraft or vessel, and is eligible for the exceptions provided in §173.156 of this part. In addition, shipments of limited quantities are not subject to subpart F (Placarding) of part 172 of this subchapter. Each package must conform to the packaging requirements of subpart B of this part and may not exceed 30 kg (66 pounds) gross weight. Except for transportation by aircraft, the following combination packagings are authorized:

(i) For flammable solids in Packing Group II, inner packagings not over 1.0 kg (2.2 pounds) net capacity each, packed in a strong outer packaging.

(ii) For flammable solids in Packing Group III, inner packagings not over 5.0 kg (11 pounds) net capacity each, packed in a strong outer packaging.

(2) For transportation by highway or rail, Charcoal briquettes (NA1361) may be packaged as a limited quantity in accordance with paragraph (b) of this section in packagings not exceeding 30 kg gross weight and are eligible for the exceptions provided in §173.156.

(c) *Consumer commodities* . Until December 31, 2013, a limited quantity package (including Charcoal briquettes (NA1361)) containing a "consumer commodity" as defined in §171.8 of this subchapter, may be renamed "Consumer commodity" and reclassed as ORM-D or, until December 31, 2012, ORM-D-AIR material and offered for transportation and transported in accordance with the applicable provisions of this subchapter in effect on October 1, 2010. For transportation by aircraft, the maximum net mass for Charcoal briquettes (NA1361) is 25 kg per package.

(d) *Limited quantities of Division 4.3*. Limited quantities of dangerous when wet solids (Division 4.3) in Packing Groups II and III are excepted from labeling requirements, unless the material is offered for transportation or transported by aircraft, and are excepted from the specification packaging requirements of this subchapter when packaged in combination packagings according to this paragraph. For transportation by aircraft, the package must also conform to applicable requirements of §173.27 of this part (*e.g.,* authorized materials, inner packaging quantity limits and closure securement) and only hazardous material authorized aboard passenger-carrying aircraft may be transported as a limited quantity. A limited quantity package that conforms to the provisions of this section is not subject to the shipping paper requirements of subpart C of part 172 of this subchapter, unless the material meets the definition of a hazardous substance, hazardous waste, marine pollutant, or is offered for transportation and transported by aircraft or vessel, and is eligible for the exceptions provided in §173.156 of this part. In addition, shipments of limited quantities are not subject to subpart F (Placarding) of part 172 of this subchapter. Each package must conform to the packaging requirements of subpart B of this part and may not exceed 30 kg (66 pounds) gross weight. Except for transportation by aircraft, the following combination packagings are authorized:

(1) For dangerous when wet solids in Packing Group II, inner packagings not over 0.5 kg (1.1 pounds) net capacity each, packed in a strong outer packaging.

(2) For dangerous when wet solids in Packing Group III, inner packagings not over 1.0 kg (2.2 pounds) net capacity each, packed in a strong outer packaging.

§173.152 Exceptions for Division 5.1 (oxidizers) and Division 5.2 (organic peroxides).

(a) *General*. Exceptions for hazardous materials shipments in the following paragraphs are permitted only if this section is referenced for the specific hazardous material in the §172.101 Table of this subchapter.

(b) *Limited quantities*. Limited quantities of oxidizers (Division 5.1) in Packing Group II and III and organic peroxides (Division 5.2) are excepted from labeling requirements, unless the material is offered for transportation or transported by aircraft, and are excepted from the specification packaging requirements of this subchapter when packaged in combination packagings according to this paragraph. For transportation by aircraft, the package must also conform to applicable requirements of §173.27 of this part (*e.g.,* authorized materials, inner packaging quantity limits and closure securement) and only hazardous material authorized aboard passenger-carrying aircraft may be transported as a limited quantity. A limited quantity package that conforms to the provisions of this section is not subject to the shipping paper requirements of subpart C of part 172 of this subchapter, unless the material meets the definition of a hazardous substance, hazardous waste, marine pollutant, or is offered for transportation and transported by aircraft or vessel, and is eligible for the exceptions provided in §173.156 of this part. In addition, shipments of limited quantities are not subject to subpart F (Placarding) of part 172 of this subchapter. Each package must conform to the packaging requirements of subpart B of this part and may not exceed 30 kg (66 pounds) gross weight. Except for transportation by aircraft, the following combination packagings are authorized:

(1) For oxidizers in Packing Group II, inner packagings not over 1.0 L (0.3 gallon) net capacity each for liquids or not over 1.0 kg (2.2 pounds) net capacity each for solids, packed in a strong outer packaging.

(2) For oxidizers in Packing Group III, inner packagings not over 5 L (1.3 gallons) net capacity each for liquids or not over 5.0 kg (11 lbs) net capacity each for solids, packed in a strong outer packaging.

(3) For organic peroxides that do not require temperature control during transportation—

(i) Except for transportation by aircraft, for Type B or C organic peroxides, inner packagings not over 25 mL (0.845 ounces) net capacity each for liquids or 100 g (3.528 ounces) net capacity for solids, packed in a strong outer packaging.

(ii) For Type D, E, or F organic peroxides, inner packagings not over 125 mL (4.22 ounces) net capacity each for liquids or 500 g (17.64 ounces) net capacity for solids, packed in a strong outer packaging.

(c) *Consumer commodities*. Until December 31, 2013, a limited quantity package containing a "consumer commodity" as defined in §171.8 of this subchapter, may be renamed "Consumer commodity" and reclassed as ORM-D or, until December 31, 2012, ORM-D-AIR material and offered for transportation and transported in accordance with the applicable provisions of this subchapter in effect on October 1, 2010.

§173.153 Exceptions for Division 6.1 (poisonous materials).

(a) *General*. Exceptions for hazardous materials shipments in the following paragraphs are permitted only if this section is referenced for the specific hazardous material in the §172.101 Table of this subchapter.

(b) *Limited quantities*. The exceptions in this paragraph do not apply to poison-by-inhalation materials. Limited quantities of poisonous material (Division 6.1) in Packing Groups II and III are excepted from the labeling requirements, unless the material is offered for transportation or transported by aircraft, and are excepted from the specification packaging requirements of this subchapter when packaged in combination packagings according to this paragraph. For transportation by aircraft, the package must also conform to applicable requirements of §173.27 of this part (*e.g.,* authorized materials, inner packaging quantity limits and closure securement) and only hazardous material authorized aboard passenger-carrying aircraft may be transported as a limited quantity. A limited quantity package that conforms to the provisions of this section is not subject to the shipping paper requirements of subpart C of part 172 of this subchapter, unless the material meets the definition of a hazardous substance, hazardous waste, marine pollutant, or is offered for transportation and transported by aircraft or vessel, and is eligible for the exceptions provided in §173.156 of this part. In addition, shipments of limited quantities are not subject to subpart F (Placarding) of part 172 of this subchapter. Each package must conform to the packaging requirements of subpart B of this part and may not exceed 30 kg (66 pounds) gross weight. Except for transportation by aircraft, the following combination packagings are authorized:

(1) For poisonous materials in Packing Group II, inner packagings not over 100 mL (3.38 ounces) each for liquids or 0.5 kg (1.1 pounds) each for solids, packed in a strong outer packaging. Inner packagings containing a liquid poisonous material which is also a drug or medicine in Packing Group II may be increased to not over 250 mL (8 ounces) each and packed in a strong outer packaging.

(2) For poisonous materials in Packing Group III, inner packagings not over 5 L (1.3 gallons) each for liquids or 5.0 kg (11 pounds) each for solids, packed in a strong outer packaging.

(c) *Consumer commodities*. Until December 31, 2013, a limited quantity package of poisonous material in Packing Group III or a drug or medicine in Packing Group II and III that is also a "consumer commodity" as defined in §171.8 of this subchapter, may be renamed "Consumer commodity" and reclassed as ORM-D or, until December 31, 2012, ORM-D-AIR material and offered for transportation and transported in accordance with the applicable provisions of this subchapter in effect on October 1, 2010.

§173.154 Exceptions for Class 8 (corrosive materials).

(a) *General*. Exceptions for hazardous materials shipments in the following paragraphs are permitted only if this section is referenced for the specific hazardous material in the §172.101 Table of this subchapter.

(b) *Limited quantities*. Limited quantities of corrosive material (Class 8) in Packing Groups II and III are excepted from labeling requirements, unless the material is offered for transportation or transported by aircraft, and are excepted from the specification packaging requirements of this subchapter when packaged in combination packagings according to this paragraph. For transportation by aircraft, the package must also conform to the applicable requirements of §173.27 of this part (*e.g.,* authorized materials, inner packaging quantity limits and closure securement) and only hazardous material authorized aboard passenger-carrying aircraft may be transported as a limited quantity. A limited quantity package that conforms to the provisions of this section is not subject to the shipping paper requirements of subpart C of part 172 of this subchapter, unless the material meets the definition of a hazardous substance, hazardous waste, marine pollutant, or is offered for transportation and transported by aircraft or vessel, and is eligible for the exceptions provided in §173.156 of this part. In addition, shipments of limited quantities are not subject to subpart F (Placarding) of part 172 of this subchapter. Each package must conform to the packaging requirements of subpart B of this part and may not exceed 30 kg (66 pounds) gross weight. Except for transportation by aircraft, the following combination packagings are authorized:

(1) For corrosive materials in Packing Group II, inner packagings not over 1.0 L (0.3 gallon) net capacity each for liquids or not over 1.0 kg (2.2 pounds) net capacity each for solids, packed in a strong outer packaging.

(2) For corrosive materials in Packing Group III, inner packagings not over 5.0 L (1.3 gallons) net capacity each for liquids or not over 5.0 kg (11 lbs) net capacity each for solids, packed in a strong outer packaging.

(c) *Consumer commodities*. Until December 31, 2013, a limited quantity package containing a "consumer commodity" as defined in §171.8 of this subchapter, may be renamed "Consumer commodity" and reclassed as ORM-D or, until December 31, 2012, ORM-D-AIR material and offered for transportation and transported in accordance with the applicable provisions of this subchapter in effect on October 1, 2010.

(d) *Materials corrosive to aluminium or steel only*. Except for a hazardous substance, a hazardous waste, or a marine pollutant, a material classed as a Class 8, Packing Group III, material solely because of its corrosive effect —

(1) On aluminium is not subject to any other requirements of this subchapter when transported by motor vehicle or rail car in a packaging constructed of materials that will not react dangerously with or be degraded by the corrosive material; or

(2) On steel is not subject to any other requirements of this subchapter when transported by motor vehicle or rail car in a bulk packaging constructed of materials that will not react dangerously with or be degraded by the corrosive material.

§173.155 Exceptions for Class 9 (miscellaneous hazardous materials).

(a) *General*. Exceptions for hazardous materials shipments in the following paragraphs are permitted only if this section is referenced for the specific hazardous material in the §172.101 Table of this subchapter.

(b) *Limited quantities of Class 9 materials*. Limited quantities of miscellaneous hazardous materials in Packing Groups II and III are excepted from labeling requirements, unless the material is offered for transportation or transported by aircraft, and are excepted from the specification packaging requirements of this subchapter when packaged in combination packagings according to this paragraph. Unless otherwise specified in paragraph (c) of this section, packages of limited quantities intended for transportation by aircraft must conform to the applicable requirements (*e.g.,* authorized materials, inner packaging quantity limits and closure securement) of §173.27 of this part. A limited quantity package that conforms to the provisions of this section is not subject to the shipping paper requirements of subpart C of part 172 of this subchapter, unless the material meets the definition of a hazardous substance, hazardous waste, marine pollutant, or is offered for transportation and transported by aircraft or vessel, and is eligible for the exceptions provided in §173.156 of this part. In addition, packages of limited quantities are not subject to subpart F (Placarding) of part 172 of this subchapter. Each package must conform to the packaging requirements of subpart B of this part and may not exceed 30 kg (66 pounds) gross weight. Except for transportation by aircraft, the following combination packagings are authorized:

(1) For miscellaneous materials in Packing Group II, inner packagings not over 1.0 L (0.3 gallon) net capacity each for liquids or not over 1.0 kg (2.2 pounds) net capacity each for solids, packed in a strong outer packaging.

(2) For miscellaneous materials in Packing Group III, inner packagings not over 5.0 L (1.3 gallons) net capacity each for liquids or not over 5.0 kg (11 lbs) net capacity each for solids, packed in a strong outer packaging.

(c) *Consumer commodities.* Until December 31, 2013, a limited quantity package containing a "consumer commodity" as defined in §171.8 of this subchapter, may be renamed "Consumer commodity" and reclassed as ORM-D or, until December 31, 2012, ORM-D-AIR material and offered for transportation and transported in accordance with the applicable provisions of this subchapter in effect on October 1, 2010.

§173.156 Exceptions for limited quantity and ORM.

(a) Exceptions for hazardous materials shipments in the following paragraphs are permitted only if this section is referenced for the specific hazardous material in the §172.101 Table or in a packaging section in this part.

(b) Packagings for limited quantity and ORM-D are specified according to hazard class in §§173.150 through 173.155 and in §173.306. In addition to other exceptions provided for limited quantity and ORM-D materials in this part:

(1) Strong outer packagings as specified in this part, marking requirements specified in subpart D of part 172 of this subchapter, and the 30 kg (66 pounds) gross weight limitation are not required for packages of limited quantity materials marked in accordance with §172.315 of this subchapter, or, until December 31, 2013, materials classed and marked as ORM-D and described as a Consumer commodity, as defined in §171.8 of this subchapter, when—

(i) Unitized in cages, carts, boxes or similar overpacks;

(ii) Offered for transportation or transported by:

(A) Rail;

(B) Private or contract motor carrier; or

(C) Common carrier in a vehicle under exclusive use for such service; and

(iii) Transported to or from a manufacturer, a distribution center, or a retail outlet, or transported to a disposal facility from one offeror.

(2) The 30 kg (66 pounds) gross weight limitation does not apply to packages of limited quantity materials marked in accordance with §172.315 of this subchapter, or, until December 31, 2013, materials classed and marked as ORM-D and described as a Consumer commodity, as defined in §171.8 of this subchapter, when offered for transportation or transported by highway or rail between a manufacturer, a distribution center, and a retail outlet provided—

(i) Inner packagings conform to the quantity limits for inner packagings specified in §§173.150(b), 173.152(b), 173.154(b), 173.155(b) and 173.306(a) and (b), as appropriate;

(ii) The inner packagings are packed into corrugated fiberboard trays to prevent them from moving freely;

(iii) The trays are placed in a fiberboard box which is banded and secured to a wooden pallet by metal, fabric, or plastic straps, to form a single palletized unit;

(iv) The package conforms to the general packaging requirements of subpart B of this part;

(v) The maximum net quantity of hazardous material permitted on one palletized unit is 250 kg (550 pounds); and

(vi) The package is properly marked in accordance with §172.315 or, until December 31, 2013, §172.316 of this subchapter.

Subpart E—Non-Bulk Packaging for Hazardous Materials Other Than Class 1 and Class 7

§173.159 Batteries, wet.

(a) Electric storage batteries, containing electrolyte acid or alkaline corrosive battery fluid (wet batteries), may not be packed with other materials except as provided in paragraphs (g) and (h) of this section and in §§173.220 and 173.222; and any battery or battery-powered device must be prepared and packaged for transport in a manner to prevent:

(1) A dangerous evolution of heat (*i.e.*, an amount of heat sufficient to be dangerous to packaging or personal safety to include charring of packaging, melting of packaging, scorching of packaging, or other evidence);

(2) Short circuits, including, but not limited to:

(i) Packaging each battery or each battery-powered device when practicable, in fully enclosed inner packagings made of non-conductive material;

(ii) Separating or packaging batteries and battery-powered devices in a manner to prevent contact with other batteries, devices or conductive materials (*e.g.*, metal) in the packagings; or

(iii) Ensuring exposed terminals are protected with non-conductive caps, non-conductive tape, or by other appropriate means; and

(3) *Damage to terminals*. If not impact resistant, the outer packaging must not be used as the sole means of protecting the battery terminals from damage or short circuiting. Batteries must be securely cushioned and packed to prevent shifting which could loosen terminal caps or reorient the terminals. Batteries contained in devices must be securely installed. Terminal protection methods include but are not limited to:

(i) Securely attaching covers of sufficient strength to protect the terminals;

(ii) Packaging the battery in a rigid plastic packaging; or

(iii) Constructing the battery with terminals that are recessed or otherwise protected so that the terminals will not be subjected to damage if the package is dropped.

(b) For transportation by aircraft:

(1) The packaging for wet batteries must incorporate an acid- or alkali-proof liner, or include a supplementary packaging with sufficient strength and adequately sealed to prevent leakage of electrolyte fluid in the event of spillage; and

(2) Any battery-powered device, equipment or vehicle must be packaged for transport in a manner to prevent unintentional activation or must have an independent means of preventing unintentional activation (*e.g.*, packaging restricts access to activation switch, switch caps or locks, recessed switches, trigger locks, temperature sensitive circuit breakers, etc.).

(c) The following specification packagings are authorized for batteries packed without other materials provided all requirements of paragraph (a) of this section, and for transportation by aircraft, paragraph (b) of this section are met:

(1) Wooden box: 4C1, 4C2, 4D, or 4F.

(2) Fiberboard box: 4G.

(3) Plywood drum: 1D.

(4) Fiber drum: 1G.

(5) Plastic drum: 1H2.

(6) Plastic jerrican: 3H2.

(7) Plastic box: 4H2.

(d) The following non-specification packagings are authorized for batteries packed without other materials provided all requirements of paragraph (a) of this section, and for transportation by aircraft, paragraph (b) of this section are met:

(1) Electric storage batteries are firmly secured to skids or pallets capable of withstanding the shocks normally incident to transportation are authorized for transportation by rail, highway, or vessel. The height of the completed unit must not exceed 1 ½ times the width of the skid or pallet. The unit must be capable of withstanding, without damage, a superimposed weight equal to two times the weight of the unit or, if the weight of the unit exceeds 907 kg (2,000 pounds), a superimposed weight of 1814 kg (4,000 pounds). Battery terminals must not be relied upon to support any part of the superimposed weight and must not short out if a conductive material is placed in direct contact with them.

(2) Electric storage batteries weighing 225 kg (500 pounds) or more, consisting of carriers' equipment, may be shipped by rail when mounted on suitable skids. Such shipments may not be offered in interchange service.

(3) One to three batteries not over 11.3 kg (25 pounds) each, packed in strong outer boxes. The maximum authorized gross weight is 34 kg (75 pounds).

(4) Not more than four batteries not over 7 kg (15 pounds) each, packed in strong outer fiberboard or wooden boxes. The maximum authorized gross weight is 30 kg (65 pounds).

(5) Not more than five batteries not over 4.5 kg (10 pounds) each, packed in strong outer fiberboard or wooden boxes. The maximum authorized gross weight is 30 kg (65 pounds).

(6) Single batteries not exceeding 34 kg (75 pounds) each, packed in 5-sided slip covers or in completely closed fiberboard boxes. Slip covers and boxes must be of solid or double-faced corrugated fiberboard of at least 91 kg (200 pounds) Mullen test strength. The slip cover or fiberboard box must fit snugly and provide inside top clearance of at least 1.3 cm (0.5 inch) above battery terminals and filler caps with reinforcement in place. Assembled for shipment, the bottom edges of the slip-cover must come to within 2.5 cm (1 inch) of the bottom of the battery. The completed package (battery and box or slip cover) must be capable of withstanding a top-to-bottom compression test of at least 225 kg (500 pounds) without damage to battery terminal caps, cell covers or filler caps.

(7) Single batteries exceeding 34 kg (75 pounds) each may be packed in completely closed fiberboard boxes. Boxes must be of double-wall corrugated fiberboard of at least 181 kg (400 pounds) test, or solid fiberboard testing at least 181 kg (400 pounds); a box may have hand holes in its ends provided that the hand holes will not materially weaken the box. Sides and ends of the box must have cushioning between the battery and walls of the box; combined thickness of cushioning material and

walls of the box must not be less than 1.3 cm (0.5 inch); and cushioning must be excelsior pads, corrugated fiberboard, or other suitable cushioning material. The bottom of the battery must be protected by a minimum of one excelsior pad or by a double-wall corrugated fiberboard pad. The top of the battery must be protected by a wood frame, corrugated trays or scored sheets of corrugated fiberboard having minimum test of 91 kg (200 pounds), or other equally effective cushioning material. Top protection must bear evenly on connectors and/or edges of the battery cover to facilitate stacking of batteries. No more than one battery may be placed in one box. The maximum authorized gross weight is 91 kg (200 pounds).

(e) When transported by highway or rail, electric storage batteries containing electrolyte or corrosive battery fluid are not subject to any other requirements of this subchapter, if all of the following are met:

(1) No other hazardous materials may be transported in the same vehicle;

(2) The batteries must be loaded or braced so as to prevent damage and short circuits in transit;

(3) Any other material loaded in the same vehicle must be blocked, braced, or otherwise secured to prevent contact with or damage to the batteries; and

(4) The transport vehicle may not carry material shipped by any person other than the shipper of the batteries.

(f) Batteries can be considered as non-spillable provided they are capable of withstanding the following two tests, without leakage of battery fluid from the battery:

(1) *Vibration test*. The battery must be rigidly clamped to the platform of a vibration machine, and a simple harmonic motion having an amplitude of 0.8 mm (0.03 inches) with a 1.6 mm (0.063 inches) maximum total excursion must be applied. The frequency must be varied at the rate of 1 Hz/min between the limits of 10 Hz to 55 Hz. The entire range of frequencies and return must be traversed in 95 ± 5 minutes for each mounting position (direction of vibrator) of the battery. The battery must be tested in three mutually perpendicular positions (to include testing with fill openings and vents, if any, in an inverted position) for equal time periods.

(2) *Pressure differential test*. Following the vibration test, the battery must be stored for six hours at 24 °C ±4 °C (75°F ±7 °F) while subjected to a pressure differential of at least 88 kPa (13 psig). The battery must be tested in three mutually perpendicular positions (to include testing with fill openings and vents, if any, in an inverted position) for at least six hours in each position.

(g) Electrolyte, acid or alkaline corrosive battery fluid, packed with batteries wet or dry, must be packed in one of the following specification packagings:

(1) In 4C1, 4C2, 4D, or 4F wooden boxes with inner receptacles of glass, not over 4.0 L (1 gallon) each with not over 8.0 L (2 gallons) total in each outside container. Inside containers must be well- cushioned and separated from batteries by a strong solid wooden partition. The completed package must conform to Packing Group III requirements.

(2) Electrolyte, acid, or alkaline corrosive battery fluid included with electric storage batteries and filling kits may be packed in strong rigid outer packagings when shipments are made by, for, or to the Departments of the Army, Navy, or Air Force of the United States. Packagings must conform to military specifications. The electrolyte, acid, or alkaline corrosive battery fluid must be packed in polyethylene bottles of not over 1.0 L (0.3 gallon) capacity each. Not more than 24 bottles, securely separated from electric storage batteries and kits, may be offered for transportation or transported in each package.

(3) In 4G fiberboard boxes with not more than 12 inside packagings of polyethylene or other material resistant to the lading, each not over 2.0 L (0.5 gallon) capacity each. Completed packages must conform to Packing Group III requirements. Inner packagings must be adequately separated from the storage battery. The maximum authorized gross weight is 29 kg (64 pounds). These packages are not authorized for transportation by aircraft.

(h) Dry batteries or battery charger devices may be packaged in 4G fiberboard boxes with inner receptacles containing battery fluid. Completed packagings must conform to Packing Group III requirements. Not more than 12 inner receptacles may be packed in one outer box. The maximum authorized gross weight is 34 kg (75 pounds).

(i) When approved by the Associate Administrator, electric storage batteries, containing electrolyte or corrosive battery fluid in a separate reservoir from which fluid is injected into the battery cells by a power device cartridge assembled with the battery, and which meet the criteria of paragraph (f) are not subject to any other requirements of this subchapter.

§173.159a Exceptions for non-spillable batteries.

(a) Exceptions for hazardous materials shipments in the following paragraphs are permitted only if this section is referenced for the specific hazardous material in the §172.101 table or in a packaging section in this part.

(b) Non-spillable batteries offered for transportation or transported in accordance with this section are subject to the incident reporting requirements. For transportation by aircraft, a telephone report in accordance with §171.15(a) is required if a fire, violent rupture, explosion or dangerous evolution of heat (*i.e.*, an amount of heat sufficient to be dangerous to packaging or personal safety to include charring of packaging, melting of packaging, scorching of packaging, or other evidence) occurs as a direct result of a non-spillable battery. For all modes of transportation, a written report in accordance with §171.16(a) is required if a fire, violent rupture, explosion or dangerous evolution of heat occurs as a direct result of a non-spillable battery.

(c) Non-spillable batteries are excepted from the packaging requirements of §173.159 under the following conditions:

(1) Non-spillable batteries must be securely packed in strong outer packagings and meet the requirements of §173.159(a). A non-spillable battery which is an integral part of and necessary for the operation of mechanical or electronic equipment must be securely fastened in the battery holder on the equipment;

(2) The battery and outer packaging must be plainly and durably marked "NONSPILLABLE" or "NONSPILLABLE BATTERY." The requirement to mark the outer package does not apply when the battery is installed in a piece of equipment that is transported unpackaged.

(d) Non-spillable batteries are excepted from all other requirements of this subchapter when offered for transportation and transported in accordance with paragraph (c) of this section and the following:

(1) At a temperature of 55 °C (131 °F), the battery must not contain any unabsorbed free-flowing liquid, and must be designed so that electrolyte will not flow from a ruptured or cracked case; and

(2) For transport by aircraft, when contained in a battery-powered device, equipment or vehicle must be prepared and packaged for transport in a manner to prevent unintentional activation in conformance with §173.159(b)(2) of this Subpart.

§173.166 Air bag inflators, air bag modules and seat-belt pretensioners.

(a) *Definitions.* An **air bag inflator** (consisting of a casing containing an igniter, a booster material, a gas generant and, in some cases, a pressure vessel (cylinder)) is a gas generator used to inflate an air bag in a supplemental restraint system in a motor vehicle. An **air bag module** is the air bag inflator plus an inflatable bag assembly. A **seat-belt pre-tensioner** contains similar hazardous materials and is used in the operation of a seat-belt restraining system in a motor vehicle.

(b) *Classification.* An air bag inflator, air bag module, or seat-belt pretensioner may be classed as Class 9 (UN3268) if:

(1) The manufacturer has submitted each design type air bag inflator, air bag module, or seat-belt pretensioner to a person approved by the Associate Administrator, in accordance with §173.56(b), for examination and testing. The submission must contain a detailed description of the inflator or pretensioner or, if more than a single inflator or pretensioner is involved, the maximum parameters of each particular inflator or pretensioner design type for which approval is sought and details on the complete package. The manufacturer must submit an application, including the test results and report recommending the shipping description and classification for each device or design type to the Associate Administrator, and must receive written notification from the Associate Administrator that the device has been approved for transportation and assigned an EX number; or,

(2) The manufacturer has submitted an application, including a classification issued by the competent authority of a foreign government to the Associate Administrator, and received written notification from the Associate Administrator that the device has been approved for transportation and assigned an EX number.

(c) *EX numbers.* When offered for transportation, the shipping paper must contain the EX number or product code for each approved inflator, module or pretensioner in association with the basic description required by §172.202(a) of this subchapter. Product codes must be traceable to the specific EX number assigned to the inflator, module or pretensioner by the Associate Administrator. The EX number or product code is not required to be marked on the outside package.

(d) *Exceptions.* (1) An air bag module or seat-belt pretensioner that has been approved by the Associate Administrator and is installed in a motor vehicle, aircraft, boat or other transport conveyance or its completed components, such as steering columns or door panels, is not subject to the requirements of this subchapter.

(2) An air bag module containing an inflator that has been previously approved for transportation is not required to be submitted for further examination or approval.

(3) An air bag module containing an inflator that has previously been approved as a Division 2.2 material is not required to be submitted for further examination to be reclassed as a Class 9 material.

(4) *Shipments for recycling.* When offered for domestic transportation by highway, rail freight, cargo vessel or cargo aircraft, a serviceable air bag module or seat-belt pretensioner removed from a motor vehicle that was manufactured as required for use in the United States may be offered for transportation and transported without compliance with the shipping paper requirement prescribed in paragraph (c) of this section. However, the word "Recycled" must be entered on the shipping paper immediately after the basic description prescribed in §172.202 of this subchapter. No more than one device is authorized in the packaging prescribed in paragraph (e)(1), (2) or (3) of this section. The device must be cushioned and secured within the package to prevent movement during transportation.

(e) *Packagings.* Rigid, outer packagings, meeting the general packaging requirements of part 173, and the packaging specification and performance requirements of part 178 of this subchapter at the Packing Group III performance level are authorized as follows. The packagings must be designed and constructed to prevent movement of the articles and inadvertent operation.

(1) 1A2, 1B2, 1G or 1H2 drums.

(2) 3A2 or 3H2 jerricans.

(3) 4C1, 4C2, 4D, 4F, 4G or 4H2 boxes.

(4) Reusable high strength plastic or metal containers or dedicated handling devices are authorized for shipment of air bag inflators, air bag modules, and seat-belt pretensioners from a manufacturing facility to the assembly facility, subject to the following conditions:

(i) The gross weight of the container or handling device may not exceed 1000 kg (2,205 pounds). The container or handling device structure must provide adequate support to allow them to be stacked at least three high with no damage to the containers or devices.

(ii) If not completely enclosed by design, the container or handling device must be covered with plastic, fiberboard, or metal. The covering must be secured to the container by banding or other comparable methods.

(iii) Internal dunnage must be sufficient to prevent shifting of the devices within the container.

(5) Packagings specified in the approval document issued by the Associate Administrator in accordance with paragraph (e) of this section are also authorized.

(f) *Labeling*. Notwithstanding the provisions of §172.402 of this subchapter, each package or handling device must display a CLASS 9 label. Additional labeling is not required when the package contains no hazardous materials other than the devices.

§173.167 Consumer commodities.

(a) Effective January 1, 2013, a "consumer commodity" (*see* §171.8 of this subchapter) when intended for transportation by aircraft may only include articles or substances of Class 2 (non-toxic aerosols only), Class 3 (Packing Group II and III only), Division 6.1 (Packing Group III only), UN3077, UN3082, and UN3175, provided such materials do not have a subsidiary risk and are authorized aboard a passenger-carrying aircraft. Friction-type closures must be secured by secondary means. Inner packagings intended to contain liquids must be capable of meeting the pressure differential requirements (75 kPa) prescribed in §173.27(c) of this part. Consumer commodities are excepted from the specification packaging requirements of this subchapter and each completed package must conform to subpart B of part. Packages of consumer commodities must also be capable of withstanding a 1.2 m drop on solid concrete in the position most likely to cause damage and a 24-hour stack test. Inner and outer packaging quantity limits for consumer commodities are as follows:

(1) Non-toxic aerosols, as defined in §171.8 of this subchapter and constructed in accordance with §173.306 of this part, in non-refillable, non-metal containers not exceeding 120 mL (4 fluid ounces) each, or in non-refillable metal containers not exceeding 820 mL (28 ounces) each, except that flammable aerosols may not exceed 500 mL (16.9 ounces) each; or

(2) Liquids, in inner packagings not exceeding 500 mL (16.9 ounces) each; or

(3) Solids, in inner packagings not exceeding 500 g (1.0 pounds) each; or

(4) Any combination thereof.

(b) Inner packagings are to be placed in an outer packaging not to exceed 30 kg (66 pounds) gross weight as prepared for shipment.

§173.171 Smokeless powder for small arms.

Smokeless powder for small arms which has been classed in Division 1.3 may be reclassed in Division 4.1, for domestic transportation by motor vehicle, rail car, vessel, or cargo-only aircraft, subject to the following conditions:

(a) The powder must be examined and approved for a Division 1.3 and Division 4.1 classification in accordance with §§173.56 and 173.58 of this part.

(b) The total quantity of smokeless powder may not exceed 45.4 kg (100 pounds) net mass in:

(1) One rail car, motor vehicle, or cargo-only aircraft; or

(2) One freight container on a vessel, not to exceed four freight containers per vessel.

(c) Only combination packagings with inner packagings not exceeding 3.6 kg (8 pounds) net mass are authorized. Inner packagings must be arranged and protected so as to prevent simultaneous ignition of the contents. The complete package must be of the same type which has been examined as required in §173.56 of this part.

(d) Inside packages that have been examined and approved by the Associate Administrator may be packaged in UN 4G fiberboard boxes meeting the Packing Group I performance level, provided all inside containers are packed to prevent shifting and the net weight of smokeless powder in any one box does not exceed 7.3 kg (16 pounds).

§173.173 Paint, paint-related material, adhesives, ink and resins.

(a) When the §172.101 Table specifies that a hazardous material be packaged under this section, the following requirements apply. Except as otherwise provided in this part, the description "Paint" is the proper shipping name for paint, lacquer, enamel, stain, shellac, varnish, liquid aluminium, liquid bronze, liquid gold, liquid wood filler, and liquid lacquer base. The description "Paint-related material" is the proper shipping name for a paint thinning, drying, reducing or removing compound. However, if a more specific description is listed in the §172.101 Table of this subchapter, that description must be used.

(b) Paint, paint-related material, adhesives, ink and resins must be packaged as follows:

(1) As prescribed in §173.202 of this part if it is a Packing Group II material or §173.203 of this part if it is a Packing Group III material; or

(2) In inner glass packagings of not over 1 L (0.3 gallon) capacity each or inner metal packagings of not over 5 L (1 gallon) each, packed in a strong outer packaging. Packages must conform to the packaging requirements of subpart B of this part but need not conform to the requirements of part 178 of this subchapter.

§173.174 Refrigerating machines.

A refrigerating machine assembled for shipment and containing 7 kg (15 pounds) or less of a flammable liquid for its operation in a strong, tight receptacle is excepted from labeling (except when offered for transportation or transported by air) and the specification packaging requirements of this subchapter. In addition, shipments are not subject to subpart F of part 172 of this subchapter (Placarding), to part 174 of this subchapter (Carriage by rail) except §174.24 (Shipping papers) and to part 177 (Carriage by highway) of this subchapter except §177.817 (Shipping papers).

§173.185 Lithium cells and batteries.

(a) *Cells and batteries*. A lithium cell or battery, including a lithium polymer cell or battery and a lithium-ion cell or battery, must conform to all of the following requirements:

(1) Be of a type proven to meet the requirements of each test in the UN Manual of Tests and Criteria (IBR; see §171.7 of this subchapter). A cell or battery and equipment containing a cell or battery that was first transported prior to January 1, 2006 and is of a type

proven to meet the criteria of Class 9 by testing in accordance with the tests in the UN Manual of Tests and Criteria, Third Revised Edition, 1999, need not be retested.

(2) Incorporate a safety venting device or otherwise be designed in a manner that will preclude a violent rupture under conditions normally incident to transportation.

(3) Be equipped with an effective means to prevent dangerous reverse current flow (e.g., diodes, fuses, etc.) if a battery contains cells or series of cells that are connected in parallel.

(4) Be packaged in combination packagings conforming to the requirements of part 178, subparts L and M, of this subchapter at the Packing Group II performance level. The lithium battery or cell must be packed in inner packagings in such a manner as to prevent short circuits, including movement which could lead to short circuits. The inner packaging must be packed within one of the following outer packagings: metal boxes (4A or 4B); wooden boxes (4C1, 4C2, 4D, or 4F); fiberboard boxes (4G); solid plastic boxes (4H2); fiber drums (1G); metal drums (1A2 or 1B2); plywood drums (1D); plastic jerricans (3H2); or metal jerricans (3A2 or 3B2).

(5) Be equipped with an effective means of preventing external short circuits.

(6) Except as provided in paragraph (d) of this section, cells and batteries with a liquid cathode containing sulfur dioxide, sulfuryl chloride or thionyl chloride may not be offered for transportation or transported if any cell has been discharged to the extent that the open circuit voltage is less than two volts or is less than 2/3 of the voltage of the fully charged cell, whichever is less.

(b) *Lithium cells or batteries packed with equipment.* Lithium cells or batteries packed with equipment may be transported as Class 9 materials if the batteries and cells meet all the requirements of paragraph (a) of this section. The equipment and the packages of cells or batteries must be further packed in a strong outer packaging. The cells or batteries must be packed in such a manner as to prevent short circuits, including movement that could lead to short circuits.

(c) *Lithium cells or batteries contained in equipment.* Lithium cells or batteries contained in equipment may be transported as Class 9 materials if the cells and batteries meet all the requirements of paragraph (a) of this section, except paragraph (a)(4) of this section, and the equipment is packed in a strong outer packaging that is waterproof or is made waterproof through the use of a liner unless the equipment is made waterproof by nature of its construction. The equipment and cells or batteries must be secured within the outer packaging and be packed so as to prevent movement, short circuits, and accidental operation during transport.

(d) *Cells and batteries, for disposal or recycling.* A lithium cell or battery offered for transportation or transported by motor vehicle to a permitted storage facility, disposal site or for purposes of recycling is excepted from the specification packaging requirements of paragraph (a)(4) of this section and the requirements of paragraphs (a)(1) and (a)(6) of this section when protected against short circuits and packed in a strong outer packaging conforming to the requirements of §§173.24 and 173.24a.

(e) *Shipments for testing (prototypes).* A lithium cell or battery is excepted from the requirements of (a)(1) of this section when transported by motor vehicle for purposes of testing. The cell or battery must be individually packed in an inner packaging, surrounded by cushioning material that is noncombustible and nonconductive. The cell or battery must be transported as a Class 9 material.

(f) A lithium cell or battery that does not comply with the provisions of this subchapter may be transported only under conditions approved by the Associate Administrator.

(g) Batteries employing a strong, impact-resistant outer casing and exceeding a gross weight of 12 kg (26.5 lbs.), and assemblies of such batteries, may be packed in strong outer packagings, in protective enclosures (for example, in fully enclosed wooden slatted crates) or on pallets. Batteries must be secured to prevent inadvertent movement, and the terminals may not support the weight of other superimposed elements. Batteries packaged in this manner are not permitted for transportation by passenger aircraft, and may be transported by cargo aircraft only if approved by the Associate Administrator prior to transportation.

§173.186 Matches.

(a) Matches must be of a type which will not ignite spontaneously or undergo marked decomposition when subjected for 8 consecutive hours to a temperature of 93°C (200°F).

(b) *Definitions.* (1) **Fusee matches** are matches the heads of which are prepared with a friction-sensitive igniter composition and a pyrotechnic composition which burns with little or no flame, but with intense heat.

(2) **Safety matches** are matches combined with or attached to the box, book or card that can be ignited by friction only on a prepared surface.

(3) **Strike anywhere matches** are matches that can be ignited by friction on a solid surface.

(4) **Wax "Vesta" matches** are matches that can be ignited by friction either on a prepared surface or on a solid surface.

(c) Safety matches and wax "Vesta" matches must be tightly packed in securely closed inner packagings to prevent accidental ignition under conditions normally incident to transportation, and further packed in outer fiber-board, wooden, or other equivalent-type packagings. These matches in outer packagings not exceeding 23 kg (50 pounds) gross weight are not subject to any other requirement (except marking) of this subchapter. These matches may be packed in the same outer packaging with materials not subject to this subchapter.

(d) Strike-anywhere matches may not be packed in the same outer packaging with any material other than safety matches or wax "Vesta" matches, which must be packed in separate inner packagings.

(e) Packagings. Strike-anywhere matches must be tightly packed in securely closed chipboard, fiberboard, wooden, or metal inner packagings to prevent accidental ignition under conditions normally incident to transportation. Each inner packaging may contain no more than 700 strike-anywhere matches and must be packed in outer steel drums (1A2), aluminium drums

(1B2), steel jerricans (3A2), wooden (4C1, 4C2), plywood (4D), reconstituted wood (4F) or fiberboard (4G) boxes, plywood (1D) or fiber (1G) drums. Gross weight of fiberboard boxes (4G) must not exceed 30 kg (66 pounds). Gross weight of other outer packagings must not exceed 45 kg (100 pounds).

§173.201 Non-bulk packagings for liquid hazardous materials in packing group I.

(a) When §172.101 of this subchapter specifies that a liquid hazardous material be packaged under this section, only non-bulk packagings prescribed in this section may be used for its transportation. Each packaging must conform to the general packaging requirements of subpart B of part 173, to the requirements of part 178 of this subchapter at the Packing Group I performance level, and to the requirements of the special provisions of Column 7 of the §172.101 Table.

(b) The following combination packagings are authorized:

Outer packagings:

Steel drum: 1A1 or 1A2
Aluminum drum: 1B1 or 1B2
Metal drum other than steel or aluminum: 1N1 or 1N2
Plywood drum: 1D
Fiber drum: 1G
Plastic drum: 1H1 or 1H2
Steel jerrican: 3A1 or 3A2
Plastic jerrican: 3H1 or 3H2
Aluminum jerrican: 3B1 or 3B2
Steel box: 4A
Aluminum box: 4B
Natural wood box: 4C1 or 4C2
Plywood box: 4D
Reconstituted wood box: 4F
Fiberboard box: 4G
Expanded plastic box: 4H1
Solid plastic box: 4H2

Inner packagings:

Glass or earthenware receptacles
Plastic receptacles
Metal receptacles
Glass ampoules

(c) Except for transportation by passenger aircraft, the following single packagings are authorized:

Steel drum: 1A1 or 1A2
Aluminum drum: 1B1 or 1B2
Metal drum other than steel, or aluminum: 1N1 or 1N2
Plastic drum: 1H1 or 1H2
Steel jerrican: 3A1 or 3A2
Plastic jerrican: 3H1 or 3H2
Aluminum jerrican: 3B1 or 3B2
Plastic receptacle in steel, aluminum, fiber or plastic drum: 6HA1, 6HB1, 6HG1, 6HH1
Plastic receptacle in steel, aluminum, wooden, plywood or fiberboard box: 6HA2, 6HB2, 6HC, 6HD2 or 6HG2
Glass, porcelain or stoneware in steel, aluminum or fiber drum: 6PA1, 6PB1 or 6PG1
Glass, porcelain or stoneware in steel, aluminum, wooden or fiberboard box: 6PA2, 6PB2, 6PC or 6PG2
Glass, porcelain or stoneware in solid or expanded plastic packaging: 6PH1 or 6PH2
Cylinders, specification or UN standard, as prescribed for any compressed gas, except 3HT and those prescribed for acetylene

§173.202 Non-bulk packagings for liquid hazardous materials in packing group II.

(a) When §172.101 of this subchapter specifies that a liquid hazardous material be packaged under this section, only non-bulk packagings prescribed in this section may be used for its transportation. Each packaging must conform to the general packaging requirements of subpart B of part 173, to the requirements of part 178 of this subchapter at the Packing Group I or II performance level (unless otherwise excepted), and to the particular requirements of the special provisions of Column 7 of the §172.101 Table.

(b) The following combination packagings are authorized:

Outer packagings:

Steel drum: 1A1 or 1A2
Aluminum drum: 1B1 or 1B2
Metal drum other than steel or aluminum: 1N1 or 1N2
Plywood drum: 1D
Fiber drum: 1G
Plastic drum: 1H1 or 1H2
Wooden barrel: 2C2
Steel jerrican: 3A1 or 3A2
Plastic jerrican: 3H1 or 3H2
Aluminum jerrican: 3B1 or 3B2
Steel box: 4A
Aluminum box: 4B
Natural wood box: 4C1 or 4C2
Plywood box: 4D
Reconstituted wood box: 4F
Fiberboard box: 4G
Expanded plastic box: 4H1
Solid plastic box: 4H2

Inner packagings:

Glass or earthenware receptacles
Plastic receptacles
Metal receptacles
Glass ampoules

(c) Except for transportation by passenger aircraft, the following single packagings are authorized:

Steel drum: 1A1 or 1A2
Aluminum drum: 1B1 or 1B2
Metal drum other than steel or aluminum: 1N1 or 1N2
Plastic drum: 1H1 or 1H2
Fiber drum: 1G (with liner)
Wooden barrel: 2C1
Steel jerrican: 3A1 or 3A2
Plastic jerrican: 3H1 or 3H2
Aluminum jerrican: 3B1 or 3B2
Plastic receptacle in steel, aluminum, fiber or plastic drum: 6HA1, 6HB1, 6HG1 or 6HH1
Plastic receptacle in steel, aluminum, wooden, plywood or fiberboard box: 6HA2, 6HB2, 6HC, 6HD2 or 6HG2
Glass, porcelain or stoneware in steel, aluminum or fiber drum: 6PA1, 6PB1 or 6PG1
Glass, porcelain or stoneware in steel, aluminum, wooden or fiberboard box: 6PA2, 6PB2, 6PC or 6PG2
Glass, porcelain or stoneware in solid or expanded plastic packaging: 6PH1 or 6PH2
Plastic receptacle in plywood drum: 6HD1
Glass, porcelain or stoneware in plywood drum or wickerwork hamper: 6PD1 or 6PD2
Cylinders, specification, as prescribed for any compressed gas, except for Specifications 8 and 3HT

§173.203 Non-bulk packagings for liquid hazardous materials in packing group III.

(a) When §172.101 of this subchapter specifies that a liquid hazardous material be packaged under this section, only non-bulk packagings prescribed in this section may be used for its transportation. Each packaging must conform to the general packaging requirements of subpart B of part 173, to the requirements of part 178 of this subchapter at the Packing Group I, II or III performance level, and to the requirements of the special provisions of Column 7 of the §172.101 Table.

(b) The following combination packagings are authorized:

Outer packagings:

Steel drum: 1A1 or 1A2
Aluminum drum: 1B1 or 1B2
Metal drum other than steel or aluminum: 1N1 or 1N2
Plywood drum: 1D
Fiber drum: 1G
Plastic drum: 1H1 or 1H2
Wooden barrel: 2C2
Steel jerrican: 3A1 or 3A2
Plastic jerrican: 3H1 or 3H2
Aluminum jerrican: 3B1 or 3B2
Steel box: 4A
Aluminum box: 4B
Natural wood box: 4C1 or 4C2
Plywood box: 4D
Reconstituted wood box: 4F
Fiberboard box: 4G
Expanded plastic box: 4H1
Solid plastic box: 4H2

Inner packagings:

Glass or earthenware receptacles
Plastic receptacles
Metal receptacles
Glass ampoules

(c) The following single packagings are authorized:

Steel drum: 1A1 or 1A2
Aluminum drum: 1B1 or 1B2
Metal drum other than steel or aluminum: 1N1
Plastic drum: 1H1 or 1H2
Fiber drum: 1G (with liner)
Wooden barrel: 2C1
Steel jerrican: 3A1 or 3A2
Plastic jerrican: 3H1 or 3H2
Aluminum jerrican: 3B1 or 3B2
Plastic receptacle in steel, aluminum, fiber or plastic drum: 6HA1, 6HB1, 6HG1 or 6HH1
Plastic receptacle in steel, aluminum, wooden, plywood or fiberboard box: 6HA2, 6HB2, 6HC, 6HD2 or 6HG2
Glass, porcelain or stoneware in steel, aluminum or fiber drum: 6PA1, 6PB1, or 6PG1
Glass, porcelain or stoneware in steel, aluminum, wooden or fiberboard box: 6PA2, 6PB2, 6PC or 6PG2
Glass, porcelain or stoneware in solid or expanded plastic packaging: 6PH1 or 6PH2
Plastic receptacle in plywood drum: 6HD1
Glass, porcelain or stoneware in plywood drum or wickerwork hamper: 6PD1 or 6PD2
Cylinders, as prescribed for any compressed gas, except for Specifications 8 and 3HT

§173.204 Non-bulk, non-specification packagings for certain hazardous materials.

When §172.101 of this subchapter specifies that a liquid or solid hazardous material be packaged under this section, any appropriate non-bulk packaging which conforms to the general packaging requirements of subpart B of part 173 may be used for its transportation. Packagings need not conform to the requirements of part 178 of this subchapter.

§173.205 Specification cylinders for liquid hazardous materials.

When §172.101 of this subchapter specifies that a hazardous material must be packaged under this section, the use of any specification or UN cylinder, except those specified for acetylene, is authorized. Cylinders used for toxic materials in Division 6.1 or 2.3 must conform to the requirements of §173.40.

§173.206 Packaging requirements for chlorosilanes.

(a) When §172.101 of this subchapter specifies that a hazardous material be packaged under this section, only non-bulk packagings prescribed in this section may be used for its transportation. Each packaging must conform to the general packaging requirements of subpart B of part 173, to the requirements of part 178 of this subchapter at the Packing Group I or II performance level (unless otherwise excepted), and to the particular requirements of the special provisions of Column (7) of the §172.101 Table.

(b) The following combination packagings are authorized:

Outer packagings:

Steel drum: 1A2
Plastic drum: 1H2
Plywood drum: 1D
Fiber drum: 1G
Steel box: 4A
Natural wood box: 4C1 or 4C2
Plywood box: 4D
Reconstituted wood box: 4F
Fiberboard box: 4G
Expanded plastic box: 4H1
Solid plastic box: 4H2

Inner packagings:

Glass or Steel receptacle

(c) Except for transportation by passenger aircraft, the following single packagings are authorized:

Steel drum: 1A1
Steel jerrican: 3A1
Plastic receptacle in steel drum: 6HA1
Cylinders (for liquids in PG I), specification or UN standard, as prescribed for any compressed gas, except Specification 3HT and those prescribed for acetylene
Cylinders (for liquids in PG II), specification, as prescribed for any compressed gas, except Specification 8 and 3HT cylinders

§173.211 Non-bulk packagings for solid hazardous materials in packing group I.

(a) When §172.101 of this subchapter specifies that a solid hazardous material be packaged under this section, only non-bulk packagings prescribed in this section may be used for its transportation. Each package must conform to the general packaging requirements of subpart B of part 173, to the requirements of part 178 of this sub-chapter at the Packing Group I performance level, and to the requirements of the special provisions of Column 7 of the §172.101 Table.

(b) The following combination packagings are authorized:

Outer packagings:

Steel drum: 1A1 or 1A2
Aluminum drum: 1B1 or 1B2
Metal drum other than steel or aluminum: 1N1 or 1N2
Plywood drum: 1D

Fiber drum: 1G
Plastic drum: 1H1 or 1H2
Wooden barrel: 2C2
Steel jerrican: 3A1 or 3A2
Plastic jerrican: 3H1 or 3H2
Aluminum jerrican: 3B1 or 3B2
Steel box: 4A
Aluminum box: 4B
Natural wood box: 4C1 or 4C2
Plywood box: 4D
Reconstituted wood box: 4F
Fiberboard box: 4G
Solid plastic box: 4H2

Inner packagings:

Glass or earthenware receptacles
Plastic receptacles
Metal receptacles
Glass ampoules

(c) Except for transportation by passenger aircraft, the following single packagings are authorized:

Steel drum: 1A1 or 1A2
Aluminum drum: 1B1 or 1B2
Metal drum other than steel or aluminum: 1N1 or 1N2
Plastic drum: 1H1 or 1H2
Fiber drum: 1G
Steel jerrican: 3A1 or 3A2
Plastic jerrican: 3H1 or 3H2
Aluminum jerrican: 3B1 or 3B2
Steel box with liner: 4A
Aluminum box with liner: 4B
Natural wood box, sift proof: 4C2
Plastic receptacle in steel, aluminum, plywood, fiber or plastic drum: 6HA1, 6HB1, 6HD1, 6HG1 or 6HH1
Glass, porcelain or stoneware in steel, aluminum, plywood or fiber drum: 6PA1, 6PB1, 6PD1 or 6PG1
Glass, porcelain or stoneware in steel, aluminum, wooden or fiberboard box: 6PA2, 6PB2, 6PC or 6PG2
Glass, porcelain or stoneware in expanded or solid plastic packaging: 6PH1 or 6PH2
Cylinders, as prescribed for any compressed gas, except for Specification 8 and 3HT

§173.212 Non-bulk packagings for solid hazardous materials in packing group II.

(a) When §172.101 of this subchapter specifies that a solid hazardous material be packaged under this section, only non-bulk packagings prescribed in this section may be used for its transportation. Each package must conform to the general packaging requirements of subpart B of part 173, to the requirements of part 178 of this subchapter at the Packing Group I or II performance level, and to the requirements of the special provisions of Column 7 of the §172.101 Table.

(b) The following combination packagings are authorized:

Outer packagings:

Steel drum: 1A1 or 1A2
Aluminum drum: 1B1 or 1B2
Metal drum other than steel or aluminum: 1N1 or 1N2
Plywood drum: 1D
Fiber drum: 1G
Plastic drum: 1H1 or 1H2
Wooden barrel: 2C2
Steel jerrican: 3A1 or 3A2
Plastic jerrican: 3H1 or 3H2
Aluminum jerrican: 3B1 or 3B2
Steel box: 4A
Aluminum box: 4B
Natural wood box: 4C1 or 4C2
Plywood box: 4D
Reconstituted wood box: 4F
Fiberboard box: 4G
Solid plastic box: 4H2

Inner packagings:

Glass or earthenware receptacles
Plastic receptacles
Metal receptacles
Glass ampoules

(c) Except for transportation by passenger aircraft, the following single packagings are authorized:

Steel drum: 1A1 or 1A2
Aluminum drum: 1B1 or 1B2
Plywood drum: 1D
Plastic drum: 1H1 or 1H2
Fiber drum: 1G
Wooden barrel: 2C1 or 2C2
Metal drum other than steel or aluminum: 1N1 or 1N2
Steel jerrican: 3A1 or 3A2
Plastic jerrican: 3H1 or 3H2
Aluminum jerrican: 3B1 or 3B2
Steel box: 4A
Steel box with liner: 4A
Aluminum box: 4B
Aluminum box with liner: 4B
Natural wood box: 4C1
Natural wood box, sift proof: 4C2
Plywood box: 4D
Reconstituted wood box: 4F
Fiberboard box: 4G
Expanded plastic box: 4H1
Solid plastic box: 4H2
Bag, woven plastic: 5H1, 5H2 or 5H3
Bag, plastic film: 5H4
Bag, textile: 5L1, 5L2 or 5L3
Bag, paper, multiwall, water resistant: 5M2
Plastic receptacle in steel, aluminum, plywood fiber or plastic drum: 6HA1, 6HB1, 6HD1, 6HG1 or 6HH1
Plastic receptacle in steel aluminum, wood, plywood or fiberboard box: 6HA2, 6HB2, 6HC, 6HD2 or 6HG2
Glass, porcelain or stoneware in steel, aluminum, plywood or fiber drum: 6PA1, 6PB1, 6PD1 or 6PG1
Glass, porcelain or stoneware in steel, aluminum, wooden or fiberboard box: 6PA2, 6PB2, 6PC or 6PG2
Glass, porcelain or stoneware in expanded or solid plastic packaging: 6PH1 or 6PH2
Cylinders, as prescribed for any compressed gas, except for Specification 8 and 3HT

§173.213 Non-bulk packagings for solid hazardous materials in packing group III.

(a) When §172.101 of this subchapter specifies that a solid hazardous material be packaged under this section, only non-bulk packagings prescribed in this section may be used for its transportation. Each package must conform to the general packaging requirements of subpart B of part 173, to the requirements of part 178 of this sub-chapter at the Packing Group I, II or III performance level, and to the requirements of the special provisions of Column 7 of the §172.101 Table.

(b) The following combination packagings are authorized:

Outer packagings:

Steel drum: 1A1 or 1A2
Aluminum drum: 1B1 or 1B2
Metal drum other than steel or aluminum: 1N1 or 1N2
Plywood drum: 1D
Fiber drum: 1G
Plastic drum: 1H1 or 1H2
Wooden barrel: 2C2
Steel jerrican: 3A1 or 3A2
Plastic jerrican: 3H1 or 3H2
Aluminum jerrican: 3B1 or 3B2
Steel box: 4A
Aluminum box: 4B
Natural wood box: 4C1 or 4C2
Plywood box: 4D
Reconstituted wood box: 4F

Fiberboard box: 4G
Solid plastic box: 4H2

Inner packagings:

Glass or earthenware receptacles
Plastic receptacles
Metal receptacles
Glass ampoules

(c) The following single packagings are authorized:

Steel drum: 1A1 or 1A2
Aluminum drum: 1B1 or 1B2
Plywood drum: 1D
Fiber drum: 1G
Plastic drum: 1H1 or 1H2
Metal drum other than steel or aluminum: 1N1 or 1N2
Wooden barrel: 2C1 or 2C2
Steel jerrican: 3A1 or 3A2
Plastic jerrican: 3H1 or 3H2
Aluminum jerrican: 3B1 or 3B2
Steel box with liner: 4A
Steel box: 4A
Aluminum box: 4B
Aluminum box with liner: 4B
Natural wood box: 4C1
Natural wood box, sift proof: 4C2
Plywood box: 4D
Reconstituted wood box: 4F
Fiberboard box: 4G
Expanded plastic box: 4H1
Solid plastic box: 4H2
Bag, woven plastic: 5H1, 5H2 or 5H3
Bag, plastic film: 5H4
Bag, textile: 5L1, 5L2 or 5L3
Bag, paper, multiwall, water resistant: 5M2
Plastic receptacle in steel, aluminum, plywood, fiber or plastic drum: 6HA1, 6HB1, 6HD1, 6HG1 or 6HH1
Plastic receptacle in steel, aluminum, wooden, plywood or fiberboard box: 6HA2, 6HB2, 6HC, 6HD2 or 6HG2
Glass, porcelain or stoneware in steel, aluminum, plywood or fiber drum: 6PA1, 6PB1, 6PD1 or 6PG1
Glass, porcelain or stoneware in steel, aluminum, wooden or fiberboard box: 6PA2, 6PB2, 6PC or 6PG2
Glass, porcelain or stoneware in expanded or solid plastic packaging: 6PH1 or 6PH2
Cylinders, as prescribed for any compressed gas, except for Specification 8 and 3HT

§173.217 Carbon dioxide, solid (dry ice).

(a) Carbon dioxide, solid (dry ice), when offered for transportation or transported by aircraft or water, must be packed in packagings designed and constructed to permit the release of carbon dioxide gas to prevent a buildup of pressure that could rupture the packagings. Packagings must conform to the general packaging requirements of subpart B of this part but need not conform to the requirements of part 178 of this subchapter.

(b) For transportation by vessel:

(1) Each transport vehicle and freight container containing solid carbon dioxide must be conspicuously marked on two sides "WARNING CO_2 SOLID (DRY ICE)."

(2) Other packagings containing solid carbon dioxide must be marked "CARBON DIOXIDE, SOLID—DO NOT STOW BELOW DECKS."

(c) For transportation by aircraft:

(1) In addition to the applicable marking requirements in subpart D of part 172, the net mass of the carbon dioxide, solid (dry ice) must be marked on the outside of the package. This provision also applies to unit load devices (ULDs) when the ULD contains dry ice and is considered the packaging.

(2) The shipper must make arrangements with the operator for each shipment.

(3) The quantity limits per package shown in Columns (9A) and (9B) of the Hazardous Materials Table in §172.101 are not applicable to dry ice being used as a refrigerant for other than hazardous materials loaded in a unit load device or other type of pallet. In such a case, the unit load device or other type of pallet must allow the venting of the carbon dioxide gas to prevent a dangerous build up of pressure, and be identified to the operator.

(4) Dry ice is excepted from the shipping paper requirements of subpart C of part 172 of this subchapter provided alternative written documentation is supplied containing the following information: proper shipping name (Dry ice or Carbon dioxide, solid), class 9, UN number 1845, the number of packages, and the net quantity of dry ice in each package. The information must be included with the description of the materials.

(5) Carbon dioxide, solid (dry ice), in quantities not exceeding 2.5 kg (5.5 pounds) per package and used as a refrigerant for the contents of the package is excepted from all other requirements of this subchapter if the requirements of paragraph (a) of this section are complied with and the package is marked "Carbon dioxide, solid" or "Dry ice", is marked with the name of the contents being cooled, and is marked with the net weight of the dry ice or an indication that the net weight is 2.5 kg (5.5 pounds) or less.

(d) Carbon dioxide, solid (dry ice), when used to refrigerate materials being shipped for diagnostic or treatment purposes (e.g., frozen medical specimens), is excepted from the shipping paper and certification requirements of this subchapter if the requirements of paragraphs (a) and (c)(2) of this section are met and the package is marked "Carbon dioxide, solid" or "Dry ice" and is marked with an indication that the material being refrigerated is being transported for diagnostic or treatment purposes.

§173.220 Internal combustion engines, self-propelled vehicles, mechanical equipment containing internal combustion engines, battery-powered equipment or machinery, fuel cell-powered equipment or machinery.

(a) *Applicability*. An internal combustion engine, self-propelled vehicle, mechanized equipment containing an internal combustion engine, a battery-powered vehicle or equipment, or a fuel cell-powered vehicle or equipment, or any combination thereof, is subject to the requirements of this subchapter when transported as cargo on a transport vehicle, vessel, or aircraft if—

(1) The engine contains a liquid or gaseous fuel. An engine may be considered as not containing fuel when the engine components and any fuel lines have been completed drained, sufficiently cleaned of residue, and purged of vapors to remove any potential hazard and the engine when held in any orientation will not release any liquid fuel;

(2) The fuel tank contains a liquid or gaseous fuel. A fuel tank may be considered as not containing fuel when the fuel tank and the fuel lines have been completed drained, sufficiently cleaned of residue, and purged of vapors to remove any potential hazard;

(3) It is equipped with a wet battery (including a non-spillable battery), a sodium battery or a lithium battery; or

(4) Except as provided in paragraph (f)(1) of this section, it contains other hazardous materials subject to the requirements of this subchapter.

(b) *Requirements.* Unless otherwise excepted in paragraph (b)(4) of this section, vehicles, engines, and equipment are subject to the following requirements:

(1) *Flammable liquid fuel.* A fuel tank containing a flammable liquid fuel must be drained and securely closed, except that up to 500 mL (17 ounces) of residual fuel may remain in the tank, engine components, or fuel lines provided they are securely closed to prevent leakage of fuel during transportation. Self-propelled vehicles containing diesel fuel are excepted from the requirement to drain the fuel tanks, provided that sufficient ullage space has been left inside the tank to allow fuel expansion without leakage, and the tank caps are securely closed.

(2) *Flammable liquefied or compressed gas fuel.*(i) For transportation by motor vehicle, rail car or vessel, fuel tanks and fuel systems containing flammable liquefied or compressed gas fuel must be securely closed. For transportation by vessel, the requirements of §§176.78(k) and 176.905 of this subchapter apply.

(ii) For transportation by aircraft:

(A) Flammable gas-powered vehicles, machines, equipment or cylinders containing the flammable gas must be completely emptied of flammable gas. Lines from vessels to gas regulators, and gas regulators themselves, must also be drained of all traces of flammable gas. To ensure that these conditions are met, gas shut-off valves must be left open and connections of lines to gas regulators must be left disconnected upon delivery of the vehicle to the operator. Shut-off valves must be closed and lines reconnected at gas regulators before loading the vehicle aboard the aircraft; or alternatively;

(B) Flammable gas powered vehicles, machines or equipment, which have cylinders (fuel tanks) that are equipped with electrically operated valves, may be transported under the following conditions:

(1) The valves must be in the closed position and in the case of electrically operated valves, power to those valves must be disconnected;

(2) After closing the valves, the vehicle, equipment or machinery must be operated until it stops from lack of fuel before being loaded aboard the aircraft;

(3) In no part of the closed system shall the pressure exceed 5% of the maximum allowable working pressure of the system or 290 psig (2000 kPa), whichever is less; and

(4) There must not be any residual liquefied gas in the system, including the fuel tank.

(3) *Truck bodies or trailers on flat cars—lammable liquid or gas powered.* T Truck bodies or trailers with automatic heating or refrigerating equipment of the flammable liquid type may be shipped with fuel tanks filled and equipment operating or inoperative, when used for the transportation of other freight and loaded on flat cars as part of a joint rail and highway movement, provided the equipment and fuel supply conform to the requirements of §177.834(l) of this subchapter.

(4) *Modal exceptions.* Quantities of flammable liquid fuel greater than 500 mL (17 ounces) may remain in the fuel tank in self-propelled vehicles and mechanical equipment only under the following conditions:

(i) For transportation by motor vehicle or rail car, the fuel tanks must be securely closed.

(ii) For transportation by vessel, the shipment must conform to §176.905 of this subchapter.

(iii) For transportation by aircraft, when carried in aircraft designed or modified for vehicle ferry operations when all the following conditions must be met:

(A) Authorization for this type operation has been given by the appropriate authority in the government of the country in which the aircraft is registered;

(B) Each vehicle is secured in an upright position;

(C) Each fuel tank is filled in a manner and only to a degree that will preclude spillage of fuel during loading, unloading, and transportation; and

(D) Each area or compartment in which a self-propelled vehicle is being transported is suitably ventilated to prevent the accumulation of fuel vapors.

(c) *Battery-powered or installed.* Batteries must be securely installed, and wet batteries must be fastened in an upright position. Batteries must be protected against a dangerous evolution of heat, short circuits, and damage to terminals in conformance with §173.159(a) and leakage; or must be removed and packaged separately under §173.159. Battery-powered vehicles, machinery or equipment including battery-powered wheelchairs and mobility aids are not subject to any other requirements of this subchapter except §173.21 of this subchapter when transported by rail, highway or vessel.

(d) *Lithium batteries.* Except as provided in §172.102, Special Provision A101 of this subchapter, vehicles, engines and machinery powered by lithium metal batteries that are transported with these batteries installed are forbidden aboard passenger-carrying aircraft. Lithium batteries contained in vehicles, engines or mechanical equipment must be securely fastened in the battery holder of the vehicle, engine or mechanical equipment and be protected in such a manner as to prevent damage and short circuits (*e.g.,* by the use of non-conductive caps that cover the terminals entirely). Lithium batteries must be of a type that have successfully passed each test in the UN Manual of Tests and Criteria as specified in §173.185 of this subchapter, unless approved by the Associate Administrator. Equipment (other than vehicles, engines or mechanical equipment) containing lithium batteries, must be described as "Lithium ion batteries contained in equipment" or "Lithium metal batteries contained in equipment," as appropriate, and transported in accordance with §173.185 and applicable special provisions.

(e) *Fuel cells.* A fuel cell must be secured and protected in a manner to prevent damage to the fuel cell. Equipment (other than vehicles, engines or mechanical equipment) such as consumer electronic devices containing fuel cells (fuel cell cartridges) must be described as "Fuel cell cartridges contained in equipment" and transported in accordance with §173.230 of this subchapter.

(f) *Other hazardous materials.* (1) Items containing hazardous materials, such as fire extinguishers, compressed gas accumulators, safety devices and other hazardous materials that are integral components of the motor vehicle, engine or mechanical equipment and that are necessary for the operation of the vehicle, engine or mechanical equipment, or for the safety of its operator or passengers, must be securely installed in the motor vehicle, engine or mechanical equipment. Such items are not otherwise subject to the requirements of this subchapter. Equipment (other than vehicles, engines or mechanical equipment) containing lithium batteries must be described as "Lithium batteries contained in equipment" and transported in accordance with §173.185 of this subchapter and applicable special provisions. Equipment (other than vehicles, engines or mechanical equipment) such as consumer electronic devices containing fuel cells (fuel cell cartridges) must be described as "Fuel cell cartridges contained in equipment" and transported in accordance with §173.230 of this subchapter.

(2) Other hazardous materials must be packaged and transported in accordance with the requirements of this subchapter.

(g) *Additional requirements for internal combustion engines and vehicles with certain electronic equipment when transported by aircraft or vessel.* When an internal combustion engine that is not installed in a vehicle or equipment is offered for transportation by aircraft or vessel, all fuel, coolant or hydraulic systems remaining in the engine must be drained as far as practicable, and all disconnected fluid pipes that previously contained fluid must be sealed with leak-proof caps that are positively retained. When offered for transportation by aircraft, vehicles equipped with theft-protection devices, installed radio communications equipment or navigational systems must have such devices, equipment or systems disabled.

(h) *Exceptions.* Except as provided in paragraph (f)(2) of this section, shipments made under the provisions of this section—

(1) Are not subject to any other requirements of this subchapter for transportation by motor vehicle or rail car; and

(2) Are not subject to the requirements of subparts D, E and F (marking, labeling and placarding, respectively) of part 172 of this subchapter or §172.604 of this subchapter (emergency response telephone number) for transportation by vessel or aircraft. For transportation by aircraft, the provisions of §173.159(b)(2) of this part as applicable, the provisions of §173.230(f), as applicable, other applicable requirements of this subchapter, including shipping papers, emergency response information, notification of pilot-in-command, general packaging requirements, and the requirements specified in §173.27 of this subchapter must be met. For transportation by vessel, additional exceptions are specified in §176.905 of this subchapter.

Subpart F—Bulk Packaging for Hazardous Materials Other Than Class 1 and Class 7

§173.240 Bulk packaging for certain low hazard solid materials.

When §172.101 of this subchapter specifies that a hazardous material be packaged under this section, only the following bulk packagings are authorized, subject to the requirements of subparts A and B of part 173 of this subchapter and the special provisions specified in Column 7 of the §172.101 Table.

(a) *Rail cars*: Class DOT 103, 104, 105, 109, 111, 112, 114, 115, or 120 tank car tanks; Class 106 or 110 multi-unit tank car tanks; and metal non-DOT specification, sift-proof tank car tanks and sift-proof closed cars.

(b) *Motor vehicles*: Specification MC 300, MC 301, MC 302, MC 303, MC 304, MC 305, MC 306, MC 307, MC 310, MC 311, MC 312, MC 330, MC 331, DOT 406, DOT 407, and DOT 412 cargo tank motor vehicles; non-DOT specification, sift-proof cargo tank motor vehicles; and sift-proof closed vehicles.

(c) *Portable tanks and closed bulk bins.* DOT 51, 56, 57 and 60 portable tanks; IMO type 1, 2 and 5, and IM 101 and IM 102 portable tanks; UN portable tanks; marine portable tanks conforming to 46 CFR part 64; and sift-proof non-DOT Specification portable tanks and closed bulk bins are authorized.

(d) *IBCs.* IBCs are authorized subject to the conditions and limitations of this section provided the IBC type is authorized according to the IBC packaging code specified for the specific hazardous material in Column (7) of the §172.101 Table of this subchapter and the IBC conforms to the requirements in subpart O of part 178 of this subchapter at the Packing Group performance level as specified in Column (5) of the §172.101 Table of this sub-chapter for the material being transported.

(1) IBCs may not be used for the following hazardous materials:

(i) Packing Group I liquids; and

(ii) Packing Group I solids that may become liquid during transportation.

(2) The following IBCs may not be used for Packing Group II and III solids that may become liquid during transportation:

(i) Wooden: 11C, 11D and 11F;

(ii) Fiberboard: 11G;

(iii) Flexible: 13H1, 13H2, 13H3, 13H4, 13H5, 13L1, 13L2, 13L3, 13L4, 13M1 and 13M2; and

(iv) Composite: 11HZ2 and 21HZ2.

(e) *Large Packagings.* Large Packagings are authorized subject to the conditions and limitations of this section provided the Large Packaging type is authorized according to the IBC packaging code specified for the specific hazardous material in Column (7) of the §172.101 Table of this subchapter and the Large Packaging conforms to the requirements in subpart Q of part 178 of this subchapter at the Packing Group performance level as specified in Column (5) of the §172.101 Table for the material being transported.

(1) Except as specifically authorized in this subchapter, Large Packagings may not be used for Packing Group I or II hazardous materials.

(2) Large Packagings with paper or fiberboard inner receptacles may not be used for solids that may become liquid in transportation.

§173.241 Bulk packagings for certain low hazard liquid and solid materials.

When §172.101 of this subchapter specifies that a hazardous material be packaged under this section, only the following bulk packagings are authorized, subject to the requirements of subparts A and B of part 173 of this subchapter and the special provisions specified in Column 7 of the §172.101 Table.

(a) *Rail cars*: Class DOT 103, 104, 105, 109, 111, 112, 114, 115, or 120 tank car tanks; Class 106 or 110 multi-unit tank car tanks and AAR Class 203W, 206W, and 211W tank car tanks.

(b) *Cargo tanks*: DOT specification MC 300, MC 301, MC 302, MC 303, MC 304, MC 305, MC 306, MC 307, MC 310, MC 311, MC 312, MC 330, MC 331, DOT 406, DOT 407, and DOT 412 cargo tank motor vehicles; and non-DOT specification cargo tank motor vehicles suitable for transport of liquids.

(c) *Portable tanks.* DOT Specification 51, 56, 57 and 60 portable tanks; IMO type 1, 2 and 5, and IM 101 and IM 102 portable tanks; UN portable tanks; marine portable tanks conforming to 46 CFR part 64; and non-DOT Specification portable tanks suitable for transport of liquids are authorized. For transportation by vessel, also see §176.340 of this subchapter. For transportation of combustible liquids by vessel, additional requirements are specified in §176.340 of this subchapter.

(d) *IBCs.* IBCs are authorized subject to the conditions and limitations of this section provided the IBC type is authorized according to the IBC packaging code specified for the specific hazardous material in Column (7) of the §172.101 Table of this subchapter and the IBC conforms to the requirements in subpart O of part 178 of this subchapter at the Packing Group performance level as specified in Column (5) of the §172.101 Table for the material being transported.

(1) IBCs may not be used for the following hazardous materials:

(i) Packing Group I liquids; and

(ii) Packing Group I solids that may become liquid during transportation.

(2) The following IBCs may not be used for Packing Group II and III solids that may become liquid during transportation:

(i) Wooden: 11C, 11D and 11F;

(ii) Fiberboard: 11G;

(iii) Flexible: 13H1, 13H2, 13H3, 13H4, 13H5, 13L1, 13L2, 13L3, 13L4, 13M1 and 13M2; and

(iv) Composite: 11HZ2 and 21HZ2.

(e) *Large Packagings.* Large Packagings are authorized subject to the conditions and limitations of this section provided the Large Packaging type is authorized according to the IBC packaging code specified for the specific hazardous material in Column (7) of the §172.101 Table of this subchapter and the Large Packaging conforms to the requirements in subpart Q of part 178 of this subchapter at the Packing Group performance level as specified in Column (5) of the §172.101 Table for the material being transported.

(1) Except as specifically authorized in this subchapter, Large Packagings may not be used for Packing Group I or II hazardous materials.

(2) Large Packagings with paper or fiberboard inner receptacles may not be used for solids that may become liquid in transportation.

§173.242 Bulk packagings for certain medium hazard liquids and solids, including solids with dual hazards.

When §172.101 of this subchapter specifies that a hazardous material be packaged under this section, only the following bulk packagings are authorized, subject to the requirements of subparts A and B of part 173 of this subchapter and the special provisions specified in Column 7 of the §172.101 Table.

(a) *Rail cars*: Class DOT 103, 104, 105, 109, 111, 112, 114, 115, or 120 tank car tanks; Class 106 or 110 multi-unit tank car tanks and AAR Class 206W tank car tanks

(b) *Cargo tanks*: Specification MC 300, MC 301, MC 302, MC 303, MC 304, MC 305, MC 306, MC 307, MC 310, MC 311, MC 312, MC 330, MC 331, DOT 406, DOT 407, and DOT 412 cargo tank motor vehicles; and non-DOT specification cargo tank motor vehicles when in compliance with §173.5a(c). Cargo tanks used to transport Class 3, Packing Group I or II, or Packing Group III with a flash point of less than 38 °C (100 °F); Class 6, Packing Group I or II; and Class 8, Packing Group I or II materials must conform to the following special requirements:

(1) Pressure relief system: Except as provided by §173.33(d), each cargo tank must be equipped with a pressure relief system meeting the requirements of §178.346-3 or §178.347-4 of this subchapter. However, pressure relief devices on MC 310, MC 311 and MC 312 cargo tanks must meet the requirements for a Specification MC 307 cargo tank (except for Class 8, Packing Group I and II). Pressure relief devices on MC 330 and MC 331 cargo tanks must meet the requirement in §178.337-9 of this subchapter.

(2) Bottom outlets: DOT 406, DOT 407 and DOT 412 must be equipped with stop-valves meeting the requirements of §178.345-11 of this subchapter; MC 304, MC 307, MC 310, MC 311, and MC 312 cargo tanks must be equipped with stop-valves capable of being remotely closed within 30 seconds of actuation by manual or mechanic means and (except for Class 8, Packing Group I and II) by a closure activated at a temperature not over 121°C (250°F); MC 330 and MC 331 cargo tanks must be equipped with internal self-closing stop-valves meeting the requirements in §178.337-11 of this subchapter.

(c) *Portable tanks*: DOT Specification 51, 56, 57 and 60 portable tanks; Specification IM 101, IM 102, and UN portable tanks when a T Code is specified in Column(7) of the §172.101 Hazardous Materials Table for a specific hazardous material; and marine portable tanks conforming to 46 CFR part 64 are authorized. DOT Specification 57 portable tanks used for the transport by vessel of Class 3, Packaging Group II materials must conform to the following:

(1) *Minimum design pressure.* Each tank must have a minimum design pressure of 62 kPa (9 psig);

(2) *Pressure relief devices.* Each tank must be equipped with at least one pressure relief device, such as a spring-loaded valve or fusible plug, conforming to the following:

(i) Each pressure relief device must communicate with the vapor space of the tank when the tank is in a normal transportation attitude. Shutoff valves may not be installed between the tank opening and any pressure relief device. Pressure relief devices must be mounted, shielded, or drained to prevent the accumulation of any material that could impair the operation or discharge capability of the device;

(ii) Frangible devices are not authorized;

(iii) No pressure relief device may open at less than 34.4 kPa (5 psig);

(iv) If a fusible device is used for relieving pressure, the device must have a minimum area of 1.25 square inches. The device must function at a temperature between 104°C. and 149°C. (220°F. and 300°F.) and at a pressure less than the design test pressure of the tank, unless this latter function is accomplished by a separate device; and

(v) No relief device may be used which would release flammable vapors under normal conditions of transportation (temperature up to and including 54°C. (130°F.).); and

(3) *Venting capacity.* The minimum venting capacity for pressure activated vents must be 6,000 cubic feet of free air per hour (measured at 101.3 kPa (14.7 psi) and 15.6°C. (60°F.)) at not more than 34.4 kPa (5 psi). The total emergency venting capacity (cu. ft./hr.) of each portable tank must be at least that determined from the following table:

Total surface area square feet [1] [2]	Cubic feet free air per hour
20	15,800
30	23,700
40	31,600
50	39,500
60	47,400
70	55,300
80	63,300
90	71,200
100	79,100
120	94,900
140	110,700
160	126,500

[1]Interpolate for intermediate sizes.

[2]Surface area excludes area of legs.

(4) Unless provided by §173.32(h)(3), an IM 101, 102 or UN portable tank with a bottom outlet and used to transport a liquid hazardous material that is a Class 3, PG I or II, or PG III with a flash point of less than 38°C (100°F); Division 5.1 PG I or II; or Division 6.1, PG I or II, must have internal valves conforming to §178.275(d)(3) of this subchapter.

(d) *IBCs.* IBCs are authorized subject to the conditions and limitations of this section provided the IBC type is authorized according to the IBC packaging code specified for the specific hazardous material in Column (7) of the §172.101 Table of this subchapter and the IBC conforms to the requirements in subpart O of part 178 of this subchapter at the Packing Group performance level as specified in Column (5) of the §172.101 Table of this sub-chapter for the material being transported.

(1) IBCs may not be used for the following hazardous materials:

(i) Packing Group I liquids; and

(ii) Packing Group I solids that may become liquid during transportation.

(2) The following IBCs may not be used for Packing Group II and III solids that may become liquid during transportation:

(i) Wooden: 11C, 11D and 11F;

(ii) Fiberboard: 11G;

(iii) Flexible: 13H1, 13H2, 13H3, 13H4, 13H5, 13L1, 13L2, 13L3, 13L4, 13M1 and 13M2; and

(iv) Composite: 11HZ2 and 21HZ2.

(e) *Large Packagings.* Large Packagings are authorized subject to the conditions and limitations of this section provided the Large Packaging type is authorized according to the IBC packaging code specified for the specific hazardous material in Column (7) of the §172.101 Table of this subchapter and the Large Packaging conforms to the requirements in subpart Q of part 178 of this subchapter at the Packing Group performance level as specified in Column (5) of the §172.101 Table for the material being transported.

(1) Except as specifically authorized in this subchapter, Large Packagings may not be used for Packing Group I or II hazardous materials.

(2) Large Packagings with paper or fiberboard inner receptacles may not be used for solids that may become liquid in transportation.

§173.243 Bulk packaging for certain high hazard liquids and dual hazard materials which pose a moderate hazard.

When §172.101 of this subchapter specifies that a hazardous material be packaged under this section, only the following bulk packagings are authorized, subject to the requirements of subparts A and B of part 173 of this subchapter and the special provisions specified in Column 7 of the §172.101 Table.

(a) *Rail cars*: Class DOT 103, 104, 105, 109, 111, 112, 114, 115, or 120 fusion-welded tank car tanks; and Class 106 or 110 multi-unit tank car tanks.

(b) *Cargo tanks.* Specification MC 304, MC 307, MC 330, MC 331 cargo tank motor vehicles; and MC 310, MC 311, MC 312, DOT 407, and DOT 412 cargo tank motor vehicles with tank design pressure of at least 172.4 kPa (25 psig). Cargo tanks used to transport Class 3 or Division 6.1 materials, or Class 8, Packing Group I or II materials must conform to the following special requirements:

(1) Pressure relief system: Except as provided by §173.33(d), each cargo tank must be equipped with a pressure relief system meeting the requirements of §178.346-3 or §178.347-4 of this subchapter. However, pressure relief devices on MC 310, MC 311 and MC 312 cargo tanks must meet the requirements for a Specification MC 307 cargo tank (except for Class 8, Packing Group I and II). Pressure relief devices on MC 330 and MC 331 cargo tanks must meet the requirement in §178.337-9 of this subchapter.

(2) Bottom outlets: DOT 407 and DOT 412 cargo tanks must be equipped with stop-valves meeting the requirements of §178.345-11 of this subchapter; MC 304, MC 307, MC 310, MC 311, and MC 312 cargo tanks must be equipped with stop-valves capable of being remotely closed within 30 seconds of actuation by manual or mechanic means and (except for Class 8, Packing Group I and II) by a closure activated at a temperature not over 121°C (250°F); MC 330 and MC 331 cargo tanks must be equipped with internal self-closing stop-valves meeting the requirements in §178.337-11 of this subchapter.

(c) *Portable tanks*. DOT Specification 51 and 60 portable tanks; UN portable tanks and IM 101 and IM 102 portable tanks when a T code is specified in Column (7) of the §172.101 Table of this subchapter for a specific hazardous material; and marine portable tanks conforming to 46 CFR part 64 with design pressure of at least 172.4 kPa (25 psig) are authorized. Unless provided by §173.32(h)(3), an IM 101, 102 or UN portable tank, with a bottom outlet, used to transport a liquid hazardous material that is a Class 3, PG I or II, or PG III with a flash point of less than 38°C (100°F); Division 5.1, PG I or II; or Division 6.1, PG I or II, must have internal valves conforming to §178.275(d)(3) of this subchapter.

(d) *IBCs*. IBCs are authorized subject to the conditions and limitations of this section provided the IBC type is authorized according to the IBC packaging code specified for the specific hazardous material in Column (7) of the §172.101 Table of this subchapter and the IBC conforms to the requirements in subpart O of part 178 of this subchapter at the Packing Group performance level as specified in Column (5) of the §172.101 Table of this sub-chapter for the material being transported.

(1) IBCs may not be used for the following hazardous materials:

(i) Packing Group I liquids; and

(ii) Packing Group I solids that may become liquid during transportation.

(2) The following IBCs may not be used for Packing Group II and III solids that may become liquid during transportation:

(i) Wooden: 11C, 11D and 11F;

(ii) Fiberboard: 11G;

(iii) Flexible: 13H1, 13H2, 13H3, 13H4, 13H5, 13L1, 13L2, 13L3, 13L4, 13M1 and 13M2; and

(iv) Composite: 11HZ2 and 21HZ2.

(e) A dual hazard material may be packaged in accordance with §173.242 if:

(1) The subsidiary hazard is Class 3 with a flash point greater than 38°C (100°F); or

(2) The subsidiary hazard is Division 6.1, Packing Group III; or

(3) The subsidiary hazard is Class 8, Packaging Group, III.

§173.244 Bulk packaging for certain pyrophoric liquids (Division 4.2), dangerous when wet (Division 4.3) materials, and poisonous liquids with inhalation hazards (Division 6.1).

When §172.101 of this subchapter specifies that a hazardous material be packaged under this section, only the following bulk packagings are authorized, subject to the requirements of subparts A and B of part 173 of this subchapter and the special provisions specified in Column 7 of the §172.101 Table.

(a) *Rail cars*: (1) Class DOT 105, 109, 112, 114, or 120 fusion-welded tank car tanks; and Class 106 or 110 multi-unit tank car tanks. For tank car tanks built prior to March 16, 2009, the following conditions apply:

(i) Division 6.1 Hazard Zone A materials must be transported in tank cars having a test pressure of 34.47 Bar (500 psig) or greater and conform to Classes 105J, 106 or 110.

(ii) Division 6.1 Hazard Zone B materials must be transported in tank cars having a test pressure of 20.68 Bar (300 psig) or greater and conform to Classes 105S, 106, 110, 112J, 114J or 120S.

(iii) Hydrogen fluoride, anhydrous must be transported in tank cars having a test pressure of 20.68 Bar (300 psig) or greater and conform to Classes 105, 112, 114 or 120.

(2) For materials poisonous by inhalation, single unit tank cars tanks built prior to March 16, 2009 and approved by the Tank Car Committee for transportation of the specified material. Except as provided in § 173.244(a)(3), tank cars built on or after March 16, 2009 used for the transportation of the PIH materials listed below, must meet the applicable authorized tank car specification listed in the following table:

Proper shipping name	Authorized tank car specification
Acetone cyanohydrin, stabilized (Note 1)	105J500I
	112J500I
Acrolein (Note 1)	105J600I
Allyl Alcohol	105J500I
	112J500I
Bromine	105J500I
Chloropicrin	105J500I
	112J500I
Chlorosulfonic acid	105J500I
	112J500I
Dimethyl sulfate	105J500I
	112J500I
Ethyl chloroformate	105J500I
	112J500I
Hexachlorocyclopentadiene	105J500I
	112J500I
Hydrocyanic acid, aquesous solution *or* Hydrogen cyanide, aqueous solution *with not more than 20% hydrogen cyanide (Note 2)*.	105J500I
	112J500I
Hydrogen cyanide, stabilized (Note 2)	105J600I
Hydrogen fluoride, anyherous	105J500I
	112J500I
Poison inhalation hazard, Zone A materials not specifically identified in this table	105J600I

Proper shipping name	Authorized tank car specification
Poison inhalation hazard, Zone B materials not specifically identified in this table	105J500I
	112J500I
Phosphorus trichloride	105J500I
	112J500I
Sulfur trioxide, stabilized	105J500I
	112J500I
Sulfuric acid, fuming	105J500I
	112J500I
Titanium tetrachloride	105J500I
	112J500I

Note 1: Each tank car must have a re-closing pressure relief device having a start-to-discharge pressure of 10.34 Bar (150 psig). Restenciling to a lower test pressure is not authorized.

Note 2: Each tank car must have a re-closing pressure relief device having a start-to-discharge pressure of 15.51 Bar (225 psig). Restenciling to a lower test pressure is not authorized.

(3) As an alternative to the authorized tank car specification listed in the table in paragraph (a)(2) of this section, a car of the same authorized tank car specification but of the next lower test pressure, as prescribed in column 5 of the table at §179.101-1 of this subchapter, may be used provided that both of the following conditions are met:

(i) The difference between the alternative and the required minimum plate thicknesses, based on the calculation prescribed in §179.100-6 of this subchapter, must be added to the alternative tank car jacket and head shield. When the jacket and head shield are made from steel with a minimum tensile strength from 70,000 p.s.i. to 80,000 p.s.i., but the required minimum plate thickness calculation is based on steel with a minimum tensile strength of 81,000 p.s.i., the thickness to be added to the jacket and head shield must be increased by a factor of 1.157. Forming allowances for heads are not required to be considered when calculating thickness differences.

(ii) The tank car jacket and head shield are manufactured from carbon steel plate as prescribed in §179.100-7(a) of this subchapter.

(b) *Cargo tanks*: Specifications MC 330 and MC 331 cargo tank motor vehicles and, except for Division 4.2 materials, MC 312 and DOT 412 cargo tank motor vehicles.

(c) *Portable tanks*: DOT 51 portable tanks and UN portable tanks that meet the requirements of this subchapter, when a T code is specified in Column (7) of the §172.101 Table of this subchapter for the specific hazardous material, are authorized. Additionally, a DOT 51 or UN portable tank used for Division 6.1 liquids, Hazard Zone A or B, must be certified and stamped to the ASME Code as specified in §178.273(b)(6) of this subchapter.

§173.245 Bulk packaging for extremely hazardous materials such as poisonous gases (Division 2.3).

When §172.101 of this subchapter specifies that a hazardous material be packaged under this section, only the following bulk packagings are authorized, subject to the requirements of subparts A and B of part 173 of this subchapter and the special provisions specified in Column 7 of the §172.101 Table.

(a) Tank car tanks and multi-unit tank car tanks, when approved by the Associate Administrator.

(b) Cargo tank motor vehicles and portable tanks, when approved by the Associate Administrator.

§173.247 Bulk packaging for certain elevated temperature materials.

When §172.101 of this subchapter specified that a hazardous material be packaged under this section, only the following bulk packagings are authorized, subject to the requirements of subparts A and B of part 173 of this subchapter and the special provisions in Column 7 of the §172.101 Table. On or after October 1, 1993, authorized packagings must meet all requirements in paragraph (g) of this section unless otherwise excepted.

(a) *Rail cars*: Class DOT 103, 104, 105, 109, 111, 112, 114, 115, or 120 tank car tanks; Class DOT 106, 110 multi-unit tank car tanks; AAR Class 203W, 206W, 211W tank car tanks; and non-DOT specification tank car tanks equivalent in structural design and accident damage resistance to specification packagings.

(b) *Cargo tanks*: Specification MC 300, MC 301, MC 302, MC 303, MC 304, MC 305, MC 306, MC 307, MC 310, MC 311, MC 312, MC 330, MC 331 cargo tank motor vehicles; DOT 406, DOT 407, DOT 412 cargo tank motor vehicles; and non-DOT specification cargo tank motor vehicles equivalent in structural design and accident damage resistance to specification packagings. A non-DOT specification cargo tank motor vehicle constructed of carbon steel which is in elevated temperature material service is excepted from §178.345-7(d)(5) of this subchapter.

(c) *Portable tanks*: DOT Specification 51, 56, 57 and 60 portable tanks; IM 101 and IM 102 portable tanks; UN portable tanks; marine portable tanks conforming to 46 CFR part 64; metal IBCs and non-specification portable tanks equivalent in structural design and accident damage resistance to specification packagings are authorized.

(d) *Crucibles*: Nonspecification crucibles designed and constructed such that the stress in the packaging does not exceed one fourth (0.25) of the ultimate strength of the packaging material at any temperature within the design temperature range. Stress is determined under a load equal to the sum of the static or working pressure in combination with the loads developed from accelerations and decelerations incident to normal transportation. For highway transportation, these forces are assumed to be "1.7g" vertical, "0.75g" longitudinal, and "0.4g" transverse, in reference to the axes of the transport vehicle. Each accelerative or decelerative load may be considered separately.

(e) *Kettles*: A kettle, for the purpose of this section, is a bulk packaging (portable tank or cargo tank) having a capacity not greater than 5678 L (1500 gallons) with an integral heating apparatus used for melting various bituminous products such as asphalt. Kettles used for the transport of asphalt or bitumen are subject to the following requirements:

(1) *Low stability kettles.* Kettles with a ratio of track-width to fully loaded center of gravity (CG) height less than 2.5 must meet all requirements of paragraph (g) of this section (track-width is the distance measured between the outer edge of the kettle tires; CG height is measured perpendicular from the road surface).

(2) *High stability kettles.* (i) Kettles with a total capacity of less than 2650 L (700 gallons) and a ratio of track-width to fully loaded CG height of 2.5 or more are excepted from all requirements of paragraph (g)(2) of this section and the rollover protection requirements of paragraph (g)(6) of this section, if closures meet the requirements of paragraph (e)(2)(iii) of this section.

(ii) Kettles with a total capacity of 2650 L (700 gallons) or more and a ratio of track-width to fully loaded CG height of 2.5 or more are excepted from the "substantially leak tight" requirements of paragraph (g)(2) of this section and the rollover protection requirements of paragraph (g)(6) of this section if closures meet the requirements of paragraph (e)(2)(iii) of this section.

(iii) Closures must be securely closed during transportation. Closures also must be designed to prevent opening and the expulsion of lading in a rollover accident.

(f) *Other bulk packagings*: Bulk packagings, other than those specified in paragraphs (a) through (e) of this section, which are used for the transport of elevated temperature materials, must conform to all requirements of paragraph (g) of this section on or after October 1, 1993.

(g) *General requirements*. Bulk packagings authorized or used for transport of elevated temperature materials must conform to the following requirements:

(1) *Pressure and vacuum control equipment.* When pressure or vacuum control equipment is required on a packaging authorized in this section, such equipment must be of a self-reclosing design, must prevent package rupture or collapse due to pressure, must prevent significant release of lading due to packaging overturn or splashing or surging during normal transport conditions, and may be external to the packaging.

(i) Pressure control equipment is not required if pressure in the packaging would increase less than 10 percent as a result of heating the lading from the lowest design operating temperature to a temperature likely to be encountered if the packaging were engulfed in a fire. When pressure control equipment is required, it must prevent rupture of the packaging from heating, including fire engulfment.

(ii) Vacuum control equipment is not required if the packaging is designed to withstand an external pressure of 100 kPa (14.5 psig) or if pressure in the packaging would decrease less than 10 percent as a result of the lading cooling from the highest design operating temperature to the lowest temperature incurred in transport. When vacuum control equipment is required, it must prevent collapse of the packaging from a cooling-induced pressure differential.

(iii) When the regulations require a reclosing pressure relief device, the lading must not render the devices inoperable (i.e. from clogging, freezing, or fouling). If the lading affects the proper operation of the device, the packaging must have:

(A) A safety relief device incorporating a frangible disc or a permanent opening, each having a maximum effective area of 22 cm^2 (3.4 $in.^2$), for transportation by highway;

(B) For transportation of asphalt by highway, a safety relief device incorporating a frangible disc or a permanent opening, each having a maximum effective area of 48 cm^2 (7.4 in^2); or

(C) For transportation by rail, a non-reclosing pressure relief device incorporating a rupture disc conforming to the requirements of §179.15 of this subchapter.

(iv) Reclosing pressure relief devices, rupture discs or permanent openings must not allow the release of lading during normal transportation conditions (i.e., due to splashing or surging).

(2) *Closures.* All openings, except permanent vent openings authorized in paragraph (g)(1)(iii) of this section, must be securely closed during transportation. Packagings must be substantially leak-tight so as not to allow any more than dripping or trickling of a non-continuous flow when overturned. Closures must be designed and constructed to withstand, without exceeding the yield strength of the packaging, twice the static loading produced by the lading in any packaging orientation and at all operating temperatures.

(3) *Strength.* Each packaging must be designed and constructed to withstand, without exceeding the yield strength of the packaging, twice the static loading produced by the lading in any orientation and at all operating temperatures.

(4) *Compatibility.* The packaging and lading must be compatible over the entire operating temperature range.

(5) *Markings.* In addition to any other markings required by this subchapter, each packaging must be durably marked in a place readily accessible for inspection in characters at least 4.8 mm (3⁄16 inch) with the manufacturer's name, date of manufacture, design temperature range, and maximum product weight (or "load limit" for tank cars) or volumetric capacity.

(6) *Accident damage protection.* For transportation by highway, external loading and unloading valves and closures must be protected from impact damage resulting from collision or overturn. Spraying equipment and the road oil application portion of packaging are excepted from this requirement.

(7) *New construction.* Specification packagings that are being manufactured for the transport of elevated temperature materials must be authorized for current construction.

(h) *Exceptions.*

(1) *General.* Packagings manufactured for elevated temperature materials service prior to October 1, 1993, which are not in full compliance with the requirements in paragraph (g) of this section, may continue in service if they meet the applicable requirements of subparts A and B of this part and meet the closure requirements in paragraph (g)(2) of this section by March 30, 1995.

(2) *Kettles.* Kettles in service prior to October 1, 1993, which are used to transport asphalt or bitumen, are excepted from specific provisions of this section as follows:

(i) Kettles with a total capacity of less than 2650 L (700 gallons), which are not in full compliance with the requirements of paragraph (g) of this section, may continue in elevated temperature material service if they meet the applicable requirements of subparts A and B of this part and if, after March 30, 1995, closures are secured during transport to resist opening in an overturn.

(ii) Kettles with a total capacity of 2650 L (700 gallons) or more, which are not in full compliance with the requirements of paragraph (g) of this section, may continue in elevated temperature material service if they meet the applicable requirements of subparts A and B of this part and if, after March 30, 1995, closures are secured during transport to resist opening in an overturn and no opening exceeds 46 cm^2 (7.1 in^2).

(3) *Molten metals and molten glass.* This section does not apply to packagings used for transportation of molten metals and molten glass by rail when movement is restricted to operating speeds less than 15 miles per hour. (See §172.203(g)(3) of this subchapter for shipping paper requirements.)

(4) *Solid elevated temperature materials.* A material which meets the definition of a solid elevated temperature material is excepted from all requirements of this subchapter except §172.325 of this subchapter.

Reserved

Subpart G—Gases; Preparation and Packaging

§173.306 Limited quantities of compressed gases.

(a) Limited quantities of compressed gases for which exceptions are permitted as noted by reference to this section in §172.101 of this subchapter are excepted from labeling, except when offered for transportation or transported by air, and, unless required as a condition of the exception, specification packaging requirements of this subchapter when packed in accordance with the following paragraphs. For transportation by aircraft, the package must also comply with the applicable requirements of §173.27 of this subchapter and only hazardous materials authorized aboard passenger-carrying aircraft may be transported as a limited quantity. In addition, shipments are not subject to subpart F (Placarding) of part 172 of this subchapter, to part 174 of this subchapter except §174.24, and to part 177 of this subchapter except §177.817. Each package may not exceed 30 kg (66 pounds) gross weight.

(1) When in containers of not more than 4 fluid ounces capacity (7.22 cubic inches or less) except cigarette lighters. Special exceptions for shipment of certain compressed gases in the ORM-D class are provided in paragraph (i) of this section.

(2) When in metal containers filled with a material that is not classed as a hazardous material to not more than 90 percent of capacity at 70°F and then charged with non-flammable nonliquefied gas. Each container must be tested to three times the pressure at 70°F and, when refilled, be retested to three times the pressure of the gas at 70°F. Also, one of the following conditions must be met:

(i) Container is not over 0.95 L (1 quart) capacity and charged to not more than 11.17 bar (482.63 kPa, 170 psig) at 21°C (70°F), and must be packed in a strong outer packaging, or

(ii) Container is not over 30 gallons capacity and charged to not more than 75 psig at 70°F.

(3) When in a metal container for the sole purpose of expelling a nonpoisonous (other than a Division 6.1 Packing Group III material) liquid, paste or powder, provided all of the following conditions are met. Special exceptions for shipment of aerosols in the ORM-D class are provided in paragraph (i) of this section.

(i) Capacity must not exceed 1L (61.0 cubic inches).

(ii) Pressure in the container must not exceed 180 psig at 130°F. If the pressure exceeds 140 psig at 130°F, but does not exceed 160 psig at 130°F, a specification DOT 2P (§178.33 of this subchapter) inside metal container must be used; if the pressure exceeds 160 psig at 130°F, a specification DOT 2Q (§178.33a of this sub-chapter) inside metal container must be used. In any event, the metal container must be capable of withstanding without bursting a pressure of one and one-half times the equilibrium pressure of the content at 130°F.

(iii) Liquid content of the material and gas must not completely fill the container at 130°F.

(iv) The container must be placed in strong outside packaging.

(v) Each container, after it is filled, must be subjected to a test performed in a hot water bath; the temperature of the bath and the duration of the test must be such that the internal pressure reaches that which would be reached at 55 °C (131 °F) (50 °C (122 °F) if the liquid phase does not exceed 95% of the capacity of the container at 50 °C (122 °F)). If the contents are sensitive to heat, the temperature of the bath must be set at between 20 °C (68 °F) and 30 °C (86 °F) but, in addition, one container in 2,000 must be tested at the higher temperature. No leakage or permanent deformation of a container may occur.

(vi) Each outside packaging must be marked "INSIDE CONTAINERS COMPLY WITH PRESCRIBED REGULATIONS."

(4) Gas samples must be transported under the following conditions:

(i) A gas sample may only be transported as non-pressurized gas when its pressure corresponding to ambient atmospheric pressure in the container is not more than 105 kPa absolute (15.22 psia).

(ii) Non-pressurized gases, toxic (or toxic and flammable) must be packed in hermetically sealed glass or metal inner packagings of not more than one L (0.3 gallons) overpacked in a strong outer packaging.

(iii) Non-pressurized gases, flammable must be packed in hermetically sealed glass or metal inner packagings of not more than 5 L (1.3 gallons) and overpacked in a strong outer packaging.

(5) For limited quantities of Division 2.2 gases with no subsidiary risk, when in a plastic container for the sole purpose of expelling a liquid, paste or powder, provided all of the following conditions are met. Special exceptions for shipment of aerosols in the ORM–D class are provided in paragraph (i) of this section.

(i) Capacity must not exceed 1 L (61.0 cubic inches).

(ii) Pressure in the container must not exceed 160 psig at 130°F. If the pressure in the container is less than 140 psig at 130°F, a non-DOT specification container may be used. If the pressure in the container exceeds 140 psig at 130°F but does not exceed 160 psig at 130°F, the container must conform to specification DOT 2S. All non-DOT specification and specification DOT 2S containers must be capable of withstanding, without bursting, a pressure of one and one-half times the equilibrium pressure of the contents at 130°F.

(iii) Liquid content of the material and gas must not completely fill the container at 130°F.

(iv) The container must be packed in strong outside packagings.

(v) Except as provided in paragraph (a)(5)(vi) of this section, each container must be subjected to a test performed in a hot water bath; the temperature of the bath and the duration of the test must be such that the internal pressure reaches that which would be reached at 55°C (131°F) or 50°C (122°F) if the liquid phase does not exceed 95% of the capacity of the container at 50°C (122°F). If the contents are sensitive to heat, or if the container is made of plastic material which softens at this test temperature, the temperature of the bath must be set at between 20°C (68°F) and 30°C (86°F) but, in addition, one container in 2,000 must be tested at the higher temperature. No leakage or permanent deformation of a container may occur except that a plastic container may be deformed through softening provided that it does not leak.

(vi) As an alternative to the hot water bath test in paragraph (a)(5)(v) of this section, testing may be performed as follows:

(A) *Pressure and leak testing before filling.* Each empty container must be subjected to a pressure equal to or in excess of the maximum expected in the filled containers at 55°C (131°F) (or 50°C (122°F) if the liquid phase does not exceed 95 percent of the capacity of the container at 50°C (122°F). This must be at least two-thirds of the design pressure of the container. If any container shows evidence of leakage at a rate equal to or greater than 3.3 x 10^{-2} mbar L/s at the test pressure, distortion or other defect, it must be rejected; and

(B) *Testing after filling.* Prior to filling, the filler must ensure that the crimping equipment is set appropriately and the specified propellant is used before filling the container. Once filled, each container must be weighed and leak tested. The leak detection equipment must be sufficiently sensitive to detect at least a leak rate of 2.0 x 10^{-3} mbar L/s at 20°C (68°F). Any filled container which shows evidence of leakage, deformation, or excessive weight must be rejected.

(vii) Each outside packaging must be marked "INSIDE CONTAINERS COMPLY WITH PRESCRIBED REGULATIONS."

(b) *Exceptions for foodstuffs, soap, biologicals, electronic tubes, and audible fire alarm systems.* Limited quantities of compressed gases (except Division 2.3 gases) for which exceptions are provided as indicated by reference to this section in §172.101 of this subchapter, when in accordance with one of the following paragraphs, are excepted from labeling, except when offered for transportation or transported by aircraft, and the specification packaging requirements of this subchapter. For transportation by aircraft, the package must comply with the applicable requirements of §173.27 of this subchapter; the net quantity per package may not exceed the quantity specified in column (9A) of the Hazardous Materials Table in §172.101 of this subchapter; and only hazardous materials authorized aboard passenger-carrying aircraft may be transported as a limited quantity. In addition, shipments are not subject to subpart F (Placarding) of part 172 of this subchapter, to part 174 of this subchapter, except §174.24, and to part 177 of this subchapter, except §177.817. Special exceptions for shipment of certain compressed gases in the ORM-D class are provided in paragraph (i) of this section.

(1) Foodstuffs or soaps in a nonrefillable metal or plastic container not exceeding 1 L (61.0 cubic inches), with soluble or emulsified compressed gas, provided the pressure in the container does not exceed 140 psig at 130°F. Plastic containers must only contain Division 2.2 non-flammable soluble or emulsified compressed gas. The metal or plastic container must be capable of withstanding, without bursting, a pressure of one and one-half times the equilibrium pressure of the contents at 130°F.

(i) Containers must be packed in strong outside packagings.

(ii) Liquid content of the material and the gas must not completely fill the container at 130°F.

(iii) Each outside packaging must be marked "INSIDE CONTAINERS COMPLY WITH PRESCRIBED REGULATIONS."

(2) Cream in refillable metal or plastic containers with soluble or emulsified compressed gas. Plastic containers must only contain Division 2.2 non-flammable soluble or emulsified compressed gas. Containers must be of such design that they will hold pressure without permanent deformation up to 375 psig and must be equipped with a device designed so as to release pressure without bursting of the container or dangerous projection of its parts at higher pressures. This exception applies to shipments offered for transportation by refrigerated motor vehicles only.

(3) Nonrefillable metal or plastic containers charged with a Division 6.1 Packing Group III or nonflammable solution containing biological products or a medical preparation which could be deteriorated by heat, and compressed gas or gases. Plastic containers must only contain 2.2 non-flammable soluble or emulsified compressed gas. The capacity of each container may not exceed 35 cubic inches (19.3 fluid ounces). The pressure in the container may not exceed 140 psig at 130°F, and the liquid content of the product and gas must not completely fill the containers at 130°F. One completed container out of each lot of 500 or less, filled for shipment, must be heated, until the pressure in the container is equivalent to equilibrium pressure of the contents at 130°F. There must be no evidence of leakage, distortion, or other defect. The container must be packed in strong outside packagings.

(4) Electronic tubes, each having a volume of not more than 30 cubic inches and charged with gas to a pressure of not more than 35 psig and packed in strong outside packagings.

(5) Audible fire alarm systems powered by a compressed gas contained in an inside metal container when shipped under the following conditions:

(i) Each inside container must have contents which are not flammable, poisonous, or corrosive as defined under this part;

(ii) Each inside container may not have a capacity exceeding 35 cubic inches (19.3 fluid ounces);

(iii) Each inside container may not have a pressure exceeding 70 psig at 70°F and the liquid portion of the gas may not completely fill the inside container at 130°F, and

(iv) Each nonrefillable inside container must be designed and fabricated with a burst pressure of not less than four times its charged pressure at 130°F. Each refillable inside container must be designed and fabricated with a burst pressure of not less than five times its charged pressure at 130°F.

(c) [Reserved]

(d) [Reserved]

(e) *Refrigerating machines.* (1) New (unused) refrigerating machines or components thereof are excepted from the specification packaging requirements of this part if they meet the following conditions. In addition, shipments are not subject to subpart F of part 172 of this subchapter, to part 174 of this subchapter except

§174.24 and to part 177 of this subchapter except §177.817.

(i) Each pressure vessel may not contain more than 5,000 pounds of Group A1 refrigerant as classified in ANSI/ASHRAE Standard 15 or not more than 50 pounds of refrigerant other than Group A1.

(ii) Machines or components having two or more charged vessels may not contain an aggregate of more than 2,000 pounds of Group I refrigerant or more than 100 pounds of refrigerant other than Group I.

(iii) Each pressure vessel must be equipped with a safety device meeting the requirements of ANSI/ASHRAE 15 (IBR, see §171.7 of this subchapter).

(iv) Each pressure vessel must be equipped with a shut-off valve at each opening except openings used for safety devices and with no other connection. These valves must be closed prior to and during transportation.

(v) Pressure vessels must be manufactured, inspected and tested in accordance with ANSI/ASHRAE 15, or when over 6 inches internal diameter, in accordance with Section VIII of the ASME Code (IBR, see §171.7 of this subchapter).

(vi) All parts subject to refrigerant pressure during shipment must be tested in accordance with ANSI/ASHRAE 15.

(vii) The liquid portion of the refrigerant, if any, may not completely fill any pressure vessel at 130°F.

(viii) The amount of refrigerant, if liquefied, may not exceed the filling density prescribed in §173.304.

(f) *Accumulators. (Articles, pressurized pneumatic or hydraulic containing non-flammable gas).* The following applies to accumulators, which are hydraulic accumulators containing nonliquefied, nonflammable gas, and nonflammable liquids or pneumatic accumulators containing nonliquefied, nonflammable gas, fabricated from materials which will not fragment upon rupture.

(1) Accumulators installed in motor vehicles, construction equipment, and assembled machinery and designed and fabricated with a burst pressure of not less than five times their charged pressure at 70°F, when shipped, are not subject to the requirements of this subchapter.

(2) Accumulators charged with limited quantities of compressed gas to not more than 200 p.s.i.g. at 70°F are excepted from labeling (except when offered for transportation by air) and the specification packaging requirements of this subchapter when shipped under the following conditions. In addition, shipments are not subject to subpart F of part 172 of this subchapter, to part 174 of this subchapter except §174.24 and to part 177 of this subchapter except §177.817.

(i) Each accumulator must be shipped as an inside packaging,

(ii) Each accumulator may not have gas space exceeding 2,500 cubic inches under stored pressure, and

(iii) Each accumulator must be tested, without evidence of failure or damage, to at least three times its charged pressure of 70°F, but not less than 120 p.s.i.g. before initial shipment and before each refilling and reshipment.

(3) Accumulators with a charging pressure exceeding 200 p.s.i.g. at 70°F are excepted from labeling (except when offered for transportation by air) and the specification packaging requirements of this subchapter when shipped under the following conditions:

(i) Each accumulator must be in compliance with the requirements stated in paragraph (f)(2),(i),(ii), and (iii) of this section, and

(ii) Each accumulator must be designed and fabricated with a burst pressure of not less than five times its charged pressure at 70°F when shipped.

(4) Accumulators intended to function as shock absorbers, struts, gas springs, pneumatic springs or other impact or energy-absorbing devices are not subject to the requirements of this subchapter provided each:

(i) Has a gas space capacity not exceeding 1.6 L and a charge pressure not exceeding 280 bar, where the product of the capacity expressed in L and charge pressure expressed in bars does not exceed 80 (for example, 0.5 L gas space and 160 bar charge pressure);

(ii) Has a minimum burst pressure of 4 times the charge pressure at 20°C for products not exceeding 0.5 L gas space capacity and 5 times the charge pressure for products greater than 0.5 L gas space capacity;

(iii) Design type has been subjected to a fire test demonstrating that the article relieves its pressure by means of a fire degradable seal or other pressure relief device, such that the article will not fragment and that the article does not rocket; and

(iv) Accumulators must be manufactured under a written quality assurance program which monitors parameters controlling burst strength, burst mode and performance in a fire situation as specified in paragraphs (f)(4)(i) through (f)(4)(iii) of this section. A copy of the quality assurance program must be maintained at each facility at which the accumulators are manufactured.

(5) Accumulators not conforming to the provisions of paragraphs (f)(1) through (f) (4) of this section, may only be transported subject to the approval of the Associate Administrator.

(g) *Water pump system tank.* Water pump system tanks charged with compressed air or limited quantities of nitrogen to not over 40 psig for single-trip shipment to installation sites are excepted from labeling (transportation by air not authorized) and the specification packaging requirements of this subchapter when shipped under the following conditions. In addition, shipments are not subject to subpart F of this subchapter, to part 174 of this subchapter except §174.24 and part 177 except §177.817.

(1) The tank must be of steel, welded with heads concave to pressure, having a rated water capacity not exceeding 120 gallons and with outside diameter not exceeding 24 inches. Safety relief devices not required.

(2) The tank must be pneumatically tested to 100 psig. Test pressure must be permanently marked on the tank.

(3) The stress at prescribed pressure must not exceed 20,000 psi using formula:

$$S = \frac{Pd}{2t}$$

where:

S = wall stress in psi:
P = prescribed pressure for the tank of at least 3 times charged pressure at 70°F or 100 psig, whichever is greater;
t = minimum wall thickness, in inches.
d = inside diameter in inches;

(4) The burst pressure must be at least 6 times the charge pressure at 70°F.

(5) Each tank must be overpacked in a strong outer packaging in accordance with §173.301(h).

(h) *Lighter refills.* (1) Lighter refills (see §171.8 of this subchapter) must not contain an ignition element but must contain a release device. Lighter refills offered for transportation under this section may not exceed 4 fluid ounces capacity (7.22 cubic inches) or contain more than 65 grams of a Division 2.1 fuel. For transportation by highway or rail, lighter refills must be tightly packed and secured against movement in strong outer packagings. For transportation by aircraft or vessel, lighter refills must be tightly packed and secured against movement in any rigid specification outer packaging authorized in Subpart L of Part 178 of this subchapter at the Packing Group II performance level.

(2) *Exceptions.* (i) For other than transportation by aircraft, special exceptions for shipment of lighter refills in the ORM-D class are provided in paragraph (i) of this section.

(ii) For highway transportation, when no more than 1,500 lighter refills covered by this paragraph are transported in one motor vehicle, the requirements of subparts C through H of part 172, and Part 177 of this subchapter do not apply. Lighter refills covered under this paragraph must be packaged in rigid, strong outer packagings meeting the general packaging requirements of subpart B of this part. Outer packagings must be plainly and durably marked on two opposing sides or ends with the words "LIGHTER REFILLS" and the number of devices contained therein in letters measuring at least 20 mm (0.79 in) in height. No person may offer for transportation or transport the lighter refills or prepare the lighter refills for shipment unless that person has been specifically informed of the requirements of this section.

(i) *Limited quantities.* (1) A limited quantity that conforms to the provisions of paragraph (a)(1), (a)(3), (a)(5), (b) or, except for transportation by aircraft, paragraph (h) of this section is excepted from labeling requirements, unless the material is offered for transportation or transported by aircraft, and the specification packaging requirements of this subchapter when packaged in combination packagings according to this paragraph. Packages must be marked in accordance with §172.315(a) or (b), as appropriate, or as authorized in paragraph (i)(2) of this section. Unless otherwise specified in paragraph (i)(2) of this section, packages of limited quantities intended for transportation by aircraft must conform to the applicable requirements (*e.g.,* authorized materials, inner packaging quantity limits and closure securement) of §173.27 of this part. A limited quantity package that conforms to the provisions of this section is not subject to the shipping paper requirements of subpart C of part 172 of this subchapter, unless the material meets the definition of a hazardous substance, hazardous waste, marine pollutant, or is offered for transportation and transported by aircraft or vessel, and is eligible for the exceptions provided in §173.156 of this part. Outside packagings conforming to this paragraph are not required to be marked "INSIDE CONTAINERS COMPLY WITH PRESCRIBED REGULATIONS." In addition, packages of limited quantities are not subject to subpart F (Placarding) of part 172 of this subchapter. Each package must conform to the packaging requirements of subpart B of this part and may not exceed 30 kg (66 pounds) gross weight.

((2) Until December 31, 2013, a limited quantity package containing a "consumer commodity" as defined in §171.8 of this subchapter may be renamed "Consumer commodity" and reclassed as ORM-D or ORM-D-AIR material and offered for transportation and transported in accordance with the applicable provisions of this subchapter in effect on October 1, 2010.

(j) Aerosols and receptacles small, containing gas with a capacity of less than 50 mL. Aerosols, as defined in §171.8 of this subchapter, and receptacles small, containing gas, with a capacity not exceeding 50 mL (1.7 oz.) and with a pressure not exceeding 970 kPa (141 psig) at 55°C (131°F), containing no hazardous materials other than a Division 2.2 gas, are not subject to the requirements of this subchapter. The pressure limit may be increased to 2,000 kPa (290 psig) at 55°C (131°F) provided the aerosols are transported in outer packages that conform to the packaging requirements of Subpart B of this part. This paragraph (j) does not apply to a self-defense spray (*e.g.,* pepper spray).

(k) *Aerosols for recycling or disposal.* Aerosols, as defined in §171.8 of this subchapter, containing a limited quantity which conforms to the provisions of paragraph (a)(3), (a)(5), (b)(1), (b)(2), or (b)(3) of this section are not subject to the 30 kg (66 pounds) gross weight limitation when transported by motor vehicle for purposes of recycling or disposal under the following conditions:

(1) The strong outer packaging and its contents must not exceed a gross weight of 500 kg (1,100 pounds);

(2) Each aerosol container must be secured with a cap to protect the valve stem or the valve stem must be removed; and

(3) The packaging must be offered for transportation or transported by—

(i) Private or contract motor carrier; or

(ii) Common carrier in a motor vehicle under exclusive use for such service.

(l) For additional exceptions, *see* §173.307.

§173.307 Exceptions for compressed gases.

(a) The following materials are not subject to the requirements of this subchapter:

(1) Carbonated beverages.

(2) Tires when inflated to pressures not greater than their rated inflation pressures. For transportation by air, tires and tire assemblies must meet the conditions in §175.8(b)(4) of this subchapter.

(3) Balls used for sports.

(4) Refrigerating machines, including dehumidifiers and air conditioners, and components thereof, such as precharged tubing containing:

(i) 12 kg (25 pounds) or less of a non-flammable, non-toxic gas;

(ii) 12 L (3 gallons) or less of ammonia solution (UN 2672);

(iii) Except when offered or transported by air, 12 kg (25 pounds) or less of a flammable, non-toxic gas;

(iv) Except when offered or transported by air or vessel, 20 kg (44 pounds) or less of a Group A1 refrigerant specified in ANSI/ASHRAE Standard 15 (IBR, see §171.7 of this subchapter); or

(v) 100 g (4 ounces) or less of a flammable, non-toxic liquefied gas.

(5) Manufactured articles or apparatuses, each containing not more than 100 mg (0.0035 ounce) of inert gas and packaged so that the quantity of inert gas per package does not exceed 1 g (0.35 ounce).

(6) Light bulbs, provided they are packaged so that the projectile effects of any rupture of the bulb will be contained within the package.

(b) [Reserved]

§173.309 Fire extinguishers.

(a) Fire extinguishers charged with a limited quantity of compressed gas to not more than 1660 kPa (241 psig) at 21 °C (70 °F) are excepted from labeling (except when offered for transportation by air) and the specification packaging requirements of this subchapter when shipped under the following conditions. In addition, shipments are not subject to subpart F of part 172 of this sub-chapter, to part 174 of this subchapter except §174.24 or to part 177 of this subchapter except §177.817.

(1) Each fire extinguisher must have contents which are nonflammable, nonpoisonous, and noncorrosive as defined under this subchapter.

(2) Each fire extinguisher must be shipped as an inner packaging.

(3) Nonspecification cylinders are authorized subject to the following conditions:

(i)The internal volume of each cylinder may not exceed 18 L (1,100 cubic inches). For fire extinguishers not exceeding 900 mL (55 cubic inches) capacity, the liquid portion of the gas plus any additional liquid or solid must not completely fill the container at 55 °C (130 °F). Fire extinguishers exceeding 900 mL (55 cubic inches) capacity may not contain any liquefied compressed gas;

(ii) Each fire extinguisher manufactured on and after January 1, 1976, must be designed and fabricated with a burst pressure of not less than six times its charged pressure at 21 °C (70 °F) when shipped;

(iii) Each fire extinguisher must be tested, without evidence of failure or damage, to at least three times its charged pressure at 21 °C (70 °F) but not less than 825 kPa (120 psig) before initial shipment, and must be marked to indicate the year of the test (within 90 days of the actual date of the original test) and with the words "MEETS DOT REQUIREMENTS." This marking is considered a certification that the fire extinguisher is manufactured in accordance with the requirements of this section. The words "This extinguisher meets all requirements of 49 CFR 173.306" may be displayed on fire extinguishers manufactured prior to January 1, 1976; and

(iv) For any subsequent shipment, each fire extinguisher must be in compliance with the retest requirements of the Occupational Safety and Health Administration Regulations of the Department of Labor, 29 CFR 1910.157.

(4) Specification 2P or 2Q (§§178.33 and 178.33a of this subchapter) inner nonrefillable metal packagings are authorized for use as fire extinguishers subject to the following conditions:

(i) The liquid portion of the gas plus any additional liquid or solid may not completely fill the packaging at 55 °C (130 °F);

(ii) Pressure in the packaging shall not exceed 1250 kPa (181 psig) at 55 °C (130 °F). If the pressure exceeds 920 kPa (141 psig) at 55 °C (130 °F), but does not exceed 1100 kPa (160 psig) at 55 °C (130 °F), a specification DOT 2P inner metal packaging must be used; if the pressure exceeds 1100 kPa (160 psig) at 55 °C (130 °F), a specification DOT 2Q inner metal packaging must be used. The metal packaging must be capable of withstanding, without bursting, a pressure of one and one-half times the equilibrium pressure of the contents at 55 °C (130 °F); and

(iii) Each completed inner packaging filled for shipment must have been heated until the pressure in the container is equivalent to the equilibrium pressure of the contents at 55 °C (130 °F) without evidence of leakage, distortion, or other defect.

(b) Specification 3A, 3AA, 3E, 3AL, 4B, 4BA, 4B240ET or 4BW (§§178.36, 178.37, 178.42, 178.46, 178.50, 178.51, 178.55 and 178.61 of this subchapter) cylinders are authorized for use as fire extinguishers.

Reserved

PART 174

Subpart C—General Handling and Loading Requirements

§174.67 Tank car unloading.

For transloading operations, the following rules must be observed:

(a) *General requirements.* (1) Unloading operations must be performed by hazmat employees properly instructed in unloading hazardous materials and made responsible for compliance with this section.

(2) Each hazmat employee who is responsible for unloading must apply the handbrake and block at least one wheel to prevent movement in any direction. If multiple tank cars are coupled together, sufficient hand brakes must be set and wheels blocked to prevent movement in both directions.

(3) Each hazmat employee who is responsible for unloading must secure access to the track to prevent entry by other rail equipment, including motorized service vehicles. This requirement may be satisfied by lining each switch providing access to the unloading area against movement and securing each switch with an effective locking device, or by using derails, portable bumper blocks, or other equipment that provides and equivalent level of safety.

(4) Each hazmat employee who is responsible for unloading must display caution signs on the track or on the tank cars to warn persons approaching the cars from the open end of the track and must be left up until after all closures are secured and the cars are in proper condition for transportation. The caution signs must be of metal or other durable material, rectangular, at 30.48 cm (12 inches) high by 38.10 cm (15 inches) wide, and bear the word "STOP." The word "STOP" must appear in letters at least 10.16 cm (4 inches) high. The letters must be white on a blue background. Additional words, such as "Tank Car Connected" or "Crew at Work," may also appear in white letters under the word "STOP."

(5) The transloading facility operator must maintain written safety procedures (such as those it may already be required to maintain pursuant to the Department of Labor's Occupational Safety and Health Administration requirements in 29 CFR 1910.119 and 1910.120) in a location where they are immediately available to hazmat employees responsible for the transloading operation.

(6) Before a manhole cover or outlet valve cap is removed from a tank car, the car must be relieved of all interior pressure by cooling the tank with water or by venting the tank by raising the safety valve or opening the dome vent at short intervals. However, if venting to relieve pressure will cause a dangerous amount of vapor to collect outside the car, venting and unloading must be deferred until the pressure is reduced by allowing the car to stand overnight, otherwise cooling the contents, or venting to a closed collection system. These precautions are not necessary when the car is equipped with a manhole cover which hinges inward or with an inner manhole cover which does not have to be removed to unload the car, and when pressure is relieved by piping vapor into a condenser or storage tank.

(b) After the pressure is released, for unloading processes that require the removal of the manhole cover, the seal must be broken and the manhole cover removed as follows:

(1) *Screw type.* The cover must be loosened by placing a bar between the manhole cover lug and knob. After two complete turns, so that the vent openings are exposed, the operation must be stopped, and if there is any sound of escaping vapor, the cover must be screwed down tightly and the interior pressure relieved as prescribed in paragraph (a)(6) of this section, before again attempting to remove the cover.

(2) *Hinged and bolted type.* All nuts must be unscrewed one complete turn, after which same precautions as prescribed for screw type cover must be observed.

(3) *Interior type.* All dirt and cinders must be carefully removed from around the cover before the yoke is unscrewed.

(c) When the car is unloaded through a bottom outlet valve, for unloading processes that require the removal of the manhole cover, the manhole cover must be adjusted as follows:

(1) *Screw type.* The cover must be put in place, but not entirely screwed down, so that air may enter the tank through the vent holes in threaded flange of the cover.

(2) *Hinged and bolted type.* A non-metallic block must be placed under one edge of the cover.

(3) *Interior type.* The screw must be tightened up in the yoke so that the cover is brought up within one-half inch of the closed position.

(d) When unloading through the bottom outlet of a car equipped with an interior manhole type cover, and in each case where unloading is done through the manhole (unless a special cover with a safety vent opening and a tight connection for the discharge outlet is used), the manhole must be protected by asbestos or metal covers against the entrance of sparks or other sources of ignition of vapor, or by being covered and surrounded with wet burlap or similar cloth material. The burlap or other cloth must be kept damp by the replacement or the application of water as needed.

(e) Seals or other substances must not be thrown into the tank and the contents may not be spilled over the car or tank.

(f) The valve rod handle or control in the dome must be operated several times to see that outlet valve in bottom of tank is on its seat before valve cap is removed.

(g) The valve cap, or the reducer when a large outlet is to be used, must be removed with a suitable wrench after the set screws are loosened and pail must be placed in position to catch any liquid that may be in the outlet chamber. If the valve cap or reducer does not unscrew easily, it may be tapped lightly with a mallet or wooden block in an upward direction. If leakage shows upon starting the removal, the cap or reducer may not be entirely unscrewed. Sufficient threads must be left engaged and sufficient time allowed to permit controlled escape of any accumulation of liquid in the outlet chamber. If the leakage stops or the rate of leakage diminishes materially, the cap or reducer may be entirely removed. If the initial rate of leakage continues, further efforts must be made to seat the outlet valve (see paragraph (f) of this section). If this fails, the cap or reducer must be screwed up tight and the tank must be unloaded through the dome. If upon removal of the outlet cap the outlet chamber is found to be blocked with frozen liquid or any other matter, the cap must be replaced immediately and a careful examination must be made to determine whether the outlet casting has been cracked. If the obstruction is not frozen liquid, the car must be unloaded through the dome. If the obstruction is frozen liquid and no crack has been found in the outlet casting, the car may, if circumstances require it, be unloaded from the bottom by removing the cap and attaching unloading connections immediately. Before opening the valve inside the tank car, steam must be applied to the outside of the outlet casting or wrap casting with burlap or other rags and hot water must be applied to melt the frozen liquid.

(h) Unloading connections must be securely attached to unloading pipes on the dome or to the bottom discharge outlets before any discharge valves are opened.

(i) Throughout the entire period of unloading and while a tank car has unloading equipment attached, the facility operator must assure that the tank car is:

(1) Attended by a designated hazmat employee who is physically present and who has an unobstructed view of the unloading operation; or

(2) Monitored by a signaling system (*e.g.,* video system, sensing equipment, or mechanical equipment) that is observed by a designated hazmat employee located either in the immediate area of the tank car or at a remote location within the facility, such as a control room. The signaling system must—

(i) Provide a level of surveillance equivalent to that provided in subparagraph (1) of this paragraph (i); and

(ii) Provide immediate notification to a designated hazmat employee of any system malfunction or other emergency so that, if warranted, responsive actions may be initiated immediately.

(j) Attendance is not required when piping is attached to a top outlet of a tank car, equipped with a protective housing required under §179.100-12 of this subchapter, for discharge of lading under the following conditions:

(1) All valves are tightly closed.

(2) The piping is not connected to hose or other unloading equipment and is fitted with a cap or plug of appropriate material and construction.

(3) The piping extends no more than 15.24 centimeters (6 inches) from the outer edge of the protective housing.

(k) In the absence of the unloader, a tank car may stand with unloading connections attached when no product is being transferred under the following conditions:

(1) The facility operator must designate a hazmat employee responsible for on-site monitoring of the transfer facility. The designated hazmat employee must be made familiar with the nature and properties of the product contained in the tank car; procedures to be followed in the event of an emergency; and, in the event of an emergency, have the ability and authority to take responsible actions.

(2) When a signaling system is used in accordance with paragraph (i) of this section, the system must be capable of alerting the designated hazmat employee in the event of an emergency and providing immediate notification of any monitoring system malfunction. If the monitoring system does not have self-monitoring capability, the designated hazmat employee must check the monitoring system hourly for proper operation.

(3) The tank car and facility shutoff valves must be secured in the closed position.

(4) Brakes must be set and wheels locked in accordance with paragraph (a)(2) of this section.

(5) Access to the track must be secured in accordance with paragraph (a)(3) of this section.

(l) As soon as a tank car is completely unloaded, all valves must be made tight by the use of a bar, wrench or other suitable tool, the unloading connections must be removed and all other closures made tight.

(m) Railroad defect cards may not be removed.

(n) If oil or gasoline has been spilled on the ground around connections, it must be covered with fresh, dry sand or dirt.

(o) All tools and implements used in connection with unloading must be kept free of oil, dirt, and grit.

PART 177

Subpart A—General Information and Regulations

§177.800 Purpose and scope of this part and responsibility for compliance and training.

(a) *Purpose and scope.* This part prescribes requirements, in addition to those contained in parts 171, 172, 173, 178 and 180 of this subchapter, that are applicable to the acceptance and transportation of hazardous materials by private, common, or contract carriers by motor vehicle.

(b) *Responsibility for compliance.* Unless this subchapter specifically provides that another person shall perform a particular duty, each carrier, including a connecting carrier, shall perform the duties specified and comply with all applicable requirements in this part and shall ensure its hazmat employees receive training in relation thereto.

(c) *Responsibility for training.* A carrier may not transport a hazardous material by motor vehicle unless each of its hazmat employees involved in that transportation is trained as required by this part and subpart H of part 172 of this subchapter.

(d) *No unnecessary delay in movement of shipments.* All shipments of hazardous materials must be transported without unnecessary delay, from and including the time of commencement of the loading of the hazardous material until its final unloading at destination.

§177.801 Unacceptable hazardous materials shipments.

No person may accept for transportation or transport by motor vehicle a forbidden material or hazardous material that is not prepared in accordance with the requirements of this subchapter.

§177.802 Inspection.

Records, equipment, packagings and containers under the control of a motor carrier, insofar as they affect safety in transportation of hazardous materials by motor vehicle, must be made available for examination and inspection by a duly authorized representative of the Department.

§177.804 Compliance with Federal Motor Carrier Safety Regulations.

(a) *General.* Motor carriers and other persons subject to this part must comply with 49 CFR part 383 and 49 CFR parts 390 through 397 (excluding §§397.3 and 397.9) to the extent those regulations apply.

(b) *Prohibition against texting.* In accordance with §392.80 of the FMCSRs a person transporting a quantity of hazardous materials requiring placarding under 49 CFR part 172 or any quantity of a material listed as a select agent or toxin in 42 CFR part 73 may not engage in, allow, or require texting while driving.

§177.810 Vehicular tunnels.

Except as regards Class 7 (radioactive) materials, nothing contained in parts 170–189 of this subchapter shall be so construed as to nullify or supersede regulations established and published under authority of State statute or municipal ordinance regarding the kind, character, or quantity of any hazardous material permitted by such regulations to be transported through any urban vehicular tunnel used for mass transportation.

§177.816 Driver training.

(a) In addition to the training requirements of §177.800, no carrier may transport, or cause to be transported, a hazardous material unless each hazmat employee who will operate a motor vehicle has been trained in the applicable requirements of 49 CFR parts 390 through 397 and the procedures necessary for the safe operation of that motor vehicle. Driver training shall include the following subjects:

(1) Pre-trip safety inspection;

(2) Use of vehicle controls and equipment, including operation of emergency equipment;

(3) Operation of vehicle, including turning, backing, braking, parking, handling, and vehicle characteristics including those that affect vehicle stability, such as effects of braking and curves, effects of speed on vehicle control, dangers associated with maneuvering through curves, dangers associated with weather or road conditions that a driver may experience (e.g., blizzards, mountainous terrain, high winds), and high center of gravity;

(4) Procedures for maneuvering tunnels, bridges, and railroad crossings;

(5) Requirements pertaining to attendance of vehicles, parking, smoking, routing, and incident reporting; and

(6) Loading and unloading of materials, including—

(i) Compatibility and segregation of cargo in a mixed load;

(ii) Package handling methods; and

(iii) Load securement.

(b) *Specialized requirements for cargo tanks and portable tanks.* In addition to the training requirement of paragraph (a) of this section, each person who operates a cargo tank or a vehicle with a portable tank with a capacity of 1,000 gallons or more must receive training applicable to the requirements of this subchapter and have the appropriate State-issued commercial driver's license required by 49 CFR part 383. Specialized training shall include the following:

(1) Operation of emergency control features of the cargo tank or portable tank;

(2) Special vehicle handling characteristics, including: high center of gravity, fluid-load subject to surge, effects of fluid-load surge on braking, characteristic differences in stability among baffled, unbaffled, and multi-compartmented tanks; and effects of partial loads on vehicle stability;

(3) Loading and unloading procedures;

(4) The properties and hazards of the material transported; and

(5) Retest and inspection requirements for cargo tanks.

(c) The training required by paragraphs (a) and (b) of this section may be satisfied by compliance with the current requirements for a Commercial Driver's License (CDL) with a tank vehicle or hazardous materials endorsement.

(d) Training required by paragraph (b) of this section must conform to the requirements of §172.704 of this subchapter with respect to frequency and recordkeeping.

§177.817 Shipping papers.

(a) *General requirements.* A person may not accept a hazardous material for transportation or transport a hazardous material by highway unless that person has received a shipping paper prepared in accordance with part 172 of this subchapter or the material is excepted from shipping paper requirements under this subchapter. A subsequent carrier may not transport a hazardous material unless it is accompanied by a shipping paper prepared in accordance with part 172 of this subchapter, except for §172.204, which is not required.

(b) *Shipper certification.* An initial carrier may not accept a hazardous material offered for transportation unless the shipping paper describing the material includes a shippers' certification which meets the requirements in §172.204 of this subchapter. Except for a hazardous waste, the certification is not required for shipments to be transported entirely by private carriage and for bulk shipments to be transported in a cargo tank supplied by the carrier.

(c) *Requirements when interlining with carriers by rail.* A motor carrier shall mark on the shipping paper required by this section, if it offers or delivers a freight container or transport vehicle to a rail carrier for further transportation;

(1) A description of the freight container or transport vehicle; and

(2) The kind of placard affixed to the freight container or transport vehicle.

(d) This subpart does not apply to a material that is excepted from shipping paper requirements as specified in §172.200 of this subchapter.

(e) *Shipping paper accessibility — accident or inspection.* A driver of a motor vehicle containing hazardous material, and each carrier using such a vehicle, shall ensure that the shipping paper required by this section is readily available to, and recognizable by, authorities in the event of accident or inspection. Specifically, the driver and the carrier shall:

(1) Clearly distinguish the shipping paper, if it is carried with other shipping papers or other papers of any kind, by either distinctively tabbing it or by having it appear first; and

(2) Store the shipping paper as follows:

(i) When the driver is at the vehicle's controls, the shipping paper shall be: (A) Within his immediate reach while he is restrained by the lap belt; and (B) either readily visible to a person entering the driver's compartment or in a holder which is mounted to the inside of the door on the driver's side of the vehicle.

(ii) When the driver is not at the vehicle's controls, the shipping paper shall be: (A) In a holder which is mounted to the inside of the door on the driver's side of the vehicle; or (B) on the driver's seat in the vehicle.

(f) *Retention of shipping papers.* Each person receiving a shipping paper required by this section must retain a copy or an electronic image thereof, that is accessible at or through its principal place of business and must make the shipping paper available, upon request, to an authorized official of a Federal, State, or local government agency at reasonable times and locations. For a hazardous waste, the shipping paper copy must be retained for three years after the material is accepted by the initial carrier. For all other hazardous materials, the shipping paper copy must be retained for one year after the material is accepted by the carrier. Each shipping paper copy must include the date of acceptance by the carrier. A motor carrier (as defined in §390.5 of subchapter B of chapter III of subtitle B) using a shipping paper without change for multiple shipments of one or more hazardous materials having the same shipping name and identification number may retain a single copy of the shipping paper, instead of a copy for each shipment made, if the carrier also retains a record of each shipment made that includes shipping name, identification number, quantity transported, and date of shipment.

§177.823 Movement of motor vehicles in emergency situations.

(a) A carrier may not move a transport vehicle containing a hazardous material unless the vehicle is marked and placarded in accordance with part 172 or as authorized in §171.12a of this subchapter, or unless, in an emergency:

(1) The vehicle is escorted by a representative of a state or local government;

(2) The carrier has permission from the Department; or

(3) Movement of the transport vehicle is necessary to protect life or property.

(b) *Disposition of contents of cargo tank when unsafe to continue.* In the event of a leak in a cargo tank of such a character as to make further transportation unsafe, the leaking vehicle should be removed from the traveled portion of the highway and every available means employed for the safe disposal of the leaking material by preventing, so far as practicable, its spread over a wide area, such as by digging trenches to drain to a hole or depression in the ground, diverting the liquid away from streams or sewers if possible, or catching the liquid in containers if practicable. Smoking, and any other source of ignition, in the vicinity of a leaking cargo tank is not permitted.

(c) *Movement of leaking cargo tanks.* A leaking cargo tank may be transported only the minimum distance necessary to reach a place where the contents of the tank or compartment may be disposed of safely. Every available means must be utilized to prevent the leakage or spillage of the liquid upon the highway.

Subpart B—Loading and Unloading

§177.834 General requirements.

(a) *Packages secured in a motor vehicle.* Any package containing any hazardous material, not permanently attached to a motor vehicle, must be secured against shifting, including relative motion between packages, within the vehicle on which it is being transported, under conditions normally incident to transportation. Packages having valves or other fittings must be loaded in a manner to minimize the likelihood of damage during transportation.

(b) Each package containing a hazardous material bearing package orientation markings prescribed in §172.312 of this subchapter must be loaded on a transport vehicle or within a freight container in accordance with such markings and must remain in the correct position indicated by the markings during transportation.

(c) *No smoking while loading or unloading.* Smoking on or about any motor vehicle while loading or unloading any Class 1 (explosive), Class 3 (flammable liquid), Class 4 (flammable solid), Class 5 (oxidizing), or Division 2.1 (flammable gas) materials is forbidden.

(d) *Keep fire away, loading and unloading.* Extreme care shall be taken in the loading or unloading of any Class 1 (explosive), Class 3 (flammable liquid), Class 4 (flammable solid), Class 5 (oxidizing), or Division 2.1 (flammable gas) materials into or from any motor vehicle to keep fire away and to prevent persons in the vicinity from smoking, lighting matches, or carrying any flame or lighted cigar, pipe, or cigarette.

(e) *Handbrake set while loading and unloading.* No hazardous material shall be loaded into or on, or unloaded from, any motor vehicle unless the handbrake be securely set and all other reasonable precautions be taken to prevent motion of the motor vehicle during such loading and unloading process.

(f) *Use of tools, loading and unloading.* No tools which are likely to damage the effectiveness of the closure of any package or other container or likely adversely to affect such package or container, shall be used for the loading or unloading of any Class 1 (explosive) material or other dangerous article.

(g) [Reserved]

(h) *Precautions concerning containers in transit; fueling road units.* Reasonable care should be taken to prevent undue rise in temperature of containers and their contents during transit. There must be no tampering with such container or the contents thereof nor any discharge of the contents of any container between point of origin and point of billed destination. Discharge of contents of any container, other than a cargo tank or IM portable tank, must not be made prior to removal from the motor vehicle. Nothing contained in this paragraph shall be so construed as to prohibit the fueling of machinery or vehicles used in road construction or maintenance.

(i) *Attendance requirements.* (1) *Loading.* A cargo tank must be attended by a qualified person at all times when it is being loaded. The person who is responsible for loading the cargo tank is also responsible for ensuring that it is so attended.

(2) *Unloading.* A motor carrier who transports hazardous materials by a cargo tank must ensure that the cargo tank is attended by a qualified person at all times during unloading. However, the carrier's obligation to ensure attendance during unloading ceases when—

(i) The carrier's obligation for transporting the materials is fulfilled;

(ii) The cargo tank has been placed upon the consignee's premises; and

(iii) The motive power has been removed from the cargo tank and removed from the premises.

(3) Except for unloading operations subject to §§177.837(d), 177.840(p), and 177.840(q), a qualified person "attends" the loading or unloading of a cargo tank if, throughout the process, he is alert and is within 7.62 m (25 feet) of the cargo tank. The qualified person attending the unloading of a cargo tank must have an unobstructed view of the cargo tank and delivery hose to the maximum extent practicable during the unloading operation.

(4) A person is "qualified" if he has been made aware of the nature of the hazardous material which is to be loaded or unloaded, he has been instructed on the procedures to be followed in emergencies, he is authorized to move the cargo tank, and he has the means to do so.

(j) Except for a cargo tank conforming to §173.29(b)(2) of this subchapter, a person may not drive a cargo tank motor vehicle containing a hazardous material regardless of quantity unless:

(1) All manhole closures are closed and secured; and

(2) All valves and other closures in liquid discharge systems are closed and free of leaks.

(k) [Reserved]

(l) *Use of cargo heaters when transporting certain hazardous material.* Transportation includes loading, carrying, and unloading.

(1) *When transporting Class 1 (explosive) materials.* A motor vehicle equipped with a cargo heater of any type may transport Class 1 (explosive) materials only if the cargo heater is rendered inoperable by: (i) Draining or removing the cargo heater fuel tank; and (ii) disconnecting the heater's power source.

(2) *When transporting certain flammable material—*

(i) *Use of combustion cargo heaters.* A motor vehicle equipped with a combustion cargo heater may be used to transport Class 3 (flammable liquid) or Division 2.1 (flammable gas) materials only if each of the following requirements are met:

(A) It is a catalytic heater.

(B) The heater's surface temperature cannot exceed 54°C (130°F)—either on a thermostatically controlled heater or on a heater without thermostatic control when the outside or ambient temperature is 16°C (61°F) or less.

(C) The heater is not ignited in a loaded vehicle.

(D) There is no flame, either on the catalyst or anywhere in the heater.

(E) The manufacturer has certified that the heater meets the requirements under paragraph (l)(2)(i) of this section by permanently marking the heater "MEETS DOT REQUIREMENTS FOR CATALYTIC HEATERS USED WITH FLAMMABLE LIQUID AND GAS."

(F) The heater is also marked "DO NOT LOAD INTO OR USE IN CARGO COMPARTMENTS CONTAINING FLAMMABLE LIQUID OR GAS IF FLAME IS VISIBLE ON CATALYST OR IN HEATER."

(G) Heater requirements under §393.77 of this title are complied with.

(ii) *Effective date for combustion heater requirements.* The requirements under paragraph (l)(2)(i) of this section govern as follows:

(A) Use of a heater manufactured after November 14, 1975, is governed by every requirement under (l)(2)(i) of this section;

(B) Use of a heater manufactured before November 15, 1975, is governed only by the requirements under (l)(2)(i) (A), (C), (D), (F) and (G) of this section until October 1, 1976; and

(C) Use of any heater after September 30, 1976, is governed by every requirement under (l)(2)(i) of this section.

(iii) *Restrictions on automatic cargo-space-heating temperature control devices.* Restrictions on these devices have two dimensions: Restrictions upon use and restrictions which apply when the device must not be used.

(A) *Use restrictions.* An automatic cargo-space-heating temperature control device may be used when transporting Class 3 (flammable liquid) or Division 2.1 (flammable gas) materials only if each of the following requirements is met:

(*1*) Electrical apparatus in the cargo compartment is nonsparking or explosion proof.

(*2*) There is no combustion apparatus in the cargo compartment.

(*3*) There is no connection for return of air from the cargo compartment to the combustion apparatus.

(*4*) The heating system will not heat any part of the cargo to more than 54°C (129°F).

(*5*) Heater requirements under §393.77 of this title are complied with.

(B) *Protection against use.* Class 3 (flammable liquid) or Division 2.1 (flammable gas) materials may be transported by a vehicle, which is equipped with an automatic cargo-space-heating temperature control device that does not meet each requirement of paragraph (l)(2)(iii) (A) of this section, only if the device is first rendered inoperable, as follows:

(*1*) Each cargo heater fuel tank, if other than LPG, must be emptied or removed.

(*2*) Each LPG fuel tank for automatic temperature control equipment must have its discharge valve closed and its fuel feed line disconnected.

(m) Tanks constructed and maintained in compliance with Spec. 106A or 110A (§§179.300, 179.301 of this subchapter) that are authorized for the shipment of hazardous materials by highway in part 173 of this subchapter must be carried in accordance with the following requirements:

(1) Tanks must be securely chocked or clamped on vehicles to prevent any shifting.

(2) Equipment suitable for handling a tank must be provided at any point where a tank is to be loaded upon or removed from a vehicle.

(3) No more than two cargo carrying vehicles may be in the same combination of vehicles.

(4) Compliance with §§174.200 and 174.204 of this subchapter for combination rail freight, highway shipments and for trailer-on-flat-car service is required.

(n) Specification 56, 57, IM 101, and IM 102 portable tanks, when loaded, may not be stacked on each other nor placed under other freight during transportation by motor vehicle.

(o) *Unloading of IM and UN portable tanks* . No person may unload an IM or UN portable tank while it remains on a transport vehicle with the motive power unit attached except under the following conditions:

(1) The unloading operation must be attended by a qualified person in accordance with the requirements in paragraph (i) of this section. The person performing unloading functions must be trained in handling emergencies that may occur during the unloading operation.

(2) Prior to unloading, the operator of the vehicle on which the portable tank is transported must ascertain that the conditions of this paragraph (o) are met.

(3) An IM or UN portable tank equipped with a bottom outlet as authorized in Column (7) of the §172.101 Table of this subchapter by assignment of a T Code in the appropriate proper shipping name entry, and that contains a liquid hazardous material of Class 3, PG I or II, or PG III with a flash point of less than 100 °F (38 °C); Division 5.1, PG I or II; or Division 6.1, PG I or II, must conform to the outlet requirements in §178.275(d)(3) of this subchapter.

§177.835 Class 1 materials.

(See also §177.834 (a) to (j).)

(a) *Engine stopped.* No Class 1 (explosive) materials shall be loaded into or on or be unloaded from any motor vehicle with the engine running.

(b) *Care in loading, unloading, or other handling of Class 1 (explosive) materials.* No bale hooks or other metal tools shall be used for the loading, unloading, or other handling of Class 1 (explosive) materials, nor shall any package or other container of Class 1 (explosive) materials, except barrels or kegs, be rolled. No packages of Class 1 (explosive) materials shall be thrown or dropped during process of loading or unloading or handling of Class 1 (explosive) materials. Special care shall be exercised to the end that packages or other containers containing Class 1 (explosive) materials shall not catch fire from sparks or hot gases from the exhaust tailpipe.

(1) Whenever tarpaulins are used for covering Class 1 (explosive) materials, they shall be secured by means of rope, wire, or other equally efficient tie downs. Class 1 (explosive) materials placards or markings required by §177.823 shall be secured, in the appropriate locations, directly to the equipment transporting the Class 1 (explosive) materials. If the vehicle is provided with placard boards, the placards must be applied to these boards.

(c) *Class 1 (explosive) materials on vehicles in combination.* Division 1.1 or 1.2 (explosive) materials may not be loaded into or carried on any vehicle or a combination of vehicles if:

(1) More than two cargo carrying vehicles are in the combination;

(2) Any full trailer in the combination has a wheel base of less than 184 inches;

(3) Any vehicle in the combination is a cargo tank which is required to be marked or placarded under §177.823; or

(4) The other vehicle in the combination contains any:

(i) Substances, explosive, n.o.s., Division 1.1A (explosive) material (Initiating explosive),

(ii) Packages of Class 7 (radioactive) materials bearing "Yellow III" labels,

(iii) Division 2.3, Hazard Zone A or Hazard Zone B materials or Division 6.1, PG I, Hazard Zone A materials, or

(iv) Hazardous materials in a portable tank or a DOT Specification 106A or 110A tank.

(d) [Reserved]

(e) *No sharp projections inside body of vehicles.* No motor vehicle transporting any kind of Class 1 (explosive) material shall have on the interior of the body in which the Class 1 (explosive) materials are contained, any inwardly projecting bolts, screws, nails, or other inwardly projecting parts likely to produce damage to any package or container of Class 1 (explosive) materials during the loading or unloading process or in transit.

(f) *Class 1 (explosive) materials vehicles, floors tight and lined.* Motor vehicles transporting Division 1.1, 1.2, or 1.3 (explosive) materials shall have tight floors; shall have that portion of the interior in contact with the load lined with either non-metallic material or non-ferrous metals, except that the lining is not required for truck load shipments loaded by the Departments of the Army, Navy or Air Force of the United States Government provided the Class 1 (explosive) materials are of such nature that they are not liable to leakage of dust, powder, or vapor which might become the cause of an explosion. The interior of the cargo space must be in good condition so that there will not be any likelihood of containers being damaged by explosed bolts, nuts, broken side panels, or floor boards, or any similar projections.

(g) No detonator assembly or booster with detonator may be transported on the same motor vehicle with any Division 1.1, 1.2 or 1.3 material (except other detonator assemblies, boosters with detonators or detonators), detonating cord Division 1.4 material or Division 1.5 material. No detonator may be transported on the same motor vehicle with any Division 1.1, 1.2 or 1.3 material (except other detonators, detonator assemblies or boosters with detonators), detonating cord Division 1.4 material or Division 1.5 material unless—

(1) It is packed in a specification MC 201 (§178.318 of this subchapter) container; or

(2) The package conforms with requirements prescribed in §173.62 of this subchapter, and its use is restricted to instances when —

(i) There is no Division 1.1, 1.2, 1.3, or 1.5 material loaded on the motor vehicle; and

(ii) A separation of 61 cm (24 inches) is maintained between each package of detonators and each package of detonating cord; or

(3) It is packed and loaded in accordance with a method approved by the Associate Administrator. One approved method requires that—

(i) The detonators are in packagings as prescribed in §173.63 of this subchapter which in turn are loaded into suitable containers or separate compartments; and

(ii) That both the detonators and the container or compartment meet the requirements of the IME Standard 22 (IBR, see §171.7 of this subchapter).

(h) *Lading within body or covered, tailgate closed.* Except as provided in paragraph (g) of this section, dealing with the transportation of liquid nitroglycerin, desensitized liquid nitroglycerin or diethylene glycol dinitrate, all of that portion of the lading of any motor vehicle which consists of Class 1 (explosive) materials shall be contained entirely within the body of the motor vehicle or within the horizontal outline thereof, without overhang or projection of any part of the load and if such motor vehicle has a tailboard or tailgate, it shall be closed and secured in place during such transportation. Every motor vehicle transporting Class 1 (explosive) materials must either have a closed body or have the body thereof covered with a tarpaulin, and in either event care must be taken to protect the load from moisture and sparks, except that subject to other provisions of these regulations, Class 1 (explosive) materials other than black powder may be transported on flat-bed vehicles if the explosive portion of the load on each vehicle is packed in fire and water resistant containers or covered with a fire and water resistant tarpaulin.

(i) *Class 1 (explosive) materials to be protected against damage by other lading.* No motor vehicle transporting any Class 1 (explosive) materials may transport as a part of its load any metal or other articles or materials likely to damage such Class 1 (explosive) material or any package in which it is contained, unless the different parts of such load be so segregated or secured in place in or on the motor vehicle and separated by bulkheads or other suitable means as to prevent such damage.

(j) *Transfer of Class 1 (explosive) materials en route.* No Division 1.1, 1.2, or 1.3 (explosive) materials shall be transferred from one container to another, or from one motor vehicle to another vehicle, or from another vehicle to a motor vehicle, on any public highway, street, or road, except in case of emergency. In such cases red electric lanterns, red emergency reflectors or red flags shall be set out in the manner prescribed for disabled or stopped motor vehicles. (See Motor Carrier Safety Regulations, part 392 of this title.) In any event, all practicable means, in addition to these hereinbefore prescribed, shall be taken to protect and warn other users of the highway against the hazard involved in any such transfer or against the hazard occasioned by the emergency making such transfer necessary.

(k) *Attendance of Class 1 (explosive) materials.* Division 1.1, 1.2, or 1.3 materials that are stored during transportation in commerce must be attended and afforded surveillance in accordance with 49 CFR 397.5. A safe haven that conforms to NFPA 498 (IBR, see §171.7 of the subchapter) constitutes a federally approved safe haven for the unattended storage of vehicles containing Division 1.1, 1.2, or 1.3 materials.

§177.837 Class 3 materials.

(See also §177.834 (a) to (j).)

(a) *Engine stopped.* Unless the engine of a cargo tank motor vehicle is to be used for the operation of a pump, Class 3 material may not be loaded into, or on, or unloaded from any cargo tank motor vehicle while the engine is running. The diesel engine of a cargo tank motor vehicle may be left running during the loading and unloading of a Class 3 material if the ambient atmospheric temperature is at or below −12°C (10°F).

(b) *Bonding and grounding containers other than cargo tanks prior to and during transfer of lading.* For containers which are not in metallic contact with each other, either metallic bonds or ground conductors shall be provided for the neutralization of possible static charges prior to and during transfer of Class 3 (flammable liquid) materials between such containers. Such bonding shall be made by first connecting an electric conductor to the container to be filled and subsequently connecting the conductor to the container from which the liquid is to come, and not in any other order. To provide against ignition of vapors by discharge of static electricity, the latter connection shall be made at a point well removed from the opening from which the Class 3 (flammable liquid) materials is to be discharged.

(c) *Bonding and grounding cargo tanks before and during transfer of lading.* (1) When a cargo tank is loaded through an open filling hole, one end of a bond wire shall be connected to the stationary system piping or integrally connected steel framing, and the other end to the shell of the cargo tank to provide a continuous electrical connection. (If bonding is to the framing, it is essential that piping and framing be electrically interconnected.) This connection must be made before any filling hole is opened, and must remain in place until after the last filling hole has been closed. Additional bond wires are not needed around All-Metal flexible or swivel joints, but are required for nonmetallic flexible connections in the stationary system piping. When a cargo tank is unloaded by a suction-piping system through an open filling hole of the cargo tank, electrical continuity shall be maintained from cargo tank to receiving tank.

(2) When a cargo tank is loaded or unloaded through a vapor-tight (not open hole) top or bottom connection, so that there is no release of vapor at a point where a spark could occur, bonding or grounding, is not required. Contact of the closed connection must be made before flow starts and must not be broken until after the flow is completed.

(3) Bonding or grounding is not required when a cargo tank is unloaded through a nonvapor-tight connection into a stationary tank provided the metallic filling connection is maintained in contact with the filling hole.

(d) *Unloading combustible liquids.* For a cargo tank unloading a material meeting the definition for combustible liquid in §173.150(f) of this subchapter, the qualified person attending the unloading operation must remain within 45.72 meters (150 feet) of the cargo tank and 7.62 meters (25 feet) of the delivery hose and must observe both the cargo tank and the receiving container at least once every five minutes during unloading operations that take more than five minutes to complete.

§177.838 Class 4 (flammable solid) materials, Class 5 (oxidizing) materials, and Division 4.2 (pyroforic liquid) materials.

(See also §177.834 (a) to (j).)

(a) *Lading within body or covered; tailgate closed; pick-up and delivery.* All of that portion of the lading of any motor vehicle transporting Class 4 (flammable solid) or Class 5 (oxidizing) materials shall be contained entirely within the body of the motor vehicle and shall be covered by such body, by tarpaulins, or other suitable means, and if such motor vehicle has a tailboard or tail-gate, it shall be closed and secured in place during such transportation: *Provided, however*, That the provisions of this paragraph need not apply to "pick-up and delivery" motor vehicles when such motor vehicles are used in no other transportation than in and about cities, towns, or villages. Shipment in water-tight bulk containers need not be covered by a tarpaulin or other means.

(b) *Articles to be kept dry.* Special care shall be taken in the loading of any motor vehicle with Class 4 (flammable solid) or Class 5 (oxidizing) materials which are likely to become hazardous to transport when wet, to keep them from being wetted during the loading process and to keep them dry during transit. Special care shall also be taken in the loading of any motor vehicle with Class 4 (flammable solid) or Class 5 (oxidizing) materials, which are likely to become more hazardous to transport by wetting, to keep them from being wetted during the loading process and to keep them dry during transit. Examples of such dangerous materials are charcoal screenings, ground, crushed, or pulverized charcoal, and lump charcoal.

(c) *Lading ventilation, precautions against spontaneous combustion.* Whenever a motor carrier has knowledge concerning the hazards of spontaneous combustion or heating of any article to be loaded on a motor vehicle, such article shall be so loaded as to afford sufficient ventilation of the load to provide reasonable assurance against fire from this cause; and in such a case the motor vehicle shall be unloaded as soon as practicable after reaching its destination. Charcoal screenings, or ground, crushed, granulated, or pulverized charcoal in bags, shall be so loaded that the bags are laid horizontally in the motor vehicle, and so piled that there will be spaces for effective air circulation, which spaces shall not be less than 10 cm (3.9 inches) wide; and air spaces shall be maintained between rows of bags. Bags shall not be piled closer than 15 cm (5.9 inches) from the top of any motor vehicle with a closed body.

(d) [Reserved]

(e) [Reserved]

(f) Nitrates, except ammonium nitrate having organic coating, must be loaded in closed or open type motor vehicles, which must be swept clean and be free of any projections capable of injuring bags when so packaged. When shipped in open typemotor vehicles, the lading must be suitably covered. Ammonium nitrate having organic coating must not be loaded in all-metal vehicles, other than those made of aluminum or aluminum alloys of the closed type.

(g) A motor vehicle may only contain 45.4 kg (100 pounds) or less net mass of material described as "Smokeless powder for small arms, Division 4.1".

(h) Division 4.2 (pyrophoric liquid) materials in cylinders. Cylinders containing Division 4.2 (pyrophoric liquid) materials, unless packed in a strong box or case and secured therein to protect valves, must be loaded with all valves and safety relief devices in the vapor space. All cylinders must be secured so that no shifting occurs in transit.

§177.839 Class 8 (corrosive) materials.

(See also §177.834 (a) through (j).)

(a) *Nitric acid.* No packaging of nitric acid of 50 percent or greater concentration may be loaded above any packaging containing any other kind of material.

(b) *Storage batteries.* All storage batteries containing any electrolyte must be so loaded, if loaded with other lading, that all such batteries will be protected against other lading falling onto or against them, and adequate means must be provided in all cases for the protection and insulation of battery terminals against short circuits.

§177.840 Class 2 (gases) materials.

(See also §177.834 (a) to (j).)

(a) *Floors or platforms essentially flat.* Cylinders containing Class 2 (gases) materials shall not be loaded onto any part of the floor or platform of any motor vehicle which is not essentially flat; cylinders containing Class 2 (gases) materials may be loaded onto any motor vehicle not having a floor or platform only if such motor vehicle be equipped with suitable racks having adequate means for securing such cylinders in place therein. Nothing contained in this section shall be so construed as to prohibit the loading of such cylinders on any motor vehicle having a floor or platform and racks as hereinbefore described.

(1) *Cylinders.* Cylinders containing Class 2 gases must be securely restrained in an upright or horizontal position, loaded in racks, or packed in boxes or crates to prevent the cylinders from being shifted, overturned or ejected from the motor vehicle under normal transportation conditions. A pressure relief device, when installed, must be in communication with the vapor space of a cylinder containing a Division 2.1 (flammable gas) material.

(2) *Cylinders for liquefied hydrogen, cryogenic liquid.* A Specification DOT-4L cylinder containing hydrogen, cryogenic liquid may only be transported on a motor vehicle as follows:

(i) The vehicle must have an open body equipped with a suitable rack or support having a means to hold the cylinder upright when subjected to an acceleration of 2 "g" in any horizontal direction;

(ii) The combined total of the hydrogen venting rates, as marked, on the cylinders transported on one motor vehicle may not exceed 60 SCF per hour;

(iii) The vehicle may not enter a tunnel; and

(iv) Highway transportation is limited to private and contract carriage and to direct movement from point of origin to destination.

(b) Portable tank containers containing Class 2 (gases) materials shall be loaded on motor vehicles only as follows:

(1) Onto a flat floor or platform of a motor vehicle.

(2) Onto a suitable frame of a motor vehicle.

(3) In either such case, such containers shall be safely and securely blocked or held down to prevent shifting relative to each other or to the supporting structure when in transit, particularly during sudden starts and stops and changes of direction of the vehicle.

(4) Requirements of paragraphs (1) and (2) of this paragraph (b) shall not be construed as prohibiting stacking of containers, provided the provisions of paragraph (3) of this paragraph (b) are fully complied with.

(c) [Reserved]

(d) *Engine to be stopped in cargo tank motor vehicles, except for transfer pump.* No Division 2.1 (flammable gas) material shall be loaded into or on or unloaded from any cargo tank motor vehicle with the engine running unless the engine is used for the operation of the transfer pump of the vehicle. Unless the delivery hose is equipped with a shut-off valve at its discharge end, the engine of the motor vehicle shall be stopped at the finish of such loading or unloading operation while the filling or discharge connections are disconnected.

(e) Chlorine cargo tank motor vehicles shall be shipped only when equipped: (1) with a gas mask of type approved by The National Institute of Occupational Safety and Health (NIOSH) Pittsburgh Research Center, U.S. Department of Health and Human Services for chlorine service; and (2) with an emergency kit for controlling leaks in fittings on the dome cover plate.

(f) A cargo tank motor vehicle used for transportation of chlorine may not be moved, coupled or uncoupled, when any loading or unloading connections are attached to the vehicle, nor may it be left without the power unit attached unless the vehicle is chocked or equivalent means are provided to prevent motion. For additional requirements, see §173.315(o) of this subchapter.

(g) Each liquid discharge valve on a cargo tank motor vehicle, other than an engine fuel line valve, must be closed during transportation except during loading and unloading.

(h) The driver of a motor vehicle transporting a Division 2.1 (flammable gas) material that is a cryogenic liquid in a package exceeding 450 L (119 gallons) of water capacity shall avoid unnecessary delays during transportation. If unforeseen conditions cause an excessive pressure rise, the driver shall manually vent the tank at a remote and safe location. For each shipment, the driver shall make a written record of the cargo tank pressure and ambient (outside) temperature:

(1) At the start of each trip,

(2) Immediately before and after any manual venting,

(3) At least once every five hours, and

(4) At the destination point.

(i) No person may transport a Division 2.1 (flammable gas) material that is a cryogenic liquid in a cargo tank motor vehicle unless the pressure of the lading is equal to or less than that used to determine the marked rated holding time (MRHT) and the one-way travel time (OWTT), marked on the cargo tank in conformance with §173.318(g) of this subchapter, is equal to or greater than the elapsed time between the start and termination of travel. This prohibition does not apply if, prior to expiration of the OWTT, the cargo tank is brought to full equilibration as specified in paragraph (j) of this section.

(j) Full equilibration of a cargo tank transporting a Division 2.1 (flammable gas) material that is a cryogenic liquid may only be done at a facility that loads or unloads a Division 2.1 (flammable gas) material that is a cryogenic liquid and must be performed and verified as follows:

(1) The temperature and pressure of the liquid must be reduced by a manually controlled release of vapor; and

(2) The pressure in the cargo tank must be measured at least ten minutes after the manual release is terminated,

(k) A carrier of carbon monoxide, cryogenic liquid must provide each driver with a self-contained air breathing apparatus that is approved by the National Institute of Occupational Safety and Health; for example, Mine Safety Appliance Co., Model 401, catalog number 461704.

(l) *Operating procedure*. Each operator of a cargo tank motor vehicle that is subject to the emergency discharge control requirements in §173.315(n) of this subchapter must carry on or within the cargo tank motor vehicle written emergency discharge control procedures for all delivery operations. The procedures must describe the cargo tank motor vehicle's emergency discharge control features and, for a passive shut-down capability, the parameters within which they are designed to function. The procedures must describe the process to be followed if a facility-provided hose is used for unloading when the cargo tank motor vehicle has a specially equipped delivery hose assembly to meet the requirements of §173.315(n)(2) of this subchapter.

(m) *Cargo tank motor vehicle safety check*. Before unloading from a cargo tank motor vehicle containing a liquefied compressed gas, the qualified person performing the function must check those components of the discharge system, including delivery hose assemblies and piping, that are readily observed during the normal course of unloading to assure that they are of sound quality, without obvious defects detectable through visual observation and audio awareness, and that connections are secure. This check must be made after the pressure in the discharge system has reached at least equilibrium with the pressure in the cargo tank. Operators need not use instruments or take extraordinary actions to check components not readily visible. No operator may unload liquefied compressed gases from a cargo tank motor vehicle with a delivery hose assembly found to have any condition identified in §180.416(g)(1) of this subchapter or with piping systems found to have any condition identified in §180.416(g)(2) of this subchapter.

(n) *Emergency shut down*. If there is an unintentional release of product to the environment during unloading of a liquefied compressed gas, the qualified person unloading the cargo tank motor vehicle must promptly shut the internal self-closing stop valve or other primary means of closure and shut down all motive and auxiliary power equipment.

(o) *Daily test of off-truck remote shut-off activation device*. For a cargo tank motor vehicle equipped with an off-truck remote means to close the internal self-closing stop valve and shut off all motive and auxiliary power equipment, an operator must successfully test the activation device within 18 hours prior to the first delivery of each day. For a wireless transmitter/receiver, the person conducting the test must be at least 45.72 m (150 feet) from the cargo tank and may have the cargo tank in his line of sight.

(p) *Unloading procedures for liquefied petroleum gas and anhydrous ammonia in metered delivery service*. An operator must use the following procedures for unloading liquefied petroleum gas or anhydrous ammonia from a cargo tank motor vehicle in metered delivery service:

(1) For a cargo tank with a capacity of 13,247.5 L (3,500 water gallons) or less, excluding delivery hose and piping, the qualified person attending the unloading operation must remain within 45.72 meters (150 feet) of the cargo tank and 7.62 meters (25 feet) of the delivery hose and must observe both the cargo tank and the receiving container at least once every five minutes when the internal self-closing stop valve is open during unloading operations that take more than five minutes to complete.

(2) For a cargo tank with a capacity greater than 13,247.5 L (3,500 water gallons), excluding delivery hose and piping, the qualified person attending the unloading operation must remain within 45.72 m (150 feet) of the cargo tank and 7.62 m (25 feet) of the delivery hose when the internal self-closing stop valve is open.

(i) Except as provided in paragraph (p)(2)(ii) of this section, the qualified person attending the unloading operation must have an unobstructed view of the cargo tank and delivery hose to the maximum extent practicable, except during short periods when it is necessary to activate controls or monitor the receiving container.

(ii) For deliveries where the qualified person attending the unloading operation cannot maintain an unobstructed view of the cargo tank, when the internal self-closing stop valve is open, the qualified person must observe both the cargo tank and the receiving container at least once every five minutes during unloading operations that take more than five minutes to complete. In addition, by the compliance dates specified in §§173.315(n)(5) and 180.405(m)(3) of this subchapter, the cargo tank motor vehicle must have an emergency discharge control capability that meets the requirements of §173.315(n)(2) or §173.315(n)(4) of this subchapter.

(q) *Unloading procedures for liquefied petroleum gas and anhydrous ammonia in other than metered delivery service*. An operator must use the following procedures for unloading liquefied petroleum gas or anhydrous ammonia from a cargo tank motor vehicle in other than metered delivery service:

(1) The qualified person attending the unloading operation must remain within 7.62 m (25 feet) of the cargo tank when the internal self-closing stop valve is open.

(2) The qualified person attending the unloading operation must have an unobstructed view of the cargo tank and delivery hose to the maximum extent practicable, except during short periods when it is necessary to activate controls or monitor the receiving container.

(r) *Unloading using facility-provided hoses.* A cargo tank motor vehicle equipped with a specially designed delivery hose assembly to meet the requirements of §173.315(n)(2) of this subchapter may be unloaded using a delivery hose assembly provided by the receiving facility under the following conditions:

(1) The qualified person monitoring unloading must visually examine the facility hose assembly for obvious defects prior to its use in the unloading operation.

(2) The qualified person monitoring unloading must remain within arm's reach of the mechanical means of closure for the internal self-closing stop valve when the internal self-closing stop valve is open except for short periods when it is necessary to activate controls or monitor the receiving container. For chlorine cargo tank motor vehicles, the qualified person must remain within arm's reach of a means to stop the flow of product except for short periods when it is necessary to activate controls or monitor the receiving container.

(3) If the facility hose is equipped with a passive means to shut off the flow of product that conforms to and is maintained to the performance standard in §173.315(n)(2) of this subchapter, the qualified person may attend the unloading operation in accordance with the attendance requirements prescribed for the material being unloaded in §177.834 of this section.

(s) *Off-truck remote shut-off activation device.* For a cargo tank motor vehicle with an off-truck remote control shut-off capability as required by §§173.315(n)(3) or (n)(4) of this subchapter, the qualified person attending the unloading operation must be in possession of the activation device at all times during the unloading process. This requirement does not apply if the activation device is part of a system that will shut off the unloading operation without human intervention in the event of a leak or separation in the hose.

(t) *Unloading without appropriate emergency discharge control equipment.* Until a cargo tank motor vehicle is equipped with emergency discharge control equipment in conformance with §§173.315(n)(2) and 180.405(m)(1) of this subchapter, the qualified person attending the unloading operation must remain within arm's reach of a means to close the internal self-closing stop valve when the internal self-closing stop valve is open except during short periods when the qualified person must activate controls or monitor the receiving container. For chlorine cargo tank motor vehicles unloaded after December 31, 1999, the qualified person must remain within arm's reach of a means to stop the flow of product except for short periods when it is necessary to activate controls or monitor the receiving container.

(u) *Unloading of chlorine cargo tank motor vehicles.* Unloading of chlorine from a cargo tank motor vehicle must be performed in compliance with Section 3 of the Chlorine Institute Pamphlet 57, "Emergency Shut-off Systems for Bulk Transfer of Chlorine" (IBR, see §171.7 of this subchapter).

(Approved by the Office of Management and Budget under control number 2137-0542)

§177.841 Division 6.1 and Division 2.3 materials.

(See also §177.834 (a) to (j).)

(a) *Arsenical compounds in bulk.* Care shall be exercised in the loading and unloading or "arsenical dust", "arsenic trioxide", and "sodium arsenate", allowable to be loaded into sift-proof, steel hopper-type or dump-type motor-vehicle bodies equipped with water-proof, dust-proof covers well secured in place on all openings, to accomplish such loading with the minimum spread of such compounds into the atmosphere by all means that are practicable; and no such loading or unloading shall be done near or adjacent to any place where there are or are likely to be, during the loading or unloading process assemblages of persons other than those engaged in the loading or unloading process, or upon any public highway or in any public place. Before any motor vehicle may be used for transporting any other articles, all detectable traces of arsenical materials must be removed therefrom by flushing with water, or by other appropriate method, and the marking removed.

(b) [Reserved]

(c) *Division 2.3 (poisonous gas) or Division 6.1 (poisonous) materials.* The transportation of a Division 2.3 (poisonous gas) or Division 6.1 (poisonous) materials is not permitted if there is any interconnection between packagings.

(d) [Reserved]

(e) A motor carrier may not transport a package:

(1) Except as provided in paragraph (e)(3) of this section, bearing or required to bear a POISON or POISON INHALATION HAZARD label or placard in the same motor vehicle with material that is marked as or known to be a foodstuffs, feed or edible material intended for consumption by humans or animals unless the poisonous material is packaged in accordance with this subchapter and is:

(i) Overpacked in a metal drum as specified in §173.25(c) of this subchapter; or

(ii) Loaded into a closed unit load device and the foodstuffs, feed, or other edible material are loaded into another closed unit load device;

(2) Bearing or required to bear a POISON, POISON GAS or POISON INHALATION HAZARD label in the driver's compartment (including a sleeper berth) of a motor vehicle; or

(3) Bearing a POISON label displaying the text "PG III," or bearing a "PG III" mark adjacent to the POISON label, with materials marked as, or known to be, foodstuffs, feed or any other edible material intended for consumption by humans or animals, unless the package containing the Division 6.1, Packing Group III material is separated in a manner that, in the event of leakage from packages under conditions normally incident to transportation, commingling of hazardous materials with foodstuffs, feed or any other edible material would not occur.

§177.842 Class 7 (radioactive) material.

(a) The number of packages of Class 7 (radioactive) materials in any transport vehicle or in any single group in any storage location must be limited so that the total transport index number does not exceed 50. The total transport index of a group of packages and overpacks is determined by adding together the transport index number on the labels on the individual packages and overpacks in the group. This provision does not apply to exclusive use shipments described in §§173.441(b), 173.457, and 173.427 of this subchapter.

(b) Packages of Class 7 (radioactive) material bearing "RADIOACTIVE YELLOW-II" or "RADIOACTIVE YELLOW-III" labels may not be placed in a transport vehicle, storage location or in any other place closer than the distances shown in the following table to any area which may be continuously occupied by any passenger, employee, or animal, nor closer than the distances shown in the table to any package containing undeveloped film (if so marked), and must conform to the following conditions:

(1) If more than one of these packages is present, the distance must be computed from the following table on the basis of the total transport index number determined by adding together the transport index number on the labels on the individual packages and overpacks in the vehicle or storeroom.

(2) Where more than one group of packages is present in any single storage location, a single group may not have a total transport index greater than 50. Each group of packages must be handled and stowed not closer than 6 m (20 feet) (measured edge to edge) to any other group. The following table is to be used in accordance with the provisions of paragraph (b) of this section:

Total transport index	Minimum separation distance in meters (feet) to nearest undeveloped film for various times of transit					Minimum distance in meters (feet) to area of persons, or minimum distance in meters (feet) from dividing partition of cargo compartments
	Up to 2 hours	2–4 hours	4–8 hours	8–12 hours	Over 12 hours	
None	0.0 (0)	0.0 (0)	0.0 (0)	0.0 (0)	0.0 (0)	0.0 (0)
0.1 to 1.0	0.3 (1)	0.6 (2)	0.9 (3)	1.2 (4)	1.5 (5)	0.3 (1)
1.1 to 5.0	0.9 (3)	1.2 (4)	1.8 (6)	2.4 (8)	3.4 (11)	0.6 (2)
5.1 to 10.0	1.2 (4)	1.8 (6)	2.7 (9)	3.4 (11)	4.6 (15)	0.9 (3)
10.1 to 20.0	1.5 (5)	2.4 (8)	3.7 (12)	4.9 (16)	6.7 (22)	1.2 (4)
20.1 to 30.0	2.1 (7)	3.0 (10)	4.6 (15)	6.1 (20)	8.8 (29)	1.5 (5)
30.1 to 40.0	2.4 (8)	3.4 (11)	5.2 (17)	6.7 (22)	10.1 (33)	1.8 (6)
40.1 to 50.0	2.7 (9)	3.7 (12)	5.8 (19)	7.3 (24)	11.0 (36)	2.1 (7)

Note: The distance in this table must be measured from the nearest point on the nearest packages of Class 7 (radioactive) material.

(c) Shipments of low specific activity materials and surface contaminated objects, as defined in §173.403 of this subchapter, must be loaded so as to avoid spillage and scattering of loose materials. Loading restrictions are set forth in §173.427 of this subchapter.

(d) Packages must be so blocked and braced that they cannot change position during conditions normally incident to transportation.

(e) Persons should not remain unnecessarily in a vehicle containing Class 7 (radioactive) materials.

(f) The number of packages of fissile Class 7 (radioactive) material in any non-exclusive use transport vehicle must be limited so that the sum of the criticality safety indices (CSIs) does not exceed 50. In loading and storage areas, fissile material packages must be grouped so that the sum of CSIs in any one group is not greater than 50; there may be more than one group of fissile material packages in a loading or storage area, so long as each group is at least 6 m (20 feet) away from all other such groups. All pertinent requirements of §§173.457 and 173.459 apply.

(g) For shipments transported under exclusive use conditions the radiation dose rate may not exceed 0.02 mSv per hour (2 mrem per hour) in any position normally occupied in the motor vehicle. For shipments transported as exclusive use under the provisions of §173.441(b) of this subchapter for packages with external radiation levels in excess of 2 mSv (200 mrem per hour) at the package surface, the motor vehicle must meet the requirements of a closed transport vehicle (see §173.403 of this subchapter). The sum of criticality safety indices (CSIs) for packages containing fissile material may not exceed 100 in an exclusive use vehicle.

§177.843 Contamination of vehicles.

(a) Each motor vehicle used for transporting Class 7 (radioactive) materials under exclusive use conditions in accordance with §173.427(b)(4) or (c) or §173.443(c) of this subchapter must be surveyed with radiation detection instruments after each use. A vehicle may not be returned to service until the radiation dose rate at every accessible surface is 0.005 mSv per hour (0.5 mrem per hour) or less and the removable (non-fixed) radioactive surface contamination is not greater than the level prescribed in §173.443(a) of this subchapter.

(b) This section does not apply to any vehicle used solely for transporting Class 7 (radioactive) material if a survey of the interior surface shows that the radiation dose rate does not exceed 0.1 mSv per hour (10 mrem per hour) at the interior surface or 0.02 mSv per hour (2 mrem per hour) at 1 meter (3.3 feet) from any interior surface. These vehicles must be stenciled with the words "For Radioactive Materials Use Only" in lettering at least 7.6 centimeters (3 inches) high in a conspicuous place, on both sides of the exterior of the vehicle. These vehicles must be kept closed at all times other than loading and unloading.

(c) In case of fire, accident, breakage, or unusual delay involving shipments of Class 7 (radioactive) material, see §§171.15, 171.16 and 177.854 of this subchapter.

(d) Each transport vehicle used to transport Division 6.2 materials must be disinfected prior to reuse if a Division 6.2 material is released from its packaging during transportation. Disinfection may be by any means effective for neutralizing the material released.

Reserved

Subpart C—Segregation and Separation Chart of Hazardous Materials

§177.848 Segregation of hazardous materials.

(a) This section applies to materials which meet one or more of the hazard classes defined in this subchapter and are:

(1) In packages that must be labeled or placarded in accordance with part 172 of this subchapter;

(2) In a compartment within a multi-compartmented cargo tank subject to the restrictions in §173.33 of this subchapter; or

(3) In a portable tank loaded in a transport vehicle or freight container.

(b) When a transport vehicle is to be transported by vessel, other than a ferry vessel, hazardous materials on or within that vehicle must be stowed and segregated in accordance with §176.83(b) of this subchapter.

(c) In addition to the provisions of paragraph (d) of this section and except as provided in §173.12(e) of this subchapter, cyanides, cyanide mixtures or solutions may not be stored, loaded and transported with acids if a mixture of the materials would generate hydrogen cyanide; Division 4.2 materials may not be stored, loaded and transported with Class 8 liquids; and Division 6.1 Packing Group I, Hazard Zone A material may not be stored, loaded and transported with Class 3 material, Class 8 liquids, and Division 4.1, 4.2, 4.3, 5.1 or 5.2 materials.

(d) Except as otherwise provided in this subchapter, hazardous materials must be stored, loaded or transported in accordance with the following table and other provisions of this section:

SEGREGATION TABLE FOR HAZARDOUS MATERIALS

Class or Division		Notes	1.1, 1.2	1.3	1.4	1.5	1.6	2.1	2.2	2.3 gas Zone A	2.3 gas Zone B	3	4.1	4.2	4.3	5.1	5.2	6.1 liquids PG I Zone A	7	8 liquids only
Explosives	1.1 and 1.2	A	*	*	*	*	*	x	x	x	x	x	x	x	x	x	x	x	x	x
Explosives	1.3		*	*	*	*	*	x		x	x	x		x	x	x	x	x		x
Explosives	1.4		*	*	*	*	*	o		o	o	o		o				o		o
Very insensitive explosives.	1.5	A	*	*	*	*	*	x	x	x	x	x	x	x	x	x	x	x	x	x
Extremely insensitive explosives.	1.6		*	*	*	*	*													
Flammable gases ..	2.1		x	x	o	x				x	o							o	o	
Non-toxic, non-flammable gases,	2.2		x			x														
Poisonous gas Zone A	2.3		x	x	o	x		x				x	x	x	x	x	x			x
Poisonous gas Zone B	2.3		x	x	o	x		o				o	o	o	o	o	o			o
Flammable liquids ..	3		x	x	o	x				x	o					o		x		
Flammable solids...	4.1		x			x				x	o							x		o
Spontaneously combustible materials.	4.2		x	x	o	x				x	o							x		x
Dangerous when wet materials.	4.3		x	x		x				x	o							x		o
Oxidizers	5.1	A	x	x		x				x	o	o						x		o
Organic peroxides ..	5.2		x	x		x				x	o							x		o
Poisonous liquids PG I Zone A.	6.1		x	x	o	x		o				x	x	x	x	x	x			x
Radioactive materials .	7		x			x		o												
Corrosive liquids ...	8		x	x	o	x				x	o		o	x	o	o	o	x		

(e) Instructions for using the segregation table for hazardous materials are as follows:

(1) The absence of any hazard class or division or a blank space in the Table indicates that no restrictions apply.

(2) The letter "X" in the Table indicates that these materials may not be loaded, transported, or stored together in the same transport vehicle or storage facility during the course of transportation.

(3) The letter "O" in the Table indicates that these materials may not be loaded, transported, or stored together in the same transport vehicle or storage facility during the course of transportation unless separated in a manner that, in the event of leakage from packages under conditions normally incident to transportation, commingling of hazardous materials would not occur. Notwithstanding the methods of separation employed, Class 8 (corrosive) liquids may not be loaded above or

SEGREGATION

adjacent to Class 4 (flammable) or Class 5 (oxidizing) materials; except that shippers may load truckload shipments of such materials together when it is known that the mixture of contents would not cause a fire or a dangerous evolution of heat or gas.

(4) The "*" in the Table indicates that segregation among different Class 1 (explosive) materials is governed by the compatibility table in paragraph (f) of this section.

(5) The note "A" in the second column of the Table means that, notwithstanding the requirements of the letter "X", ammonium nitrate (UN1942) and ammonium nitrate fertilizer may be loaded or stored with Division 1.1 (explosive) or Division 1.5 materials.

(6) When the §172.101 Table or §172.402 of this subchapter requires a package to bear a subsidiary hazard label, segregation appropriate to the subsidiary hazard must be applied when that segregation is more restrictive than that required by the primary hazard. However, hazardous materials of the same class may be stowed together without regard to segregation required for any secondary hazard if the materials are not capable of reacting dangerously with each other and causing combustion or dangerous evolution of heat, evolution of flammable, poisonous, or asphyxiant gases, or formation of corrosive or unstable materials.

(f) Class 1 (explosive) materials shall not be loaded, transported, or stored together, except as provided in this section, and in accordance with the following Table:

COMPATIBILITY TABLE FOR CLASS 1 (EXPLOSIVE) MATERIALS.

Compatibility Group	A	B	C	D	E	F	G	H	J	K	L	N	S
A		X	X	X	X	X	X	X	X	X	X	X	X
B	X		X	$X_{(4)}$	X	X	X	X	X	X	X	X	4/5
C	X	X		2	2	X	6	X	X	X	X	3	4/5
D	X	$X_{(4)}$	2		2	X	6	X	X	X	X	3	4/5
E	X	X	2	2		X	6	X	X	X	X	3	4/5
F	X	X	X	X	X		X	X	X	X	X	X	4/5
G	X	X	6	6	6	X		X	X	X	X	X	4/5
H	X	X	X	X	X	X	X		X	X	X	X	4/5
J	X	X	X	X	X	X	X	X		X	X	X	4/5
K	X	X	X	X	X	X	X	X	X		X	X	4/5
L	X	X	X	X	X	X	X	X	X	X	1	X	X
N	X	X	3	3	3	X	X	X	X	X	X		4/5
S	X	4/5	4/5	4/5	4/5	4/5	4/5	4/5	4/5	4/5	X	4/5	

(g) Instructions for using the compatibility table for Class 1 (explosive) materials are as follows:

(1) A blank space in the Table indicates that no restrictions apply.

(2) The letter "X" in the Table indicates that explosives of different compatibility groups may not be carried on the same transport vehicle.

(3) The numbers in the Table mean the following:

(i) "1" means an explosive from compatibility group L shall only be carried on the same transport vehicle with an identical explosive.

(ii) "2" means any combination of explosives from compatibility groups C, D, or E is assigned to compatibility group E.

(iii) "3" means any combination of explosives from compatibility groups C, D, or E with those in compatibility group N is assigned to compatibility group D.

(iv) "4" means 'see §177.835(g)' when transporting detonators.

(v) "5" means Division 1.4S fireworks may not be loaded on the same transport vehicle with Division 1.1 or 1.2 (explosive) materials.

(vi) "6" means explosive articles in compatibility group G, other than fireworks and those requiring special handling, may be loaded, transported and stored with other explosive articles of compatibility groups C, D and E, provided that explosive substances (such as those not contained in articles) are not carried in the same transport vehicle.

(h) Except as provided in paragraph (i) of this section, explosives of the same compatibility group but of different divisions may be transported together provided that the whole shipment is transported as though its entire contents were of the lower numerical division (i.e., Division 1.1 being lower than Division 1.2). For example, a mixed shipment of Division 1.2 (explosive) materials and Division 1.4 (explosive) materials, both of compatibility group D, must be transported as Division 1.2 (explosive) materials.

(i) When Division 1.5 materials, compatibility group D, are transported in the same freight container as Division 1.2 (explosive) materials, compatibility group D, the shipment must be transported as Division 1.1 (explosive) materials, compatibility group D.

Subpart D—Vehicles and Shipments in Transit; Accidents

§177.854 Disabled vehicles and broken or leaking packages; repairs.

(a) *Care of lading, hazardous materials.* Whenever for any cause other than necessary traffic stops any motor vehicle transporting any hazardous material is stopped upon the traveled portion of any highway or shoulder thereof, special care shall be taken to guard the vehicle and its load or to take such steps as may be necessary to provide against hazard. Special effort shall be made to remove the motor vehicle to a place where the hazards of the materials being transported may be provided against. See §§392.22, 392.24, and 392.25 of this title for warning devices required to be displayed on the highway.

(b) *Disposition of containers found broken or leaking in transit.* When leaks occur in packages or containers during the course of transportation, subsequent to initial loading, disposition of such package or container shall be made by the safest practical means afforded under paragraphs (c), (d), and (e) of this section.

(c) *Repairing or overpacking packages.* (1) Packages may be repaired when safe and practicable, such repairing to be in accordance with the best and safest practice known and available.

(2) Packages of hazardous materials that are damaged or found leaking during transportation, and hazardous materials that have spilled or leaked during transportation, may be forwarded to destination or returned to the shipper in a salvage drum in accordance with the requirements of §173.3(c) of this subchapter.

(d) *Transportation of repaired packages.* Any package repaired in accordance with the requirements of paragraph (c)(1) of this section may be transported to the nearest place at which it may safely be disposed of only in compliance with the following requirements:

(1) The package must be safe for transportation.

(2) The repair of the package must be adequate to prevent contamination of or hazardous admixture with other lading transported on the same motor vehicle therewith.

(3) If the carrier is not himself the shipper, the consignee's name and address must be plainly marked on the repaired package.

(e) *Disposition of unsafe broken packages.* In the event any leaking package or container cannot be safely and adequately repaired for transportation or transported, it shall be stored pending proper disposition in the safest and most expeditious manner possible.

(f) *Stopped vehicles; other dangerous articles.* Whenever any motor vehicle transporting Class 3 (flammable liquid), Class 4 (flammable solid), Class 5 (oxidizing), Class 8 (corrosive), Class 2 (gases), or Division 6.1 (poisonous) materials, is stopped for any cause other than necessary traffic stops upon the traveled portion of any highway or a shoulder next thereto, the following requirements shall be complied with during the period of such stop:

(1) For motor vehicles other than cargo tank motor vehicles used for the transportation of Class 3 (flammable liquid) or Division 2.1 (flammable gas) materials and not transporting Division 1.1, 1.2, or 1.3 (explosive) materials, warning devices must be set out in the manner prescribed in §392.22 of this title.

(2) For cargo tanks used for the transportation of Class 3 (flammable liquid) or Division 2.1 (flammable gas) materials, whether loaded or empty, and vehicles transporting Division 1.1, 1.2, or 1.3 (explosive) materials, warning devices must be set out in the manner prescribed by §392.25 of this title.

(g) *Repair and maintenance of vehicles containing certain hazardous materials.*— (1) *General.* No person may use heat, flame or spark producing devices to repair or maintain the cargo or fuel containment system of a motor vehicle required to be placarded, other than COMBUSTIBLE, in accordance with subpart F of part 172 of this subchapter. As used in this section, "containment system" includes all vehicle components intended physically to contain cargo or fuel during loading or filling, transport, or unloading.(1) *General.* No person may use heat, flame or spark producing devices to repair or maintain the cargo or fuel containment system of a motor vehicle required to be placarded, other than COMBUSTIBLE, in accordance with subpart F of part 172 of this subchapter. As used in this section, "containment system" includes all vehicle components intended physically to contain cargo or fuel during loading or filling, transport, or unloading.

(2) *Repair and maintenance inside a building.* No person may perform repair or maintenance on a motor vehicle subject to paragraph (g)(1) of this section inside a building unless:

(i) The motor vehicle's cargo and fuel containment systems are closed (except as necessary to maintain or repair the vehicle's motor) and do not show any indication of leakage;

(ii) A means is provided, and a person capable to operate the motor vehicle is available, to immediately remove the motor vehicle if necessary in an emergency;

(iii) The motor vehicle is removed from the enclosed area upon completion of repair or maintenance work; and

(iv) For motor vehicles loaded with Division 1.1, 1.2, or 1.3 (explosive), Class 3 (flammable liquid), or Division 2.1 (flammable gas) materials, all sources of spark, flame or glowing heat within the area of enclosure (including any heating system drawing air there from) are extinguished, made inoperable or rendered explosion-proof by a suitable method. *Exception:* Electrical equipment on the vehicle, necessary to accomplish the maintenance function, may remain operational.

(h) *No repair with flame unless gas-free.* No repair of a cargo tank used for the transportation of any Class 3 (flammable liquid) or Division 6.1 (poisonous liquid) material, or any compartment thereof, or of any container for fuel of whatever nature may be repaired by any method employing a flame, arc, or other means of welding, unless the tank or compartment shall first have been made gas-free.

Reserved

PART 178

§178.1 Purpose and scope.

This part prescribes the manufacturing and testing specifications for packaging and containers used for the transportation of hazardous materials in commerce.

§178.2 Applicability and responsibility.

(a) *Applicability.* (1) The requirements of this part apply to packagings manufactured—

(i) To a DOT specification, regardless of country of manufacture; or

(ii) To a UN standard, for packagings manufactured within the United States. For UN standard packagings manufactured outside the United States, see §173.24(d)(2) of this subchapter. For UN standard packagings for which standards are not prescribed in this part, see §178.3(b).

(2) A manufacturer of a packaging subject to the requirements of this part is primarily responsible for compliance with the requirements of this part. However, any person who performs a function prescribed in this part shall perform that function in accordance with this part.

(b) *Specification markings.* When this part requires that a packaging be marked with a DOT specification or UN standard marking, marking of the packaging with the appropriate DOT or UN markings is the certification that—

(1) Except as otherwise provided in this section, all requirements of the DOT specification or UN standard, including performance tests, are met; and

(2) All functions performed by, or on behalf of, the person whose name or symbol appears as part of the marking conform to requirements specified in this part.

(c) *Notification.* (1) Except as specifically provided in §§178.337–18 and 178.345–10 of this part, the manufacturer or other person certifying compliance with the requirements of this part, and each subsequent distributor of that packaging must:

(i) Notify each person to whom that packaging is transferred—

(A) Of all requirements in this part not met at the time of transfer, and

(B) With information specifying the type(s) and dimensions of the closures, including gaskets and any other components needed to ensure that the packaging is capable of successfully passing the applicable performance tests. This information must include any procedures to be followed, including closure instructions for inner packagings and receptacles, to effectively assemble and close the packaging for the purpose of preventing leakage in transportation. Closure instructions must provide for a consistent and repeatable means of closure that is sufficient to ensure the packaging is closed in the same manner as it was tested. For packagings sold or represented as being in conformance with the requirements of this subchapter applicable to transportation by aircraft, this information must include relevant guidance to ensure that the packaging, as prepared for transportation, will withstand the pressure differential requirements in §173.27 of this subchapter.

(ii) Retain copies of each written notification for the amount of time that aligns with the packaging's periodic retest date, *i.e.*, every 12 months for single or composite packagings and every 24 months for combination packagings; and

(iii) Make copies of all written notifications available for inspection by a representative of the Department.

(2) The notification required in accordance with this paragraph (c) may be in writing or by electronic means, including e-mailed transmission or transmission on a CD or similar device. If a manufacturer or subsequent distributor of the packaging utilizes electronic means to make the required notifications, the notification must be specific to the packaging in question and must be in a form that can be printed in hard copy by the person receiving the notification.

(d) Except as provided in paragraph (c) of this section, a packaging not conforming to the applicable specifications or standards in this part may not be marked to indicate such conformance.

(e) *Definitions.* For the purpose of this part—

Manufacturer means the person whose name and address or symbol appears as part of the specification markings required by this part or, for a packaging marked with the symbol of an approval agency, the person on whose behalf the approval agency certifies the packaging.

Specification markings mean the packaging identification markings required by this part including, where applicable, the name and address or symbol of the packaging manufacturer or approval agency.

(f) No packaging may be manufactured or marked to a packaging specification that was in effect on September 30, 1991, and that was removed from this part 178 by a rule published in the *Federal Register* on December 21, 1990 and effective October 1, 1991.

§178.3 Marking of packagings.

(a) Each packaging represented as manufactured to a DOT specification or a UN standard must be marked on a non-removable component of the packaging with specification markings conforming to the applicable specification, and with the following:

(1) In an unobstructed area, with letters, and numerals identifying the standards or specification (e.g. UN 1A1, DOT 4B240ET, etc.).

(2) Unless otherwise specified in this part, with the name and address or symbol of the packaging manufacturer or, where specifically authorized, the symbol of the approval agency certifying compliance with a UN standard. Symbols, if used, must be registered with the Associate Administrator. Duplicative symbols are not authorized.

(3) The markings must be stamped, embossed, burned, printed or otherwise marked on the packaging to provide adequate accessibility, permanency, contrast, and legibility so as to be readily apparent and understood.

(4) Unless otherwise specified, letters and numerals must be at least 12.0 mm (0.47 inches) in height except

that for packagings of less than or equal to 30 L (7.9 gallons) capacity for liquids or 30 kg (66 pounds) capacity for solids the height must be at least 6.0 mm (0.2 inches). For packagings having a capacity of 5 L (1 gallon) or 5 kg (11 pounds) or less, letters and numerals must be of an appropriate size.

(5) For packages with a gross mass of more than 30 kg (66 pounds), the markings or a duplicate thereof must appear on the top or on a side of the packaging.

(b) A UN standard packaging for which the UN standard is set forth in this part may be marked with the United Nations symbol and other specification markings only if it fully conforms to the requirements of this part. A UN standard packaging for which the UN standard is not set forth in this part may be marked with the United Nations symbol and other specification markings for that standard as provided in the ICAO Technical Instructions or the IMDG Code subject to the following conditions:

(1) The U.S. manufacturer must establish that the packaging conforms to the applicable provisions of the ICAO Technical Instructions (IBR, see §171.7 of this subchapter) or the IMDG Code (IBR, see §171.7 of this subchapter), respectively.

(2) If an indication of the name of the manufacturer or other identification of the packaging as specified by the competent authority is required, the name and address or symbol of the manufacturer or the approval agency certifying compliance with the UN standard must be entered. Symbols, if used, must be registered with the Associate Administrator.

(3) The letters "USA" must be used to indicate the State authorizing the allocation of the specification marks if the packaging is manufactured in the United States.

(c) Where a packaging conforms to more than one UN standard or DOT specification, the packaging may bear more than one marking, provided the packaging meets all the requirements of each standard or specification. Where more than one marking appears on a packaging, each marking must appear in its entirety.

(d) No person may mark or otherwise certify a packaging or container as meeting the requirements of a manufacturing special permit unless that person is the holder of or a party to that special permit, an agent of the holder or party for the purpose of marking or certification, or a third party tester.

Subpart L—Non-Bulk Performance-Oriented Packaging Standards

§178.500 Purpose, scope and definitions.

(a) This subpart prescribes certain requirements for non-bulk packagings for hazardous materials. Standards for these packagings are based on the UN Recommendations.

(b) Terms used in this subpart are defined in §171.8 of this subchapter.

§178.502 Identification codes for packagings.

(a) Identification codes for designating kinds of packagings consist of the following:

(1) A numeral indicating the kind of packaging, as follows:

(i) "1" means a drum.

(ii) "2" means a wooden barrel.

(iii) "3" means a jerrican.

(iv) "4" means a box.

(v) "5" means a bag.

(vi) "6" means a composite packaging.

(vii) "7" means a pressure receptacle.

(2) A capital letter indicating the material of construction, as follows:

(i) "A" means steel (all types and surface treatments).

(ii) "B" means aluminum.

(iii) "C" means natural wood.

(iv) "D" means plywood.

(v) "F" means reconstituted wood.

(vi) "G" means fiberboard.

(vii) "H" means plastic.

(viii) "L" means textile.

(ix) "M" means paper, multi-wall.

(x) "N" means metal (other than steel or aluminum).

(xi) "P" means glass, porcelain or stoneware.

(3) A numeral indicating the category of packaging within the kind to which the packaging belongs. For example, for steel drums ("1A"), "1" indicates a non-removable head drum (i.e., "1A1") and "2" indicates a removable head drum (i.e., "1A2").

(b) For composite packagings, two capital letters are used in sequence in the second position of the code, the first indicating the material of the inner receptacle and the second, that of the outer packaging. For example, a plastic receptacle in a steel drum is designated "6HA1".

(c) For combination packagings, only the code number for the outer packaging is used.

(d) Identification codes are set forth in the standards for packagings in §§178.504 through 178.523 of this subpart.

Note to §178.502: Plastics materials include other polymeric materials such as rubber.

§178.503 Marking of packagings.

(a) A manufacturer must mark every packaging that is represented as manufactured to meet a UN standard with the marks specified in this section. The markings must be durable, legible and placed in a location and of such a size relative to the packaging as to be readily visible, as specified in §178.3(a). Except as otherwise provided in this section, every reusable packaging liable to undergo a reconditioning process which might obliterate the packaging marks must bear the marks specified in paragraphs (a)(1) through (a)(6) and (a)(9) of this section in a permanent form (e.g. embossed) able to withstand the reconditioning process. A marking may be applied in a single line or in multiple lines provided the correct sequence is used. As illustrated by the examples in paragraph (e) of this section, the following information must be presented in the correct sequence. Slash marks should be used to separate this information. A packaging conforming to a UN standard must be marked as follows:

(1) Except as provided in paragraph (e)(1)(ii) of this section, the United Nations symbol as illustrated in paragraph (e)(1)(i) of this section (for embossed metal receptacles, the letters "UN") may be applied in place of the symbol;

(2) A packaging identification code designating the type of packaging, the material of construction and, when appropriate, the category of packaging under §§178.504 through 178.523 of this subpart within the type to which the packaging belongs. The letter "V" must follow the packaging identification code on packagings tested in accordance with §178.601(g)(2); for example, "4GV". The letter "W" must follow the packaging identification code on packagings when required by an approval under the provisions of §178.601(h) of this part;

(3) A letter identifying the performance standard under which the packaging design type has been successfully tested, as follows:

(i) X—for packagings meeting Packing Group I, II and III tests;

(ii) Y—for packagings meeting Packing Group II and III tests; or

(iii) Z—for packagings only meeting Packing Group III tests;

(4) A designation of the specific gravity or mass for which the packaging design type has been tested, as follows:

(i) For packagings without inner packagings intended to contain liquids, the designation shall be the specific gravity rounded down to the first decimal but may be omitted when the specific gravity does not exceed 1.2; and

(ii) For packagings intended to contain solids or inner packagings, the designation shall be the maximum gross mass in kilograms;

(5)(i) For single and composite packagings intended to contain liquids, the test pressure in kilopascals rounded down to the nearest 10 kPa of the hydrostatic pressure test that the packaging design type has successfully passed;

(ii) For packagings intended to contain solids or inner packagings, the letter "S";

(6) The last two digits of the year of manufacture. Packagings of types 1H and 3H shall also be marked with the month of manufacture in any appropriate manner; this may be marked on the packaging in a different place from the remainder of the markings;

(7) The state authorizing allocation of the mark. The letters 'USA' indicate that the packaging is manufactured and marked in the United States in compliance with the provisions of this subchapter;

(8) The name and address or symbol of the manufacturer or the approval agency certifying compliance with subpart L and subpart M of this part. Symbols, if used must be registered with the Associate Administrator;

(9) For metal or plastic drums or jerricans intended for reuse or reconditioning as single packagings or the outer packagings of a composite packaging, the thickness of the packaging material, expressed in millimeters (rounded to the nearest 0.1 mm), as follows:

(i) Metal drums or jerricans must be marked with the nominal thickness of the metal used in the body. The marked nominal thickness must not exceed the minimum thickness of the steel used by more than the thickness tolerance stated in ISO 3574 (IBR, see §171.7 of this subchapter). (See appendix C of this part.) The unit of measure is not required to be marked. When the nominal thickness of either head of a metal drum is thinner than that of the body, the nominal thickness of the top head, body, and bottom head must be marked (*e.g.*, "1.0-1.2-1.0" or "0.9-1.0-1.0").

(ii) Plastic drums or jerricans must be marked with the minimum thickness of the packaging material. Minimum thicknesses of plastic must be as determined in accordance with §173.28(b)(4). The unit of measure is not required to be marked;

(10) In addition to the markings prescribed in paragraphs (a)(1) through (a)(9) of this section, every new metal drum having a capacity greater than 100 L must bear the marks described in paragraphs (a)(1) through (a)(6), and (a)(9)(i) of this section, in a permanent form, on the bottom. The markings on the top head or side of these packagings need not be permanent, and need not include the thickness mark described in paragraph (a)(9) of this section. This marking indicates a drum's characteristics at the time it was manufactured, and the information in paragraphs (a)(1) through (a)(6) of this section that is marked on the top head or side must be the same as the information in paragraphs (a)(1) through (a)(6) of this section permanently marked by the original manufacturer on the bottom of the drum; and

(11) Rated capacity of the packaging expressed in liters may be marked.

(b) For a packaging with a removable head, the markings may not be applied only to the removable head.

(c) *Marking of reconditioned packagings.* (1) If a packaging is reconditioned, it shall be marked by the reconditioner near the marks required in paragraphs (a)(1) through (6) of this section with the following additional information:

(i) The name of the country in which the reconditioning was performed (in the United States, use the letters "USA");

(ii) The name and address or symbol of the reconditioner. Symbols, if used, must be registered with the Associate Administrator;

(iii) The last two digits of the year of reconditioning;

(iv) The letter "R"; and

(v) For every packaging successfully passing a leakproofness test, the additional letter "L".

(2) When, after reconditioning, the markings required by paragraph (a)(1) through (a)(5) of this section no longer appear on the top head or the side of the metal drum, the reconditioner must apply them in a durable form followed by the markings in paragraph (c)(1) of this section. These markings may identify a different performance capability than that for which the original design type had been tested and marked, but may not identify

a greater performance capability. The markings applied in accordance with this paragraph may be different from those which are permanently marked on the bottom of a drum in accordance with paragraph (a)(10) of this section.

(d) *Marking of remanufactured packagings.* For remanufactured metal drums, if there is no change to the packaging type and no replacement or removal of integral structural components, the required markings need not be permanent (e.g., embossed). Every other remanufactured drum must bear the marks required in paragraphs (a)(1) through (a)(6) of this section in a permanent form (e.g., embossed) on the top head or side. If the metal thickness marking required in paragraph (a)(9)(i) of this section does not appear on the bottom of the drum, or if it is no longer valid, the remanufacturer also must mark this information in permanent form.

(e) The following are examples of symbols and required markings:

(1)(i) The United Nations symbol is:

(ii) The circle that surrounds the letters "u" and "n" may have small breaks provided the following provisions are met:

(A) The total gap space does not exceed 15 percent of the circumference of the circle;

(B) There are no more than four gaps in the circle;

(C) The spacing between gaps is separated by no less than 20 percent of the circumference of the circle (72 degrees); and

(D) The letters "u" and "n" appear exactly as depicted in §178.503(e)(1)(i) with no gaps.

(2) Examples of markings for a new packaging are as follows:

(i) For a fiberboard box designed to contain an inner packaging:

(as in §178.503(a)(1) through (a)(8) of this subpart).

(ii) For a steel drum designed to contain liquids:

(as in §178.503(a)(1) through (a)(9) of this subpart).

(iii) For a steel drum to transport solids or inner packagings:

(as in §178.503(a)(1) through (a)(8) of this subpart).

(3) Examples of markings for reconditioned packagings are as follows:

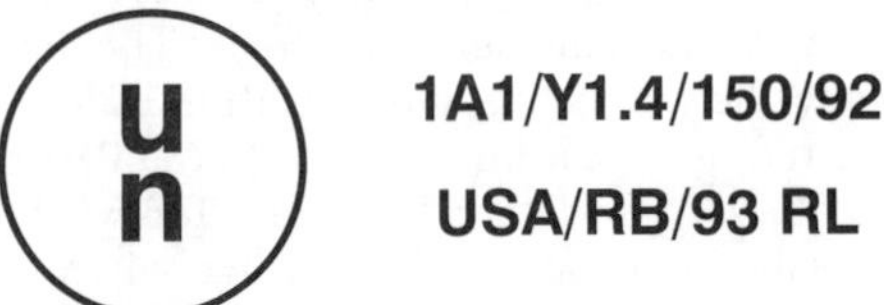

(as in §178.503(c)(1)).

(f) A manufacturer must mark every UN specification package represented as manufactured to meet the requirements of §178.609 for packaging of infectious substances with the marks specified in this section. The markings must be durable, legible, and must be readily visible, as specified in §178.3(a). An infectious substance packaging that successfully passes the tests conforming to the UN standard must be marked as follows:

(1) The United Nations symbol as illustrated in paragraph (e) of this section.

(2) The code designating the type of packaging and material of construction according to the identification codes for packagings specified in §178.502.

(3) The text "CLASS 6.2".

(4) The last two digits of the year of manufacture of the packaging.

(5) The country authorizing the allocation of the mark. The letters "USA" indicate the packaging is manufactured and marked in the United States in compliance with the provisions of this subchapter.

(6) The name and address or symbol of the manufacturer or the approval agency certifying compliance with subparts L and M of this part. Symbols, if used, must be registered with the Associate Administrator for Hazardous Materials Safety.

(7) For packagings meeting the requirements of §178.609(i)(3), the letter "U" must be inserted immediately following the marking designating the type of packaging and material required in paragraph (f)(2) of this section.

Subpart M—Testing of Non-Bulk Packagings and Packages

§178.602 Preparation of packagings and packages for testing.

(a) Except as otherwise provided in this subchapter, each packaging and package must be closed in preparation for testing and tests must be carried out in the same manner as if prepared for transportation, including inner packagings in the case of combination packagings.

(b) For the drop and stacking test, inner and single-unit receptacles other than bags must be filled to not less than 95% of maximum capacity (see §171.8 of this subchapter) in the case of solids and not less than 98% of maximum in the case of liquids. Bags containing solids shall be filled to the maximum mass at which they may be used. The material to be transported in the packagings may be replaced by a non-hazardous material, except for chemical compatibility testing or where this would invalidate the results of the tests.

(c) If the material to be transported is replaced for test purposes by a non-hazardous material, the material used must be of the same or higher specific gravity as the material to be carried, and its other physical properties (grain, size, viscosity) which might influence the results of the required tests must correspond as closely as possible to those of the hazardous material to be transported. Water may also be used for the liquid drop test under the conditions specified in §178.603(e) of this subpart. It is permissible to use additives, such as bags of lead shot, to achieve the requisite total package mass, so long as they are placed so that the test results are not affected.

(d) Paper or fiberboard packagings must be conditioned for at least 24 hours immediately prior to testing in an atmosphere maintained—

(1) At 50 percent ± percent relative humidity, and at a temperature of 23°C ± 2°C (73°F ± 4°F). Average values should fall within these limits. Short-term fluctuations and measurement limitations may cause individual measurements to vary by up to ± 5 percent relative humidity without significant impairment of test reproducibility;

(2) At 65 percent ± 2 percent relative humidity, and at a temperature of 20°C ± 2°C (68°F ± 4°F), or 27°C ± 2°C (81°F ± 4°F). Average values should fall within these limits. Short-term fluctuations and measurement limitations may cause individual measurements to vary by up to ± 5 percent relative humidity without significant impairment of test reproducibility; or

(3) For testing at periodic intervals only (i.e., other than initial design qualification testing), at ambient conditions.

(e) Except as otherwise provided, each packaging must be closed in preparation for testing in the same manner as if prepared for actual shipment. All closures must be installed using proper techniques and torques.

(f) Bung-type barrels made of natural wood must be left filled with water for at least 24 hours before the tests.

Reserved

PART 180

Subpart D—Qualification and Maintenance of Intermediate Bulk Containers

§180.352 Requirements for retest and inspection of IBCs.

(a) *General*. Each IBC constructed in accordance with a UN standard for which a test or inspection specified in paragraphs (b)(1), (b)(2) and (b)(3) of this section is required may not be filled and offered for transportation or transported until the test or inspection has been successfully completed. This paragraph does not apply to any IBC filled prior to the test or inspection due date. The requirements in this section do not apply to DOT 56 and 57 portable tanks.

(b) *Test and inspections for metal, rigid plastic, and composite IBCs*. Each IBC is subject to the following test and inspections:

(1) Each IBC intended to contain solids that are loaded or discharged under pressure or intended to contain liquids must be tested in accordance with the leak-proofness test prescribed in §178.813 of this subchapter prior to its first use in transportation and every 2.5 years thereafter, starting from the date of manufacture or the date of a repair conforming to paragraph (d)(1) of this section. For this test, the IBC is not required to have its closures fitted.

(2) An external visual inspection must be conducted initially after production and every 2.5 years starting from the date of manufacture or the date of a repair conforming to paragraph (d)(1) of this section to ensure that:

(i) The IBC is marked in accordance with requirements in §178.703 of this subchapter. Missing or damaged markings, or markings difficult to read must be restored or returned to original condition.

(ii) Service equipment is fully functional and free from damage which may cause failure. Missing, broken, or damaged parts must be repaired or replaced.

(iii) The IBC is capable of withstanding the applicable design qualification tests. The IBC must be externally inspected for cracks, warpage, corrosion or any other damage which might render the IBC unsafe for transportation. An IBC found with such defects must be removed from service or repaired in accordance with paragraph (d) of this section. The inner receptacle of a composite IBC must be removed from the outer IBC body for inspection unless the inner receptacle is bonded to the outer body or unless the outer body is constructed in such a way (e.g., a welded or riveted cage) that removal of the inner receptacle is not possible without impairing the integrity of the outer body. Defective inner receptacles must be replaced in accordance with paragraph (d) of this section or the entire IBC must be removed from service. For metal IBCs, thermal insulation must be removed to the extent necessary for proper examination of the IBC body.

(3) Each metal, rigid plastic and composite IBC must be internally inspected at least every five years to ensure that the IBC is free from damage and to ensure that the IBC is capable of withstanding the applicable design qualification tests.

(i) The IBC must be internally inspected for cracks, warpage, and corrosion or any other defect that might render the IBC unsafe for transportation. An IBC found with such defects must be removed from hazardous materials service until restored to the original design type of the IBC.

(ii) Metal IBCs must be inspected to ensure the minimum wall thickness requirements in §178.705(c)(1)(iv) of this subchapter are met. Metal IBCs not conforming to minimum wall thickness requirements must be removed from hazardous materials service.

(c) *Visual inspection for flexible, fiberboard, or wooden IBCs*. Each IBC must be visually inspected prior to first use and permitted reuse, by the person who places hazardous materials in the IBC, to ensure that:

(1) The IBC is marked in accordance with requirements in §178.703 of this subchapter. Additional marking allowed for each design type may be present. Required markings that are missing, damaged or difficult to read must be restored or returned to original condition.

(2) Proper construction and design specifications have been met.

(i) Each flexible IBC must be inspected to ensure that:

(A) Lifting straps if used are securely fastened to the IBC in accordance with the design type.

(B) Seams are free from defects in stitching, heat sealing or gluing which would render the IBC unsafe for transportation of hazardous materials. All stitched seam-ends must be secure.

(C) Fabric used to construct the IBC is free from cuts, tears and punctures. Additionally, fabric must be free from scoring which may render the IBC unsafe for transport.

(ii) Each fiberboard intermediate bulk container must be inspected to ensure that:

(A) Fluting or corrugated fiberboard is firmly glued to facings.

(B) Seams are creased and free from scoring, cuts, and scratches.

(C) Joints are appropriately overlapped and glued, stitched, taped or stapled as prescribed by the design. Where staples are used, the joints must be inspected for protruding staple-ends which could puncture or abrade the inner liner. All such ends must be protected before the IBC is authorized for hazardous materials service.

(iii) Each wooden IBC must be inspected to ensure that:

(A) End joints are secured in the manner prescribed by the design.

(B) IBC walls are free from defects in wood. Inner protrusions which could puncture or abrade the liner must be covered.

(d) *Requirements applicable to repair of IBCs*. (1) Except for flexible and fiberboard IBCs and the bodies of rigid plastic and composite IBCs, damaged IBCs may be repaired and the inner receptacles of composite packagings may be replaced and returned to service provided:

(i) The repaired IBC conforms to the original design type, is capable of withstanding the applicable design qualification tests, and is retested and inspected in accordance with the applicable requirements of this section;

(ii) An IBC intended to contain liquids or solids that are loaded or discharged under pressure is subjected to a leakproofness test as specified in §178.813 of this sub-chapter and is marked with the date of the test; and

(iii) The IBC is subjected to the internal and external inspection requirements as specified in paragraph (b) of this section.

(iv) The person performing the tests and inspections after the repair must durably mark the IBC near the manufacturer's UN design type marking to show the following:

(A) The country in which the tests and inspections were performed;

(B) The name or authorized symbol of the person performing the tests and inspections; and

(C) The date (month, year) of the tests and inspections.

(v) Retests and inspections performed in accordance with paragraphs (d)(1)(i) and (ii) of this section may be used to satisfy the requirements for the 2.5 and five year periodic tests and inspections required by paragraph (b) of this section, as applicable.

(2) Except for flexible and fiberboard IBCs, the structural equipment of an IBC may be repaired and returned to service provided:

(i) The repaired IBC conforms to the original design type and is capable of withstanding the applicable design qualification tests; and

(ii) The IBC is subjected to the internal and external inspection requirements as specified in paragraph (b) of this section.

(3) Service equipment may be replaced provided:

(i) The repaired IBC conforms to the original design type and is capable of withstanding the applicable design qualification tests;

(ii) The IBC is subjected to the external visual inspection requirements as specified in paragraph (b) of this section; and

(iii) The proper functioning and leak tightness of the service equipment, if applicable, is verified.

(e) *Requirements applicable to routine maintenance of IBCs*. Except for routine maintenance of metal, rigid plastics and composite IBCs performed by the owner of the IBC, whose State and name or authorized symbol is durably marked on the IBC, the party performing the routine maintenance shall durably mark the IBC near the manufacturer's UN design type marking to show the following:

(1) The county in which the routine maintenance was carried out; and

(2) The name or authorized symbol of the party performing the routine maintenance.

(f) *Retest date*. The date of the most recent periodic retest must be marked as provided in §178.703(b) of this subchapter.

(g) *Record retention*. (1) The owner or lessee of the IBC must keep records of periodic retests, initial and periodic inspections, and tests performed on the IBC if it has been repaired or remanufactured.

(2) Records must include design types and packaging specifications, test and inspection dates, name and address of test and inspection facilities, names or name of any persons conducting test or inspections, and test or inspection specifics and results.

(3) Records must be kept for each packaging at each location where periodic tests are conducted, until such tests are successfully performed again or for at least 2.5 years from the date of the last test. These records must be made available for inspection by a representative of the Department on request.

Subpart E—Qualification and Maintenance of Cargo Tanks

§180.407 Requirements for test and inspection of specification cargo tanks.

(a) *General.* (1) A cargo tank constructed in accordance with a DOT specification for which a test or inspection specified in this section has become due, may not be filled and offered for transportation or transported until the test or inspection has been successfully completed. This paragraph does not apply to any cargo tank filled prior to the test or inspection due date.

(2) Except during a pressure test, a cargo tank may not be subjected to a pressure greater than its design pressure or MAWP.

(3) A person witnessing or performing a test or inspection specified in this section must meet the minimum qualifications prescribed in §180.409.

(4) Each cargo tank must be evaluated in accordance with the acceptable results of tests and inspections prescribed in §180.411.

(5) Each cargo tank which has successfully passed a test or inspection specified in this section must be marked in accordance with §180.415.

(6) A cargo tank which fails a prescribed test or inspection must:

(i) Be repaired and retested in accordance with §180.413; or

(ii) Be removed from hazardous materials service and the specification plate removed, obliterated or covered in a secure manner.

(b) *Conditions requiring test and inspection of cargo tanks.* Without regard to any other test or inspection requirements, a specification cargo tank must be tested and inspected in accordance with this section prior to further use if:

(1) The cargo tank shows evidence of dents, cuts, gouges, corroded or abraded areas, leakage, or any other condition that might render it unsafe for hazardous materials service. At a minimum, any area of a cargo tank showing evidence of dents, cuts, digs, gouges, or corroded or abraded areas must be thickness tested in accordance with the procedures set forth in paragraphs (i)(2), (i)(3), (i)(5), and (i)(6) of this section and evaluated in accordance with the criteria prescribed in §180.411. Any signs of leakage must be repaired in accordance with §180.413. The suitability of any repair affecting the structural integrity of the cargo tank must be determined either by the testing required in the applicable manufacturing specification or in paragraph (g)(1)(iv) of this section.

(2) The cargo tank has sustained damage to an extent that may adversely affect its lading retention capability. A damaged cargo tank must be pressure tested in accordance with the procedures set forth in paragraph (g) of this section.

(3) The cargo tank has been out of hazardous materials transportation service for a period of one year or more. Each cargo tank that has been out of hazardous materials transportation service for a period of one year or more must be pressure tested in accordance with §180.407(g) prior to further use.

(4) [Reserved]

(5) The Department so requires based on the existence of probable cause that the cargo tank is in an unsafe operating condition.

(c) *Periodic test and inspection.* Each specification cargo tank must be tested and inspected as specified in the following table by an inspector meeting the qualifications in §180.409. The retest date shall be determined from the specified interval identified in the following table from the most recent inspection or the CTMV certification date.

COMPLIANCE DATES—INSPECTIONS AND TEST UNDER §180.407(C)

Test or Inspection (cargo tank specification, configuration, and service)	Date by which first test must be completed (see note 1)	Interval period after first test
External Visual Inspection:		
All cargo tanks designed to be loaded by vacuum with full opening rear heads	September 1, 1991	6 months.
All other cargo tanks	September 1, 1991	1 year.
Internal Visual Inspection:		
All insulated cargo tanks, except MC 330, MC 331, MC 338 (see Note 4)	September 1, 1991	1 year.
All cargo tanks transporting lading corrosive to the tank	September 1, 1991	1 year.
All other cargo tanks except MC 338	September 1, 1995	5 years.
Lining Inspection:		
All lined cargo tanks transporting lading corrosive to the tank	September 1, 1991	1 year.
Leakage Test:		
MC 330 and MC 331 cargo tanks in chlorine service	September 1, 1991	2 years.
All other cargo tanks except MC 338	September 1, 1991	1 year.
Pressure Test:		
(Hydrostatic or pneumatic) (See Notes 2 and 3)	....	....
All cargo tanks which are insulated with no manhole or insulated and lined, except MC 338	September 1, 1991	1 year.
All cargo tanks designed to be loaded by vacuum with full opening rear heads	September 1, 1992	2 years.
MC 330 and MC 331 cargo tanks in chlorine service	September 1, 1992	2 years.
All other cargo tanks	September 1, 1995	5 years.
Thickness Test:		
All unlined cargo tanks transporting material corrosive to the tank, except MC 338	September 1, 1992	2 years.

Note 1: If a cargo tank is subject to an applicable inspection or test requirement under the regulations in effect on December 30, 1990, and the due date (as specified by a requirement in effect on December 30, 1990) for completing the required inspection or test occurs before the compliance date listed in Table I, the earlier date applies.

Note 2: Pressure testing is not required for MC 330 and MC 331 cargo tanks in dedicated sodium metal service.

Note 3: Pressure testing is not required for uninsulated lined cargo tanks with a design pressure or MAWP 15 psig or less, which receive an external visual inspection and lining inspection at least once each year.

Note 4: Insulated cargo tanks equipped with manholes or inspection openings may perform either an internal visual inspection in conjunction with the external visual inspection or a hydrostatic or pneumatic pressure-test of the cargo tank.

(d) *External visual inspection and testing*. The following applies to the external visual inspection and testing of cargo tanks:

(1) Where insulation precludes a complete external visual inspection as required by paragraphs (d)(2) through (d)(6) of this section, the cargo tank also must be given an internal visual inspection in accordance with paragraph (e) of this section. If external visual inspection is precluded because any part of the cargo tank wall is externally lined, coated, or designed to prevent an external visual inspection, those areas of the cargo tank must be internally inspected. If internal visual inspection is precluded because the cargo tank is lined, coated, or designed so as to prevent access for internal inspection, the tank must be hydrostatically or pneumatically tested in accordance with paragraph (g)(1)(iv) of this section. Those items able to be externally inspected must be externally inspected and noted in the inspection report.

(2) The external visual inspection and testing must include as a minimum the following:

(i) The tank shell and heads must be inspected for corroded or abraded areas, dents, distortions, defects in welds and any other conditions, including leakage, that might render the tank unsafe for transportation service;

(ii) The piping, valves, and gaskets must be carefully inspected for corroded areas, defects in welds, and other conditions, including leakage, that might render the tank unsafe for transportation service;

(iii) All devices for tightening manhole covers must be operative and there must be no evidence of leakage at manhole covers or gaskets;

(iv) All emergency devices and valves including self-closing stop-valves, excess flow valves and remote closure devices must be free from corrosion, distortion, erosion and any external damage that will prevent safe operation. Remote closure devices and self-closing stop valves must be functioned to demonstrate proper operation;

(v) Missing bolts, nuts and fusible links or elements must be replaced, and loose bolts and nuts must be tightened;

(vi) All markings on the cargo tank required by parts 172, 178 and 180 of this subchapter must be legible;

(vii) [Reserved]

(viii) All major appurtenances and structural attachments on the cargo tank including, but not limited to, suspension system attachments, connecting structures, and those elements of the upper coupler (fifth wheel) assembly that can be inspected without dismantling the upper coupler (fifth wheel) assembly must be inspected for any corrosion or damage which might prevent safe operation;

(ix) For cargo tanks transporting lading corrosive to the tank, areas covered by the upper coupler (fifth wheel) assembly must be inspected at least once in each two year period for corroded and abraded areas, dents, distortions, defects in welds, and any other condition that might render the tank unsafe for transportation service. The upper coupler (fifth wheel) assembly must be removed from the cargo tank for this inspection.

(3) All reclosing pressure relief valves must be externally inspected for any corrosion or damage which might prevent safe operation. All reclosing pressure relief valves on cargo tanks carrying lading corrosive to the valve must be removed from the cargo tank for inspection and testing. Each reclosing pressure relief valve required to be removed and tested must open at no less than the required set pressure and no more than 110 percent of the required set pressure, and must reseat to a leak-tight condition at no less than 90 percent of the start-to-discharge pressure or the pressure prescribed for the applicable cargo tank specification.

(4) Ring stiffeners or other appurtenances, installed on cargo tanks constructed of mild steel or high-strength, low-alloy steel, that create air cavities adjacent to the tank shell that do not allow for external visual inspection must be thickness tested in accordance with paragraphs (i)(2) and (i)(3) of this section, at least once every 2 years. At least four symmetrically distributed readings must be taken to establish an average thickness for the ring stiffener or appurtenance. If any thickness reading is less than the average thickness by more than 10%, thickness testing in accordance with paragraphs (i)(2) and (i)(3) of this section must be conducted from the inside of the cargo tank on the area of the tank wall covered by the appurtenance or ring stiffener.

(5) Corroded or abraded areas of the cargo tank wall must be thickness tested in accordance with the procedures set forth in paragraphs (i)(2), (i)(3), (i)(5) and (i)(6) of this section.

(6) The gaskets on any full opening rear head must be:

(i) Visually inspected for cracks or splits caused by weather or wear; and

(ii) Replaced if cuts or cracks which are likely to cause leakage, or are of a depth one-half inch or more, are found.

(7) The inspector must record the results of the external visual examination as specified in §180.417(b).

(e) *Internal visual inspection*. (1) When the cargo tank is not equipped with a manhole or inspection opening, or the cargo tank design precludes an internal inspection, the tank shall be hydrostatically or pneumatically tested in accordance with 180.407(c) and (g).

(2) The internal visual inspection must include as a minimum the following:

(i) The tank shell and heads must be inspected for corroded and abraded areas, dents, distortions, defects in welds, and any other condition that might render the tank unsafe for transportation service.

(ii) Tank liners must be inspected as specified in §180.407(f).

(3) Corroded or abraded areas of the cargo tank wall must be thickness tested in accordance with paragraphs (i)(2), (i)(3), (i)(5) and (i)(6) of this section.

(4) The inspector must record the results of the internal visual inspection as specified in §180.417(b).

(f) *Lining inspection*. The integrity of the lining on all lined cargo tanks, when lining is required by this subchapter, must be verified at least once each year as follows:

(1) Rubber (elastomeric) lining must be tested for holes as follows:

(i) Equipment shall consist of:

(A) A high frequency spark tester capable of producing sufficient voltage to ensure proper calibration;

(B) A probe with an "L" shaped 2.4 mm (0.09 inch) diameter wire with up to a 30.5 cm (12-inch) bottom leg (end bent to a 12.7 mm (0.5 inch) radius), or equally sensitive probe; and

(C) A steel calibration coupon 30.5 cm X 30.5 cm (12 inches X 12 inches) covered with the same material and thickness as that to be tested. The material on the coupon shall have a test hole to the metal substrate made by puncturing the material with a 22 gauge hypodermic needle or comparable piercing tool.

(ii) The probe must be passed over the surface of the calibration coupon in a constant uninterrupted manner until the hole is found. The hole is detected by the white or light blue spark formed. (A sound lining causes a dark blue or purple spark.) The voltage must be adjusted to the lowest setting that will produce a minimum 12.7 mm (0.5 inch) spark measured from the top of the lining to the probe. To assure that the setting on the probe has not changed, the spark tester must be calibrated periodically using the test calibration coupon, and the same power source, probe, and cable length.

(iii) After calibration, the probe must be passed over the lining in an uninterrupted stroke.

(iv) Holes that are found must be repaired using equipment and procedures prescribed by the lining manufacturer or lining installer.

(2) Linings made of other than rubber (elastomeric material) must be tested using equipment and procedures prescribed by the lining manufacturer or lining installer.

(3) Degraded or defective areas of the the cargo tank liner must be removed and the cargo tank wall below the defect must be inspected. Corroded areas of the tank wall must be thickness tested in accordance with paragraphs (i)(2), (i)(3), (i)(5) and (i)(6) of this section.

(4) The inspector must record the results of the lining inspection as specified in §180.417(b).

(g) *Pressure test.* All components of the cargo tank wall, as defined in §178.320(a) of this subchapter, must be pressure tested as prescribed by this paragraph.

(1) *Test Procedure* - (i) As part of the pressure test, the inspector must perform an external and internal visual inspection, except that on an MC 338 cargo tank, or a cargo tank not equipped with a manhole or inspection opening, an internal inspection is not required.

(ii) All self-closing pressure relief valves, including emergency relief vents and normal vents, must be removed from the cargo tank for inspection and testing.

(A) Each self-closing pressure relief valve that is an emergency relief vent must open at no less than the required set pressure and no more than 110 percent of the required set pressure, and must reseat to a leak-tight condition at no less than 90 percent of the start-to-discharge pressure or the pressure prescribed for the applicable cargo tank specification.

(B) Normal vents (1 psig vents) must be tested according to the testing criteria established by the valve manufacturer.

(C) Self-closing pressure relief devices not tested or failing the tests in this paragraph (g)(1)(ii) must be repaired or replaced.

(iii) Except for cargo tanks carrying lading corrosive to the tank, areas covered by the upper coupler (fifth wheel) assembly must be inspected for corroded and abraded areas, dents, distortions, defects in welds, and any other condition that might render the tank unsafe for transportation service. The upper coupler (fifth wheel) assembly must be removed from the cargo tank for this inspection.

(iv) Each cargo tank must be tested hydrostatically or pneumatically to the internal pressure specified in the following table. At no time during the pressure test may a cargo tank be subject to pressures that exceed those identified in the following table:

Specification	Test pressure
MC 300, 301, 302, 303, 305, 306	20.7 kPa (3 psig) or design pressure, whichever is greater.
MC 304, 307	275.8 kPa (40 psig) or 1.5 times the design pressure, whichever is greater.
MC 310, 311, 312	20.7 kPa (3 psig) or 1.5 times the design pressure, whichever is greater.
MC 330, 331	1.5 times either the MAWP or the re-rated pressure, whichever is applicable.
MC 338	1.25 times either the MAWP or the re-rated pressure, whichever is applicable.
DOT 406	34.5 kPa (5 psig) or 1.5 times the MAWP, whichever is greater.
DOT 407	275.8 kPa (40 psig) or 1.5 times the MAWP, whichever is greater.
DOT 412	1.5 times the MAWP.

(v) [Reserved]

(vi) Each cargo tank of a multi-tank cargo tank motor vehicle must be tested with the adjacent cargo tanks empty and at atmospheric pressure.

(vii) All closures except pressure relief devices must be in place during the test. All prescribed loading and unloading venting devices rated at less than test pressure may be removed during the test. If retained, the devices must be rendered inoperative by clamps, plugs, or other equally effective restraining devices. Restraining devices may not prevent detection of leaks or damage the venting devices and must be removed immediately after the test is completed.

(viii) *Hydrostatic test method.* Each cargo tank, including its domes, must be filled with water or other liquid having similar viscosity, at a temperature not exceeding 100 °F. The cargo tank must then be pressurized to not less than the pressure specified in paragraph (g)(1)(iv) of this section. The cargo tank, including its closures, must hold the prescribed test pressure for at least 10 minutes during which time it shall be inspected for leakage, bulging or any other defect.

(ix) *Pneumatic test method.* Pneumatic testing may involve higher risk than hydrostatic testing. Therefore, suitable safeguards must be provided to protect personnel and facilities should failure occur during the test. The cargo tank must be pressurized with air or an insert gas. The pneumatic test pressure in the cargo tank must be reached by gradually increasing the pressure to one-half of the test pressure. Thereafter, the pressure must be increased in steps of approximately one-tenth of the test pressure until the required test pressure has been reached. The test pressure must be held for at least 5 minutes. The pressure must then be reduced to the MAWP, which must be maintained during the time the entire cargo tank surface is inspected. During the inspection, a suitable method must be used for detecting the existence of leaks. This method must consist either of coating the entire surface of all joints under pressure with a solution of soap and water, or using other equally sensitive methods.

(2) When testing an insulated cargo tank, the insulation and jacketing need not be removed unless it is otherwise impossible to reach test pressure and maintain a condition of pressure equilibrium after test pressure is reached, or the vacuum integrity cannot be maintained in the insulation space. If an MC 338 cargo tank used for the transportation of a flammable gas or oxygen, refrigerated liquid is opened for any reason, the cleanliness must be verified prior to closure using the procedures contained in §178.338-15 of this subchapter.

(3) Each MC 330 and MC 331 cargo tank constructed of quenched and tempered steel in accordance with Part UHT in Section VIII of the ASME Code (IBR, see §171.7 of this subchapter), or constructed of other than quenched and tempered steel but without postweld heat treatment, used for the transportation of anhydrous ammonia or any other hazardous materials that may cause corrosion stress cracking, must be internally inspected by the wet fluorescent magnetic particle method immediately prior to and in conjunction with the performance of the pressure test prescribed in this section. Each MC 330 and MC 331 cargo tank constructed of quenched and tempered steel in accordance with Part UHT in Section VIII of the ASME Code and used for the transportation of liquefied petroleum gas must be internally inspected by the wet fluorescent magnetic particle method immediately prior to and in conjunction with the performance of the pressure test prescribed in this section. The wet fluorescent magnetic particle inspection must be in accordance with Section V of the ASME Code and CGA Technical Bulletin TB-2 (IBR, see §171.7 of this subchapter). This paragraph does not apply to cargo tanks that do not have manholes. (See §180.417(c) for reporting requirements.)

(4) All pressure bearing portions of a cargo tank heating system employing a medium such as, but not limited to, steam or hot water for heating the lading must be hydrostatically pressure tested at least once every 5 years. The test pressure must be at least the maximum system design operating pressure and must be maintained for five minutes. A heating system employing flues for heating the lading must be tested to ensure against lading leakage into the flues or into the atmosphere.

(5) *Exceptions.* (i) Pressure testing is not required for MC 330 and MC 331 cargo tanks in dedicated sodium metal service.

(ii) Pressure testing is not required for uninsulated lined cargo tanks, with a design pressure or MAWP of 15 psig or less, which receive an external visual inspection and a lining inspection at least once each year.

(6) *Acceptance criteria.* A cargo tank that leaks, fails to retain test pressure or pneumatic inspection pressure, shows distortion, excessive permanent expansion, or other evidence of weakness that might render the cargo tank unsafe for transportation service, may not be returned to service, except as follows: A cargo tank with a heating system which does not hold pressure may remain in service as an unheated cargo tank if:

(i) The heating system remains in place and is structurally sound and no lading may leak into the heating system, and

(ii) The specification plate heating system information is changed to indicate that the cargo tank has no working heating system.

(7) The inspector must record the results of the pressure test as specified in §180.417(b).

(h) *Leakage test.* The following requirements apply to cargo tanks requiring a leakage test:

(1) Each cargo tank must be tested for leaks in accordance with paragraph (c) of this section. The leakage test must include testing product piping with all valves and accessories in place and operative, except that any venting devices set to discharge at less than the leakage test pressure must be removed or rendered inoperative during the test. All internal or external self-closing stop valves must be tested for leak tightness. Each cargo tank of a multi-cargo tank motor vehicle must be tested with adjacent cargo tanks empty and at atmospheric pressure. Test pressure must be maintained for at least 5 minutes. Cargo tanks in liquefied compressed gas service must be externally inspected for leaks during the leakage test. Suitable safeguards must be provided to protect personnel should a failure occur. Cargo tanks may be leakage tested with hazardous materials contained in the cargo tank during the test. Leakage test pressure must be no less than 80% of MAWP marked on the specification plate except as follows:

(i) A cargo tank with an MAWP of 690 kPa (100 psig) or more may be leakage tested at its maximum normal operating pressure provided it is in dedicated service or services; or

(ii) An MC 330 or MC 331 cargo tank in dedicated liquefied petroleum gas service may be leakage tested at not less than 414 kPa (60 psig).

(iii) An operator of a specification MC 330 or MC 331 cargo tank, and a nonspecification cargo tank authorized under §173.315(k) of this subchapter, equipped with a meter may check leak tightness of the internal self-closing stop valve by conducting a meter creep test. (See Appendix B to this part.)

(iv) An MC 330 or MC 331 cargo tank in dedicated service for anhydrous ammonia may be leakage tested at not less than 414 kPa (60 psig).

(v) A non-specification cargo tank required by §173.8(d) of this subchapter to be leakage tested, must be leakage tested at not less than 16.6 kPa (2.4 psig), or as specified in paragraph (h)(2) of this section.

(2) Cargo tanks used to transport petroleum distillate fuels that are equipped with vapor collection equipment may be leak tested in accordance with the Environmental Protection Agency's "Method 27—Determination of Vapor Tightness of Gasoline Delivery Tank Using Pressure-Vacuum Test," as set forth in Appendix A to 40 CFR part 60. Test methods and procedures and maximum allowable pressure and vacuum changes are in 40 CFR 63.425(e)(1). The hydrostatic test alternative, using liquid in Environmental Protection Agency's "Method 27—Determination of Vapor Tightness of Gasoline Delivery Tank Using Pressure-Vacuum Test," may not be used to satisfy the leak testing requirements of this paragraph. The test must be conducted using air.

Editor's Note: Method 27 is located in this Guide after the Appendixes in part 180.

(3) A cargo tank that fails to retain leakage test pressure may not be returned to service as a specification cargo tank, except under conditions specified in §180.411(d).

(4) After July 1, 2000, Registered Inspectors of specification MC 330 and MC 331 cargo tanks, and nonspecification cargo tanks authorized under §173.315(k) of this subchapter must visually inspect the delivery hose assembly and piping system while the assembly is under leakage test pressure utilizing the rejection criteria listed in §180.416(g). Delivery hose assemblies not permanently attached to the cargo tank motor vehicle may be inspected separately from the cargo tank motor vehicle. In addition to a written record of the inspection prepared in accordance with §180.417(b), the Registered Inspector conducting the test must note the hose identification number, the date of the test, and the condition of the hose assembly and piping system tested.

(5) The inspector must record the results of the leakage test as specified in §180.417(b).

(i) *Thickness testing.* (1) The shell and head thickness of all unlined cargo tanks used for the transportation of materials corrosive to the tank must be measured at least once every 2 years, except that cargo tanks measuring less than the sum of the minimum prescribed thickness, plus one-fifth of the original corrosion allowance, must be tested annually.

(2) Measurements must be made using a device capable of accurately measuring thickness to within +/– 0.002 of an inch.

(3) Any person performing thickness testing must be trained in the proper use of the thickness testing device used in accordance with the manufacturer's instruction.

(4) Thickness testing must be performed in the following areas, as a minimum:

(i) Areas of the tank shell and heads and shell and head area around any piping that retains lading;

(ii) Areas of high shell stress such as the bottom center of the tank;

(iii) Areas near openings;

(iv) Areas around weld joints;

(v) Areas around shell reinforcements;

(vi) Areas around appurtenance attachments;

(vii) Areas near upper coupler (fifth wheel) assembly attachments;

(viii) Areas near suspension system attachments and connecting structures;

(ix) Known thin areas in the tank shell and nominal liquid level lines; and

(x) Connecting structures joining multiple cargo tanks of carbon steel in a self-supporting cargo tank motor vehicle.

(5) Minimum thicknesses for MC 300, MC 301, MC 302, MC 303, MC 304, MC 305, MC 306, MC 307, MC 310, MC 311, and MC 312 cargo tanks are determined based on the definition of minimum thickness found in §178.320(a) of this subchapter. The following Tables I and II identify the "In-Service Minimum Thickness" values to be used to determine the minimum thickness for the referenced cargo tanks. The column headed "Minimum Manufactured Thickness" indicates the minimum values required for new construction of DOT 400 series cargo tanks, found in Tables I and II of §§178.346–2, 178.347–2, and 178.348–2 of this subchapter. In-Service Minimum Thicknesses for MC 300, MC 301, MC 302, MC 303, MC 304, MC 305, MC 306, MC 307, MC 310, MC 311, and MC 312 cargo tanks are based on 90 percent of the manufactured thickness specified in the DOT specification, rounded to three places.

TABLE I.—IN SERVICE MINIMUM THICKNESS FOR MC 300, MC 303, MC 304, MC 306, MC 307, MC 310, MC 311, AND MC 312 SPECIFICATION CARGO TANKS CONSTRUCTED OF STEEL AND STEEL ALLOYS

Minimum manufactured thickness (US gauge or inches)	Nominal decimal equivalent for (inches)	In-service minimum thickness reference (inches)
19	0.0418	0.038
18	0.0478	0.043
17	0.0538	0.048
16	0.0598	0.054
15	0.0673	0.061
14	0.0747	0.067
13	0.0897	0.081
12	0.1046	0.094
11	0.1196	0.108
10	0.1345	0.121
9	0.1495	0.135
8	0.1644	0.148
7	0.1793	0.161
3/16	0.1875	0.169
1/4	0.2500	0.225
5/16	0.3125	0.281
3/8	0.3750	0.338

TABLE II.—IN SERVICE MINIMUM THICKNESS FOR MC 301, MC 302, MC 304, MC 305, MC 306, MC 307, MC 311, AND MC 312 SPECIFICATION CARGO TANKS CONSTRUCTED OF ALUMINUM AND ALUMINUM ALLOYS

Minimum manufactured thickness	In-service minimum thickness (inches)
0.078	0.070
0.087	0.078
0.096	0.086
0.109	0.098
0.130	0.117
0.141	0.127
0.151	0.136
0.172	0.155
0.173	0.156
0.194	0.175
0.216	0.194
0.237	0.213
0.270	0.243
0.360	0.324
0.450	0.405
0.540	0.486

(6) An owner of a cargo tank that no longer conforms with the minimum thickness prescribed for the design as manufactured may use the cargo tank to transport

authorized materials at reduced maximum weight of lading or reduced maximum working pressure, or combinations thereof, provided the following conditions are met:

(i) A Design Certifying Engineer must certify that the cargo tank design and thickness are appropriate for the reduced loading conditions by issuance of a revised manufacturer's certificate, and

(ii) The cargo tank motor vehicle's nameplate must reflect the revised service limits.

(7) An owner of a cargo tank that no longer conforms with the minimum thickness prescribed for the specification may not return the cargo tank to hazardous materials service. The tank's specification plate must be removed, obliterated or covered in a secure manner.

(8) The inspector must record the results of the thickness test as specified in §180.417(b).

(9) For MC 331 cargo tanks constructed before October 1, 2003, minimum thickness shall be determined by the thickness indicated on the U1A form minus any corrosion allowance. For MC 331 cargo tanks constructed after October 1, 2003, the minimum thickness will be the value indicated on the specification plate. If no corrosion allowance is indicated on the U1A form then the thickness of the tank shall be the thickness of the material of construction indicated on the U1A form with no corrosion allowance.

(10) For 400-series cargo tanks, minimum thickness is calculated according to tables in each applicable section of this subchapter for that specification: §178.346–2 for DOT 406 cargo tanks, §178.347–2 for DOT 407 cargo tanks, and §178.348–2 for DOT 412 cargo tanks.

§180.415 Test and inspection markings.

(a) Each cargo tank successfully completing the test and inspection requirements contained in §180.407 must be marked as specified in this section.

(b) Each cargo tank must be durably and legibly marked, in English, with the date (month and year) and the type of test or inspection performed, subject to the following provisions:

(1) The date must be readily identifiable with the applicable test or inspection.

(2) The markings must be in letters and numbers at least 32 mm (1.25 inches) high, near the specification plate or anywhere on the front head.

(3) The type of test or inspection may be abbreviated as follows:

(i) V for external visual inspection and test;

(ii) I for internal visual inspection;

(iii) P for pressure test;

(iv) L for lining inspection;

(v) T for thickness test; and

(vi) K for leakage test for a cargo tank tested under §180.407, except §180.407(h)(2); and

(vii) K-EPA27 for a cargo tank tested under §180.407(h)(2) after October 1, 2004.

Examples to paragraph (b). The markings "10-99 P, V, L" represent that in October 1999 a cargo tank passed the prescribed pressure test, external visual inspection and test, and the lining inspection. The markings "2-00 K-EPA27" represent that in February 2000 a cargo tank passed the leakage test under §180.407(h)(2). The markings "2-00 K, K-EPA27" represent that in February 2000 a cargo tank passed the leakage test under both §180.407(h)(1) and under EPA Method 27 in 180.407(h)(2).

(c) For a cargo tank motor vehicle composed of multiple cargo tanks constructed to the same specification, which are tested and inspected at the same time, one set of test and inspection markings may be used to satisfy the requirements of this section. For a cargo tank motor vehicle composed of multiple cargo tanks constructed to different specifications, which are tested and inspected at different intervals, the test and inspection markings must appear in the order of the cargo tank's corresponding location, from front to rear.

§180.416 Discharge system inspection and maintenance program for cargo tanks transporting liquefied compressed gases.

(a) *Applicability*. This section is applicable to an operator using specification MC 330, MC 331, and nonspecification cargo tanks authorized under §173.315(k) of this subchapter for transportation of liquefied compressed gases other than carbon dioxide. Paragraphs (b), (c), (d)(1), (d)(5), (e), (f), and (g)(1) of this section, applicable to delivery hose assemblies, apply only to hose assemblies installed or carried on the cargo tank.

(b) *Hose identification*. By July 1, 2000, the operator must assure that each delivery hose assembly is permanently marked with a unique identification number and maximum working pressure.

(c) *Post-delivery hose check*. After each unloading, the operator must visually check that portion of the delivery hose assembly deployed during the unloading.

(d) *Monthly inspections and tests*. (1) The operator must visually inspect each delivery hose assembly at least once each calendar month the delivery hose assembly is in service.

(2) The operator must visually inspect the piping system at least once each calendar month the cargo tank is in service. The inspection must include fusible elements and all components of the piping system, including bolts, connections, and seals.

(3) At least once each calendar month a cargo tank is in service, the operator must actuate all emergency discharge control devices designed to close the internal self-closing stop valve to assure that all linkages operate as designed. Appendix A to this part outlines acceptable procedures that may be used for this test.

(4) The operator of a cargo tank must check the internal self-closing stop valve in the liquid discharge opening for leakage through the valve at least once each calendar month the cargo tank is in service. On cargo tanks equipped with a meter, the meter creep test as outlined in Appendix B to this part or a test providing equivalent accuracy is acceptable. For cargo tanks that are not equipped with a meter, Appendix B to this part outlines one acceptable method that may be used to check internal self-closing stop valves for closure.

(5) After July 1, 2000, the operator must note each inspection in a record. That record must include the inspection date, the name of the person performing the inspection, the hose assembly identification number, the company name, the date the hose was assembled and tested, and an indication that the delivery hose assembly and piping system passed or failed the tests and

inspections. A copy of each test and inspection record must be retained by the operator at its principal place of business or where the vehicle is housed or maintained until the next test of the same type is successfully completed.

(e) *Annual hose leakage test.* The owner of a delivery hose assembly that is not permanently attached to a cargo tank motor vehicle must ensure that the hose assembly is annually tested in accordance with §180.407(h)(4).

(f) *New or repaired delivery hose assemblies.* Each operator of a cargo tank must ensure each new and repaired delivery hose assembly is tested at a minimum of 120 percent of the hose maximum working pressure.

(1) The operator must visually examine the delivery hose assembly while it is under pressure.

(2) Upon successful completion of the pressure test and inspection, the operator must assure that the delivery hose assembly is permanently marked with the month and year of the test.

(3) After July 1, 2000, the operator must complete a record documenting the test and inspection, including the date, the signature of the inspector, the hose owner, the hose identification number, the date of original delivery hose assembly and test, notes of any defects observed and repairs made, and an indication that the delivery hose assembly passed or failed the tests and inspections. A copy of each test and inspection record must be retained by the operator at its principal place of business or where the vehicle is housed or maintained until the next test of the same type is successfully completed.

(g) *Rejection criteria.* (1) No operator may use a delivery hose assembly determined to have any condition identified below for unloading liquefied compressed gases. An operator may remove and replace damaged sections or correct defects discovered. Repaired hose assemblies may be placed back in service if retested successfully in accordance with paragraph (f) of this section.

(i) Damage to the hose cover that exposes the reinforcement.

(ii) Wire braid reinforcement that has been kinked or flattened so as to permanently deform the wire braid.

(iii) Soft spots when not under pressure, bulging under pressure, or loose outer covering.

(iv) Damaged, slipping, or excessively worn hose couplings.

(v) Loose or missing bolts or fastenings on bolted hose coupling assemblies.

(2) No operator may use a cargo tank with a piping system found to have any condition identified in this paragraph (g)(2) for unloading liquefied compressed gases.

(i) Any external leak identifiable without the use of instruments.

(ii) Bolts that are loose, missing, or severely corroded.

(iii) Manual stop valves that will not actuate.

(iv) Rubber hose flexible connectors with any condition outlined in paragraph (g)(1) of this section.

(v) Stainless steel flexible connectors with damaged reinforcement braid.

(vi) Internal self-closing stop valves that fail to close or that permit leakage through the valve detectable without the use of instruments.

(vii) Pipes or joints that are severely corroded.

§180.417 Reporting and record retention requirements.

(a) *Vehicle certification.* (1) Each owner of a specification cargo tank must retain the manufacturer's certificate, the manufacturer's ASME U1A data report, where applicable, and related papers certifying that the specification cargo tank identified in the documents was manufactured and tested in accordance with the applicable specification. This would include any certification of emergency discharge control systems required by §173.315(n) of this subchapter or §180.405(m). The owner must retain the documents throughout his ownership of the specification cargo tank and for one year thereafter. In the event of a change in ownership, the prior owner must retain non-fading photo copies of these documents for one year.

(2) Each motor carrier who uses a specification cargo tank motor vehicle must obtain a copy of the manufacturer's certificate and related papers or the alternative report authorized by paragraph (a)(3) (i) or (ii) of this section and retain the documents as specified in this paragraph (a)(2). A motor carrier who is not the owner of a cargo tank motor vehicle must also retain a copy of the vehicle certification report for as long as the cargo tank motor vehicle is used by that carrier and for one year thereafter. The information required by this section must be maintained at the company's principal place of business or at the location where the vehicle is housed or maintained. The provisions of this section do not apply to a motor carrier who leases a cargo tank for less than 30 days.

(3) *DOT Specification cargo tanks manufactured before September 1, 1995—*

(i) *Non-ASME Code stamped cargo tanks—* If an owner does not have a manufacturer's certificate for a cargo tank and he wishes to certify it as a specification cargo tank, the owner must perform appropriate tests and inspections, under the direct supervision of a Registered Inspector, to determine if the cargo tank conforms with the applicable specification. Both the owner and the Registered Inspector must certify that the cargo tank fully conforms to the applicable specification. The owner must retain the certificate, as specified in this section.

(ii) *ASME Code Stamped cargo tanks.* If the owner does not have the manufacturer's certificate required by the specification and the manufacturer's data report required by the ASME, the owner may contact the National Board for a copy of the manufacturer's data report, if the cargo tank was registered with the National Board, or copy the information contained on the cargo tank's identification and ASME Code plates. Additionally, both the owner and the Registered Inspector must certify that the cargo tank fully conforms to the specification. The owner must retain such documents, as specified in this section.

(b) *Test or inspection reporting*. Each person performing a test or inspection as specified in §180.407 must prepare a written report, in English, in accordance with this paragraph.

(1) Each test or inspection report must include the following information:

(i) Owner's and manufacturer's unique serial number for the cargo tank;

(ii) Name of cargo tank manufacturer;

(iii) Cargo tank DOT or MC specification number;

(iv) MAWP of the cargo tank;

(v) Minimum thickness of the cargo tank shell and heads when the cargo tank is thickness tested in accordance with §180.407(d)(5), §180.407(e)(3), §180.407(f)(3), or §180.407(i);

(vi) Indication of whether the cargo tank is lined, insulated, or both; and

(vii) Indication of special service of the cargo tank (*e.g.*, transports material corrosive to the tank, dedicated service, etc.)

(2) Each test or inspection report must include the following specific information as appropriate for each individual type of test or inspection:

(i) Type of test or inspection performed;

(ii) Date of test or inspection (month and year);

(iii) Listing of all items tested or inspected, including information about pressure relief devices that are removed, inspected and tested or replaced, when applicable (type of device, set to discharge pressure, pressure at which device opened, pressure at which device reseated, and a statement of disposition of the device (*e.g.*, reinstalled, repaired, or replaced)); information regarding the inspection of upper coupler assemblies, when applicable (visually examined in place, or removed for examination); and, information regarding leakage and pressure testing, when applicable (pneumatic or hydrostatic testing method, identification of the fluid used for the test, test pressure, and holding time of test);

(iv) Location of defects found and method of repair;

(v) ASME or National Board Certificate of Authorization number of facility performing repairs, if applicable;

(vi) Name and address of person performing test;

(vii) Registration number of the facility or person performing the test;

(viii) Continued qualification statement, such as "cargo tank meets the requirements of the DOT specification identified on this report" or "cargo tank fails to meet the requirements of the DOT specification identified on this report";

(ix) DOT registration number of the registered inspector; and

(x) Dated signature of the registered inspector and the cargo tank owner.

(3) The owner and the motor carrier, if not the owner, must each retain a copy of the test and inspection reports until the next test or inspection of the same type is successfully completed. This requirement does not apply to a motor carrier leasing a cargo tank for fewer than 30 days.

(c) *Additional requirements for Specification MC 330 and MC 331 cargo tanks.* (1) After completion of the pressure test specified in §180.407(g)(3), each motor carrier operating a Specification MC 330 or MC 331 cargo tank in anhydrous ammonia, liquefied petroleum gas, or any other service that may cause stress corrosion cracking, must make a written report containing the following information:

(i) Carrier's name, address of principal place of business, and telephone number;

(ii) Complete identification plate data required by Specification MC 330 or MC 331, including data required by ASME Code;

(iii) Carrier's equipment number;

(iv) A statement indicating whether or not the tank was stress relieved after fabrication;

(v) Name and address of the person performing the test and the date of the test;

(vi) A statement of the nature and severity of any defects found. In particular, information must be furnished to indicate the location of defects detected, such as in weld, heat-affected zone, the liquid phase, the vapor phase, or the head-to-shell seam. If no defect or damage was discovered, that fact must be reported;

(vii) A statement indicating the methods employed to make repairs, who made the repairs, and the date they were completed. Also, a statement of whether or not the tank was stress relieved after repairs and, if so, whether full or local stress relieving was performed;

(viii) A statement of the disposition of the cargo tank, such as "cargo tank scrapped" or "cargo tank returned to service"; and

(ix) A statement of whether or not the cargo tank is used in anhydrous ammonia, liquefied petroleum gas, or any other service that may cause stress corrosion cracking. Also, if the cargo tank has been used in anhydrous ammonia service since the last report, a statement indicating whether each shipment of ammonia was certified by its shipper as containing 0.2 percent water by weight.

(2) A copy of the report must be retained by the carrier at its principal place of business during the period the cargo tank is in the carrier's service and for one year thereafter. Upon a written request to, and with the approval of, the Field Administrator, Regional Service Center, Federal Motor Carrier Safety Administration for the region in which a motor carrier has its principal place of business, the carrier may maintain the reports at a regional or terminal office.

(3) The requirement in paragraph (c)(1) of this section does not apply to a motor carrier leasing a cargo tank for less than 30 days.

(d) *Supplying certificates and reports*. Each person offering a DOT-specification cargo tank for sale or lease must provide the purchaser or lessee a copy of the cargo tank certificate of compliance, records of repair, modification, stretching, or rebarrelling; and the most recent inspection and test reports made under this section. Copies of such reports must be provided to the lessee if the cargo tank is leased for more than 30 days.

Subpart G—Qualification and Maintenance of Portable Tanks

§180.605 Requirements for periodic testing, inspection and repair of portable tanks.

(a) A portable tank constructed in accordance with a DOT specification for which a test or inspection specified in this subpart has become due, must be tested or inspected prior to being returned for transportation.

(b) *Conditions requiring test and inspection of portable tanks.* Without regard to any other test or inspection requirements, a Specification or UN portable tank must be tested and inspected in accordance with this section prior to further use if any of the following conditions exist:

(1) The portable tank shows evidence of dents, corroded or abraded areas, leakage, or any other condition that might render it unsafe for transportation service.

(2) The portable tank has been in an accident and has been damaged to an extent that may adversely affect its ability to retain the hazardous material.

(3) The portable tank has been out of hazardous materials transportationservice for a period of one year or more.

(4) The portable tank has been modified from its original design specification.

(5) The portable tank is in an unsafe operating condition based on the existence of probable cause.

(c) *Schedule for periodic inspections and tests.* Each Specification portable tank must be tested and inspected in accordance with the following schedule:

(1) Each IM or UN portable tank must be given an initial inspection and test before being placed into service, a periodic inspection and test at least once every 5 years, and an intermediate periodic inspection and test at least every 2.5 years following the initial inspection and the last 5 year periodic inspection and test.

(2) Each Specification 51 portable tank must be given a periodic inspection and test at least once every five years.

(3) Each Specification 56 or 57 portable tank must be given a periodic inspection and test at least once every 2.5 years.

(4) Each Specification 60 portable tank must be given a periodic inspectionand test at the end of the first 4-year period after the original test; at least once every 2 years thereafter up to a total of 12 years of service; and at least once annually thereafter. Retesting is not required on a rubber-lined tank except before each relining.

(d) *Intermediate periodic inspection and test.* For IM and UN portable tanks the intermediate 2.5 year periodic inspection and test must include at least an internal and external examination of the portable tank and its fittings taking into account the hazardous materials intended to be transported; a leakage test; and a test of the satisfactory operation of all service equipment. Sheathing, thermal insulation, etc. need only be removed to the extent required for reliable appraisal of the condition of the portable tank. For portable tanks intended for the transportation of a single hazardous material, the internal examination may be waived if it is leakage tested in accordance with the procedures in paragraph (h) of this section prior to each filling, or if approved by the Associate Administrator. Portable tanks used for dedicated transportation of refrigerated liquefied gases that are not fitted with inspection openings are excepted from the internal inspection requirement.

(e) *Periodic inspection and test.* The 5 year periodic inspection and test must include an internal and external examination and, unless excepted, a pressure test as specified in this section. Sheathing, thermal insulation, etc. need only to be removed to the extent required for reliable appraisal of the condition of the portable tank. Except for DOT Specification 56 and 57 portable tanks, reclosing pressure relief devices must be removed from the tank and tested separately unless they can be tested while installed on the portable tank. For portable tanks where the shell and equipment have been pressure-tested separately, after assembly they must be subjected together to a leakage test and effectively tested and inspected for corrosion. Portable tanks used for the transportation of refrigerated, liquefied gases are excepted from the requirement for internal inspection and the hydraulic pressure test during the 5-year periodic inspection and test, if the portable tanks were pressure tested to a minimum test pressure of 1.3 times the design pressure using an inert gas as prescribed in §178.338-16(a) and (b) of this subchapter before putting the portable tank into service initially and after any exceptional inspections and tests specified in paragraph (f) of this section.

(f) *Exceptional inspection and test.* The exceptional inspection and test is necessary when a portable tank shows evidence of damaged or corroded areas, or leakage, or other conditions that indicate a deficiency that could affectthe integrity of the portable tank. The extent of the exceptional inspection and test must depend on the amount of damage or deterioration of the portable tank. It must include at least the inspection and a pressure test according to paragraph (e) of this section. Pressure relief devices need not be tested or replaced unless there is reason to believe the relief devices have been affected by the damage or deterioration.

(g) *Internal and external examination.* The internal and external examinations must ensure that:

(1) The shell is inspected for pitting, corrosion, or abrasions, dents, distortions, defects in welds or any other conditions, including leakage, that might render the portable tank unsafe for transportation;

(2) The piping, valves, and gaskets are inspected for corroded areas, defects, and other conditions, including leakage, that might render the portable tank unsafe for filling, discharge or transportation;

(3) Devices for tightening manhole covers are operative and there is no leakage at manhole covers or gaskets;

(4) Missing or loose bolts or nuts on any flanged connection or blank flange are replaced or tightened;

(5) All emergency devices and valves are free from corrosion, distortion and any damage or defect that could prevent their normal operation. Remote closure devices and self-closing stop-valves must be operated to demonstrate proper operation;

(6) Required markings on the portable tank are legible and in accordance with the applicable requirements; and

(7) The framework, the supports and the arrangements for lifting the portable tank are in satisfactory condition.

(h) *Pressure test procedures for specification 51, 56, 57, 60, IM or UN portable tanks* . (1) Each Specification 57 portable tank must be leak tested by a minimum sustained air pressure of at least 3 psig applied to the entire tank. Each Specification 51 or 56 portable tank must be tested by a minimum pressure (air or hydrostatic) of at least 2 psig or at least one and one-half times the design pressure (maximum allowable working pressure, or rerated pressure) of the tank, whichever is greater. The leakage test for portable tanks used for refrigerated liquefied gas must be performed at 90% of MAWP. Leakage tests for all other portable tanks must be at a pressure of at least 25% of MAWP. During each air pressure test, the entire surface of all joints under pressure must be coated with or immersed in a solution of soap and water, heavy oil, or other material suitable for the purpose of detecting leaks. The pressure must be held for a period of time sufficiently long to assure detection of leaks, but in no case less than five minutes. During the air or hydrostatic test, relief devices may be removed, but all the closure fittings must be in place and the relief device openings plugged. Lagging need not be removed from a lagged tank if it is possible to maintain the required test pressure at constant temperature with the tank disconnected from the source of pressure.

(2) Each Specification 60 portable tank must be retested by completely filling the tank with water or other liquid having a similar viscosity, the temperature of the liquid must not exceed 37.7 °C (100 °F) during the test, and applying a pressure of 60 psig. The portable tank must be capable of holding the prescribed pressure for at least 10 minutes without leakage, evidence of impending failure, or failure. All closures shall be in place while the test is made and the pressure shall be gauged at the top of the tank. Safety devices and/or vents shall be plugged during this test.

(3) Each Specification IM or UN portable tank, except for UN portable tanks used for non-refrigerated and refrigerated liquefied gases, and all piping, valves and accessories, except pressure relief devices, must be hydrostatically tested with water, or other liquid of similar density and viscosity, to a pressure not less than 150% of its maximum allowable working pressure. UN portable tanks used for the transportation of non-refrigerated liquefied gases must be hydrostatically tested with water, or other liquid of similar density and viscosity, to a pressure not less than 130% of its maximum allowable working pressure. UN portable tanks used for the transportation of refrigerated liquefied gases may be tested hydrostatically or pneumatically using an inert gas to a pressure not less than 1.3 times the design pressure. For pneumatic testing, due regard for protection of all personnel must be taken because of the potential hazard involved in such a test. The pneumatic test pressure in the portable tank must be reached by gradually increasing the pressure to one-half of the test pressure. Thereafter, the test pressure must be increased in steps of approximately one-tenth of the test pressure until the required test pressure has been reached. The pressure must then be reduced to a value equal to four-fifths of the test pressure and held for a sufficient time to permit inspection of the portable tank for leaks. The minimum test pressure for a portable tank is determined on the basis of the hazardous materials that are intended to be transported in the portable tanks. For liquid, solid and non-refrigerated liquefied gases, the minimum test pressure for specific hazardous materials are specified in the applicable T Codes assigned to a particular hazardous material in the §172.101 Table of this subchapter. While under pressure the tank shall be inspected for leakage, distortion, or any other condition which might render the tank unsafe for service. A portable tank fails to meet the requirements of the pressure test if, during the test, there is permanent distortion of the tank exceeding that permitted by the applicable specification; if there is any leakage; or if there are any deficiencies that would render the portable tank unsafe for transportation. Any portable tank that fails must be rejected and may not be used again for the transportation of a hazardous material unless the tank is adequately repaired, and, thereafter, a successful test is conducted in accordance with the requirements of this paragraph. An approval agency shall witness the hydrostatic or pneumatic test. Any damage or deficiency that might render the portable tank unsafe for service shall be repaired to the satisfaction of the witnessing approval agency. The repaired tank must be retested to the original pressure test requirements. Upon successful completion of the hydrostatic or pneumatic test, as applicable, the witnessing approval agency shall apply its name, identifying mark or identifying number in accordance with paragraph (k) of this section.

(i) *Rejection criteria.* When evidence of any unsafe condition is discovered, the portable tank may not be returned to service until it has been repaired and the pressure test is repeated and passed.

(j) *Repair.* The repair of a portable tank is authorized, provided such repairs are made in accordance with the requirements prescribed in the specification for the tank's original design and construction. In addition to any other provisions of the specification, no portable tank may be repaired so as to cause leakage or cracks or so as to increase the likelihood of leakage or cracks near areas of stress concentration due to cooling metal shrinkage in welding operations, sharp fillets, reversal of stresses, or otherwise. No field welding may be done except to non-pressure parts. Any cutting, burning or welding operations on the shell of an IM or UN portable tank must be done with the approval of the approval agency and be done in accordance with the requirements of this subchapter, taking into account the pressure vessel code used for the construction of the shell. A pressure testto the original test pressure must be performed after the work is completed.

(k) *Inspection and test markings.* (1) Each IM or UN portable tank must be durably and legibly marked, in English, with the date (month and year) of the last pressure test, the identification markings of the approval agency witnessing the test, when required, and the date of the last visual inspection. The marking must be placed on or near the metal identification plate, in letters and numerals of not less than 3 mm (0.118 inches) high when on the metal identification plate, and 12 mm (0.47 inches) high when on the portable tank.

(2) Each Specification DOT 51, 56, 57 or 60 portable tank must be durably and legibly marked, in English, with the date (month and year) of the most recent periodic retest. The marking must be placed on or near the metal certification plate and must be in accordance with §178.3 of this subchapter. The letters and numerals must not be less than 3 mm (0.118 inches) high when on the metal certification plate, and 12 mm (0.47 inches) high when on the portable tank, except that a portable tank manufactured under a previously authorized specification may continue to be marked with smaller markings if originally authorized under that specification (for example, DOT Specification 57 portable tanks).

(l) *Record retention.* The owner of each portable tank or his authorized agent shall retain a written record of the date and results of all required inspections and tests, including an ASME manufacturer's date report, if applicable, and the name and address of the person performing the inspection or test, in accordance with the applicable specification. The manufacturer's data report, including a certificate(s) signed by the manufacturer, and the authorized design approval agency, as applicable, indicating compliance with the applicable specification of the portable tank, must be retained in the files of the owner, or his authorized agent, during the time that such portable tank is used for such service, except for Specifications 56 and 57 portable tanks.

Reserved

PART 397

Subpart A—General

§397.1 Application of the rules in this part.

(a) The rules in this part apply to each motor carrier engaged in the transportation of hazardous materials by a motor vehicle which must be marked or placarded in accordance with §177.823 of this title and to—

(1) Each officer or employee of the motor carrier who performs supervisory duties related to the transportation of hazardous materials; and

(2) Each person who operates or who is in charge of a motor vehicle containing hazardous materials.

(b) Each person designated in paragraph (a) of this section must know and obey the rules in this part.

§397.2 Compliance with Federal motor carrier safety regulations.

A motor carrier or other person to whom this part is applicable must comply with the rules in Part 390 through 397, inclusive, of this subchapter when he/she is transporting hazardous materials by a motor vehicle which must be marked or placarded in accordance with §177.823 of this title.

§397.3 State and local laws, ordinances and regulations.

Every motor vehicle containing hazardous materials must be driven and parked in compliance with the laws, ordinances, and regulations of the jurisdiction in which it is being operated, unless they are at variance with specific regulations of the Department of Transportation which are applicable to the operation of that vehicle and which impose a more stringent obligation or restraint.

§397.5 Attendance and surveillance of motor vehicles.

(a) Except as provided in paragraph (b) of this section, a motor vehicle which contains a Division 1.1, 1.2, or 1.3 (explosive) material must be attended at all times by its driver or a qualified representative of the motor carrier that operates it.

(b) The rules in paragraph (a) of this section do not apply to a motor vehicle which contains Division 1.1, 1.2, or 1.3 material if all the following conditions exist—

(1) The vehicle is located on the property of a motor carrier, on the property of a shipper or consignee of the explosives, in a safe haven, or, in the case of a vehicle containing 50 pounds or less of a Division 1.1, 1.2, or 1.3 material, on a construction or survey site; and

(2) The lawful bailee of the explosives is aware of the nature of the explosives the vehicle contains and has been instructed in the procedures which must be followed in emergencies; and

(3) The vehicle is within the bailee's unobstructed field of view or is located in a safe haven.

(c) A motor vehicle which contains hazardous materials other than Division 1.1, 1.2, or 1.3, materials, and which is located on a public street or highway, or the shoulder of a public highway, must be attended by its driver. However, the vehicle need not be attended while its driver is performing duties which are incident and necessary to the driver's duties as the operator of the vehicle.

(d) For purposes of this section—

(1) A motor vehicle is attended when the person in charge of the vehicle is on the vehicle, awake, and not in a sleeper berth, or is within 100 feet of the vehicle and has it within his/her unobstructed field of view.

(2) A qualified representative of a motor carrier is a person who—

(i) Has been designated by the carrier to attend the vehicle;

(ii) Is aware of the nature of the hazardous materials contained in the vehicle he/she attends;

(iii) Has been instructed in the procedures he/she must follow in emergencies; and

(iv) Is authorized to move the vehicle and has the means and ability to do so.

(3) A safe haven in an area specifically approved in writing by local, State, or Federal governmental authorities for the parking of unattended vehicles containing Division 1.1, 1.2, or 1.3 materials.

(e) The rules in this section do not relieve the driver from any obligation imposed by law relating to the placing of warning devices when a motor vehicle is stopped on a public street or highway.

§397.7 Parking.

(a) A motor vehicle which contains Division 1.1, 1.2, or 1.3 materials must not be parked under any of the following circumstances—

(1) On or within 5 feet of the traveled portion of a public street or highway;

(2) On private property (including premises of a fueling or eating facility) without the knowledge and consent of the person who is in charge of the property and who is aware of the nature of the hazardous materials the vehicle contains; or

(3) Within 300 feet of a bridge, tunnel, dwelling, or place where people work, congregate, or assemble, except for brief periods when the necessities of operation require the vehicle to be parked and make it impracticable to park the vehicle in any other place.

(b) A motor vehicle which contains hazardous materials other than Division 1.1, 1.2, or 1.3 materials must not be parked on or within five feet of the traveled portion of public street or highway except for brief periods when the necessities of operation require the vehicle to be parked and make it impracticable to park the vehicle in any other place.

§397.9 [Removed and Reserved]

§397.11 Fires.

(a) A motor vehicle containing hazardous materials must not be operated near an open fire unless its driver has first taken precautions to ascertain that the vehicle can safely pass the fire without stopping.

(b) A motor vehicle containing hazardous materials must not be parked within 300 feet of an open fire.

§397.13 Smoking.

No person may smoke or carry a lighted cigarette, cigar, or pipe on or within 25 feet of—

(a) A motor vehicle which contains Class 1 materials, Class 5 materials, or flammable materials classified as Division 2.1, Class 3, Divisions 4.1 and 4.2; or

(b) An empty tank motor vehicle which has been used to transport Class 3, flammable materials or Division 2.1 flammable gases, which, when so used, was required to be marked or placarded in accordance with the rules in §177.823 of this title.

§397.15 Fueling.

When a motor vehicle which contains hazardous materials is being fueled—

(a) Its engine must not be operating; and

(b) A person must be in control of the fueling process at the point where the fuel tank is filled.

§397.17 Tires.

(a) A driver must examine each tire on a motor vehicle at the beginning of each trip and each time the vehicle is parked.

(b) If, as the result of an examination pursuant to paragraph (a) of this section, or otherwise, a tire is found to be flat, leaking, or improperly inflated, the driver must cause the tire to be repaired, replaced, or properly inflated before the vehicle is driven. However, the vehicle may be driven to the nearest safe place to perform the required repair, replacement, or inflation.

(c) If, as the result of an examination pursuant to paragraph (a) of this section, or otherwise, a tire is found to be overheated, the driver shall immediately cause the overheated tire to be removed and placed at a safe distance from the vehicle. The driver shall not operate the vehicle until the cause of the overheating is corrected.

(d) Compliance with the rules in this section does not relieve a driver from the duty to comply with the rules in §§397.5 and 397.7.

§397.19 Instructions and documents.

(a) A motor carrier that transports Division 1.1, 1.2, or 1.3 (explosive) materials must furnish the driver of each motor vehicle in which the explosives are transported with the following documents:

(1) A copy of the rules in this part;

(2) [Reserved]

(3) A document containing instructions on procedures to be followed in the event of accident or delay. The documents must include the names and telephone numbers of persons (including representatives of carriers or shippers) to be contracted, the nature of the explosives being transported, and the precautions to be taken in emergencies such as fires, accidents, or leakages.

(b) A driver who receives documents in accordance with paragraph (a) of this section must sign a receipt for them. The motor carrier shall maintain the receipt for a period of one year from the date of signature.

(c) A driver of a motor vehicle which contains Division 1.1, 1.2, or 1.3 materials must be in possession of, be familiar with, and be in compliance with

(1) The documents specified in paragraph (a) of this section;

(2) The documents specified in §177.817 of this title; and

(3) The written route plan specified in §397.67.

PLAIN LANGUAGE EXPLANATIONS

The following plain language explanations are provided to help you understand the regulations. In the plain language explanations, regulatory references are provided so you can go to and read the actual regulations. This book does not contain all of the Hazardous Materials Regulations. Most of the regulatory references to the Hazardous Materials Regulations can be found in the regulations section at the front of this book.

TRAINING

Hazmat training for each hazmat employee is one of the most important steps to take to ensure the safe movement of hazardous materials. The hazmat employer is responsible for training each hazmat employee.

Training

A hazmat employee is an individual, including a self-employed individual, who, during the course of employment performs any function subject to the Hazardous Materials Regulations. Before any hazmat employee performs any function subject to the Hazardous Materials Regulations the hazmat employee must be trained, tested and certified by the hazmat employer. (*172.702*)

An employee may perform hazmat job functions before completing hazmat training, provided:

- The employee does so under the direct supervision of a properly trained and knowledgeable hazmat employee; **and**
- The hazmat training is completed within 90 days of employment or change in job function. (*172.704(c)(1)*)

Training Content

Hazmat employee training must include the following types of training.

1. General awareness/familiarization

   This training must provide familiarity with the regulations and enable the employee to recognize and identify hazardous materials. (*172.704(a)(1)*)

2. Function-specific

   This training must provide specific information on the regulations and special permits/exemptions that are applicable to the functions performed by the employee. (*172.704(a)(2)*)

3. Safety

   This training must provide information on hazmat emergency response information; measures to protect the employee from hazards to which they may be exposed, including measures implemented to protect the employee from exposure; and methods and procedures for avoiding accidents. (*172.704(a)(3)*)

   Employees who repair, modify, recondition, or test packagings may be excepted from safety training. (*172.704(e)*)

4. Security awareness

   This training must provide an awareness of security risks associated with hazardous materials transportation and methods designed to enhance transportation security. This training must also include a component covering how to recognize and respond to possible security threats. (*172.704(a)(4)*)

5. In-depth security

   Each hazmat employee of a person/company that is required to have a security plan must be trained on the security plan and its implementation if they handle materials covered by the plan, perform a function related to the materials covered by the plan, or are responsible for implementing the plan. This training must include company security objectives, organizational security structure, specific security procedures, specific security duties and responsibilities for each employee, and specific actions to be taken in the event of a security breach. (*172.704(a)(5)*)

6. Modal specific

   In addition to the above five types of training, modal specific training requirements may be required for the individual modes of transportation (air, rail, highway, or vessel). (*172.700(c)*)

   For example, by highway, drivers must also be trained on the safe operation of the motor vehicle in which they operate, or intend to operate, and the applicable requirements of the Federal Motor Carrier Safety Regulations. (*177.816*)

Recurrent Training

Recurrent training is required once every three years. (*172.704*)

However, if a new regulation is adopted or an existing regulation is changed, that relates to a function performed by a hazmat employee, that employee must be instructed on the new or revised regulations. This training must be completed before the employee performs the function and before the employee's three year recurrent training. The employee only needs to be instructed on the new or revised requirements. Testing and recordkeeping for the new or revised requirements are not required until the employee's three year recurrent training.

For in-depth security training, if the employer's security plan is revised during the three-year recurrent training cycle, the hazmat employee must be trained within 90 days of implementation of the revised security plan. (*172.704(c)(2)*)

Training Record

Hazmat training for each hazmat employee must be documented by the employer. A record of current hazmat training, include the preceding three years, must be created and retained by the hazmat employer for each hazmat employee. The hazmat employee's training record must include:

- The employee's name;
- The most recent completion date of the employee's training;
- A description, copy, or location of the training materials used;
- The name and address of the person providing the training; and
- Certification that the employee has been trained and tested.

This record must be retained for as long as the hazmat employee is employed by the hazmat employer and for 90 days thereafter. (*172.704(d)*)

SECURITY

SECURITY

Security Plans

Applicability

Each person/company that offers for transportation in commerce or transports in commerce one or more of the following must develop and adhere to a security plan for hazardous materials. Used below, "large bulk quantity" means a quantity greater than 3,000 Kg (6,614 pounds) for solids or 3,000 liters (792 gallons) for liquids and gases in a single packaging.

1. Any quantity of a Division 1.1, 1.2, or 1.3 material;
2. A quantity of a Division 1.4, 1.5, or 1.6 material requiring placarding;
3. A large bulk quantity of Division 2.1 material;
4. A large bulk quantity of Division 2.2 material with a subsidiary hazard of 5.1;
5. Any quantity of a material poisonous by inhalation;
6. A large bulk quantity of a Class 3 material meeting the criteria for Packing Group I or II;
7. A quantity of a desensitized explosives meeting the definition of a Division 4.1 or Class 3 material requiring placarding in accordance with §172.504(c);
8. A large bulk quantity of a Division 4.2 material meeting the criteria for Packing Group I or II;
9. Any quantity of a Division 4.3 material;
10. A large bulk quantity of a Division 5.1 material in Packing Groups I and II; perchlorates; or ammonium nitrate, ammonium nitrate fertilizers, or ammonium nitrate emulsions, suspensions, or gels;
11. Any quantity of organic peroxide, Type B, liquid or solid, temperature controlled;
12. A large bulk quantity of Division 6.1 material;
13. A select agent or toxin regulated by the Centers for Disease Control and Prevention under 42 CFR part 73 or the United States Department of Agriculture under 9 CFR part 121;
14. A quantity of uranium hexafluoride requiring placarding under §172.505(b);
15. International Atomic Energy Agency (IAEA) Code of Conduct Category 1 and 2 materials including Highway Route Controlled quantities as defined in 49 CFR 173.403 or known radionuclides in forms listed as RAM-QC by the Nuclear Regulatory Commission;
16. A large bulk quantity of Class 8 material meeting the criteria for Packing Group I.

Exceptions

The transportation activities of a farmer, who generates less than $500,000 annually in gross receipts from the sale of agricultural commodities or products, are not subject to the security plan requirements if such activities are:

1. Conducted by highway or rail;
2. In direct support of their farming operations; and
3. Conducted within a 150-mile radius of those operations. (*172.800*)

The security plan requirements do not apply to combustible liquids. Combustible liquids that are not a hazardous substance, hazardous waste, or marine pollutant, in non-bulk packaging are not subject to the Hazardous Materials Regulations (this includes security plans). In addition, the security plan requirements in Part 172, Subpart I are not listed as required compliance for combustible liquids in bulk packaging or combustible liquids that are also a hazardous substance, hazardous waste, or marine pollutant. (*173.150(f)*)

Plan content

Each plan must include an assessment of possible transportation security risks for shipments of hazardous materials including site-specific or location-specific risks associated with facilities at which the materials are prepared for transport, stored, or unloaded incidental to movement, and appropriate measures to address them. Specific measures may vary depending upon the level of threat. At a minimum, a security plan must consist of the following elements:

- *Personnel security* — Measures to confirm information provided by job applicants hired for positions that involve access to and handling of hazardous materials covered by the security plan. These measures must be consistent with applicable Federal and State laws and requirements concerning employment practices and individual privacy.
- *Unauthorized access* — Measures to address the assessed risk that unauthorized persons may gain access to hazmat covered by the security plan or transport conveyances being prepared for transportation of materials covered by the security plan.
- *En route security* — Measures to address the assessed security risks of shipments of hazmat covered by the security plan en route from origin to destination, including shipments stored incidental to movement. (*172.802*)

In addition, a security plan must also include:

- Identification by job title of the senior management official responsible for overall development and implementation of the security plan.
- Security duties for each position or department that is responsible for implementing the plan or a portion of the plan and the process of notifying employees when specific elements of the plan must be implemented.
- A plan for training hazmat employees in accordance with §172.704 on security awareness and in-depth security training.

Sensitive security information

Security plans must be marked as Sensitive Security Information (SSI) as required by 49 CFR 15.13. Security plans on paper must include the protective marking “SENSITIVE SECURITY INFORMATION” at the top and the distribution limitation warning at the bottom of:

- The front and back cover, including a binder cover or folder;
- Any title page; and
- Each page of the document.

The distribution limitation warning must read as follows:

WARNING: This record contains Sensitive Security Information that is controlled under 49 CFR parts 15 and 1520. No part of this record may be disclosed to persons without a “need to know”, as defined in 49 CFR parts 15 and 1520, except with the written permission of the Administrator of the Transportation Security Administration or the Secretary of Transportation. Unauthorized release may result in civil penalty or other action. For U.S. government agencies, public disclosure is governed by 5 U.S.C. 552 and 49 CFR parts 15 and 1520.

Non-paper security plans must be clearly and conspicuously marked with the protective marking and distribution limitation warning so that a viewer or listener is reasonably likely to see or hear them when obtaining access to the security plan. (*15.13*)

Plan retention/revisions

The security plan, including the risk assessment, must be in writing and must be retained as long as it remains in effect. Copies must be made available to those employees who are responsible for implementing it, consistent with personnel security clearance or background investigation restrictions and demonstrated need to know. (*172.802*)

The security plan must be reviewed at least annually and revised and/or updated as necessary to reflect changes in circumstances. When the plan is updated or revised, all employees responsible for implementing it must be notified and all copies of the plan must be maintained as of the date of the most recent revision. (*172.802*)

A copy of the security plan or an electronic file must be maintained at the employer’s principal place of business and must be made available upon request, at a reasonable time and location, to an authorized DOT or DHS official. (*172.802*)

Additional rail requirements

Each rail carrier transporting one or more of the following materials is subject to additional safety and security planning requirements:

- A highway route-controlled quantity of a Class 7 (radioactive) material;
- More than 2,268 kg (5,000 pounds) in a single carload of Division 1.1, 1.2, or 1.3 explosives; or
- A quantity of a material poisonous by inhalation in a single bulk packaging. (*172.820*)

The additional safety and security planning requirements include:

- Compiling commodity data from the previous year;
- Analyzing the safety and security risks for the transportation route;
- Identifying practicable alternative routes over which the carrier has authority to operate;
- Selecting the route to be used based on the analysis of the above two route requirements;
- Completing the above requirements by September 1, 2009 and in subsequent years no later than the end of the calendar year following the year to which the analyses apply. (*172.820*)

Each rail carrier must ensure the safety and security plan it develops and implements includes all of the following:

1. A procedure under which the rail carrier must consult with offerors and consignees in order to develop measures for minimizing the duration of any storage of the material incidental to movement.
2. Measures to prevent unauthorized access to the materials during storage or delays in transit.
3. Measures to mitigate risk to population centers associated with in-transit storage.
4. Measure to be taken in the event of an escalating threat level for materials stored in transit.
5. Procedures for notifying the consignee in the event of a significant delay during transportation. (*172.820*)

Each rail carrier must maintain the above information so it is accessible at or through its principal place of business and retain this information for a minimum of two years. (*172.820*)

Security Training

In-depth security training

Each hazmat employee of a person/company that is required to have a security plan must be trained on the security plan and its implementation. This training must include company security objectives, organizational security structure, specific security procedures, specific security duties and employee responsibilities, and actions to be taken in the event of a security breach. (*172.704(a)(5)*)

Security awareness training

Every hazmat employee must receive security awareness training. This training must provide an awareness of security risks associated with hazardous materials transportation and methods designed to enhance transportation security. This training must also include a component covering how to recognize and respond to possible security threats. (*172.704(a)(4)*)

Reserved

CLASSIFICATION

The first step in shipping a material is determining what it is and if it is subject to the Hazardous Materials Regulations. Is it a hazardous material?

The regulations define a hazardous material as a material which is "capable of posing an unreasonable risk to health, safety, and property when transported in commerce." A material is considered "hazardous" if it meets one or more of the hazard class definitions found in the Hazardous Materials Regulations, and/or is a hazardous substance, hazardous waste, marine pollutant, or elevated-temperature material.

To determine if the material is a hazardous material you need to know:

1. The hazard classifications in the Hazardous Materials Regulations;
2. Criteria for hazardous substances, hazardous wastes, marine pollutants, and elevated temperature materials; and
3. The hazard(s) of the material.

Hazard Classifications

The regulations define nine hazard classes. Some of the classes are further subdivided into divisions. There is also a category of hazardous materials known as "Other Regulated Material" or "ORM-D."

A material is considered "hazardous" if it meets one or more of the Hazardous Materials Regulations hazard class definitions listed below.

Class No.	Division No.(if any)	Name of class or division	49 CFR reference for definitions
None	....	Forbidden Materials	173.21
None	....	Forbidden Explosives	173.54
1	1.1	Explosives (with a mass explosion hazard)	173.50
1	1.2	Explosives (with a projection hazard)	173.50
1	1.3	Explosives (with predominantly a fire hazard)	173.50
1	1.4	Explosives (with no significant blast hazard)	173.50
1	1.5	Very insensitive explosives; blasting agents	173.50
1	1.6	Extremely insensitive detonating substances	173.50
2	2.1	Flammable gas	173.115
2	2.2	Non-flammable compressed gas	173.115
2	2.3	Poisonous gas	173.115
3	....	Flammable and combustible liquid	173.120
4	4.1	Flammable solid	173.124
4	4.2	Spontaneously combustible material	173.124
4	4.3	Dangerous when wet material	173.124
5	5.1	Oxidizer	173.127
5	5.2	Organic peroxide	173.128
6	6.1	Poisonous materials	173.132
6	6.2	Infectious substance (Etiologic agent)	173.134
7	....	Radioactive material	173.403
8	....	Corrosive material	173.136
9	....	Miscellaneous hazardous material	173.140
None	....	Other regulated material: ORM-D	173.144

Some hazardous materials are further subdivided into packing groups. The packing group indicates the degree of danger presented by the hazardous material.

Packing Group I — great danger
Packing Group II — medium danger
Packing Group III — minor danger

Packing groups are assigned to most hazardous materials, except Class 2, Class 7, ORM-D materials, and some Division 6.2 and Class 9 materials.

Other Criteria

A material may be regulated as a hazardous material even though it does not meet the definition of any of the first eight hazard classes. These materials fall into Class 9 (miscellaneous). A Class 9 material does not meet the definition of any other hazard classification (1-8), but is regulated as a hazardous material because it is a hazardous substance, hazardous waste, marine pollutant, or elevated-temperature material.

A **Hazardous Substance** is listed in Appendix A to §172.101 and is in a quantity in one package that meets or exceeds the reportable quantity (RQ). (*171.8*)

A **Hazardous Waste** is a material subject to the waste manifest requirements of the Environmental Protection Agency as specified in 40 CFR 262. (*171.8*)

A **Marine Pollutant** is listed in Appendix B to §172.101 and is packaged in a concentration which equals or exceeds 10% for marine pollutants and 1% for severe marine pollutants (*171.8*). Marine pollutants are always regulated by vessel. They are only regulated by air, rail or highway when in bulk packagings. (*171.4*)

An **Elevated Temperature Material** is a material in a bulk package that is:

- In a liquid phase at a temperature at or above 100°C (212°F); or
- In a liquid phase with a flash point at or above 37.8°C (100°F) that is intentionally heated above its flash point; or
- In a solid phase and at a temperature at or above 240°C (464°F). (*171.8*)

A material that meets the criteria of one of the first eight hazard classes (1-8) may also meet the definition of a hazardous substance, hazardous waste, marine pollutant, or elevated-temperature material. These materials would be regulated with additional requirements specifically for the additional hazard(s) they present.

Hazard of the Material

You must know the hazard or hazards of the material you want to ship. For some materials this is easy since the material and its hazard may be well known. For example, Gasoline is a well known flammable liquid.

For other materials, the hazard may not be known or there may be some question as to what the hazard is. These materials need to be tested to determine their hazard or hazards. This is often true for new products, mixtures and solutions.

An important part of hazard classification is packing groups. For some materials the hazard class or division may be known, such as Class 3, but to properly package and ship the material you need to know the packing group. Testing may be required to determine if it is a packing group I, II, or III material.

Another part of hazard classification is determining if the material is a hazardous substance, hazardous waste, marine pollutant, or elevated-temperature material. Always check to see if your material meets any of these definitions.

If a material has more than one hazard, refer to §173.2a to determine the primary hazard for the material. In some situations additional requirements are mandated for certain subsidiary hazards, such as poison inhalation hazard.

If the hazard of a material is uncertain and must be determined by testing, a sample of the material may be shipped for testing by using a tentative proper shipping name, hazard class, identification number, and packing group based on the shipper's knowledge of the material and the criteria in the regulations. (*172.101(c)(11)*)

Once the hazard classification of the material is known, you can go to the §172.101 Hazardous Materials Table to select a description/proper shipping name that best describes the hazardous material.

HAZARDOUS MATERIALS TABLE

Once you know the hazard(s) of your material you can go to the §172.101 Hazardous Materials Table and pick a description/proper shipping name to describe your material.

The Hazardous Materials Table does not contain the names of all the materials that are hazardous; but it does contain all the descriptions/proper shipping names that can be used to describe a hazardous material. You can only use the descriptions/names that are listed in the Table. You must select the description/name that best describes the hazardous material.

The Hazardous Materials Table provides the majority of the information needed to properly prepare a hazardous material for shipment. This information is presented in the 10 columns of the Hazardous Materials Table.

Symbols	Hazardous materials descriptions and proper shipping names	Hazard class or Division	Identification Numbers	PG	Label Codes	Special Provisions (§172.102)	(8) Packaging (§173.***)			(9) Quantity limitations (see §§173.27 and 175.75)		(10) Vessel stowage	
							Exceptions	Non-bulk	Bulk	Passenger aircraft/rail	Cargo aircraft only	Location	Other
(1)	(2)	(3)	(4)	(5)	(6)	(7)	(8A)	(8B)	(8C)	(9A)	(9B)	(10A)	(10B)
	Aerosols, *poison, (each not exceeding 1 L capacity)*	2.2	UN1950		2.2, 6.1		306	None	None	Forbidden	Forbidden	A	48, 87, 126
I	**Air bag inflators,** *or* **Air bag modules,** *or* **Seat-belt pretensioners**	1.4G	UN0503	II	1.4G	161	None	62	None	Forbidden	75 kg	02	
	Air bag inflators, *or* **Air bag modules,** *or* **Seat-belt pretensioners**	9	UN3268	III	9	160	166	166	166	25 kg	100 kg	A	
	Air, compressed	2.2	UN1002		2.2	78	306, 307	302	302	75 kg	150 kg	A	
	Air, refrigerated liquid, *(cryogenic liquid)*	2.2	UN1003		2.2, 5.1	T75, TP5, TP22	320	316	318, 319	Forbidden	Forbidden	D	51

The columns from left to right are:

Column 1: Symbols

Column 1 uses six symbols to identify hazardous materials which have special shipping conditions - such as restrictions for air, domestic, international, or water vessel transport

Symbol	Meaning
+	Fixes the proper shipping name, hazard class, and packing group without regard to whether the material meets the definition of that class or packing group, or meets any other hazard class definition.
A	Restricts the application of the requirements to materials offered for transportation by aircraft - unless the material is a hazardous substance or hazardous waste.
D	Identifies proper shipping names which are appropriate for domestic transportation, but which may be inappropriate for international transportation.
G	Identifies proper shipping names for which one or more technical names must be entered in parentheses, in association with the basic description.
I	Identifies proper shipping names which are appropriate for international transportation. An alternate proper shipping name may be selected when only domestic transportation is involved.
W	Restricts the application of the requirements to materials offered for transportation by vessel - unless the material is a hazardous substance or hazardous waste.

Column 2: Hazardous Materials Descriptions and Proper Shipping Names

Column 2 provides descriptions/proper shipping names that can be used to describe hazardous materials. Only the names shown in Roman type (not italics) are proper shipping names. Although the words in italics are not part of the proper shipping name they may be used in addition to the proper shipping name. Punctuation marks are not part of the proper shipping name, but may be used in addition to the proper shipping name.

To determine the proper shipping name, locate the material's technical name in Column 2 of the table. A technical name is a recognized chemical name or microbiological name currently used in scientific and technical handbooks, journals, and texts, such as Acetone or Sodium Peroxide.

HAZARDOUS MATERIALS TABLE

If you cannot find the material's technical name, select a generic or n.o.s. name that most accurately describes the material.

Examples: A material not listed by its technical name which meets the Class 3 (flammable liquid) material classification may be best described by Flammable liquid, n.o.s. If the material is an alcohol which is not listed by name, Alcohol, n.o.s. may be more accurate.

Next, make certain that the hazard class shown in Column 3 for the proper shipping name corresponds to the classification of the material.

Example: Compounds, cleaning liquid has two entries in the table. One entry is listed as a Class 8 material, the other as a Class 3 material.

When selecting a proper shipping name be sure to watch for an I or D in Column 1 of the Table. The hazard class for the same shipping name can be different for domestic and international entries.

I, Ammonia anhydrous, 2.3, UN1005
D, Ammonia anhydrous, 2.2, UN1005

Certain n.o.s. and generic proper shipping names in column 2 of the table are required to be supplemented with technical names. The letter "G" in column 1 identifies instances in which you are required to enter a technical name or names in parentheses with the basic description.

Example: UN1993, Flammable liquids, n.o.s. (benzene), 3, II

Proper shipping names may be singular or plural, and may be written in either upper or lower case letters. Proper shipping names may be spelled as they appear in the Table or in the same manner as in the IMDG Code or the ICAO Technical Instructions.

Example: Sulfuric acid (Hazmat Table) or Sulphuric acid (IMDG)

The word "*or*" in italics indicates that either description in Roman type may be used as the proper shipping name.

Example: Lithium hydroxide, monohydrate *or* Lithium hydroxide, solid

When one entry references another by the use of the word "see", and both names are in Roman type, either name may be used as the proper shipping name.

Example: Ethyl alcohol, *see* Ethanol

When the proper shipping name includes a concentration range as part of the description, the actual concentration may be used in place of the range.

Hazardous wastes must be identified by the most appropriate proper shipping name. If the shipping name does not include the word "waste" it must be included before the shipping name.

Example: Waste Acetone

A mixture or solution not identified specifically by name - that consists of a single predominant hazardous material identified in the table by technical name and one or more hazardous and/or non-hazardous material must be described using the proper shipping name of the hazardous material and the qualifying word “mixture” or “solution”, as appropriate. In §172.101(c)(10) some restrictions apply, such as the hazard class, packing group, or subsidiary hazard of the mixture or solution must be the same as the hazardous material identified in the table.

Example: The proper shipping name for a solution of Brucine and a non-regulated material could be Brucine solution.

When a material meets the definition of a hazard class or packing group other than that shown in Columns 3 and 5, respectively, or does not meet the subsidiary hazard(s) shown in Column 6, the material must be described by a more appropriate proper shipping name - one that lists the correct hazard class, packing group, or subsidiary hazard(s) of the material.

However, if the proper shipping name is preceded by a plus (+) in Column 1 the proper shipping name, hazard class, and packing group are fixed, even if the material does not meet the definition of the class, packing group or any other hazard class.

Column 3: Hazard Class or Division

Column 3 lists the hazard class or division which corresponds to the proper shipping name. When the material is too hazardous to be transported, the word “Forbidden” will be shown. (This prohibition does not apply if the material is diluted, stabilized, or incorporated in a device and is classified according to the hazard class definitions.)

Since the hazard class or division will impact how a material is packaged and labeled, it is important that the one listed for the selected proper shipping name matches the material being transported. This becomes an issue when more than one hazard class or division is listed for a given proper shipping name.

Example: Paint related material is listed as both a Class 8 (corrosive) material and a Class 3 (flammable liquid) material. If you look across the table, you will notice that the labels, special provisions, authorized packagings, and quantity limitations differ between the two entries.

If the proper shipping name is preceded by a plus (+) in Column 1 the proper shipping name, hazard class, and packing group are fixed even if the material does not meet the definition of the class, packing group or any other hazard class.

Column 4: Identification Numbers

Column 4 provides the material’s UN, NA, or ID identification number. Numbers preceded by the letters “UN” are appropriate for both international and domestic transportation. Those preceded by “NA” are for domestic transportation and not for international transportation - except to and from Canada. Those preceded by “ID” are for use with the ICAO Technical Instructions for air transport.

The NA9000 series of identification numbers are used with proper shipping names which are not appropriately covered by the international regulations, or not appropriately addressed by the international standards for emergency response information purposes.

Column 5: PG (Packing Group)

Column 5 lists packing groups which correspond to the proper shipping name and hazard class of the material. No packing groups are assigned to Class 2, Class 7, ORM-D materials, and some Division 6.2 and Class 9 materials.

When more than one packing group is listed for a given proper shipping name, the correct one must be determined using the criteria detailed in Subpart D of Part 173.

HAZARDOUS MATERIALS TABLE

If a material is a hazardous waste or a hazardous substance, and the proper shipping name is preceded in Column 1 by the letter "A" or "W", the packing group is modified to read "III" when offered for transportation or transported by a mode for which its transportation is not otherwise regulated.

If the proper shipping name is preceded by a plus (+) in Column 1 the proper shipping name, hazard class, and packing group are fixed even if the material does not meet the definition of the class, packing group or any other hazard class.

Column 6: Label Codes

Column 6 identifies the label codes which represent the hazard warning label(s) that must be applied to the material's packaging - unless the material is excepted from the labeling requirements. When more than one label code is listed, the first code shown indicates the material's primary hazard. Additional label codes indicate subsidiary hazards.

If a material has more than one hazard, all applicable subsidiary labels may not be listed in Column 6. In such cases, consult §172.402.

Column 7: Special Provisions

Column 7 contains special provisions or instructions specific to the hazardous material. The codes listed in this column are defined in Section 172.102. Provisions referenced by a number only are multi-modal, and may apply to both bulk and non-bulk packagings. Provisions coded by letters are applicable as follows

Code	Applies to
A	Transportation by aircraft
B	Bulk packagings, other than UN or IM portable tanks or Intermediate bulk containers
IB	Intermediate bulk containers
IP	Intermediate bulk containers
N	Non-bulk packagings
R	Transportation by rail
T	UN or IM portable tanks
TP	UN or IM portable tanks
W	Transportation by water

Column 8: Packaging

Column 8 contains three columns of packaging authorizations: exceptions (8A), non-bulk (8B), and bulk (8C). The numerical references in these columns are to sections within Part 173 which list the applicable packagings or packaging exceptions. If the word "None" is listed, that type of packaging or exception is not authorized - except as may be provided by special provisions in Column 7.

Column 9: Quantity Limitations

Column 9 lists quantity limitations for passenger-carrying aircraft or rails cars (9A), and cargo aircraft only (9B). The limits listed are the maximum quantities that can be offered for transport in a single packaging. If the word "Forbidden" is listed the material may not be offered or transported in the applicable mode.

Column 10: Vessel Stowage

Column 10 identifies the authorized storage locations on board both passenger and cargo vessels (10A), and specifies other stowage requirements for specific hazardous materials (10B).

PACKAGING

The Hazardous Materials Regulations have very specific requirements for all hazmat packagings. These requirements are designed to ensure that the packaging is appropriate for the material and that it can withstand the conditions normally encountered in transport. The responsibility for packaging a hazardous material rests with the individual who offers the material for transport.

Packaging can be any container authorized by the regulations to contain a hazardous material. This includes boxes, drums, cylinders, portable tanks, intermediate bulk containers, and cargo tanks.

Once a description/proper shipping name is selected in the Hazardous Materials Table packaging can be determined by going to Column 8. Column 8 is divided into three columns: 8A-Exceptions, 8B-Non-bulk and 8C-Bulk. Column 8A-Exceptions should be checked for any exceptions to the regulations. Columns 8B-Non-bulk and 8C-Bulk should be checked to determine what packagings are authorized for the material. All the numbers listed in these columns reference sections in Part 173. If "202" is listed you would go to §173.202 to find information on the packaging. Any special provisions listed in Column 7 of the Table should also be checked for packaging applications.

In addition to the packaging requirements referenced in the Hazardous Materials Table, all packagings must be in compliance with the general packaging requirements found in 173.24 and 173.24a or 173.24b. These requirements include acceptable condition for transport, compatibility, mixed contents, closures, venting, filling limits, etc.

Terms

The following key terms will help in the understanding of packaging.

Bulk packagings, other than vessels or barges, and including transport vehicles or freight containers, are packagings in which hazardous materials are loaded with no intermediate form of containment and which have:

- A maximum capacity greater than 450 L (119 gals) as a receptacle for a liquid;
- A maximum net mass greater than 400 kg (882 lbs) and a maximum capacity greater than 450 L (119 gals), as a receptacle for a solid; or
- A water capacity greater than 454 kg (1,000 lbs), as a receptacle for a gas.

A Large Packaging in which hazardous materials are loaded with an intermediate form of containment, such as one or more articles or inner packagings, is also a bulk packaging. (*171.8*)

Combination packaging consists of one or more inner packagings secured in a non-bulk packaging. It does not include a composite packaging. (*171.8*)

Composite packaging consists of an outer packaging and an inner receptacle forming an integral packaging. Once assembled it remains an integrated single unit. (*171.8*)

Large packaging is a packaging that:

- Consists of an outer packaging that contains articles or inner packagings;
- Is designated for mechanical handling;
- Exceeds 400 kg net mass or 450 liters (118.9 gallons) capacity;
- Has a volume of not more than 3 cubic meters (m^3); and
- Conforms to the construction, testing and marking requirements specified in subparts P and Q of part 178. (*171.8*)

Non-bulk packagings have:

- A maximum capacity of 450 L (119 gals) or less as a receptacle for a liquid;
- A maximum net mass of 400 kg (882 lbs) or less and a maximum capacity of 450 L (119 gals) or less as a receptacle for a solid; or
- A water capacity of 454 kg (1,000 lbs) or less as a receptacle for a gas. (*171.8*)

Overpack is an enclosure that is used by a single consignor to provide protection or convenience in handling of a package or to consolidate two or more packages. (*171.8*)

Package is a packaging with its contents. (*171.8*)

PACKAGING

Packaging is a receptacle and any other components needed to meet the packaging requirements. (*171.8*)

Packing group indicates the degree of danger presented by the hazardous material.

Packing Group I — great danger
Packing Group II — medium danger
Packing Group III — minor danger (*171.8*)

Salvage packaging is a special packaging conforming to §173.3. Damaged, defective, leaking packages or hazardous materials that have spilled or leaked are placed into this packaging for transport for recovery or disposal. (*171.8*)

Manufacturer's UN Packaging Marking

Most non-bulk packagings must meet the standards in the UN Recommendations. The Manufacturer's UN packaging marking is placed on every packaging that has been manufactured and tested to meet a UN standard. (*178.503*) To understand the marking you must know what the codes mean.

Packaging codes

Codes that contain only one capital letter indicate single packagings, and can be deciphered as follows:

The first numeral designates the kind of packaging:

1 = Drum
2 = Wooden barrel
3 = Jerrican
4 = Box
5 = Bag
6 = Composite packaging
7 = Pressure receptacle (*178.502*)

The letter indicates the material of construction:

A = Steel (all types and surface treatments)
B = Aluminum
C = Natural wood
D = Plywood
F = Reconstituted wood
G = Fiberboard
H = Plastic
L = Textile
M = Paper, multi-wall
N = Metal (other than steel and aluminum)
P = Glass, porcelain, or stoneware (*178.502*)

A second numeral indicates the category of packaging within the packaging:

1 = Non-removable head (for drums)
2 = Removable head (for drums) (*178.502*)

Examples: "1A1" indicates a steel drum with a non-removable head.
"4D" indicates a plywood box.

Codes that contain two capital letters indicate composite packagings. The first letter designates the material for the inner receptacle, and the second, the material for the outer packaging. (178.502)

Example: "6HA1" would be a plastic receptacle in a steel drum with a non-removable head.

Performance levels

X — for packagings meeting Packing Group I, II and III tests.
Y — for packagings meeting Packing Group II and III tests.
Z — for packagings meeting only Packing Group III tests. (*178.503*)

Solids or inner packaging

S = packagings intended to contain solids or inner packagings. (*178.503*)

Content and sequence

The manufacturer's marking must include, in the following order:

UN symbol;
Packaging code;
Performance level;
Gross mass or specific gravity;
Solids or inner packagings, or hydrostatic test pressure;
Year of manufacture;
Country of authorization and marking; and
Manufacturer symbol or name and address.

Minimum thickness is required for some packagings intended for reuse or reconditioning. (*178.503*)

Examples:

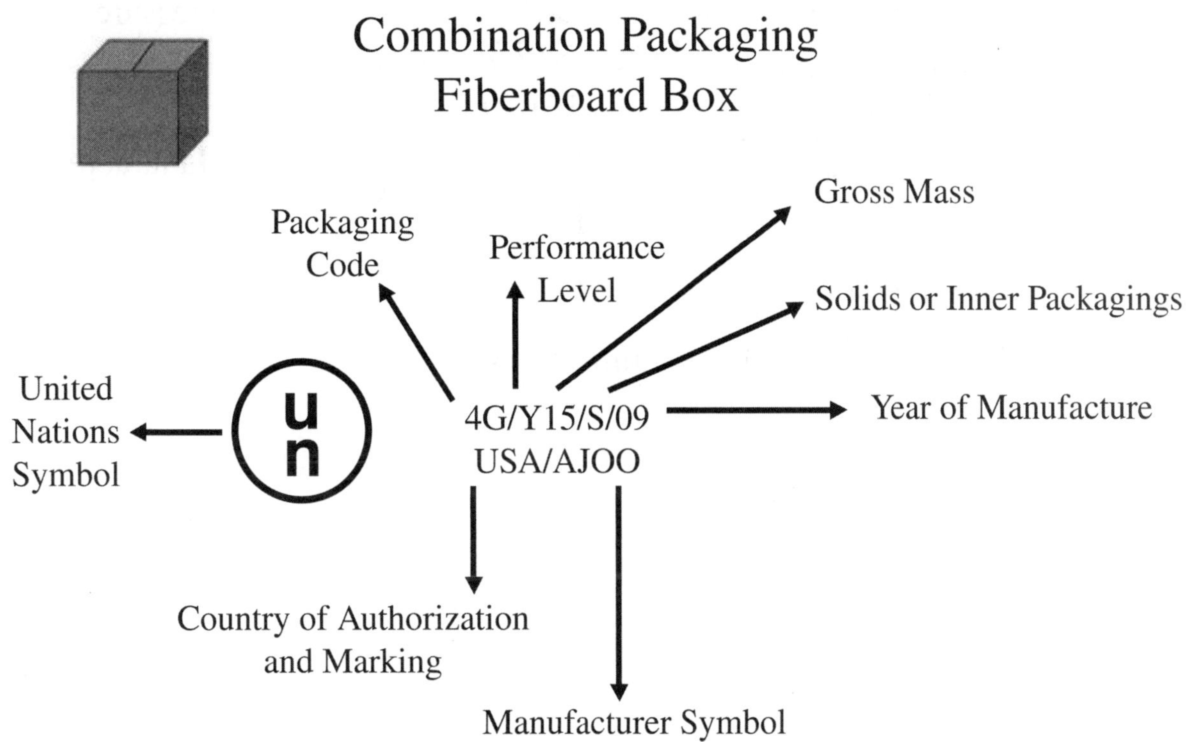

PACKAGING

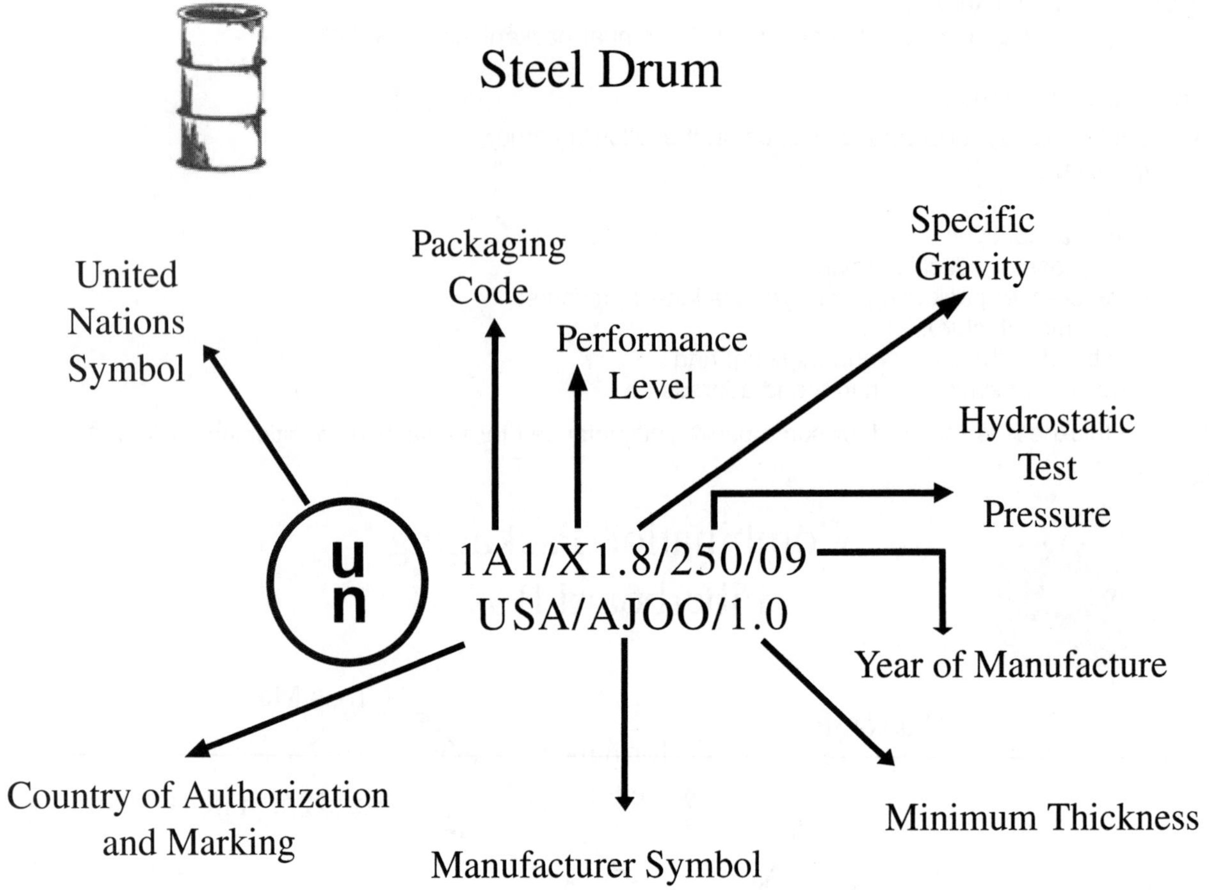

Steel Drum
United Nations Symbol
Packaging Code
Performance Level
Specific Gravity
Hydrostatic Test Pressure
u n
1A1/X1.8/250/09
USA/AJOO/1.0
Year of Manufacture
Country of Authorization and Marking
Manufacturer Symbol
Minimum Thickness

SHIPPING PAPERS

A "shipping paper" is a document used to identify the freight being offered for transportation. This term covers any shipping order, bill of lading, waybill, manifest or other document serving a similar purpose.

A "waste manifest" is a shipping paper prepared on a prescribed form on which all hazardous wastes must be identified. The original hazardous waste manifest is required to accompany each shipment from the point of pickup to its final destination and appropriate copies of the document must be properly distributed. Copies are provided for the generator, transporter and treatment, storage or disposal (TSD) facility.

It is the responsibility of the shipper to properly prepare the shipping paper tendered with the shipment of hazardous materials to the initial carrier. (*172.200*) The carrier, in turn, must be certain that the shipping paper is properly prepared prior to accepting the shipment. (*177.817*)

With a few exceptions, a shipping paper is required to accompany each shipment of hazardous materials during transportation. This requirement may be met by a photocopy of the shipping paper or by a carrier prepared shipping paper such as a freight bill or waybill. For shipments of hazardous waste the original shipper prepared hazardous waste manifest is the only authorized documentation. (*40 CFR 262.20*)

Exceptions

Shipping papers are not required for a material, unless it is a hazardous substance, hazardous waste, or marine pollutant, that:

- Has "A" in Col. (1) of the hazmat table, except when offered/transported by air;
- Has "W" in Col. (1) of the hazmat table, except when offered/transported by water;
- Is a limited quantity, except when offered/transported by air or water;
- Before January 1, 2014, an ORM-D material when offered for transport by highway or rail; or
- Is a Category B infectious substance prepared in accordance with §173.199. (*172.200*)

In addition, shipping papers are not required for materials transported as Materials of Trade. (*173.6*)

Shipping Paper Retention

Hazardous materials shipping papers must be retained by the **carrier** for **one year** after the material is accepted (*177.817*). Hazardous materials shipping papers must be retained by the **shipper** for **two years** after the material is accepted by the initial carrier *(172.201)*. **Hazardous waste** manifests (shipping papers) must be retained for **three years** after the material is accepted by the initial carrier *(172.205)*.

The hazardous materials shipping paper that is retained may be a paper copy or an electronic image. The date of acceptance by the carrier must be included on the retained shipping paper. (*172.201, 177.817*)

Display on Papers

Hazardous materials are required to be indicated on shipping papers in one of three ways when non-hazardous materials are also listed.

1. The hazardous materials must be listed **first**, before any non-hazardous materials, or
2. the hazardous materials must be entered in a **color** that clearly contrasts with non-hazardous materials entries (hazardous materials entries may be highlighted only on reproductions of the shipping paper), or
3. the hazardous materials must be identified with an "**X**" in a column captioned "HM" placed before the basic shipping description. (*172.201*)

Description

Each hazardous material that is offered for transport must be clearly described on the shipping paper using the applicable information from the Hazardous Materials Table.

This shipping description must include the material's:

Identification number;
Proper shipping name;
Hazard class or division number;
Subsidiary Hazard Class(es) or Division number(s) entered in parentheses, if required;
Packing group, if any;

Total quantity (mass or volume or activity for Class 7 or net explosive mass for Class 1) and unit of measurement; and

Number and type of packages. (*172.202*)

Example: 2 drums, UN1717, Acetyl chloride, 3, (8), II , 180 lbs.

Subsidiary hazard Classes or Divisions must be included, in parenthesis, when subsidiary labels are required. Unless excepted, a subsidiary label is required when indicated in Column 6 (Label Codes) of the Hazardous Materials Table or by §172.402.

Example: UN1463, Chromium trioxide, anhydrous, 5.1, (6.1, 8), II

The total quantity is **not** required for hazardous material packaging containing only residue (i.e., one that has not been cleaned and purged or refilled with a non-hazardous material), for cylinders, and for bulk packagings. However, some indication of total quantity must be shown for cylinders and bulk packagings (such as 10 cylinders or 1 cargo tank). (*172.202*)

The first five (5) items — often referred to as the material's **basic description** — must be shown in sequence, with no additional information interspersed unless authorized by the regulations. The identification number must include the letters "UN" or "NA", as appropriate. The packing group must be shown in Roman numerals and may be preceded by the letters "PG". (*172.202*)

Examples: UN1203, Gasoline, 3, PG II
UN2359, Diallylamine, 3, (6.1, 8), II

The basic description may be shown in an alternate sequence, with the proper shipping name listed first. This sequence may continue to be used until January 1, 2013. (*171.14(e)*)

Examples: Gasoline, 3, UN1203, PG II
Diallylamine, 3, (6.1, 8), UN2359, II

Technical Names

Proper shipping names identified by the letter "G" in Column (1) of the Hazardous Materials Table must include technical names in parentheses. Most n.o.s. and other generic proper shipping names must have the technical name of the hazardous material entered in parentheses in association with the basic description. The regulations allow that if a "technical name" is required, it may be entered between the proper shipping name and the hazard class or following the basic description.

Organic peroxides which may qualify for more than one generic listing, depending on concentration, must include with the technical name the actual concentration being shipped or the concentration range for the appropriate generic listing.

A Division 6.2 material with the ID number UN2814 or UN2900 that is suspected to contain an unknown Category A infectious substance must have the words "suspected Category A infectious substance" entered in parentheses in place of the technical name.

If a hazardous material is a mixture or solution the technical names of at least two components most predominately contributing to the hazards of the mixture or solution must be entered. If a material contains two or more hazardous substances only the names of the two substances with the lowest RQ's must be entered. For hazardous waste, the waste code (such as D001) if appropriate, may be used to identity the hazardous substance. (*172.203*)

Examples: UN1760, Corrosive liquids, n.o.s. (Caprylyl chloride), 8, II
UN2924, Flammable liquid, corrosive, n.o.s., 3, (8), II (methanol, potassium hydroxide)

Reportable Quantity (RQ)

If the material is a hazardous substance, the letters "**RQ**" must appear either before or after the basic description. The letters "RQ" may be entered in an HM Column on a shipping paper in place of the "X". (*172.203*)

Limited Quantity

When a shipping paper is required for a material being shipped as a limited quantity, the words "Limited Quantity" or the abbreviation "Ltd. Qty" must be entered following the basic description. (*172.203*)

Elevated Temperature Material

The fact that a material is an elevated temperature material must be indicated in the proper shipping name (such as "Molten" or "Elevated temperature") or the word "HOT" must immediately precede the proper shipping name. (*172.203*)

Poisons

For all materials meeting the poisonous by inhalation criteria, the words "Poison-Inhalation Hazard" or "Toxic-Inhalation Hazard" and the words "Zone A", "Zone B", "Zone C", or "Zone D", for gases or "Zone A" or "Zone B" for liquids, as appropriate, must be added on the shipping paper immediately following the shipping description. For anhydrous ammonia transported within the U.S., only the words "Inhalation Hazard" must be added in association with the shipping description. (*172.203*)

Marine Pollutant

If a material is a marine pollutant the words "Marine Pollutant" must be entered in association with the basic shipping description, unless the proper shipping name indicates that it is a marine pollutant. (*172.203*)

Special Permits

For a shipment of hazardous materials made under a special permit the shipping paper must bear the notation "DOT-SP" followed by the number issued for the special permit. The notation must be clearly associated with the description to which the special permit applies. (*172.203*)

Empty Packagings

The description on the shipping paper for a packaging containing only the residue of a hazardous material may include the words "RESIDUE: Last Contained" in association with the basic description of the hazardous material last contained in the packaging. (*172.203*)

Additional Information

A shipping paper may contain additional information not inconsistent with the required hazardous materials information. This information must follow all the information required by the hazardous materials regulations. (*172.201*)

Prohibited Entries

No shipping paper entry shall identify a material as hazardous by hazard class or identification number unless the material described is a regulated hazardous material. (*172.202*)

Shipper's Certification

When a shipment is offered to the initial carrier the shipping paper must bear the prescribed shipper's certification. This certification indicates that the hazardous material is being offered in compliance with the Hazardous Materials Regulations. (*172.204*)

No certification is required for the return of an empty tank car which previously contained a hazardous material and which has not been cleaned or purged. (*172.204)*

SHIPPING PAPERS

Except for hazardous waste, a shipper's certification is not required for the highway transport of hazardous materials when:

- Transported in a cargo tank supplied by the carrier, or
- Transported by the shipper as a private carrier, unless the material is to be reshipped or transferred from one carrier to another. (*172.204*)

The following two shipper's certifications are authorized for use. (*172.204*)

"This is to certify that the above-named materials are properly classified, described, packaged, marked and labeled, and are in proper condition for transportation according to the applicable regulations of the Department of Transportation."

"I hereby declare that the contents of this consignment are fully and accurately described above by the proper shipping name, and are classified, packaged, marked and labeled/placarded, and are in all respects in proper condition for transport according to applicable international and national governmental regulations."

For shipments by air the following certification is authorized. (*172.204*)

"I hereby certify that the contents of this consignment are fully and accurately described above by the proper shipping name and are classified, packaged, marked and labeled, and are in proper condition for carriage by air according to applicable national governmental regulations."

Note: In the above certification, the word "packed" may be used instead of the word "packaged" until October 1, 2010.

On any shipment moving by air, the following additional certification must appear. (*172.204*)

"This shipment is within the limitations prescribed for passenger aircraft/cargo aircraft only (delete nonapplicable)."

Shipper's certifications must be printed (manually or mechanically) on the shipping paper. The certification must be legibly signed by an authorized representative of the shipper. The signature may be applied manually, by typewriter, or other mechanical means (*172.204*). On hazardous waste manifests the signature must be handwritten. (*172.205*)

Emergency Response Telephone Number

In most cases, an emergency response telephone number must be entered on the shipping paper. It can be immediately following the description of each hazardous material or if the number applies to every hazardous material entered on the shipping paper, entered once on the shipping paper in a clearly visible location and identified as an emergency response information number. (*172.604*)

The emergency response phone number must be monitored at all times the hazardous material is in transportation by a person who has knowledge of the hazardous material and has comprehensive emergency response and incident mitigation information for the material or has immediate access to such a person. *(172.604)*

The emergency response telephone number must be:

- The number of the person offering the hazardous material for transport, when that person is also the emergency response information provider; or
- The number of an agency or organization (emergency response information telephone service provider) capable of and accepting responsibility for providing the response information. The person registered with the emergency response information telephone service must ensure that the service has the current information on the material before it is offered for transport. *(172.604)*

When the person offering hazardous materials for transport is also the emergency response telephone information provider they must enter their name on the shipping paper near the emergency response telephone number. In addition, when the telephone number of an emergency response information (ERI) telephone service provider is used, the name of the person who is registered with the ERI service provider, or the contract number or other unique identifier assigned by the ERI provider must be placed on the shipping paper near the emergency response telephone number. However, the above requirements do not apply if the name or identifier is entered elsewhere on the shipping paper in a prominent, readily identifiable, and clearly visible manner to be easily and quickly found. *(172.604)*

The emergency response telephone number requirements do not apply to:

- Limited quantities;

- Materials described with the shipping names listed in §172.604(d)(2); or
- Vehicles or containers containing lading that has been fumigated and display the FUMIGANT marking, unless other hazardous material is present. *(172.604)*

Emergency Response Information

Most hazardous material shipments (except those that do not require shipping papers or an ORM-D material) must have emergency response information on or in association with the shipping paper. If the information is in association with the shipping paper it may be in the form of the Emergency Response Guidebook, a Material Safety Data Sheet, or any other form that provides all the information required by the regulations. (*172.602*)

The following emergency response information is required:

1. basic description and technical name of the hazardous material;
2. immediate hazards to health;
3. fire and explosion risks;
4. precautions to be taken immediately;
5. immediate methods for handling fires;
6. initial methods for handling spills or leaks in the absence of fire; and
7. preliminary first aid measures. (*172.602*)

This information must be on or kept with the shipping papers, away from the packages containing the hazardous material and in a location immediately accessible in the event of an incident. (*172.602*)

Reserved

MARKING

Markings provide important information about the contents of a packaging, freight container or transport vehicle and help warn of the hazards posed by that material during transport. Markings give additional information, not provided by labels or placards, about the hazardous material in a package or vehicle.

Applicability

The marking requirements apply to non-bulk and bulk packagings transported by rail, air, vessel, and highway. Certain requirements specifically apply to transport vehicles and freight containers.

Responsibility

The individual who prepares non-bulk packages of hazardous material for transport is responsible for marking the package. This responsibility includes:

- Checking that any relevant markings already displayed are in the correct location and are in accordance with the regulations.
- Removing or obliterating any markings which are not applicable or which may reduce the effectiveness of the required markings.
- Applying any new markings in accordance with the regulations. (*172.300*)

In most cases, the responsibility for marking bulk packagings, freight containers, and transport vehicles rests with the individual initiating the shipment. The carrier is responsible for replacing identification number markings that are lost, damaged, or destroyed during transit.

Prohibited Markings

No packaging may be marked with a proper shipping name or identification number unless the packaging contains the identified material or its residue. (*172.303*)

Marking Specifications

To withstand the conditions normally encountered during transportation, all markings must be:

- Durable,
- In English,
- Printed on, or affixed to, the surface of a package, or on a label, tag, or sign,
- Displayed on a background of sharply-contrasting color,
- Unobscured by labels or attachments, and
- Located away from any other markings — such as advertising — which could substantially reduce their effectiveness. (*172.304, 172.308*)

Manufacturer/Specification Packaging Marking

In addition to the markings that will be covered in the remainder of this section, most authorized packaging for hazardous materials must be marked with a UN packaging marking (for non-bulk packaging) or a specification packaging marking (such as DOT 406). For more information on these markings see the plain language explanation Packaging.

4G/Y145/S/83
USA/RA

MARKING

Non-Bulk Markings

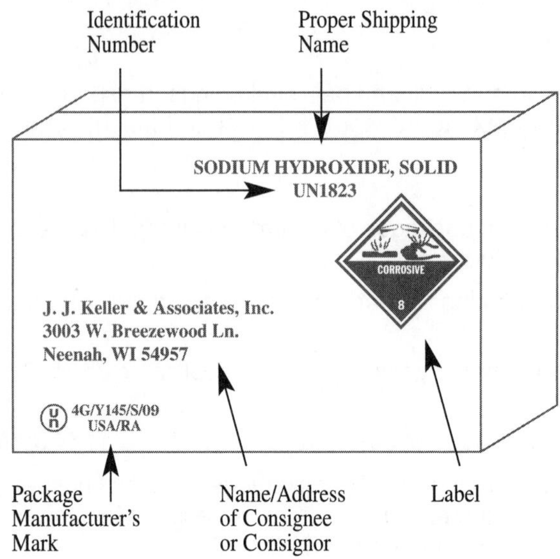

Most non-bulk packagings must be "marked" with the following information:

- Description/proper shipping name;
- Technical name(s), when required;
- Identification number (preceded by "UN" or "NA" or "ID", as appropriate);
- Consignee's or consignor's name and address;
- DOT-SP, when required. (*172.301*)

Description/proper shipping name

The description/proper shipping name as shown in Column (2) of the Hazardous Materials Table must be marked on the non-bulk package. (*172.301*)

Abbreviations are usually not allowed in a proper shipping name marking. Two specific exceptions include:

- "ORM" in place of "Other Regulated Material;" and
- Abbreviations which appear as part of the authorized description in Column (2) of the Hazardous Materials Table.

Technical name(s)

Any package that contains hazardous materials described by a proper shipping name preceded by the symbol "G" in Column (1) of the Hazardous Materials Table must be marked with the technical name in parenthesis in association with the proper shipping name. However, a technical name should not be marked on the outer package of a Division 6.2 material. (*172.301*)

Example: Flammable liquids, n.o.s. (ethanol)

Identification number

The identification number shown in Column (3) of the Hazardous Materials Table, for the description/proper shipping name being used must be marked on the package. The appropriate "UN" or "NA" or "ID" prefix must be included. (*172.301*)

Examples: UN1263 NA1993 ID8000

Name and address

The consignee's or consignor's name and address must be marked on the package unless:

- The package is transported by highway only, and will not be transferred from one motor carrier to another; or
- The package is transported as part of a carload lot, truckload lot, or freight container load and the entire contents of the rail car, truck, or freight container are shipped from one consignor to one consignee. (*172.301*)

Special permit packagings

DOT-SP (followed by the special permit number) is required to be plainly and durably marked on the packaging if it is authorized for use under a special permit. (*172.302*)

Example: DOT-SP 01234

Hazardous substances

Non-bulk packagings which contain a reportable quantity of a hazardous substance must be marked with the letters "RQ" in association with the proper shipping name. (*172.324*)

Example: RQ, ENVIRONMENTALLY HAZARDOUS SUBSTANCE, SOLID, N.O.S. (DIAZI-NON), UN3077

PCB's

Polychlorinated biphenyls (PCB's) are hazard substances which must be transported in compliance with the Hazardous Materials Regulations and special additional requirements of the Environmental Protection Agency (EPA).

The Hazardous Materials Regulations require the proper shipping name, identification number and RQ to be marked on each non-bulk package.

The EPA regulations require the package to be marked with the large PCB mark illustrated to the right (minimum size 6 x 6 inches). (*40 CFR 761.45(a)*)

The EPA regulations also require a vehicle to be marked on each side and each end with this marking, if transporting PCB containers that contain more than 45 kg (99.4 pounds) of liquid PCB's in concentrations of 50 ppm to 500 ppm. (*40 CFR 761.40*)

Large Quantities of Non-Bulk Packages

A transport vehicle or freight container containing only a single hazardous material in non-bulk packages must be marked, on each side and each end with the identification number specified for the material, subject to the following:

- Each package must be marked with the same proper shipping name and identification number;
- The aggregate gross weight of the hazardous material is 4,000 kg (8,820 lb) or more;
- All of the hazardous material is loaded at one loading facility; and
- The transport vehicle or freight container contains no other material, hazardous or non-hazardous. (*172.301*)

The above requirement does not apply to Class 1 materials, Class 7 materials, or non-bulk packagings for which identification numbers are not required, such as ORM-D materials. (*172.301*)

Large Quantities of Non-Bulk Poison Inhalation Hazards

A transport vehicle or freight container loaded at one loading facility with 1,000 kg (2,205 lb) or more of non-bulk packages containing a material poisonous by inhalation (in Hazard Zone A or B), having the same shipping name and identification number, must be marked with the identification number specified for the material, on each side and each end. (*172.313*)

If the transport vehicle or freight container contains more than one inhalation hazard material that meets the above identification number marking requirement, it must be marked with the identification number for only one material. That one identification number is determined by the following:

- For different materials in the same hazard zone, the identification number of the material having the greatest aggregate gross weight.
- For different materials in both Hazard Zone A or B, the identification number for the Hazard Zone A material. (*172.313*)

Bulk Markings

Unless specifically excepted, all bulk packagings of hazardous materials must be marked with the identification number(s) of the contents. These numbers, unless otherwise provided, must be marked:

- On each end and each side of a packaging having a capacity of 3,785 L (1,000 gal) or more.
- On two opposite sides of a packaging with a capacity of less than 3,785 L (1,000 gal).
- On each end and each side of a tube-trailer motor vehicle. (*172.302*)

CARGO TANK

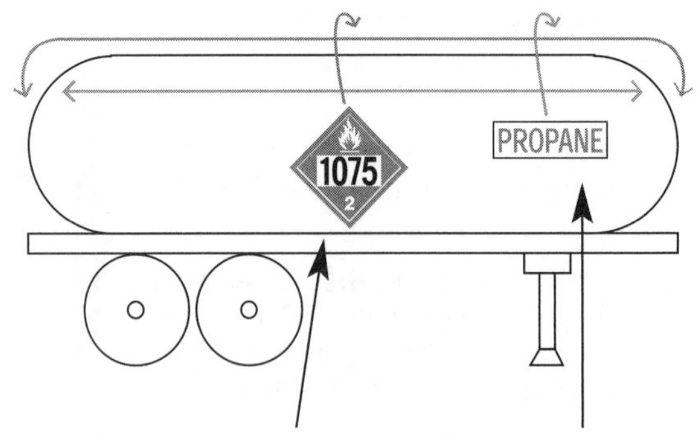

Identification number on placard, orange panel, or white square-on-point configuration. Display on all 4 sides.

Common name or proper shipping name for Class 2 materials. Display on all 4 sides.

There are three ways in which identification numbers may be displayed:

- On an orange panel 

- Across the center of a primary hazard placard 

- On a white square-on-point configuration (diamond-shaped) — with the same outside dimensions as a placard (If the material is in a class that does not allow identification numbers on placards or require placards)(*172.332*)

If the identification number markings on a portable tank or cargo tank are not visible, the transport vehicle or freight container used to transport the tank must also be marked with the numbers, on each side and each end. (*172.326, 172.328*)

When a tank is permanently installed within an enclosed cargo body of a transport vehicle or freight container, the identification number marking need only be displayed on the sides and ends of the tank that are visible when the enclosed cargo body is opened or entered. (*172.328*)

Identification numbers are not required on the ends of portable tanks and cargo tanks having more than one compartment if hazardous materials having different identification number are transported in the compartments. The identification numbers on the sides of the tank must be displayed in the same sequence as the compartments containing the materials they identify. (*172.336*)

When a bulk packaging is labeled instead of placarded, the identification number markings may be displayed on the bulk packaging in the same manner as would be required for a non-bulk packaging. (172.336)

Special permit packagings

DOT-SP (followed by the special permit number) is required to be marked on the packaging if it is authorized for use under a special permit. (*172.302*)

Portable tanks

In addition to displaying the applicable identification number (as described above), portable tanks must be marked with:

- Material's proper shipping name — on two opposite sides. For water transport the size of the shipping name must be 65 mm or more.

MARKING

- Owner or lessee's name (*172.326*)

IBCs

When an IBC is labeled instead of placarded, the IBC may display the proper shipping name and identification number markings in 25 mm (one inch) letters/numbers. These smaller markings may be used in place of the identification number on an orange panel or placard.*(172.514)*

Cargo tanks

In addition to the identification number, cargo tanks (except for certain nurse tanks) which are used to transport gases (Class 2 material) must be marked on each side and each end with the proper shipping name or appropriate common name of the gas. (*172.328*)

Examples: CARBON DIOXIDE, REFRIGERATED LIQUID or REFRIGERANT GAS

Tank cars

In addition to the identification number, tank cars — when required by a special provision to the Hazardous Materials Table or by Section 172.330 — must be marked on each side with the proper shipping name or appropriate common name. (*172.330*)

In addition to the identification number marking, multi-unit tank car tanks must be marked on two opposing sides with the proper shipping name or appropriate common name.

Shipping name change

If the proper shipping name for a hazardous material in a bulk packaging has been changed, the bulk packaging may not need to be remarked if:

- The bulk packaging was marked before October 1, 1991, in conformance with the Hazardous Materials Regulations in effect on September 30, 1991; and
- The marking contains the same key words as the current proper shipping name in the Section 172.101 Hazardous Materials Table. (*172.302*)

Example: A tank car marked "ANHYDROUS AMMONIA" does not need to be remarked "ANHYDROUS AMMONIA, LIQUEFIED"

Petroleum sour crude oil

A bulk packaging used to transport petroleum crude oil containing hydrogen sulfide (sour crude oil) in sufficient concentrations that the vapors may present an inhalation hazard must be marked with the GHS toxic pictogram or a warning statement such as "Danger Possible Hydrogen Sulfide Inhalation Hazard." The marking must be displayed at each location (manhole, loading head) where exposure to hydrogen sulfide vapors may occur.

Additional Marking Requirements

Additional markings may be required depending on the hazardous material being transported or the type of package used.

Hazardous wastes

Proper shipping names marked on non-bulk packages of hazardous waste are not required to include the word "waste" if the EPA marking shown to the right is displayed in accordance with 40 CFR Section 262.32. (*172.301*)

Liquids in combination packagings

Combination packagings that have inner packagings which contain liquid hazardous materials, or single packaging fitted with vents, or open cryogenic receptacles must be marked with orientation arrows. These arrows must be displayed on two opposite vertical sides of the packaging — with the arrows pointing in the correct upward direction. (*172.312*)

Orientation arrows are not required on packages which contain:

- Inner packagings of cylinders;
- Flammable liquids in inner packagings of one liter or less prepared as a limited quantity or consumer commodity, except for transport by air;
- Flammable liquids in inner packagings of 120 mL (4 fluid oz) or less prepared as a limited quantity or consumer commodity, when packed with sufficient absorption material between the inner and outer packagings to completely absorb the liquid contents, when transported by air.
- Liquids contained in manufactured articles (e.g., alcohol or mercury in thermometers) which are leak-tight in all orientations;
- Hermetically sealed inner packagings. (*172.312*)
- Liquid infectious substances in primary receptacles not exceeding 50 mL (1.7 oz).

Inhalation hazards

Packagings which contain materials poisonous-by-inhalation must be marked “INHALATION HAZARD” in association with the required labels or placards or the proper shipping name. The INHALATION HAZARD marking is not required if the package bears a poison gas or poison inhalation hazard label (or placard), displaying the words “Inhalation Hazard.” (*172.313*)

This marking on bulk packages must be on two opposite sides, and have:

- A minimum width of 6 mm (0.24 inches) and a minimum height of 100 mm (3.9 inches) for rail cars.
- A minimum width of 4 mm (0.16 inch) and a minimum height of 25 mm (1.0 inch) for portable tanks with capacities of less than 3,785 L (1,000 gallons) and intermediate bulk containers.
- A minimum width of 6 mm (0.24 inches) and a minimum height of 50 mm (2 inches) for cargo tanks and other bulk packagings. (*172.302(b)*)

Poisons

Non-bulk plastic outer packagings used as single or composite packagings for Division 6.1 materials must be permanently marked with the word “POISON” — either by embossing or other durable means. This marking must be at least 6.3 mm (0.25 in) in height and located within 150 mm (6 in) of the closure of the packaging. (*172.313*)

A package containing a Division 6.1 (poison) material in Packing Group III may be marked “PG III” adjacent to the POISON label. (*172.313*)

Limited quantity

A package containing a limited quantity of hazardous materials is not required to be marked with the proper shipping name and identification number when marked with the square-on-point limited quantity marking or for packages meeting the requirements for air transport, marked with the square-on-point “Y” limited quantity marking. (*172.315*)

The width of the border forming the square-on-point marking must be at least 2 mm. Each side of the marking must be at least 100 mm. If package size requires a reduced size marking, then each side of the marking may be reduced to no less than 50 mm. For cargo transport units transported by water, each side of the marking must have a minimum size of 250 mm. (*172.315*)

MARKING

Before January 1, 2013, for transport by aircraft, a limited quantity properly marked as an ORM-D-AIR is not required to be marked with the square-on-point "Y" limited quantity marking.

Before January 1, 2014, a limited quantity properly marked as an ORM-D is not required to be marked with the square-on-point limited quantity marking. (*172.315*)

ORM-Ds

Packages that contain a material classed as a consumer commodity (ORM-D) must be marked on at least one side or end with the ORM-D designation — immediately following or below the proper shipping name. If the shipment is to be by air, the marking must be ORM-D-AIR. (*172.316*)

Both the ORM-D and ORM-D-AIR markings must be within a rectangle that is approximately 6.3 mm (0.25 in) larger on each side than the designation. (*172.316*)

Before January 1, 2013, an air shipment that meets the requirements for an ORM-D may be marked with ORM-D-AIR. Before January 1, 2014, packages of ORM-D material may be marked with the ORM-D marking. Starting on the above January dates, only the appropriate square-on-point limited quantity marking may be used and the ORM-D markings must not be used.

Keep away from heat

For transportation by aircraft, packages containing self-reactive substances of Division 4.1 or organic peroxides of Division 5.2 must be marked with the KEEP AWAY FROM HEAT handling mark. (*172.317*)

Marine pollutants

Non-bulk packagings which contain marine pollutants and are offered for transportation by vessel must be marked with:

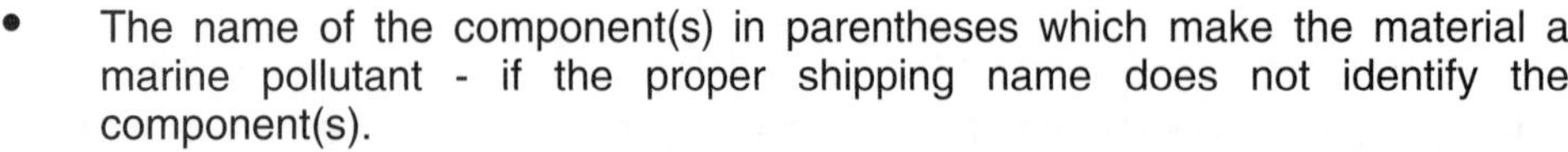

- The name of the component(s) in parentheses which make the material a marine pollutant - if the proper shipping name does not identify the component(s).
- The marine pollutant mark, shown below, in association with the required hazard warning label(s) or in the absence of labels, in association with the proper shipping name. (*172.322*)

The marine pollutant marking is not required on single packagings or combination packagings where each single package or each inner packaging of combination packagings has:

- A net quantity of 5 L (1.3 gallons) or less for liquids; or
- A net mass of 5 kg (11 pounds) or less for solids.

The marine pollutant marking is not required on a combination packaging containing a marine pollutant, other than a severe marine pollutant, in inner packagings each of which contains:

- 5 L (1.3 gallons) or less net capacity for liquids; or
- 5 kg (11 pounds) or less net capacity for solids. *(172.322)*

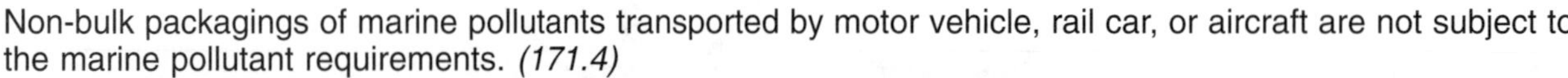

Non-bulk packagings of marine pollutants transported by motor vehicle, rail car, or aircraft are not subject to the marine pollutant requirements. *(171.4)*

Bulk packagings with a capacity of 3,785 L (1,000 gallons) or more and which contain marine pollutants must be marked on each side and each end with the marine pollutant marking. A bulk packaging with a capacity of less than 3,785 L (1,000 gallons) must be marked on two opposing sides or ends with the marine pollutant marking. (*172.322*)

Except for transportation by vessel, a bulk packaging, freight container, or transport vehicle that bears a label or placard is not required to display the marine pollutant marking. (*172.322*)

A package of limited quantity material that is marked with a square-on-point limited quantity marking is not required to display the marine pollutant marking. (*172.322*)

The symbol and border must be black on a white background, or be of a contrasting color to the surface on which the marking is displayed. Each side of the marking must be at least 100 mm (3.9 inches) for non-bulk packages and bulk packages with a capacity of less than 3,785 L (1,000 gallons). For all other bulk packages each side of the marking must be at least 250 mm (9.8 inches). (*172.322*)

Elevated temperature materials

Most bulk packagings which contain an elevated temperature material must be marked on two opposing sides with the word "HOT" — in black or white Gothic lettering on a contrasting background. This marking must be displayed on the packaging itself or in black lettering on a white square-on-point configuration that is the same size as a placard. (*172.325*)

Bulk packagings that contain molten aluminum or molten sulfur must be marked "MOLTEN ALUMINUM" or "MOLTEN SULFUR" as appropriate, instead of with the word "HOT". (*172.325*)

For bulk packagings that must be marked, HOT, MOLTEN ALUMINUM or MOLTEN SULFUR the letters must be at least:

- 100 mm (3.9 inches) for rail cars
- 25 mm (1 inch) for portable tanks with capacities of less than 3,785 L (1,000 gallons)
- 50 mm (2 inches) for cargo tanks and other bulk packagings

The regulations also allow the combining of the "HOT" marking and the identification number marking on the same white square-on-point configuration. In this case, the identification number must be displayed in the center of the white square-on-point configuration and the word "HOT" must be displayed in the upper corner. The word "HOT" must be in black letters having a height of at least 50 mm (2 inches). (*172.325*)

Explosive materials

A package of Class 1 material must be marked with the EX-number for each substance, article or device contained therein unless it is excepted from this marking as specified in section 172.320. In some situations no marking is required, for others a national stock number or product code may be used. (*172.320*)

When more than five different Class 1 materials are packed in the same package, no more than five of the numbers, codes, or a combination thereof are required to be marked on the package. (*172.320*)

Radioactive materials

Each package containing Class 7 (radioactive) materials must also be marked with:

- Gross mass if the gross mass of the package exceeds 50 kilograms (110 pounds)
- TYPE A or TYPE B, as appropriate
- For TYPE B(U) or B(M), the appropriate radiation symbol according to Appendix B to Part 172
- USA in conjunction with the specification marking, if the package is destined for export. (*172.310*)

Regulated medical waste

A bulk packaging containing a regulated medical waste must be marked with the BIOHAZARD marking, visible from the direction it faces:

- on two opposing sides or ends if the packaging has a capacity of less than 1,000 gallons (3,785 L), or
- on each end and each side if the packaging has a capacity of 1,000 gallons (3,785 L) or more.

The BIOHAZARD marking must be at least 6 inches (152.4 mm) on each side. This marking may be displayed on a white square-on-point configuration. (*172.323*)

BIOHAZARD

Non-odorized LPG

No person may offer or transport certain specification cylinders, a portable tank, cargo tank, or tank car containing liquefied petroleum gas (LPG) that is unodorized unless the words "NON-ODORIZED" or "NOT ODORIZED" appear near the proper shipping name. This marking must appear on two opposing sides on portable tanks, cargo tanks, and tank cars. (*172.301, 172.326, 172.328, 172.330*)

Overpacks

Each overpack must be marked with the proper shipping names and identification numbers of the materials contained within, unless the markings on the inside packages are visible. (*173.25*)

Each overpack that contains packages subject to the orientation marking requirements must be marked with orientation arrows on two opposite vertical sides, with the arrows pointing in the correct upward direction. (*172.312*)

Each overpack must be marked with the word "OVERPACK" indicating that the inner packages comply with prescribed specifications (when specification packagings are required), unless the specification markings on the inside packages are visible. (*173.25*)

Each overpack containing packages of limited quantity materials or ORM-D materials must be marked "OVERPACK", unless all required markings are visible through the overpack. (*173.25*)

Each overpack containing excepted quantities must be marked with the required markings for the inside packages, unless all the required markings are visible through the overpack. Overpacks of excepted quantities are not required to be marked "OVERPACK." (*173.25*)

Fumigant marking

Each truck body or trailer, rail car, or freight container containing lading that has been fumigated or treated with any material, or is undergoing fumigation must have a FUMIGANT marking displayed so that it can be seen by any person attempting to enter the vehicle or container. (*173.9*)

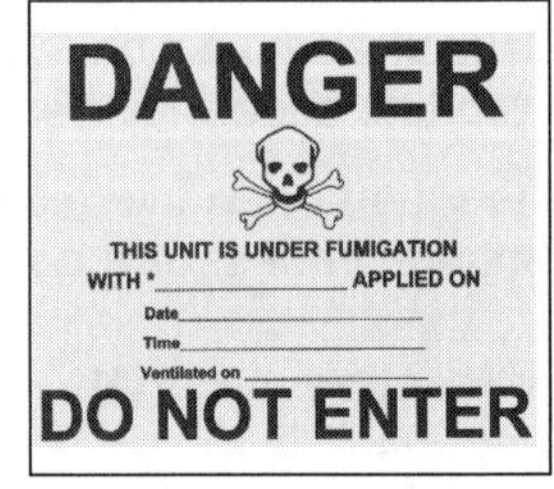

The printing on the FUMIGANT marking must be red or black. The size of the white background must be at least 30 cm (11.8 inches) wide and at least 25 cm (9.8 inches) high. The technical name of the fumigant, must be entered in the appropriate place (*) on the marking.

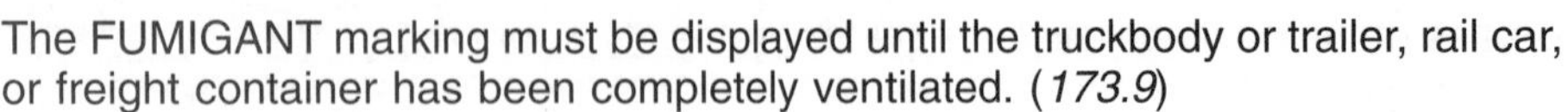

The FUMIGANT marking must be displayed until the truckbody or trailer, rail car, or freight container has been completely ventilated. (*173.9*)

Small quantities

When transported by highway or rail, packages containing small quantities of materials meeting the requirements in Section 173.4 must be marked with the statement "This package conforms to 49 CFR 173.4 for domestic highway or rail transport only". (*173.4*)

Excepted quantities

When transported by air or vessel, packages containing excepted quantities of materials meeting the requirements in Section 173.4a must be marked with the excepted quantities marking. The "*" must be replace with the hazard class or division of each hazardous material in the package. The "**" must be replaced with the name of the shipper or consignee, if not shown elsewhere on the package. (*173.4a*)

For illustrations of markings see the color charts at the back of the book.

Reserved

LABELING

Labels are printed on or affixed to packages containing hazardous materials. Labels are color- and symbol-coded to provide easy and immediate warning of the hazardous materials inside the package. The regulations have specific requirements for design, use, placement, prohibitions, and exceptions for labels.

Responsibility

The labeling of packages of hazardous materials is the responsibility of the shipper and packages must be properly labeled at the time they are offered for transport. (*172.400*)

Although the shipper is responsible for the actual labeling, the carrier also has labeling responsibilities. A carrier must only accept and transport packages that have been properly labeled. (*177.801*)

Applicability

The labeling requirements apply primarily to non-bulk packagings. The following must also display labels if they are not placarded:

- Bulk packagings with a volumetric capacity of less than 18m^3 (640 feet^3), other than cargo tanks, portable tanks or tank cars.
- Portable tanks of less than 3,785 L (1,000 gallon) capacity.
- DOT Specification 106 or 110 multi-unit tank car tanks.
- Overpacks, freight containers, or unit load devices of less than 18m^3 (640 feet^3), which contain a package for which labels are required. (*172.400*)

Exceptions

Not all non-bulk packages of hazardous materials will require labels. Labels are not required on the following:

- A cylinder or Dewar flask (§173.320) containing Division 2.1, 2.2, or 2.3 gas that is not overpacked, and is durably and legibly marked in accordance with CGA C-7, Appendix A.

- A package or unit of military explosives — including ammunition — shipped by or on behalf of the U.S. Department of Defense (DOD) when:
  - (1) in freight container-load, car-load or truck-load shipments, if loaded and unloaded by the shipper or DOD; or
  - (2) in unitized or palletized break-bulk shipments by cargo vessel under charter to DOD, if at least one required label is displayed on each unitized or palletized load.
- A package of hazardous materials — other than ammunition — that is loaded and unloaded under the supervision of DOD personnel and is escorted by DOD personnel in a separate vehicle.
- A compressed gas cylinder which is permanently mounted in or on a transport vehicle.
- A freight container, aircraft unit load device, or portable tank which is placarded in accordance with the regulations or identified as provided in the ICAO Technical Instructions.
- An overpack or unit load device in or on which labels that represent the hazard(s) inside are visible.
- A package of low specific-activity radioactive material when being transported in a conveyance assigned for the exclusive use of the consignor under Section 173.427(a)(6).
- A package containing a material classed as ORM-D. (*172.400a*)
- A package containing a combustible liquid. *(173.150(f))*

Limited Quantities

There are also exceptions to labeling for limited quantities in the following sections:

- Section 173.306 — Class 2 (gases)
- Section 173.150 — Class 3 (flammable and combustible liquids)

LABELING

- Section 173.151 — Class 4 (flammable solids)
- Section 173.152 — Class 5 (oxidizers and organic peroxides)
- Section 173.153 — Division 6.1 (poisons)
- Section 173.421 — Class 7 (radioactives)
- Section 173.154 — Class 8 (corrosives)
- Section 173.155 — Class 9 (miscellaneous)

Prohibited Labeling

No package may be labeled with a label specified in the Hazardous Materials Regulations unless the package contains the hazardous material and the label represents the hazard of the material in the package. (*172.401*)

Packages may not be offered for transportation or transported with a marking or label that could be confused with or conflicts with the labels that are required unless the package is labeled in conformance with:

- The United Nations recommendations;
- The International Maritime Dangerous Goods Code;
- The International Civil Aviation Organization's Technical Instructions;
- Canada's Transport Dangerous Goods Regulations; or
- The Globally Harmonized System of Classification and Labeling of Chemicals.(*172.401*)

General Requirements

Once a material has been classified and a proper shipping name has been selected, determining the appropriate labels is a fairly easy process.

1. Locate the selected proper shipping name in Column (2) of the Hazardous Materials Table (Section 172.101).
2. Refer to Column (6) of the Table for the appropriate label code(s).
3. The first label code listed indicates the material's primary hazard. Any additional label codes indicate subsidiary hazards. (*172.101*)
4. Using the label codes from Column (6) in the Table, find the name of the label(s) required to be on the package in the label substitution table in Section 172.101(g).

Except for the label code 6.1, the label codes are the same as the hazard classes or divisions and only one label is possible. For example, a 2.3 label code is a 2.3 (poison gas) label, a 3 label code is a Class 3 (flammable liquid) label, and a 5.1 label code is a 5.1 (oxidizer) label.

For the label code 6.1, there are two possible labels. If the material has an inhalation hazard, Zone A or B, a Poison Inhalation Hazard label is required. If the material does not have an inhalation hazard, Zone A or B, a Poison label is required.

Each package containing a hazardous material must be labeled with the label(s) prescribed in Column (6) of the Hazardous Materials Table. (*172.400*)

If the material has more than one hazard, all applicable subsidiary labels may not be listed in the Table (such as generic or n.o.s. shipping names). If this is the case, subsidiary labels must be determined according to section 172.402. (*172.101*)

A Division 6.1 subsidiary label is not required on a package containing a Class 8 material which has a subsidiary hazard of Division 6.1, if the toxicity of the material is based solely on the corrosive destruction of tissue rather than systemic poisoning. Also, a Division 4.1 subsidiary label is not required on a package bearing a Division 4.2 label. (*172.400a*)

Placement

A label on a package of hazardous materials must be clearly visible and not obscured by markings or attachments. Each label must be printed on or affixed to a background of contrasting color, or must have a dotted or solid line outer border. (*172.406*)

The label must be printed on or affixed to a surface (other than the bottom) of the package containing the hazardous material. If possible, the label should be located on the same surface of the package and near the proper shipping name marking. (*172.406*)

However, a label may be printed on or placed on a securely affixed tag, or may be affixed by other suitable means to:

- A package that contains no radioactive material and which has dimensions less than those of the required label; or
- A cylinder; or
- A package which has such an irregular surface that a label cannot be satisfactorily affixed. (*172.406*)

When primary and subsidiary labels are required they must be displayed next to each other, within 6 inches of one another. (*172.406*) Except as discussed in the following section, duplicate labeling is not required on a package. This means that only one of each different required label would have to be on the package.

NON-BULK PACKAGE

Duplicate Labeling

Generally, only one of each different required label must be displayed on a package. However, duplicate labels must be displayed on at least two sides or two ends (other than the bottom) of the following:

- Each package or overpack having a volume of 1.8 m^3 (64 $feet^3$) or more;
- Each non-bulk package containing a radioactive material;
- Each DOT 106 or 110 multi-unit tank car tank. Labels must be displayed on each end;
- Each portable tank of less than 3,785 L (1000 gallons) capacity; and
- Each freight container or aircraft unit load device having a volume of 1.8 m^3 (64 cubic feet) or more, but less than 18 m^3 (640 cubic feet). One of each required label must be displayed on or near the closure. (*172.406*)

Consolidated Packaging

When two or more packages containing compatible hazardous materials are placed within the same outside container or overpack, the outside container or overpack must be labeled for each class of material contained within, unless labels on the inside packages are visible and represent each hazardous material in the container or overpack. (*172.404*)

Modifications

Text is not required on a primary or subsidiary label for Classes 1, 2, 3, 4, 5, 6, and 8. However, the label must meet the other specifications in the regulations. (*172.405*)

The OXIDIZER label may be modified for packages of hazardous materials with the proper shipping name "Oxygen, refrigerated liquid" or "Oxygen, compressed". The word "OXIDIZER" may be replaced with "OXYGEN" and the Division number "5.1" with "2". The word "OXYGEN" must appear on the label. (*172.405*)

LABELING

The POISON label may be modified by displaying the text "TOXIC" instead of "POISON". (*172.430*)

The POISON label may be modified for packages of hazardous materials in Division 6.1, PG III. The text "PG III" may be displayed instead of "POISON" or "TOXIC" below the midline of the label. (*172.405*)

Specifications

Labels must meet the following specifications.

- Must be durable and weather-resistant. Must be able to withstand a 30-day exposure to conditions incident to transportation.
- Each diamond-shaped label must be a minimum size of 100 mm (3.9 inches) on each side — with each side having a solid-line inner border 5.0 to 6.3 mm (0.2 to 0.25 inches) from the edge.
- The printing, inner border, and symbol on each label must be as shown in the regulations.
- Specifications for color must be as prescribed in Appendix A to Part 172.
- The specified color must extend to the edge of the label (with a few exceptions), as prescribed for each label.
- The symbol, text, numbers, and border of a label must be black — except that white may be used on a label with a one color background of green, red, or blue. White must be used for the text and class number for the CORROSIVE label.
- The hazard class or division number must be at least 6.3 mm (0.25 inches) and not greater than 12.7 mm (0.5 inches).
- When text is displayed, in must be in letters measuring at least 7.6 mm (0.3 inches) in height, with a few exceptions.
- Labels may contain form identification information — including the name of the maker — provided that such information is printed outside of the solid-line inner border in no larger than 10-point type. (*172.407*)

Hazard class or division numbers must be displayed in the lower corner of primary and subsidiary labels. (*172.402*)

Special Labels

The CARGO AIRCRAFT ONLY label must be displayed on packages that can only be transported by cargo aircraft. It must be a rectangle at least 110 mm (4.3 inches) in height and 120 mm (4.7 inches) in width. (*172.407*)

The CARGO AIRCRAFT ONLY label conforming to the requirements in Section 172.448 on December 31, 2008 may continue to be used until January 1, 2013. The word "DANGER" must be at least 12.7 mm (0.5 inches) in height. (*172.407*)

Packages which contain the residue of radioactive material can be excepted from most of the regulations if they meet certain requirements in Section 173.428. One of those requirements is that labels that were previously applied must be removed or obliterated and the EMPTY label must be applied. The EMPTY label must be 6 inches by 6 inches. (*172.450*)

Radioactive labels are unique in that the regulations require that specific information be printed (manually or mechanically) on the labels. This information (CONTENTS, ACTIVITY, and TRANSPORT INDEX) is specific to each individual package. (*172.403*)

For illustrations of labels see the color charts at the back of the book.

Reserved

PLACARDING

Hazardous materials placards correspond very closely with the shape, color and design of the hazardous materials warning labels. However, placards are much larger than labels. Placards alert persons to the potential dangers associated with the particular hazardous material contained in a motor vehicle, rail car, freight container, cargo tanks, and portable tanks.

Applicability

The placarding requirements apply to each person who offers for transport or transports hazardous materials. The placarding requirements do not apply to the following materials:

- Infectious substances (Division 6.2).
- Materials classed as ORM-D.
- Materials authorized to be transported as a limited quantity (when properly identified on the shipping papers or marked in accordance with section 172.315).
- Materials packaged as "small quantities" (under the provisions of Sections 173.4, 173.4a, or 173.4b).
- Materials prepared in accordance with Section 173.13 (Exceptions for Classes 3, 8, 9, and Division 4.1, 4.2, 4.3, 5.1, 6.1).
- Combustible liquids in non-bulk packagings. (*172.500*)

Placards are used to identify the hazard of any quantity of materials contained in bulk packagings, freight containers, unit load devices, transport vehicles, or rail cars.

Responsibility

The responsibility for affixing or supplying placards varies according to the mode of transport and the type of packaging used to transport the hazardous material.

Bulk packagings

The person who offers a bulk packaging containing hazardous material for transportation is required to affix the required placards prior to or at the time the packaging is offered for transportation. (*172.514*)

Freight containers and aircraft unit load devices

Each person who offers for transportation and each person who loads and transports a hazardous material in a freight container or aircraft unit load device is required to affix the placards specified. (*172.512*)

Transport vehicles

When a hazardous material is offered for transport by highway, the individual offering the material must provide the carrier with the required placards — prior to, or at the same time, the material is offered for transport — unless the appropriate placards are already affixed to the vehicle. (*172.506*)

According to Section 177.823, the carrier may not move the vehicle until he or she has affixed the required placards — unless it is an emergency situation and one of the following three conditions are met:

- The vehicle is escorted by a representative of state or local government.
- The carrier has received permission from DOT to move the vehicle.
- Movement of the vehicle is necessary to protect life and property.

Rail cars

When a hazardous material is offered for transport by rail, the individual offering the material must affix the required placard(s) to the rail car — unless the car is already properly placarded. The rail carrier may not accept a rail car for transport unless the required placards are affixed. (*172.508*)

Prohibited Placarding

No person may affix a placard to a packaging, freight container, unit load device, motor vehicle or rail car unless the material being offered for transportation is a hazardous material, the placard applied represents the hazard of the material being offered, and the placards conform to the requirements in the regulations. (*172.502*)

PLACARDING

Signs, advertisements, slogans (such as Drive Safely), or other devices that could be confused with a prescribed placard are prohibited. (*172.502*)

Permissive Placarding

Placards may be displayed for a hazardous material, even when not required, if the placarding conforms to the regulations. (*172.502*)

A bulk packaging, freight container, unit load device, transport vehicle or rail car placarded according to Canada's Transport Dangerous Goods (TDG) Regulations, the International Maritime Dangerous Goods (IMDG) Code, or the United Nations (UN) Recommendations is allowed.

Placarding Tables

Placarding requirements vary according to the category of the material (hazard class, division, packing group or description) being transported and the type of packaging (bulk or non-bulk) containing the material. Each bulk packaging, freight container, unit load device, transport vehicle, or rail car containing hazardous material must be placarded on each side and each end (with some exceptions) with the type of placards specified in Table 1 or Table 2 of Section 172.504.

To determine what placards are required, you must know:

- The type of packaging (bulk or non-bulk) containing the hazardous material(s).
- The hazard category (class, division, packing group, or description) and subsidiary hazard(s) of the hazardous material(s) present.
- The weight of non-bulk packages in each hazard category.

The most dangerous categories of hazardous materials are located in Table 1. Any quantity of material falling within the categories listed in Table 1 must be placarded. (*172.504*)

TABLE 1

Category of material (Hazard class or division number and additional description, as appropriate)	Placard name	Placard design section reference (§)
1.1	EXPLOSIVES 1.1	172.522
1.2	EXPLOSIVES 1.2	172.522
1.3	EXPLOSIVES 1.3	172.522
2.3	POISON GAS	172.540
4.3	DANGEROUS WHEN WET	172.548
5.2 (Organic peroxide, Type B, liquid or solid, temperature controlled)	ORGANIC PEROXIDE	172.552
6.1 (material poisonous by inhalation (see §171.8 of this subchapter))	POISON INHALATION HAZARD	172.555
7 (Radioactive Yellow III label only)	RADIOACTIVE1	172.556

[1] RADIOACTIVE placard also required for exclusive use shipments of low specific activity material and surface contaminated objects transported in accordance with §173.427(a) of this subchapter.

The remaining hazard categories are assigned to Table 2. Any quantity of these materials also must be placarded, but the regulations provide some exceptions under certain conditions. (*172.504*)

TABLE 2

Category of material (Hazard class or division number and additional description, as appropriate)	Placard name	Placard design section reference (§)
1.4	EXPLOSIVES 1.4	172.523
1.5	EXPLOSIVES 1.5	172.524
1.6	EXPLOSIVES 1.6	172.525
2.1	FLAMMABLE GAS	172.532
2.2	NON-FLAMMABLE GAS	172.528
3	FLAMMABLE	172.542
Combustible liquid	COMBUSTIBLE	172.544
4.1	FLAMMABLE SOLID	172.546
4.2	SPONTANEOUSLY COMBUSTIBLE	172.547
5.1	OXIDIZER	172.550
5.2 (Other than organic peroxide, Type B, liquid or solid, temperature controlled)	ORGANIC PEROXIDE	172.552
6.1 (Other than material poisonous by inhalation)	POISON	172.554
6.2	(None)	
8	CORROSIVE	172.558
9	CLASS 9 (see §172.504(f)(9))	172.560
ORM-D	(None)	

Placarding Exceptions

Table 2

A transport vehicle or freight container which contains less than 454 kg (1,001 lb) aggregate gross weight of hazardous materials in non-bulk packages covered by Table 2 is not required to display placards. (This exception does not apply to bulk packages or materials with subsidiary hazards that must be placarded according to Section 172.505.) (*172.504*)

Dangerous placard

If a transport vehicle, rail car, freight container, or unit load device contains non-bulk packagings of two or more categories of Table 2 materials, the DANGEROUS placard may be displayed instead of the separate placards specified in Table 2. (*172.504*)

When 1,000 kg (2,205 lb) or more of one hazard category is loaded at one facility, on one vehicle, rail car, freight container, or unit load device, the DANGEROUS placard cannot be used for that one hazard category; instead the placard specified in Table 2 must be displayed. (*172.504*)

If there are three or more different categories of Table 2 materials in one vehicle, rail car, freight container, or unit load device, and one material is over the 1,000 kg (2,205 lb) and requires an individual class placard, the DANGEROUS placard may still be used for the other categories of Table 2 material that fall below the 1,000 kg (2,205 lb) limit. (*172.504*)

Residue

Except for materials subject to §172.505, placarding for subsidiary hazards, non-bulk packagings that contain only the residue of a Table 2 material do not have to be included when determining the placards for a transport vehicle, rail car, freight container, or unit load device. (*172.504*)

Class 1

When more than one division placard is required for Class 1 materials on a transport vehicle, rail car, freight container, or unit load device, only the placard representing the lowest division number must be displayed.

The EXPLOSIVES 1.4 placard is not required for those Division 1.4 Compatibility Group S (1.4S) materials that are not required to be labeled 1.4S. (*172.504*)

PLACARDING

For shipments of Class 1 (explosive) materials by aircraft or vessel, the applicable compatibility group letter must be displayed on the required placards. (*172.504*)

Gases

A NON-FLAMMABLE GAS placard is not required on a motor vehicle containing a non-flammable gas if the vehicle contains flammable gas or oxygen and is placarded FLAMMABLE GAS or OXYGEN, as required.

The OXYGEN placard may be used for domestic shipments of oxygen, compressed, or oxygen, refrigerated liquid, in place of a NON-FLAMMABLE GAS placard. (*172.504*)

Flammables

A COMBUSTIBLE placard is not required on a cargo tank, portable tank, or compartmented tank car containing both flammable and combustible liquids when placarded FLAMMABLE. (*172.504*)

Oxidizers

OXIDIZER placards are not required for Division 5.1 materials on freight containers, unit load devices, transport vehicles, or rail cars which also contain Division 1.1 or 1.2 materials and which are placarded with EXPLOSIVES 1.1 or 1.2 placards, as required. (*172.504*)

(For transportation by transport vehicle or rail car only) An OXIDIZER placard is not required for Division 5.1 materials on a transport vehicle, rail car, or freight container that also contains Division 1.5 materials and is placarded with EXPLOSIVES 1.5 placards, as required. (*172.504*)

Poisons

For domestic transportation, a POISON placard is not required on a transport vehicle or freight container required to display a POISON INHALATION HAZARD or POISON GAS placard. (*172.504*)

Also for domestic transportation, a POISON INHALATION HAZARD placard is not required on a transport vehicle or freight container that is already placarded with the POISON GAS placard. *(172.504)*

For Division 6.1, Packing Group III materials, a POISON placard may be modified to display the text "PG III" below the mid line of the placard in place of the word "POISON." *(172.504)*

Class 9

CLASS 9 placards are not required for domestic transportation. (*172.504*)

Freight containers/unit load devices

A motor vehicle transporting freight containers or aircraft unit load devices that are not required to be placarded is not required to display placards. (*172.512*)

Freight containers and unit load devices being transported for delivery to a consignee immediately after an air or water shipment are also allowed to use the exception for less than 454 kg (1,001 lb) of Table 2 materials.

A freight container or aircraft unit load device that is only transported by air and is prepared according to Part 5, Chapter 2, Section 2.7 of the ICAO Technical Instructions is not required to be placarded. (*172.512*)

Rail

A rail car loaded with a transport vehicle or freight container that is not required to be placarded is not required to display placards. (*172.504*)

Subsidiary Placards

There are two types of placards:

- Primary placards are used to indicate the primary hazard of the material.
- Subsidiary placards indicate additional hazards the material possesses.

Generally, only the primary placards are displayed. Section 172.505 of the regulations requires placards to be displayed for certain subsidiary hazards. Other subsidiary placards are permitted to be displayed but are not required. (*172.505*)

Placard Specifications

To be in compliance, all placards must meet the following specifications:

- Placards must be designed as specified in Part 172, Subpart F.
- Specifications for color must be as prescribed in Appendix A to Part 172.
- Placards must measure at least 250 mm (9.84 inches) on each side, and have a solid-line inner border approximately 12.7 mm (0.5 inches) from each edge.
- Other than for the RADIOACTIVE or DANGEROUS placard, text indicating the hazard (e.g., FLAMMABLE) is not required. Text is not required on the OXYGEN placard only when an identification number is displayed.
- Placards may contain form identification information — including the name of the maker — provided that such information is printed outside of the solid-line inner border in no larger than 10-point type.
- Reflective and retro-reflective material may be used on a placard provided color, strength and durability requirements are maintained.
- The material used for placards may be any plastic, metal, or other material capable of withstanding, without deterioration or a substantial reduction in effectiveness, a 30-day exposure to open-weather conditions. (*172.519*)

Tagboard placards must:

- Be of a quality at least equal to that designated commercially as white tagboard.
- Have a weight of at least 80 kg (176 lbs) per ream of 610 mm x 910 mm (24 in x 36 in), waterproofing materials included.
- Be able to pass a 414 kPa (60 p.s.i.) Mullen test. (*172.519*)

Placard Placement

Motor vehicles and rail cars

Each placard affixed to a motor vehicle or rail car must be readily visible from the direction it faces — except from the direction of another vehicle or rail car to which the vehicle or rail car is coupled. (*172.516*)

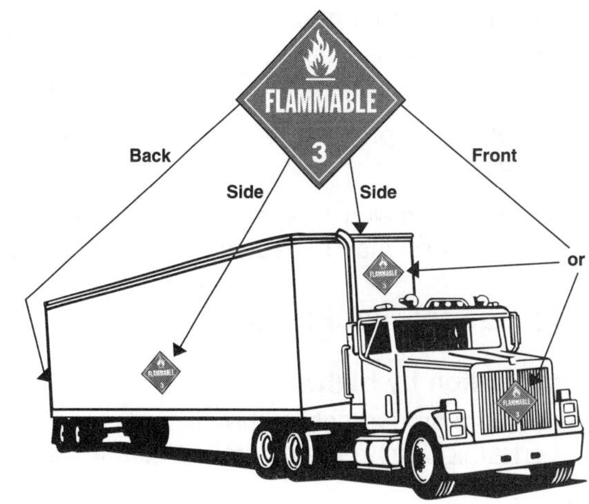

The placard placement for the front of a vehicle may be on the front of the truck-tractor instead of, or in addition to, the placard on the front of the cargo-carrying body (e.g., trailer or semitrailer). (*172.516*)

Placards displayed on freight containers or portable tanks loaded on (not enclosed in) a motor vehicle or rail car may be used to meet the placarding requirements for the motor vehicle or rail car. (*172.516*)

In addition, each placard on a transport vehicle, bulk packaging, unit load device, or freight container must be:

- Located clear of any appurtenances and devices (e.g., ladders and pipes).
- Away from any markings — such as advertising — which might substantially reduce its effectiveness. (A minimum distance of 76 mm (3 inches) is required.)
- Affixed to a background of contrasting color, or must have a dotted or solid line outer border which contrasts with the background color.
- Maintained by the carrier so that its effectiveness will not be reduced in any way.
- Placed, as far as practicable, so that dirt or water is not directed to it from the wheels of the vehicle.
- Securely attached or affixed or placed in a placard holder.

- Displayed with the word(s) or identification number (when authorized) shown horizontally, reading left to right. (*172.516*)

Bulk packagings

In most situations, a bulk packaging containing hazardous materials that requires placards must be placarded on each side and each end (*172.514*). Each bulk packaging that previously contained a hazardous material requiring placards must remain placarded when emptied unless:

- Sufficiently cleaned of residue and purged of vapors to remove any potential hazard, or
- Refilled with a material requiring different placards or no placards (provided that any residue remaining in the packaging is no longer hazardous), or
- Contains the residue of a hazardous substance in Class 9 in a quantity less than the reportable quantity, and conforms to §173.29(b)(1). (*172.514*)

The following bulk packagings may be placarded on only two opposite sides or may be labeled instead of placarded:

- A portable tank having a capacity of less than 3,785 liters (1,000 gallons);
- A DOT 106 or 110 multi-unit tank car tank;
- A bulk packaging other than a portable tank, cargo tank, or tank car (such as a bulk bag or box) with a volumetric capacity of less than 18 cubic meters (640 cubic feet);
- An intermediate bulk container (IBC); and
- A large packaging. (*172.514*)

Freight containers and aircraft unit load devices

A freight container or unit load device with a capacity less than 18 cubic meters (640 cubic feet) is required to display one placard as specified in Section 172.504. This requirement does not apply if the freight container or aircraft unit load device is:

- Labeled according to Subpart E of the regulations.
- Contains radioactive materials requiring the RADIOACTIVE YELLOW III label and is placarded with one RADIOACTIVE placard and labeled according to Subpart E.
- Identified as prepared according to Part 5, Chapter 2, Section 2.7 of the ICAO Technical Instructions. (*172.512*)

Freight containers with a capacity of less than 640 cubic feet that are not transported by air do not have to be placarded — they may be labeled instead. (*172.512*)

White Square Background

Transportation by highway

A square background behind the placard is required when a motor vehicle is used to transport a highway route-controlled quantity of radioactive material. (*172.507*)

Transportation by rail

A white square background behind the placard is required when the following materials are transported by rail:

- EXPLOSIVES 1.1 or 1.2
- POISON GAS or POISON INHALATION HAZARD
- FLAMMABLE GAS — When transported in DOT 113 tank cars (*172.510*)

ID Number Display

Identification numbers may be displayed on most placards. Identification numbers must not be displayed on DANGEROUS, subsidiary, RADIOACTIVE, EXPLOSIVES 1.1, 1.2, 1.3, 1.4, 1.5, or 1.6 placards. (*172.334*)

White Bottom Combustible

A COMBUSTIBLE placard used to display an identification number must have a white bottom for rail transportation. This placard may be used for highway transportation. (*172.332*)

For illustrations of placards see the color charts at the back of the book.

Reserved

REGISTRATION

A National Registration Program, including an annual fee, is in place for persons who offer for transport or transport certain hazardous materials. The annual fee funds a nation-wide emergency response training and planning grant program for states, Indian tribes and local communities.

Who Must Register

Any person who offers for transport, or transports in foreign, interstate or intrastate commerce, any of the following, is subject to the Pipeline and Hazardous Materials Safety Administration's (PHMSA) hazardous materials registration and fee requirements:

- Any highway route-controlled quantity of a Class 7 (radioactive) material;
- More than 25 kg (55 pounds) of a Division 1.1, 1.2, 1.3 (explosive) material;
- More than one L (1.06 quarts) per package of a material extremely toxic by inhalation (i.e.,"material poisonous by inhalation," as defined in §171.8, that meets the criteria for "hazard zone A," as specified in §173.116(a) or §173.133(a));
- A hazardous material in a bulk packaging having a capacity equal to or greater than 13,248 L (3,500 gallons) for liquids or gases, or more than 13.24 cubic meters (468 cubic feet) for solids;
- A shipment in other than a bulk packaging (being offered or loaded at one loading facility using one transport vehicle) of 2,268 kg (5,000 pounds) gross weight or more of one class of hazardous materials for which placarding is required; or
- A quantity of hazardous material that requires placarding. Except farmers in direct support of farming operations. (*107.601*)

Exceptions

The following are excepted from the registration and fee requirements:

- An agency of the federal government;
- A state agency;
- An agency of a political subdivision of a state;
- An Indian tribe;
- An employee of any of the above agencies;
- A hazmat employee — including the owner/operator of a motor vehicle that transports hazardous materials and is leased to a registered motor carrier under a 30-day or longer lease (or an equivalent contractual relationship);
- A person domiciled outside the U.S., who offers from locations outside the U.S., hazardous materials for transportation in commerce, provided the country in which the person is domiciled does not require U.S. persons to register or pay a fee. (*107.606*)

Registration Form

Each person subject to the program must submit a complete and accurate "Hazardous Materials Registration Statement" on DOT Form F 5800.2 and the required fee to PHMSA by June 30 of each year. A registration year begins on July 1 and ends on June 30 of the following year. (*107.608*)

Fee

Each person must pay an annual registration fee. The amount of the fee will vary, depending on the person's business classification (small business, not-for-profit, or other than small business and not-for-profit) and the registration year. See §107.612 for fee specifics.

Recordkeeping

Each person subject to the registration program must maintain the following at his or her principal place of business for three (3) years from the date of the certificate's issuance:

- A copy of the registration statement filed with PHMSA, and

REGISTRATION

- The Certificate of Registration issued by PHMSA. (*107.620*)

Motor carriers and persons who transport hazardous materials by vessel that are subject to the registration requirements must carry, on board **each** vehicle that is transporting a hazardous material requiring registration:

- A copy of the carriers' current Certificate of Registration; or
- Another document bearing the registration number identified as the "U.S. DOT Hazmat Reg. No.". (*107.620*)

Information on the registration program and its requirements can be obtained by calling PHMSA at (202) 366-4109.

HIGHWAY TRANSPORT

The transportation of hazardous materials by highway, as with other modes of transport, requires compliance with additional requirements specific to the mode of transport. These additional highway requirements are addressed below.

Accepting a Shipment

A motor carrier may not accept for transportation or transport any hazardous material classed as "Forbidden" in the 172.101 Hazardous Materials Table or hazardous materials not properly classed, described, packaged, marked, labeled or not in a condition for transport as required by the Hazardous Materials Regulations. (*171.2*)

A motor carrier may not transport any hazardous materials that require a Hazardous Materials Safety Permit, unless the carrier holds a safety permit issued by the Federal Motor Carrier Safety Administration. (*385.404*)

No person may offer a motor carrier any hazardous materials that require a Hazardous Materials Safety Permit, unless the carrier holds a safety permit issued by the Federal Motor Carrier Safety Administration. (*173.22*)

CDL Hazardous Materials Endorsement

The driver of a commercial motor vehicle that is required to be placarded for hazardous materials or contains any quantity of a material listed as a select agent or toxin in 42 CFR part 73 must have a hazardous materials endorsement on their Commercial Driver's License. (*383.93*)

Papers

Shipping papers

Hazardous materials shipments must be accompanied by proper shipping papers, such as bills of lading, hazardous waste manifests, etc. During the course of the trip, the driver is responsible for maintaining the shipping papers according to requirements, so that they are easily accessible to authorities in the event of an incident, accident or inspection.

- If the hazardous material shipping paper is carried with any other papers, it must be clearly distinguished, either by tabbing it or having it appear first.
- When the driver is at the controls, the shipping papers must be within immediate reach when the driver is restrained by the seat belt.
- The shipping papers must be readily visible to someone entering the driver's compartment, or in a holder mounted on the inside of the door on the driver's side.
- If the driver is not in the vehicle, the shipping papers must be either in the holder on the door or on the driver's seat. (*177.817*)

Each carrier receiving a shipping paper must retain a copy or an electronic image of the shipping paper for one year after the material is accepted. Each shipping paper copy must include the date of acceptance by the carrier. Hazardous waste shipping papers must be retained for three years after the material is accepted by the initial carrier. (*177.817*)

Emergency response information

Most hazardous material shipments (except those that do not require shipping papers or an ORM-D material) must have emergency response information on or in association with the shipping paper. If the information is in association with the shipping paper it may be in the form of the Emergency Response Guidebook, a Material Safety Data Sheet, or any other form that provides all the information required in §172.602. (*172.602*)

In most cases, an emergency response telephone number must be entered on the shipping paper. It can be immediately following the description of each hazardous material or if the number applies to every hazardous material entered on the shipping paper, entered once on the shipping paper in a clearly visible location. (*172.604*)

When the person offering hazardous materials for transport is also the emergency response telephone information provider they must enter their name on the shipping paper near the emergency response telephone number. In addition, when the telephone number of an emergency response information (ERI) telephone service provider is used, the name of the person who is registered with the ERI service provider, or the contract number or other unique identifier assigned by the ERI provider must be placed on the shipping paper near the emergency response telephone number. However, the above requirements do not apply if the name or identifier is entered elsewhere on the shipping paper in a prominent, readily identifiable, and clearly visible manner to be easily and quickly found. *(172.604)*

Proof of registration

Motor carriers subject to the federal registration requirements must carry, on board **each** vehicle that is transporting a hazardous material requiring registration:

- A copy of the carriers' current Certificate of Registration; or
- Another document bearing the registration number identified as the "U.S. DOT Hazmat Reg. No.". (*107.620*)

Class 1 (explosive) shipments

Some special documents are required to be carried by drivers of vehicles transporting Division 1.1, 1.2, or 1.3 (explosive) materials:

- A copy of Part 397, Transportation of Hazardous Materials; Driving and Parking Rules (Federal Motor Carrier Safety Regulations).
- A document containing instructions on what to do in the event of an accident or delay in the shipment. This information must also include the name of the Class 1 materials hauled and names and phone numbers of all persons to be contacted in the event of accident or delay.
- Proper shipping papers.
- A written route plan for the movement of Class 1 materials that complies with the requirements. (*397.19*)

Hazardous waste shipment

A driver **must not** accept a shipment of hazardous waste unless it is accompanied by a properly prepared Uniform Hazardous Waste Manifest.

Delivery of hazardous wastes **must be** made ONLY to the facility or alternate facility designated on the hazardous waste manifest. If delivery can not be made, the driver should contact his dispatcher or other designated official, immediately.

Loading/Unloading

Securing packages

Any package containing any hazardous material that is not permanently attached to a vehicle must be secured against any shifting, including motion between packages, within the vehicle during normal transportation conditions. Packages having valves or other fittings must be loaded in a manner to minimize the likelihood of damage during transportation. (*177.834*)

Each package bearing orientation arrow markings must be loaded on a vehicle or within a freight container in accordance with the arrow markings. The package must remain in this position during transportation. (*177.834*)

No smoking

Smoking on or near any vehicle while loading or unloading any Class 1 (explosive), Class 3 (flammable liquid), Class 4 (flammable solid), Class 5 (oxidizer), or Division 2.1 (flammable gas) materials is forbidden. Further, care should be taken to keep all fire sources — matches and smoking materials in particular — away from vehicles hauling any of the above materials. (*177.834*)

Set handbrake

During the loading and unloading of any hazardous materials shipment the handbrake on the vehicle must be set, and all precautions taken to prevent movement of the vehicle. (*177.834*)

Tools

Any tools used in loading or unloading hazardous material must be used with care not to damage the closures on any packages or containers, or to harm packages of Class 1 (explosive) material and other hazardous materials in any way. (*177.834*)

Attending vehicles

There are attendance requirements for **cargo tanks** that are being loaded and unloaded with hazardous materials. Such a tank must be attended at all times during loading and unloading by a qualified person. The person who is responsible for loading the cargo is also responsible for seeing that the vehicle is attended. However, the carrier's obligation to oversee unloading ceases when all these conditions are met:

- The carrier's transportation obligation is completed.
- The cargo tank is placed on the consignee's premises.
- Motive power is removed from the cargo tank and the premises. (*177.834*)

A person qualified to attend the cargo tank during loading and/or unloading must be:

- Alert.
- Within 7.62 m (25 feet) of the cargo tank.
- Have an unobstructed view of the cargo tank and delivery hose to the maximum extent practicable.
- Aware of the nature of the hazardous material.
- Instructed in emergency procedures.
- Authorized to and capable of moving the cargo tank if necessary. (*177.834*)

Segregation

The regulations contain segregation requirements for highway transport that indicate which hazardous materials may not be loaded, transported or stored together.

Materials which are in packages that must be labeled or placarded, in a compartment within a multi-compartmented cargo tank, or in a portable tank loaded in a transport vehicle or freight container are subject to the segregation requirements. (*177.848*)

In addition to the following Table and §173.12(e), cyanides or cyanide mixtures or solutions may not be stored, loaded and transported with acids if a mixture of the materials would generate hydrogen cyanide; Division 4.2 materials may not be stored, loaded and transported with Class 8 liquids; and Division 6.1 Packing Group I, Hazard Zone A material may not be stored, loaded and transported with Class 3 material, Class 8 liquids, and Division 4.1, 4.2, 4.3, 5.1 or 5.2 materials. (*177.848*)

Except as mentioned in the previous paragraph hazardous materials may not be loaded, transported, or stored together for highway transport, except as provided in the following Table. (*177.848*)

SEGREGATION TABLE FOR HAZARDOUS MATERIALS

Class or division		Notes	1.1 1.2	1.3	1.4	1.5	1.6	2.1	2.2	2.3 gas Zone A	2.3 gas Zone B	3	4.1	4.2	4.3	5.1	5.2	6.1 liquids PG I Zone A	7	8 liquids only
Explosives	1.1 and 1.2	A	*	*	*	*	*	X	X	X	X	X	X	X	X	X	X	X	X	X
Explosives	1.3		*	*	*	*	*	X		X	X	X		X	X	X	X	X		X
Explosives	1.4		*	*	*	*	*	O		O	O	O		O				O		O
Very insensitive explosives.	1.5	A	*	*	*	*	*	X	X	X	X	X	X	X	X	X	X	X	X	X
Extremely insensitive explosives.	1.6		*	*	*	*	*													
Flammable gases	2.1		X	X	O	X				X	O							O	O	
Non-toxic, non-flammable gases.	2.2		X			X														
Poisonous gas Zone A	2.3		X	X	O	X		X				X	X	X	X	X	X			X
Poisonous gas Zone B	2.3		X	X	O	X		O				O	O	O	O	O	O			O
Flammable liquids	3		X	X	O	X				X	O					O		X		

HIGHWAY TRANSPORT

SEGREGATION TABLE FOR HAZARDOUS MATERIALS, Continued

Class or division		Notes	1.1 1.2	1.3	1.4	1.5	1.6	2.1	2.2	2.3 gas Zone A	2.3 gas Zone B	3	4.1	4.2	4.3	5.1	5.2	6.1 liquids PG I Zone A	7	8 liquids only
Flammable solids.	4.1		X			X				X	O							X		O
Spontaneously combustible materials.	4.2		X	X	O	X				X	O							X		X
Dangerous when wet materials.	4.3		X	X		X				X	O							X		O
Oxidizers	5.1	A	X	X		X				X	O	O						X		O
Organic peroxides	5.2		X	X		X				X	O							X		O
Poisonous liquids PG I Zone A.	6.1		X	X	O	X		O				X	X	X	X	X	X			X
Radioactive materials . .	7		X			X		O												
Corrosive liquids	8		X	X	O	X				X	O		O	X	O	O	O	X		

Instructions for using the segregation table are as follows:

The absence of any hazard class or division or a blank space in the Table indicates that no restrictions apply.

The letter "X" in the Table indicates that these materials may not be loaded, transported, or stored together in the same transport vehicle or storage facility during the course of transportation.

The letter "O" in the Table indicates that these materials may not be loaded, transported, or stored together in the same transport vehicle or storage facility during the course of transportation unless separated in a manner that, in the event of leakage from packages under conditions normally incident to transportation, commingling of hazardous materials would not occur. Notwithstanding the methods of separation employed, Class 8 (corrosive) liquids may not be loaded above or adjacent to Class 4 (flammable) or Class 5 (oxidizing) materials; except that shippers may load truckload shipments of such materials together when it is known that the mixture of contents would not cause a fire or a dangerous evolution of heat or gas.

The "*" in the Table indicates that segregation among different Class 1 (explosive) materials is governed by the compatibility table for Class 1 (explosive) materials.

The note "A" in the second column of the Table means that, notwithstanding the requirements of the letter "X", ammonium nitrate (UN1942) and ammonium nitrate fertilizer may be loaded or stored with Division 1.1 (explosive) or Division 1.5 materials.

When the §172.101 Table or §172.402 requires a package to bear a subsidiary hazard label, segregation appropriate to the subsidiary hazard must be applied when that segregation is more restrictive than that required by the primary hazard. However, hazardous materials of the same class may be stowed together without regard to segregation required any secondary hazard if the materials are not capable of reacting dangerously with each other and causing combustion or dangerous evolution of heat, evolution of flammable, poisonous, or asphyxiant gases, or formation of corrosive or unstable materials. (*177.848*)

Placards

The Hazardous Materials Regulations require most vehicles hauling hazardous goods to be placarded. **The shipper is responsible for providing the appropriate placards** to the motor carrier for a shipment. **The carrier is responsible for applying them correctly to the vehicle** and maintaining them during transport (*172.506*). In addition, carriers are responsible for any placarding necessitated by aggregate shipments which collect at their terminals. Large freight containers (640 cubic feet or more) must be placarded by the shipper.

Placards must be affixed on all four sides of the vehicle, trailer or cargo carrier. (*172.504*) The front placard may be on the front of the tractor or the cargo body. (*172.516*) Placards must be removed from any vehicle not carrying hazardous materials. (*172.502*)

On the Road

Railroad crossings

Any marked or placarded vehicle (except Divisions 1.5, 1.6, 4.2, 6.2, and Class 9), any cargo tank motor vehicle, loaded or empty, used to transport any hazardous material, or a vehicle carrying any amount of chlorine, must stop at railroad crossings. (*392.10*)

Stops must be made within 50 feet of the crossing, but no closer than 15 feet. When you determine it is safe to cross the tracks, you may do so, but do not shift gears while crossing the tracks. (*392.10*)

Stops need not be made at:

- Streetcar crossings or industrial switching tracks within a business district.
- Crossings where a police officer or flagman is directing traffic.
- Crossings which are controlled by a stop-and-go traffic light which is green.
- Abandoned rail lines and industrial or spur line crossings clearly marked "exempt". (*392.10*)

Tunnels

Unless there is no other practicable route, marked or placarded shipments of hazardous materials should not be driven through tunnels. Operating convenience cannot be used as a determining factor in such decisions. (*177.810*)

In addition, the provisions of the Hazardous Materials Regulations do not supersede state or local laws and ordinances which may be more restrictive concerning hazardous materials and urban vehicular tunnels used for mass transportation. (*177.810*)

Routing

A motor carrier transporting hazardous materials required to be marked or placarded shall operate the vehicle over routes which do not go through or near heavily populated areas, places where crowds are assembled, tunnels, narrow streets, or alleys, except when:

- There is no practicable alternative;
- It is necessary to reach a terminal, points of loading or unloading, facilities for food, fuel, rest, repairs, or a safe haven; or
- A deviation is required by emergency conditions. (*397.67*)

Operating convenience is not a basis for determining if a route can be used. A motor carrier shall also comply with the routing designations of States or Indian tribes as authorized by federal regulations. (*397.67*)

Attending vehicles

Any marked or placarded vehicle containing hazardous materials which is on a public street or highway or the shoulder of any such road must be attended by the driver. Except when transporting Division 1.1, 1.2, or 1.3 materials, the vehicle does not have to be attended when the driver is performing duties necessary to the operation of the vehicle. (*397.5*)

What exactly does "attended" mean? In terms of the regulations a motor vehicle is attended when the person in charge is on the vehicle and awake (cannot be in the sleeper berth) or is within 100 feet of the vehicle and has an unobstructed view of it. (*397.5*)

Fueling

When a marked or placarded vehicle is being fueled, the engine must be shut off and a person must be in control of the fueling process at the point where the fuel tank is filled. (*397.15*)

Parking

Marked or placarded vehicles containing hazardous materials should not be parked on or within five feet of the traveled portion of any roadway. If the vehicle does not contain Division 1.1, 1.2, or 1.3 material, it may be stopped for brief periods when operational necessity requires parking the vehicle, and it would be impractical to stop elsewhere. Further restrictions apply to vehicles hauling Division 1.1, 1.2, or 1.3 (explosive) materials. Standard warning devices are to be set out as required by law when a vehicle is stopped along a roadway. (*397.7*)

HIGHWAY TRANSPORT

Emergency carrier information contact

If a transport vehicle (semi-trailer or freight container-on-chassis) contains hazardous materials that require shipping papers and the vehicle is separated from its motive power and parked at a location [other than a facility operated by the consignor or consignee or a facility subject to the emergency response information requirements in §172.602(c)(2)] the carrier must:

- Mark the vehicle with the telephone number of the motor carrier on the front exterior near the brake hose and electrical connections, or on a label, tag, or sign attached to the vehicle at the brake hose or electrical connection; or
- Have the shipping paper and emergency response information readily available on the transport vehicle. (*172.606*)

These requirements do not apply if the vehicle is marked on an orange panel, a placard, or a plain white square-on-point configuration with the identification number of each hazardous material contained within. The identification number(s) must be visible on the outside of the vehicle. (*172.606*)

Carriers must instruct drivers of vehicles transporting hazardous materials that require shipping papers to contact them (e.g., by telephone or radio) in the event of an incident involving the hazardous material. (*172.606*)

Tire checks

The driver of any vehicle which must be marked or placarded because it contains hazardous materials, must examine each tire on their vehicle at the beginning of each trip and each time the vehicle is parked. (*397.17*)

If any defect is found in a tire, it should be repaired or replaced immediately. The vehicle may, however, be driven a short distance to the nearest safe place for repair. (*397.17*)

If a hot tire is found, it must be removed from the vehicle immediately and placed at a safe distance from the vehicle. Such a vehicle may not be operated until the cause of the overheating is corrected. (*397.17*)

Texting

A person transporting a quantity of hazardous materials that requires placarding or any quantity of a material listed as a select agent or toxin in 42 CFR part 73 must not engage in, allow, or require texting while driving. (*177.804*)

Smoking

No person may smoke or carry any lighted smoking materials on or within 25 feet of marked or placarded vehicles containing any Class 1 (explosive) materials, Class 5 (oxidizer) materials or flammable materials classified as Division 2.1, Class 3, Divisions 4.1 and 4.2, or any empty tank vehicle that has been used to transport Class 3 (flammable) liquid or Division 2.1 (flammable gas) materials. (*397.13*)

Fires

A marked or placarded vehicle containing hazardous materials should not be driven near an open fire; unless careful precautions have been taken to be sure the vehicle can completely pass the fire without stopping. In addition, a marked or placarded vehicle containing hazardous materials should not be parked within 300 feet of any open fire. (*397.11*)

Damaged packages

Packages of hazardous materials that are damaged or found leaking during transportation, and hazardous materials that have spilled or leaked during transportation, may be forwarded to their destination or returned to the shipper in a salvage drum in accordance with the regulations. (*177.854*)

Packages may be repaired in accordance with the best and safest practice known and available. (*177.854*)

Any package repaired in accordance with the requirements in the regulations may be transported to the nearest place where it may safely be disposed. This may be done only if the following requirements are met:

- The package must be safe for transportation.
- The repair of the package must be adequate to prevent contamination of other lading or producing a hazardous mixture with other lading transported on the same motor vehicle.

- If the carrier is not the shipper, the consignee's name and address must be plainly marked on the repaired package. (*177.854*)

In the event any leaking package or container cannot be safely and adequately repaired for transportation or transported, it shall be stored pending proper disposition in the safest and most expeditious manner possible. (*177.854*)

Vehicle maintenance

No person may use heat, flame or spark producing devices to repair or maintain the cargo or fuel containment system of a motor vehicle required to be placarded, other than COMBUSTIBLE. The containment system includes all vehicle components intended physically to contain cargo or fuel during loading or filling, transport, or unloading. (*177.854*)

Vehicle emergency movement

As a rule, a driver may not move a vehicle that is not properly marked and placarded, if required by the regulations. However, in certain emergency situations, a vehicle not properly marked and placarded may be moved if at least one of the following conditions is met:

- Vehicle is escorted by a state or local government representative.
- Carrier has received permission from the Department of Transportation.
- Movement is necessary to protect life and property. (*177.823*)

Hazard warning signals

It is recommended that flame producing signals not be used when transporting hazardous materials of any type. The use of flame producing signals is specifically prohibited for any vehicle transporting Division 1.1, 1.2 or 1.3 (explosive) materials or any cargo tank motor vehicle used to transport Class 3 (flammable liquids) or Division 2.1 (flammable gas) materials, loaded or empty. In place of flame producing signals, emergency reflective triangles, red electric lanterns, or red emergency reflectors should be used. Emergency signals shall be used as required in 392.22 of the Motor Carrier Safety Regulations. (*392.25*)

Reserved

INCIDENT REPORTING

Whenever there is a spillage, discharge, or leakage of hazardous materials, including hazardous wastes and hazardous substances or there is direct involvement of the hazardous materials arising out of an accident, special reports must be made in accordance with the following:

Immediate Notice of Certain Hazardous Materials Incidents

Immediate notice to the National Response Center must be made as soon as practical, but no later than 12 hours after the occurrence of any hazardous materials incident. The notice must be made by the person in physical possession of the hazardous material. (*171.15*)

Immediate notice is required whenever any of the following occur during transportation in commerce. This includes loading, unloading, and temporary storage.

(a) A person is killed.

(b) A person receives an injury requiring hospitalization.

(c) An evacuation of the general public occurs lasting one or more hours.

(d) A major transportation artery or facility is closed or shut down for one hour or more.

(e) The operational flight pattern or routine of an aircraft is altered.

(f) Fire, breakage, spillage or suspected radioactive contamination involving a shipment of radioactive materials.

(g) Fire, breakage, spillage or suspected contamination involving a shipment of infectious substances, other than a regulated medical waste.

(h) A release of a marine pollutant occurs in a quantity exceeding 450 L (119 gal.) for liquids or 400 kg (882 lbs.) for solids.

(i) During transport by aircraft, a fire, violent rupture, explosion or dangerous evolution of heat occurs as a direct result of a battery or a battery-powered device.

(j) In the judgement of the person in possession of the material there exists a situation which should be reported even though it does not meet one of the specific criteria listed above. (*171.15*)

NATIONAL RESPONSE CENTER
Toll-free: (800) 424-8802
Toll call: (202) 267-2675
Online: *http://www.nrc.uscg.mil*

Each immediate notice must provide the following information:

(a) Name of reporter.

(b) Name and address of person represented by reporter.

(c) Phone number where reporter can be contacted.

(d) Date, time and location of incident.

(e) Extent of injuries, if any.

(f) Class or division, proper shipping name and quantity of hazardous materials involved, if such information is available.

(g) Type of incident, nature of hazardous materials involvement, and whether or not there is a continuing danger to life at the scene. (*171.15*)

Detailed Hazardous Materials Incident Reports

Each person in physical possession of a hazardous material at the time of an incident must submit a Hazardous Materials Incident Report. A written Hazardous Materials Incident Report must be made within 30 days of discovery of an incident arising out of the transportation, loading, unloading or temporary storage of hazardous materials as follows:

(a) As a follow-up to any incident requiring an immediate notice as required by §171.15;

(b) As the result of an unintentional release of hazardous materials or any quantity of hazardous waste;

(c) A specification cargo tank (1,000 gallons or more) suffers structural damage to the lading retention system or damage that requires repair to a system intended to protect the lading retention system, even though there is no release of material;

(d) An undeclared shipment is discovered; or

(e) A fire, violent rupture, explosion or dangerous evolution of heat occurs as a direct result of a battery or a battery-powered device. (*171.16*)

Except when an immediate notice is required (§171.15), Hazardous Materials Incident Reports are not required for the following incidents:

1. A release of a minimal amount of material from—
   - A vent, for materials for which venting is authorized;
   - The routine operation of a seal, pump, compressor, or valve; or
   - Connection or disconnection of loading or unloading lines, provided that the release does not result in property damage.
2. An unintentional release of material when—
   - The material is properly classed as ORM-D or is a Packing Group III material in Class or Division 3, 4, 5, 6.1, 8, or 9;
   - Each package has a capacity of less than 20 liters (5.2 gallons) for liquids or less than 30 kg (66 pounds) for solids;
   - The total aggregate release is less than 20 liters (5.2 gallons) for liquids or less than 30 kg (66 pounds) for solids; and
   - The material is not offered for transportation or transported by aircraft, is not a hazardous waste or is not an undeclared hazardous material.
3. An undeclared hazardous material discovered in an air passenger's checked or carry-on baggage during the airport screening process. (*171.16*)

The incident report must be updated within one year of the date of occurrence of the incident whenever:

1. A death results from injury caused by the material;
2. There was a misidentification of the material or package information on a prior report;
3. Damage, loss or related cost that was not known when the initial report was filed becomes known;
4. Damage, loss or related cost changes by $25,000 or more, or 10% of the prior total estimate, whichever is greater. (*171.16*)

The incident report shall be made on the prescribed form DOT F 5800.1. (*171.16*)

The written incident report should be sent to:

Information Systems Manager, PHH-60
Pipeline and Hazardous Materials Safety Administration
Department of Transportation
East Building, 1200 New Jersey Ave., SE.
Washington, D.C. 20590-0001

In place of the written incident report an electronic incident report may be submitted at *http://hazmat.dot.gov.*

In filing a Hazardous Materials Incident Report, you should endeavor to provide all of the information required to complete the report and to provide as much information as possible about the incident. These reports are analyzed by DOT to discover unsafe or inefficient packagings' of hazardous materials, and to make such changes in the hazardous materials regulations as experience indicates are necessary for safety.

A copy of the Hazardous Materials Incident Report shall be retained, for a period of two years, at the reporting person's principal place of business. If the report is maintained at other than the reporting person's principal place of business, the report must be made available at the reporting person's principal place of business within 24 hours of a request by an authorized official. (*171.16*)

The filing of the reports mentioned above does not relieve you from the responsibility to report spillages of hazardous materials, or spillages of other materials that may create a pollution problem to local, state and federal agencies concerned with environmental controls as required in the area where any such spillage occurs.

Hazardous Substance Discharge Notification

Certain releases of hazardous substances must be reported to the National Response Center at (toll free) (800) 424-8802 or (toll) (202) 267-2675. Only spills of hazardous substances which equal or exceed the designated reportable quantity are required to be reported, however, it is recommended that all spills be reported. The report must be made as soon as a person has knowledge of the discharge. (*40 CFR 302.6*)

Releases of reportable quantities of an extremely hazardous substance listed in Title 40, Section 355.42, when they occur in transportation or storage incidental to transportation must be immediately reported to the local emergency planning committee if available or to the state emergency planning commission, or as a last resort by dialing 911 or calling the operator. (*40 CFR 355.42*)

Reserved

PUBLIC RELATIONS IN EMERGENCIES

Introduction

The transportation of hazardous materials is becoming increasingly complex with the introduction of new products, increased transportation demands and more hazardous materials than ever before being transported at any given time. Incidents have been few, but increased interest by enforcement agencies, news media and the public in the potential dangers involved in the transportation of hazardous materials has resulted in some citations for violations of the regulations and in bad publicity for the industry.

This section explains how public relations planning can be made an integral part of a company's action program. It is intended to assist in handling emergencies and communicating necessary information to each other and to the public. It suggests plans of action which can be used to help insure against bad publicity, exaggerations, distortions and misunderstandings.

Responsibility

When an emergency occurs, there are three major areas of responsibility for the company's management and employees:

- Personnel (care for the injured and protect against further injury)
- Property (confine the disaster to the smallest possible area and protect property and equipment)
- Public (the basic facts must be given to an interested and anxious public as quickly and accurately and as completely as possible)

With regard to Personnel, one public relations responsibility is to find out who is involved and get full information on them. Contact relatives before releasing names to the press. Where Property is concerned, such as equipment involved in an incident, refer to specific models in reporting on the incident.

In getting the story to the Public, be sure to include safety measures taken, the previous good safety record of your firm and of the employees involved. Be sure to note in your disaster reporting any uncontrollable factors which might have contributed to the cause of the emergency (poor visibility, gas leak in terminal area, etc.). And follow up as more information becomes available with calls to the news media regarding your company's efforts to clean up after the incident.

A Written Plan

Don't rely on anyone to remember what to do in a hazardous materials emergency. Put the plans in writing. Give copies to everyone who might be involved, directly or indirectly, in the plan's execution (don't overlook telephone operators).

Have on hand an alphabetical listing of hazardous materials with instructions on how to deal with them in emergencies (such as the Emergency Response Guidebook). These instructions, besides telling you what to do, can provide the factual basis for your press releases. This is a very important point since it is virtually impossible for an individual to know about all the dangerous commodities that may be handled. More than 15,000 are listed and classified as hazardous materials and there are thousands of shipments of these materials made daily in the United States.

Emergency Information

Include in the emergency instructions the suggestion that each terminal appoint an emergency chief for each shift. It would be this person's responsibility to pass on information to reporters as fast as he can get accurate data. During periods of delay in assembling facts, the information chief should reassure the press and electronic media representatives that information is coming as quickly as possible.

Reporters and photographers may want to visit the scene of a fire or explosion. The decision as to when to permit them to do so must be based fundamentally on safety considerations. As soon as the key person determines it is safe for a small number of reporters to visit the scene, the information chief should arrange to escort them on a tour, pointing out action the company has taken to contain the disaster.

PUBLIC RELATIONS IN EMERGENCIES

Importance of Attitude

Chances are that newspaper reporters, photographers and electronic media representatives will arrive at the scene of the disaster expecting to be met with denials, flat refusals of information, thin excuses for non-cooperation or an arbitrary prohibition against photographs. They have come to have that attitude after bitter experiences.

The attitude of the company representative who deals with the media will make a big difference in the handling of news about your emergency. To quote one public relations manual, "Never try to cover up, no matter what the circumstances. It's regrettable, but not a public disgrace to have an accident. To have one in print after you have denied it is highly embarrassing to a company...Remember, the newspapers have free access to hospitals, coroners, fire and police departments and to your own employees. They will get the news. See that they get it correctly."

Just as bad as covering up is speculating or spreading rumor unfounded in fact. Only established facts should be given. Never make guesses. It is much better to say, "the cause has not been determined" or "due to safety and security reasons, we cannot allow visitors to approach the scene of the accident as yet."

For the best possible treatment by the media, be forthrightly honest, openly reasonable, willing to cooperate and ready to explain delays. Recognition and proper handling of these public relations responsibilities will enhance the reputation of your company as a responsible organization in the community where the emergency occurs.

One Company's Procedure

Many companies already have a specific order of action for emergency procedures. Here is one firm's recommended course of action for the emergency information chief following an incident. (First, division management and other specified company personnel are notified.)

1. Attend first to the three P's - personnel, property and public.
2. Inform the public through the news media.
3. Report all available facts concerning the incident and assure news personnel that additional information will be forthcoming as soon as possible.
4. Allow news personnel to visit the scene as soon as the area is safe. Be sure they are accompanied by a company representative.
5. Make sure the news personnel understand the scope of the emergency, as it is common for them to exaggerate the seriousness of it.
6. Inform families of injured personnel first, before releasing their names to the press.
7. Make telephone facilities available to the news personnel when possible.
8. Explain safety procedures and safeguards that have been established to protect against emergencies.
9. Government officials will usually investigate. Cooperate fully and be sure they are assisted in every way.

Positive Actions in Between

Work with firefighters and police departments in your community before emergencies force it. Not all policemen and firemen are familiar with how to handle hazardous materials. Go over your emergency action plans with them and with policemen, and have them tour the facility so they will be familiar with it should an emergency ever occur. This action will also make them aware of your company's desire to cooperate, to be a good citizen, and will help when your reputation is on the line in an emergency.

MATERIALS OF TRADE

Definition

Materials of Trade (MOTs) are hazardous materials that are carried on a motor vehicle:

- For the purpose of protecting the health and safety of the motor vehicle operator or passengers (such as insect repellant or self-contained breathing apparatus);
- For the purpose of supporting the operation or maintenance of a motor vehicle, including its auxiliary equipment (such as a spare battery, gasoline, or engine starting fluid); or
- By a private motor carrier (including vehicles operated by a rail carrier) in direct support of a principal business that is other than transportation by motor vehicle (such as lawn care, plumbing, welding, or painting operations). (*173.6*)

Hazard and Size Limitations

A material of trade (not including self-reactive material, poisonous by inhalation material, or hazardous waste) is limited to:

- A Class 3, 8, 9, Division 4.1, 5.1, 5.2, 6.1, or ORM-D material contained in a packaging having a gross mass or capacity not over—
  - 0.5 kg (1 pound) or 0.5 L (1 pint) for Packing Group I material,
  - 30 kg (66 pounds) or 30 L (8 gallons) for a Packing Group II , III , or ORM-D material,
  - 1500 L (400 gallons) for a diluted mixture, not to exceed 2 percent concentration, of a Class 9 material;
- A Division 2.1 or 2.2 material in a cylinder with a gross weight not over 100 kg (220 pounds);
- A non-liquefied Division 2.2 material with no subsidiary hazard in a permanently mounted tank manufactured to the ASME Code at not more than 70 gallons water capacity; or
- A Division 4.3 material in Packing Group II or III contained in a packaging having a gross capacity not exceeding 30 mL (1 ounce).
- A Division 6.2 material, other than a Category A infectious substance, contained in human or animal samples being transported for research, diagnosis, investigational activities, or disease treatment or prevention or is a biological product or regulated medical waste.
  - The material must be contained in a combination packaging. For liquids the inner packaging must be leakproof, and the outer packaging must contain sufficient absorbent material to absorb the entire contents of the inner packaging.
  - For sharps, the inner packaging must be constructed of a rigid material resistant to punctures and securely closed to prevent leaks or punctures, and the outer packaging must be securely closed to prevent leaks or punctures.
  - For solids, liquids, and sharps, the outer packaging must be a strong, tight packaging securely closed and secured against movement, including relative motion between packages, within the vehicle on which it is being transported.
  - For other than regulated medical waste, the amount of Division 6.2 material in a combination packaging must be in:
    1. One or more inner packagings, each of which may not contain more than 0.5 kg (1.1 lbs) or 0.5 L (17 ounces), and an outer packaging containing not more than 4 kg (8.8 lbs) or 4 L (1 gal); or
    2. A single inner packaging containing not more than 16 kg (35.2 lbs) or 16 L (4.2 gal) in a single outer packaging.
  - For regulated medical waste, a combination packaging must consist of one or more inner packagings, each of which may not contain more than 4 kg (8.8 lbs) or 4 L (1 gal), and an outer packaging containing not more than 16 kg (35.2 lbs) or 16 L (4.2 gal).

Gross Weight

The gross weight of all MOTs on a motor vehicle may not exceed 200 kg (440 pounds), not including 1,500 L (400 gallons) or less of a diluted mixture of Class 9 material, as mentioned above. (*173.6*)

MATERIALS OF TRADE

Markings

A non-bulk packaging, other than a cylinder, must be marked with a common name or proper shipping name to identify the material it contains (such as spray paint or Isopropyl Alcohol).

The letters "RQ" must be included if the packaging contains a reportable quantity of a hazardous substance.

A bulk packaging containing a diluted mixture (≤2%) of a Class 9 material must be marked on two opposing sides with the identification number. (*173.6*)

Packaging

Packaging for MOTs must be leak tight for liquids and gases, sift proof for solids, and be securely closed, secured against shifting, and protected against damage. Each material must be packaged in the manufacturer's original packaging or a packaging of equal or greater strength and integrity. Packaging for gasoline must be made of metal or plastic and conform to the Hazardous Materials Regulations or OSHA regulations.

Outer packagings are not required for receptacles (e.g. cans and bottles) that are secured against shifting in cages, carts, bins, boxes or compartments. (*173.6*)

Cylinders

A cylinder or other pressure vessel containing a Division 2.1 or 2.2 material must conform to the marking, labeling, packaging, qualification, maintenance, and use requirements of the Hazardous Materials Regulations, except that outer packagings are not required. Manifolding of cylinders is authorized provided all valves are tightly closed. (*173.6*)

Drivers

The operator of a motor vehicle that contains MOTs must be informed of the presence of the material and must be informed of the MOTs requirements in Section 173.6 of the Hazardous Materials Regulations. (*173.6*)

Exceptions

The MOTs requirements do not require:

- Shipping papers;
- Emergency response information;
- Placarding;
- Formal hazmat training; or
- Retention of training records.

LIMITED QUANTITIES

Definition

The definition for a limited quantity is "when specified as such in a section applicable to a particular material, means the maximum amount of a hazardous material for which there is a specific labeling or packaging exception." (*171.8*)

Limited quantities are smaller amounts of hazardous materials in non-bulk combination packages. A limited quantity exception can only be used for a hazardous material if Column 8A of the Hazardous Materials Table references the specific section in Part 173 that contains the exception.

The sections in Part 173 that contain the limited quantity requirements for the different hazard classes and divisions are:

173.306 - Class 2 (gases)
173.150 - Class 3 (flammable and combustible liquids)
173.151 - Class 4 (flammable solids)
173.152 - Class 5 (oxidizers and organic peroxides)
173.153 - Division 6.1 (poisons)
173.421 - Class 7 (radioactives)
173.154 - Class 8 (corrosives)
173.155 - Class 9 (miscellaneous)

Although not identified as limited quantities, Class 1 has some similar exceptions listed in Section 173.63 for certain less hazardous explosive materials.

Packaging

In each instance, the limited quantity packaging must conform to the specifications, quantity and weight limits specified for each hazard class/division and packing group. These specifications will be listed in the section in Part 173 (listed in Column 8A of the Hazardous Materials Table) that is applicable to the material.

Except for Class 1 and 7, each combination package may not exceed 66 pounds (30 kg) gross weight. The size of inner packagings that are allowed to be used varies depending on the hazard/division and packing group of the material. For example, Class 8 materials in packing group III must be in inner packagings not over 1.3 gallons (5 L) for liquids or not over 11 pounds (5 kg) for solids. These inner packagings must be packed in a strong outer packaging.

Markings

A package containing a limited quantity of hazardous materials is not required to be marked with the proper shipping name and identification number when marked with the square-on-point limited quantity marking or for packages meeting the requirements for air transport, marked with the square-on-point "Y" limited quantity marking. (*172.315*)

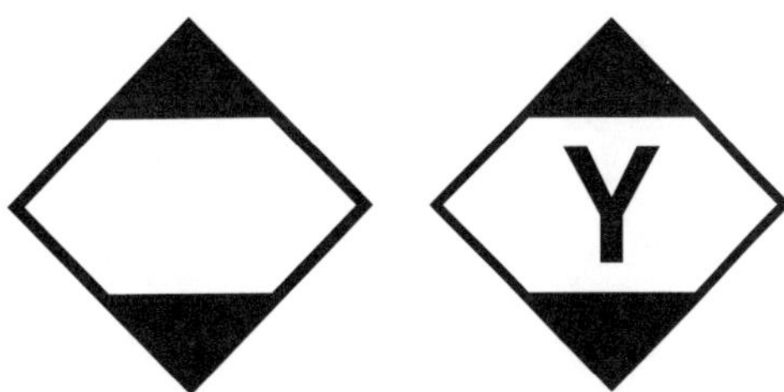

The width of the border forming the square-on-point marking must be at least 2 mm. Each side of the marking must be at least 100 mm. If package size requires a reduced size marking, then each side of the marking may be reduced to no less than 50 mm. For cargo transport units transported by water, each side of the marking must have a minimum size of 250 mm. (*172.315*)

LIMITED QUANTITIES

Labels

Labels are not required for most limited quantities of hazardous materials. However, labels are required for a limited quantity transported by aircraft and for poison-by-inhalation materials. (*173.150 - 173.155*)

Shipping Papers

Shipping papers are required for a limited quantity of a hazardous material that is a hazardous substance, hazardous waste, marine pollutant, or is offered/transported by air or water. When a shipping paper is required the words "Limited Quantity" or the abbreviation "Ltd Qty" must follow the basic description for the hazardous material. (*172.203*)

Training

If you offer or transport limited quantities, your employees are performing functions subject to the Hazardous Materials Regulations and they must receive hazmat training. Before these hazmat employees perform any function they must be trained, tested and certified as required by the Hazardous Materials Regulations. (*172.704*)

Exceptions

Listing an emergency response telephone number on the shipping papers is not required for limited quantities. (*172.604(c)*)

Most limited quantities are excepted from the specification packaging, shipping papers, labeling, and placarding requirements. For examples of this see §173.150 through §173.155.

Be sure to read the exceptions section in Part 173 (listed in Column 8A of the Hazardous Materials Table) that is applicable to your material. This section will specify the exceptions allowed and the requirements that must be met to offer and transport the material as a limited quantity.

Compliance Dates

Before January 1, 2013, for transport by aircraft, a limited quantity properly marked as an ORM-D-AIR is not required to be marked with the square-on-point "Y" limited quantity marking. (*172.315*)

Before January 1, 2014, a limited quantity properly marked as an ORM-D is not required to be marked with the square-on-point limited quantity marking. (*172.315*)

CONSUMER COMMODITIES

Definition

The official definition for a consumer commodity is "a material that is packaged and distributed in a form intended or suitable for sale through retail sales agencies or instrumentalities for consumption by individuals for purposes of personal care or household use. This term also includes drugs and medicines." (*171.8*)

A consumer commodity (ORM-D) is a small amount of hazardous materials that can be purchased at grocery, hardware, and discount stores. This includes products such as bleach, aerosols, drain cleaners, lighter fluid, cosmetics, and small arms ammunition.

Before a hazardous material can even be considered to be a consumer commodity, it must first meet the limited quantity requirements for the material. If the material meets these requirements and is packaged as a limited quantity, then the regulations in the section referenced in Part 173 (listed in Column 8A of the Hazardous Materials Table) must be checked to see if the material is allowed to be a consumer commodity. Not all limited quantities are allowed to be reclassified as ORM-D. If the material is allowed to be a consumer commodity and it meets the definition of a consumer commodity, then the material may be renamed "consumer commodity" and reclassified as ORM-D.

The ORM-D classification and markings are being phased out of the hazardous materials regulations. They are being replaced with the new limited quantity requirements and markings.

Before January 1, 2013, an air shipment that meets the requirements for an ORM-D may be marked with ORM-D-AIR. Before January 1, 2014, packages of ORM-D material may be marked with the ORM-D marking. Starting on the above January dates, only the appropriate square-on-point limited quantity marking may be used and the ORM-D markings must not be used. *(172.316)*

Packaging

Before a hazardous material can be considered a consumer commodity it must first meet the limited quantity packaging requirements for the material. In each instance, the limited quantity packaging must conform to the specifications, quantity and weight limits specified for each hazard class/division and packing group. These specifications will be listed in the section in Part 173 (listed in Column 8A of the Hazardous Materials Table) that is applicable to the material.

Markings

Packages that contain a material classed as a consumer commodity (ORM-D) must be marked on at least one side or end with the ORM-D designation - immediately following or below the proper shipping name. If the shipment is to be by air, the marking must be ORM-D-AIR. (*172.316*)

Both the ORM-D and ORM-D-AIR markings must be within a rectangle that is approximately 6.3 mm (0.25 in) larger on each side than the designation. (*172.316*)

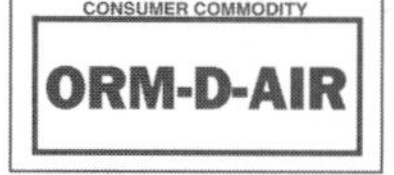

Some shippers mark packages of consumer commodities with the limited quantity marking and the ORM-D marking. This dual marking is not prohibited by the regulations. Interpretations from PHMSA acknowledge that a material packaged as a limited quantity and properly classed as ORM-D may display both the limited quantity marking and ORM-D marking.

Before January 1, 2013, an air shipment that meets the requirements for an ORM-D may be marked with ORM-D-AIR. Before January 1, 2014, packages of ORM-D material may be marked with the ORM-D marking. Starting on the above January dates, only the appropriate square-on-point limited quantity marking may be used and the ORM-D markings must not be used. *(172.316)*

Training

If you offer or transport consumer commodities, your employees are performing functions subject to the Hazardous Materials Regulations and they must receive hazmat training. Before these hazmat employees perform any function they must be trained, tested and certified as required by the Hazardous Materials Regulations. (*172.704*)

CONSUMER COMMODITIES

Exceptions

Consumer commodities are generally excepted from specification packaging, labeling, placarding, shipping paper, and emergency response information requirements. Some of these exceptions are not allowed if transport is by aircraft or the material is a hazardous substance, hazardous waste, or marine pollutant.

Be sure to read the exceptions section in Part 173 (listed in Column 8A of the Hazardous Materials Table) that is applicable to your material. This section will specify the exceptions allowed and the requirements that must be met to offer and transport the material as a consumer commodity.

Compliance Dates

The ORM-D classification and markings are being phased out of the hazardous materials regulations. They are being replaced with the new limited quantity requirements and markings.

Before January 1, 2013, an air shipment that meets the requirements for an ORM-D may be marked with ORM-D-AIR. Before January 1, 2014, packages of ORM-D material may be marked with the ORM-D marking. Starting on the above January dates, only the appropriate square-on-point limited quantity marking may be used and the ORM-D markings must not be used. *(172.316)*

HAZARDOUS WASTES

Identification of Hazardous Waste

A Hazardous Waste is defined by the Environmental Protection Agency (EPA) under the Resource Conservation and Recovery Act (RCRA) as a waste that is dangerous or potentially harmful to our health or the environment. Hazardous wastes can be liquids, solids, gases, or sludges. They can be discarded commercial products, like cleaning fluids or pesticides, or the by-products of manufacturing processes.

The EPA has identified more than 700 specific wastes which are hazardous and assigned each an EPA Hazardous Waste Number. These appear in lists such as the F-list (non-specific source wastes), K-list (source-specific wastes), and the P-list and U-list (discarded commercial chemical products). However, other wastes not listed by EPA may still exhibit one or more of the hazardous characteristics. If your waste is not an EPA-listed hazardous waste, it is important to sample and analyze the waste, or identify the waste based on process knowledge to determine if it exhibits a hazardous characteristic.

There are four characteristic classifications for all hazardous wastes and specific testing procedures are prescribed for determination of classification by the generator.

These characteristics, with a simple definition of each, are:

1. Ignitability: a liquid, solid, semi-solid, or contained gaseous material which has a primary hazard of fire ignition or flammability; or an oxidizer as defined in 49 CFR 173.127. Any waste material which, after prescribed testing, exhibits the characteristic of ignitability, but is not listed as an identified waste is assigned the EPA Hazardous Waste Number D001.
2. Corrosivity: a material whose primary hazard in transportation is that of corrosion. Any waste material which, after prescribed testing, exhibits the characteristics of corrosivity, but is not listed as an identified waste, is assigned the EPA Hazardous Waste Number D002.
3. Reactivity: a material which is normally unstable or which reacts with water, heat sources, or other materials to generate toxic vapors or fumes; or is potentially explosive. Any waste material which, after prescribed testing, exhibits the characteristic of reactivity, but is not listed as an identified waste, is assigned the EPA Hazardous Waste Number D003.
4. Toxicity: a material which has a primary hazard in transportation of having adverse effects on human health or environmental quality such as poisons. Any waste material which, after prescribed testing, exhibits the characteristic of toxicity, is assigned the EPA Hazardous Waste Number, D004 - D043, which corresponds to the identified contaminant which makes it toxic. These identification numbers and contaminants are listed in 40 CFR 261.24.

It is the responsibility of the generator to identify the hazardous waste on a properly prepared waste manifest. Transporters are not permitted to accept a shipment of hazardous waste unless the manifest has been properly completed and signed by the generator.

EPA Transporter Number

Any carrier engaged in the transportation of any hazardous waste must obtain an EPA identification number. (*40 CFR 263.11*)

Waste Manifest

The term "manifest" is used to identify the shipping paper on which all hazardous wastes must be listed and described. A properly prepared uniform hazardous waste manifest with a prescribed number of copies is required to accompany each hazardous waste shipment from point of origin to final destination. In addition to manifest requirements specified in Sections 263.20 through 263.22 of 40 CFR, transporters must also comply with the shipping paper requirements of Sections 172.200 through 172.205 and 177.817 of 49 CFR.

EPA has mandated a nationally uniform hazardous waste manifest (EPA Form 8700-22) which is required to accompany all hazardous waste shipments. If more than four waste materials are included in one shipment, a continuation sheet (EPA Form 8700-22A) must be prepared.

HAZARDOUS WASTES

Manifest tracking system

EPA has established a manifest tracking system to monitor the flow of hazardous waste from "cradle to grave." This system consists of the following procedures:

1. A transporter may not accept a hazardous waste shipment from a generator unless it is accompanied by a properly prepared and signed hazardous waste manifest. The generator's signature must be entered manually. (*40 CFR 262*) This differs from the DOT hazardous materials shipping paper requirements (*49 CFR 172.204*) which permit the shipper's signature to be entered by mechanical means.
2. The transporter must sign and date the manifest before leaving the generator's facility. If more than one transporter will be used, the generator must supply additional copies of the manifest for each of the transporters. It is the transporter's responsibility to assure that the manifest accompanies the shipment of hazardous waste. (*40 CFR 263.20*)
3. Upon delivery to either the designated disposal facility or to another transporter, a handwritten signature must be obtained on the manifest from the consignee to document the delivery. (*40 CFR 263.20*)
4. The transporter must keep one copy of the completed manifest, and provide the disposal facility or interline carrier with the remaining copies of the manifest. (*40 CFR 263.22*)
5. Each transporter must retain a copy of the manifest in its files for a period of three years from the date the hazardous waste was accepted by the initial transporter. (*40 CFR 263.22, 49 CFR 172.205*)
6. The disposal facility must within 30 days after delivery mail one copy of the manifest to the generator to complete the paperwork chain and retain a copy for itself (*40 CFR 264.71*). If the generator does not receive the completed manifest within 35 days of the date the waste was accepted by the initial transporter, the generator must contact the transporter and/or the disposal facility. If after 45 days of the date the waste was accepted by the initial transporter, the generator still has not received the completed manifest from the disposal facility, the generator must notify EPA and file an Exception Report. (*40 CFR 262.42*)

Manifest requirements

Federal EPA regulations require the following information to appear on the uniform waste manifest:

1. Generator's U.S. EPA identification number;
2. Total number of pages (i.e. page 1 of 2);
3. Emergency response phone number;
4. Manifest tracking number;
5. Generator's name, address, phone number, and site address, if applicable;
6. First transporter's name and U.S. EPA identification number;
7. Second transporter's name and U.S. EPA identification number, if applicable;
8. Designated facility's name, address, phone number, and U.S. EPA identification number;
9. U.S. DOT description (shipping name, hazard class/division, ID number, packing group);
10. Number and type of containers using the appropriate abbreviation:

    BA = Burlap, cloth, paper, or plastic bags
    CF = Fiber or plastic boxes, cartons, cases
    CM = Metal boxes, cartons, cases (including roll-offs)
    CW = Wooden boxes, cartons, cases
    CY = Cylinders
    DF = Fiberboard or plastic drums, barrels, kegs
    DM = Metal drums, barrels, kegs
    DT = Dump truck
    DW = Wooden drums, barrels, kegs
    HG = Hopper or gondola cars
    TC = Tank cars
    TP = Portable tanks
    TT = Cargo tanks (tank trucks);
11. Total quantity;

HAZARDOUS WASTES

12. Units of measure using the appropriate abbreviation:

    G = Gallons (liquid only)
    K = Kilograms
    L = Liters (liquid only)
    M = Metric tons (1,000 kg)
    N = Cubic meters
    P = Pounds
    T = Tons (2,000 lbs)
    Y = Cubic yards;

13. Waste codes;
14. Special handling instructions;
15. Generator/Offeror certifications.

Pictured below is an example of the Uniform Hazardous Waste Manifest:

Please print or type. (Form designed for use on elite (12-pitch) typewriter.) Form Approved. OMB No. 2050-0039

UNIFORM HAZARDOUS WASTE MANIFEST | 1. Generator ID Number | 2. Page 1 of | 3. Emergency Response Phone | **4. Manifest Tracking Number**

GENERATOR

5. Generator's Name and Mailing Address | Generator's Site Address (if different than mailing address)
Generator's Phone:

6. Transporter 1 Company Name | U.S. EPA ID Number

7. Transporter 2 Company Name | U.S. EPA ID Number

8. Designated Facility Name and Site Address | U.S. EPA ID Number
Facility's Phone:

9a. HM	9b. U.S. DOT Description (including Proper Shipping Name, Hazard Class, ID Number, and Packing Group (if any))	10. Containers		11. Total Quantity	12. Unit Wt./Vol.	13. Waste Codes
		No.	Type			
	1.					
	2.					
	3.					
	4.					

14. Special Handling Instructions and Additional Information

15. **GENERATOR'S/OFFEROR'S CERTIFICATION:** I hereby declare that the contents of this consignment are fully and accurately described above by the proper shipping name, and are classified, packaged, marked and labeled/placarded, and are in all respects in proper condition for transport according to applicable international and national governmental regulations. If export shipment and I am the Primary Exporter, I certify that the contents of this consignment conform to the terms of the attached EPA Acknowledgment of Consent.
I certify that the waste minimization statement identified in 40 CFR 262.27(a) (if I am a large quantity generator) or (b) (if I am a small quantity generator) is true.

Generator's/Offeror's Printed/Typed Name | Signature | Month | Day | Year

INT'L

16. International Shipments ☐ Import to U.S. ☐ Export from U.S. Port of entry/exit:
Transporter signature (for exports only): Date leaving U.S.:

TR ANSPORTER

17. Transporter Acknowledgment of Receipt of Materials
Transporter 1 Printed/Typed Name | Signature | Month | Day | Year
Transporter 2 Printed/Typed Name | Signature | Month | Day | Year

DESIGNATED FACILITY

18. Discrepancy
18a. Discrepancy Indication Space ☐ Quantity ☐ Type ☐ Residue ☐ Partial Rejection ☐ Full Rejection
Manifest Reference Number:
18b. Alternate Facility (or Generator) | U.S. EPA ID Number
Facility's Phone:
18c. Signature of Alternate Facility (or Generator) | Month | Day | Year

19. Hazardous Waste Report Management Method Codes (i.e., codes for hazardous waste treatment, disposal, and recycling systems)
1. | 2. | 3. | 4.

20. Designated Facility Owner or Operator: Certification of receipt of hazardous materials covered by the manifest except as noted in Item 18a
Printed/Typed Name | Signature | Month | Day | Year

EPA Form 8700-22 (Rev. 3-05) Previous editions are obsolete. DESIGNATED FACILITY TO DESTINATION STATE (IF REQUIRED)

SAMPLE

HAZARDOUS WASTES

Package Marking

It is the responsibility of the generator to use specification packaging for DOT regulated wastes. The generator is also responsible for marking and labeling each package in compliance with the requirements of 49 CFR. The information provided in this section is intended to provide assistance to transporter personnel in recognizing properly prepared packages and determining regulatory compliance prior to transporting.

Each package or container of hazardous waste, except portable tanks, cargo tanks, or bulk containers, offered for transportation by highway must be marked and labeled as follows:

1. The proper shipping name of the hazardous waste;
2. The identification number from 49 CFR 172.101, if the material is regulated by DOT;
3. The name and address of the generator or TSD facility unless: (a) the package is transported by highway and will not be transferred from one transporter to another, or (b) the package is part of a truckload or freight container load from one generator to one TSD facility;
4. A package of 119 gallons or less must be marked with the following wording and information:

   HAZARDOUS WASTE - Federal Law Prohibits Improper Disposal. If found, contact the nearest police or public safety authority or the US Environmental Protection Agency.
   Generator's Name and Address ________________
   Generator's EPA Identification Number ________________
   Manifest Tracking Number ________________
   (40 CFR 262.32)

NOTE: When this marking is affixed to a package, the word "Waste" is not required to be used in the proper shipping name on the package.

5. The DOT hazard class label(s) when required *(49 CFR 172.400)*.

Vehicle placards must be displayed when transporting hazardous wastes which require placarding under DOT regulations (*49 CFR 172.504*).

NOTE: When transporting packages containing over 1 liter of liquids identified as "toxic by inhalation", additional labeling, marking, placarding, and shipping paper requirements are prescribed in 49 CFR Part 172.

Transfer Facility

A transporter may store manifested shipments of hazardous wastes in Department of Transportation specification containers at a transfer facility for a period of ten days or less without being subject to EPA storage requirements in 40 CFR Parts 264, 265, 268, and 270. *(40 CFR 263.12)*

Discharges

In the event of a discharge of hazardous waste during transportation, the transporter must take appropriate action to protect human health and the environment.

If a discharge of hazardous waste occurs during transportation and an official (state or local government or federal agency) acting within the scope of his or her official responsibilities determines that immediate removal of the waste is necessary to protect human health or the environment, that official may authorize the removal of the waste by transporters who do not have an EPA identification number and without the preparation of a manifest.

The transporter must immediately notify the **National Response Center at: (800) 424-8802 or (202) 267-2675** in the event of a discharge of a hazardous waste that meets any of the criteria in 49 CFR 171.15. (*40 CFR 263.30*)

It is the transporter's responsibility to clean up any hazardous waste discharge that occurs during transportation or take such action as may be required or approved by federal, state, or local officials so that the discharge no longer presents a hazard to human health or the environment. *(40 CFR 263.31)*

EPA Identified Hazardous Wastes

EPA has identified more than 700 materials and industrial by-products that pose a threat to human health or the environment. Generators of wastes that are not specifically identified and listed must analyze the material to determine if it meets the characteristics of a hazardous waste. If so, all the applicable requirements for manifesting, packaging, labeling, marking, and placarding must be followed.

HAZARDOUS WASTES

Three lists of hazardous wastes are included in 40 CFR Part 261. They identify wastes that are from specific sources, non-specific sources, and discarded commercial products. These lists include the EPA Hazardous Waste Number for each waste, the waste name, and the hazard code. One or more of the following hazard codes will be the basis for the material being listed in one of the tables:

Ignitable Wastes.......... (I)
Corrosive Wastes .... (C)
Reactive Wastes .... (R)
Toxicity Characteristic Wastes.......... (E)
Acute Hazardous Wastes.......... (H)
Toxic Wastes .... (T)

NOTE: Certain constituents listed in 40 CFR Appendix VII to Part 261 cause the wastes to be identified as Toxicity Characteristic (E); or Toxic (T).

In addition to those wastes listed by the EPA, hazardous materials identified by the Department of Transportation in the Hazardous Materials Table *(49 CFR 172.101)* that are intended for disposal, and which meet one of the EPA hazard categories are hazardous wastes. These will be designated on the hazardous waste manifest by the word "Waste" immediately preceding the proper shipping name of the material as it appears in the table.

Use of the Hazardous Waste Tables

To determine the proper shipping name for wastes to be transported, you must first know how to identify them. There are three lists of Hazardous Wastes in Part 261, Title 40 CFR; they are: (1) Section 261.31 Hazardous wastes from non-specific sources; (2) Section 261.32 Hazardous wastes from specific sources; and (3) Section 261.33 Discarded commercial chemical products, off-specification species, container residues, and spill residues thereof (2 separate lists are included in this section). Of these, which one would you use for shipment of your waste?

In Section 261.33 there are two separate lists. Both lists identify specific chemicals. The difference between the two is that one lists acutely hazardous wastes (identified as "P" wastes) while the other lists toxic chemicals (identified as "U" wastes). These two lists are utilized when the discarded material meets the definition and characteristics of the virgin material.

Example: There are several drums of Acetone collecting dust in your stockroom. Some of the drums have been opened and a small quantity of the Acetone has been used. For whatever reason, your company no longer needs the material, nor wants it taking up space. Since the material is unchanged characteristically from when it was new, you would go to Section 261.33 and see if you can find it on one of the lists of specific chemicals. You will not find it on the list of "P" wastes; however, Acetone is listed as U002 on the list of "U" wastes. There is no need to look further.

Section 261.31 deals with hazardous wastes from non-specific sources or Waste Streams (identified as "F" wastes). These wastes are the by-products of a variety of manufacturing processes and treatments. This would be the section to turn to when considering how to identify your spent solvents for transportation.

Example: Instead of discarding your Acetone as in the previous example, you utilize it in a manufacturing process. The by-product is a mixture of things, one of which is the spent non-halogenated solvent-Acetone. The material you now have is not virgin Acetone and can not be discarded as such. You would utilize the list in Section 261.31 where you would find Acetone listed in the waste stream F003.

Section 261.32 identifies hazardous wastes from specific sources. These specific source wastes are identified as "K" wastes and range from "Ammonia still lime sludge from coking operations" to various Wastewater treatment sludges.

HAZARDOUS WASTES

If, by now, you have utilized all the lists and still have not identified your wastes for transportation, there is one final step to take. Subpart C of Part 261 deals with identification of materials according to the characteristic (identified as "D" wastes) they exhibit. Materials not identified in the aforementioned lists, but exhibit one or more of the characteristics must be identified as required in Subpart C of Part 261.

ID CROSS REFERENCE

The identification number cross reference index to proper shipping names in the §172.101 Table is useful in determining a shipping name when only an identification number is known. It is also useful in checking if an identification number is correct for a specific shipping description. This listing is for information purposes only, the §172.101 Hazardous Materials Table should be consulted for authorized shipping names and identification numbers.

IDENTIFICATION NUMBER CROSS REFERENCE INDEX TO PROPER SHIPPING NAMES IN §172.101

NOTE: This listing is provided for information purposes only.

(1)-Identification Number	(2)-Description
UN0004	Ammonium picrate
UN0005	Cartridges for weapons
UN0006	Cartridges for weapons
UN0007	Cartridges for weapons
UN0009	Ammunition, incendiary
UN0010	Ammunition, incendiary
UN0012	Cartridges for weapons, inert projectile or Cartridges, small arms
UN0014	Cartridges for weapons, blank or Cartridges, small arms, blank
UN0015	Ammunition, smoke
UN0016	Ammunition, smoke
UN0018	Ammunition, tear-producing
UN0019	Ammunition, tear-producing
UN0020	Ammunition, toxic
UN0021	Ammunition, toxic
NA0027	Black powder for small arms
UN0027	Black powder or Gunpowder
UN0028	Black powder, compressed or Gunpowder, compressed or Black powder, in pellets or Gunpowder, in pellets
UN0029	Detonators, non-electric
UN0030	Detonators, electric
UN0033	Bombs
UN0034	Bombs
UN0035	Bombs
UN0037	Bombs, photo-flash
UN0038	Bombs, photo-flash
UN0039	Bombs, photo-flash
UN0042	Boosters
UN0043	Bursters
UN0044	Primers, cap type
UN0048	Charges, demolition
UN0049	Cartridges, flash
UN0050	Cartridges, flash
UN0054	Cartridges, signal
UN0055	Cases, cartridge, empty with primer
UN0056	Charges, depth
UN0059	Charges, shaped
UN0060	Charges, supplementary explosive
UN0065	Cord, detonating
UN0066	Cord, igniter
UN0070	Cutters, cable, explosive
UN0072	Cyclotrimethylenetrinitramine, wetted or Cyclonite, wetted or Hexogen, wetted or RDX, wetted
UN0073	Detonators for ammunition
UN0074	Diazodinitrophenol, wetted
UN0075	Diethyleneglycol dinitrate, desensitized
UN0076	Dinitrophenol
UN0077	Dinitrophenolates
UN0078	Dinitroresorcinol
UN0079	Hexanitrodiphenylamine or Dipicrylamine or Hexyl
UN0081	Explosive, blasting, type A
UN0082	Explosive, blasting, type B
UN0083	Explosive, blasting, type C
UN0084	Explosive, blasting, type D
UN0092	Flares, surface
UN0093	Flares, aerial
UN0094	Flash powder
UN0099	Fracturing devices, explosive
UN0101	Fuse, non-detonating
UN0102	Cord detonating or Fuse detonating
UN0103	Fuse, igniter
UN0104	Cord, detonating, mild effect or Fuse, detonating, mild effect
UN0105	Fuse, safety
UN0106	Fuzes, detonating
UN0107	Fuzes, detonating
UN0110	Grenades, practice
UN0113	Guanyl nitrosaminoguanylidene hydrazine, wetted
UN0114	Guanyl nitrosaminoguanyltetrazene, wetted or Tetrazene, wetted
UN0118	Hexolite or Hexotol
UN0121	Igniters
NA0124	Jet perforating guns, charged oil well, with detonator
UN0124	Jet perforating guns, charged
UN0129	Lead azide, wetted
UN0130	Lead styphnate, wetted or Lead trinitroresorcinate, wetted
UN0131	Lighters, fuse
UN0132	Deflagrating metal salts of aromatic nitroderivatives, n.o.s
UN0133	Mannitol hexanitrate, wetted or Nitromannite, wetted
UN0135	Mercury fulminate, wetted
UN0136	Mines
UN0137	Mines
UN0138	Mines
UN0143	Nitroglycerin, desensitized
UN0144	Nitroglycerin, solution in alcohol
UN0146	Nitrostarch
UN0147	Nitro urea
UN0150	Pentaerythrite tetranitrate wetted or Pentaerythritol tetranitrate wetted or PETN, wetted, or Pentaerythrite tetranitrate, or Pentaerythritol tetranitrate, or PETN, desensitized
UN0151	Pentolite
UN0153	Trinitroaniline or Picramide
UN0154	Trinitrophenol or Picric acid
UN0155	Trinitrochlorobenzene or Picryl chloride
UN0159	Powder cake, wetted or Powder paste, wetted
UN0160	Powder, smokeless
UN0161	Powder, smokeless
UN0167	Projectiles
UN0168	Projectiles
UN0169	Projectiles
UN0171	Ammunition, illuminating
UN0173	Release devices, explosive
UN0174	Rivets, explosive
UN0180	Rockets
UN0181	Rockets
UN0182	Rockets
UN0183	Rockets
UN0186	Rocket motors
UN0190	Samples, explosive
UN0191	Signal devices, hand
UN0192	Signals, railway track, explosive
UN0193	Signals, railway track, explosive
UN0194	Signals, distress
UN0195	Signals, distress
UN0196	Signals, smoke
UN0197	Signals, smoke
UN0204	Sounding devices, explosive
UN0207	Tetranitroaniline
UN0208	Trinitrophenylmethylnitramine or Tetryl
UN0209	Trinitrotoluene or TNT
UN0212	Tracers for ammunition
UN0213	Trinitroanisole
UN0214	Trinitrobenzene
UN0215	Trinitrobenzoic acid
UN0216	Trinitro-m-cresol
UN0217	Trinitronaphthalene
UN0218	Trinitrophenetole
UN0219	Trinitroresorcinol or Styphnic acid
UN0220	Urea nitrate
UN0221	Warheads, torpedo
UN0222	Ammonium nitrate
UN0224	Barium azide
UN0225	Boosters with detonator
UN0226	Cyclotetramethylenetetranitramine, wetted or HMX, wetted or Octogen, wetted
UN0234	Sodium dinitro-o-cresolate
UN0235	Sodium picramate
UN0236	Zirconium picramate
UN0237	Charges, shaped, flexible, linear
UN0238	Rockets, line-throwing
UN0240	Rockets, line-throwing
UN0241	Explosive, blasting, type E
UN0242	Charges, propelling, for cannon
UN0243	Ammunition, incendiary, white phosphorus
UN0244	Ammunition, incendiary, white phosphorus
UN0245	Ammunition, smoke, white phosphorus
UN0246	Ammunition, smoke, white phosphorus

ID CROSS REFERENCE

(1)-Identification Number	(2)-Description
UN0247	Ammunition, incendiary
UN0248	Contrivances, water-activated
UN0249	Contrivances, water-activated
UN0250	Rocket motors with hypergolic liquids
UN0254	Ammunition, illuminating
UN0255	Detonators, electric
UN0257	Fuzes, detonating
UN0266	Octolite or Octol
UN0267	Detonators, non-electric
UN0268	Boosters with detonator
UN0271	Charges, propelling
UN0272	Charges, propelling
UN0275	Cartridges, power device
NA0276	Model rocket motor
UN0276	Cartridges, power device
UN0277	Cartridges, oil well
UN0278	Cartridges, oil well
UN0279	Charges, propelling, for cannon
UN0280	Rocket motors
UN0281	Rocket motors
UN0282	Nitroguanidine or Picrite
UN0283	Boosters
UN0284	Grenades
UN0285	Grenades
UN0286	Warheads, rocket
UN0287	Warheads, rocket
UN0288	Charges, shaped, flexible, linear
UN0289	Cord, detonating
UN0290	Cord, detonating or Fuse, detonating
UN0291	Bombs
UN0292	Grenades
UN0293	Grenades
UN0294	Mines
UN0295	Rockets
UN0296	Sounding devices, explosive
UN0297	Ammunition, illuminating
UN0299	Bombs, photo-flash
UN0300	Ammunition, incendiary
UN0301	Ammunition, tear-producing
UN0303	Ammunition, smoke
UN0305	Flash powder
UN0306	Tracers for ammunition
UN0312	Cartridges, signal
UN0313	Signals, smoke
UN0314	Igniters
UN0315	Igniters
UN0316	Fuzes, igniting
UN0317	Fuzes, igniting
UN0318	Grenades, practice
UN0319	Primers, tubular
UN0320	Primers, tubular
UN0321	Cartridges for weapons
UN0322	Rocket motors with hypergolic liquids
NA0323	Model rocket motor
UN0323	Cartridges, power device
UN0324	Projectiles
UN0325	Igniters
UN0326	Cartridges for weapons, blank
UN0327	Cartridges for weapons, blank or Cartridges, small arms, blank
UN0328	Cartridges for weapons, inert projectile
UN0329	Torpedoes
UN0330	Torpedoes
NA0331	Ammonium nitrate-fuel oil mixture
UN0331	Explosive, blasting, type B or Agent blasting, Type B
UN0332	Explosive, blasting, type E or Agent blasting, Type E
UN0333	Fireworks
UN0334	Fireworks
UN0335	Fireworks
UN0336	Fireworks
NA0337	Toy caps
UN0337	Fireworks
UN0338	Cartridges for weapons, blank or Cartridges, small arms, blank
UN0339	Cartridges for weapons, inert projectile or Cartridges, small arms
UN0340	Nitrocellulose
UN0341	Nitrocellulose
UN0342	Nitrocellulose, wetted
UN0343	Nitrocellulose, plasticized
UN0344	Projectiles
UN0345	Projectiles
UN0346	Projectiles
UN0347	Projectiles
UN0348	Cartridges for weapons
UN0349	Articles, explosive, n.o.s.
UN0350	Articles, explosive, n.o.s.
UN0351	Articles, explosive, n.o.s.
UN0352	Articles, explosive, n.o.s.
UN0353	Articles, explosive, n.o.s.
UN0354	Articles, explosive, n.o.s.
UN0355	Articles, explosive, n.o.s.
UN0356	Articles, explosive, n.o.s.
UN0357	Substances, explosive, n.o.s.
UN0358	Substances, explosive, n.o.s.
UN0359	Substances, explosive, n.o.s.
UN0360	Detonator assemblies, non-electric
UN0361	Detonator assemblies, non-electric
UN0362	Ammunition, practice
UN0363	Ammunition, proof
UN0364	Detonators for ammunition
UN0365	Detonators for ammunition
UN0366	Detonators for ammunition
UN0367	Fuzes, detonating
UN0368	Fuzes, igniting
UN0369	Warheads, rocket
UN0370	Warheads, rocket
UN0371	Warheads, rocket
UN0372	Grenades, practice
UN0373	Signal devices, hand
UN0374	Sounding devices, explosive
UN0375	Sounding devices, explosive
UN0376	Primers, tubular
UN0377	Primers, cap type
UN0378	Primers, cap type
UN0379	Cases, cartridges, empty with primer
UN0380	Articles, pyrophoric
UN0381	Cartridges, power device
UN0382	Components, explosive train, n.o.s.
UN0383	Components, explosive train, n.o.s.
UN0384	Components, explosive train, n.o.s.
UN0385	5-Nitrobenzotriazol
UN0386	Trinitrobenzenesulfonic acid
UN0387	Trinitrofluorenone
UN0388	Trinitrotoluene and Trinitrobenzene mixtures or TNT and trinitrobenzene mixtures or TNT and hexanitrostilbene mixtures or Trinitrotoluene and hexanitrostilnene mixtures
UN0389	Trinitrotoluene mixtures containing Trinitrobenzene and Hexanitrostilbene or TNT mixtures containing trinitrobenzene and hexanitrostilbene
UN0390	Tritonal
UN0391	RDX and HMX mixtures, wetted, or RDX and HMX mixtures, desensitized
UN0392	Hexanitrostilbene
UN0393	Hexotonal
UN0394	Trinitroresorcinol, wetted or Styphnic acid, wetted
UN0395	Rocket motors, liquid fueled
UN0396	Rocket motors, liquid fueled
UN0397	Rockets, liquid fueled
UN0398	Rockets, liquid fueled
UN0399	Bombs with flammable liquid
UN0400	Bombs with flammable liquid
UN0401	Dipicryl sulfide
UN0402	Ammonium perchlorate
UN0403	Flares, aerial
UN0404	Flares, aerial
UN0405	Cartridges, signal
UN0406	Dinitrosobenzene
UN0407	Tetrazol-1-acetic acid
UN0408	Fuzes, detonating
UN0409	Fuzes, detonating
UN0410	Fuzes, detonating
UN0411	Pentaerythrite tetranitrate or Pentaerythritol tetranitrate or PETN
UN0412	Cartridges for weapons
UN0413	Cartridges for weapons, blank
UN0414	Charges, propelling, for cannon
UN0415	Charges, propelling
UN0417	Cartridges for weapons, inert projectile or Cartridges, small arms
UN0418	Flares, surface
UN0419	Flares, surface
UN0420	Flares, aerial
UN0421	Flares, aerial
UN0424	Projectiles
UN0425	Projectiles
UN0426	Projectiles
UN0427	Projectiles
UN0428	Articles, pyrotechnic
UN0429	Articles, pyrotechnic
UN0430	Articles, pyrotechnic
UN0431	Articles, pyrotechnic
UN0432	Articles, pyrotechnic
UN0433	Powder cake, wetted or Powder paste, wetted
UN0434	Projectiles
UN0435	Projectiles
UN0436	Rockets

(1)-Identification Number	(2)-Description
UN0437	Rockets
UN0438	Rockets
UN0439	Charges, shaped
UN0440	Charges, shaped
UN0441	Charges, shaped
UN0442	Charges, explosive, commercial
UN0443	Charges, explosive, commercial
UN0444	Charges, explosive, commercial
UN0445	Charges, explosive, commercial
UN0446	Cases, combustible, empty, without primer
UN0447	Cases, combustible, empty, without primer
UN0448	5-Mercaptotetrazol-1-acetic acid
UN0449	Torpedoes, liquid fueled
UN0450	Torpedoes, liquid fueled
UN0451	Torpedoes
UN0452	Grenades practice
UN0453	Rockets, line-throwing
UN0454	Igniters
UN0455	Detonators, non-electric
UN0456	Detonators, electric
UN0457	Charges, bursting, plastics bonded
UN0458	Charges, bursting, plastics bonded
UN0459	Charges, bursting, plastics bonded
UN0460	Charges, bursting, plastics bonded
UN0461	Components, explosive train, n.o.s.
UN0462	Articles, explosive, n.o.s.
UN0463	Articles, explosive, n.o.s.
UN0464	Articles, explosive, n.o.s.
UN0465	Articles, explosive, n.o.s.
UN0466	Articles, explosive, n.o.s.
UN0467	Articles, explosive, n.o.s.
UN0468	Articles, explosive, n.o.s.
UN0469	Articles, explosive, n.o.s.
UN0470	Articles, explosive, n.o.s.
UN0471	Articles, explosive, n.o.s.
UN0472	Articles, explosive, n.o.s.
UN0473	Substances, explosive, n.o.s.
UN0474	Substances, explosive, n.o.s.
UN0475	Substances, explosive, n.o.s.
UN0476	Substances, explosive, n.o.s.
UN0477	Substances, explosive, n.o.s.
UN0478	Substances, explosive, n.o.s.
UN0479	Substances, explosives, n.o.s.
UN0480	Substances, explosives, n.o.s.
UN0481	Substances, explosive, n.o.s.
UN0482	Substances, explosive, very insensitive, n.o.s., or Substances, EVI, n.o.s.
UN0483	Cyclotrimethylenetrinitramine, desensitize or Cyclonite, desensitized or Hexogen, desensitized or RDX, desensitized
UN0484	Cyclotetramethylenetetranitramine, desensitized or Octogen, desensitized or HMX, desensitized
UN0485	Substances, explosive, n.o.s.
UN0486	Articles, explosive, extremely insensitive or Articles, EEI
UN0487	Signals, smoke
UN0488	Ammunition, practice
UN0489	Dinitroglycoluril or Dingu
UN0490	Nitrotriazolone or NTO
UN0491	Charges, propelling
UN0492	Signals, railway track, explosive
UN0493	Signals, railway track, explosive
NA0494	Jet perforating guns, charged oil well, with detonator
UN0494	Jet perforating guns, charged
UN0495	Propellant, liquid
UN0496	Octonal
UN0497	Propellant, liquid
UN0498	Propellant, solid
UN0499	Propellant, solid
UN0500	Detonator assemblies, non-electric
UN0501	Propellant, solid
UN0502	Rockets
UN0503	Air bag inflators, or Air bag modules, or Seat-belt pretensioners
UN0504	1H-Tetrazole
UN0505	Signals, distress
UN0506	Signals, distress
UN0507	Signals, smoke
UN0508	1-Hydroxybenzotriazole, anhydrous
UN0509	Powder, smokeless
UN1001	Acetylene, dissolved
UN1002	Air, compressed
UN1003	Air, refrigerated liquid
UN1005	Ammonia, anhydrous
UN1006	Argon
UN1008	Boron trifluoride
UN1009	Bromotrifluoromethane or Refrigerant gas R 13B1
UN1010	Butadienes, stabilized or Butadienes and Hydrocarbon mixture, stabilized
UN1011	Butane
UN1012	Butylene
UN1013	Carbon dioxide
UN1016	Carbon monoxide, compressed
UN1017	Chlorine
UN1018	Chlorodifluoromethane or Refrigerant gas R22
UN1020	Chloropentafluoroethane or Refrigerant gas R115
UN1021	1-Chloro-1,2,2,2-tetrafluoroethane or Refrigerant gas R 124
UN1022	Chlorotrifluoromethane or Refrigerant gas R 13
UN1023	Coal gas, compressed
UN1026	Cyanogen
UN1027	Cyclopropane
UN1028	Dichlorodifluoromethane or Refrigerant gas R 12
UN1029	Dichlorofluoromethane or Refrigerant gas R 21
UN1030	1, 1-Difluoroethane or Refrigerant gas R 152a
UN1032	Dimethylamine, anhydrous
UN1033	Dimethyl ether
UN1035	Ethane
UN1036	Ethylamine
UN1037	Ethyl chloride
UN1038	Ethylene, refrigerated liquid
UN1039	Ethyl methyl ether
UN1040	Ethylene oxide or Ethylene oxide with nitrogen
UN1041	Ethylene oxide and Carbon dioxide mixtures
UN1043	Fertilizer ammoniating solution
UN1044	Fire extinguishers
UN1045	Fluorine, compressed
UN1046	Helium
UN1048	Hydrogen bromide, anhydrous
UN1049	Hydrogen, compressed
UN1050	Hydrogen chloride, anhydrous
UN1051	Hydrogen cyanide, stabilized
UN1052	Hydrogen fluoride, anhydrous
UN1053	Hydrogen sulfide
UN1055	Isobutylene
UN1056	Krypton, compressed
NA1057	Lighters
UN1057	Lighters or Lighter refills
UN1058	Liquefied gases
UN1060	Methyl acetylene and propadiene mixtures, stabilized
UN1061	Methylamine, anhydrous
UN1062	Methyl bromide
UN1063	Methyl chloride or Refrigerant gas R 40
UN1064	Methyl mercaptan
UN1065	Neon, compressed
UN1066	Nitrogen, compressed
UN1067	Dinitrogen tetroxide
UN1069	Nitrosyl chloride
UN1070	Nitrous oxide
UN1071	Oil gas, compressed
UN1072	Oxygen, compressed
UN1073	Oxygen, refrigerated liquid
UN1075	Petroleum gases, liquefied or Liquefied petroleum gas
UN1076	Phosgene
UN1077	Propylene
UN1078	Refrigerant gases, n.o.s.
UN1079	Sulfur dioxide
UN1080	Sulfur hexafluoride
UN1081	Tetrafluoroethylene, stabilized
UN1082	Trifluorochloroethylene, stabilized
UN1083	Trimethylamine, anhydrous
UN1085	Vinyl bromide, stabilized
UN1086	Vinyl chloride, stabilized
UN1087	Vinyl methyl ether, stabilized
UN1088	Acetal
UN1089	Acetaldehyde
UN1090	Acetone
UN1091	Acetone oils
UN1092	Acrolein, stabilized
UN1093	Acrylonitrile, stabilized
UN1098	Allyl alcohol
UN1099	Allyl bromide
UN1100	Allyl chloride
UN1104	Amyl acetates
UN1105	Pentanols
UN1106	Amylamines
UN1107	Amyl chlorides
UN1108	1-Pentene
UN1109	Amyl formates
UN1110	n-Amyl methyl ketone
UN1111	Amyl mercaptans
UN1112	Amyl nitrate
UN1113	Amyl nitrites
UN1114	Benzene
UN1120	Butanols
UN1123	Butyl acetates

ID CROSS REFERENCE

(1)-Identification Number	(2)-Description
UN1125	n-Butylamine
UN1126	1-Bromobutane
UN1127	Chlorobutanes
UN1128	n-Butyl formate
UN1129	Butyraldehyde
UN1130	Camphor oil
UN1131	Carbon disulfide
UN1133	Adhesives
UN1134	Chlorobenzene
UN1135	Ethylene chlorohydrin
UN1136	Coal tar distillates, flammable
UN1139	Coating solution
UN1143	Crotonaldehyde *or* Crotonaldehyde, stabilized
UN1144	Crotonylene
UN1145	Cyclohexane
UN1146	Cyclopentane
UN1147	Decahydronaphthalene
UN1148	Diacetone alcohol
UN1149	Dibutyl ethers
UN1150	1,2-Dichloroethylene
UN1152	Dichloropentanes
UN1153	Ethylene glycol diethyl ether
UN1154	Diethylamine
UN1155	Diethyl ether or Ethyl ether
UN1156	Diethyl ketone
UN1157	Diisobutyl ketone
UN1158	Diisopropylamine
UN1159	Diisopropyl ether
UN1160	Dimethylamine solution
UN1161	Dimethyl carbonate
UN1162	Dimethyldichlorosilane
UN1163	Dimethylhydrazine, unsymmetrical
UN1164	Dimethyl sulfide
UN1165	Dioxane
UN1166	Dioxolane
UN1167	Divinyl ether, stabilized
UN1169	Extracts, aromatic, liquid
UN1170	Ethanol or Ethyl alcohol or Ethanol solutions or Ethyl alcohol solutions
UN1171	Ethylene glycol monoethyl ether
UN1172	Ethylene glycol monoethyl ether acetate
UN1173	Ethyl acetate
UN1175	Ethylbenzene
UN1176	Ethyl borate
UN1177	2-Ethylbutyl acetate
UN1178	2-Ethylbutyraldehyde
UN1179	Ethyl butyl ether
UN1180	Ethyl butyrate
UN1181	Ethyl chloroacetate
UN1182	Ethyl chloroformate
UN1183	Ethyldichlorosilane
UN1184	Ethylene dichloride
UN1185	Ethyleneimine, stabilized
UN1188	Ethylene glycol monomethyl ether
UN1189	Ethylene glycol monomethyl ether acetate
UN1190	Ethyl formate
UN1191	Octyl aldehydes
UN1192	Ethyl lactate
UN1193	Ethyl methyl ketone or Methyl ethyl ketone
UN1194	Ethyl nitrite solutions
UN1195	Ethyl propionate
UN1196	Ethyltrichlorosilane
UN1197	Extracts, flavoring, liquid
UN1198	Formaldehyde solutions, flammable
UN1199	Furaldehydes
UN1201	Fusel oil
UN1202	Gas oil or Diesel fuel or Heating oil, light
NA1203	Gasohol
UN1203	Gasoline
UN1204	Nitroglycerin solution in alcohol
UN1206	Heptanes
UN1207	Hexaldehyde
UN1208	Hexanes
UN1210	Printing ink or Printing ink related material
UN1212	Isobutanol or isobutyl alcohol
UN1213	Isobutyl acetate
UN1214	Isobutylamine
UN1216	Isooctenes
UN1218	Isoprene, stabilized
UN1219	Isopropanol or isopropyl alcohol
UN1220	Isopropyl acetate
UN1221	Isopropylamine
UN1222	Isopropyl nitrate
UN1223	Kerosene
UN1224	Ketones, liquid, n.o.s.
UN1228	Mercaptans, liquid, flammable, toxic, n.o.s. or Mercaptan mixtures, liquid, flammable, toxic, n.o.s.
UN1229	Mesityl oxide
UN1230	Methanol
UN1231	Methyl acetate
UN1233	Methylamyl acetate
UN1234	Methylal
UN1235	Methylamine, aqueous solution
UN1237	Methyl butyrate
UN1238	Methyl chloroformate
UN1239	Methyl chloromethyl ether
UN1242	Methyldichlorosilane
UN1243	Methyl formate
UN1244	Methylhydrazine
UN1245	Methyl isobutyl ketone
UN1246	Methyl isopropenyl ketone, stabilized
UN1247	Methyl methacrylate monomer, stabilized
UN1248	Methyl propionate
UN1249	Methyl propyl ketone
UN1250	Methyltrichlorosilane
UN1251	Methyl vinyl ketone, stabilized
UN1259	Nickel carbonyl
UN1261	Nitromethane
UN1262	Octanes
UN1263	Paint
UN1263	Paint related material
UN1264	Paraldehyde
UN1265	Pentanes
UN1266	Perfumery products
UN1267	Petroleum crude oil
UN1268	Petroleum distillates, n.o.s. or Petroleum products, n.o.s.
NA1270	Petroleum oil
UN1272	Pine oil
UN1274	n-Propanol or Propyl alcohol, normal
UN1275	Propionaldehyde
UN1276	n-Propyl acetate
UN1277	Propylamine
UN1278	1-Chloropropane
UN1278	Propyl chloride
UN1279	1,2-Dichloropropane
UN1280	Propylene oxide
UN1281	Propyl formates
UN1282	Pyridine
UN1286	Rosin oil
UN1287	Rubber solution
UN1288	Shale oil
UN1289	Sodium methylate solutions
UN1292	Tetraethyl silicate
UN1293	Tinctures, medicinal
UN1294	Toluene
UN1295	Trichlorosilane
UN1296	Triethylamine
UN1297	Trimethylamine, aqueous solutions
UN1298	Trimethylchlorosilane
UN1299	Turpentine
UN1300	Turpentine substitute
UN1301	Vinyl acetate, stabilized
UN1302	Vinyl ethyl ether, stabilized
UN1303	Vinylidene chloride, stabilized
UN1304	Vinyl isobutyl ether, stabilized
UN1305	Vinyltrichlorosilane, stabilized
UN1306	Wood preservatives, liquid
UN1307	Xylenes
UN1308	Zirconium suspended in a liquid
UN1309	Aluminum powder, coated
UN1310	Ammonium picrate, wetted
UN1312	Borneol
UN1313	Calcium resinate
UN1314	Calcium resinate, fused
UN1318	Cobalt resinate, precipitated
UN1320	Dinitrophenol, wetted
UN1321	Dinitrophenolates, wetted
UN1322	Dinitroresorcinol, wetted
UN1323	Ferrocerium
UN1324	Films, nitrocellulose base
NA1325	Fusee
UN1325	Flammable solids, organic, n.o.s.
UN1326	Hafnium powder, wetted
UN1328	Hexamethylenetetramine
UN1330	Manganese resinate
UN1331	Matches, strike anywhere
UN1332	Metaldehyde
UN1333	Cerium
UN1334	Naphthalene, crude or Naphthalene, refined
UN1336	Nitroguanidine, wetted or Picrite, wetted
UN1337	Nitrostarch, wetted
UN1338	Phosphorus, amorphous
UN1339	Phosphorus heptasulfide
UN1340	Phosphorus pentasulfide
UN1341	Phosphorus sesquisulfide
UN1343	Phosphorus trisulfide
UN1344	Trinitrophenol, wetted or Picric acid, wetted
UN1345	Rubber scrap or shoddy

(1)-Identification Number	(2)-Description
UN1346	Silicon powder, amorphous
UN1347	Silver picrate, wetted
UN1348	Sodium dinitro-o-cresolate, wetted
UN1349	Sodium picramate, wetted
NA1350	Sulfur
UN1350	Sulfur
UN1352	Titanium powder, wetted
UN1353	Fibers or Fabrics inpregnated with weakly nitrated nitrocellulose, n.o.s.
UN1354	Trinitrobenzene, wetted
UN1355	Trinitrobenzoic acid, wetted
UN1356	Trinitrotoluene, wetted or TNT, wetted
UN1357	Urea nitrate, wetted
UN1358	Zirconium powder, wetted
UN1360	Calcium phosphide
NA1361	Charcoal
UN1361	Carbon
UN1362	Carbon, activated
UN1363	Copra
UN1364	Cotton waste, oily
NA1365	Cotton
UN1365	Cotton, wet
UN1369	p-Nitrosodimethylaniline
UN1370	Dimethylzinc
UN1372	Fibers, animal or Fibers, vegetable
UN1373	Fibers or Fabrics, animal or vegetable or Synthetic, n.o.s.
UN1374	Fish meal, unstablized or Fish scrap, unstabilized
UN1376	Iron oxide, spent, or Iron sponge, spent
UN1378	Metal catalyst, wetted
UN1379	Paper, unsaturated oil treated
UN1380	Pentaborane
UN1381	Phosphorus, white dry or Phosphorus, white, under water or Phosphorus white, in solution or Phosphorus, yellow dry or Phosphorus, yellow, under water or Phosphorus, yellow, in solution
UN1382	Potassium sulfide, anhydrous or Potassium sulfide
UN1383	Pyrophoric metals, n.o.s., or Pyrophoric alloys, n.o.s.
UN1384	Sodium dithionite or Sodium hydrosulfite
UN1385	Sodium sulfide, anhydrous or Sodium sulfide
UN1386	Seed cake
UN1387	Wool waste, wet
UN1389	Alkali metal amalgam, liquid
UN1390	Alkali metal amides
UN1391	Alkali metal dispersions, or Alkaline earth metal dispersions
UN1392	Alkaline earth metal amalgams, liquid
UN1393	Alkaline earth metal alloys, n.o.s.
UN1394	Aluminum carbide
UN1395	Aluminum ferrosilicon powder
UN1396	Aluminum powder, uncoated
UN1397	Aluminum phosphide
UN1398	Aluminum silicon powder, uncoated
UN1400	Barium
UN1401	Calcium
UN1402	Calcium carbide
UN1403	Calcium cyanamide
UN1404	Calcium hydride
UN1405	Calcium silicide
UN1407	Cesium or Caesium
UN1408	Ferrosilicon
UN1409	Metal hydrides, water reactive, n.o.s.
UN1410	Lithium aluminum hydride
UN1411	Lithium aluminum hydride, ethereal
UN1413	Lithium borohydride
UN1414	Lithium hydride
UN1415	Lithium
UN1417	Lithium silicon
UN1418	Magnesium, powder or Magnesium alloys, powder
UN1419	Magnesium aluminum phosphide
UN1420	Potassium, metal alloys, liquid
UN1421	Alkali metal alloys, liquid, n.o.s.
UN1422	Potassium sodium alloys, liquid
UN1423	Rubidium
UN1426	Sodium borohydride
UN1427	Sodium hydride
UN1428	Sodium
UN1431	Sodium methylate
UN1432	Sodium phosphide
UN1433	Stannic phosphide
UN1435	Zinc ashes
UN1436	Zinc powder or Zinc dust
UN1437	Zirconium hydride
UN1438	Aluminum nitrate
UN1439	Ammonium dichromate
UN1442	Ammonium perchlorate
UN1444	Ammonium persulfate
UN1445	Barium chlorate, solid
UN1446	Barium nitrate
UN1447	Barium perchlorate, solid
UN1448	Barium permanganate
UN1449	Barium peroxide
UN1450	Bromates, inorganic, n.o.s.
UN1451	Cesium nitrate or Caesium nitrate
UN1452	Calcium chlorate
UN1453	Calcium chlorite
UN1454	Calcium nitrate
UN1455	Calcium perchlorate
UN1456	Calcium permanganate
UN1457	Calcium peroxide
UN1458	Chlorate and borate mixtures
UN1459	Chlorate and magnesium chloride mixture solid
UN1461	Chlorates, inorganic, n.o.s.
UN1462	Chlorites, inorganic, n.o.s.
UN1463	Chromium trioxide, anhydrous
UN1465	Didymium nitrate
UN1466	Ferric nitrate
UN1467	Guanidine nitrate
UN1469	Lead nitrate
UN1470	Lead perchlorate, solid
UN1471	Lithium hypochlorite, dry or Lithium hypochlorite mixture
UN1472	Lithium peroxide
UN1473	Magnesium bromate
UN1474	Magnesium nitrate
UN1475	Magnesium perchlorate
UN1476	Magnesium peroxide
UN1477	Nitrates, inorganic, n.o.s.
UN1479	Oxidizing, solid, n.o.s.
UN1481	Perchlorates, inorganic, n.o.s.
UN1482	Permanganates, inorganic, n.o.s.
UN1483	Peroxides, inorganic, n.o.s.
UN1484	Potassium bromate
UN1485	Potassium chlorate
UN1486	Potassium nitrate
UN1487	Potassium nitrate and sodium nitrite mixtures
UN1488	Potassium nitrite
UN1489	Potassium perchlorate
UN1490	Potassium permanganate
UN1491	Potassium peroxide
UN1492	Potassium persulfate
UN1493	Silver nitrate
UN1494	Sodium bromate
UN1495	Sodium chlorate
UN1496	Sodium chlorite
UN1498	Sodium nitrate
UN1499	Sodium nitrate and potassium nitrate mixtures
UN1500	Sodium nitrite
UN1502	Sodium perchlorate
UN1503	Sodium permanganate
UN1504	Sodium peroxide
UN1505	Sodium persulfate
UN1506	Strontium chlorate
UN1507	Strontium nitrate
UN1508	Strontium perchlorate
UN1509	Strontium peroxide
UN1510	Tetranitromethane
UN1511	Urea hydrogen peroxide
UN1512	Zinc ammonium nitrite
UN1513	Zinc chlorate
UN1514	Zinc nitrate
UN1515	Zinc permanganate
UN1516	Zinc peroxide
UN1517	Zirconium picramate, wetted
UN1541	Acetone cyanohydrin, stabilized
UN1544	Alkaloids, solid, n.o.s. or Alkaloid salts, solid, n.o.s.
UN1545	Allyl isothiocyanate, stabilized
UN1546	Ammonium arsenate
UN1547	Aniline
UN1548	Aniline hydrochloride
UN1549	Antimony compounds, inorganic, solid, n.o.s.
UN1550	Antimony lactate
UN1551	Antimony potassium tartrate
UN1553	Arsenic acid, liquid
UN1554	Arsenic acid, solid
UN1555	Arsenic bromide
NA1556	Methyldichloroarsine
UN1556	Arsenic compounds, liquid, n.o.s.
UN1557	Arsenic compounds, solid, n.o.s.
UN1558	Arsenic
UN1559	Arsenic pentoxide
UN1560	Arsenic trichloride
UN1561	Arsenic trioxide
UN1562	Arsenical dust
UN1564	Barium compounds, n.o.s.
UN1565	Barium cyanide
UN1566	Beryllium compounds, n.o.s.
UN1567	Beryllium, powder

ID CROSS REFERENCE

(1)-Identification Number	(2)-Description
UN1569	Bromoacetone
UN1570	Brucine
UN1571	Barium azide, wetted
UN1572	Cacodylic acid
UN1573	Calcium arsenate
UN1574	Calcium arsenate and calcium arsenite, mixtures, solid
UN1575	Calcium cyanide
UN1577	Chlorodinitrobenzenes, liquid
UN1578	Chloronitrobenzenes, solid
UN1579	4-Chloro-o-toluidine hydrochloride, solid
UN1580	Chloropicrin
UN1581	Chloropicrin and methyl bromide mixtures
UN1582	Chloropicrin and methyl chloride mixtures
UN1583	Chloropicrin mixtures, n.o.s.
UN1585	Copper acetoarsenite
UN1586	Copper arsenite
UN1587	Copper cyanide
UN1588	Cyanides, inorganic, solid, n.o.s.
UN1589	Cyanogen chloride, stabilized
UN1590	Dichloroanilines, liquid
UN1591	o-Dichlorobenzene
UN1593	Dichloromethane
UN1594	Diethyl sulfate
UN1595	Dimethyl sulfate
UN1596	Dinitroanilines
UN1597	Dinitrobenzenes, liquid
UN1598	Dinitro-o-cresol
UN1599	Dinitrophenol solutions
UN1600	Dinitrotoluenes, molten
UN1601	Disinfectants, solid, toxic, n.o.s.
UN1602	Dyes, liquid, toxic, n.o.s. or Dye intermediates, liquid, toxic, n.o.s.
UN1603	Ethyl bromoacetate
UN1604	Ethylenediamine
UN1605	Ethylene dibromide
UN1606	Ferric arsenate
UN1607	Ferric arsenite
UN1608	Ferrous arsenate
UN1611	Hexaethyl tetraphosphate
UN1612	Hexaethyl tetraphosphate and compressed gas mixtures
NA1613	Hydrocyanic acid, aqueous solutions
UN1613	Hydrocyanic acid, aqueous solutions or Hydrogen cyanide, aqueous solutions
UN1614	Hydrogen cyanide, stabilized
UN1616	Lead acetate
UN1617	Lead arsenates
UN1618	Lead arsenites
UN1620	Lead cyanide
UN1621	London purple
UN1622	Magnesium arsenate
UN1623	Mercuric arsenate
UN1624	Mercuric chloride
UN1625	Mercuric nitrate
UN1626	Mercuric potassium cyanide
UN1627	Mercurous nitrate
UN1629	Mercury acetate
UN1630	Mercury ammonium chloride
UN1631	Mercury benzoate
UN1634	Mercury bromides
UN1636	Mercury cyanide
UN1637	Mercury gluconate
UN1638	Mercury iodide
UN1639	Mercury nucleate
UN1640	Mercury oleate
UN1641	Mercury oxide
UN1642	Mercury oxycyanide, desensitized
UN1643	Mercury potassium iodide
UN1644	Mercury salicylate
UN1645	Mercury sulfates
UN1646	Mercury thiocyanate
UN1647	Methyl bromide and ethylene dibromide mixtures, liquid
UN1648	Acetonitrile
UN1649	Motor fuel anti-knock mixtures
UN1650	beta-Naphthylamine, solid
UN1651	Naphthylthiourea
UN1652	Naphthylurea
UN1653	Nickel cyanide
UN1654	Nicotine
UN1655	Nicotine compounds, solid, n.o.s. or Nicotine preparations, solid, n.o.s.
UN1656	Nicotine hydrochloride liquid or solution
UN1657	Nicotine salicylate
UN1658	Nicotine sulfate solution
UN1659	Nicotine tartrate
UN1660	Nitric oxide, compressed
UN1661	Nitroanilines
UN1662	Nitrobenzene
UN1663	Nitrophenols
UN1664	Nitrotoluenes, liquid
UN1665	Nitroxylenes, liquid
UN1669	Pentachloroethane
UN1670	Perchloromethyl mercaptan
UN1671	Phenol, solid
UN1672	Phenylcarbylamine chloride
UN1673	Phenylenediamines
UN1674	Phenylmercuric acetate
UN1677	Potassium arsenate
UN1678	Potassium arsenite
UN1679	Potassium cuprocyanide
UN1680	Potassium cyanide, solid
UN1683	Silver arsenite
UN1684	Silver cyanide
UN1685	Sodium arsenate
UN1686	Sodium arsenite, aqueous solutions
UN1687	Sodium azide
UN1688	Sodium cacodylate
UN1689	Sodium cyanide, solid
UN1690	Sodium fluoride, solid
UN1691	Strontium arsenite
UN1692	Strychnine or Strychnine salts
NA1693	Tear gas devices
UN1693	Tear gas substances, liquid, n.o.s.
UN1694	Bromobenzyl cyanides, liquid
UN1695	Chloroacetone, stabilized
UN1697	Chloroacetophenone, solid
UN1698	Diphenylamine chloroarsine
UN1699	Diphenylchloroarsine, liquid
UN1700	Tear gas candles
UN1701	Xylyl bromide, liquid
UN1702	1,1,2,2-Tetrachloroethane
UN1704	Tetraethyl dithiopyrophosphate
UN1707	Thallium compounds, n.o.s.
UN1708	Toluidines, liquid
UN1709	2,4-Toluylenediamine, solid or 2,4-Toluenediamine, solid
UN1710	Trichloroethylene
UN1711	Xylidines, liquid
UN1712	Zinc arsenate or Zinc arsenite or Zinc arsenate and zinc arsenite mixtures
UN1713	Zinc cyanide
UN1714	Zinc phosphide
UN1715	Acetic anhydride
UN1716	Acetyl bromide
UN1717	Acetyl chloride
UN1718	Butyl acid phosphate
UN1719	Caustic alkali liquids, n.o.s.
UN1722	Allyl chloroformate
UN1723	Allyl iodide
UN1724	Allyltrichlorosilane, stabilized
UN1725	Aluminum bromide, anhydrous
UN1726	Aluminum chloride, anhydrous
UN1727	Ammonium hydrogendifluoride, solid
UN1728	Amyltrichlorosilane
UN1729	Anisoyl chloride
UN1730	Antimony pentachloride, liquid
UN1731	Antimony pentachloride, solutions
UN1732	Antimony pentafluoride
UN1733	Antimony trichloride, liquid
UN1733	Antimony trichloride, solid
UN1736	Benzoyl chloride
UN1737	Benzyl bromide
UN1738	Benzyl chloride
UN1739	Benzyl chloroformate
UN1740	Hydrogendifluoride, solid, n.o.s.
UN1741	Boron trichloride
UN1742	Boron trifluoride acetic acid complex, liquid
UN1743	Boron trifluoride propionic acid complex, liquid
UN1744	Bromine or Bromine solutions
UN1745	Bromine pentafluoride
UN1746	Bromine trifluoride
UN1747	Butyltrichlorosilane
UN1748	Calcium hypochlorite, dry or Calcium hypochlorite mixtures, dry
UN1749	Chlorine trifluoride
UN1750	Chloroacetic acid, solution
UN1751	Chloroacetic acid, solid
UN1752	Chloroacetyl chloride
UN1753	Chlorophenyltrichlorosilane
UN1754	Chlorosulfonic acid
UN1755	Chromic acid solution
UN1756	Chromic fluoride, solid
UN1757	Chromic fluoride, solution
UN1758	Chromium oxychloride
NA1759	Ferrous chloride, solid
UN1759	Corrosive solids, n.o.s.
NA1760	Chemical kit
NA1760	Compounds, cleaning liquid
NA1760	Compounds, tree killing, liquid or Compounds, weed killing, liquid
NA1760	Ferrous chloride, solution

(1)-Identification Number	(2)-Description
UN1760	Corrosive liquids, n.o.s.
UN1761	Cupriethylenediamine solution
UN1762	Cyclohexenyltrichlorosilane
UN1763	Cyclohexyltrichlorosilane
UN1764	Dichloroacetic acid
UN1765	Dichloroacetyl chloride
UN1766	Dichlorophenyltrichlorosilane
UN1767	Diethyldichlorosilane
UN1768	Difluorophosphoric acid, anhydrous
UN1769	Diphenyldichlorosilane
UN1770	Diphenylmethyl bromide
UN1771	Dodecyltrichlorosilane
UN1773	Ferric chloride, anhydrous
UN1774	Fire extinguisher charges
UN1775	Fluoroboric acid
UN1776	Fluorophosphoric acid anhydrous
UN1777	Fluorosulfonic acid
UN1778	Fluorosilicic acid
UN1779	Formic acid
UN1780	Fumaryl chloride
UN1781	Hexadecyltrichlorosilane
UN1782	Hexafluorophosphoric acid
UN1783	Hexamethylenediamine solution
UN1784	Hexyltrichlorosilane
UN1786	Hydrofluoric acid and Sulfuric acid mixtures
UN1787	Hydriodic acid
UN1788	Hydrobromic acid
UN1789	Hydrochloric acid
UN1790	Hydrofluoric acid
UN1791	Hypochlorite solutions
UN1792	Iodine monochloride
UN1793	Isopropyl acid phosphate
UN1794	Lead sulfate
UN1796	Nitrating acid mixtures
UN1798	Nitrohydrochloric acid
UN1799	Nonyltrichlorosilane
UN1800	Octadecyltrichlorosilane
UN1801	Octyltrichlorosilane
UN1802	Perchloric acid
UN1803	Phenolsulfonic acid, liquid
UN1804	Phenyltrichlorosilane
UN1805	Phosphoric acid solution
UN1806	Phosphorus pentachloride
UN1807	Phosphorus pentoxide
UN1808	Phosphorus tribromide
UN1809	Phosphorus trichloride
UN1810	Phosphorus oxychloride
UN1811	Potassium hydrogendifluoride solid
UN1812	Potassium fluoride, solid
UN1813	Potassium hydroxide, solid
UN1814	Potassium hydroxide, solution
UN1815	Propionyl chloride
UN1816	Propyltrichlorosilane
UN1817	Pyrosulfuryl chloride
UN1818	Silicon tetrachloride
UN1819	Sodium aluminate, solution
UN1823	Sodium hydroxide, solid
UN1824	Sodium hydroxide solution
UN1825	Sodium monoxide
UN1826	Nitrating acid mixtures spent
UN1826	Nitrating acid mixtures, spent
UN1827	Stannic chloride, anhydrous
UN1828	Sulfur chlorides
UN1829	Sulfur trioxide, stabilized
UN1830	Sulfuric acid
UN1831	Sulfuric acid, fuming
UN1832	Sulfuric acid, spent
UN1833	Sulfurous acid
UN1834	Sulfuryl chloride
UN1835	Tetramethylammonium hydroxide solution
UN1836	Thionyl chloride
UN1837	Thiophosphoryl chloride
UN1838	Titanium tetrachloride
UN1839	Trichloroacetic acid
UN1840	Zinc chloride, solution
UN1841	Acetaldehyde ammonia
UN1843	Ammonium dinitro-o-cresolate, solid
UN1845	Carbon dioxide, solid or Dry ice
UN1846	Carbon tetrachloride
UN1847	Potassium sulfide, hydrated
UN1848	Propionic acid
UN1849	Sodium sulfide, hydrated
UN1851	Medicine, liquid, toxic, n.o.s.
UN1854	Barium alloys, pyrophoric
UN1855	Calcium, pyrophoric or Calcium alloys, pyrophoric
UN1856	Rags, oily
UN1857	Textile waste, wet
UN1858	Hexafluoropropylene, compressed or Refrigerant gas R 1216
UN1859	Silicon tetrafluoride
UN1860	Vinyl fluoride, stabilized
UN1862	Ethyl crotonate
UN1863	Fuel, aviation, turbine engine
UN1865	n-Propyl nitrate
UN1866	Resin solution
UN1868	Decaborane
UN1869	Magnesium or Magnesium alloys
UN1870	Potassium borohydride
UN1871	Titanium hydride
UN1872	Lead dioxide
UN1873	Perchloric acid
UN1884	Barium oxide
UN1885	Benzidine
UN1886	Benzylidene chloride
UN1887	Bromochloromethane
UN1888	Chloroform
UN1889	Cyanogen bromide
UN1891	Ethyl bromide
UN1892	Ethyldichloroarsine
UN1894	Phenylmercuric hydroxide
UN1895	Phenylmercuric nitrate
UN1897	Tetrachloroethylene
UN1898	Acetyl iodide
UN1902	Diisooctyl acid phosphate
UN1903	Disinfectant, liquid, corrosive n.o.s. or Disinfectants, liquid, corrosive n.o.s.
UN1905	Selenic acid
UN1906	Sludge, acid
UN1907	Soda lime
UN1908	Chlorite solution
UN1910	Calcium oxide
NA1911	Diborane mixtures
UN1911	Diborane
UN1912	Methyl chloride and methylene chloride mixtures
UN1913	Neon, refrigerated liquid
UN1914	Butyl propionates
UN1915	Cyclohexanone
UN1916	2,2'-Dichlorodiethyl ether
UN1917	Ethyl acrylate, stabilized
UN1918	Isopropylbenzene
UN1919	Methyl acrylate, stabilized
UN1920	Nonanes
UN1921	Propyleneimine, stabilized
UN1922	Pyrrolidine
UN1923	Calcium dithionite or Calcium hydrosulfite
UN1928	Methyl magnesium bromide, in ethyl ether
UN1929	Potassium dithionite or Potassium hydrosulfite
UN1931	Zinc dithionite or Zinc hydrosulfite
UN1932	Zirconium scrap
UN1935	Cyanide solutions, n.o.s.
UN1938	Bromoacetic acid solution
UN1939	Phosphorus oxybromide
UN1940	Thioglycolic acid
UN1941	Dibromodifluoromethane
UN1942	Ammonium nitrate
UN1944	Matches, safety
UN1945	Matches, wax, Vesta
UN1950	Aerosols
UN1950	Aerosols, flammable, n.o.s.
UN1951	Argon, refrigerated liquid
UN1952	Ethylene oxide and carbon dioxide mixtures
UN1953	Compressed gas, toxic, flammable, n.o.s.
NA1954	Refrigerant gases, n.o.s. or Dispersant gases, n.o.s.
UN1954	Compressed gas, flammable, n.o.s.
NA1955	Organic phosphate, mixed with compressed gas or Organic phosphate compound, mixed with compressed gas or Organic phosphorus compound, mixed with compressed gas
UN1955	Compressed gas, toxic, n.o.s.
UN1956	Compressed gas, n.o.s.
UN1957	Deuterium, compressed
UN1958	1,2-Dichloro-1,1,2,2-tetrafluoroethane or Refrigerant gas R 114
UN1959	1,1-Difluoroethylene or Refrigerant gas R 1132a
NA1961	Ethane-Propane mixture, refrigerated liquid
UN1961	Ethane, refrigerated liquid
UN1962	Ethylene
UN1963	Helium, refrigerated liquid
UN1964	Hydrocarbon gas mixture, compressed, n.o.s.
UN1965	Hydrocarbon gas mixture, liquefied, n.o.s.
UN1966	Hydrogen, refrigerated liquid
NA1967	Parathion and compressed gas mixture
UN1967	Insecticide gases, toxic, n.o.s.
UN1968	Insecticide gases, n.o.s.
UN1969	Isobutane

ID CROSS REFERENCE

ID CROSS REFERENCE

(1)-Identification Number	(2)-Description
UN1970	Krypton, refrigerated liquid
UN1971	Methane, compressed or Natural gas, compressed
UN1972	Methane, refrigerated liquid or Natural gas, refrigerated liquid
UN1973	Chlorodifluoromethane and chloropentafluoroethane mixture or Refrigerant gas R502
UN1974	Chlorodifluorobromomethane or Refrigerant gas R12B1
UN1975	Nitric oxide and dinitrogen tetroxide mixtures or Nitric oxide and nitrogen dioxide mixtures
UN1976	Octafluorocyclobutane or Refrigerant gas RC 318
UN1977	Nitrogen, refrigerated liquid
UN1978	Propane
UN1982	Tetrafluoromethane or Refrigerant gas R 14
UN1983	1-Chloro-2,2,2-trifluoroethane or refrigerant gas R 133a
UN1984	Trifluoromethane or Refrigerant gas R 23
UN1986	Alcohols, flammable, toxic, n.o.s.
NA1987	Denatured alcohol
UN1987	Alcohols, n.o.s.
UN1988	Aldehydes, flammable, toxic, n.o.s.
UN1989	Aldehydes, n.o.s.
UN1990	Benzaldehyde
UN1991	Chloroprene, stabilized
UN1992	Flammable liquids, toxic, n.o.s.
NA1993	Combustible liquid, n.o.s.
NA1993	Compounds, cleaning liquid
NA1993	Compounds, tree killing, liquid or Compounds, weed killing, liquid
NA1993	Diesel fuel
NA1993	Fuel oil
UN1993	Flammable liquids, n.o.s.
UN1994	Iron pentacarbonyl
NA1999	Asphalt
UN1999	Tars, liquid
UN2000	Celluloid
UN2001	Cobalt naphthenates, powder
UN2002	Celluloid, scrap
UN2004	Magnesium diamide
UN2006	Plastics, nitrocellulose-based, self-heating, n.o.s.
UN2008	Zirconium powder, dry
UN2009	Zirconium, dry
UN2010	Magnesium hydride
UN2011	Magnesium phosphide
UN2012	Potassium phosphide
UN2013	Strontium phosphide
UN2014	Hydrogen peroxide, aqueous solutions
UN2015	Hydrogen peroxide, stabilized or Hydrogen peroxide aqueous solutions, stabilized
UN2016	Ammunition, toxic, non-explosive
UN2017	Ammunition, tear-producing, non-explosive
UN2018	Chloroanilines, solid
UN2019	Chloroanilines, liquid
UN2020	Chlorophenols, solid
UN2021	Chlorophenols, liquid
UN2022	Cresylic acid
UN2023	Epichlorohydrin
UN2024	Mercury compound, liquid, n.o.s.
UN2025	Mercury compound, solid, n.o.s.
UN2026	Phenylmercuric compounds, n.o.s.
UN2027	Sodium arsenite, solid
UN2028	Bombs, smoke, non-explosive
UN2029	Hydrazine, anhydrous
UN2030	Hydrazine aqueous solution
UN2031	Nitric acid
UN2032	Nitric acid, red fuming
UN2033	Potassium monoxide
UN2034	Hydrogen and Methane mixtures, compressed
UN2035	1,1,1,-Trifluoroethane or Refrigerant gas, R 143a
UN2036	Xenon, compressed
UN2037	Gas cartridges
UN2037	Receptacles, small, containing gas or gas cartridges
UN2038	Dinitrotoluenes
UN2044	2,2-Dimethylpropane
UN2045	Isobutyraldehyde or isobutyl aldehyde
UN2046	Cymenes
UN2047	Dichloropropenes
UN2048	Dicyclopentadiene
UN2049	Diethylbenzene
UN2050	Diisobutylene, isomeric compounds
UN2051	2-Dimethylaminoethanol
UN2052	Dipentene
UN2053	Methyl isobutyl carbinol
UN2054	Morpholine
UN2055	Styrene monomer, stabilized
UN2056	Tetrahydrofuran
UN2057	Tripropylene
UN2058	Valeraldehyde
UN2059	Nitrocellulose, solution, flammable
UN2067	Ammonium nitrate based fertilizer
UN2071	Ammonium nitrate based fertilizer
UN2073	Ammonia solutions
UN2074	Acrylamide, solid
UN2075	Chloral, anhydrous, stabilized
UN2076	Cresols, liquid
UN2077	alpha-Naphthylamine
UN2078	Toluene diisocyanate
UN2079	Diethylenetriamine
UN2186	Hydrogen chloride, refrigerated liquid
UN2187	Carbon dioxide, refrigerated liquid
UN2188	Arsine
UN2189	Dichlorosilane
UN2190	Oxygen difluoride, compressed
UN2191	Sulfuryl fluoride
UN2192	Germane
UN2193	Hexafluoroethane or Refrigerant gas R 116
UN2194	Selenium hexafluoride
UN2195	Tellurium hexafluoride
UN2196	Tungsten hexafluoride
UN2197	Hydrogen iodide, anhydrous
UN2198	Phosphorus pentafluoride
UN2199	Phosphine
UN2200	Propadiene, stabilized
UN2201	Nitrous oxide, refrigerated liquid
UN2202	Hydrogen selenide, anhydrous
UN2203	Silane
UN2204	Carbonyl sulfide
UN2205	Adiponitrile
UN2206	Isocyanates, toxic, n.o.s. or Isocyanate, solutions, toxic, n.o.s.
UN2208	Calcium hypochlorite mixtures, dry
UN2209	Formaldehyde solutions
UN2210	Maneb or Maneb preparations
UN2211	Polymeric beads expandable
NA2212	Asbestos
UN2212	Blue Asbestos or Brown Asbestos
UN2213	Paraformaldehyde
UN2214	Phthalic anhydride
UN2215	Maleic anhydride
UN2215	Maleic anhydride, molten
UN2216	Fish meal, stabilized or Fish scrap, stabilized
UN2217	Seed cake
UN2218	Acrylic acid, stabilized
UN2219	Allyl glycidyl ether
UN2222	Anisole
UN2224	Benzonitrile
UN2225	Benzene sulfonyl chloride
UN2226	Benzotrichloride
UN2227	n-Butyl methacrylate, stabilized
UN2232	2-Chloroethanal
UN2233	Chloroanisidines
UN2234	Chlorobenzotrifluorides
UN2235	Chlorobenzyl chlorides, liquid
UN2236	3-Chloro-4-methylphenyl isocyanate, liquid
UN2237	Chloronitroanilines
UN2238	Chlorotoluenes
UN2239	Chlorotoluidines, solid
UN2240	Chromosulfuric acid
UN2241	Cycloheptane
UN2242	Cycloheptene
UN2243	Cyclohexyl acetate
UN2244	Cyclopentanol
UN2245	Cyclopentanone
UN2246	Cyclopentene
UN2247	n-Decane
UN2248	Di-n-butylamine
UN2249	Dichlorodimethyl ether, symmetrical
UN2250	Dichlorophenyl isocyanates
UN2251	Bicyclo[2,2,1]hepta-2,5-diene, stabilized or 2,5-Norbornadiene, stabilized
UN2252	1,2-Dimethoxyethane
UN2253	N,N-Dimethylaniline
UN2254	Matches, fusee
UN2256	Cyclohexene
UN2257	Potassium
UN2258	1,2-Propylenediamine
UN2259	Triethylenetetramine
UN2260	Tripropylamine
UN2261	Xylenols, solid
UN2262	Dimethylcarbamoyl chloride
UN2263	Dimethylcyclohexanes
UN2264	N,N-Dimethylcyclohexylamine
UN2265	N,N-Dimethylformamide
UN2266	Dimethyl-N-propylamine
UN2267	Dimethyl thiophosphoryl chloride
UN2269	3,3'-Iminodipropylamine
UN2270	Ethylamine, aqueous solution

(1)-Identification Number	(2)-Description
UN2271	Ethyl amyl ketone
UN2272	N-Ethylaniline
UN2273	2-Ethylaniline
UN2274	N-Ethyl-N-benzylaniline
UN2275	2-Ethylbutanol
UN2276	2-Ethylhexylamine
UN2277	Ethyl methacrylate, stabilized
UN2278	n-Heptene
UN2279	Hexachlorobutadiene
UN2280	Hexamethylenediamine, solid
UN2281	Hexamethylene diisocyanate
UN2282	Hexanols
UN2283	Isobutyl methacrylate, stabilized
UN2284	Isobutyronitrile
UN2285	Isocyanatobenzotrifluorides
UN2286	Pentamethylheptane
UN2287	Isoheptenes
UN2288	Isohexenes
UN2289	Isophoronediamine
UN2290	Isophorone diisocyanate
UN2291	Lead compounds, soluble, n.o.s.
UN2293	4-Methoxy-4-methylpentan-2-one
UN2294	N-Methylaniline
UN2295	Methyl chloroacetate
UN2296	Methylcyclohexane
UN2297	Methylcyclohexanone
UN2298	Methylcyclopentane
UN2299	Methyl dichloroacetate
UN2300	2-Methyl-5-ethylpyridine
UN2301	2-Methylfuran
UN2302	5-Methylhexan-2-one
UN2303	Isopropenylbenzene
UN2304	Naphthalene, molten
UN2305	Nitrobenzenesulfonic acid
UN2306	Nitrobenzotrifluorides, liquid
UN2307	3-Nitro-4-chlorobenzotrifluoride
UN2308	Nitrosylsulfuric acid, liquid
UN2309	Octadiene
UN2310	Pentane-2,4-dione
UN2311	Phenetidines
UN2312	Phenol, molten
UN2313	Picolines
UN2315	Polychlorinated biphenyls, liquid
UN2316	Sodium cuprocyanide, solid
UN2317	Sodium cuprocyanide, solution
UN2318	Sodium hydrosulfide
UN2319	Terpene hydrocarbons, n.o.s.
UN2320	Tetraethylenepentamine
UN2321	Trichlorobenzenes, liquid
UN2322	Trichlorobutene
UN2323	Triethyl phosphite
UN2324	Triisobutylene
UN2325	1,3,5-Trimethylbenzene
UN2326	Trimethylcyclohexylamine
UN2327	Trimethylhexamethylenediamines
UN2328	Trimethylhexamethylene diisocyanate
UN2329	Trimethyl phosphite
UN2330	Undecane
UN2331	Zinc chloride, anhydrous
UN2332	Acetaldehyde oxime
UN2333	Allyl acetate
UN2334	Allylamine

(1)-Identification Number	(2)-Description
UN2335	Allyl ethyl ether
UN2336	Allyl formate
UN2337	Phenyl mercaptan
UN2338	Benzotrifluoride
UN2339	2-Bromobutane
UN2340	2-Bromoethyl ethyl ether
UN2341	1-Bromo-3-methylbutane
UN2342	Bromomethylpropanes
UN2343	2-Bromopentane
UN2344	Bromopropanes
UN2345	3-Bromopropyne
UN2346	Butanedione
UN2347	Butyl mercaptans
UN2348	Butyl acrylates, stabilized
UN2350	Butyl methyl ether
UN2351	Butyl nitrites
UN2352	Butyl vinyl ether, stabilized
UN2353	Butyryl chloride
UN2354	Chloromethyl ethyl ether
UN2356	2-Chloropropane
UN2357	Cyclohexylamine
UN2358	Cyclooctatetraene
UN2359	Diallylamine
UN2360	Diallylether
UN2361	Diisobutylamine
UN2362	1,1-Dichloroethane
UN2363	Ethyl mercaptan
UN2364	n-Propyl benzene
UN2366	Diethyl carbonate
UN2367	alpha-Methylvaleraldehyde
UN2368	alpha-Pinene
UN2370	1-Hexene
UN2371	Isopentenes
UN2372	1,2-Di-(dimethylamino)ethane
UN2373	Diethoxymethane
UN2374	3,3-Diethoxypropene
UN2375	Diethyl sulfide
UN2376	2,3-Dihydropyran
UN2377	1,1-Dimethoxyethane
UN2378	2-Dimethylaminoacetonitrile
UN2379	1,3-Dimethylbutylamine
UN2380	Dimethyldiethoxysilane
UN2381	Dimethyl disulfide
UN2382	Dimethylhydrazine, symmetrical
UN2383	Dipropylamine
UN2384	Di-n-propyl ether
UN2385	Ethyl isobutyrate
UN2386	1-Ethylpiperidine
UN2387	Fluorobenzene
UN2388	Fluorotoluenes
UN2389	Furan
UN2390	2-Iodobutane
UN2391	Iodomethylpropanes
UN2392	Iodopropanes
UN2393	Isobutyl formate
UN2394	Isobutyl propionate
UN2395	Isobutyryl chloride
UN2396	Methacrylaldehyde, stabilized
UN2397	3-Methylbutan-2-one
UN2398	Methyl tert-butyl ether
UN2399	1-Methylpiperidine
UN2400	Methyl isovalerate
UN2401	Piperidine

(1)-Identification Number	(2)-Description
UN2402	Propanethiols
UN2403	Isopropenyl acetate
UN2404	Propionitrile
UN2405	Isopropyl butyrate
UN2406	Isopropyl isobutyrate
UN2407	Isopropyl chloroformate
UN2409	Isopropyl propionate
UN2410	1,2,3,6-Tetrahydropyridine
UN2411	Butyronitrile
UN2412	Tetrahydrothiophene
UN2413	Tetrapropylorthotitanate
UN2414	Thiophene
UN2416	Trimethyl borate
UN2417	Carbonyl fluoride
UN2418	Sulfur tetrafluoride
UN2419	Bromotrifluoroethylene
UN2420	Hexafluoroacetone
UN2421	Nitrogen trioxide
UN2422	Octafluorobut-2-ene or Refrigerant gas R 1318
UN2424	Octafluoropropane or Refrigerant gas R 218
UN2426	Ammonium nitrate, liquid
UN2427	Potassium chlorate, aqueous solution
UN2428	Sodium chlorate, aqueous solution
UN2429	Calcium chlorate aqueous solution
UN2430	Alkylphenols, solid, n.o.s.
UN2431	Anisidines
UN2432	N,N-Diethylaniline
UN2433	Chloronitrotoluenes, liquid
UN2434	Dibenzyldichlorosilane
UN2435	Ethylphenyldichlorosilane
UN2436	Thioacetic acid
UN2437	Methylphenyldichlorosilane
UN2438	Trimethylacetyl chloride
UN2439	Sodium hydrogendifluoride
UN2440	Stannic chloride pentahydrate
UN2441	Titanium trichloride, pyrophoric or Titanium trichloride mixtures, pyrophoric
UN2442	Trichloroacetyl chloride
UN2443	Vanadium oxytrichloride
UN2444	Vanadium tetrachloride
UN2446	Nitrocresols, solid
UN2447	Phosphorus white, molten
NA2448	Sulfur, molten
UN2448	Sulfur, molten
UN2451	Nitrogen trifluoride
UN2452	Ethylacetylene, stabilized
UN2453	Ethyl fluoride or Refrigerant gas R161
UN2454	Methyl fluoride or Refrigerant gas R 41
UN2456	2-Chloropropene
UN2457	2,3-Dimethylbutane
UN2458	Hexadiene
UN2459	2-Methyl-1-butene
UN2460	2-Methyl-2-butene
UN2461	Methylpentadienes
UN2463	Aluminum hydride
UN2464	Beryllium nitrate
UN2465	Dichloroisocyanuric acid, dry or Dichloroisocyanuric acid salts
UN2466	Potassium superoxide

ID CROSS REFERENCE

(1)-Identification Number	(2)-Description
UN2468	Trichloroisocyanuric acid, dry
UN2469	Zinc bromate
UN2470	Phenylacetonitrile, liquid
UN2471	Osmium tetroxide
UN2473	Sodium arsanilate
UN2474	Thiophosgene
UN2475	Vanadium trichloride
UN2477	Methyl isothiocyanate
UN2478	Isocyanates, flammable, toxic, n.o.s. or Isocyanate solutions, flammable, toxic, n.o.s.
UN2480	Methyl isocyanate
UN2481	Ethyl isocyanate
UN2482	n-Propyl isocyanate
UN2483	Isopropyl isocyanate
UN2484	tert-Butyl isocyanate
UN2485	n-Butyl isocyanate
UN2486	Isobutyl isocyanate
UN2487	Phenyl isocyanate
UN2488	Cyclohexyl isocyanate
UN2490	Dichloroisopropyl ether
UN2491	Ethanolamine or Ethanolamine solutions
UN2493	Hexamethyleneimine
UN2495	Iodine pentafluoride
UN2496	Propionic anhydride
UN2498	1,2,3,6-Tetrahydrobenzaldehyde
UN2501	Tris-(1-aziridinyl)phosphine oxide, solution
UN2502	Valeryl chloride
UN2503	Zirconium tetrachloride
UN2504	Tetrabromoethane
UN2505	Ammonium fluoride
UN2506	Ammonium hydrogen sulfate
UN2507	Chloroplatinic acid, solid
UN2508	Molybdenum pentachloride
UN2509	Potassium hydrogen sulfate
UN2511	2-Chloropropionic acid
UN2512	Aminophenols
UN2513	Bromoacetyl bromide
UN2514	Bromobenzene
UN2515	Bromoform
UN2516	Carbon tetrabromide
UN2517	1-Chloro-1,1-difluoroethane or Refrigerant gas R 142b
UN2518	1,5,9-Cyclododecatriene
UN2520	Cyclooctadienes
UN2521	Diketene, stabilized
UN2522	2-Dimethylaminoethyl methacrylate
UN2524	Ethyl orthoformate
UN2525	Ethyl oxalate
UN2526	Furfurylamine
UN2527	Isobutyl acrylate, inhibited
UN2528	Isobutyl isobutyrate
UN2529	Isobutyric acid
UN2531	Methacrylic acid, stabilized
UN2533	Methyl trichloroacetate
UN2534	Methylchlorosilane
UN2535	4-Methylmorpholine or n-methlymorpholine
UN2536	Methyltetrahydrofuran
UN2538	Nitronaphthalene
UN2541	Terpinolene
UN2542	Tributylamine
UN2545	Hafnium powder, dry
UN2546	Titanium powder, dry
UN2547	Sodium superoxide
UN2548	Chlorine pentafluoride
UN2552	Hexafluoroacetone hydrate, liquid
UN2554	Methyl allyl chloride
UN2555	Nitrocellulose with water
UN2556	Nitrocellulose with alcohol
UN2557	Nitrocellulose mixture with or without plasticizer or Nitrocellulose mixture with or without pigment
UN2558	Epibromohydrin
UN2560	2-Methylpentan-2-ol
UN2561	3-Methyl-1-butene
UN2564	Trichloroacetic acid, solution
UN2565	Dicyclohexylamine
UN2567	Sodium pentachlorophenate
UN2570	Cadmium compounds
UN2571	Alkylsulfuric acids
UN2572	Phenylhydrazine
UN2573	Thallium chlorate
UN2574	Tricresyl phosphate
UN2576	Phosphorus oxybromide, molten
UN2577	Phenylacetyl chloride
UN2578	Phosphorus trioxide
UN2579	Piperazine
UN2580	Aluminum bromide, solution
UN2581	Aluminum chloride, solution
UN2582	Ferric chloride, solution
UN2583	Alkyl sulfonic acids, solid or Aryl sulfonic acids, solid
UN2584	Alkyl sulfonic acids, liquid or Aryl sulfonic acids, liquid
UN2585	Alkyl sulfonic acids, solid or Aryl sulfonic acids, solid
UN2586	Alkyl sulfonic acids, liquid or Aryl sulfonic acids, liquid
UN2587	Benzoquinone
UN2588	Pesticides, solid, toxic, n.o.s.
UN2589	Vinyl chloroacetate
UN2590	White asbestos
UN2591	Xenon, refrigerated liquid
UN2599	Chlorotrifluoromethane and trifluoromethane azeotropic mixture or Refrigerant gas R 503
UN2601	Cyclobutane
UN2602	Dichlorodifluoromethane and difluoroethane azeotropic mixture or Refrigerant gas R 500
UN2603	Cycloheptatriene
UN2604	Boron trifluoride diethyl etherate
UN2605	Methoxymethyl isocyanate
UN2606	Methyl orthosilicate
UN2607	Acrolein dimer, stabilized
UN2608	Nitropropanes
UN2609	Triallyl borate
UN2610	Triallylamine
UN2611	Propylene chlorohydrin
UN2612	Methyl propyl ether
UN2614	Methallyl alcohol
UN2615	Ethyl propyl ether
UN2616	Triisopropyl borate
UN2617	Methylcyclohexanols
UN2618	Vinyl toluenes, stabilized
UN2619	Benzyldimethylamine
UN2620	Amyl butyrates
UN2621	Acetyl methyl carbinol
UN2622	Glycidaldehyde
UN2623	Firelighters, solid
UN2624	Magnesium silicide
UN2626	Chloric acid aqueous solution
UN2627	Nitrites, inorganic, n.o.s.
UN2628	Potassium fluoroacetate
UN2629	Sodium fluoroacetate
UN2630	Selenates or Selenites
UN2642	Fluoroacetic acid
UN2643	Methyl bromoacetate
UN2644	Methyl iodide
UN2645	Phenacyl bromide
UN2646	Hexachlorocyclopentadiene
UN2647	Malononitrile
UN2648	1,2-Dibromobutan-3-one
UN2649	1,3-Dichloroacetone
UN2650	1,1-Dichloro-1-nitroethane
UN2651	4,4'-Diaminodiphenyl methane
UN2653	Benzyl iodide
UN2655	Potassium fluorosilicate
UN2656	Quinoline
UN2657	Selenium disulfide
UN2659	Sodium chloroacetate
UN2660	Nitrotoluidines (mono)
UN2661	Hexachloroacetone
UN2664	Dibromomethane
UN2667	Butyltoluenes
UN2668	Chloroacetonitrile
UN2669	Chlorocresols solution
UN2670	Cyanuric chloride
UN2671	Aminopyridines
UN2672	Ammonia solution
UN2673	2-Amino-4-chlorophenol
UN2674	Sodium fluorosilicate
UN2676	Stibine
UN2677	Rubidium hydroxide solution
UN2678	Rubidium hydroxide
UN2679	Lithium hydroxide, solution
UN2680	Lithium hydroxide
UN2681	Caesium hydroxide solution
UN2682	Caesium hydroxide
UN2683	Ammonium sulfide solution
UN2684	3-Diethylamino-propylamine
UN2685	N,N-Diethylethylenediamine
UN2686	2-Diethylaminoethanol
UN2687	Dicyclohexylammonium nitrite
UN2688	1-Bromo-3-chloropropane
UN2689	Glycerol alpha-monochlorohydrin
UN2690	N-n-Butyl imidazole
UN2691	Phosphorus pentabromide
UN2692	Boron tribromide
UN2693	Bisulfites, aqueous solutions, n.o.s.
UN2698	Tetrahydrophthalic anhydrides
UN2699	Trifluoroacetic acid
UN2705	1-Pentol
UN2707	Dimethyldioxanes
UN2709	Butyl benzenes
UN2710	Dipropyl ketone
UN2713	Acridine
UN2714	Zinc resinate
UN2715	Aluminum resinate
UN2716	1,4-Butynediol

(1)-Identification Number	(2)-Description
UN2717	Camphor
UN2719	Barium bromate
UN2720	Chromium nitrate
UN2721	Copper chlorate
UN2722	Lithium nitrate
UN2723	Magnesium chlorate
UN2724	Manganese nitrate
UN2725	Nickel nitrate
UN2726	Nickel nitrite
UN2727	Thallium nitrate
UN2728	Zirconium nitrate
UN2729	Hexachlorobenzene
UN2730	Nitroanisole, liquid
UN2732	Nitrobromobenzenes, liquid
UN2733	Amine, flammable, corrosive, n.o.s. or Polyamines, flammable, corrosive, n.o.s.
UN2734	Amine, liquid, corrosive, flammable, n.o.s. or Polyamines, liquid, corrosive, flammable, n.o.s.
UN2735	Amines, liquid, corrosive, n.o.s. or Polyamines, liquid, corrosive, n.o.s.
UN2738	N-Butylaniline
UN2739	Butyric anhydride
UN2740	n-Propyl chloroformate
UN2741	Barium hypochlorite
UN2742	Chloroformates, toxic, corrosive, flammable, n.o.s.
UN2743	n-Butyl chloroformate
UN2744	Cyclobutyl chloroformate
UN2745	Chloromethyl chloroformate
UN2746	Phenyl chloroformate
UN2747	tert-Butylcyclohexylchloroformate
UN2748	2-Ethylhexyl chloroformate
UN2749	Tetramethylsilane
UN2750	1,3-Dichloropropanol-2
UN2751	Diethylthiophosphoryl chloride
UN2752	1,2-Epoxy-3-ethoxypropane
UN2753	N-Ethylbenzyltoluidines liquid
UN2754	N-Ethyltoluidines
UN2757	Carbamate pesticides, solid, toxic
UN2758	Carbamate pesticides, liquid, flammable, toxic
UN2759	Arsenical pesticides, solid, toxic
UN2760	Arsenical pesticides, liquid, flammable, toxic
UN2761	Organochlorine, pesticides, solid, toxic
UN2762	Organochlorine pesticides liquid, flammable, toxic
UN2763	Triazine pesticides, solid, toxic
UN2764	Triazine pesticides, liquid, flammable, toxic
UN2771	Thiocarbamate pesticides, solid, toxic
UN2772	Thiocarbamate pesticide, liquid, flammable, toxic
UN2775	Copper based pesticides, solid, toxic
UN2776	Copper based pesticides, liquid, flammable, toxic
UN2777	Mercury based pesticides, solid, toxic
UN2778	Mercury based pesticides, liquid, flammable, toxic
UN2779	Substituted nitrophenol pesticides, solid, toxic
UN2780	Substituted nitrophenol pesticides, liquid, flammable, toxic
UN2781	Bipyridilium pesticides, solid, toxic
UN2782	Bipyridilium pesticides, liquid, flammable, toxic
UN2783	Organophosphorus pesticides, solid, toxic
UN2784	Organophosphorus pesticides, liquid, flammable, toxic
UN2785	4-Thiapentanal
UN2786	Organotin pesticides, solid, toxic
UN2787	Organotin pesticides, liquid, flammable, toxic
UN2788	Organotin compounds, liquid, n.o.s.
UN2789	Acetic acid, glacial or Acetic acid solution
UN2790	Acetic acid solution
UN2793	Ferrous metal borings or Ferrous metal shavings or Ferrous metal turnings or Ferrous metal cuttings
UN2794	Batteries, wet, filled with acid
UN2795	Batteries, wet, filled with alkali
UN2796	Battery fluid, acid
UN2796	Sulfuric acid
UN2797	Battery fluid, alkali
UN2798	Phenyl phosphorus dichloride
UN2799	Phenyl phosphorus thiodichloride
UN2800	Batteries, wet, non-spillable
UN2801	Dyes, liquid, corrosive n.o.s. or Dye intermediates, liquid, corrosive, n.o.s.
UN2802	Copper chloride
UN2803	Gallium
UN2805	Lithium hydride, fused solid
UN2806	Lithium nitride
UN2809	Mercury
NA2810	Compounds, tree killing, liquid or Compounds, weed killing, liquid
UN2810	Toxic, liquids, organic, n.o.s.
UN2811	Toxic, solids, organic, n.o.s.
UN2812	Sodium aluminate, solid
UN2813	Water-reactive solid, n.o.s.
UN2814	Infectious substances, affecting humans
UN2815	N-Aminoethylpiperazine
UN2817	Ammonium hydrogendifluoride, solution
UN2818	Ammonium polysulfide, solution
UN2819	Amyl acid phosphate
UN2820	Butyric acid
UN2821	Phenol solutions
UN2822	2-Chloropyridine
UN2823	Crotonic acid, solid
UN2826	Ethyl chlorothioformate
UN2829	Caproic acid
UN2830	Lithium ferrosilicon
UN2831	1,1,1-Trichloroethane
UN2834	Phosphorous acid
UN2835	Sodium aluminum hydride
UN2837	Bisulfate, aqueous solution
UN2838	Vinyl butyrate, stabilized
UN2839	Aldol
UN2840	Butyraldoxime
UN2841	Di-n-amylamine
UN2842	Nitroethane
UN2844	Calcium manganese silicon
NA2845	Ethyl phosphonous dichloride, anhydrous
NA2845	Methyl phosphonous dichloride
UN2845	Pyrophoric liquids, organic, n.o.s.
UN2846	Pyrophoric solids, organic, n.o.s.
UN2849	3-Chloropropanol-1
UN2850	Propylene tetramer
UN2851	Boron trifluoride dihydrate
UN2852	Dipicryl sulfide, wetted
UN2853	Magnesium fluorosilicate
UN2854	Ammonium fluorosilicate
UN2855	Zinc fluorosilicate
UN2856	Fluorosilicates, n.o.s.
UN2857	Refrigerating machines
UN2858	Zirconium, dry
UN2859	Ammonium metavanadate
UN2861	Ammonium polyvanadate
UN2862	Vanadium pentoxide
UN2863	Sodium ammonium vanadate
UN2864	Potassium metavanadate
UN2865	Hydroxylamine sulfate
UN2869	Titanium trichloride mixtures
UN2870	Aluminum borohydride or Aluminum borohydride in devices
UN2871	Antimony powder
UN2872	Dibromochloropropane
UN2873	Dibutylaminoethanol
UN2874	Furfuryl alcohol
UN2875	Hexachlorophene
UN2876	Resorcinol
UN2878	Titanium sponge granules or Titanium sponge powders
UN2879	Selenium oxychloride
UN2880	Calcium hypochlorite, hydrated or Calcium hypochlorite, hydrated mixtures
UN2881	Metal catalyst, dry
UN2900	Infectious substances, affecting animals
UN2901	Bromine chloride
UN2902	Pesticides, liquid, toxic, n.o.s.
UN2903	Pesticides, liquid, toxic, flammable, n.o.s.
UN2904	Chlorophenolates, liquid or Phenolates, liquid
UN2905	Chlorophenolates, solid or Phenolates, solid
UN2907	Isosorbide dinitrate mixture
UN2908	Radioactive material, excepted package-empty packaging
UN2909	Radioactive material, excepted package-articles manufactured from natural uranium or depleted uranium or natural thorium
UN2910	Radioactive material, excepted package-limited quantity of material
UN2911	Radioactive material, excepted package-instruments or articles
UN2912	Radioactive material, low specific activity (LSA-I)
UN2913	Radioactive material, surface contaminated objects (SCO-I or SCO-II)
UN2915	Radioactive material, Type A package
UN2916	Radioactive material, Type B(U) package
UN2917	Radioactive material, Type B(M) package

ID CROSS REFERENCE

(1)-Identification Number	(2)-Description
UN2919	Radioactive material, transported under special arrangement
UN2920	Corrosive liquids, flammable, n.o.s.
UN2921	Corrosive solids, flammable, n.o.s.
UN2922	Corrosive liquids, toxic, n.o.s.
UN2923	Corrosive solids, toxic, n.o.s.
UN2924	Flammable liquids, corrosive, n.o.s.
UN2925	Flammable solids, corrosive, organic, n.o.s.
UN2926	Flammable solids, toxic, organic, n.o.s.
NA2927	Ethyl phosphonothioic dichloride, anhydrous
NA2927	Ethyl phosphorodichloridate
UN2927	Toxic liquids, corrosive, organic, n.o.s.
UN2928	Toxic solids, corrosive, organic n.o.s.
UN2929	Toxic liquids, flammable, organic, n.o.s.
UN2930	Toxic solids, flammable, organic, n.o.s.
UN2931	Vanadyl sulfate
UN2933	Methyl 2-chloropropionate
UN2934	Isopropyl 2-chloropropionate
UN2935	Ethyl 2-chloropropionate
UN2936	Thiolactic acid
UN2937	alpha-Methylbenzyl alcohol, liquid
UN2940	9-Phosphabicyclononanes or Cyclooctadiene phosphines
UN2941	Fluoroanilines
UN2942	2-Trifluoromethylaniline
UN2943	Tetrahydrofurfurylamine
UN2945	N-Methylbutylamine
UN2946	2-Amino-5-diethylaminopentane
UN2947	Isopropyl chloroacetate
UN2948	3-Trifluoromethylaniline
UN2949	Sodium hydrosulfide
UN2950	Magnesium granules, coated
UN2956	5-tert-Butyl-2,4,6-trinitro-m-xylene or Musk xylene
UN2965	Boron trifluoride dimethyl etherate
UN2966	Thioglycol
UN2967	Sulfamic acid
UN2968	Maneb stabilized or Maneb preparations, stabilized
UN2969	Castor beans or Castor meal or Castor pomace or Castor flake
UN2977	Radioactive material, uranium hexafluoride, fissile
UN2978	Radioactive material, uranium hexafluoride
UN2983	Ethylene oxide and propylene oxide mixtures
UN2984	Hydrogen peroxide, aqueous solutions
UN2985	Chlorosilanes, flammable, corrosive, n.o.s.
UN2986	Chlorosilanes, corrosive, flammable, n.o.s.
UN2987	Chlorosilanes, corrosive, n.o.s.
UN2988	Chlorosilanes, water-reactive, flammable, corrosive, n.o.s.
UN2989	Lead phosphite, dibasic
UN2990	Life-saving appliances, self inflating
UN2991	Carbamate pesticides, liquid, toxic, flammable
UN2992	Carbamate pesticides, liquid, toxic
UN2993	Arsenical pesticides, liquid, toxic, flammable
UN2994	Arsenical pesticides, liquid, toxic
UN2995	Organochlorine pesticides, liquid, toxic, flammable.
UN2996	Organochlorine pesticides, liquid, toxic
UN2997	Triazine pesticides, liquid, toxic, flammable
UN2998	Triazine pesticides, liquid, toxic
UN3002	Phenyl urea pesticides, liquid, toxic
UN3005	Thiocarbamate pesticide, liquid, toxic, flammable
UN3006	Thiocarbamate pesticide, liquid, toxic
UN3009	Copper based pesticides, liquid, toxic, flammable
UN3010	Copper based pesticides, liquid, toxic
UN3011	Mercury based pesticides, liquid, toxic, flammable
UN3012	Mercury based pesticides, liquid, toxic
UN3013	Substituted nitrophenol pesticides, liquid, toxic, flammable
UN3014	Substituted nitrophenol pesticides, liquid, toxic
UN3015	Bipyridilium pesticides, liquid, toxic, flammable
UN3016	Bipyridilium pesticides, liquid, toxic
UN3017	Organophosphorus pesticides, liquid, toxic, flammable
UN3018	Organophosphorus pesticides, liquid, toxic
UN3019	Organotin pesticides, liquid, toxic, flammable
UN3020	Organotin pesticides, liquid, toxic
UN3021	Pesticides, liquid, flammable, toxic
UN3022	1,2-Butylene oxide, stabilized
UN3023	2-Methly-2-heptanethiol
UN3024	Coumarin derivative pesticides, liquid, flammable, toxic
UN3025	Coumarin derivative pesticides, liquid, toxic, flammable
UN3026	Coumarin derivative pesticides, liquid, toxic
UN3027	Coumarin derivative pesticides, solid, toxic
UN3028	Batteries, dry, containing potassium hydroxide solid
UN3048	Aluminum phosphide pesticides
UN3054	Cyclohexyl mercaptan
UN3055	2-(2-Aminoethoxy) ethanol
UN3056	n-Heptaldehyde
UN3057	Trifluoroacetyl chloride
UN3064	Nitroglycerin, solution in alcohol
UN3065	Alcoholic beverages
UN3066	Paint or Paint related material
UN3070	Ethylene oxide and dichlorodifluoromethane mixture
UN3071	Mercaptans, liquid, toxic, flammable, n.o.s. or Mercaptan mixtures, liquid, toxic, flammable, n.o.s.
UN3072	Life-saving appliances, not self inflating
UN3073	Vinylpyridines, stabilized
NA3077	Hazardous waste, solid, n.o.s.
NA3077	Other regulated substances, solid, n.o.s.
UN3077	Environmentally hazardous substance, solid, n.o.s.
UN3078	Cerium
UN3079	Methacrylonitrile, stabilized
UN3080	Isocyanates, toxic, flammable, n.o.s. or Isocyanate solutions, toxic, flammable, n.o.s.
NA3082	Hazardous waste, liquid, n.o.s.
NA3082	Other regulated substances, liquid, n.o.s.
UN3082	Environmentally hazardous substance, liquid, n.o.s.
UN3083	Perchloryl fluoride
UN3084	Corrosive solids, oxidizing, n.o.s.
UN3085	Oxidizing solid, corrosive, n.o.s.
UN3086	Toxic solids, oxidizing, n.o.s.
UN3087	Oxidizing solid, toxic, n.o.s.
UN3088	Self-heating solid, organic, n.o.s.
UN3089	Metal powders, flammable, n.o.s.
UN3090	Lithium battery
UN3091	Lithium batteries, contained in equipment
UN3091	Lithium batteries, packed with equipment
UN3092	1-Methoxy-2-proponal
UN3093	Corrosive liquids, oxidizing, n.o.s.
UN3094	Corrosive liquids, water-reactive, n.o.s.
UN3095	Corrosive solids, self-heating, n.o.s.
UN3096	Corrosive solids, water-reactive, n.o.s.
UN3097	Flammable solid, oxidizing, n.o.s.
UN3098	Oxidizing liquid, corrosive, n.o.s.
UN3099	Oxidizing liquid, toxic, n.o.s.
UN3100	Oxidizing solid, self-heating, n.o.s.
UN3101	Organic peroxide type B, liquid
UN3102	Organic peroxide type B, solid
UN3103	Organic peroxide type C, liquid
UN3104	Organic peroxide type C, solid
UN3105	Organic peroxide type D, liquid
UN3106	Organic peroxide type D, solid
UN3107	Organic peroxide type E, liquid
UN3108	Organic peroxide type E, solid
UN3109	Organic peroxide type F, liquid
UN3110	Organic peroxide type F, solid
UN3111	Organic peroxide type B, liquid, temperature controlled
UN3112	Organic peroxide type B, solid, temperature controlled
UN3113	Organic peroxide type C, liquid, temperature controlled
UN3114	Organic peroxide type C, solid, temperature controlled
UN3115	Organic peroxide type D, liquid, temperature controlled
UN3116	Organic peroxide type D, solid, temperature controlled
UN3117	Organic peroxide type E, liquid, temperature controlled
UN3118	Organic peroxide type E, solid, temperature controlled
UN3119	Organic peroxide type F, liquid, temperature controlled
UN3120	Organic peroxide type F, solid, temperature controlled
UN3121	Oxidizing solid, water-reactive, n.o.s.
UN3122	Toxic liquids, oxidizing, n.o.s.
UN3123	Toxic liquids, water-reactive, n.o.s.

(1)-Identification Number	(2)-Description
UN3124	Toxic solids, self-heating, n.o.s.
UN3125	Toxic solids, water-reactive, n.o.s.
UN3126	Self-heating solid, corrosive, organic, n.o.s.
UN3127	Self-heating solid, oxidizing, n.o.s.
UN3128	Self-heating solid, toxic, organic, n.o.s.
UN3129	Water-reactive liquid, corrosive, n.o.s.
UN3130	Water-reactive liquid, toxic, n.o.s.
UN3131	Water-reactive solid, corrosive, n.o.s.
UN3132	Water-reactive solid, flammable, n.o.s.
UN3133	Water-reactive solid, oxidizing, n.o.s.
UN3134	Water-reactive solid, toxic, n.o.s.
UN3135	Water-reactive solid, self-heating, n.o.s.
UN3136	Trifluoromethane, refrigerated liquid
UN3137	Oxidizing, solid, flammable, n.o.s.
UN3138	Ethylene, acetylene and propylene in mixture, refrigerated liquid
UN3139	Oxidizing, liquid, n.o.s.
UN3140	Alkaloids, liquid, n.o.s. or Alkaloid salts, liquid, n.o.s.
UN3141	Antimony compounds, inorganic, liquid, n.o.s.
UN3142	Disinfectants, liquid, toxic, n.o.s
UN3143	Dyes, solid, toxic, n.o.s. or Dye intermediates, solid, toxic, n.o.s.
UN3144	Nicotine compounds, liquid, n.o.s. or Nicotine preparations, liquid, n.o.s.
UN3145	Alkylphenols, liquid, n.o.s.
UN3146	Organotin compounds, solid, n.o.s.
UN3147	Dyes, solid, corrosive, n.o.s. or Dye intermediates, solid, corrosive, n.o.s.
UN3148	Water-reactive, liquid, n.o.s.
UN3149	Hydrogen peroxide and peroxyacetic acid mixtures, stabilized
UN3150	Devices, small, hydrocarbon gas powered or Hydrocarbon gas refills for small devices
UN3151	Polyhalogenated biphenyls, liquid or Polyhalogenated terphenyls liquid
UN3152	Polyhalogenated biphenyls, solid or Polyhalogenated terphenyls, solid
UN3153	Perfluoro(methyl vinyl ether)
UN3154	Perfluoro(ethyl vinyl ether)
UN3155	Pentachlorophenol
UN3156	Compressed gas, oxidizing, n.o.s.
UN3157	Liquefied gas, oxidizing, n.o.s.
UN3158	Gas, refrigerated liquid, n.o.s.
UN3159	1,1,1,2-Tetrafluoroethane or Refrigerant gas R 134a
UN3160	Liquefied gas, toxic, flammable, n.o.s.
UN3161	Liquefied gas, flammable, n.o.s.
UN3162	Liquefied gas, toxic, n.o.s.
UN3163	Liquefied gas, n.o.s.
UN3164	Articles, pressurized pneumatic or hydraulic
UN3165	Aircraft hydraulic power unit fuel tank
UN3166	Engines, internal combustion or Engines, fuel cell
UN3166	Vehicle, flammable gas powered or Vehicle, fuel cell, flammable gas powered
UN3166	Vehicle, flammable liquid powered or Vehicle, fuel cell, flammable liquid powered
UN3167	Gas sample, non-pressurized, flammable, n.o.s.
UN3168	Gas sample, non-pressurized, toxic, flammable, n.o.s.
UN3169	Gas sample, non-pressurized, toxic, n.o.s.
UN3170	Aluminum smelting by-products or Aluminum remelting by-products
UN3171	Battery-powered vehicle or Battery-powered equipment
UN3172	Toxins, extracted from living sources, liquid, n.o.s.
UN3174	Titanium disulphide
UN3175	Solids containing flammable liquid, n.o.s.
UN3176	Flammable solid, organic, molten, n.o.s.
NA3178	Smokeless powder for small arms
UN3178	Flammable solid, inorganic, n.o.s.
UN3179	Flammable solid, toxic, inorganic, n.o.s.
UN3180	Flammable solid, corrosive, inorganic, n.o.s.
UN3181	Metal salts of organic compounds, flammable, n.o.s.
UN3182	Metal hydrides, flammable, n.o.s.
UN3183	Self-heating liquid, organic, n.o.s.
UN3184	Self-heating liquid, toxic, organic, n.o.s.
UN3185	Self-heating liquid, corrosive, organic, n.o.s.
UN3186	Self-heating liquid, inorganic, n.o.s.
UN3187	Self-heating liquid, toxic, inorganic, n.o.s.
UN3188	Self-heating liquid, corrosive, inorganic, n.o.s.
UN3189	Metal powder, self-heating, n.o.s.
UN3190	Self-heating solid, inorganic, n.o.s.
UN3191	Self-heating solid, toxic, inorganic, n.o.s.
UN3192	Self-heating solid, corrosive, inorganic, n.o.s.
UN3194	Pyrophoric liquid, inorganic, n.o.s.
UN3200	Pyrophoric solid, inorganic, n.o.s.
UN3205	Alkaline earth metal alcoholates, n.o.s.
UN3206	Alkali metal alcoholates, self-heating, corrosive, n.o.s.
UN3208	Metallic substance, water-reactive, n.o.s.
UN3209	Metallic substance, water-reactive, self-heating, n.o.s.
UN3210	Chlorates, inorganic, aqueous solution, n.o.s.
UN3211	Perchlorates, inorganic, aqueous solution, n.o.s.
UN3212	Hypochlorites, inorganic, n.o.s.
UN3213	Bromates, inorganic, aqueous solution, n.o.s.
UN3214	Permanganates, inorganic, aqueous solution, n.o.s.
UN3215	Persulfates, inorganic, n.o.s.
UN3216	Persulfates, inorganic, aqueous solution, n.o.s.
UN3218	Nitrates, inorganic, aqueous solution, n.o.s.
UN3219	Nitrites, inorganic, aqueous solution, n.o.s.
UN3220	Pentafluoroethane or Refrigerant gas R 125
UN3221	Self-reactive liquid type B
UN3222	Self-reactive solid type B
UN3223	Self-reactive liquid type C
UN3224	Self-reactive solid type C
UN3225	Self-reactive liquid type D
UN3226	Self-reactive solid type D
UN3227	Self-reactive liquid type E
UN3228	Self-reactive solid type E
UN3229	Self-reactive liquid type F
UN3230	Self-reactive solid type F
UN3231	Self-reactive liquid type B, temperature controlled
UN3232	Self-reactive solid type B, temperature controlled
UN3233	Self-reactive liquid type C, temperature controlled
UN3234	Self-reactive solid type C, temperature controlled
UN3235	Self-reactive liquid type D, temperature controlled
UN3236	Self-reactive solid type D, temperature controlled
UN3237	Self-reactive liquid type E, temperature controlled
UN3238	Self-reactive solid type E, temperature controlled
UN3239	Self-reactive liquid type F, temperature controlled
UN3240	Self-reactive solid type F, temperature controlled
UN3241	2-Bromo-2-nitropropane-1,3-diol
UN3242	Azodicarbonamide
UN3243	Solids containing toxic liquid, n.o.s.
UN3244	Solids containing corrosive liquid, n.o.s.
UN3246	Methanesulfonyl chloride
UN3247	Sodium peroxoborate, anhydrous
UN3248	Medicine, liquid, flammable, toxic, n.o.s.
UN3249	Medicine, solid, toxic, n.o.s.
UN3250	Chloroacetic acid, molten
UN3251	Isosorbide-5-mononitrate
UN3252	Difluoromethane or Refrigerant gas R 32
UN3253	Disodium trioxosilicate
UN3254	Tributylphosphane
UN3255	tert-Butyl hypochlorite
UN3256	Elevated temperature liquid, flammable, n.o.s.
UN3257	Elevated temperature liquid, n.o.s.
UN3258	Elevated temperature solid, n.o.s.
UN3259	Amines, solid, corrosive, n.o.s. or, Polyamines, solid, corrosive n.o.s.
UN3260	Corrosive solid, acidic, inorganic, n.o.s.
UN3261	Corrosive solid, acidic, organic, n.o.s.
UN3262	Corrosive solid, basic, inorganic, n.o.s.
UN3263	Corrosive solid, basic, organic, n.o.s.
UN3264	Corrosive liquid, acidic, inorganic, n.o.s.
UN3265	Corrosive liquid, acidic, organic, n.o.s.
UN3266	Corrosive liquid, basic, inorganic, n.o.s.

ID CROSS REFERENCE

(1)-Identification Number	(2)-Description
UN3267	Corrosive liquid, basic, organic, n.o.s.
UN3268	Air bag inflators or Air bag modules or Seat-belt pre-tensioners
UN3269	Polyester resin kit
UN3270	Nitrocellulose membrane filters
UN3271	Ethers, n.o.s.
UN3272	Esters, n.o.s.
UN3273	Nitriles, flammable, toxic, n.o.s.
UN3274	Alcoholates solution, n.o.s.
UN3275	Nitriles, toxic, flammable, n.o.s.
UN3276	Nitriles, toxic, liquid, n.o.s.
UN3277	Chloroformates, toxic, corrosive, n.o.s.
UN3278	Organophosphorus compound, toxic, liquid, n.o.s.
UN3279	Organophosphorus compound, toxic, flammable, n.o.s.
UN3280	Organoarsenic compound, liquid, n.o.s.
UN3281	Metal carbonyls, liquid, n.o.s.
UN3282	Organometallic compound, toxic, liquid, n.o.s.
UN3283	Selenium compound, solid, n.o.s.
UN3284	Tellurium compound, solid, n.o.s.
UN3285	Vanadium compound, n.o.s.
UN3286	Flammable liquid, toxic, corrosive, n.o.s.
UN3287	Toxic liquid, inorganic, n.o.s.
UN3288	Toxic solid, inorganic, n.o.s.
UN3289	Toxic liquid, corrosive, inorganic, n.o.s.
UN3290	Toxic solid, corrosive, inorganic, n.o.s.
UN3291	Regulated medical waste, n.o.s. or Clinical waste, unspecified, n.o.s. or (BIO) Medical waste, n.o.s or Biomedical waste, n.o.s. or Medical waste, n.o.s.
UN3292	Batteries, containing sodium
UN3292	Cells, containing sodium
UN3293	Hydrazine, aqueous solution
UN3294	Hydrogen cyanide, solution in alcohol
UN3295	Hydrocarbons, liquid, n.o.s.
UN3296	Heptafluoropropane or Refrigerant gas R 227
UN3297	Ethylene oxide and chlorotetra-fluoroethane mixture
UN3298	Ethylene oxide and pentafluoroethane mixture
UN3299	Ethylene oxide and tetrafluorethane mixture
UN3300	Ethylene oxide and carbon dioxide mixture
UN3301	Corrosive liquid, self-heating, n.o.s.
UN3302	2-Dimethylaminoethyl acrylate
UN3303	Compressed gas, toxic, oxidizing, n.o.s.
UN3304	Compressed gas, toxic, corrosive, n.o.s.
UN3305	Compressed gas, toxic, flammable, corrosive, n.o.s.
UN3306	Compressed gas, toxic, oxidizing, corrosive, n.o.s.
UN3307	Liquified gas, toxic, oxidizing, n.o.s.
UN3308	Liquified gas, toxic, corrosive, n.o.s.
UN3309	Liquified gas, toxic, flammable, corrosive, n.o.s.
UN3310	Liquified gas, toxic, oxidizing, corrosive, n.o.s.
UN3311	Gas, refrigerated liquid, oxidizing, n.o.s.
UN3312	Gas, refrigerated liguid, flammable, n.o.s.
UN3313	Organic pigments, self-heating
UN3314	Plastic molding compound
UN3316	Chemical kits
UN3316	First aid kits
UN3317	2-Amino-4,6-Dinitrophenol, wetted
UN3318	Ammonia solution
UN3319	Nitroglycerin mixture, desensitized, solid, n.o.s.
UN3320	Sodium borohydride and sodium hydroxide solution
UN3321	Radioactive material, low specific activity (LSA-II)
UN3322	Radioactive material, low specific activity (LSA-III)
UN3327	Radioactive material, Type A package, fissile
UN3328	Radioactive material, Type B(U) package, fissile
UN3329	Radioactive material, Type B(M) package, fissile
UN3331	Radioactive material, transported under special arrangement, fissile
UN3332	Radioactive material, Type A package, special form
UN3333	Radioactive material, Type A package, special form, fissile
NA3334	Self-defense spray, non-pressurized
UN3334	Aviation regulated liquid, n.o.s.
UN3335	Aviation regulated solid, n.o.s.
UN3336	Mercaptans, liquid, flammable, n.o.s. or Mercaptan mixture, liquid, flammable, n.o.s.
UN3337	Refrigerant gas R404A
UN3338	Refrigerant gas R407A
UN3339	Refrigerant gas R407B
UN3340	Refrigerant gas R407C
UN3341	Thiourea dioxide
UN3342	Xanthates
UN3343	Nitroglycerin mixture, desensitized, liquid, flammable, n.o.s.
UN3344	Pentaerythrite tetranitrate mixture, desensitized solid, n.o.s. or Pentaerythritol tetranitrate mixtrure, desensitized, solid, n.o.s. or PETN mixture, desensitized, solid, n.o.s.
UN3345	Phenoxyacetic acid derivative pesticide, solid, toxic
UN3346	Phenoxyacetic acid derivative pesticide, liquid, flammable, toxic
UN3347	Phenoxyacetic acid derivative pesticide, liquid, toxic, flammable
UN3348	Phenoxyacetic acid derivative pesticide, liquid, toxic
UN3349	Pyrethroid pesticide, solid, toxic
UN3350	Pyrethroid pesticide, liquid, flammable, toxic
UN3351	Pyrethroid pesticide, liquid, toxic, flammable
UN3352	Pyrethroid pesticide, liquid, toxic
UN3354	Insecticide gases, flammable, n.o.s.
UN3355	Insecticide gases toxic, flammable, n.o.s.
NA3356	Oxygen generator, chemical, spent
UN3356	Oxygen generator, chemical
UN3357	Nitroglycerin mixture, desensitized, liquid, n.o.s.
UN3358	Refrigerating machines
UN3360	Fibers, vegetable, dry
UN3361	Chlorosilanes, toxic, corrosive, n.o.s.
UN3362	Chlorosilanes, toxic, corrosive, flammable, n.o.s.
UN3363	Dangerous goods in machinery or Dangerous goods in apparatus
UN3364	Trinitrophenol (picric acid), wetted
UN3365	Trinitrochlorobenzene (picryl chloride), wetted
UN3366	Trinitrotoluene (TNT), wetted
UN3367	Trinitrobenzene, wetted
UN3368	Trinitrobenzoic acid, wetted
UN3369	Sodium dinitro-o-cresolate, wetted
UN3370	Urea nitrate, wetted
UN3371	2-Methylbutanal
UN3373	Biological substance, Category B
UN3375	Ammonium nitrate emulsion or Ammonium nitrate suspension or Ammonium nitrate gel
UN3376	4-Nitrophenylhydrazine
UN3377	Sodium perborate monohydrate
UN3378	Sodium carbonate peroxyhydrate
UN3379	Desensitized explosive, liquid, n.o.s.
UN3380	Desensitized explosive, solid, n.o.s.
UN3381	Toxic by inhalation liquid, n.o.s.
UN3382	Toxic by inhalation liquid, n.o.s.
UN3383	Toxic by inhalation liquid, flammable, n.o.s.
UN3384	Toxic by inhalation liquid, flammable, n.o.s.
UN3385	Toxic by inhalation liquid, water-reactive, n.o.s.
UN3386	Toxic by inhalation liquid, water-reactive, n.o.s.
UN3387	Toxic by inhalation liquid, oxidizing, n.o.s.
UN3388	Toxic by inhalation liquid, oxidizing, n.o.s.
UN3389	Toxic by inhalation liquid, corrosive, n.o.s.
UN3390	Toxic by inhalation liquid, corrosive, n.o.s.
UN3391	Organometallic substance, solid, pyrophoric
UN3392	Organometallic substance, liquid, pyrophoric
UN3393	Organometallic substance, solid, pyrophoric, water-reactive
UN3394	Organometallic substance, liquid, pyrophoric, water-reactive
UN3395	Organometallic substance, solid, water-reactive
UN3396	Organometallic substance, solid, water-reactive, flammable
UN3397	Organometallic substance, solid, water-reactive, self-heating
UN3398	Organometallic substance, liquid, water-reactive
UN3399	Organometallic substance, liquid, water-reactive, flammable
UN3400	Organometallic substance, solid, self-heating
UN3401	Alkali metal amalgam, solid

(1)-Identification Number	(2)-Description
UN3402	Alkaline earth metal amalgams, solid
UN3403	Potassium, metal alloys, solid
UN3404	Potassium sodium alloys, solid
UN3405	Barium chlorate, solution
UN3406	Barium perchlorate, solution
UN3407	Chlorate and magnesium chloride mixture solution
UN3408	Lead perchlorate, solution
UN3409	Chloronitrobenzenes, liquid
UN3410	4-Chloro-o-toluidine hydrochloride, solution
UN3411	beta-Naphthylamine solution
UN3412	Formic acid
UN3413	Potassium cyanide solution
UN3414	Sodium cyanide solution
UN3415	Sodium fluoride solution
UN3416	Chloroacetophenone, liquid
UN3417	Xylyl bromide, solid
UN3418	2,4-Toluylenediamine solution or 2,4-Toluenediamine solution
UN3419	Boron trifluoride acetic acid complex, solid
UN3420	Boron trifluoride propionic acid complex, solid
UN3421	Potassium hydrogendifluoride solution
UN3422	Potassium fluoride solution
UN3423	Tetramethylammonium hydroxide, solid
UN3424	Ammonium dinitro-o-cresolate solution
UN3425	Bromoacetic acid solid
UN3426	Acrylamide solution
UN3427	Chlorobenzyl chlorides, solid
UN3428	3-Chloro-4-methylphenyl isocyanate, solid
UN3429	Chlorotoluidines, liquid
UN3430	Xylenols, liquid
UN3431	Nitrobenzotrifluorides, solid
UN3432	Polychlorinated biphenyls, solid
UN3434	Nitrocresols, liquid
UN3436	Hexafluoroacetone hydrate, solid
UN3437	Chlorocresols, solid
UN3438	alpha-Methylbenzyl alcohol, solid
UN3439	Nitriles, toxic, solid, n.o.s.
UN3440	Selenium compound, liquid, n.o.s.
UN3441	Chlorodinitrobenzenes, solid
UN3442	Dichloroanilines, solid
UN3443	Dinitrobenzenes, solid
UN3444	Nicotine hydrochloride, solid
UN3445	Nicotine sulphate, solid
UN3446	Nitrotoluenes, solid
UN3447	Nitroxylenes, solid
UN3448	Tear gas substance, solid, n.o.s.
UN3449	Bromobenzyl cyanides, solid
UN3450	Diphenylchloroarsine, solid
UN3451	Toluidines, solid
UN3452	Xylidines, solid
UN3453	Phosphoric acid, solid
UN3454	Dinitrotoluenes, solid
UN3455	Cresols, solid
UN3456	Nitrosylsulphuric acid, solid
UN3457	Chloronitrotoluenes, solid
UN3458	Nitroanisoles, solid
UN3459	Nitrobromobenzenes, solid
UN3460	N-Ethylbenzyltoluidines, solid
UN3462	Toxins, extracted from living sources, solid, n.o.s.
UN3463	Propionic acid
UN3464	Organophosphorus compound, toxic, solid, n.o.s.
UN3465	Organoarsenic compound, solid, n.o.s.
UN3466	Metal carbonyls, solid, n.o.s.
UN3467	Organometallic compound, toxic, solid, n.o.s.
UN3468	Hydrogen in a metal hydride storage system or Hydrogen in a metal hydride storage system contained in equipment of Hydrogen in a metal hydride storage system packed with equipment
UN3469	Paint, flammable, corrosive
UN3469	Paint related material, flammable, corrosive
UN3470	Paint, corrosive, flammable
UN3470	Paint related material corrosive, flammable
UN3471	Hydrogendifluoride solution, n.o.s
UN3472	Crotonic acid, liquid
UN3473	Fuel cell cartridges or Fuel cell cartridges contained in equipment or Fuel cell cartridges packed with equipment
UN3474	1-Hydroxybenzotriazole, monohydrate
UN3475	Ethanol and gasoline mixture or Ethanol and motor spirit mixture or Ethanol and petrol mixture
UN3476	Fuel cell cartridges or Fuel cell cartridges contained in equipment or Fuel cell cartridges packed with equipment
UN3477	Fuel cell cartridges or Fuel cell cartridges contained in equipment or Fuel cell cartridges packed with equipment
UN3478	Fuel cell cartridges or Fuel cell cartridges contained in equipment or Fuel cell cartridges packed with equipment
UN3479	Fuel cell cartridges or Fuel cell cartridges contained in equipment or Fuel cell cartridges packed with equipment
UN3482	Alkali metal dispersions, flammable or Alkaline earth metal dispersions, flammable
UN3483	Motor fuel anti-knock mixture, flammabie
UN3484	Hydrazine aqueous solution, flammable
UN3485	Calcium hypochlorite, dry, corrosive or Calcium hypochlorite mixtures, dry, corrosive
UN3486	Calcium hypochlorite mixture, dry, corrosive
UN3487	Calcium hypochlorite, hydrated, corrosive or Calcium hypochlorite, hydrated mixture, corrosive
UN3488	Toxic by inhalation, flammable, corrosive, n.o.s.
UN3489	Toxic by inhalation liquid, flammable, corrosive, n.o.s.
UN3490	Toxic by inhalation liquid, water-reactive, flammable, n.o.s.
UN3491	Toxic by inhalation liquid, water-reactive, flammable, n.o.s
UN3492	Toxic by inhalation liquid, corrosive, flammable, n.o.s.
UN3493	Toxic by inhalation liquid, corrosive, flammable, n.o.s.
UN3494	Petroleum sour crude oil, flammable, toxic
UN3495	Iodine
UN3496	Batteries, nickel-metal hydride
ID8000	Consumer commodity
NA9035	Gas identification set
NA9191	Chlorine dioxide, hydrate, frozen
NA9202	Carbon monoxide, refrigerated liquid
NA9206	Methyl phosphonic dichloride
NA9260	Aluminum, molten
NA9263	Chloropivaloyl chloride
NA9264	3,5-Dichloro-2,4,6-trifluoropyridine
NA9269	Trimethoxysilane

ID CROSS REFERENCE

NOTES

NOTES

NOTES

NOTES

NOTES

NOTES

NOTES

NOTES

NOTES